C₄ Plant Biology

This is a volume in the
PHYSIOLOGICAL ECOLOGY series
Edited by Harold A. Mooney

A complete list of books in this series appears at the end of the volume.

C_4 Plant Biology

Edited by

Rowan F. Sage
Department of Botany
University of Toronto
Toronto, Ontario, Canada

Russell K. Monson
Department of E.P.O. Biology
University of Colorado
Boulder, Colorado

Academic Press

San Diego   London   Boston   New York   Sydney   Tokyo   Toronto

Cover photo courtesy of Ron Dengler.

This book is printed on acid-free paper.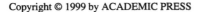

Copyright © 1999 by ACADEMIC PRESS

All Rights Reserved.
No part of this publication may be reproduced or transmitted in any form or by any means, electronic or mechanical, including photocopy, recording, or any information storage and retrieval system, without permission in writing from the publisher.

Academic Press
a division of Harcourt Brace & Company
525 B Street, Suite 1900, San Diego, California 92101-4495, USA
http://www.apnet.com

Academic Press Limited
24-28 Oval Road, London NW1 7DX, UK
http://www.hbuk.co.uk/ap/

Library of Congress Catalog Card Number: 98-87080

International Standard Book Number: 0-12-614440-0

Printed and bound by CPI Group (UK) Ltd, Croydon, CR0 4YY

Transferred to Digital Print 2011

Contents

Contributors xi
Preface xiii

Part I
Perspectives

1. Why C_4 Photosynthesis?
Rowan F. Sage

    I. Introduction   3
    II. The Problem with Rubisco   3
    III. How C_4 Photosynthesis Solves the Rubisco Problem   7
    IV. Significant Variations on the C_4 Theme   10
    V. The Consequences of the Evolution of C_4 Photosynthesis   11
    VI. C_4 Photosynthesis in the Future   12
    VII. Summary   14
        References   14

2. C_4 Photosynthesis: A Historical Overview
Marshall D. Hatch

    I. Introduction   17
    II. The Prediscovery Scene (Before 1965)   18
    III. Discovery: Radiotracer Evidence (1954–1967)   19
    IV. Mechanism, Functions, and Recognition (1965–1970)   22
    V. Selected Aspects of the C_4 Story (1970–)   25
    VI. Summary   40
        References   40

Part II
Structure–Function of the C_4 Syndrome

3. The Biochemistry of C_4 Photosynthesis
Ryuzi Kanai and Gerald E. Edwards

    I. Introduction   49

II. CO_2 Concentration and Rubisco Activity in Bundle Sheath Cells   60
III. Energetics of C_4 Photosynthesis   62
IV. Coordination of the Two Cell Types in C_4 Photosynthesis   68
V. Intracellular Transport of Metabolites   74
VI. Summary   78
References   80

4. Regulation of the C_4 Pathway
Richard C. Leegood and Robert P. Walker

I. Introduction   89
II. An Overview of the Regulation of the C_4 Cycle   89
III. Regulation of the C_4 Cycle   95
IV. Interactions between the C_4 Cycle and Mitochondrial Metabolism   109
V. Regulation of the Benson–Calvin Cycle in C_4 Plants   111
VI. Regulation of Product Synthesis in Leaves of C_4 Plants   114
VII. Summary   120
References   121

5. Leaf Structure and Development in C_4 Plants
Nancy G. Dengler and Timothy Nelson

I. Introduction   133
II. Kranz Anatomy and Biochemical Compartmentation   134
III. Development of the C_4 Syndrome   153
IV. Summary   163
References   164

6. Modeling C_4 Photosynthesis
Susanne von Caemmerer and Robert T. Furbank

I. Introduction   173
II. Basic Model Equations   174
III. Analysis of the Model   184
IV. Summary   205
References   205

Part III
Ecology of C_4 Photosynthesis

7. Environmental Responses
Steve P. Long

I. Introduction   215
II. Light   216
III. Nitrogen   221

IV. Water   226
   V. Temperature   231
   VI. Summary   240
      References   242

8. Success of C₄ Photosynthesis in the Field: Lessons from Communities Dominated by C₄ Plants
Alan K. Knapp and Ernesto Medina

   I. Introduction   251
   II. The C_4-Dominated Tallgrass Prairies of North America   252
   III. The C_4-Dominated Neotropical Savannas   264
   IV. Summary   276
      References   277

9. C₄ Plants and Herbivory
Scott A. Heckathorn, Samuel J. McNaughton, and James S. Coleman

   I. Introduction   285
   II. Unique Features of C_4 Species That May Affect Herbivores and Plant Responses to Herbivory   286
   III. How Important Are C_4 Characteristics to Herbivory Tolerance and Resistance?   295
   IV. Summary   306
      References   307

10. The Biogeography of C₄ Photosynthesis: Patterns and Controlling Factors
Rowan F. Sage, David A. Wedin, and Meirong Li

   I. Introduction   313
   II. The Global Distribution of C_4 Photosynthesis   314
   III. Factors Controlling the Distribution of C_4 Species   331
   IV. C_4 Plants in the Future   343
   V. Summary   350
      References   356

Part IV
The Evolution of C₄ Photosynthesis

11. The Origins of C₄ Genes and Evolutionary Pattern in the C₄ Metabolic Phenotype
Russell K. Monson

   I. Introduction   377
   II. The Evolution of C_4 Genes   378
   III. The Evolution of C_4 Metabolism as Evidenced in C_3–C_4 Intermediates   389

IV. The Evolution of C_4 Metabolism in Relation to Ecological Factors and Plant Growth Form   399
V. Summary   403
References   404

12. Phylogenetic Aspects of the Evolution of C_4 Photosynthesis
Elizabeth A. Kellogg

I. Introduction   411
II. Phylogenetic Pattern   412
III. Number and Time of C_4 Origins   429
IV. Evolution of C_4 Photosynthesis   433
V. Summary   439
References   439

13. Paleorecords of C_4 Plants and Ecosystems
Thure E. Cerling

I. Introduction   445
II. Paleorecords   448
III. Summary   464
References   465

Part V
C_4 Plants and Humanity

14. Agronomic Implications of C_4 Photosynthesis
R. Harold Brown

I. Introduction   473
II. Adaptation and Characterization of C_4 Crop Plants   475
III. Productivity of C_4 Crop Plants   479
IV. Crop Quality   487
V. Weeds   491
VI. C_3–C_4 Mixtures   494
VII. Turf   499
VIII. Summary   501
References   503

15. C_4 Plants and the Development of Human Societies
Nikolaas J. van der Merwe and Hartmut Tschauner

I. Introduction   509
II. C_4 Plants, Carbon Isotopes, and Human Evolution   510
III. The Emergence of Food Production   514
IV. Millets in China   517
V. C_4 Plants in Africa   520

VI. Maize in the Americas　525
　　　VII. The Past 500 Years　534
　　　VIII. Summary　537
　　　　　References　538

16. The Taxonomic Distribution of C_4 Photosynthesis
Rowan F. Sage, Meirong Li, and Russell K. Monson

　　　I. Introduction　551
　　　II. Methodology　552
　　　III. Lists of C_4 Taxa　555
　　　IV. Summary　580
　　　　　References　581

Index　585
Previous Volumes in Series　597

Contributors

Numbers in parentheses indicate the pages on which the authors' contributions begin.

R. Harold Brown (473), Department of Crop and Soil Sciences, University of Georgia, Athens, Georgia 30602

Thure E. Cerling (445), Department of Geology and Geophysics, University of Utah, Salt Lake City, Utah 84112

James S. Coleman (285), Department of Biology, Syracuse University, Syracuse, New York, 13244; and Desert Research Institute, Reno, Nevada 89512

Nancy G. Dengler (133), Department of Botany, University of Toronto, Toronto, Ontario M5S 3B2, Canada

Gerald E. Edwards (49), Department of Botany, Washington State University, Pullman, Washington 99164

Robert T. Furbank (173), Division of Plant Industry, Commonwealth Scientific and Industrial Research Organization, Canberra 2601, Australia

Marshall D. Hatch (17), CSIRO Plant Industry, Canberra 2601, Australia

Scott A. Heckathorn (285), Department of Biology, Syracuse University, Syracuse, New York, 13244

Ryuzi Kanai (49), Department of Biochemistry and Molecular Biology, Saitama University, Urawa 388, Japan

Elizabeth A. Kellogg (411), Department of Biology, University of Missouri, St. Louis, Missouri 68121

Alan K. Knapp (251), Division of Biology, Kansas State University, Manhattan, Kansas 66506

Richard C. Leegood (89), Robert Hill Institute and Department of Animal and Plant Science, University of Sheffield, Sheffield S10 2UQ, United Kingdom

Meirong Li (313, 551), Department of Botany, University of Toronto, Toronto, Ontario M5S 3B2, Canada

Steve P. Long (215), Department of Biological and Chemical Sciences, John Tabor Laboratories, University of Essex, Colchester CO4 2SQ, United Kingdom

Samuel J. McNaughton (285), Department of Biology, Syracuse University, Syracuse, New York 13244

Ernesto Medina (251), Centro de Ecologia, Instituto Venezolano de Investigaciones Cientificas, Caracas, Venezuela

Russell K. Monson (377, 551), Department of E.P.O. Biology, University of Colorado, Boulder, Colorado 80309

Timothy Nelson (133), Biology Department, Yale University, New Haven, Connecticut 06520

Rowan F. Sage (3, 313, 551), Department of Botany, University of Toronto, Toronto, Ontario M5S 3B2, Canada

Hartmut Tschauner (509), Department of Anthropology, Peabody Museum, Harvard University, Cambridge, Massachusetts 02138

Nikolaas J. van der Merwe (509), Department of Anthropology, Peabody Museum, Harvard University, Cambridge, Massachusetts 02138

Susanne von Caemmerer (173), Research School of Biological Sciences, Australian National University, Canberra 2601, Australia

Robert P. Walker (89), Robert Hill Institute and Department of Animal and Plant Science, University of Sheffield, Sheffield S10 2UQ, United Kingdom

David A. Wedin (313), School of Natural Resources, University of Nebraska, Lincoln, Nebraska 68583

Preface

The spinach is a C_3 plant
It has a Nobel pathway
Its compensation point is high
From early morn till noon-day.
It has carboxydismutase
But has no malate transferase;
The spinach is a regal plant
But lacks the Hatch–Slack pathway.

F. A. Smith, H. Beevers, and others
Sung to the tune of "O Tannenbaum"
Reprinted from *Photosynthesis and Photorespiration*
(1971; John Wiley & Sons, Inc., New York)

The eclectic mixture of seasonal frivolity and glee of scientific discovery was obvious when 14 esteemed scientists took the floor to sing this musical postscript to the Conference on Photosynthesis and Photorespiration in Canberra, Australia, in early December 1970. The occasion marked the close of one of the most influential conferences in the 20th century dealing with the topic of photosynthesis. For the first time, scientists from around the world had come together to synthesize the biochemistry, physiology, anatomy, systematics, ecology, and economic significance of plants possessing a unique pathway of photosynthetic CO_2 assimilation. As one revisits the proceedings from this meeting (Hatch *et al.*, 1971), it is obvious, even to those of us who were too young to attend the sessions, that the integrative spirit of this group of scientists was seminal to the wave of understanding of C_4 photosynthesis that has been achieved in the almost 30 years since. This book, although the product of a couple of C_4 "baby-boomers," was conceived and created as a salute to the seminal sessions of that 1970 meeting in Canberra and as means of restating the integrative necessity of understanding C_4 photosynthesis.

Three modes of photosynthesis predominate in terrestrial plants: the C_3 mode, which is employed by most higher plant species; the Crassulacean acid metabolism (CAM) mode, employed by 20,000 or more succulents and epiphytes; and the C_4 mode, employed by approximately 8000 of the estimated 250,000 higher plant species. Although far fewer species use the C_4 pathway, their ecological and economic significance is substantial. C_4 plants dominate all tropical and subtropical grasslands, most temperate grasslands, and most disturbed landscapes in warmer regions of the world. Major C_4 crops include maize, sorghum, millets, and amaranths, and 8 of the world's 10 most invasive weeds possess C_4 photosynthesis. In addition to its current economic significance, recent work indicates that the appearance of C_4 species in the vast grassland ecosystems of eastern Africa and southwestern Asia in the past 10–20 million years greatly influenced evolutionary patterns in many faunal lineages, including *Homo sapiens*. The spread and domestication of C_4 species in that region in the more recent past have had major impacts on the timing and development of human societies. The anthropological and ecological significance of C_4 plants may increase in the future, as C_4-dominated savannas are thought to significantly influence the long-term carbon dynamics of the soil and atmosphere. The responses of C_4 savannas to future increases in atmospheric CO_2 concentrations and climate change lie at the foundation of any attempts to understand and predict dynamics in the global carbon cycle. Because of these issues, an improved understanding of the biology of C_4 photosynthesis will be required by more than the traditional audience of crop scientists, plant physiologists, and plant ecologists. Land managers, paleoecologists, community and ecosystem ecologists, systematists, and anthropologists are some of the specialists who could be well served by a comprehensive volume summarizing our current knowledge of C_4 plant biology.

This book has been produced with the aim of providing a broad overview of the subject of C_4 photosynthesis, while retaining enough scientific depth to engage those scientists who specialize in C_4 photosynthesis and to further catalyze the integration that was begun at the Canberra conference. In Chapter 1, Sage provides a brief overview of why the CO_2 concentrating mechanism, which is a hallmark of C_4 photosynthesis, may exist, focusing on the evolutionary constraints imposed by more than 3 billion years of photosynthetic existence in the C_3 mode. In Chapter 2, Hal Hatch, one of the discoverers of C_4 photosynthesis, provides a firsthand account of the events surrounding that incipient recognition that not all plants assimilate CO_2 in the same way as *Chlorella* or spinach. The personal recollections that Hal offers and the broad historical perspective that he provides to the more recent discoveries associated with C_4 photosynthesis are invaluable parables to anyone wishing to understand the development of important scientific disciplines. In the chapters following Hatch's historical overview, the book takes on a loose hierarchy of scale, beginning at the level of organelles and cells and progressing to communities and ecosystems. In chapters of the second section, including those by Ryuzi Kanai and Gerald Edwards, Richard Leegood and Robert Walker, Nancy Dengler and Timothy Nelson, and Susanne von Caemmerer and Robert Furbank, strong themes of coordinated structure and function are developed. In chapters of the third section, including those by Steve Long, Alan Knapp and Ernesto Medina, Scott Heckathorn, Samuel McNaughton, and James Coleman, and Rowan Sage, David Wedin, and Meirong Li, the dominant theme is that of ecological performance and its translation into the geographic distribution of C_4 plants. In chapters of the fourth section, including those by Russell Monson, Elizabeth Kellogg, and Thure Cerling, C_4 evolutionary patterns are considered. These patterns include the evolutionary patterns within the biochemistry of C_4 photosynthesis, within the phylogenetic record of C_4 photosynthesis, and within the fossil record of C_4 photosynthesis. In the fifth group of chapters, the topic of C_4 photosynthesis in relation to human societies is developed. This section includes chapters by Harold Brown and by Nicolaas van der Merwe and Hartmut Tschauner, who focus on the relevance of C_4 photosynthesis to agriculture and the role of C_4 photosynthesis in the development of agrarian human societies. The final chapter of the book, by Rowan Sage, Meirong Li, and Russell Monson, is devoted to the known systematic distribution of C_4 photosynthesis, including a comprehensive list of C_4 taxa.

The production of this book required the energy and wisdom of many collaborators. We especially acknowledge the encouragement of Jim Ehleringer and Bob Pearcy, who initially approached us with the idea of putting together a book like this. Our editor at Academic Press, Chuck Crumly, was both encouraging and patient as we worked through the complexities of soliciting and editing the various chapter manuscripts. We thank the many reviewers of the chapters for their valuable input. We especially thank our families for their patience, support, and understanding during long hours of writing and editing. Above all, we thank those authors who contributed chapters to the book. These individuals represent some of the most enthusiastic students of C_4 photosynthesis, and those that are carrying the spirit of the Canberra conference forward into the next 30 years of research.

ROWAN F. SAGE
RUSSELL K. MONSON

Reference

Hatch, M. D., Osmond, C. B., and Slatyer, R. O., eds. (1971). Photosynthesis and photorespiration. John Wiley and Sons, Inc., New York.

I

Perspectives

1

Why C₄ Photosynthesis?

Rowan F. Sage

I. Introduction

The net acquisition of carbon by photosynthetic organisms is catalyzed by ribulose-1,5-bisphosphate carboxylase/oxygenase (Rubisco) through the carboxylation of RuBP, forming two phosphoglyceric acid (PGA) molecules. In addition to RuBP carboxylation, Rubisco catalyzes a second reaction, the oxygenation of RuBP, producing phosphoglycolate (Fig. 1). Oxygenation is considered a wasteful side reaction of Rubisco because it uses active sites that otherwise would be used for carboxylation, it consumes RuBP, and the recovery of carbon in phosphogylcolate consumes ATP and reducing equivalents while releasing previously fixed CO_2 (Sharkey, 1985). Oxygenation may be inevitable, given similarities in the reaction sequence for oxygenation and carboxylation of RuBP (Andrews *et al.*, 1987). From an evolutionary standpoint, oxygenation could be considered a design flaw that reduces Rubisco performance under a specific set of conditions. Should those conditions ever arise and persist, then opportunities may exist for alternative physiological modes to evolve. C₄ photosynthesis appears to be one such alternative, appearing in response to the prehistoric advent of atmospheric conditions that allowed for significant oxygenase activity and photorespiration (Ehleringer *et al.*, 1991).

II. The Problem with Rubisco

Rubisco evolved early in the history of life, more than 3 billion years ago (Hayes, 1994). The CO_2 content of the atmosphere at this time was orders

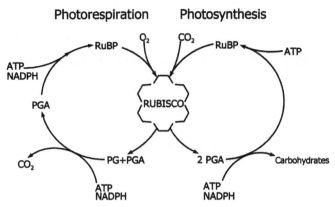

Figure 1 A schematic of the photorespiratory cycle and photosynthesis. Photosynthesis occurs when RuBP is carboxylated by Rubisco, and the products (two phosphoglyceric acid molecules; PGA) are processed into carbohydrates and used to regenerate RuBP in reaction sequences requiring ATP and NADPH. Photorespiration begins with the oxygenation of RuBP to form one phosphoglycolate (PG) and PGA, in a side reaction catalyzed by Rubisco. Processing the phosphoglycolate to PGA and eventually RuBP requires ATP and reducing power (indicated by NADPH).

of magnitude greater than now, and O_2 was rare (Fig. 2A; Kasting, 1987; Holland, 1994). In this environment, oxygenase activity was uncommon, probably less than one oxygenation per billion carboxylations (Fig. 2C). Atmospheric O_2 level remained low and was unable to support an oxygenation rate that was more than 1% of the carboxylation rate until approximately 2 billion years ago. At this time, atmospheric O_2 began to rise, eventually surpassing 200 mbar (20%) about 0.6 billion years ago (Kasting, 1987; Berner and Canfield, 1989; Holland, 1994). Atmospheric CO_2 level continuously declined prior to 1 billion years ago, yet at the advent of the first land plants some 450 million years ago, atmospheric CO_2 was still high enough to saturate Rubisco and minimize oxygenase activity (Fig. 2C). Coal-forming forests of the Carboniferous period (360 to 280 million years ago) contributed to a reduction in CO_2 partial pressures to less than 500 mbar and a rise in O_2 partial pressures to more than 300 mbar (Berner and Canfield, 1989; Berner, 1994). As a consequence, Rubisco oxygenase activity is predicted to have become significant (>20% of the carboxylation rate at 30°C) for the first time in nonstressed plants at the prevailing atmospheric conditions (Fig. 2B,D). After the Carboniferous period, CO_2 levels are modeled to have risen to more than five times current levels for about 200 million years, and the rate of RuBP oxygenation was again a small percentage of the carboxylation rate. Over the past 100 million years, atmospheric CO_2 levels declined from more than 1,000 μbar, eventually

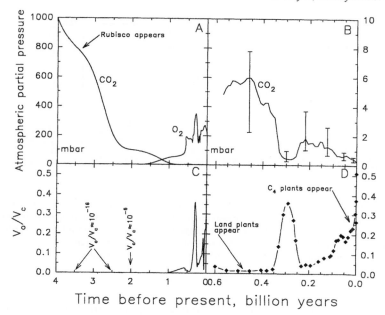

Figure 2 The modeled change in atmospheric CO_2 and O_2 partial pressures over (A) the past 4 billion years (adapted from Kasting, 1987, and Berner and Canfield, 1989); and (B) the past 600 million years (according to Berner, 1994). C and D present the modeled change in Rubisco oxygenase activity (v_o) to carboxylase activity (v_c) at 30°C for the corresponding CO_2 and O_2 levels presented in A and B, respectively, calculated assuming a spinach C_3-type Rubisco according to Jordan and Ogren, 1984.

falling below 200 μbar during the Pleistocene epoch (2 to 0.01 million years ago). In the last 15 million years, Rubisco oxygenase activity is modeled to have risen above 20% of carboxylase activity at 30°C, eventually surpassing 40% of carboxylase activity at the low CO_2 levels (180 μbar) experienced during the late Pleistocene (Fig. 2D). It is only after atmospheric CO_2 levels are low enough to allow the rate of RuBP oxygenation to exceed 20% to 30% of the carboxylation potential that C_4 plants appear in the fossil record (Cerling, Chapter 13). No evidence exists for C_4 photosynthesis during the Carboniferous (Cerling, Chapter 13).

Above 200 mbar oxygen, CO_2 partial pressures of less than 500 μbar pose two problems. First, as a substrate for Rubisco carboxylation, CO_2 availability becomes strongly limiting, reducing the turnover of the enzyme *in vivo* and imposing a limitation on photosynthesis by reducing the capacity of Rubisco to consume RuBP (Fig. 3A; Sage, 1995). Second, O_2 competition becomes significant at warmer temperature (>30°C), and this causes a high rate of photorespiration (Fig. 3B; Sharkey, 1988; Sage, 1995). Whereas a

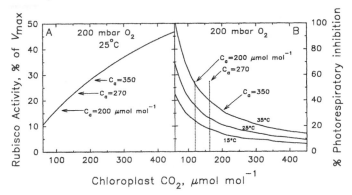

Figure 3 The modeled response of (A) Rubisco activity (as a percent of V_{max}) as a function of stromal CO_2 concentration, and (B) the percent of photorespiratory inhibition of photosynthesis ($0.5v_o/v_c \times 100\%$). Modeled according to Sage (1995) using equations from Farquhar and von Caemmerer (1982). C_a indicates atmospheric CO_2 content corresponding to indicated chloroplast CO_2 concentrations. (Note: at sea level, μmol mol$^{-1} \cong \mu$bar.)

CO_2-substrate deficiency occurs only when the capacity for Rubisco to consume RuBP limits the rate of CO_2 assimilation in plants, photorespiration is inhibitory regardless of whether Rubisco capacity or the capacity of the leaf to regenerate RuBP limits photosynthesis (Sharkey, 1985). At current CO_2 levels, photorespiration can reduce photosynthesis by more than 40% at warmer temperatures (Sharkey, 1988; Ehleringer et al., 1991).

The rise in photorespiratory potential triggered by the increase in the atmospheric O_2:CO_2 over the past 50 million years created high evolutionary pressure for dealing with the consequences of Rubisco oxygenation (Ehleringer et al., 1991). In all higher photosynthetic organisms, however, an elaborate biochemical edifice was already built around Rubisco, such that substantial barriers likely prevented the evolution of a novel carboxylase to replace Rubisco. Not only would the new carboxylase be required, but the accompanying biochemistry to regenerate acceptor molecules and process photosynthetic products would likely have to change as well. Such new photosynthetic systems would then have to compete against preexisting Rubisco-based systems, which though inefficient, would have had the advantage of working reasonably well and of being integrated into the associated cellular, organismal, and ecological systems. To a large degree, the potential for RuBP oxygenation represents a systematic constraint around which evolution must work, an evolutionary spandrel *sensu* Gould and Lewontin (1979) (Somerville et al., 1983). In a situation similar to that of a restoration architect who is constrained by preexisting structures, evolution is constrained by preexisting enzymes, genes, and regulatory systems in dealing with novel challenges (Gould and Lewontin, 1979). In plants and algae, the evolutionary response to declining atmospheric CO_2 levels was to modify existing leaf physiology to create CO_2 concentrating systems ("CO_2 pumps") that were coupled to preexisting Rubisco-based biochemistry. In land plants, the most elaborate and successful of these modifications is C_4 photosynthesis.

III. How C_4 Photosynthesis Solves the Rubisco Problem

The options for dealing with photorespiration are limited in plants employing only C_3 photosynthesis. This can be demonstrated using Eq. 1, which describes the relationship between Rubisco kinetic parameters and the ratio of photorespiratory CO_2 release to photosynthetic CO_2 fixation (Andrews and Lorimer, 1987; Sharkey, 1988).

$$\frac{\text{Photorespiration}}{\text{Photosynthesis}} = \frac{0.5 v_o}{v_c} = 0.5 \left(\frac{1}{S} \times \frac{O}{C} \right) \quad (1)$$

The term v_o is the rate of RuBP oxygenation, v_c is the rate of RuBP carboxylation, S is the specificity of Rubisco for CO_2 relative to O_2, C is the CO_2 concentration in the chloroplast stroma, and O is the O_2 concentration in the stroma. According to Eq. 1, photorespiration can be reduced by changing Rubisco properties to increase S, increasing C, or reducing O. There has been an increase in S from less than 10 mol CO_2 per mol O_2 in primitive bacteria to near 80 in C_3 plants (Table I). In C_3 plants, S shows at most

Table I Range of Rubisco Specificity for CO_2 Relative to O_2 from Various Classes of Organisms[a,b]

Organism type	Rubisco specificity factor
Photosynthetic bacteria (single subunit)	9 to 15
Cyanobacteria	40 to 60
Green algae	50 to 70
C_4 plants	55 to 85
C_3 plants	75 to 85

[a] Values reported here summarize work from the early 1980s. Other groups report specificity factors that are 20% higher, on average, because of different assumptions concerning the pK for the CO_2 to bicarbonate equilibrium (Andrews and Lorimer, 1987).

[b] It has been reported that thermophilic red algae have specificity factors as high as 238 (Uemura et al., 1997). Although this represents an improvement over terrestrial plants, these algae have a very low Rubisco turnover rate, indicating the range of Rubisco specificity in C_3 plants may reflect a balance between photorespiration potential and turnover capacity in the terrestrial environment.

Data from Pierce, J. (1988). Prospects for manipulating the substrate specificity of ribulose bisphosphate carboxylase/oxygenase. *Physiol. Plant* **72**, 690–698.

modest variation (25% or less) and no clear relationship with habitat (Kane et al., 1994; Kent and Tomany, 1995). The fact that in C_3 plants, S clusters around 80 (or 100, depending on the methodology used, see Table I footnote), despite wide ranges in habitat, indicates that evolutionary change in Rubisco specificity may have reached a plateau, at least for terrestrial plants.

No evidence exists that C_3 plants reduce stromal O_2 concentration in order to reduce photorespiration. Theoretically, this would be difficult because gaseous diffusion occurs in response to molar concentration differences. At 210,000 μmol O_2 mol^{-1} air (210 mbar), atmospheric O_2 counteracts any attempt at O_2 depletion by plants. For example, reduction of O_2 in the stroma to 200,000 μmol mol^{-1} would have only a small effect on photorespiration, but would create an inward diffusion gradient for O_2 that is 50 times greater than the maximum CO_2 diffusion gradient (200 μmol mol^{-1}) normally encountered by C_3 plants.

Increasing CO_2 concentration is much easier to accomplish, simply by opening stomates or speeding CO_2 diffusion through the mesophyll cells. Increasing stomatal conductance comes at a high cost in terms of transpiration and reduced water use efficiency, however, and can only raise intercellular CO_2 levels to near atmospheric (Farquhar and Sharkey, 1982). For example, a 25% increase in conductance from 0.4 to 0.5 mol m^{-2} sec^{-1} in a C_3 leaf assimilating CO_2 at 30 μmol m^{-2} sec^{-1} in air (365 μbar CO_2) would raise the transpiration rate by 25% if leaf temperature remained the same. In contrast, intercellular CO_2 levels would increase 5–10% depending on the response of photosynthesis to this conductance change (modeled using standard gas exchange equations and CO_2 responses found in *Chenopodium album*, Sage et al., 1990). Similarly, increasing mesophyll transfer conductance involves high nitrogen investment in carbonic anhydrase (Evans and von Caemmerer, 1996); but again, this can only raise stromal CO_2 levels to near intercellular values.

The only way to increase chloroplast CO_2 above ambient is through expenditure of energy. In aquatic environments of alkaline pH, the high availability of bicarbonate in the surrounding medium is an important opportunity for CO_2 enrichment because the charged HCO_3^{-1} ion can be actively accumulated within membrane-bound compartments (Osmond et al., 1982). Many, if not most, algae actively accumulate bicarbonate, which is converted to CO_2 in the interior of algal cells by carbonic anhydrase (Coleman, 1991). In terrestrial environments, by contrast, this strategy is not a viable option because the aerial environment only provides carbon as freely permeable CO_2.

To most effectively overcome the problems that Rubisco experiences in terrestrial atmospheres, two primary requirements must be met. First, Rubisco should be localized inside a diffusion barrier that restricts efflux of CO_2. Otherwise, any attempt to concentrate CO_2 would result in rapid counter flux against the concentrating system. Second, some way must be developed to move CO_2 into the compartment in which Rubisco is localized. Two systems have appeared to meet these requirements. In some terrestrial C_3 plants, the photorespiratory enzyme glycine decarboxylase is localized in the bundle-sheath tissue or some functional equivalent, to which is transported one of the products of photorespiration (glycine) (Monson, Chapter 11). Following glycine decarboxylation in the bundle-sheath tissues, CO_2 concentration is locally increased, suppressing the rate of RuBP oxygenation in the bundle-sheath cells. Species employing glycine transport loops to reduce photorespiratory CO_2 loss exhibit CO_2 compensation points than can be less than half that of typical C_3 species (Monson, Chapter 11). Many glycine transporters are closely related to C_4 species, leading to the suggestion that compartmentalization of glycine decarboxylation is an important step in the evolution of C_4 photosynthesis (Ehleringer and Monson, 1993).

The most effective means for overcoming photorespiration in terrestrial settings is the C_4 mechanism of CO_2 concentration, involving the carboxylation of PEP and the shuttling of four and three carbon acids between two specialized compartments—the mesophyll tissue, where initial fixation of inorganic carbon occurs, and an inner ring of "bundle-sheath" cells surrounding the vascular bundle (either the bundle sheath proper or, in many species, a functionally similar mestome sheath; Dengler and Nelson, Chapter 5). Unlike C_3 leaves, in which anatomy is modified for high conductance of CO_2 into the chloroplast stroma of mesophyll cells, in C_4 plants, Rubisco is localized in the bundle-sheath tissue, in which CO_2 is released by decarboxylation of four carbon acids (Hatch, Chapter 2; Kanai and Edwards, Chapter 3). As a result of these features, CO_2 concentrations are raised from an intercellular level in the mesophyll tissue of 100 to 200 μmol mol^{-1} to between 1,000 and 3,000 μmol mol^{-1} in the bundle-sheath tissue (von Caemmerer and Furbank, Chapter 6, this volume). Photorespiration is greatly reduced, and CO_2 concentrations in the bundle sheath are maintained at values near what is required to saturate Rubisco.

The enzymes incorporated into the C_4 pathway all have important roles in C_3 plants (Hatch, 1987). PEP carboxylase, for example, is important in stomatal function, pH balance, nitrogen assimilation, carbohydrate metabolism, and osmotic regulation (Cockburn, 1983; Chollet *et al.*, 1996). Thus, no new enzymes appear to have been required in the evolution of C_4 metabolism. Instead, expression of existing biochemistry is modified so that activities of certain enzymes (PEP carboxylase, pyruvate,P_i dikinase, decarboxylating enzymes) are enhanced, whereas for others such as Rubisco, changes in the spatial pattern of enzyme expression occur (Kanai and Edwards, Chapter 3; Leegood and Walker, Chapter 4; Monson, Chapter 11).

Compared to ecologically similar C_3 species, C_4 plants generally exhibit higher photosynthesis rates at low CO_2 and elevated temperature and have higher efficiencies of light, water, and nitrogen use in warm to hot environments (Long, Chapter 7; Knapp and Medina, Chapter 8). These improvements in photosynthetic performance appear to have greatly enhanced the fitness of C_4 grasses in low latitudes, salinized soils, and temperate regions with hot summers and some growth season precipitation, such that C_4 species often dominate these landscapes (Sage, Wedin, and Li, Chapter 10). The C_4 syndrome was so successful that ancient C_3 grasses that once dominated the open tropics are suggested to have been largely replaced by C_4 lineages (Renvoize and Clayton, 1992). Significantly, the rise of most grassland biomes of the world is associated with the geographical expansion of the C_4 syndrome (Cerling, Chapter 13; Renvoize and Clayton, 1992).

IV. Significant Variations on the C_4 Theme

C_4 photosynthesis evolved independently at least 31 times and is now represented by 8,000 to 10,000 species classified in 18 diverse taxonomic families (Kellogg, Chapter 12; Sage, Li, and Monson, Chapter 16). Important variations exist on the basic C_4 theme, reflecting differences in evolutionary origin of the various C_4 species (Dengler and Nelson, Chapter 5; Kellogg, Chapter 12). The primary difference between the varying groups of C_4 species is the decarboxylation step in the bundle sheath. Three major decarboxylation types have been identified: those using NADP-malic enzyme (NADP-ME subtype), those using NAD-malic enzyme (NAD-ME subtype), and those using PEP carboxykinase (PCK subtype) (Hatch, Chapter 2; Kanai and Edwards, Chapter 3). Whereas most C_4 species primarily use only one of the decarboxylating enzymes, a number of species employ a second enzyme in a supporting role; for example, PCK species usually employ NAD-ME in the bundle sheath to generate NADH from C_4 acids imported from the mesophyll. This NADH is then oxidized in the mitochondria to generate ATP for the PCK reaction (Leegood and Walker, Chapter 4). Associated with the type of enzyme used for decarboxylation are a number of additional features that affect whole plant performance. In species using NADP-ME, the import of malate from the mesophyll allows for NADPH formation in the bundle-sheath cells during the decarboxylating step. As a consequence, bundle-sheath chloroplasts in the NADP-ME species require less photosystem II (PSII) activity and therefore less O_2 is evolved in the bundle sheath (Kanai and Edwards, Chapter 3). However, to fully reduce the triose–phosphates generated by CO_2 fixation, NADP-ME subtypes must transport PGA to the mesophyll, where PSII activity is localized (Leegood and Walker, Chapter 4).

Significant correlations in subtype distribution have been observed in response to aridity and human domestication. NAD-ME species tend to dominate the C_4 flora at the arid end of the C_4 distribution range, whereas NADP-ME species predominate at the moist end (Knapp and Medina, Chapter 8; Sage, Wedin, and Li, Chapter 10; Hattersley, 1992; Schulze et al., 1996). In agricultural settings, the large majority of C_4 row crops, pasture plants, and weeds are NADP-ME (Brown, Chapter 14). The mechanistic reasons causing these trends are not apparent because differences in physiological performance associated with the subtypes have not been linked to differential responses to aridity or productive potential. NAD-ME species have, on average, lower quantum yields than NADP-ME and PCK species (Ehleringer and Pearcy, 1983), and these differences may have significance in situations in which water availability affects canopy development and the degree to which leaves are self-shaded.

V. The Consequences of the Evolution of C_4 Photosynthesis

At a minimum, the evolutionary rise and diversification of C_4 photosynthesis in the last 15 to 30 million years enhanced productivity and survival of herbaceous plants in the hot, often dry or nutrient-poor soils of the tropics and subtropics (Hatch, Chapter 2; Brown, Chapter 14). In these environments, C_4 photosynthesis may enable survival in sites too harsh for C_3 vegetation. For example, Schulze et al. (1996) observed that along aridity gradients in Namibia, the presence of the C_4 pathway allows for plant growth in locations that are too xeric to support C_3 plants—at least under the atmospheric CO_2 conditions of today and the recent geological past. Abiotic stress such as drought is not a prerequisite for C_4 dominance, however, because C_4 plants are important in virtually any warm terrestrial habitat, including wetlands, where ecological disturbance minimizes competition from woody C_3 vegetation (Knapp and Medina, Chapter 8; Sage, Wedin, and Li, Chapter 10).

In a larger sense, expansion of C_4-dominated grasslands in the late-Miocene to early Pliocene initiated a series of evolutionary responses in grazing fauna and their carnivore specialists, thereby contributing to the great assemblage of mammals that were once common on the world's savannas (Cerling et al., 1997). McFadden and Cerling (1994), for example, discuss relationships between the expansion of C_4 grasslands during the Pliocene and the evolutionary diversification and success of modern horse lineages. Our own species may well owe its existence to C_4 plants (Cerling, Chapter 13; van der Merwe and Tschauner, Chapter 15). Hominids arose in the mixed C_4 grass–C_3 woodlands of eastern and southern Africa during a period of oscillating climate and atmospheric CO_2 concentration (Raymo,

1991; Stanley, 1995). East Africa was particularly affected by increasing drought, which, in combination with low CO_2, allowed for widespread expansion of C_4-dominated savannas and grasslands (Cerling et al., 1997), which in turn would have created a new suite of evolutionary challenges for hominids of the region (Stanley, 1995). Unlike C_3 woodlands, which provide protective cover and a range of plant foods for hominids to exploit, C_4 grasslands provide little cover and few plant resources. Perennial C_4 grasses generally do not produce roots and tubers for human consumption, and their seeds are small and shatter easily, reducing gathering potential. Grazing animals effectively convert primary productivity of C_4 plants into a rich food resource that would benefit any hominid line able to overcome the challenges of the open grassland and hunt the C_4 herbivores. Thus, a scenario for human evolution is that expansion of C_4 grasslands created a new "niche" that favored many of the traits (large brains, complex social organization, sophisticated weaponry) that distinguishes our evolutionary line from all others (Stanley, 1995).

For modern humans, C_4 plants are major contributors to food production in both the industrialized and developing nations (Brown, Chapter 14). Historically, exploitation of C_4 plants has contributed to the success and expansion of major civilizations. The rise of pre-Columbian city-states in the Americas was due in no small part to maize agriculture (van der Merwe and Tschauner, Chapter 15). Moreover, the adoption of C_4 crops often created revolutions in human society with far-reaching social impacts. The spread of sugarcane to the Caribbean islands in the 1600s, for example, provided Europe with relatively cheap sugar, altering the diets, social customs, and economies of Western Europe. To grow the cane, large numbers of Africans were kidnapped and sold into slavery in the Caribbean, contributing in large part to the social matrix characteristic of the region today (Hobhouse, 1992).

Significantly, most animal protein consumed by humans is dependent on C_4 productivity, because major grass forage species in warm climates are C_4 and grain supplements in animal feed are mainly derived from C_4 row crops (Brown, Chapter 14). For various reasons, C_4 grasses support substantially higher levels of herbivory on average than most C_3-dominated ecosystems (Heckathorn et al., Chapter 9). Because of this, humans will continue to rely heavily on C_4 forage to meet the protein demands of future generations.

VI. C_4 Photosynthesis in the Future

With human modifications of the global environment, an important question is, "How will C_4 plants fare in coming centuries?" Atmospheric

CO_2 enrichment generally favors C_3 species over C_4 species on nutrient enriched soils (Sage, Wedin, and Li, Chapter 10). The overall effect of CO_2 on the C_3/C_4 balance in natural landscapes is uncertain, however, because nutrient limitations restrict the ability of C_3 species to exploit high CO_2. Morever, elevated CO_2 can enhance C_4 performance by restricting stomatal apertures and promoting water savings. This can be important in grasslands, where high CO_2 enhances water relations and growth of C_4 grasses at the expense of C_3 grasses during dry episodes (Owensby *et al.*, 1996). C_4-dominated grasslands are often characterized by high variation in water availability, ranging between wet years, when precipitation is similar to mesic forests, to dry years, when precipitation mimics desert patterns (Knapp and Medina, Chapter 8).

Effects of warming are also difficult to predict because the seasonal timing is as important as the degree of warming. If warming is greater in winter, for example, a cool growing season favorable to C_3 plants may lengthen, whereas the hot summer C_4 season may be attenuated by earlier drought (Sage, Wedin, and Li, Chapter 10). This may be occurring in Alberta, Canada, where C_3 grasses have expanded at the expense of their C_4 neighbors in response to milder winter and springtime temperatures of recent decades (Peat, 1997). Alternatively, summer heating may enable C_4 species to migrate to higher latitudes and altitudes, as long as growth season moisture is available. How CO_2 enrichment and global warming interact with the timing and variation of precipitation may be more important to the change in C_3/C_4 dynamics than direct photosynthetic responses to CO_2 or temperature.

Regardless of how C_4 species respond to atmospheric and climate change, the most important factor in the near future likely will be how C_4 species respond to human activities, which now affect every ecosystem where C_4 plants grow. In this regard, the future of C_4 plants will likely depend on human decisions for the landscape, namely whether forest, pasture, or urbanized uses are implemented. In the tropics, the recent spread of pastures at the expense of forests (Reiners *et al.*, 1994; Dale, 1996) represents the greatest shift in C_3/C_4 ranges in recent history. Further deforestation is predicted, with management for cattle pastures leading to near permanent replacement of C_3 with C_4 landscapes. Conversion of forests to C_4 grasslands has substantial effects well beyond the affected regions—surface albedo, evapotranspiration, carbon sequestration, and trace gas flux to the atmosphere are altered, with potential impacts on the global climate system (Reiners *et al.*, 1994; Dale, 1996).

Despite these educated speculations, there will always be substantial uncertainty about the future of C_4 vegetation. Threshold effects, biological invasions, and species evolution could alter ecological trajectories in unforeseen directions. The most difficult item to predict, however, will likely be management decisions, which will respond to "serendipity and human fickleness" (van der Merwe and Tschauner, Chapter 15), items that no futurist has ever had much success in predicting.

VII. Summary

At the center of photosynthetic carbon fixation is the carboxylation of RuBP by Rubisco, an ancient enzyme originating near the beginning of life. In addition to RuBP carboxylation, Rubisco catalyzes RuBP oxygenation, leading in turn to wasteful photorepiratory metabolism. For most of the history of life on Earth, photorespiration was insignificant because atmospheric CO_2 levels were high enough to suppress RuBP oxygenation. During this time, an elaborate Rubisco-based photosynthetic system predominated in all oxygenic photosynthetic organisms. In recent geological periods, atmospheric conditions changed with the rise of O_2 partial pressures to 210 mbar and the reduction of CO_2 partial pressures to below 500 μbar. To a large degree, plants appeared constrained by their history of Rubisco-based photosynthesis. Other than modifying the characteristics of Rubisco to increase CO_2 specificity, the primary solution to the photorespiratory problem appears to have been to couple a carbon-concentrating system onto Rubisco-based carbon metabolism. In terrestrial plants, the carbon concentrating solution was the C_4 system, based on preexisting biochemistry commonly used in pH balance, intercellular transport, and osmotic adjustment. The consequence of this evolutionary development is just now being appreciated, as the rise of many faunal groups, including humans, has been linked to the C_4 solution to photorespiratory inhibition.

Acknowledgments

I thank Jarmila Pittermann, Russ Monson, and Gerry Edwards for comments and advice in the preparation of this chapter.

References

Andrews, T. J. and Lorimer, G. H. (1987). Rubisco: structure, mechanisms, and prospects for improvement. *In* "The Biochemistry of Plants" Vol. 10 (M. D. Hatch and N. K. Boardman, eds.), pp. 131–218. Academic Press, New York.

Berner, R. A. (1994). 3Geocarb II: A revised model of atmospheric CO_2 over Phanerozoic time. *Amer. J. Sci.* **291**, 339–376.

Berner, R. A., and Canfield, D. E. (1989). A new model for atmospheric oxygen over Phanerozoic time. *Amer. J. Sci.* **289**, 333–361.

Cerling, T. E., Harris, J. M., MacFadden, B. J., Leakey, M. G., Quade, J., Eisenmann, V. and Ehleringer, J. R. (1997). Global vegetation change through the Miocene/Pliocene boundary. *Nature* **389**, 153–158.

Chollet, R., Vidal, J. and O'Leary, M. H. (1996). Phosphoenolpyruvate carboxylase: A ubiquitous, highly regulated enzyme in plants. *Annu. Rev. Plant Physiol. Plant Mol. Biol.* **47**, 273–298.

Cockburn, W. (1983). Stomatal mechanism as the basis of the evolution of CAM and C_4 photosynthesis. *Plant Cell Environ.* **6**, 275–279.

Coleman, J. R. (1991). The molecular and biochemical analyses of CO_2-concentrating mechanisms in cyanobactria and microalgae. *Plant Cell Environ.* **14**, 861–867.

Dale, V. H. (1997). The relationship between land-use change and climate change. *Ecol. Appl.* **7**, 753–769.

Ehleringer, J. and Pearcy, R. W. (1983). Variation in quantum yield for CO_2 uptake among C_3 and C_4 plants. *Plant Physiol.* **73**, 555–559.

Ehleringer, J. R. and Monson, R. K. (1993). Evolutionary and ecological aspects of photosynthetic pathway variation. *Annu. Rev. Ecol. Syst.* **24**, 411–439.

Ehleringer, J. R., Sage, R. F., Flanagan, L. B. and Pearcy, R. W. (1991). Climate change and the evolution of C_4 photosynthesis. *Trends Ecol. Evol.* **6**, 95–99.

Evans, J. R. and von Caemmerer, S. (1996). Carbon dioxide diffusion inside leaves. *Plant Physiol.* **110**, 339–346.

Farquhar, G. D. and Sharkey, T. D. (1982). Stomatal conductance and photosynthesis. *Annu. Rev. Plant Physiol.* **33**, 317–345.

Gould, S. J. and Lewontin, R. C. (1979). The spandrels of San Marco and the Panglossian paridigm: A critique of the adaptationist programme. *Proc. R. Soc. Lond. B* **205**, 581–598.

Hatch, M. D. (1987). C_4 photosynthesis: A unique blend of modified biochemistry, anatomy and ultrastructure. *Biochim. Biophys. Acta* **895**, 81–106.

Hattersley, P. W. (1992). C_4 photosynthetic pathway variation in grasses (Poaceae): Its significance for arid and semi-arid lands. In "Desertified Grasslands: Their Biology and Management" (G. P. Chapman, ed.) Linnean Society Symposium Series No. 13. Academic Press, San Diego.

Hayes, J. M. (1994). Global methanothophy at the Archean–Proterozoic transition. In " Early Life on Earth" (S. Bengston, ed.), pp. 220–236. Columbia University Press, New York.

Hobhouse, H. (1992). "Seeds of Change." Revised Ed., Papermac, London.

Holland, H. D. (1994). Early Proterozoic atmospheric change. In " Early Life on Earth" (S. Bengston, ed.), pp. 237–244. Columbia University Press, New York.

Jordan, D. B. and Ogren, W. L. (1984) The CO_2/O_2 specificity of ribulose-1,5-bisphosphate carboxylase/oxygenase—dependence on ribulosebisphosphate concentration, pH and temperature. *Planta* **161**, 308–313.

Kasting, J. F. (1987). Theoretical constraints of oxygen and carbon dioxide concentrations in the precambrian atmosphere. *Precambrian Res.* **34**, 205–299.

Kane, H. J., Viil, J., Entsch, B., Paul, K., Morell, M. K. and Andrews, T. J. (1994). An improved method for measuring the CO_2/O_2 specificity of riboulosebisphosphate carboxylase–oxygenase. *Aust. J. Plant Physiol.* **21**, 449–461.

Kent, S. S. and Tomany, M. J. (1995). The differential of the ribulose 1,5-bisphosphate carboxylase/oxygenase specificity factor among higher plants and the potential for biomass enhancement. *Plant Physiol. Biochem.* **33**, 71–80.

MacFadden, B. J. and Cerling, T. E. (1994). Fossil horses, carbon isotopes and global change. *Trends Ecol. Evol.* **9**, 481–486.

Osmond, C. B., Winter, K. and Ziegler, H. (1982). Functional significance of different pathways of CO_2 fixation in photosynthesis. In "Encyclopaedia of Plant Physiology, New Series, Physiological Plant Ecology II" Vol. 12B (O. L. Lange, P. S. Nobel, B. Osmond, and H. Zeigler, eds.), pp. 480–547. Springer-Verlag, Berlin.

Owensby, C. E., Ham, J. M., Knapp, A., Rice, C. W., Coyne, P. I. and Auen, L. M. (1996). Ecosystem-level responses of tallgrass prairie to elevated CO_2. In "Carbon Dioxide and Terrestrial Ecosystems" (G. W. Kock and H. A. Mooney, eds.), pp. 147–162. Academic Press, San Diego.

Peat, H. C. L. (1997). Dynamics of C_3 and C_4 Productivity in Northern Mixed Grass Prairies. M. Sc. Thesis. University of Toronto, Ontario, Canada.

Pierce, J. (1988). Prospects for manipulating the substrate specificity of ribulose bisophosphate carboxylase/oxygenase. *Physiol. Plant.* **72**, 690–698.

Raymo, M. E. (1992). Global climate change: A three million year perspective. In "Start of a Glacial" NATO SAI series 1: Global Environmental Change, Vol. 3, (G. J. Kukla and E. Went, eds.), pp. 208–223. Springer-Verlag, Berlin.

Reiners, W. A., Bouwman, A. F., Parsons, W. F. and Keller, M. (1994). Tropical rain forest conversion to pasture changes in vegetation and soil properties. *Ecol. Appl.* **4**, 363–377.

Renvoize, S. A. and Clayton, W. D. (1992). Classification and evolution of the grasses. In "Grass Evolution and Domestication" (G. P. Chapman, ed.), pp. 3–37. Cambridge University Press, Cambridge.

Sage, R. F. (1995). Was low atmospheric CO_2 during the Pleistocene a limiting factor for the origin of agriculture. *Global Change Biol.* **1**, 93–106.

Sage, R. F., Sharkey, T. D. and Pearcy, R. W. (1990). The effect of leaf nitrogen and temperature on the CO_2 response of photosynthesis in the C_3 dicot *Chenopodium album* L. *Aust. J. Plant Physiol.* **17**, 135–148.

Schulze, E.-D., Ellis, R., Schulze, W., Trimborn, P. and Ziegler, H. (1996). Diversity, metabolic types and $\delta^{13}C$ carbon isotope ratios in the grass flora of Namibia in relation to growth form, precipitation and habitat conditions. *Oecologia* **106**, 352–369.

Sharkey, T. D. (1985). O_2-insensitive photosynthesis in C_3 plants. Its occurrence and a possible explanation. *Plant Physiol.* **78**, 71–75.

Sharkey, T. D. (1988). Estimating the rate of photorespiration in leaves. *Physiol. Plant.* **73**, 147–152.

Somerville, C., Fitchen, J., Somerville, S., McIntosh, L. and Nargang, F. (1983). Enhancement of net photosynthesis by genetic manipulation of photorespiration and RuBP carboxylase/oxygenase. In "Advances in Gene Technology: Molecular Genetics of Plants and Animals" (K. Downey, R. W. Voellmy, F. Ahmad, and J. Schultz, eds.). pp. 296–309. Academic Press, San Diego.

Stanley, S. M. (1995). Climatic forcing and the origin of the human genus. In "Effects of Past Global Change of Life" pp. 233–243. Board on Earth Sciences and Resources Commission on Geosciences, Environment, and Resources, National Research Council, National Academic Press, Washington, D. C.

Uemura, K., Anwaruzzaman, M., Miyachi, S, Yokota, A. (1997). Ribulose-1,5-bisphosphate carboxylase/oxygenase from thermophilic red algae with a strong specificity for CO_2 fixation. *Biochem. Biophys. Res. Comm.* **233**, 568–571.

2

C_4 Photosynthesis: A Historical Overview

Marshall D. Hatch

I. Introduction

The editors have asked me to give an account of my recollections of the events surrounding the discovery of C_4 photosynthesis and what I see to be the key developments in the elucidation of the process in the intervening years. I do this by first considering in a more or less chronological way the scientific scene and critical events prior to and surrounding the discovery of this process. This is followed by an account of the key developments leading to the initial resolution of its mechanism and likely function, and finally, the general recognition of the C_4 process. This would take us to about 1970–1971. It was at that time that an international meeting was held in Canberra, Australia, to consider the latest developments in photosynthesis and photorespiration (proceedings, Hatch *et al.*, 1971). That meeting was a watershed for the recognition and acceptance of the C_4 option for photosynthesis by the broader plant research community; I mention it frequently.

Rather than continue this simple chronological catalog of events beyond 1970, I thought it would be more useful and readable to cover subsequent developments as a series of case histories of selected topics. In considering these topics I hope that most of the seminal discoveries and events critical to the development of the C_4 story will be highlighted. Generally, my remarks are confined to research and events that occurred before 1990. The bias is toward biochemistry because that is my prime interest and I am sure that there is undue emphasis on the work from my laboratory, partly reflecting my failing memory. I tried to resist delving into the kind of detail that would be more appropriate to the following chapters. As a result, important work will have been overlooked or failed to make the cut, and I aplogize to those concerned.

II. The Prediscovery Scene (Before 1965)

By the early 1960s the mechanism of the photosynthetic process operating in the algae *Chlorella* and *Scenedemus* was essentially resolved (Calvin and Bassham, 1962), and a similar process had been demonstrated in some higher plants. At that time there was no reason to doubt that this process, termed the *Calvin cycle* or more commonly now, the *photosynthetic carbon reduction (PCR) cycle*, would account for CO_2 assimilation in all autotrophic organisms. However, being wise after the fact, we can now see that a keen and observant reader of the plant biology literature at that time might reasonably have suspected that a different photosynthetic process could be operating in one particular group of higher plants. This was a group of grasses, including species like maize and sorghum all in the Panicoideae subfamily, that shared a range of unique or unusual anatomical, ultrastructural, and physiological features all related in one way or another to the photosynthetic process (Table I). The first of these unusual features identified was the specialized Kranz leaf anatomy, featuring two distinct types of photosynthetic cells (Haberlandt, 1884). For convenience, biochemists at least refer to these two cell types as *mesophyll* and *bundle sheath cells*, although technically, these terms are not always anatomically correct (see Chapter 5). Later, Rhoades and Carvalho (1944) noted that, in these particular species with Kranz anatomy, starch accumulated only in the bundle sheath

Table I Unique or Unusual Features Shared by Some Tropical Grasses Now Known to Be C_4

Feature	Reference
Specialised Kranz anatomy	Haberlandt, 1884
Bifunctional chloroplasts/cells	Rhoades and Carvalho, 1944
Dimorphic chloroplasts	Hodge *et al.* 1955
High water-use efficiency	Shantz and Piemeisel, 1927
Low CO_2 compensation point	Meidner, 1962; Moss, 1962
High growth rates	Loomis and Williams, 1963
Leaf photosynthesis	
High maximum rates and light saturation	Hesketh and Moss, 1963; Hesketh, 1963
High temperature optima	Murata and Iyama, 1963; El-Sharkaway and Hesketh, 1964
Low postillumination CO_2 burst	Tregunna, Krotkov, and Nelson, 1964

chloroplasts and speculated that carbon may be assimilated in mesophyll cells but stored in bundle sheath cells. Then, with the development of the electron microscope, Hodge et al. (1955) showed that the bundle sheath chloroplasts of maize had an unusual internal membrane structure with single unappressed thylakoid membranes and an almost total absence of grana.

Earlier, Shantz and Piemeisel (1927) had conducted a wide survey of species for water-use efficiency. What went unnoticed at that time was that the plants they examined fell into two distinct groups with regard to this parameter. For grass species, a low efficiency group lost between 600 and 900 g of water per gram of dry matter produced compared with another group, including several species of the Panicoideae, which lost water in the range of only 260 to 340 g per gram of dry matter produced. Much later, there was evidence that some of these particular grasses had a high potential for rapid growth (Loomis and Williams, 1963). However, most critical were a series of observations made about photosynthesis itself, largely made possible by the recent development of gas exchange technology. For instance, representatives of this group of grasses showed very low CO_2 compensation points (Meidner, 1962; Moss, 1962) and a very low postillumination CO_2 burst (Tregunna et al., 1964). These features were coupled with unusually high leaf photosynthesis rates, only saturating near full sunlight (Hesketh, 1963; Hesketh and Moss, 1963) and high temperature optima for photosynthesis (Murata and Iyama, 1963; El-Sharkaway and Hesketh, 1964). However, this remarkable coincidence of unusual or unique features in this subgroup of grasses went largely unnoticed at that time, and so did not provide the clue leading to the recognition of the C_4 process. Not until several years later were these features recognized as being invariably linked with the operation of C_4 photosynthesis. The discovery of this process began another way.

III. Discovery: Radiotracer Evidence (1954–1967)

Although PGA was the major early labeled product of $^{14}CO_2$ assimilation in the species studied by Calvin and collaborators, they recognized that significant label appeared in dicarboxylic acids such as malate under some conditions (see Calvin and Bassham, 1962). The C_4 story begins with a brief unreferenced comment in the 1954 Annual Report of the Hawaiian Sugar Planters Association Experiment Station (HSPAES), based on labeling from $^{14}CO_2$, that "it seems clear that in sugarcane phosphoglycerate is not the major carrier of newly fixed carbon that it is found to be in some algae." Over the next 6 years there appeared further brief reports in the HSPAES Annual Reports (1956–1960) identifying malate and aspartate as major early labeled products. A third labeled compound, originally identified as phosphomalate, was later shown to be 3-phosphoglycerate (PGA). These results were also mentioned in an abstract of a report to the annual meeting of the Hawaiian Academy of Science by Kortschak, Hartt, and Burr. Later, these data were briefly mentioned again by Burr in a paper on the use of radioisotopes in the Hawaiian sugar plantations at a Pacific Science Congress. These proceedings were published (Burr, 1962).

It was many years before we were to learn that, about that time, a young Russian scientist had reported his studies on the labeling patterns obtained when maize leaves assimilated $^{14}CO_2$ (Karpilov, 1960). These studies, reported in detail in the proceedings of a Russian agricultural research institute, clearly showed malate and aspartate as major early labeled products; according to the translation he made the conservative conclusion that the results were "not characteristic of other plant species." A joint paper 3 years later seemed to confuse the picture by discussing artifactual effects of killing leaves in boiling ethanol (Tarchevskii and Karpilov, 1963). He did not publish again on the mechanism of photosynthesis in maize for several years (Karpilov, 1969). It was only about that time that Roger Slack and I first became aware of Karpilov's earlier work and, needless to say, the Hawaiian and Russian workers would have been unaware of each other's work throughout most of this period.

In the early part of the 1960s Roger Slack and I were working in the laboratory of the Colonial Sugar Refining Company in Brisbane, Australia, on aspects of carbohydrate metabolism in sugarcane. That laboratory maintained regular contact with the Hawaiian laboratory. As a result, we were aware of their work on $^{14}CO_2$ labeling, briefly mentioned in their Annual Reports, and had often discussed its possible implications. It was not until 1965 that they finally published their results in a detailed and accessible form (Kortschak et al., 1965). The kinetics of radiolabeling they reported were clearly consistent with malate and aspartate being labeled from $^{14}CO_2$ before PGA and showed the rapid labeling of PCR cycle intermediates and products only after longer exposure to $^{14}CO_2$. They concluded that, "In sugarcane carbon assimilation proceeds by a path qualitatively different from many other plants."

At that time Roger Slack and I were both at the end of current research projects. So, over a glass or two of beer at a scientific meeting in Hobart later that year, we decided to follow up this Hawaiian work to see if we could understand what it all meant. Our results on time-course labeling from $^{14}CO_2$ (Hatch and Slack, 1966) were very similar to those reported by the Hawaiian group. These results were also similar to those reported for maize by Karpilov (1960) but, as already mentioned, another 3 years were to pass before we were to learn about that work. We extended the studies on sugarcane in several ways. For instance, we showed (1) that the chemically unstable dicarboxylic acid, oxaloacetate, was labeled at the same time as aspartate and malate, (2) that C_4 acids were initially labeled in the C-4 carboxyl carbon, and (3) by pulse–chase labeling that this C-4 carboxyl gives rise to the C-1 carboxyl of PGA and then hexose phosphates in a way consistent with the operation of the PCR cycle (Hatch and Slack, 1966).

From these studies we developed a simple working model for the path of carbon assimilation for photosynthesis in sugarcane (Fig. 1). This model proposed the carboxylation of a 3-carbon compound, either phosphoenolpyruvate (PEP) or pyruvate, giving a C_4 dicarboxylic acid (oxaloacetate, malate, or aspartate) with the fixed carbon in the C-4 carboxyl. Following the transfer of this C-4 carbon to provide the C-1 carboxyl of PGA, it was proposed that the remaining 3-carbon compound could act as a precursor for the regeneration of the primary CO_2 acceptor. At that time, we favored the idea that the transfer of the C-4 carbon to PGA proceeded by a transcarboxylation reaction (Hatch and Slack, 1966). This, of course, turned out to be incorrect; the reason for discounting the more obvious decarboxylation–refixation mechanism at that time was the likely inefficiency of this process through diffusive loss of CO_2. Unfortunately, our erroneous conclusion that these grasses, like sugarcane and maize, contained only low levels of Rubisco (Slack and Hatch, 1967; discussed in more detail later in this

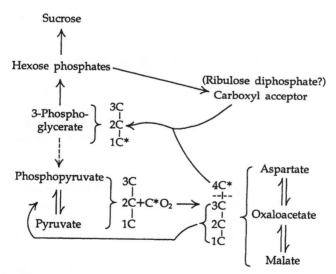

Figure 1 Model for C_4 photosynthesis based on the kinetics of labelling and the distribution of label in intermediates and products after assimilation of $^{14}CO_2$ by sugarcane leaves [Reprinted with permission from Hatch, M. D., and Slack, C. R. (1966). Photosynthesis by sugar cane leaves. A new carboxylation reaction and the pathway of sugar formation. *Biochem. J.* **101**, 103–111].

chapter) appeared to add support to our proposal of a transcarboxylation reaction.

In a following paper (Hatch et al., 1967) we confirmed that similar novel labeling of C_4 acids occured in sugarcane with leaves of different age and when light and CO_2 were varied. We also made the exciting discovery that this distinctive early labeling of C_4 acids from $^{14}CO_2$ occurred in a number of other grass species from different tribes. Of course, many other plants tested showed classical PGA-dominated labeling consistent with the operation of the conventional PCR cycle (Hatch et al., 1967). Significantly, a sedge (*Cyperus*, family Cyperaceae) also showed classical C_4 acid labeling. In the same year, Osmond (1967) reported similar labeling in a dicotyledonous species, *Atriplex spongiosa*. By this time we had convinced ourselves, at least, that a substantially different process was operating to assimilate CO_2 in these species. General recognition was to take a few more years.

About that time we named this process the *C_4 dicarboxylic acid pathway of photosynthesis*, and this became abbreviated to C_4 pathway or C_4 photosynthesis. It is interesting to note that these latter terms were widely used by the time of the international meeting on photosynthesis held in Canberra in 1970 (see Hatch et al., 1971), along with the terms C_3 photosynthesis and C_4 and C_3 plants. However, the only publication I could find preceding that meeting where such terms were used was Osmond et al. (1969) who referred to C_3-type and C_4-type plants and photosynthesis.

IV. Mechanism, Functions, and Recognition (1965–1970)

The model for C_4 photosynthesis based on our initial radiotracer studies (Fig. 1) provided the basis for a variety of predictions about the possible enzymes involved. There followed a cyclic process in which enzymes were identified leading to more predictions. A search for the primary carboxylating enzyme revealed two contenders, PEP carboxylase and NADP malic enzyme (Slack and Hatch, 1967). Both of these enzymes were at least 20 times more active in the species showing C_4 acid labeling than in other species. We favored PEP carboxylase for this role because of the low activity of NADP malic enzyme in the carboxylating direction relative to photosynthesis. Only later was the decarboxylating function of NADP malic enzyme recognized.

With PEP as the likely CO_2 acceptor, it seemed that these C_4 species should be capable of converting pyruvate to PEP. After ruling out pyruvate kinase as an option, we ultimately showed that PEP formation in these species was due to a new and unique enzyme pyruvate,P_i dikinase (Hatch and Slack, 1968), and that this enzyme was light–dark regulated (Slack,

1968; Hatch and Slack, 1969a). The reaction is shown in the following equation:

$$\text{Pyruvate} + \text{ATP} + P_i \rightleftarrows \text{PEP} + \text{AMP} + PP_i.$$

This discovery, together with the resolution of the unusual features of this reaction and the regulation of the enzyme involved, is considered in more detail in Section IV.C. With the resolution of this mechanism it became clear that the products AMP and PP_i must be rapidly metabolized. This soon led to the discovery of very high activities of adenylate kinase and pyrophosphatase in C_4 leaves, with activities at least 40 times higher than those in leaves of C_3 plants (Hatch et al., 1969).

With oxaloacetic acid (OAA) identified as the product of the primary carboxylation reaction there was clearly a need for a reaction to reduce this acid to malate, the major C_4 acid detected in $^{14}CO_2$ labeling experiments. After reasoning that chloroplast-generated NADPH was the most likely reductant, we subsequently discovered the then unique NADP-specific malate dehydrogenase (Hatch and Slack, 1969b). It was reasonable to think that this could be a C_4-specific enzyme, so we were surprised at the time to find the enzyme was also present in leaves of C_3 plants; however, it was up to 10 times more active in the C_4 plants we were studying at that time. As discussed in more detail later in this chapter, this enzyme also turned out to be light–dark regulated (Johnson and Hatch, 1970).

The years 1969 and 1970 saw the resolution of some major problems and questions associated with the operation of the C_4 pathway. At the International Botanical Congress held in Seattle in the summer of 1969, Roger Slack and I met for the first time others who were looking at some of these problems—in particular, the role of Rubisco in C_4 plants and the inter- and intracellular locations of this and other photosynthetic enzymes. Especially important was our contact with Olle Björkman, who told us that they had found substantial Rubisco activity in C_4 leaves, about half that in C_3 leaves, and that this activity was confined to bundle sheath cells. These studies, then in press (Björkman and Gauhl, 1969), also explained our lower Rubisco activities (see previously mentioned material) by showing that few bundle sheath cells are broken with the blending procedure we used. It turned out that this provided an excellent method for isolating intact bundle sheath cells, and this procedure was refined and first used for studies of bundle sheath cell metabolism in Clanton Black's laboratory (Edwards et al., 1970; Edwards and Black, 1971). In fact, much of our current knowledge of bundle sheath cell metabolism was generated using these so-called bundle sheath cell strand preparations (see Hatch, 1987, and more recent papers).

Complementing these studies from Björkman and Gauhl were our studies showing by nonaqueous fractionation of leaves that pyruvate,P_i dikinase,

NADP malate dehydrogenase, and most of the adenylate kinase and pyophosphatase in maize leaves was located in mesophyll chloroplasts. It was a surprise to us at that time that these chloroplast contained little or no Rubisco; it was confined to bundle sheath chloroplasts along with NADP malic enzyme and most of the other PCR cycle enzymes (Slack, 1969; Slack et al., 1969).

These data, taken together with the data of Björkman and Gauhl (1969) and also Berry et al. (1970), provided the final clues to how the chloroplasts in the two types of cells must act cooperatively to finally assimilate CO_2. The clear inference was that CO_2 fixed into malate in mesophyll cells is released again in bundle sheath cells via NADP malic enzyme and then refixed via Rubisco and the PCR cycle. The rest fell into place with the decarboxylation product, pyruvate, cycling back to mesophyll cells, where it is converted to PEP via pyruvate,P_i dikinase in the mesophyll chloroplasts. Figure 2 shows this view of the mechanism of C_4 photosynthesis presented (Hatch, 1971a) at an international meeting on photosynthesis held in Canberra in late 1970 (see proceedings, Hatch et al., 1971). By the time of this meeting it could be said for the first time that the C_4 pathway was not only discovered, but also generally recognized and accepted by the wider plant research community.

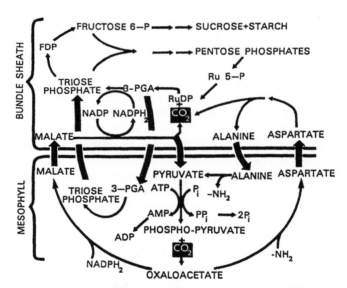

Figure 2 A scheme for C_4 photosynthesis in species like sugarcane and maize: a 1970 view [From Hatch, M. D. (1971a). Mechanism and function of C_4 photosynthesis. In "Photosynthesis and Photorespiration" (M. D. Hatch, C. B. Osmond, and R. O. Slayter, eds.), pp. 139–152. Wiley-Interscience, New York].

V. Selected Aspects of the C_4 Story (1970–   )

A. The Biochemical Subgroups of C_4

The C_4 species initially examined in some detail happened to be NADP-malic enzyme type (NADP-ME-type). As just indicated, by 1970 it was possible to propose a detailed scheme to account for photosynthesis in this group of species (Hatch, 1971a) and this interpretation has remained essentially unaltered (see Kanai and Edwards, Chapter 3, this volume). However, it was already clear by that time that there were other species with C_4-like radiolabeling and anatomical features but with little of the decarboxylating enzyme, NADP malic enzyme. Thus, there were apparently at least two biochemically distinct groups of C_4 plants. This became even more evident when it was shown (Andrews *et al.*, 1971; Hatch, 1971a) that the species lacking NADP malic enzyme also contained lower levels of NADP malate dehydrogenase but much higher levels of aspartate and alanine aminotransferases (Table II). So the questions were, how were C_4 acids decarboxylated in these species, and was aspartate more directly involved in the process?

Soon after, this problem was partly resolved when Edwards *et al.* (1971) reported high levels of PEP carboxykinase in some but not all species lacking high NADP malic enzyme. This raised interesting problems because the substrate, oxaloacetate, occurs in such low levels in leaves and because of the energetic implications of the ATP requirement. In fact, it was to take more than 20 years to largely resolve the mechanism of photosynthesis in this C_4 group, generally called phosphoenolpyruvate carboxykinase (PCK)-type species (Burnell and Hatch, 1988; Carnal *et al.*, 1993).

The immediate outcome of this discovery was that there must be at least three groups of C_4 species because there were several species lacking high levels of NADP malic enzyme and PEP carboxykinase. The resolution to that problem came when, by chance, I heard at a meeting that plant

Table II Enzyme-Based Subdivision of C_4 Species into Two Groups: 1970

	Activity (μmol min^{-1} mg^{-1} Chl)[a]			
Subgroup	NADP malic enzyme	NADP malate dehydrogenase	Aspartate aminotransferase	Alanine aminotransferase
1	10–12	9–14	4–5	0.3–0.7
2	0.1–0.8	0.2–4	22–41	16–32

[a] The range of activities for a number of species are shown. Data from Andrews, T. J., Johnson, H. S., Slack, C. R., and Hatch, M. D. (1971). Malic enzyme and aminotransferases in relation to 3-phosphoglycerate formation in plants with the C_4 dicarboxylic acid pathway of photosynthesis. *Phytochemistry* **10**, 2005–2013.

mitochondria from various nonleaf tissues had been shown to contain significant activities of an NAD-specific malic enzyme. Our testing for this activity among those species with low activities of the two identified decarboxylases soon revealed levels some 50 times higher than the activities present in leaves of C_3 plants or, indeed, other groups of C_4 plants (Hatch and Kagawa, 1974). Subsequent surveys showed that most, if not all, C_4 species fit into one of these three groups defined by the principal decarboxylating enzyme (Gutierrez et al., 1974; Hatch et al., 1975). These groups were named NADP-ME-type, PEP-carboxykinase-type (PCK-type), and NAD-malic enzyme-type (NAD-ME-type) (Hatch et al., 1975), and the mechanism proposed at that time to account for C_4 acid decarboxylation in NAD-ME-types (Hatch and Kagawa, 1974) has been fully supported by subsequent studies (see Hatch, 1987).

One should specially acknowledge the contributions of Gerry Edwards and colleagues during the 1970s to aspects of the resolution of this three-subgroup division of C_4 species. These included surveys of a large number of C_4 species representing the three subgroups for the inter- and intracellular distribution of enzymes and the metabolic and photosystem activities associated with mesophyll and bundle sheath cells (see Edwards et al., 1976; Edwards and Huber, 1981; Edwards and Walker, 1983). Steve Huber was a key collaborator in many of these studies.

The resolution of the details of C_4 acid decarboxylation in PCK-type species turned out to be much more difficult. For some time, efforts were confused by the erroneous conclusion that PEP carboxykinase was located in the bundle sheath chloroplasts. Such a location left no problem regarding the source of ATP to drive the reaction (i.e., oxaloacetate + ATP → PEP + CO_2 + ADP). However, when this enzyme was clearly shown to be cytosolic (Ku et al., 1980; Chapman and Hatch, 1983), the critical question to answer was the source of the ATP. Resolution of this problem came with a series of studies (Burnell and Hatch, 1988; Hatch et al., 1988; Carnal et al., 1993) showing that aspartate decarboxylation via oxaloacetate and PEP carboxykinase was linked with the phosphorylation-coupled respiration of malate in bundle sheath mitochondria. The important thing to note was that this malate oxidation occurred via NAD malic enzyme, so that this reaction would contribute at least 25% of the CO_2 released in bundle sheath cells, and more if less than 3 ATP molecules are generated for each NADH oxidised in mitochondria (see Kanai and Edwards, Chapter 3, this volume).

I suspect that the photosynthetic process in most C_4 species will be adequately described by one or other of these three mechanisms, described in detail in an earlier review (Hatch, 1987; also see Kanai and Edwards, Chapter 3, this volume). These schemes could be termed "classical" NADP-ME-type, NAD-ME-type, and PCK-type mechanisms. However, one should add a cautionary note. Although it is convenient to keep these groupings of C_4 plants and the associated metabolic schemes, it should be acknowledged that significant variations on these themes may occur. The only certain way of identifying such deviations from the classical pathways is by detailed pulse–chase analysis of $^{14}CO_2$ fixation, and this has been done for only a relatively few species. Inferential evidence could be provided by analysis of the activities and intercellular location of enzymes and photoreactions. An example of such a deviant species is the NADP-ME-type dicot *Flaveria bidentis*. This species has high aspartate and alanine aminotransferase activities, more typical of the other two groups of C_4 plants, combined with substantial NADP-MDH and photosystem-2 activity in bundle sheath cells (Meister *et al.*, 1996). These features are not typical of the grass species on which the classical NADP-ME-type scheme was based (see Hatch, 1987) and suggest that aspartate as well as malate may contribute to the flux of C_4 acids to bundle sheath cells. This conclusion was supported by radiotracer analysis of carbon flux through C_4 acids during steady-state leaf photosynthesis (Meister *et al.*, 1996). Whether this is common in dicots of this group is not clear. At least based on analysis of the levels and distribution of enzyme and photosystem-2 activities, aspartate may contribute to carbon flux in some NADP-ME-type grass species as well (Edwards and Walker, 1983).

B. The CO_2-Concentrating Function and Cell Permeability

As already mentioned, by 1970 it was clear that, for the NADP-ME-type C_4 plants studied at that time, malate is decarboxylated in bundle sheath cells and the released CO_2 refixed via Rubisco and the associated PCR cycle. During the international meeting held in Canberra in late 1970 it was suggested that the special reactions of the C_4 pathway may serve to concentrate CO_2 in bundle sheath cells (Björkman, 1971; Hatch, 1971a). It is interesting to note that although this speculation was correct, the perceived purpose of concentrating CO_2 to overcome an apparently very low CO_2 affinity of Rubisco was not. This idea of concentrating CO_2 received support from the later evidence that large pools of CO_2/HCO_3 develop in C_4 leaves in the light (Hatch, 1971b; Furbank and Hatch, 1987). However, it was not until the discovery of the oxygenase activity of Rubisco, the competitive nature of this reaction and its link with photorespiration, that the full implications of concentrating CO_2 in bundle sheath cells of C_4 plants were realized (Chollet and Ogren, 1975; Hatch and Osmond, 1976).

By the time of these reviews in the mid-1970s, it was clearly understood that high bundle sheath CO_2 would suppress the Rubisco oxygenase reaction and reduce or eliminate photorespiration. This provided the basis for explaining some of the C_4-specific features listed in Table I, such as the low CO_2 compensation point, lack of a postillumination burst of CO_2, and the high potential for photosynthesis. Later, the apparently low photorespiration in C_4 plants also provided an explanation for why the quantum yields for C_4 plants are higher than those for C_3 plants at higher temperatures but not at lower temperatures (Ehleringer and Pearcy, 1983; Jordan and Ogren, 1984).

The plant cell wall–membrane interface normally offers little resistance to the diffusion of gases like CO_2 and O_2. Following acceptance of the decarboxylation–refixation mechanism of C_4 photosynthesis, the next key question was how might this operate without unacceptable diffusive loss of CO_2? With the recognition of the additional energy cost of $2ATP/CO_2$ fixed for the C_4-mediated transfer of CO_2 to bundle sheath cells (Hatch and Slack, 1970), it followed that substantial leakage of CO_2 would reduce photosynthetic efficiency. However, quantum requirement measurements at limiting light indicated that such losses were not large (Ehleringer and Björkman, 1977; Ehleringer and Pearcy, 1983). The quantitative relationship between the extent of CO_2 leakage and the quantum requirement for C_4 photosynthesis has been considered by Farquhar (1983) and more recently by Furbank et al. (1990).

Laetsch (1971) was one of the first to note that most C_4 grass species contained a suberin lamella imbedded in the cell wall between mesophyll and bundle sheath cells and to pose the question of whether this might be responsible for limiting the loss of CO_2 from bundle sheath cells. There is presumptive evidence that such a suberin layer may limit the movement of gases like CO_2 (see Hatch and Osmond, 1976). However, it soon became apparent that this could not be the only explanation, because a number of C_4 grass species lack this suberin layer (Hattersley and Browning, 1981) as do apparently all C_4 dicot species. The dilemma is that these species seem to be almost as efficient at retaining CO_2 as judged by quantum yield measurements or direct measurement of the CO_2 permeability of bundle sheath cells (see later this chapter). The only clue at this stage is that species with a suberin lamella generally have bundle sheath chloroplasts and mitochondria arranged in a centrifugal position within the cell, that is, adjacent to the mesophyll cells. By contrast, in those species lacking this layer in the mesophyll–bundle sheath cell wall, the chloroplasts and mitochondria are invariably arranged in a centripetal position; that is, against the inner wall. Hattersley and Browning (1981) have suggested that this arrangement may substitute for suberin as a means of limiting diffusive loss of CO_2. However, we are no closer now to an experimental resolution of the question of how CO_2 leakage is minimized in these species than we were in the early 1980s.

It has proved difficult to devise procedures for measuring either the permeability of the mesophyll–bundle sheath interface to CO_2, the actual CO_2 concentration in bundle sheath cells, or the leak of stored CO_2 from bundle sheath cells. These have been key questions since the early 1970s.

Not until relatively recently have estimates been made of the permeability of the mesophyll–bundle sheath interface to CO_2 (see Jenkins et al., 1989; Brown and Byrd, 1993), the possible cell CO_2 concentrations (Jenkins et al., 1989; Dai et al., 1995), and the percentage of CO_2 leaking from bundle sheath cells (Farquhar, 1983; Henderson et al., 1992; Hatch et al., 1995). Estimates of the bundle sheath CO_2 concentration under normal atmospheric conditions suggest a value of about 50 μM, which is more than 10 times the mesophyll concentration. However, concentrations may be lower in developing or senescing leaves (Dai et al., 1995). These analyses involve procedures that are indirect or technically difficult, and it seems unlikely that the last word has been said about the measurement of these various parameters (see following chapters).

Another interesting aspect of this CO_2 concentrating mechanism is the compromises and trade-offs that had to be made in optimizing cell permeability. As previously discussed (Hatch, 1987), one compromise was between making cells permeable enough to permit the essential fluxes of metabolites to sustain the C_4 process without allowing unacceptable rates of CO_2 leakage. The second compromise centers around the need to develop a cell that minimizes leakage of CO_2 without unduly limiting the escape of photosystem-2–generated O_2 from bundle sheath cells. For one model of bundle sheath CO_2/O_2 exchanges during photosynthesis, the computed bundle sheath $[O_2]$ required to give an efflux rate matching O_2 production was nearly twice the atmospheric O_2 concentration (Jenkins et al., 1989). However, smaller gradients, and hence a lower bundle sheath O_2 concentration, would be required if the resistance to gas diffusion is lower (see Dai et al., 1995). Of course, for those NADP-ME-type species with low photosystem-2 activity in bundle sheath cells, $[O_2]$ in these cells would presumably be similar to air levels.

The critical intercellular flux of metabolites, referred to previously, deserves some further comment. As the C_4 story unfolded, the apparent need for rapid flux of C_4 acids, C_3 compounds, and other metabolites between cells was initially of great concern. The flux rates required were much greater than those commonly recorded for plant cells (see Hatch, 1987). Laetsch (1971) was first to note the profusion of plasmodesmata linking mesophyll and bundle sheath cells in C_4 leaves and to link this with the demands for metabolite flux. A quantitative model based on plasmodesmatal-mediated diffusion of metabolites supported the plausibility of such a system (Osmond and Smith, 1976; Hatch and Osmond, 1976). A range of subsequent studies with isolated bundle sheath cells, prepared as strands surrounding vascular tissue, clearly demonstrated that these cells are highly permeable to a range of metabolites (Edwards et al., 1970; Hatch and Kagawa, 1976; see Hatch, 1987). Later, a procedure was developed to directly measure the permeability of isolated bundle sheath cells to metabolites (Weiner et al., 1988) giving permeability coefficients averaging about 3 μmol min^{-1} mg^{-1} Chl for a 1 mM gradient. This would suggest that gradients of only about 2 to 3 mM between mesophyll and bundle sheath would be sufficient to give fluxes matching maximum photosynthesis rates. Measurement of cell metabolite concentrations gave somewhat higher apparent concentration gradients (Stitt and Heldt, 1985).

C. Pyruvate,P$_i$ Dikinase and Other Light–Dark Regulated Enzymes

1. Pyruvate,Pi Dikinase As already mentioned, we had predicted that C$_4$ leaves should be able to convert pyruvate rapidly to PEP, and the subsequent discovery of pyruvate,P$_i$ dikinase (Hatch and Slack, 1968) was critical to the further development of our ideas on the mechanism of C$_4$ photosynthesis. It is interesting to recount the role good fortune played in this important discovery and the subsequent resolution of the mechanism and regulation of this enzyme. It is also difficult to think of another enzyme with more unique features connected with its mechanism and regulation. Our first attempts at detecting ATP-dependent conversion of pyruvate to PEP yielded activities of only about 2% of that required to support maximum rates of photosynthesis. Of course, we did not know then how much the odds were stacked against us. For instance, it transpired that pyruvate,P$_i$ dikinase required a thiol and Mg^{2+} for stability, NH$_4^+$ was an essential cofactor for catalysis, and P$_i$ was a cosubstrate (see Edwards et al., 1985). Furthermore, it was one of those odd enzymes that is inactivated in the cold and, in addition, we were unaware that the enzyme was rapidly inactivated in leaves kept in low light or darkness (see later this chapter).

Fortunately, our plants were growing in the phytotron next to our laboratory so that the enzyme was probably only partially dark-inactivated if immediately extracted. More important, we were inadvertently adding NH$_4^+$ and P$_i$ with the coupling enzyme PEP carboxylase, which had been eluted from a column with phosphate and then stored in (NH$_4$)$_2$SO$_4$. There is little doubt that without these accidental additions we would not have detected this unique activity at that time, and this surely would have slowed the resolution of the C$_4$ process enormously. As these matters were recognized and other problems overcome, activities increased severalfold to levels matching the photosynthesis rate (Hatch and Slack, 1968, 1969a).

Also of historic interest is the fact that we initially believed that this PEP synthesizing activity in C$_4$ leaves was catalyzed by an enzyme (PEP synthase) that had just previously been described in bacteria by Cooper and Kornberg (1965). In this reaction, ATP phosphorylates pyruvate but the products are AMP + P$_i$ rather than ADP. Our early analysis with the partially purified leaf enzyme suggested a similar reaction because we were inadvertently adding P$_i$, and the true product, PP$_i$, was being hydrolyzed by a contaminating pyrophosphatase. Of course, as finally resolved, the reaction in leaves turned out to be essentially unique in the annals of enzymology in that two substrates, pyruvate and P_i, are phosphorylated from one molecule of ATP (Hatch and Slack, 1968; equation in Section IV).

No less unique was the remarkable mechanism of dark–light regulation of pyruvate,P_i dikinase. Roger Slack discovered this dark–light effect on this enzyme while pretreating maize leaves in dark or light to modify the density of chloroplasts (Slack, 1968). Later, we showed that the extracted dark-inactivated enzyme could be reactivated by incubation with P_i (Hatch and Slack, 1969a). This followed from the observation that the dark-inactivated enzyme was commonly seen to partially activate in the standard assay system. A search of the reaction components revealed P_i to be the critical activating factor. In the course of testing reaction components we also discovered that the enzyme in Sephadex G-25–treated leaf extracts was inactivated when ADP was added. Significantly, this inactive enzyme was also activated by subsequent treatment with P_i (Hatch and Slack, 1969a). All this finally led, many years later, to the elucidation of the remarkable mechanism for inactivation–activation of pyruvate,P_i dikinase (see Burnell and Hatch, 1985). The key features of this mechanism include: (1) the unprecedented ADP-mediated phosphorylation of a threonine residue to inactivate the enzyme; (2) the fact that a prerequisite for this inactivation is the phosphorylation of a catalytic site histidine, thus explaining the requirement for ATP as well as ADP for inactivation; (3) reactivation by an unprecedented phosphorylitic removal of the threonine phosphate; and (4) the rare involvement of a single protein catalyzing these unique and mechanistically different processes of activation and inactivation.

2. *NADP Malate Dehydrogenase* As already mentioned, NADP-MDH was one of the two new enzymes discovered during the course of elucidating the C_4 pathway (Hatch and Slack, 1969b). Soon after, this enzyme was shown to be dark–light regulated in leaves and, like several PCR-cycle enzymes, this apparently involved interconversion between an active dithiol form of the enzyme and an inactive disulfide form (Johnson and Hatch, 1970). It is interesting to recall that some of the subsequent developments in the elucidation of this regulatory process. In 1977, we showed that a low-molecular-weight protein factor was required for both the thiol-mediated activation and the O_2-mediated inactivation of the enzyme (Kagawa and Hatch, 1977). Later it became obvious that this protein was, in fact, thioredoxin-m (Jacquot *et al.*, 1984), one of the family of thioredoxins involved in this type of thiol redox process. Not surprisingly, the degree of activation of NADP-MDH in leaves was related to the redox state of the electron transport system (Hatch, 1977; Nakamoto and Edwards, 1986). However, the actual mechanism was quite complex. The prevailing NADPH/NADP ratio had a strong influence through effects on the activation and inactivation processes as well as via an inhibitory effect of NADP on the reaction itself (Ashton and Hatch, 1983; Rebeille and Hatch, 1986). As a result, a high degree of activation of NADP-MDH was only possible with a prevailing high ratio of NADPH to NADP. Subsequently, it was shown that activation requires the reduction of two disulfide bridges and the precise location of these in the amino acid sequence has been determined (see Chapter 4 for recent developments).

3. PEP Carboxylase The identification of PEP carboxylase as the primary carboxylating enzyme of C_4 photosynthesis (Slack and Hatch, 1967) followed soon after the initial radiotracer studies. However, the full appreciation of the regulatory processes operating on this enzyme came much later. Uedan and Sugiyama (1976) were first to purify the C_4 enzyme, and they elaborated on early reports of cooperativity with respect to PEP binding and activation by glucose 6-phosphate. The enzyme was also inhibited by its product oxaloacetate (Lowe and Slack, 1971) and by malate and aspartate (Huber and Edwards, 1975). The latter paper provided some resolution of a variety of conflicting observations on the inhibitory effects of these C_4 acids. However, it was difficult to see how these effects could account for the complete suppression of PEP carboxylase in the dark, although the likely need for such down regulation was apparent. Surprisingly, it took another 10 years for the first clues to emerge that such regulation occurred (Karabourniotis *et al.*, 1985; Huber and Sugiyama, 1986; Doncaster and Leegood, 1987) and that this was based on a phosphorylation mechanism (Nimmo *et al.*, 1987; Jiao and Chollet, 1988). How could these effects of light be missed for so long? It was largely the result of assaying PEP carboxylase at high pH with excess of PEP and the activator glucose 6-phosphate. Under these conditions there are only small differences in activity between the "light" and "dark" forms of the enzyme. Subsequent developments in the regulation of PEP carboxylase are considered in Chapter 4.

D. Metabolite Transport

By 1975, the schemes drawn up to account for C_4 photosynthesis implied the operation of a variety of metabolite transport processes into and from both chloroplasts and mitochondria (Hatch *et al.*, 1975). These were remarkable not only for their variety but also for the rapid rates required. Most were stoichiometrically linked to the photosynthetic process, therefore requiring fluxes 5 to 10 times those previously described for metabolite transport in chloroplasts or mitochondria. The question was, were these unique transport systems or simply modifications of those operating in leaves of C_3 plants? Huber and Edwards (1977a,b) were the first to directly address this question demonstrating transport of pyruvate and PEP in mesophyll chloroplasts. Subsequent studies suggested that most of these transport processes were similar to those already described in C_3 leaves (see Hatch, 1987). However, it should be emphasized that many of these transport systems have not been studied in detail.

There are apparently two unusual transport systems, and the case of sodium-dependent pyruvate transport is particularly interesting. Many years ago Brownell and Crossland (1972) showed that sodium was an essential micronutrient for the growth of several C_4 species but not C_3 species. Subsequent attempts to explain this remarkable observation by identifying the site of action of sodium met with only limited success (Brownell, 1979). Then in 1987, in the course of studying the uptake of pyruvate by mesophyll chloroplasts, Ohnishi and Kanai (1987) reported that sodium was essential for this process, possibly via a Na^+/pyruvate symport mechanism. Only later was the possible link between these observations made (Boag et al., 1988). Subsequent studies clearly confirmed that this Na^+-dependent pyruvate uptake was almost certainly the basis for the Na dependency for growth of C_4 plants. This depended on showing that in certain C_4 species, including maize, that did not require Na^+ for growth (Brownell, 1979), pyruvate was taken up by a process driven by a proton gradient instead of a sodium gradient (Aoki et al., 1992).

The other exception is the phosphate translocator of C_4 mesophyll chloroplasts, which has been modified to effectively transport PEP as well as P_i, PGA, and triosephosphates. Presumptive evidence that PEP was being moved on the same translocator as these other compounds (Huber and Edwards, 1977b; Day and Hatch, 1981) has been confirmed by a variety of subsequent studies (Gross et al., 1990). The current status of metabolite transport associated with C_4 photosynthesis is considered in Chapter 3.

E. C_3–C_4 Intermediates

During the 1975 Steenbock Symposium in Madison, I recall wondering with Harold Brown why so few plants with features intermediate between C_3 and C_4 species had been found. At that meeting, Harold Brown reported his analysis of the apparently intermediate nature of the species *Panicum milioides* (Brown and Brown, 1975). I should mention here that, in the following years, Harold Brown went on to make many critical contributions in areas relating the physiology and biochemistry of both C_4 species and C_3–C_4 intermediates. At that time, only one other species with such intermediate features was known, *Mullogo verticillata* (Kennedy and Laetsch, 1974). It seemed that if C_4 plants evolved from C_3 plants in a multistep process, then some of these intermediate stages should have stabilized and survived. In addition, there was the possibility of viable progeny of crosses between C_3 and C_4 species, already demonstrated to be a practical possibility by Björkman (1976).

It turned out that there were more C_3–C_4 intermediates but not many more. Currently, only about 24 species, representing eight different genera in six families, are known (Rawsthorne, 1992). Significantly, all but two of the genera also include C_3 and C_4 species. In the 10 years from 1975 there was a period of collecting physiological and biochemical statistics on the growing number of intermediates. By the time of a review of this field in 1987 by Edwards and Ku (1987), it was clear that the great majority of these species assimilated CO_2 essentially via the C_3 mode, but generally had a somewhat lower CO_2 compensation point and, apparently, reduced photorespiration. Edwards and Ku considered how this might be the result of reassimilation of photorespiratory CO_2 and the advantage of releasing this CO_2 from mitochondria, which were specially abundant in bundle sheath cells and surrounded by chloroplasts. About that time, others directly demonstrated that a greater percentage of photorespiratory CO_2 is refixed in C_3–C_4 intermediates relative to C_3 plants (Bauwe *et al.* 1987; Hunt *et al.* 1987).

The critical discovery that made sense of all this came soon after when Hylton *et al.* (1988) showed that, in the C_3–C_4 intermediates, glycine decarboxylase was exclusively located in the mitochondria of the bundle sheath cells. This clearly provided the opportunity for elevating the concentration of CO_2 in bundle sheath cells and therefore reducing the contribution of the oxygenase reaction for at least part of the Rubisco complement in these leaves. Such a stable anatomical/biochemical arrangement would provide a foundation for the next evolutionary step towards C_4 photosynthesis. In fact, some other C_3–C_4 intermediates had indeed advanced more closely to C_4, especially among *Flavaria* (Monson *et al.*, 1986, 1987).

F. Carbon Isotope Discrimination in C_4

The fact that C_4 plants discriminate less than C_3 plants against the heavier isotopes of carbon during CO_2 assimilation was an important element in the development of the C_4 story. The characteristically higher ratio of ^{13}C to ^{12}C of C_4 plants has been widely used to identify C_4 and C_3 species in broad-ranging surveys (Smith and Brown, 1973; see Farquhar *et al.*, 1989). Differences in this ratio have had a range of other uses, including assessing the degree of C_3–C_4 intermediacy of species (Edwards and Ku, 1987), providing evidence for expansion of C_4 plants in geological time (Cerling, Chapter 11, this volume), and assessing the extent of CO_2 leakage during C_4 photosynthesis (Farquhar, 1983; Henderson *et al.*, 1992).

Bender (1968) was first to recognize that higher plants fall into two distinct groups on the basis of the ratio of ^{13}C to ^{12}C in their organic carbon and that this was related to the operation of C_4 or C_3 photosynthesis. This discovery arose through an interest in carbon-14 dating for archeological purposes and early observations that corn cobs and kernals have a higher

^{13}C to ^{12}C ratio than tissues of a wide variety of plant species (see Bender, 1968). Bender went on to survey a number of grass species and showed the clear link between the higher ^{13}C to ^{12}C ratio and the taxonomic group of grasses we had previously identified as C_4 (see Hatch et al., 1967). Of course, plants also discriminate against ^{14}C and, as Bender (1968) recognized, this had implications for carbon-14 dating in that a new correction would have to be applied for material originating from C_4 plants to avoid an error of about 200 years.

Later, this difference in ^{13}C to ^{12}C ratio was shown to hold for C_3 and C_4 dicots (Tregunna et al., 1970) and for the various organic constituents of C_3 and C_4 plants (Whelan et al., 1970). However, some algae and gymnosperms also showed higher ratios similar to C_4 plants (Smith and Epstein, 1971). The bases for these differences in carbon isotope ratios between plants fixing CO_2 via C_3 and C_4 pathways, and also via crassulacean acid metabolism (CAM), have been examined together with the effects of varying environmental conditions (see Farquhar et al., 1989).

G. Random Recollections, Anecdotes, and Epilog

1. Controversies and Conflicts Conflicts and disagreements are bound to arise in any field of research. As frustrating as they may be at the time, they usually turn out, in retrospect, to benefit the field as the controversy concentrates minds and efforts on the inadequacy of the prevailing evidence. C_4 photosynthesis has had its share. I have already mentioned the "low Rubisco-transcarboxylase" story where we got off on the wrong track, and also the confusion about the location of PEP carboxykinase. There were some others that deserve a brief mention.

As I mentioned already, by the time of the meeting held in Canberra in late 1970, there was quite strong support coming from several laboratories for the interpretations put forward at that time to account for C_4 photosynthesis in NADP-ME-type grasses at least (Hatch, 1971a). I thought there was reasonably widespread acceptance of these interpretations at that time. However, in the following 3 years a number of papers appeared from one laboratory leading to claims that, in sugarcane and some other NADP-ME-type species, PEP carboxylase was largely confined to nongreen tissues, Rubisco and the PCR-cycle were located in mesophyll cells, and bundle sheath cells functioned only for synthesis and storage of starch (see Coombs and Baldry, 1972). This radically different interpretation of C_4 photosynthesis, appearing in high-profile journals, created considerable confusion, especially among the onlookers of the developing C_4 pathway story. There followed soon after a number of papers (see Hatch and Osmond, 1976) clearly supporting the original interpretations of the C_4 mechanism. Especially important among these was the paper from Ku et al. (1974) that was a direct response to these claims, using the same species as the Coombs group. I went to the International Botanical Congress in Leningrad in 1975 expecting fireworks, but by then the controversy was over.

For some time around 1970 and for the next few years there was a not-insignificant controversy about the levels of photosystem-2 in the bundle sheath chloroplasts of NADP-ME-type species like maize, sorghum, and sugarcane. Decarboxylation of malate via NADP malic enzyme should provide about half the total NADPH necessary for the subsequent reduction of PGA formed in bundle sheath cells of these species (see Kanai and Edwards, Chapter 3). The question was, where was the remaining NADPH generated and how much PGA might need to be shuttled to mesophyll cells for reduction? Initial determinations suggested low photosystem-2 levels in bundle sheath cell chloroplasts of species like maize and sorghum and presumably, a low capacity for light-dependent NADP reduction (Downton et al., 1970; Anderson et al., 1971). This was linked with the lack or deficiency of grana in these chloroplasts. The situation was then confused by reports that these bundle sheath chloroplasts contained substantial photosystem-2 activity and could photoreduce NADP if plastocyanin was added to link the two photosystems (Bishop et al., 1971; Smillie et al., 1972). These authors proposed that these chloroplasts could be competent for NADP reduction *in vivo* and that a linking factor may have been lost during isolation. However, various subsequent analyses indicated that the capacity of these bundle sheath cells for light-dependent NADP reduction was indeed limited, and it was also clear that the chloroplasts were at least highly deficient in other photosystem-2 associated activities (see Edwards et al., 1976; Chapman et al., 1980). Notably, however, these and later studies also revealed that some other grass species belonging to this NADP-ME-type group do show significant bundle sheath photosystem-2 activity. Furthermore, as recently shown, the bundle sheath chloroplasts of some NADP-ME-type dicotyledonous C_4 species have a substantial capacity for light-dependent NADP reduction, and this is associated with the increased involvement of aspartate in the transfer of carbon to bundle sheath cells (Meister et al., 1996).

Since the mid-1970s there have been conflicting views on whether C_4 plants have distinct advantages over C_3 plants in terms of potential growth rates and dry matter production. By the earlier 1970s the collective results of a variety of studies on plant and crop productivity strongly suggested that C_4 plants were capable of higher rates of primary productivity than C_3 plants. This conclusion was questioned by Gifford (1974) whose analyses led to the conclusion that, although C_4 plants had a higher potential for photosynthesis, there was no apparent difference in growth rates between C_3 and C_4 crop plants. He argued that any differences were likely due to the longer growing season for C_4 plants giving higher annual yield. Later, Monteith (1978) argued strongly against this conclusion and questioned the basis of the earlier analysis. His analysis came up with a range for the four highest recorded short-term growth rates for C_3 plants of 34–39 g m^{-2} day^{-1} compared with 50–54 g m^{-2} day^{-1} for C_4 plants.

There followed occasional reports of C_4 plants setting new growth records until the whole question was raised again by Snaydon (1991). He concluded that, "when direct comparisons are made of the productivity of C_3 and C_4 species, there is no consistent difference in productivity between the two groups." What was difficult to reconcile with this conclusion was the fact that in the wide range of data analyzed in that study, 11 of the 12 most productive species in the sample were C_4. Furthermore, the average for the top 10 C_4 species was about 72 tonnes ha^{-1}year^{-1} compared with 37 tonnes ha^{-1}year^{-1} for the top 10 C_3 species. Similar differences were seen when the maximum rates of short-term dry matter production were compared (see Loomis and Gerakis, 1975; Monteith, 1978). As these authors point out, this advantage of C_4 is most evident when the C_4 species are grown in tropical latitudes and disappears at higher latitudes. Clearly, to assess potential growth and productivity of C_3 and C_4 species one should take data only for plants growing under near optimal environmental conditions. Under these conditions, the results clearly show that C_4 photosynthesis provides plants with the potential for higher productivity. Furthermore, this potential will not be evident in all C_4 plants, many of which have evolved to exploit the advantages of the C_4 process for survival under arid or salty conditions rather than for rapid growth.

2. People and Events: Early Days Some random recollections of people and their exploits might be in order at this point. I have already related the most fortunate circumstances surrounding our entry into the C_4 business. I was also most fortunate to have formed the liaison with Roger Slack to persue this work. Compared with that exciting first 5 years up to 1970, with new discoveries every few months, the rest seems like hard work. We were quite isolated from the mainstream plant science community for much of that period, which, in retrospect, I believe was a good thing. This allowed us to focus on the task at hand without the unproductive distractions that might otherwise have been generated. However, this only works up to a point and we were both very much enlightened in many ways by the contacts we made on an overseas trip in mid-1969 that included the International Botanical Congress in Seattle. Among those we met in Seattle was Olle Björkman, who set the record straight for us on Rubisco in C_4 plants (see Section IV). Later, Olle Björkman contributed a great deal to the developing C_4 story with a mammoth study of the progeny of crosses between C_3 and C_4 *Atriplex* species and a wide range of comparative studies on the physiology of C_3 and C_4 species (see Björkman, 1976). He was the epitome of concentration when in the middle of a research project; I can vouch for the fact that he was liable to ignore his mail for weeks on end.

Roger Slack was a delightful colleague as well as a thinker and a "doer." Besides sharing opposite sides of the same bench we had in common bad tempers and incessant smoking. Roger smoked a pipe and I smoked everything that burnt, in the laboratory as well of course—how things have changed! On at least two occasions Roger set himself alight by failing to extinguish a used match before putting it back into the same box (an odd habit), thus igniting the remaining matches just as he got the box back in his pocket.

We kept in contact with Hugo Kortschak (see Section III, this chapter) during the late 1960s and beyond. The last time was at the Rank Prize ceremony in Britain in 1981. I don't recall discussing with him why they postponed publication of their work for so long. However, I gathered from Andy Benson that, in part, this may have been due to a less than enthusiastic response when Burr and Hartt spoke of their results during contacts with Calvin and colleagues in the mid to late 1950s. Andy Benson recalled some of these visits at a recent symposium on C_4 photosynthesis held in Canberra. Apparently, the Hawaiians were discouraged by the reaction that their data could be interpreted as a simple extension of the significant C_4 acid labeling commonly seen in *Chlorella* and other species under certain conditions.

I met the Russian Yuri Karpilov at the 1975 International Botanical Congress held in Leningrad (now St. Petersburg), the one who 15 years earlier had described the labeling patterns resulting from $^{14}CO_2$ assimilation by maize leaves (see Section III). He was kind enough to invite me out to dinner; we conversed through an interpreter and consumed a great deal of vodka between reminiscences about fate and good fortune in science. I was shocked to hear that he was killed soon after in a bicycle accident in Moscow.

Clanton Black was an important early player in the C_4 field, together with Gerry Edwards, who was originally on a postdoctoral fellowship in his laboratory. Clanton and colleagues wrote an important paper correlating various unusual physiological features of the plants distinguished by using C_4 photosynthesis as well as pointing out that they featured prominently in the list of the world's worst weeds (Black et al., 1969). He was also involved with Gerry Edwards in pioneering the separation of mesophyll and bundle sheath cells from *Digitaria* leaves (see Section IV) and the analysis of the metabolic activities of those cells (see Black, 1973). All those who attended the international meeting on photosynthesis in Canberra at the end of 1970 will remember Clanton Black's masterly demonstration of the preparation of these cells. There was a lively discussion at the time about whether the cigar ash he dropped in the mortar while grinding the leaves was critical to the success of the operation. Gerry Edwards has remained a major contributor to the developing C_4 story over the intervening years. A remarkable amount of important work came out of his small basement laboratory in Madison. His research has spanned from the biochemistry of the C_4 process, its regulation, the photochemistry of C_4 plants, through to a variety physiological studies, including more recent work on photorespiration in C_4 and related matters.

"Mac" Leatsch and Barry Osmond both made critical contributions in the early development of the C_4 story. As mentioned already, Barry was especially concerned with the biochemical and physiological aspects of photorespiration in C_4 (see Hatch *et al.*, 1971) as well as the intercellular movement of metabolites (Osmond, 1971; Osmond, and Smith, 1976). He was one of the important contributors to the developing broad picture of C_4 interrelationships. Laetsch was particularly concerned with structure–function studies associated with the specialized anatomy and chloroplast structure of C_4 leaves. He contacted us in late 1967 with information on anatomy and ultrastructure for the newly discovered C_4 species and also suggestions and questions (see Laetsch and Price, 1968). We remained in regular contact over the next few years. His contributions on the anatomical and ultrastructural features of C_4 plants were an important component of the 1970 photosynthesis meeting held in Canberra, and he was a major contributor to the free-flowing discussions. Leatsch was also physically formidable, distinguishing himself at one stage of a midconference excursion by removing, single-handed, a large post from the roadside to allow the bus we were traveling in to complete a U-turn. He then put the post back.

3. *Epilog* Many of the major questions raised by the early 1970s have now been largely resolved. Among the more important of these were the nature of the biochemical options for C_4 photosynthesis, the mechanism of concentrating CO_2 in bundle sheath cells and its advantages, the nature of the intercellular transport of metabolites, the mechanism of dark–light regulation of pyruvate,P_i dikinase and NADP malate dehydrogenase, the taxonomic diversity of the C_4 process, and the basis of the different physiological and performance features characteristic of C_4 plants.

Much of the effort is now on developmental aspects and the associated regulation of gene expression. Molecular techniques will also aid in understanding the regulatory processes operating *in vivo* and such aspects as the evolution of the process. What is also needed is for people with the breadth of skills, interests, and focus of Paul Hattersely (see Hattersely, 1992) to continue to collate and make functional and evolutionary sense of the vast amount of taxonomic, structural, physiological, and ecological data relating to the C_4 process and its biochemical subtypes. Interrelated with this is the question of the origins and significance of the C_3–C_4 intermediate species (see Monson, Chapter 11, this volume). I would also like to know what the rates of CO_2 leakage are for different C_4 species. Estimates so far vary widely (see Section V.B.). There is also the question of how CO_2 is effectively retained in those species lacking a suberin lamella in the bundle sheath cell wall. Another question concerns the relationship, if any, between the varying quantum yields of different C_4 species and differences in the CO_2 leak rate.

What remarkable good fortune it was to become, by chance, associated with this extraordinary C_4 process and to be able to continue this liaison over the years. The so-called C_4 syndrome has added a new dimension to the basics of photosynthesis, plant physiology, and plant productivity. This whole field of plant biology would have been so much less exciting and interesting if the Cretaceous decline in atmospheric carbon dioxide, which apparently provided the pressure for C_4 evolution (see Ehleringer et al., 1991), had never occurred.

VI. Summary

With the advantage of hindsight, an account of the scientific scene leading to the discovery of the C_4 photosynthetic process is given. The events surrounding the discovery are described, together with the key developments leading to the initial resolution of the mechanism and function of this process. The subsequent highlights of the developing C_4 story are then considered in a series of "case histories" dealing with particular aspects of the process. In particular, I identify critical papers that initiated key developments and discuss where ideas originated. There are some personal reminiscences and also some comments on key figures in the early development of the C_4 process.

References

Anderson, J. M., Woo, K. C., and Boardman, N. K. (1971). Photochemical systems in mesophyll and bundle sheath chloroplasts of C_4 plants. *Biochim. Biophys. Acta* **245**, 398–408.

Andrews, T. J., Johnson, H. S., Slack, C. R., and Hatch, M. D. (1971). Malic enzyme and aminotransferases in relation to 3-phosphoglycerate formation in plants with the C_4 dicarboxylic acid pathway of photosynthesis. *Phytochemistry* **10**, 2005–2013.

Aoki, N., Ohnishi, J., and Kanai, P. (1992). Two different mechanisms for transport of pyruvate into mesophyll chloroplasts of C_4 plants—a comparative study. *Plant Cell Physiol.* **33**, 805–809.

Ashton, A. R., and Hatch, M. D. (1983). Regulation of C_4 photosynthesis: Regulation of activation and inactivation of NADP malate dehydrogenase by NADP and NADPH. *Arch. Biochem. Biophys.* **227**, 416–424.

Bauwe, H., Keerberg, O., Bassuner, R., Parnik, T., and Bassuner, B. (1987). Reassimilation of carbondioxide by *Flaveria* (Asteraceae) species representing different types of photosynthesis. *Planta* **172**, 214–218.

Bender, M. (1968). Mass spectrometric studies of carbon 13 variations in corn and other grasses. *Radiocarbon.* **10**, 468–472.

Berry, J. A., Downton, W. J. S., and Tregunna, E. B. (1970). The photosynthetic carbon metabolism of *Zea mays* and *Gomphrena globosa:* The location of the CO_2 fixation and carboxyl transfer reactions. *Can. J. Bot.* **48,** 777-786.
Bishop, D. G., Andersen, K. S. and Smillie, R. M. (1971). Incomplete membrane-bound photosynthetic electron transfer pathway in agranal chloroplasts. *Biochem. Biophys. Res. Commun.* **42,** 74-81.
Björkman, O. (1971). Comparative photosynthetic CO_2 exchange in higher plants. *In* "Photosynthesis and Photorespiration" (M. D. Hatch, C. B. Osmond, and R. O. Slatyer, eds.), pp. 18-32 Wiley-Interscience, New York.
Björkman, O. (1976). Adaptive and genetic aspects of C_4 photosynthesis. *In* "CO_2 Metabolism and Plant Productivity" (R. H. Burris and C. C. Black, eds.), pp. 287-309. University Park Press, Baltimore.
Björkman, O., and Gauhl, E. (1969). Carboxydismutase activity in plants with and without β-carboxylation photosynthesis. *Planta* **88,** 197-203.
Black, C. C. (1973). Photosynthetic carbon fixation in relation to net CO2 uptake. *Annu. Rev. Plant Physiol.* **24,** 253-286.
Black, C. C., Chen. T. M., and Brown, R.H. (1969). Biochemical basis for plant competition. *Weed Sci.* **17,** 338-344.
Boag, T. S., Brownell, P. F., and Grof, C. P. (1988). The essentiality of sodium resolved? *Life Sci. Adv.* **7,** 169-170.
Brown, R. H., and Brown, W. V. (1975). Photosynthetic characteristics of *Panicum milioides*, a species with reduced photorespiration. *Crop Sci.* **15,** 681-685.
Brown, R. H., and Byrd, G. T. (1993). Estimation of bundle sheath cell conductance in C_4 species and O_2 insensitivity of photosynthesis. *Plant Physiol.* **103,** 1183-1188.
Brownell, P. F. (1979). Sodium as an essential micronutrient element for plants and its possible role in metabolism. *Adv. Bot. Res.* **7,** 117-224.
Brownell, P. F., and Crossland, C. J. (1972). The requirement of Na^+ as a micronutrient by species having the C_4 decarboxylic acid pathway of photosynthesis. *Plant Physiol.* **49,** 794-797.
Burnell, J. N., and Hatch, M. D. (1985). Light-dark modulation of leaf pyruvate,P_i dikinase. *Trends Biochem. Sci.* **10,** 288-291.
Burnell, J. N., and Hatch, M. D. (1988). Photosynthesis in PEP carboxykinase-type C_4 plants: Pathways of C_4 acid decarboxylation in bundle sheath cells of *Urochloa panicoides*. *Arch. Biochem. Biophys.* **260,** 187-199.
Burr, G. O. (1962). The use of radioisotopes by Hawaiian sugar plantations. *Int. J. Appl. Rad. Iso.* **13,** 365-374.
Calvin, M., and Bassham, J. A. (1962). "The Photosynthesis of Carbon Compounds." Benjamin, New York.
Carnal, N. W., Agostino, A., and Hatch, M. D. (1993). Photosynthesis in phosphoenolpyruvate carboxykinase-type C_4 plants: Mechanism and regulation of C_4 acid decarboxylation in bundle sheath cells. *Arch. Biochem. Biophys.* **306,** 360-367.
Chapman, K. S. R., and Hatch, M. D. (1983). Intracellular location of phosphoenolpyruvate carboxykinase and other C_4 photosynthetic enzymes in mesophyll and bundle sheath protoplasts of *Panicum maximum*. *Plant Sci. Lett.* **29,** 145-154.
Chapman, K. S. R., Berry, J. A., and Hatch, M. D. (1980). Photosynthetic metabolism in the bundle sheath cells of *Zea mays:* Sources of ATP and NADPH. *Arch. Biochem. Biophys.* **202,** 330-341.
Chollet, R., and Ogren, W. L. (1975). Regulation of photorespiration in C_3 and C_4 species. *Botan. Rev.* **41,** 137-179.
Coombs, J., and Baldry, C. W. (1972). C_4-pathway in *Pennisetum purpureum*. *Nature* **238,** 268-270.
Cooper, R. A., and Kornberg, H. L. (1965). Net conversion of pyruvate to phosphoenolpyruvate in *Escherichia coli*. *Biochim. Biophys. Acta* **104,** 618-621.

Dai, Z., Ku, M. S. B., and Edwards, G. E. (1995). C_4 photosynthesis: Effect of leaf development on the CO_2 concentrating mechanism and photorespiration in maize. *Plant Physiol.* **107**, 815–825.

Day, D. A., and Hatch, M. D. (1981). Transport of 3-phosphoglyceric acid, phosphoenolpyruvate and inorganic phosphate in maize mesophyll chloroplasts and the effect of 3-phosphoglycerate on malate and phosphoenolpyruvate production. *Arch. Biochem. Biophys.* **211**, 743–749.

Doncaster, H. D., and Leegood, R. C. (1987). Regulation of phosphoenolpyruvate carboxylase activity in maize leaves. *Plant Physiol.* **84**, 82–87.

Downton, W. J. S., Berry, J. A., and Tregunna, E. B. (1970). C_4 photosynthesis: Noncyclic electron flow and grana development in bundle sheath chloroplasts. *Zeit. Pflanzenphysiol.* **63**, 194–198.

Edwards, G. E., and Black, C. C. (1971). Isolation of mesophyll and bundle sheath cells from *Digitaria sanguinalis* (L) Scop. leaves and a scanning microscope study. *Plant Physiol.* **47**, 149–156.

Edwards, G. E., and Huber, S. C. (1981). The C_4 pathway. In "The Biochemistry of Plants," Vol. 8. "Photosynthesis" (M. D. Hatch and N. K. Boardman, eds.), pp. 237–281. Academic Press, New York.

Edwards, G. E., and Ku, M. S. B. (1987). Biochemistry of C_3-C_4 intermediates. In "Biochemistry of Plants," Vol. 10. "Photosynthesis" (M. D. Hatch and N. K. Boardman, eds.), pp. 275–325. Academic Press, New York.

Edwards, G. E., and Walker, D. A. (1983). "C_3, C_4: Mechanisms and Cellular and Environmental Regulation of Photosynthesis." Blackwell, Oxford.

Edwards, G. E., Lee, S. S., Chen, T. M., and Black, C.C. (1970). Carboxylation reactions and photosynthesis of carbon compounds in isolated mesophyll and bundle sheath cells of *Digitaria sanguinalis*. *Biochem. Biophys. Res. Commun.* **39**, 389–395.

Edwards, G. E., Kanai, R., and Black, C. C. (1971). Phosphoenolpyruvate carboxykinase in leaves of certain plants which fix CO_2 by the C_4 dicarboxylic cycle of photosynthesis. *Biochem. Biophys. Res. Commun.* **45**, 278–285.

Edwards, G. E., Huber, S. C., Ku, S. B., Rathnam, C. K., Gutierrez, M., and Mayne, B. C. (1976). Variation in photochemical activities of C_4 plant in relation to CO_2 fixation. In "CO_2 Metabolism and Plant Productivity" (R. H. Burris and C. C. Black, eds.), pp. 83–112. University Park Press, Baltimore.

Edwards, G. E., Nakamoto, H., Burnell, J. N., and Hatch, M. D. (1985). Pyruvate, Pi dikinase, and NADP malate dehydrogenase in C_4 photosynthesis. *Annu. Rev. Plant Physiol.* **36**, 255–286.

Ehleringer, J., and Björkman, O. (1977). Quantum yield for CO_2 uptake in C_3 and C_4 plants: Dependence on temperature, CO_2 and O_2 concentration. *Plant Physiol.* **59**, 86–90.

Ehleringer, J., and Pearcy, R.W. (1983). Variation in quantum yield for CO_2 uptake in C_3 and C_4 plants. *Plant Physiol.* **73**, 555–559.

Ehleringer, J., Sage, R. F., Flanagan, L. B., and Pearcy, R. W. (1991). Climate change and the evolution of C_4 photosynthesis. *Trends Ecol. Evol.* **6**, 95–99.

El-Sharkaway, M. A., and Hesketh, J.D. (1964). Effect of temperature and water deficit on leaf photosynthetic rates of different species. *Crop Sci.* **4**, 514–518.

Farquhar, G. D. (1983). On the nature of carbon isotope discrimination of C_4 species. *Aust. J. Plant Physiol.* **10**, 205–226.

Farquhar, G. D., Ehleringer, J. R., and Hubick, K. T. (1989). Carbon isotope discrimination and photosynthesis. *Annu. Rev. Plant Physiol. Plant Mol. Biol.* **40**, 503–537.

Furbank, R. T., and Hatch, M. D. (1987). Mechanism of C_4 photosynthesis: The size and composition of the inorganic carbon pool in bundle sheath cells. *Plant Physiol.* **85**, 958–964.

Furbank, R.T., Jenkins, C. L. D., and Hatch, M. D. (1990). C_4 photosynthesis: Quantum requirement, C_4 acid overcycling and Q cycle involvement. *Aust. J. Plant Physiol.* **17**, 1–7.

Gifford, R. M. (1974). A comparison of potential photosynthesis productivity and yield of plant species with differing photosynthetic metabolism. *Aust. J. Plant Physiol.* **1**, 107–117.

Gross, A., Bruckner, G., Heldt, H. W., and Flugge, U. (1990). Comparison of the kinetic properties, inhibition and labeling of the phosphate translocator from maize and spinach mesophyll chloroplasts. *Planta* **180**, 262–271.

Gutierrez, M., Gracen, V. E., and Edwards, G. E. (1974). Biochemical and cytological relationships in C_4 plants. *Planta* **19**, 279–300.

Haberlandt, G. (1884). "Physiological Plant Anatomy" (Transl. M. Drummond). Macmillan, London.

Hatch, M. D. (1971a). Mechanism and function of C_4 photosynthesis. *In* "Photosynthesis and Photorespiration" (M. D. Hatch, C. B. Osmond, and R. O. Slatyer, eds.), pp. 139–152. Wiley-Interscience, New York.

Hatch, M. D. (1971b). The C_4 pathway of photosynthesis: Evidence for an intermediate pool of CO_2 and the identity of the donor C_4 dicarboxylic acid. *Biochem. J.* **125**, 425–432.

Hatch, M. D. (1977). Light–dark mediated regulation of NADP malate dehydrogenase in isolated chloroplasts from *Z. mays*. *In* "Photosynthetic Organelles." *Plant Cell Physiol.* (Special issue, S. Miyachi, S. Katoh, Y. Fujita, J. Shibata, eds.) pp. 311–314.

Hatch, M. D. (1987). C_4 photosynthesis: A unique blend of modified biochemistry, anatomy and ultrastructure. *Biochim. Biophys. Acta* **895**, 81–106.

Hatch, M. D., and Kagawa, T. (1974). Activity, location and role of NAD malic enzyme in leaves with C_4 pathway photosynthesis. *Aust. J. Plant Physiol.* **1**, 357–369.

Hatch, M. D., and Kagawa, T. (1976). Photosynthetic activities of isolated bundle sheath cells in relation to differing mechanisms of C_4 photosynthesis. *Arch. Biochem. Biophys.* **175**, 39–53.

Hatch, M. D., and Osmond, C. B. (1976). Compartmentation and transport in C_4 photosynthesis. *In* "Transport in Plants" (U. Huber and C. R. Stocking, eds.), Vol. 3. pp. 145–184. Springer-Verlag, Berlin.

Hatch, M. D., and Slack, C.R. (1966). Photosynthesis by sugar cane leaves: A new carboxylation reaction and the pathway of sugar formation. *Biochem. J.* **101**, 103–111.

Hatch, M. D., and Slack, C. R. (1968). A new enzyme for the interconversion of pyruvate and phosphopyruvate and its role in the C_4 dicarboxylic acid pathway of photosynthesis. *Biochem. J.* **106**, 141–146.

Hatch, M. D., and Slack, C. R. (1969a). Studies on the mechanism of activation and inactivation of pyruvate, phosphate dikinase. *Biochem. J.* **112**, 549–558.

Hatch, M. D., and Slack, C.R. (1969b). NADP-specific malate dehydrogenase and glycerate kinase in leaves and their location in chloroplasts. *Biochem. Biophys. Res. Commun.* **34**, 589–593.

Hatch, M. D., and Slack, C. R. (1970). The C_4 dicarboxylic acid pathway of photosynthesis. *In* "Progress in Phytochemistry" (L. Reinhold and Y. Liwschitz, eds.), Vol. 2., pp. 35–106. Wiley-Interscience, London.

Hatch, M. D., Slack, C. R., and Johnson, H. S. (1967). Further studies on a new pathway of photosynthetic CO_2 fixation in sugarcane and its occurrence in other plant species. *Biochem. J.* **102**, 417–422.

Hatch, M. D., Slack, C. R., and Bull, T. A. (1969). Light induced changes in the content of some enzymes of the C_4 dicarboxylic acid pathway of photosynthesis. *Phytochemistry* **8**, 697–706.

Hatch, M.D., Osmond, C. B., and Slatyer, R. O. (eds.) (1971). "Photosynthesis and Photorespiration." Wiley-Interscience, New York.

Hatch, M. D., Kagawa, T., and Craig, S. (1975). Subdivision of C_4 pathway species based on differing C_4 acid decarboxylating systems and ultrastructural features. *Aust. J. Plant Physiol.* **2**, 111–128.

Hatch, M. D., Agostino, A., and Burnell, J. N. (1988). Photosynthesis in phosphoenolpyruvate carboxykinase-type C_4 plants: Activity and role of mitochondria in bundle sheath cells. *Arch. Biochem. Biophys.* **261**, 357–367.

Hatch, M. D., Agostino, A., and Jenkins, C. L. D. (1995). Measurement of the leakage of CO_2 from bundle sheath cells of leaves during C_4 photosynthesis. *Plant Physiol.* **108**, 173–181.

Hattersley, P. W. (1992). Diversification of photosynthesis. In "Grass Evolution and Domestication" (G. P. Chapman, ed.), pp. 38–116. Cambridge University Press, Cambridge, U.K.

Hattersley, P. W., and Browning, A. J. (1981). Occurrence of the suberised lamella in leaves of grasses of different photosynthetic types. *Protoplasma* **109**, 371–401.

Henderson, S. A., von Caemmerer, S., and Farquhar, G. D. (1992). Short-term measurements of carbon isotope discrimination in C_4 species. *Aust. J. Plant Physiol.* **19**, 263–285.

Hesketh, J. D. (1963). Limitations to photosynthesis responsible for differences among species. *Crop Sci.* **3**, 493–496.

Hesketh, J. D., and Moss, D. N. (1963). Variations in the response of photosynthesis to light. *Crop Sci.* **3**, 107–110.

Hodge, A. J., McLean, J. D., and Mercer, F. V. (1955). Ultrastructure of the lamellae and grana in the chloroplasts of *Zea mays* L. *J. Biophys. Biochem. Cytol.* **25**, 605–614.

Huber, S. C., and Edwards, G. E. (1975). Inhibition of phosphoenolpyruvate carboxylase from C_4 plants by malate and aspartate. *Canad. J. Bot.* **53**, 1925–1933.

Huber, S. C., and Edwards, G. E. (1977a). Transport in C_4 mesophyll chloroplasts. Characterization of the pyruvate carrier. *Biochim. Biophys. Acta* **462**, 583–602.

Huber, S. C., and Edwards, G. E. (1977b). Transport in C_4 mesophyll chloroplasts: Evidence for an exchange of inorganic phosphate and phosphoenolpyruvate. *Biochim. Biophys. Acta.* **462**, 603–612.

Huber, S. C., and Sugiyama, T. (1986). Changes in sensitivity to effectors of maize leaf phosphoenolpyruvate carboxylase during light–dark transitions. *Plant Physiol.* **81**, 674–677.

Hunt, S., Smith, A. M., and Woolhouse, H. W. (1987). Evidence for a light-dependent system for reassimilation of photorespiratory CO_2 in the C_3–C_4 intermediate *Moricandia arvensis*. *Planta* **171**, 227–234.

Hylton, C. M., Rawsthorne, S., Smith, A. M., Jones, D. A., and Woolhouse, H.W. (1988). Glycine decarboxylase is confined to the bundle sheath cells of leaves of C_3–C_4 intermediate species. *Planta* **175**, 452–459.

Jacquot, J-P., Gadal, P., Nishizawa, A. N., Yee, B. C., Crawford, N. A., and Buchanan, B. B. (1984). Enzyme regulation in C_4 photosynthesis. *Arch. Biochem. Biophys.* **228**, 170–178.

Jiao, J. A., and Chollet, R. (1988). Light/dark regulation of maize leaf phosphoenolpyruvate carboxylase by *in vivo* phosphorylation. *Arch. Biochem. Biophys.* **216**, 409–417.

Jenkins, C. L. D., Furbank, R. T., and Hatch, M.D. (1989). Mechanisms of C_4 photosynthesis: A model describing the inorganic carbon pool in bundle sheath cells. *Plant Physiol.* **91**, 1372–1381.

Johnson, H. S., and Hatch, M. D. (1970). Properties and regulation of leaf NADP-malate dehydrogenase and malic enzyme in plants with the C_4 dicarboxylic acid pathway of photosynthesis. *Biochem. J.* **119**, 273–280.

Jordan, D. B., and Ogren, W. L. (1984). The CO_2 specificity of ribulose 1,5-bisphosphate carboxylase/oxygenase. *Planta* **161**, 308–313.

Kagawa, T., and Hatch, M. D. (1977). Regulation of C_4 photosynthesis: Characterization of a protein factor mediating the activation and inactivation of NADP malate dehydrogenase. *Arch. Biochem. Biophys.* **184**, 290–297.

Karabourniotis, G., Manetas, Y., and Gavalas, N.A. (1985). Detecting photoactivation in phosphoenolpyruvate carboxylase in C_4 plants. *Plant Physiol.* **77**, 300–302.

Karpilov, Yu, S. (1960). The distribution of radioactive carbon 14 amongst the products of photosynthesis of maize. *Trudy Kazansk Sel'shokhoz Institute* **41**(1), 15–24.

Karpilov, Yu, S. (1969). Peculiarities of the functions and structure of the photosynthetic apparatus in some species of plants of tropical origin. *Proc. Moldavian Inst. Irrig. Veg. Res.* **11**, 1–34.

Kennedy, R. A., and Laetsch, W. M. (1974). Plant species intermediate for C_3, C_4 photosynthesis. *Science* **184**, 1087-1089.

Kortschak, H. P., Hartt, C. E., and Burr, G. O. (1965). Carbon dioxide fixation in sugar cane leaves. *Plant Physiol.* **40**, 209-213.

Ku, S. B., Gutierrez, M., and Edwards, G. E. (1974). Localisation of the C_4 and C_3 pathways of photosynthesis in the leaves of *Pennisetum purpureum* and other C_4 species. *Planta* **119**, 267-278.

Ku, M. S. B., Spalding, M. H., and Edwards, G.E. (1980). Intracellular localisation of phosphoenolpyruvate carboxykinase in leaves of C_4 and CAM plants. *Plant Sci. Lett.* **19**, 1-8.

Laetsch, W. M. (1971). Chloroplast structural relationships in leaves of C_4 plants. *In* "Photosynthesis and Photorespiration" (M. D. Hatch, C. B. Osmond, and R. O. Slatyer, eds.), pp. 323-348. Wiley-Interscience, New York.

Laetsch, W. M., and Price, I. (1968). Development of the dimorphic chloroplasts of sugar cane. *Amer. J. Bot.* **56**, 77-87.

Loomis, R. S., and Gerakis, P. A. (1975). Productivity of agricultural systems. *In* "Photosynthesis and Productivity in Different Environments" (J. D. Cooper, ed.), pp. 145-172. Cambridge University Press.

Loomis, R. S., and Williams, W. A. (1963). Maximum crop productivity—an estimate. *Crop Sci.* **3**, 67-72.

Lowe, J., and Slack, C.R. (1971). Inhibition of maize leaf phosphoenolpyruvate carboxylase by oxaloacetate. *Biochim. Biophys. Acta.* **235**, 207-209.

Meidner, H. (1962). The minimum intercellular space CO_2 concentration of maize leaves and its influence on stomatal movement. *J. Exp. Bot.* **13**, 284-293.

Meister, M., Agostino, A., and Hatch, M.D. (1996). The roles of malate and aspartate in C_4 photosynthetic metabolism of *Flaveria bidentis* (L). *Planta.* **199**, 262-269.

Monson, R. K., Moore, B. D., Ku, M. S. B., and Edwards, G.E. (1986). Co-function of C_3 and C_4 photosynthetic pathways in C_3, C_4 and C_3-C_4 intermediate *Flaveria* species. *Planta* **168**, 493-502.

Monson, R. K., Schuster, W. S., and Ku, M. S. B. (1987). Photosynthesis *in Flaveria brownii* A. M. Powell. a C_4-like C_3-C_4 intermediate. *Plant Physiol.* **85**, 1063-1067.

Monteith, J. L. (1978). Reassessment of maximum growth rates for C_3 and C_4 crops. *J. Exp. Agric.* **14**, 1-5.

Moss, D. N. (1962). The limiting carbon dioxide concentration for photosynthesis. *Nature* **193**, 587.

Murata, Y., and Iyama, J. (1963). Studies on the photosynthesis of forage crops. II. Influence of air temperature. *Proc. Crop Sci. Soc. Japan* **31**, 315-322.

Nakamoto, H., and Edwards, G.E. (1986). Light activation of pyruvate,P_i dikinase and NADP malate dehydrogenase in mesophyll protoplasts of maize. *Plant Physiol.* **82**, 312-315.

Nimmo, G. A., McNaughton, G. L., Fewson, C. A., Wilkins, M. B., and Nimmo, H.G. (1987). Changes in the kinetics properties and phosphorylation state of phosphoenolpyruvate carboxylase in *Zea mays* leaves in response to light and dark. *FEBS Lett.* **213**, 18-22.

Ohnishi, J., and Kanai, R. (1987). Pyruvate uptake by mesophyll and bundle sheath chloroplasts of a C_4 plant, *Panicum miliaceum. Plant Cell Physiol.* **28**, 1-10.

Osmond, C. B. (1967). β-Carboxylation during photosynthesis in Atriplex. *Biochim. Biophys. Acta.* **141**, 197-199.

Osmond, C. B. (1971). Metabolite transport in C_4 photosynthesis. *Aust. J. Biol. Sci.* **24**, 159-163.

Osmond, C. B., and Smith, F. A. (1976). Symplastic transport of metabolites during C_4 photosynthesis. *In* "Intercellular Communication in Plants" (B. E. S. Gunning and R. W. Robards, eds.), pp. 229-241. Springer, Berlin.

Osmond, C. B., Troughton, J. H., and Goodchild, D. J. (1969). Physiological, biochemical and structural studies of photosynthesis and photorespiration in two species of *Atriplex. Zeit. Pflanzenphysiol.* **61**, 218-237.

Rawsthorne, S. (1992). C_3–C_4 intermediate photosynthesis: Linking physiology to gene expression. *Plant J.* **2**, 267–274.

Rebeille, F., and Hatch, M. D. (1986). Regulation of NADP-malate dehydrogenase in C_4 plants: Relationships among enzyme activity, NADPH to NADP ratios and thioredoxin state in intact maize mesophyll chloroplasts. *Arch. Biochem. Biophys.* **249**, 171–179.

Rhoades, M. M., and Carvalho, A. (1944). The function and structure of the parenchyma sheath plastids of maize leaf. *Bull Torrey Botanical Club* **7**, 335–346.

Shantz, H. L., and Piemeisel, L. N. (1927). The water requirement of plants at Akron Colorado. *J. Agric. Res.* **34**, 1093–1189.

Slack, C. R. (1968). The photoactivation of phosphopyruvate synthase in leaves of *Amaranthus palmerii*. *Biochem. Biophys. Res. Commun.* **30**, 483–488.

Slack, C. R. (1969). Localisation of certain photosynthetic enzymes in mesophyll and parenchyma sheath chloroplasts of maize and *Amaranthus palmeri*. *Phytochemistry* **8**, 1387–1391.

Slack, C. R., and Hatch, M. D. (1967). Comparative studies on the activities of carboxylases and other enzymes in relation to the new pathway of photosynthetic CO_2 fixation in tropical grasses. *Biochem. J.* **103**, 660–665.

Slack, C. R., Hatch, M. D., and Goodchild, D. J. (1969). Distribution of enzymes in mesophyll and parenchyma sheath chloroplasts of maize leaves in relation to the C_4 dicarboxylic acid pathway of photosynthesis. *Biochem. J.* **114**, 489–498.

Smillie, R. M., Andersen, K. S., Tobin, N. F., Entsch, B., and Bishop, D. G. (1972). NADP photoreduction from water by agranal chloroplasts isolated from bundle sheath cells of maize. *Plant Physiol.* **49**, 471–475.

Smith, B. N., and Brown, W. V. (1973). The Kranz syndrome in Gramineae as indicated by carbon isotope ratios. *Am. J. Bot.* **60**, 505–513.

Smith, B. N., and Epstein, S. (1971). Two categories of $^{13}C/^{12}C$ ratios for higher plants. *Plant Physiol.* **47**, 380–384.

Snaydon, R. W. (1991). The productivity of C_3 and C_4 plants: A reassessment. *Functional Ecol.* **5**, 321–330.

Stitt, M., and Heldt, H.W. (1985). Generation and maintenance of concentration gradients between mesophyll and bundle sheath in maize leaves. *Biochim. Biophys. Acta* **808**, 400–414.

Tarchevskii, I. A., and Karpilov, Yu. S. (1963). On the nature of products of short-term photosynthesis. *Soviet Plant Physiol.* **10**, 183–184.

Tregunna, E. B., Krotkov, G., and Nelson, C. D. (1964). Further evidence on the effects of light on respiration during photosynthesis. *Can. J. Bot.* **42**, 989–997.

Tregunna, E. B., Smith, B. N., Berry, J. A., and Downton, W. J. S. (1970). Some methods for studying the photosynthetic taxonomy of Angiosperms. *Can. J. Bot.* **48**, 1209–1214.

Uedan, K., and Sugiyama, T. (1976). Purification and characterisation of phosphoenolpyruvate carboxylase from maize leaves. *Plant Physiol.* **57**, 906–910.

Weiner, H., Burnell, J. N., Woodrow, I. E., Heldt, H. W., and Hatch, M.D. (1988). Metabolite diffusion into bundle sheath cells from C_4 plants. Relation to C_4 photosynthesis and plasmodesmatal function. *Plant Physiol.* **88**, 815–822.

Whelan, T., Sackett, W. M., and Benedict, C. R. (1970). Carbon isotope discrimination in a plant possessing the C_4 dicarboxylic acid pathway. *Biochem. Biophys. Res. Commun.* **41**, 1205–1210.

II

Structure–Function of the C$_4$ Syndrome

3

The Biochemistry of C_4 Photosynthesis

Ryuzi Kanai and Gerald E. Edwards

I. Introduction

C_4 photosynthesis consists of the coordinated function of two cell types in the leaves, usually designated mesophyll cells (MC) and bundle sheath cells (BSC), because enzymes of the C_4 pathway are located separately in these morphologically distinct cell types. In C_4 leaves, atmospheric CO_2 enters through stomata and is first accessible to MC, where it is fixed by phosphoenolpyruvate (PEP) carboxylase to form oxaloacetate, and then malate and aspartate. These C_4 dicarboxylic acids are transported to BSC where they are decarboxylated, and the released CO_2 refixed by ribulose-1,5-bisphosphate (RuBP) carboxylase (Rubisco) and assimilated through the enzymes of the photosynthetic carbon reduction (PCR) cycle to form sucrose and starch. Although anatomic differentiation is apparent in BSC, they are functionally similar to C_3 MC in carbon assimilation except for the presence of enzymes concerned with decarboxylation of C_4 acids.

The physiological significance of separate but coordinate function of the two cell types in C_4 photosynthesis is the specialization of MC for generation of a high concentration of CO_2 in BSC in order to reduce the oxygenase activity of Rubisco and consequential reduction of photorespiration. Without consideration of a possible positive function of photorespiration in C_3 plants (*cf.*, Osmond and Grace, 1995), it is clear that C_4 plants have the capacity to perform effective photosynthesis under conditions in which RuBP oxygenase activity is restricted. C_4 photosynthesis can be visualized as a mechanism to provide Rubisco with near saturating CO_2 when C_4 plants can afford a high stomatal conductance, or to provide sufficient CO_2 for survival and growth when stomatal conductance is low.

During the evolution of C_4 photosynthesis from C_3 plants, the MC developed a high level of carbonic anhydrase (CA) and PEP carboxylase for initial CO_2 fixation in the cytoplasm, and pyruvate, orthophosphate (P_i) dikinase in the chloroplasts for provision of PEP, the HCO_3^- acceptor. It is equally important that the synthesis of some key photosynthetic enzymes in carbon metabolism of C_3 photosynthesis is repressed in MC of C_4 plants. This includes Rubisco and phosphoribulokinase of the PCR cycle in MC chloroplasts, and enzymes of glycine decarboxylation in the photosynthetic carbon oxidation pathway (PCO cycle) in MC mitochondria. Differences in the C_4 pathway in three subgroups are illustrated in the first section of this chapter through highlighting differentiation in photosynthetic functions of MC and BSC. This is followed by concise information on the enzyme reactions and properties of the respective enzymes. In the second section, the CO_2 concentration in BSC and the activity of Rubisco, which is exclusively localized in these cells, are discussed. In the third section, the energetics of C_4 photosynthesis is dealt with, including the theoretic maximum efficiency and *in vivo* energy requirements of C_4 plants. Although there is evidence for cooperation between the two cell types in C_4 leaves from *in vivo* studies, cell separation techniques have allowed studies with isolated cell types as well as with intact organelles. These have been critical in understanding the division of labor and coordination of the two cell types in the intercellular and intracellular transport of metabolites. These are discussed in Sections IV and V. For previous reviews on the biochemistry of C_4 photosynthesis see Edwards and Walker (1983), Hatch (1987), and Leegood and Osmond (1990).

A. The Three C_4 Subgroups

C_4 plants have been separated into three subgroups based on differences in the enzymes of the decarboxylation step in BSC. These are the NADP-malic enzyme (NADP-ME), NAD-malic enzyme (NAD-ME), and PEP carboxykinase (PEP-CK) types. Each C_4 type shows not only morphologic differentiation in their arrangement of bundle sheath chloroplasts and ultrastructure, but also further biochemical differences between MC and BSC, and in the method of transport of metabolites between the cells (Gutierrez *et al.*, 1974b; Hatch *et al.*, 1975).

Biochemical pathways of the three C_4 subgroups are summarized in Figs. 1, 2, and 3. Common to all C_4 plants is the initial fixation of HCO_3^- by PEP carboxylase to form oxaloacetate in the MC cytoplasm. As atmospheric CO_2 enters the MC via stomata, carbonic anhydrase in the MC cytoplasm helps to equilibrate the CO_2 to HCO_3^-. Malate and aspartate are formed from oxaloacetate in MC. As determined by $^{14}CO_2$-fixation experiments, the main initial product is malate in C_4 species of the NADP-ME type, whereas aspartate is the major product in the NAD-ME and PEP-CK types.

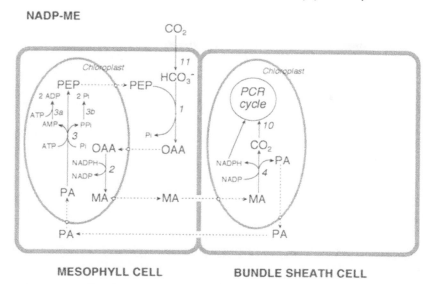

Figure 1 Subgroup of C_4 pathway: NADP-ME (malic enzyme) type. Compound abbreviations for Figs. 1–3: Ala, alanine; Asp, aspartate; MA, malate; OAA, oxaloacetate; PA, pyruvate; PEP, phosphoenolpyruvate; Pi, orthophosphate; PPi, pyrophosphate. Enzyme abbreviations: 1, PEP carboxylase; 2, NADP-malate dehydrogenase; 3, pyruvate phosphate dikinase; 3a, adenylate kinase; 3b, pyrophosphatase; 4, NADP-malic enzyme; 5, NAD-malic enzyme; 6, PEP carboxykinase; 7, NAD-malate dehydrogenase; 8, alanine aminotransferase; 9, aspartate aminotransferase; 10, RuBP carboxylase; 11, carbonic anhydrase; 12, respiratory electron transport system.

1. NADP-ME Type In NADP-ME C_4 species, bundle sheath chloroplasts of C_4 grasses are usually arranged in a centrifugal position relative to the vascular bundle, and have thylakoid membranes with reduced grana stacking. As is evident from Fig. 1, chloroplasts in MC and BSC play a critical role in the C_4 pathway. Oxaloacetate, formed by PEP carboxylase in the cytoplasm, is transported to MC chloroplasts, where most of the oxaloacetate is reduced to malate (% is species dependent) by NADP-specific malate dehydrogenase and the remainder is converted to aspartate by aspartate aminotransferase. These acids are exported from MC to BSC, presumably through plasmodesmata, which are abundant at the interface of the two cell types. In BSC chloroplasts, malate is decarboxylated by NADP-malic enzyme to feed CO_2 and reduced NADP to the PCR cycle. The other product of decarboxylation, pyruvate, is returned to MC chloroplasts, where it is phosphorylated by pyruvate,P_i dikinase to form PEP, the acceptor of inorganic carbon. Decarboxylation through NADP-malic enzyme may also occur via aspartate being metabolized to malate in BSC through aspartate aminotransferase and malate dehydrogenase; alternatively in some NADP-

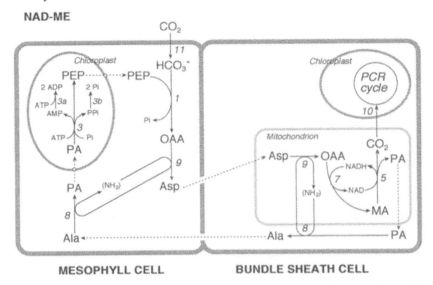

Figure 2 Subgroup of C_4 pathway: NAD-ME (malic enzyme) type. (See Fig. 1 legend for compounds and enzymes.)

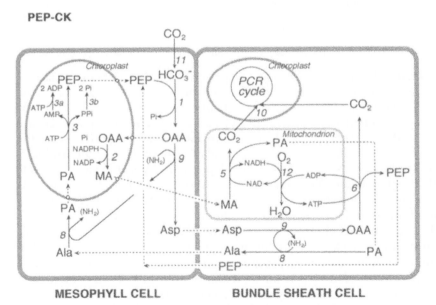

Figure 3 Subgroup of C_4 pathway: PEP-CK (carboxykinase) type. (See Fig. 1 legend for compounds and enzymes.)

ME species, PEP carboxykinase may serve as a secondary decarboxylase (Gutierrez *et al.*, 1974b; Walker *et al.*, 1997).

2. NAD-ME Type Bundle sheath chloroplasts of NAD-ME C_4 species have thylakoid membranes with developed grana stackings. Both chloroplasts and mitochondria are located together in a centripetal position relative to the vascular bundle, except in some grass species of *Panicum* and *Eragrostis* (Ohsugi *et al.*, 1982; Prendergast *et al.*, 1986). The main initial product of $^{14}CO_2$-fixation is aspartate via aspartate aminotransferase in the MC cytoplasm. The aspartate is transported to BSC mitochondria, where it is deaminated by aspartate aminotransferase. The product oxaloacetate is reduced to malate by NAD-malate dehydrogenase and then the malate is decarboxylated by NAD-ME to feed CO_2 to bundle sheath chloroplasts. Thus, bundle sheath mitochondria play a decisive role in this C_4 subtype, as illustrated in Fig. 2. The decarboxylation product, pyruvate, is converted to alanine, which is shuttled to the MC where it is used for resynthesis of PEP; alanine aminotransferases in the cytoplasm of MC and BSC have a key role in this process.

3. PEP-CK Type Bundle sheath chloroplasts of PEP-CK types have well-developed grana stacks. The chloroplasts are arranged evenly or in a centrifugal position in BSC of this C_4 subgroup. PEP carboxykinase in the bundle sheath cytoplasm is the main decarboxylation enzyme, but BSC mitochondria also possess appreciable activity of NAD-malic enzyme. Although aspartate is the main initial product of $^{14}CO_2$-fixation through the high aspartate aminotransferase activity in the MC cytoplasm, some malate is formed in MC chloroplasts. As shown in Fig. 3, aspartate transported from MC cytoplasm to BSC is deaminated and decarboxylated by PEP-CK, whereas malate transported to BSC mitochondria is decarboxylated by NAD-ME resulting in both decarboxylases feeding CO_2 to BSC chloroplasts. The NADH formed by NAD-ME is oxidized through the mitochondrial electron transport chain to produce ATP by oxidative phosphorylation. The ATP is exported to the cytoplasm, where it is used for the PEP-CK reaction. Of the two decarboxylation products, pyruvate may return to MC chloroplasts through alanine, as noted in the NAD-ME type species. PEP is suggested to return directly to the MC cytoplasm, because only low activity of pyruvate kinase is detectable in BSC. Relatively low activity of pyruvate,P_i dikinase in PEP-CK C_4 plants compared with the other C_4 types (*cf.*, Table I) may also reflect the return of PEP. Mechanisms to balance distribution of nitrogen and phosphate between MC and BSC remain to be explored.

B. Enzymes of the C_4 Pathway: Reaction and Properties

After proposing the C_4-dicarboxylic acid pathway of photosynthesis in 1966, Hatch and Slack identified many of the enzymes of the C_4 pathway and showed their activities were sufficient to account for *in vivo* photosyn-

Table I Summary of Enzyme Activities of the C_4 Pathway and Location in the Leaf

	Intercellular and intracellular enzyme location[a]	NADP-ME	NAD-ME	PEP-CK
		Enzyme activity in whole leaf extract (μmol min^{-1} mg^{-1} chlorophyll)		
PEP carboxylase[b,c]	M cyt	13~24	12~25	17~27
Pyruvate,Pi dikinase[b]	M chlt	4~8	4~9	2~4
Adenylate kinase[d]	M chlt >B	41~87	36~70	
Pyrophosphatase[d]	M chlt >B	37~57	52~74	
NADP-malate dehydrogenase[b]	M chlt >B	10~17	1~2	2~5
NADP-malic enzyme[b]	B chlt	10~16	<1	<1
NAD-malic enzyme[b]	B mit	<1	5~18	1~3
PEP carboxykinase[b]	B cyt	<1	<1	6~17
Aspartate aminotransferase[b]	M chlt >B	5~9		
	M cyt >B mit, cyt		27~46	44~60
Alanine aminotransferase[b]	M cyt =B cyt	3~8	30~63	38~45
RuBP carboxylase[c,e,f]	B chlt	1~4	1~3	1~4
Carbonic anhydrase[a]	M cyt	35~68	79~89	28

For values listed as <1, the activity was less than 1 or not detectable.

[a] M, B, main localization in mesophyll or bundle sheath cell, respectively; M > B, more in mesophyll cell; M < B; more in bundle sheath cell; M = B, equally distributed; chlt, chloroplasts; cyt, cytoplasm; mit, mitochondria.
[b] Hatch, 1987.
[c] Gutierrez et al., 1974a.
[d] Hatch and Burnell, 1990.
[e] Hatch and Osmond, 1976.
[f] The lower activities in the range given are likely underestimates due to inactivation or loss of enzyme during extraction procedures because they are below rates of leaf photosynthesis.

thetic rates (ca. 3–5 μmol CO_2 min^{-1} [mg Chl]$^{-1}$; see Hatch and Osmond, 1976). Subsequent studies established the distribution between MC and BSC and their intracellular localization. The enzyme activities on a chlorophyll basis in whole leaf extract of C_4 plants and their compartmentation are summarized in Table I.

1. PEP Carboxylase PEP carboxylase (EC 4.1.1.31) from C_4 leaves consists of a homotetramer with 110 kDa subunits. The purified enzyme from maize leaves has molecular activity (molar activity per minute per mol of enzyme:

$$\begin{array}{c} CH_2 \\ | \\ CO\text{-}PO_3H_2 \\ | \\ COOH \end{array} \quad \begin{array}{c} H^*CO_3^- \\ \xrightarrow{\quad Mg^{2+} \quad} \end{array} \quad \begin{array}{c} ^*COOH \\ | \\ CH_2 \\ | \\ C=O \\ | \\ COOH \end{array} \quad + \quad H_3PO_4 \quad \Delta G^{o'} = -31.8 \text{ kJ} \cdot \text{mol}^{-1}$$

phosphoenol-pyruvate (PEP)   oxaloacetate (OAA)

(Reaction I)

mol min^{-1} [mol enzyme]$^{-1}$) of 9920 at pH 7.0 and 22°C and K_m values for HCO_3^- and PEP are 0.02 mM and 1–2 mM, respectively (Uedan and Sugiyama, 1976). The catalytic mechanism starts with binding of metal^{2+}, PEP and HCO_3^- in this order to the active site having Lys, His, and Arg residues. The chemical steps are summarized as follows: (1) phosphate transfer from PEP to form carboxyphosphate and enolate of pyruvate, (2) carboxyphosphate decomposes to form enzyme-bound CO_2 and phosphate, (3) CO_2 combines with the metal-stabilized enolate, and then (4) the products oxaloacetate and phosphate are released from the enzyme (for details, *cf.*, Chollet *et al.*, 1996).

Regulation of enzyme activity by metabolic control has been a subject of study especially at nearly physiological assay conditions, including positive and negative effectors such as glucose 6-phosphate and malate, respectively. In addition, light activation and diurnal changes in kinetic properties were shown to be via phosphorylation–dephosphorylation at a Ser residue near the C terminal of the protein by a specific protein kinase–phosphatase system. Interestingly, the phosphorylated enzyme, which occurs in the light in C_4 plants (and the dark phase of CAM plants), is the active form that is more sensitive to positive effectors and less sensitive to negative effectors (Carter *et al.*, 1996; Chollet *et al.*, 1996).

2. NADP-Malate Dehydrogenase Malate dehydrogenase specific to NADP (EC 1.1.1.82), found by Hatch and Slack (1969a) in the chloroplasts of

$$\begin{array}{c} COOH \\ | \\ CH_2 \\ | \\ C=O \\ | \\ COOH \end{array} + NADPH+H^+ \underset{Mg^{2+}}{\longleftrightarrow} \begin{array}{c} COOH \\ | \\ CH_2 \\ | \\ CHOH \\ | \\ COOH \end{array} + NADP^+ \quad \Delta G° = +30 \text{ kJ·mol}^{-1}$$

oxaloacetate (OAA)   malate (MA)

(Reaction II)

some C_4 and C_3 plants, was shown to be activated in the light (Johnson and Hatch, 1970). The enzyme purified from maize leaves is a homodimer with 43 kDa subunits with molecular activity of 60,500 at pH 8.5 and 25°C. The K_m values for NADPH, oxaloacetate, NADP$^+$, and malate are 24, 56, and 73 μM and 32 mM, respectively (Kagawa and Bruno, 1988). Light activation is mediated by the ferredoxin–thioredoxin m system, which reduces a disulphide group on the enzyme (Edwards *et al.*, 1985; Droux *et al.*, 1987). Further modulation of the activation state occurs through a high NADPH–NADP$^+$ ratio in MC chloroplasts in the light (Rebeille and Hatch, 1986).

3. Pyruvate,P_i Dikinase The enzyme activity first reported as "PEP synthetase" by Hatch and Slack in 1967 was identified as pyruvate, orthophosphate

$$\underset{\substack{\text{pyruvate}\\\text{(PA)}}}{\underset{\substack{|\\\text{COOH}}}{\overset{\text{CH}_3}{\underset{|}{\text{C}=\text{O}}}}} + \text{H}_3\text{PO}_4 + \text{ATP} \underset{\text{Mg}^{2+}}{\longleftrightarrow} \underset{\substack{\text{phosphoenol-}\\\text{pyruvate (PEP)}}}{\underset{\substack{|\\\text{COOH}}}{\overset{\text{CH}_2}{\underset{|}{\text{CO-PO}_3\text{H}_2}}}} + \underset{\substack{\text{pyrophosphate}\\\text{(PPi)}}}{\text{H}_2\text{O}_3\text{P-O-PO}_3\text{H}_2} + \text{AMP} \qquad \Delta G^{0'} = -13 \text{ kJ}\cdot\text{mol}^{-1}$$

(Reaction III)

dikinase (EC 2.7.9.1), a new key enzyme in the C_4 pathway, which is also activated by illumination (Hatch and Slack, 1969b). The purified enzyme from maize leaves is a homotetramer with 94 kDa subunits, having a molecular activity of 2,600 at pH 7.5 and 22°C; K_m values for pyruvate, P_i, ATP, PEP, pyrophosphate and AMP are 250, 1,500, 15, 140, 40, and <10 μM, respectively (Sugiyama, 1973). Although the reaction itself is reversible, it proceeds to form PEP *in vivo* because high activity of pyrophosphatase and adenylate kinase are present in the same compartment, the MC chloroplasts (*cf.*, Figs. 1–3). The reaction mechanism includes phosphorylation of P_i to form pyrophosphate and His residues at the active site of the enzyme with γ- and β-P of ATP, respectively, and then pyruvate reacts with the His-P residue to form PEP. As for light–dark regulation, the active enzyme having His-P is inactivated by phosphorylation of a specific Thr residue with β-P of ADP, and it is reactivated by phosphorolysis of Thr-P to form pyrophosphate. Interestingly, a single regulatory protein mediates both phosphorylation and dephosphorylation of the Thr residue (for details, *cf.* Edwards *et al.*, 1985).

4. Enzymes of C_4 Acid Decarboxylation   Although C_4 decarboxylating enzymes had already been discovered in plant tissues, their role in C_4 photosynthesis was revealed because of high activities in the BSC of each C_4 subgroup [by Slack and Hatch (1967) for NADP-malic enzyme (EC 1.1.1.40, Reaction IV); by Edwards *et al.* (1971) for PEP carboxykinase (EC 4.1.1.49, Reaction V); and by Hatch and Kagawa (1974) for NAD-malic enzyme (EC 1.1.1.39, Reaction VI)]. The inorganic carbon product of these three decarboxylation enzymes was proved to be CO_2, and not HCO_3^-, after some debates; HCO_3^- is rather inhibitory to the decarboxylation reactions (Jenkins *et al.*, 1987).

NADP-malic enzyme in C_4 plants is located in BSC chloroplasts, whereas

$$\underset{\substack{\text{malate}\\\text{(MA)}}}{\underset{\substack{|\\\text{COOH}}}{\overset{\text{COOH}}{\underset{|}{\underset{|}{\overset{|}{\text{CH}_2}}\text{CHOH}}}}} + \text{NADP}^+ \underset{\text{Mg}^{2+},\text{Mn}^{2+}}{\longleftrightarrow} \text{CO}_2 + \underset{\substack{\text{pyruvate}\\\text{(PA)}}}{\underset{\substack{|\\\text{COOH}}}{\overset{\text{CH}_3}{\underset{|}{\text{C}=\text{O}}}}} + \text{NADPH+H}^+ \qquad \Delta G^{0'} = -1.3 \text{ kJ}\cdot\text{mol}^{-1}$$

(Reaction IV)

that in CAM plants is located in the MC cytoplasm (cf., Edwards and Andreo, 1992). The enzyme purified from sugarcane leaves has a molecular activity of 17,750 at pH 8.0 and 30°C; K_m values for malate and $NADP^+$ are 120 and 5 μM, respectively. The major form of the enzyme is a homotetramer with 62 kDa subunits at pH 8.0, but a homodimer at pH 7.0 (Iglesias and Andreo, 1990). The former is considered the active form in the light as it has higher V_{max} and lower K_m ($NADP^+$) and higher affinity for Mg^{2+} compared with the latter. A possible light–dark regulation by the ferredoxin–thioredoxin system was also suggested from the thiol–disulfide interchange of the maize enzyme (Drincovich and Andreo, 1994). Although the enzyme activity is reversible (K_m for CO_2, 1.1 mM), the ratio of the decarboxylation–carboxylation reaction was about 10/1 at pH 8.1 and 0.6 mM CO_2 (Jenkins et al., 1987).

$$\begin{array}{l} \text{COOH} \\ | \\ \text{CH}_2 \\ | \\ \text{C=O} \\ | \\ \text{COOH} \end{array} + \text{ATP} \underset{Mn^{2+}}{\xleftrightarrow{\hspace{1cm}}} CO_2 + \begin{array}{l} \text{CH}_2 \\ | \\ \text{CO-PO}_3\text{H}_2 \\ | \\ \text{COOH} \end{array} + \text{ADP} \quad \Delta G^{0'} = 0 \text{ kJ} \cdot \text{mol}^{-1}$$

oxaloacetate (OAA)   phosphoenol-pyruvate (PEP)

(Reaction V)

The localization of PEP carboxykinase in BSC cytoplasm was confirmed after some misinterpretation (Watanabe et al., 1984). The purified enzyme from three C_4 species was reported as a hexamer with 68 kDa subunits having a specific activity of 36–51 U (molecular activity ca. 14,700–20,800) at pH 7.6 and 25°C, an absolute requirement for Mn^{2+}; and K_m values for oxaloacetate, ATP, and CO_2 of 12–25 μM, 16–25 μM, and 2.5 mM, respectively (Burnell, 1986). Later work, however, showed that these purifications yielded a truncated form of the subunit at the N-terminal end due to rapid proteolysis. The native form, isolated from five C_4 species, has 67–71 kDa subunits (Walker and Leegood, 1996). The enzyme having the larger subunit, for example, from *Panicum maximum*, is phosphorylated–dephosphorylated by dark–light treatment of the leaves *in vivo*, but its role in regulation of the enzyme activity remains to be studied.

$$\begin{array}{l} \text{COOH} \\ | \\ \text{CH}_2 \\ | \\ \text{CHOH} \\ | \\ \text{COOH} \end{array} + \text{NAD}^+ \underset{Mn^{2+}}{\xleftrightarrow{\hspace{1cm}}} CO_2 + \begin{array}{l} \text{CH}_2 \\ | \\ \text{C=O} \\ | \\ \text{COOH} \end{array} + \text{NADH} + \text{H}^+ \quad \Delta G^{0'} = -1.3 \text{ kJ} \cdot \text{mol}^{-1}$$

malate (MA)   pyruvate (PA)

(Reaction VI)

NAD-malic enzyme located in the BSC mitochondria is the C_4 acid decarboxylase functioning in C_4 photosynthesis in NAD-ME type species; it also has a role in PEP-CK type C_4 plants (see Figs. 2 and 3). Different isoforms are widely reported in various tissues of plant species, even among the C_4 species (Hatch et al., 1974; Artus and Edwards, 1985). The purified enzyme from *Eleusine coracana* is a homooctamer, having 63 kDa subunits with a molecular activity of 60,500 at pH 7.2 and 31°C (Murata et al., 1989). Mn^{2+} is absolutely required for activity and acetyl CoA, CoA, and fructose 1,6-bisphosphate (FuBP) are potent activators. K_m values for malate and NAD^+ in the presence of FuBP are 2.2 and 0.63 mM, respectively. Antiserum raised against the enzyme from *E. coracana*, a monocot, does not inhibit the enzyme activity of a C_4 dicot, *Amaranthus edulis*. The purified enzyme from *Amaranthus hypocondriacus* is a heterotetramer composed of two 65 kD α-subunits, which are responsible for catalytic activity, and two 60 kD β-subunits (Long et al., 1994).

5. Aminotransferases By the work of Hatch and associates in the 1970s, aspartate aminotransferase (EC 2.6.1.1, Reaction VII) and alanine aminotransferase (EC 2.6.1.2, Reaction VIII) were proved to be essential in the C_4 pathway of NAD-ME and PEP-CK types (Hatch, 1987). These enzymes exist in both MC and BSC mainly as cytoplasmic and/or mitochondrial isoenzymes.

```
COOH      +   COOH     <-------->   COOH      +   COOH
 |             |                     |             |
CH₂           CH₂                   CH₂           CH₂
 |             |                     |             |
HCNH₂         CH₂                   C=O           CH₂
 |             |                     |             |
COOH          C=O                   COOH          HCNH₂
               |                                   |
              COOH                                COOH

aspartate   2-oxoglutarate       oxaloacetate    glutamate
 (Asp)        (2OG)                 (OAA)          (Glu)

            (Reaction VII)
```

A main aspartate aminotransferase (AspAT) isoform located in the MC cytoplasm was first purified from *P. maximum* (PEP-CK type). It is a homodimer with 42 kDa subunits having molecular activity of 18,200 at pH 8.0 and 25°C (Numazawa et al., 1989). Of the three AspAT isoforms in *Panicum miliaceum* and *E. coracana* (NAD-ME type), two major isoforms are specific to C_4 plants and are localized in the MC cytoplasm (cAspAT) and BSC mitochondria (mAspAT). The third minor component is a plastidic isoform (pAspAT) that is similar to that of C_3 plants (Taniguchi and Sugiyama, 1990; Taniguchi et al., 1996). All isoforms have been purified and consist of a homodimer with approximately 40 kDa subunits. Although the isoelectric points of the proteins and cross-reactivity against antisera are different, kinetic properties are similar: K_m values for aspartate, 2-oxoglutarate, glutamate, and oxaloacetate are 1.3–3.0, 0.07–0.20, 8–32, 0.023–0.085 mM, respectively.

Three isoforms of alanine aminotransferase (AlaAT) exist in *P. miliaceum*

$$\underset{\substack{\text{alanine}\\\text{(Ala)}}}{\begin{array}{c}CH_3\\|\\HCNH_2\\|\\COOH\end{array}} + \underset{\substack{\text{2-oxoglutarate}\\\text{(2OG)}}}{\begin{array}{c}COOH\\|\\CH_2\\|\\CH_2\\|\\C=O\\|\\COOH\end{array}} \longleftrightarrow \underset{\substack{\text{pyruvate}\\\text{(PA)}}}{\begin{array}{c}CH_3\\|\\C=O\\|\\COOH\end{array}} + \underset{\substack{\text{glutamate}\\\text{(Glu)}}}{\begin{array}{c}COOH\\|\\CH_2\\|\\CH_2\\|\\HCNH_2\\|\\COOH\end{array}}$$

(Reaction VIII)

(NAD-ME type) leaves; the major forms occurring in the MC and BSC cytoplasm have been purified (cAlaAT). The enzyme is a homodimer with 50 kDa subunits having a molecular activity of 39,500 at pH 7.5 and 25°C. K_m values for alanine, 2-oxoglutarate, pyruvate and glutamate are 6.7, 0.15, 0.33 and 5.0 mM, respectively. The cAlaAT seems to be functional in C_4 photosynthesis, as the protein increases in parallel with other C_4 enzymes during greening of seedlings (Son et al., 1991).

6. RuBP Carboxylase–Oxygenase Ribulose-1,5-bisphosphate carboxylase—oxygenase (Rubisco, EC 4.1.1.39) of C_4 plants, like that of C_3 plants, consists

$$\underset{\substack{\text{ribulose 1,5-bisphosphate}\\\text{(RuBP)}}}{\begin{array}{c}CH_2O\text{-}PO_3H_2\\|\\C=O\\|\\HCOH\\|\\HCOH\\|\\CH_2O\text{-}PO_3H_2\end{array}} + {}^*CO_2 \xrightarrow{Mg^{2+}} \underset{\text{2x3-phosphoglycerate (PGA)}}{\begin{array}{c}CH_2O\text{-}PO_3H_2\\|\\HCOH\\|\\{}^*COOH\end{array} + \begin{array}{c}COOH\\|\\HCOH\\|\\CH_2O\text{-}PO_3H_2\end{array}} \quad \Delta G^{o\prime} = -35.2 \text{ kJ} \cdot \text{mol}^{-1}$$

$$+ \; {}^*O_2 \longrightarrow \underset{\text{phosphoglycolate}}{\begin{array}{c}CH_2O\text{-}PO_3H_2\\|\\CO^*OH\end{array}} + \underset{\text{3-phosphoglycerate}}{\begin{array}{c}COOH\\|\\HCOH\\|\\CH_2O\text{-}PO_3H_2\end{array}}$$

(Reaction IX)

of eight large subunits of 53.5 kDa and eight small subunits of 13 kDa (L_8S_8). Kinetic properties show some difference between C_3 and C_4 plants (Yeoh et al., 1980; 1981). The range of $K_m(CO_2)$ values for the enzyme among C_3 species (12–25 μM) is lower than those in C_4 (28–34 μM), although variation of $K_m(RuBP)$ values (15–82 μM in C_4) is not related to the photosynthetic pathway of higher plants. Among the C_4 grasses, the $K_m(CO_2)$ values of Rubisco in PEP-CK type (28–41, mean 35 μM) are significantly lower than those in NAD- and NADP-ME types (41–63, mean 53 μM). Molecular activities of some C_4 Rubiscos (2280–4020, mean of 3240 mol CO_2 mol enzyme^{-1} min^{-1}) are about twofold higher than those from C_3 higher plant species (Seemann et al., 1984). These higher V_{max} values of Rubisco in C_4 plants accompanied by higher $K_m(CO_2)$ may be an evolutionary change that allows a high activity per unit Rubisco protein under high levels of CO_2 in BSC in the light. However, there are no clear differences between the C_4 and C_3 Rubisco in their specificity factor (S_{rel}), that is, the relative specificity to react with CO_2 versus O_2, which is based on V_{max} values and Michaelis constants for the two gases. An improved measurement of the specificity factor of Rubisco resulted in a value of 79 mol/mol for maize leaves, which is marginally smaller than those of five C_3 higher plant species (82–90 mol/mol) (Kane et al., 1994).

7. Carbonic Anhydrase Carbonic anhydrase (CA: carbonate hydratase, EC 4.2.1.1), which catalyses the first step in C_4 photosynthesis, is located in the

$$CO_2 + H_2O \longleftrightarrow H_2CO_3 \longleftrightarrow H^+ + HCO_3^-$$
(Reaction X)

mesophyll cell cytoplasm (Gutierrez et al., 1974a; Ku and Edwards, 1975), where it equilibrates the aerial CO_2 to HCO_3^-. Some activity may also be associated with the MC plasmamembrane (Utsunomiya and Muto, 1993). HCO_3^- is the form of inorganic carbon used as a substrate in the PEP carboxylase reaction. In the absence of CA it is presumed that the rate of nonenzymatic equilibration of atmospheric CO_2 to HCO^{3-} would be rate limiting for C_4 photosynthesis, although this has not been proved. Its virtual absence in BSC is understandable because the inorganic carbon substrate used by Rubisco is CO_2, which is also the direct product of the three decarboxylation enzymes (Hatch and Burnell, 1990). Maize CA has been purified, but its molecular and kinetic properties were not reported (Burnell, 1990). Antibodies from maize CA cross-reacted with a 24–25 kDa subunit in a crude extract of C_4 plants by Western blot analysis. Studies with several C_4 leaf CA showed a $K_m(CO_2)$ of 0.8–2.8 mM at 0°C and maximum activity at 35 mM CO_2, which was far higher than the maximum photosynthetic rates of C_4 plants. However, their activities measured under conditions considered to exist *in vivo*, pH 7.2 and 4 μM CO_2, were 28–89 μmol CO_2 hydration min^{-1} (mg Chl)$^{-1}$, just sufficient for C_4 photosynthesis (Hatch and Burnell, 1990).

II. CO_2 Concentration and Rubisco Activity in Bundle Sheath Cells

As mentioned in the introduction, the C_4 pathway of photosynthesis evolved to concentrate CO_2 around Rubisco in BSC, which allows much more efficient CO_2 assimilation than in C_3 plants under limiting CO_2 in the atmosphere. In C_3 plants, PCO cycle activity is a consequence of the kinetic properties of Rubisco; O_2 competes with CO_2 to produce phosphoglycolate, which is metabolized in the glycolate pathway to glycerate.

In C_4 photosynthesis, the level of CO_2 in BSC is influenced by a variety of factors, particularly (1) the intercellular CO_2 partial pressure, (2) the solubility of CO_2 as influenced by temperature, (3) the capacity for transport of C_4 acids from MC through plasmodesmata (*cf.*, Osmond and Smith, 1976), (4) the reactions of decarboxylation followed by (5) refixation via Rubisco in the BSC, and (6) leakage of CO_2 (and HCO_3^-) from BSC to MC by passive diffusion, which depends on diffusive resistance of the BSC cell wall (*cf.*, Section IIIB).

Various models have been used to estimate the size of the CO_2 pool in BSC during C_4 photosynthesis (see He and Edwards, 1996; Chapter 6 by von Caemmerer and Furbank, this volume). Estimates indicate the C_4 cycle can raise the levels of CO_2 in the BSC to 30–75 μM (*ca.* 3 to 8 times higher than in C_3 plants), resulting in PCO cycle activity as low as 3% of net CO_2 uptake (Jenkins *et al.*, 1989; Dai *et al.*, 1993; He and Edwards, 1996).

The capacity of a C_4 plant to increase the supply of CO_2 to Rubisco is illustrated when comparing the response of photosynthesis in maize (C_4) to that of wheat (C_3) with varying intercellular levels of CO_2 (Fig. 4). The higher initial slope and lower CO_2 compensation point (intercept on the

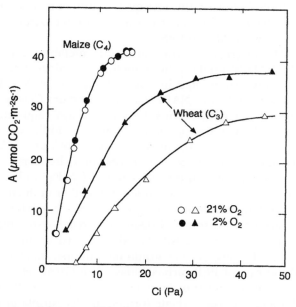

Figure 4 Response of the rate of CO_2-fixation in maize and wheat to varying intercellular levels of CO_2. At atmospheric level of CO_2 (33 Pa), the corresponding intercellular level of CO_2 (Ci) was approximately 16 Pa in maize and 21 Pa in wheat. (From Dai, Edwards, Ku, unpublished).

x axis) in the C_4 plant is a consequence of the CO_2 concentrating mechanism. In C_4 plants, the large increase in efficiency of Rubisco to function in CO_2 fixation through the C_4 system allows for a lower investment in Rubisco protein than in C_3 plants (Ku et al., 1979; Sage et al., 1987). For example, maize has about half the Rubisco content of wheat (Edwards, 1986).

III. Energetics of C_4 Photosynthesis

A. Minimum Energy Requirements in the Three C_4 Subgroups

The minimum energy requirements for CO_2 fixation to the level of triose phosphate in C_3 plants are 3 ATP and 2 NADPH per CO_2 in the absence of photorespiration. The same is true for the refixation of CO_2 generated by C_4 acid decarboxylation in BSC of C_4 plants by Rubisco and subsequent PCR cycle enzymes, because the oxygenase activity of Rubisco is negligible due to high CO_2 concentration in BSC in the light. In C_4 leaves, however, extra energy is required for regeneration of PEP, the acceptor for the initial CO_2 fixation, from pyruvate through pyruvate,P_i dikinase and adenylate kinase in mesophyll chloroplasts; it costs 2 ATP per CO_2 fixed by PEP carboxylase. Differentiation of the energy requirements among C_4 subgroups takes place by the difference in decarboxylation enzymes and transport metabolites between MC and BSC.

In the NADP-ME type (cf., Fig. 1), primary shuttle metabolites are malate/pyruvate, and the NADPH used for the reduction of oxaloacetate to malate in MC chloroplasts is regenerated in BSC chloroplasts by NADP-ME for the PCR cycle; thus, 5 ATP and 2 NADPH are required per CO_2 assimilated. The NAD-ME type (cf., Fig. 2) retains the same energy requirement; the primary shuttle metabolites are aspartate/alanine, no additional energy is required for transamination reactions in MC and BSC, and NADH produced by NAD-malic enzyme is recycled in BSC mitochondria.

The situation is complicated in the PEP-CK type (cf., Fig. 3) because PEP carboxykinase and NAD-malic enzyme function together in the decarboxylation step and the relative stochiometries are uncertain. NADH is generated by malate decarboxylation through NAD-malic enzyme. Assuming a maximum of 3 ATP are generated per NADH oxidized through mitochondrial phosphorylation, NAD malic enzyme would have to operate at a minimum of one-third of the rate of PEP carboxykinase (see Carnal et al., 1993, in which interregulation of the two decarboxylases is discussed). With this stochiometry, 25% of the oxaloacetate formed in MC cytoplasm would be reduced to malate in MC chloroplasts (requirement of 0.25 NADPH per CO_2 fixed) and 75% converted to aspartate. In this case, the ATP requirement to form PEP in mesophyll chloroplasts is only 0.5 ATP per CO_2

fixed because of direct return of PEP (three-fourths of the decarboxylation products via PEP-CK) from BSC. With this stoichiometry, the minimum energy requirements of the PEP-CK type are estimated to be 3.5 ATP and 2.25 NADPH per CO_2 assimilated. However, adopting the recent estimate of mitochondrial ATP generation of 2.5 ATP per NADH oxidized (Hinkle *et al.*, 1991), the energy requirements in the PEP-CK type will be increased to 3.57 ATP and 2.29 NADPH, and NAD-malic enzyme would need to operate at 40% of the rate of PEP carboxykinase. Another possibility for ATP supply to PEP carboxykinase is through the oxidation of triose phosphate to 3-phosphoglycerate in the cytoplasm. In fact, isolated BSC of PEP-CK species decarboxylate oxaloacetate when supplied with triose phosphate, NAD^+ and ADP (Burnell and Hatch, 1988a).

B. *In Vivo* Energy Requirements in C_4 Plants

For the estimation of energy requirements of C_4 photosynthesis *in vivo* it is necessary to evaluate several factors, such as the leakage of CO_2 from BSC, costs of CO_2 assimilation to form carbohydrates and nitrate assimilation, and level of photorespiration, if any. The magnitude of leakiness of BSC has been estimated using techniques such as ^{14}C labeling (Hatch *et al.*, 1995), carbon isotope discrimination (Evans *et al.*,1986; Henderson *et al.*, 1992), O_2 sensitivity of photosynthesis (He and Edwards, 1996) and application of mathematical models (*cf.*, He and Edwards, 1996; Chapter 6 by von Caemmerer and Furbank, this volume), with values as a percentage of the rate of C_4 cycle ranging from 8% to 50%. Using a leakage value of 20% as a reasonable estimate, for every 5 CO_2 delivered to the BSC by the C_4 cycle, 1 CO_2 would leak, resulting in a C_4 cycle expense of 2.5 ATP per net CO_2 fixed by PCR cycle. Besides the C_4 and PCR cycles, other demands for assimilatory power are for the synthesis of carbohydrate and assimilation of nitrate, namely the cost of conversion from triose phosphate to carbohydrates (sucrose and starch) and from nitrate to glutamate, respectively. It is apparent that these do not add much to the total assimilatory power requirements per CO_2 fixed in mature leaves (Furbank *et al.*, 1990; Edwards and Baker, 1993).

Although photorespiration via the PCO cycle is strongly suppressed during C_4 photosynthesis, it is not completely eliminated. The extent of photorespiration may vary with leaf age, environmental conditions that affect intercellular CO_2 levels, and species (see Marek and Stewart, 1983; de Veau and Burris, 1989; Dai *et al.*, 1993, 1995, 1996). In maize, PCO cycle activity measured by incorporation of $^{18}O_2$ into glycolate under normal atmospheric levels of CO_2 and O_2 indicated that the velocity of oxygenase (v_o) was about 3% of the net rate of CO_2 assimilation (A) in mature leaves of 3-month-old plants, and 11% in leaves of 9-day-old seedlings (as opposed to 54% in wheat, a C_3 plant; de Veau and Burris, 1989).

Considering the previously mentioned factors, an estimate of the energy requirements for mature leaves of maize, an NADP-ME species, under normal levels of CO_2 and O_2 is shown in Table II. This analysis shows a total of 5.7 ATP and 2 NADPH per CO_2 fixed. About 90% of the reductive power is accounted for by the reduction of 3-phosphoglycerate to triose phosphate. The ratio of the true rate of O_2 evolution (J_{O_2}) to net CO_2 uptake (A) in this case is 1.1. This value is remarkably consistent with the J_{O_2}/A values determined by oxygen isotope exchange (Badger, 1985) and by chlorophyll fluorescence analysis (Genty et al., 1989; Krall and Edwards, 1992; Oberhuber and Edwards, 1993; Oberhuber et al., 1993). The relationship between J_{O_2} and A in maize under varying temperature, light intensity, and CO_2 concentrations fits well the theoretical energy requirements for C_4 photosynthesis (Edwards and Baker, 1993). However, under low temperature stress in the field there is evidence that additional electron sinks may be induced (Baker et al., 1995).

The relatively low J_{O_2}/A ratio in C_4 plants compared to C_3 plants suggests that respiratory processes during C_4 photosynthesis are low (Badger 1985; Edwards and Baker, 1993). From measurements of J_{O_2} by O_2 isotope analysis on three monocot species of each C_4 subgroup at ambient external CO_2 levels, the average rate of O_2 uptake for NADP-ME was only 4% of J_{O_2}, for NAD-ME 9% of J_{O_2}, and for PEP-CK 22% of J_{O_2} (Furbank and Badger, 1982). Photorespiration and the associated O_2 consumption during C_4 photosynthesis could occur by PCO cycle activity, pseudocyclic electron flow (linear

Table II Estimate of Energy Required in NADP-malic Enzyme Type C_4 Species (e.g., Maize) per CO_2 Fixed Considering Bundle Sheath Leakage of CO_2, Carbohydrate Synthesis, a Low Level of PCO Cycle Activity, and Nitrate Assimilation

Function	ATP	NADPH
1. C_3 pathway per net CO_2 fixed	3	2
2. C_4 pathway, allowing for 25% overcycling	2.5	0
3. ATP per C in triose-P converted to carbohydrate (sucrose)	0.08	0
4. Rubisco oxygenase[a] where v_o is 3% of the net rate of CO_2 fixation (A)	0.14	0.09
5. Nitrate assimilation[b]	0.02	0.11
Total	5.7	2.2

This analysis of energy requirements assumes no dark-type mitochondrial respiration is occuring in the light.

[a] The activity of Rubisco oxygenase used is that from measurements on the rates of incorporation of $^{18}O_2$ into glycolate and CO_2 fixation in leaves of 3-month-old maize plants (de Veau and Burris, 1989). The ATP and NADPH requirement was calculated considering that the true rate of O_2 evolution with respect to Rubisco (J_{O_2}) = $v_c + v_o$, that the net rate of CO_2 fixation with respect to Rubisco (A) = $v_c -$ 0.5 v_o, and that for each O_2 reaction with RuBP approximately 3 ATP and 2 NADPH are consumed in the PCO cycle and conversion of the products to RuBP (Krall and Edwards, 1992).

[b] Based on a C/N ratio in maize of 35/1, assuming all nitrate is assimilated in leaves in the day, and that carbon loss by respiration in the dark is 20% of carbon gain in the light (see Edwards and Baker, 1993).

flow with O_2 as electron acceptor), or normal mitochondrial respiration. First, all C_4 plants examined have lower rates of O_2 uptake than C_3 plants due to the low RuBP oxygenase activity. The differences between the three C_4 subgroups in O_2 uptake may be explained by differences in the mechanism of generation of the additional ATP required in C_4 photosynthesis above what is produced from linear electron flow to NADP. The BSC chloroplasts of NADP-ME species are specialized for production of ATP by cyclic photophosphorylation which contributes additional ATP without contributing to O_2 uptake (see Edwards and Baker, 1993). The very low rate of O_2 uptake (4% of J_{O_2}) in NADP-ME species may be accounted for by low PCO cycle activity (Table II). The lack of production of O_2 in BSC by PSII activity may contribute to low PCO cycle activity in this subgroup. Neither NAD-ME species nor PEP-CK species have chloroplasts that are specialized in cyclic photophosphorylation. The higher rate of O_2 uptake in NAD-ME species compared to NADP-ME species may be due to production of additional ATP for the C_4 cycle via pseudocyclic electron flow and/ or a higher RuBP oxygenase activity in BSC. In PEP-CK species, additional ATP for the C_4 cycle is provided via BSC mitochondrial respiration, which contributes to O_2 uptake. According to the minimum energy requirements for PEP-CK species (*cf.*, Section III.A), the rate of O_2 uptake associated with respiratory production of ATP (using NADH produced via NAD-malic enzyme) would be about 12.5% of J_{O_2}, which, along with RuBP oxygenase activity, may account for this subgroup having the highest rates of O_2 uptake (22% of J_{O_2}).

C. Energy Requirements of C_4 Photosynthesis Compared to Maximum Quantum Yield

Based on the biochemistry of C_4 photosynthesis, the requirements for assimilatory power have been defined (*cf.*, Section III.A). Knowing the energy requirements per CO_2 fixed, and considering the photochemical efficiency for production of ATP and NADPH using absorbed quanta, the theoretical quantum yield for CO_2 fixation (or O_2 evolved) can be calculated. The maximum experimental quantum yield can be determined under limiting light and compared to the theoretical values (Ehleringer and Björkman, 1977; Monson *et al.*, 1982; Ehleringer and Pearcy, 1983; Furbank *et al.*, 1990; Edwards and Baker, 1993; Lal and Edwards, 1995). Theoretical considerations include the degree of engagement of the Q-cycle, the H^+/ ATP ratio for ATP synthase, the degree of overcycling of the C_4 cycle, and the extent to which cyclic versus pseudocyclic electron flow contributes additional ATP.

Calculation of the theoretical maximum quantum yield versus the experimental quantum yield can be illustrated for maize, an NADP-ME species.

The energy requirements for MC and BSC are shown in Fig. 5, in which NADPH production is assumed to be produced photochemically only in the MC chloroplasts.

If the Q-cycle is fully engaged (3 $H^+/\bar{e}$) and 4 H^+ are required per ATP synthesized (Rumberg and Berry, 1995), then 3 ATP would be synthesized per 2 NADPH generated by whole chain electron flow. In this case, production of the 2.1 NADPH in MC chloroplasts per CO_2 fixed would require 8.4 quanta be used for whole chain transfer of 4.2 $\bar{e}$, which would generate 3.15 ATP according to the previously mentioned stoichiometry. If the additional 0.15 ATP needed in MC were generated by pseudocyclic electron

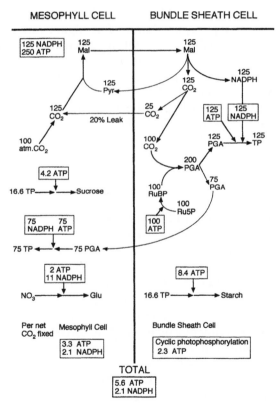

Figure 5 Illustration of energy requirements of mesophyll and bundle sheath cells of NADP-ME C_4 species. Note that 125 TP generated in BSC + 75 TP generated in MC equals a total of 200 TP; 166.6 TP are needed for regeneration of RuBP and the remainder used for synthesis of sucrose and starch. Abbreviations: TP, triose-P; Mal, malate; Pyr, pyruvate; PGA, 3-phosphoglycerate; Glu, glutamate. (See Table II for overall summary of ATP and NADPH requirements.)

transport with the same efficiencies, this would require an additional 0.4 quanta. Production of 2.3 ATP by cyclic electron flow in BSC chloroplasts per CO_2 fixed would require an additional 4.6 quanta (assuming 2 H^+ generated by cyclic electron flow per quanta and 4 H^+ used per ATP produced). Overall, a minimum of 13.4 quanta are required in this example. This ratio will increase due to PCO cycle activity and the expense can be calculated as follows, assuming each O_2 reacting with RuBP consumes approximately 2 NADPH (cf., Krall and Edwards, 1992). Considering $J_{O_2(\text{Rubisco})} = v_c + v_o$, and $A = v_c - 0.5\ v_o$, the increase in quantum requirement (QR) due to RuBP oxygenase activity per net CO_2 fixed can be calculated.

$$\text{Increase in QR due to PCO cycle activity} = \frac{[J_{O_2(\text{Rubisco})}\ (8)]}{A} - 8$$

$$= \frac{[v_c\ (8\ \text{quanta}/CO_2) + v_o\ (8\ \text{quanta}/O_2)]}{(v_c - 0.5\ v_o)} - 8$$

For example, in mature leaves of maize, v_o was found to be 3% of A (de Veau and Burris 1989), which would require an additional 0.35 quanta per CO_2 fixed. In young leaves of maize, de Veau and Burris (1989) found v_o was 11% of A, which would require an additional 1.32 quanta. In summary, the minimum quantum requirement would be 13.4 without PCO cycle activity, and 13.8 and 14.7 per CO_2 fixed, with v_o equal to 3% of A, and with v_o equal to 11% of A, respectively. The maximum efficiency of transfer of excitons from light harvesting chlorophyll to the reaction centers under limiting light must also be considered; assuming an efficiency of 85% (Krause and Weis, 1991), then the quantum requirement on the basis of absorbed quanta in these examples would be 15.8, 16.2, and 17.3, respectively (i.e., maximum quantum yields of 0.057–0.063). For example, in maize (NADP-ME), the measured minimum quantum requirement is approximately 17 quanta/CO_2 fixed—that is, a maximum quantum yield of 0.059 (Monson et al., 1982; Ehleringer and Pearcy, 1983; Dai et al., 1993). In general, experimental values of the maximum quantum yields in C_4 plants under limiting light are about half of the theoretical maximum of 0.125 quanta per CO_2 for generation of 2 NADPH per CO_2 fixed.

The maximum quantum yields measured in C_3 and C_4 plants are similar under current atmospheric CO_2 at approximately 25°C, with the C_3 plants having an additional investment in PCO cycle activity and the C_4 plants having an additional investment in the C_4 pathway. Higher temperatures at atmospheric levels of CO_2, or subatmospheric levels of CO_2 at moderate temperatures, decrease the maximum yield in C_3 plants to values below that of C_4 plants due to increased RuBP oxygenase activity (see Leegood and Edwards, 1996; Chapter 7, this volume).

It is well known that the maximum quantum yield in C_3 plants is higher than that of C_4 plants with sufficient CO_2 enrichment to minimize PCO cycle activity (*ca.* 3 times atmospheric levels) because C_4 photosynthesis has the additional energy requirements for the C_4 cycle. However, under extremely high CO_2 levels (10%), the maximum quantum yields per O_2 evolved in the C_4 plants maize (NADP-ME) and *P. miliaceum* (NAD-ME) are as high as 0.105, which is the same as in C_3 plants under high CO_2 (Lal and Edwards, 1996). This suggests that in C_4 plants under very high CO_2 there is direct fixation of atmospheric CO_2 in the BSC through the PCR cycle, that the C_4 pathway is inhibited, and that all the assimilatory power is used to support reactions of the C_3 pathway as in C_3 plants.

IV. Coordination of the Two Cell Types in C_4 Photosynthesis

A. Separation of the Two Cell Types and Enzyme Distribution

Following the discovery of C_4 photosynthesis and its association with two photosynthetic cell types, there were questions not only about the mechanism, but controversy also developed over the localization of key enzymes. Roger Slack was the first to separate chloroplasts of the two cell types with nonaqueous mixtures of hexane and carbon tetrachloride, and he demonstrated differential location of some enzymes in the C_4 pathway as well as ^{14}C-intermediates of the pathway (Slack, 1969; Slack *et al.*, 1969). Using aqueous media, brief grindings repeated with C_4 leaf segments were later devised for the separation of MC and BSC contents in the form of partial leaf extracts. The "differential" grinding methods in the early 1970s brought about controversy over the localization of some enzymes, such as PEP carboxylase and Rubisco in the two cell types (see Chapter 2 by Hatch, this volume). However, careful handling of the procedure was subsequently effective to obtain intact MC and BSC chloroplasts and bundle sheath strands (e.g., Jenkins and Russ, 1984; Jenkins and Boag, 1985; and Sheen and Bogorad, 1985, respectively).

In a limited number of C_4 species such as *Digitaria sanguinalis* and other species of the genus, intact MC and bundle sheath strands were separated by simple maceration of the leaf segments with a mortar and pestle and subsequent filtration with nylon nets (Edwards and Black, 1971). Treatment of C_4 leaf segments with digestive enzymes such as pectinase and cellulase proved to be a very effective method for separating the two cell types from various species in a pure and intact state, namely mesophyll protoplasts and bundle sheath strands (Kanai and Edwards, 1973a,b). Improvements in commercial enzymes and testing protocols and the use of young seedlings has allowed the enzymatic method to be applicable to a variety of C_4 plants, especially for isolating intact organelles from protoplasts (Edwards *et al.*, 1979; Moore *et al.*, 1984; Watanabe *et al.*, 1984). A disadvantage of the method is the limited quantity of organelles in spite of the high degree of intactness.

Distribution of enzyme activities related to the C_4 pathway, PCR and PCO cycles, and other enzymes of primary carbon metabolism are listed in Table III, based on data from some representative C_4 species of the three C_4 subgroups. Very low activity in a specific cell type may indicate either a low degree of contamination by the other cell type or actual presence of the enzyme—for example, PEP carboxylase, pyruvate,P_i dikinase, and NADP-malate dehydrogenase in BSC, or Rubisco, phosphoribulokinase, and malic enzymes in MC. As for the photorespiratory glycolate pathway, BSCs possess most of the enzymes except for the absence of glycerate kinase, whereas MCs lack Rubisco in the chloroplasts and enzymes of glycine decarboxylation in the mitochondria. Carbonic anhydrase is localized in MC; little activity occurs in BSC chloroplasts in contrast to high activities in chloroplasts of C_3 plants where the PCR cycle is located. Nitrate reductase and nitrite reductase are localized in MC cytoplasm and chloroplasts, respectively, in most C_4 species (Rathnam and Edwards, 1976). Cytoplasmic and chloroplastic isoforms of glutamine synthetase are distributed evenly in the two cell types; but ferredoxin-dependent glutamate synthase (Fd-GOGAT) is mostly located in BSC chloroplasts in maize leaf (Harel *et al.*, 1977; Becker *et al.*, 1993). Thus, nitrate assimilation is restricted to MC, whereas recycling of ammonia produced by the photorespiratory pathway may occur in BSC. *In situ* immunolocalization studies have also been a valuable tool in demonstrating inter- and intracellular location of certain key enzymes in C_4 photosynthesis (e.g., Hattersley *et al.*, 1977; Wang *et al.*, 1992, 1993).

B. Intercellular Transport of Metabolites

Because of differential location of the enzymes, operation of the C_4 pathway of photosynthesis necessitates a rapid bidirectional movement of various metabolites between MC and BSC. The intercellular transport of metabolites includes C_3/C_4 acids, as illustrated in Figs. 1–3, C_3 compounds in the reductive phase of the PCR cycle, and glycerate of the photorespiratory PCO cycle. As the enzymes of the reductive phase of the PCR cycle are distributed in both cell types, MC chloroplasts are able to share this function, especially in many C_4 species of the NADP-ME type, where the BSC chloroplasts have reduced grana and are deficient in Photosystem II activity. The glycerate formed by the glycolate pathway in the BSC is apparently transported and converted to 3-phosphoglycerate in MC due to exclusive localization of glycerate kinase in MC chloroplasts (Usuda and Edwards, 1980a,b).

C. Evidence for Coordination of Two Cell Types

In addition to the enzyme distribution evidence, there is a large body of deductive evidence based on a combination of time-course (during

Table III Intra- and Intercellular Localization of Enzymes in the Mesophyll Cell and Bundle Sheath Cell of C4 Plants

	MC/BSC enzyme activity: μmol hour^{-1} (mg Chl)$^{-1}$		
	NADP-ME type *Zea mays*	PEP-CK type *Panicum maximum* (*Urochloa panicoides*)	NAD-ME type *Eleusine indica* (*Panicum miliaceum*)
C$_4$ pathway			
Carbonic anhydrase[a,b]	98/2	(98/2)[c]	(93/7)
PEP carboxylase[d,e]	864/14	2590/3	2400/10
Pyruvate,Pi dikinase[b]	188/15	(66/1)	82/1
Adenylate kinase[a,f]	56/10		
Pyrophosphatase[a,f]	49/16		
NADP-malate dehydrogenase[b,d]	805/<1	72/25	175/51
NADP-malic enzyme[b,d]	<1/1690	<1/85	7/28
PEP carboxykinase[d]	<1/<1	<1/542	<1/1
NAD-malic enzyme[d]	<1/126	62/263	11/554
Aspartate aminotransferase[d]	324/70	(1150/753)	1400/205
Alanine aminotransferase[d]	33/53	(612/335)	942/579
PCR cycle			
RuBP carboxylase[b,d]	<1/389	4/249	1/395
Phosphoribulokinase[b,d]	<1/2940	(75/1310)	24/2450
Phosphoriboisomerase[d]	75/1500		
PGA kinase[b]	1290/2450	(2110/701)	2350/1090
NADP-triose-P dehydrogenase[b,d]	705/1400	(240/457)	250/477
Triose-P isomerase[b]	526/545	(277/481)	209/813

Photorespiratory glycolate pathway			
Glycolate oxidase[b]	1.2/10		1.2/22
Hydroxypyruvate reductase[b]	20/184	(1/26)	40/526
Catalase ($\times 10^{-3}$)[b]	9/60	(15/330)	11/60
Glycerate kinase[g]	7.5/<1	(11/46)	(22/2)
Others (sucrose synthesis, glycolysis, oxidative pentose-P pathway, respiration)			
UDPG pyrophosphorylase[g]	797/285		(392/443)
Sucrose-P synthase[g]	56/3.2		
Phosphoglyceromutase[b]	160/23	95/35	150/115
Enolase[b]	203/18	181/100	174/95
Glucose-6-P dehydrogenase[b]	20/9	(37/18)	(24/34)
6-Phosphogluconate dehydrogenase[b]	10/8	(28/18)	11/38
Cytochrome oxidase[b]	33/146	(78/149)	46/117

For values listed as <1, the activity was less than 1 or not detectable.
[a] Relative value.
[b] Ku and Edwards, 1975.
[c] R. Kanai, unpublished results.
[d] Kanai and Edwards, 1973a.
[e] Gutierrez et al., 1974a.
[f] Slack et al., 1969.
[g] Usuda and Edwards, 1980a.

feeding of $^{14}CO_2$) and pulse-chase data ($^{14}CO_2$ pulse and $^{12}CO_2$ chase) under steady state photosynthesis. All of this information supports cooperative interaction between cells as the only reasonable explanation for C_4 photosynthesis and allows precise deductions about how it occurs (*cf.*, Hatch, 1987; see also Chapter 2 by Hatch, this volume).

One of the most direct visual proofs for metabolite transport from MC to BSC is an autoradiograph of a transverse leaf section of *Atriplex spongiosa*, an NAD-ME C_4 dicot, after 2 sec of $^{14}CO_2$-fixation in the light (Fig. 6). This shows most of the label in BSC and the remaining label in MC and hypodermal cells (Osmond, 1971). Within such a short time of $^{14}CO_2$-photosynthesis, about 90% of the label was found in C_4 acids; mostly aspartate, as would be expected for this subtype. The microautoradiography clearly indicates rapid transport of these C_4 acids from MC cytoplasm to BSC.

A unique approach to separate functions of the two cell types during C_4 photosynthesis *in vivo* was a light-enhanced dark (LED) $^{14}CO_2$-fixation experiment using intact leaves of maize (Samejima and Miyachi, 1971, 1978, or *Cynodon dactylon* (Black *et al.*, 1973). Leaves were illuminated in CO_2-free air, followed by dark $^{14}CO_2$-fixation, and then reillumination in CO_2-free air. Preillumination of leaves significantly enhanced the subsequent dark $^{14}CO_2$-fixation into malate and aspartate, but no label was transferred to 3-phosphoglycerate and sugar phosphates. The transfer of ^{14}C- to the PCR cycle intermediates and synthesis of sucrose occurred only when the leaves were reilluminated. This experiment apparently separates the function of MC and BSC in time, and demonstrates that light is required

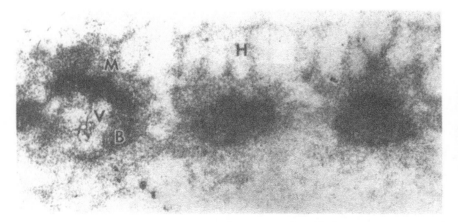

Figure 6 Microautoradiograph of a transverse section of *Atriplex spongiosa* leaf tissue after 2 second $^{14}CO_2$-photosynthesis. M, mesophyll cells; B, bundle sheath cells; V, vascular bundle; H, hypodermis. [Reprinted with permission from C. B. Osmond (1971). Metabolite transport in C_4 photosynthesis. *Aust. J. Biol. Sci.* **24**, 159–163, with permission.]

for use of the C_4 acids as donors of CO_2 to the PCR cycle in BSC. Interestingly, when a ^{14}C-bicarbonate solution was supplied directly to BSC by vacuum infiltration through vascular tissue of cut maize leaves, the main initial photosynthetic products in the light were 3-phosphoglycerate and sugar phosphates. This indicates that CO_2 directly supplied to BSC by the vascular tissue is fixed by the PCR cycle, whereas atmospheric CO_2 is fixed into C_4 acids by the C_4 pathway. These experiments provide further evidence for the photosynthetic function of MC and BSC and the cooperation required in light-mediated processes.

In general, intercellular transport of metabolites is mediated by apoplastic and/or symplastic processes. Although the former is via the cell wall and entails movement across the two plasma membranes, the latter occurs between the cytoplasm of two cells through the plasmodesmata. Osmond calculated the flux of C_4 acids through the plasmodesmata, which occupy approximately 3% of the cell wall interface between MC and BSC. He suggested that the rapid movement of metabolites between the two cell types may be accomplished by a diffusive process in the symplasm, with approximately 10 mM concentration gradient; thus it is not necessary to postulate a special mechanism for active transport (for detail, *cf.*, Osmond and Smith, 1976). Electron micrographs of C_4 leaves show many plasmodesmata at the interface of MC and BSC and, in many species, a thick, suberized layer in the BSC cell wall (Laetsch, 1971; Evert *et al.*, 1977).

Actual estimations of concentration gradients of the intermediates in leaves of maize and *Amaranthus edulis* were made by partial mechanical separation of MC and BSC followed by enzymatic assay of the metabolites (Leegood, 1985; Stitt and Heldt, 1985; Leegood and von Caemmerer, 1988). For example, in photosynthesizing maize leaves, gradients of malate and triose phosphates from MC to BSC were 18 and 10 mM, respectively, whereas that of 3-phosphoglycerate from BSC to MC was 9 mM (Stitt and Heldt, 1985). However, the estimated concentration of pyruvate between MC and BSC was similar. This unexpected result can be explained by active accumulation of pyruvate in MC chloroplasts in the light (*cf.*, Section V.A.3), which allows a concentration gradient of pyruvate to be maintained from the cytosol of BSC to the cytosol of MC. Large concentration gradients guarantee rapid intercellular transport of each metabolite.

Isolated bundle sheath strands were obtained from several C_4 species and used for studying the permeability of compounds of different molecular sizes (Weiner *et al.*, 1988; Valle *et al.*, 1989). The BSC in these preparations retained their structural integrity and transport into BSC via severed plasmodesmata, which were originally connected to MC. The size exclusion limit into BSC was estimated to be approximately 900–1000 kDa by inhibition of alanine aminotransferase using inhibitor dyes having various molecular weights and by observations on plasmolysis of the bundle sheath strands with polyethyleneglycol (PEG) of various molecular sizes. The apparent diffusion rates of the compounds with smaller molecular weight were in the range of 2 to 5 mmol min^{-1} (mg Chl)$^{-1}$ per mmolar gradient between bundle sheath strands and the suspending medium. This suggests that various metabolites of low molecular weight can readily diffuse between MC and BSC through plasmodesmata.

V. Intracellular Transport of Metabolites

Compared with C_3 plants, C_4 species, by necessity, have a more extensive and diverse intracellular transport of metabolites, namely the unique transport of C_3 and C_4 acids associated with intercellular transport, PEP and phosphorylated C_3 compounds in the PCR cycle, and glycerate in the PCO cycle. Differentiation in the transport properties of respective organelles in MC and BSC is also expected because of compartmentation of key enzymes and the unique feature of intercellular coordination. Using differential centrifugation in sucrose density gradients, chloroplasts, mitochondria and peroxisomes have been separated from mesophyll and bundle sheath protoplasts isolated from some C_4 species (Gutierrez *et al.*, 1975; Moore *et al.*, 1984; Watanabe *et al.*, 1984; Ohnishi *et al.*, 1985). However, direct studies on carrier-mediated mechanisms of metabolite transport and their kinetic properties have been performed mainly with MC chloroplasts. This is largely because of technical limitations in isolating sufficient quantities of intact organelles from BSC.

A. Mesophyll Chloroplasts

1. Translocators of C_4 Acids In NADP-ME (and to some extent PEP-CK) C_4 plants, the initial CO_2-fixation product in MC, oxaloacetate, is transported from the cytoplasm to the chloroplasts by a specific translocator having a high affinity for oxaloacetate (K_m: ca. 45 μM) but very low affinity for malate (Hatch *et al.*, 1984). Malate (and some aspartate in NADP-ME–type plants) formed in mesophyll chloroplasts is exported to the cytoplasm by a C_4 dicarboxylic acid translocator (K_m for malate: 0.5 mM, Day and Hatch, 1981), which is similar to that of C_3 chloroplasts. Purification and molecular cloning of these C_4 acid translocators can be anticipated for C_4 mesophyll chloroplasts, because a malate/2-oxoglutarate translocator from the chloroplast envelope of spinach has already been cloned (Weber *et al.*, 1995).

2. Translocator of P_i/Triose Phosphate and PEP Because pyruvate,P_i dikinase is localized in the chloroplasts and PEP is used by PEP carboxylase in the cytoplasm of MC, transport of pyruvate and PEP was logically discovered using mesophyll chloroplasts of *Digitaria sanguinalis*, an NADP-ME C_4 plant

(Huber and Edwards, 1977a,b). Although pyruvate uptake was mediated by a new translocator in the envelope of C_4 mesophyll chloroplasts, PEP was transported by a phosphate translocator having the extra capacity to carry PEP in addition to P_i, 3-phosphoglycerate and triose phosphate, which are tranported by the phosphate/triose–phosphate translocator in C_3 chloroplasts. In *P. miliaceum*, both MC and BSC chloroplasts are able to transport PEP (Ohnishi *et al.*, 1989). The P_i translocator of maize MC chloroplasts possesses a higher affinity for all transported substrates compared to that of C_3 chloroplasts: 50 times higher for PEP and 2-phosphoglycerate (compounds with the phosphate group at the C-2 position), and 6 times higher for P_i, triose phosphates and 3-phosphoglycerate (compounds with phosphate group at C-3 position) (Gross *et al.*, 1990). Among the phosphate translocator cDNAs cloned recently from several C_3 and C_4 species, those from *Flaveria trinervia* (C_4) and *F. pringlei* (C_3) showed 94% homology of amino acids in the mature protein. A computer-aided molecular modeling of the translocators suggests that minor changes in amino acids at the translocation pore might be sufficient to extend high substrate specificity to PEP in the C_4 phosphate translocator (Fischer *et al.*, 1994).

3. Active Pyruvate Transport After the study of pyruvate tranport in *D. sanguinalis* in MC in the dark (Huber and Edwards, 1977a), pyruvate uptake in maize and *P. miliaceum* MC chloroplasts was shown to be promoted by light (Flügge *et al.*, 1985; Ohnishi and Kanai, 1987a, respectively). In illuminated chloroplasts of *P. miliaceum*, the initial rate of transport was 7–10 times higher and the amount of pyruvate accumulated was 10–30 times higher than in the dark. Good correlations between the light-dependent pyruvate uptake and stromal alkalization of illuminated chloroplasts suggested that active transport is primarily driven by the pH gradient, which is formed across the envelope (Ohnishi and Kanai, 1987b). In maize MC chloroplasts, in fact, generation of an artificial pH gradient by a pH shift from 7 to 6 in the suspension medium (a H^+-jump) induced a 5- to 10-fold enhancement of pyruvate uptake in the dark (Ohnishi and Kanai, 1990). In *P. miliaceum*, however, addition of 10 mM Na^+ into the medium (a Na^+-jump) induced a similar enhancement of pyruvate uptake in the dark (Ohnishi and Kanai, 1987c). In terms of the pyruvate uptake induced by these cation jumps in the dark, C_4 plants can be divided into two groups: a H^+ type and a Na^+ type. NADP-ME type C_4 species of Arundineleae and Andropogoneae in the Gramineae belong to the former group, whereas most of the other C_4 monocots and all dicots thus far studied belong to the latter (Aoki *et al.*, 1992).

In the H^+ type, the sole source of driving force for the active pyruvate uptake in MC chloroplasts is the H^+ gradient formed across the envelope and H^+ and pyruvate are cotransported in a one-to-one ratio in the light

(Aoki et al., 1994). Moreover, the pH gradient is maintained by co-export of H^+ and PEP^{3-} via the phosphate translocator as illustrated in Fig. 7 (Aoki and Kanai, 1995).

In the Na^+ type, the driving force for the active pyruvate transport in MC chloroplasts was originally considered to be by a Na^+ gradient generated by light dependent H^+ uptake with Na^+ efflux, because Na^+ and pyruvate were apparently cotransported in the light and darkness (Ohnishi et al., 1990). In this case, light (or ATP) dependent Na^+ efflux or Na^+/H^+ antiport across the envelope would be expected. However, in a Na^+ jump experiment in the dark, there is no evidence for measurable H^+ efflux, nor change in ATP content in the chloroplast. On the contrary, addition of K-pyruvate or Na-pyruvate decreases the stromal pH only in the light with the Na-pyruvate being most effective (Fig. 8).

Furthermore, illumination and Na^+-jump treatments were cooperative in promoting the initial rate of pyruvate uptake in MC chloroplasts of *P. miliaceum* suspended in medium at pH 7.8. Thus, in the Na^+ type, pyruvate uptake may occur by Na^+ dependent acceleration of pyruvate-H^+ cotransport (Aoki and Kanai, 1997).

B. Bundle Sheath Chloroplasts

The BSC chloroplasts have a phosphate transporter (Flugge and Heldt, 1991) and a glycolate transporter (Ohnishi and Kanai, 1988) that have

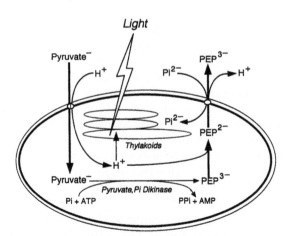

Figure 7 Scheme of H^+ mobilization accompanied with pyruvate import and phosphoenolpyruvate export in the light in mesophyll chloroplasts of C_4 species having H^+ type pyruvate transport. [Reprinted with permission from N. Aoki and R. Kanai (1995). The role of phosphoenolpyruvate in proton/pyruvate cotransport into mesophyll chloroplasts of maize. *Plant Cell Physiol.* **36**, 187–189, with permission.]

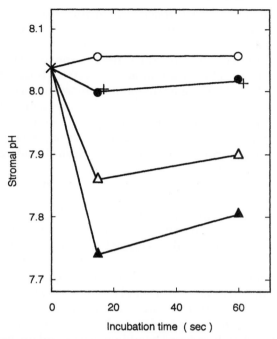

Figure 8 Changes in stromal pH of *Panicum miliaceum* mesophyll chloroplasts in the light by the addition of K- or Na-pyruvate. Intact chloroplasts isolated from mesophyll protoplasts of *P. miliaceum* were suspended in the medium of pH 7.8 and preincubated for 5 min in the light (500 μmol quanta $m^{-2} s^{-1}$). Stromal pH was calculated from the uptake of [^{14}C]-5,5-dimethyloxazolidine-2,4-dione by silicone-oil filtering centrifugation. Addition at 0 time: (+) water, (●) 5 mM Na-gluconate, (○) 5 mM K-gluconate, (△) 5 mM K-pyruvate, (▲) 5 mM Na-pyruvate. There were no significant changes in stromal pH in darkness by these additives (*cf.* Aoki and Kanai, unpublished, 1997).

similar properties to these transporters in C_3 chloroplasts although the metabolic exchanges during C_4 photosynthesis are different due to the coordinated functions between MC and BSC chloroplasts. During CO_2 fixation by RuBP carboxylase, BSC chloroplasts export some of the product PGA in exchange for triose-P or P_i. This occurs by the exported PGA being converted to triose-P in MC chloroplasts and re-entering the BSC chloroplast, or the triose-P being metabolized to sucrose, releasing P_i, which is taken up by the BSC chloroplast. Whereas C_3 chloroplasts export glycolate in exchange for glycerate during photorespiration, in C_4 plants the glycolate synthesized in BSC during photorespiration is exported, metabolized to glycerate and the glycerate imported by MC chloroplasts, where it is converted to PGA (*cf.* Section IV). This transport is mediated by a glycolate/glycerate transporter in BS and MC chloroplasts (Ohnishi and Kanai, 1988).

BSC chloroplasts of NADP-ME species like maize also transport malate and pyruvate. Intact BSC chloroplasts isolated from maize leaves are capable of high rates of malate decarboxylation and $^{14}CO_2$-assimilation only in the presence of 3-phosphoglycerate and/or triose phosphate, with significant enhancement of the rates by the addition of aspartate (Boag and Jenkins, 1985; Taniguchi, 1986). The former compounds are required for the generation of intermediates of the PCR cycle as well as for recycling of NADPH/ $NADP^+$ in the decarboxylation reaction. As aspartate itself is not metabolized, nor does it influence the decarboxylation reaction, the enhancement effect suggests a malate translocator, which differs from C_4 acid translocators in C_3 and C_4 MC chloroplasts.

These differences in the characteristics of malate transport in MC and BSC chloroplasts of maize are illustrated in Fig. 9. The initial rate and final level of ^{14}C-malate uptake in maize MC chloroplasts (Fig. 9A) were the highest in the absence of aspartate (−Asp in the figure). The rate was reduced by adding aspartate together with malate (+Asp) and even more by preincubation with aspartate for 5 min before malate addition (pre+Asp). However, ^{14}C-malate uptake into maize BSC chloroplasts (Fig. 9B) was enhanced by adding aspartate (+Asp), and even more so by preincubation with aspartate (pre+Asp). In contrast, ^{14}C-aspartate uptake with/without malate was essentially the same in MC and BSC chloroplasts of maize; namely, the rate was slightly reduced by adding malate and further inhibited by preincubation with malate (data not shown). Interestingly, addition of pyruvate had an enhancement effect on ^{14}C-malate uptake in maize BSC chloroplasts similar to that of aspartate (Fig. 9C). These results indicate that the envelope of BSC chloroplasts possesses a new malate transport system in addition to the C_4 acid and/or aspartate translocators that have previously been found in C_3 and C_4 MC chloroplasts (Day and Hatch, 1981; Werner-Washburne and Keegstra; 1985). Although there is evidence that BSC chloroplasts can transport pyruvate (Taniguchi, 1986; Ohnishi and Kanai, 1987a), the mechanism of this transport relative to that in MC chloroplasts remains to be fully characterized.

VI. Summary

It is clear that since the late 1960s the biochemistry of C_4 photosynthesis in three C_4 subgroups has become well defined and functional characteristics of many of the key translocators have been elucidated. However, much work remains to be done on isolation of translocators and description of their physical and molecular properties. The two photosynthetic cells, MC and BSC, must cooperate in production of ATP and NADPH to meet the energy requirements of the C_4 cycle and PCR cycle; the relative contribu-

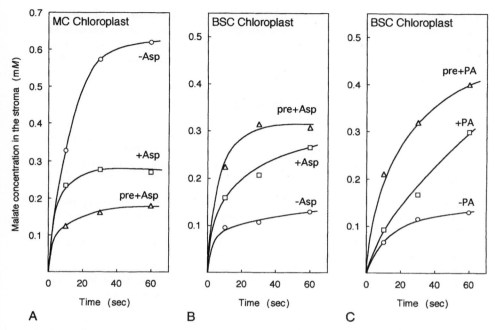

Figure 9 ^{14}C-Malate uptake by mesophyll and bundle sheath chloroplasts isolated from maize leaves: effects of aspartate and pyruvate. BSC chloroplasts from maize leaves were isolated according to Jensen and Boag (1985). Intactness of the BSC chloroplasts was more than 80%, whereas that of mesophyll chloroplasts was more than 90%; estimation of intactness was made with phase-contrast microscopy and an NADP-triose phosphate dehydrogenase activity test (instead of ferricyanide test) resulted in similar percentage intactness. Uptake of ^{14}C-malate (0.5 mM, in the medium) was measured by silicone-oil filtering centrifugation at 5°C and pH 8.0 after preincubation in the light (300 μmol quanta m^{-2} s^{-1}). A, B: Malate uptake by mesophyll and BSC chloroplasts, respectively, without (-Asp), with (+Asp), and after 5 min preincubation with (pre+Asp) 0.5 mM aspartate. The initial uptake rates (in mmol [mgChl]$^{-1}$ hour^{-1}) at 10 sec in mesophyll chloroplasts were 3.9, 2.8, and 1.5, respectively, whereas those in BSC chloroplasts were 1, 1.6, and 2.3, respectively. C: Effect of pyruvate on malate uptake without (-PA), with (+PA), and after preincubation with (pre+PA) 5 mM pyruvate; the initial uptake rates were 1, 1.4, and 3.2, respectively (Taniguchi and Kanai, unpublished).

tions of the two chloroplast types and degree of flexibility in shared production of assimilatory power is not clear. Although metabolic control and balance within and between the C_4 and PCR cycle are required, they are poorly understood. One can presently only speculate on how the two cycles are coordinated under a range of environmental conditions to provide a sufficient level of CO_2 to the BSC to minimize RuBP oxygenase activity and at the same time avoid excessive overcycling and leakage of CO_2 from BSC. Molecular research supported by solid plant biochemistry (Hatch, 1986) continues to be important in considering the origin of enzymes and translocators essential to C_4 photosynthesis. Finally, the relevance of the variations on the biochemistry of C_4 photosynthesis and Kranz type leaf anatomy to survival and performance in different habitats remains to be elucidated.

References

Aoki, N. and Kanai, R. (1995). The role of phosphoenolpyruvate in proton/pyruvate cotransport into mesophyll chloroplasts of maize. *Plant Cell Physiol.* **36,** 187-189.
Aoki, N. and Kanai, R. (1997). Reappraisal of the role of sodium in the light-dependent active transport of pyruvate into mesophyll chloroplasts of C_4 plants. *Plant Cell Physiol.* **38,** 1217-1225.
Aoki, N., Ohnishi, J. and Kanai, R. (1992). Two different mechanisms for transport of pyruvate into mesophyll chloroplasts of C_4 plants—a comparative study. *Plant Cell Physiol.* **33,** 805-809.
Aoki, N., Ohnishi, J. and Kanai, R. (1994). Proton/pyruvate cotransport into mesophyll chloroplasts of C_4 plants. *Plant Cell Physiol.* **35,** 801-806.
Artus, N. and G. E. Edwards. (1985). NAD-malic enzyme from plants. *FEBS Lett.* **182,** 225-233.
Badger, M. R. (1985). Photosynthetic oxygen exchange. *Ann. Rev. Plant. Physiol.* **36,** 27-53.
Baker, N. R., Oxborough, K. and Andrews, J. R. (1995). Operation of alternative electron acceptor to CO_2 in maize crops during periods of low temperatures. In "Photosynthesis Research. Xth International Photosynthesis Congress" (P. Mathis, ed.), Vol. IV, pp. 771-776, Kluwer Academic Publishers, Dordrecht.
Becker, T. W., Perrot-Rechenmann, C., Suzuki, A. and Hirel, B. (1993). Subcellular and immunocytochemical localization of the enzymes involved in ammonia assimilation in mesophyll and bundle-sheath cells of maize leaves. *Planta* **191,** 129-136.
Black, C. C. Campbell, W. H., Chen, T. M. and Dittrich, P. (1973). The monocotyledons: Their evolution and comparative biology III. Pathways of carbon metabolism related to net carbon dioxide assimilation by monocotyledons. *Quart. Rev. Biol.* **48,** 299-313.
Boag, S. and Jenkins, C. L. D. (1985). CO_2 assimilation and malate decarboxylation by isolated bundle sheath chloroplasts from *Zea mays*. *Plant Physiol.* **80,** 165-170.
Burnell, J. N. (1986). Purification and properties of phosphoenolpyruvate carboxykinase from C_4 plants. *Aust. J. Plant Physiol.* **13,** 577-587.
Burnell, J. N. (1990). Immunological study of carbonic anhydrase in C_3 and C_4 plants using antibodies to maize cytosolic and spinach chloroplastic carbonic anhydrase. *Plant Cell Physiol.* **31,** 423-427.
Burnell, J. N. and Hatch, M. D. (1988a). Photosynthesis in phosphoenolpyruvate carboxykinase-type C_4 plants: Photosynthetic activities of isolated bundle sheath cells from *Urochloa panicoides*. *Arch. Biochem. Biophys.* **260,** 177-186.
Burnell, J. N. and Hatch, M. D. (1988b). Photosynthesis in phosphoenolpyruvate carboxykinase-type C_4 plants: Pathways of C_4 acid decarboxylation in bundle sheath cells of *Urochloa panicoides*. *Arch. Biochem. Biophys.* **260,** 187-199.
Carnal, N. W., Agostino, A., and Hatch, M. D. (1993). Photosynthesis in phosphoenolpyruvate carboxykinase-type C_4 plants: Mechanism and regulation of C_4 acid decarboxylation in bundle sheath cells. *Arch. Biochem. Biophys.* **306,** 360-367.
Carter, P. J., Fewson, C. A., Nimmo, G. A., Nimmo, H. G. and Wilkins, M. B. (1996). Roles of circadian rhythms, light and temperature in regulation of phosphoenolpyruvate carboxylase in Crassulacean acid metabolism. In "Crassulacean Acid Metabolism" (K. Winter and J. A. C. Smith, eds), pp. 46-52. Springer-Verlag, Berlin.

Chollet, R., Vidal, J. and O'Leary, M. H. (1996). Phosphoenolpyruvate carboxylase: A ubiquitous, highly regulated enzyme in plants. *Annu. Rev. Plant Physiol. Mol. Biol.* **47,** 273–298.

Dai, Z., Ku, M. S. B. and Edwards, G. E. (1993). C_4 Photosynthesis: The CO_2-concentrating mechanism and photorespiration. *Plant Physiol.* **103,** 83–90.

Dai, Z., Ku, M. S. B. and Edwards, G. E. (1995). C_4 Photosynthesis: The effects of leaf development on the CO_2-concentrating mechanism and photorespiration in maize. *Plant Physiol.* **107,** 815–825.

Dai, Z. Ku, M. S. B., and Edwards, G. E. (1996). Oxygen sensitivity of photosynthesis in C_3, C_4, and C_3-C intermediate species of *Flaveria*. *Planta* **198,** 563–571.

Day, D. A. and Hatch, M. D. (1981). Transport of 3-phosphoglyceric acid, phosphoenolpyruvate, and inorganic phosphate in maize mesophyll chloroplasts, and the effect of 3-phosphoglyceric acid on malate and phosphoenolpyruvate production. *Arch. Biochem. Biophys.* **211,** 743–749.

de Veau, E. J. and Burris, J. E. (1989). Photorespiratory rates in wheat and maize as determined by ^{18}O-labelling. *Plant Physiol.* **90,** 500–511.

Drincovich, M. F. and Andreo, C. S. (1994). Redox regulation of maize NADP-malic enzyme by thiol-disulfide interchange: effect of reduced thioredoxin on activity. *Biochim. Biophys. Acta* **1206,** 10–16.

Droux, M, Miginiac-Maslow, M., Jacquot, J.-P., Gadal, P., Crawford, N. A., Kosower, N. S. and Buchanan, B. B. (1987). Ferredoxin–thioredoxin reductase: A catalytically active dithiol group links photoreduced ferredoxin to thioredoxin functional in photosynthetic enzyme regulation. *Arch. Biochem. Biophys.* **256,** 372–380.

Edwards, G. E. (1986). Carbon fixation and partitioning in the leaf. *In* "Regulation of Carbon and Nitrogen Reduction and Utilization in Maize" (J. C. Shannon, D. P. Knievel and C. D. Boyer, eds.) pp. 51–65, The American Society of Plant Physiologists, Beltsville, Maryland.

Edwards, G. E. and Andreo, C. S. (1992). NADP-malic enzyme in plants. *Phytochemistry.* **31,** 1845–1857.

Edwards, G. E. and Baker, N. R. (1993). Can CO_2 assimilation in maize leaves be predicted accurately from chlorophyll fluorescence analysis? *Photosyn. Res.* **37,** 89–102.

Edwards, G. E. and Black, C. C., Jr. (1971). Isolation of mesophyll cells and bundle sheath cells from *Digitaria sanguinalis* (L.) Scop. leaves and a scanning microscopy study of the internal leaf cell morphology. *Plant Physiol.* **47,** 149–156.

Edwards, G. E. and Walker, D. A. (1983). C_3, C_4: Mechanisms, and cellular and environmental regulation, of photosynthesis. Blackwell Scientific Publ., Oxford, U.K. pp. 542.

Edwards, G. E., Kanai, R. and Black, C. C., Jr. (1971). Phosphoenolpyruvate carboxykinase in leaves of certain plants which fix CO_2 by the C_4-dicarboxylic acid cycle of photosynthesis. *Biochem. Biophys. Res. Commun.* **45,** 278–285.

Edwards, G. E., Lilley, R. M., Craig, S. and Hatch, M. D. (1979). Isolation of intact and functional chloroplasts from mesophyll and bundle sheath protoplasts of the C_4 plant *Panicum miliaceum*. *Plant Physiol.* **63,** 821–827.

Edwards, G. E., Nakamoto, H., Burnell, J. N. and Hatch, M. D. (1985). Pyruvate, Pi dikinase and NADP-malate dehydrogenase in C_4 photosynthesis: Properties and mechanism of light/dark regulation. *Annu. Rev. Plant Physiol.* **36,** 255–286.

Ehleringer, J. and Björkman, O. (1977). Quantum yields for CO_2 uptake in C_3 and C_4 plants. Dependence of temperature, CO_2, and O_2 concentration. *Plant Physiol.* **59,** 86–90.

Ehleringer, J. and Pearcy, R. W. (1983). Variation in quantum yield for CO_2 uptake among C_3 and C_4 plants. *Plant Physiol.* **73,** 555–559.

Evans, J. R., Sharkey, T. D., Berry, J. A. and Farquhar, G. D. (1986). Carbon isotope discrimination measured concurrently with gas-exchange to investigate CO_2 diffusion in leaves of higher plants. *Aust. J. Plant Physiol.* **13,** 281–292.

Evert, R. F., Escherich, W, and Heyser, W. (1977). Leaf structure in relation to solute transport and phloem leading in *Zea mays* L. *Planta* **138**, 279–294.

Fischer, K., Arbinger, B., Kammerer, B. Busch, C., Brink, S., Wallmeier, H., Sauer, N., Eckerskorn, C. and Flügge, U.-I. (1994). Cloning and *in vivo* expression of functional triose phosphate/phosphate translocators from C_3- and C_4-plants: evidence for the putative participation of specific amino acid residues in the recognition of phosphoenolpyruvate. *Plant J.* **5**, 215–226.

Flügge, U. I. and Heldt, H. W. (1991). Metabolite translocators of the chloroplast envelope. *Annu. Rev. Plant Physiol. Plant Mol. Biol.* **42**, 129–144.

Flügge, U. I., Stitt, M. and Heldt, H. W. (1985). Light-driven uptake of pyruvate into mesophyll chloroplasts from maize. *FEBS Lett.* **183**, 335–339.

Furbank, R. T. and Badger, M. R. (1982). Photosynthetic oxygen exchange in attached leaves of C_4 monocotyledons. *Aust. J. Plant Physiol.* **9**, 553–558.

Furbank, R. T., Jenkins, C. L. D. and Hatch, M. D. (1990). C_4 photosynthesis: Quantum requirement, C_4 acid overcycling and Q-cycle involvement. *Aust. J. Plant Physiol.* **17**, 1–7.

Genty, B., Briantais, J.-M, and Baker, N. R. (1989). The relationship between the quantum yield of photosynthetic electron transport and quenching of chlorophyll fluorescence. *Biochim. Biophys. Acta* **990**, 87–92.

Gross, A., Brückner, G., Heldt, H. W. and Flügge, U.-I. (1990). Comparison of the kinetic properties, inhibition and labelling of the phosphate translocators from maize and spinach mesophyll chloroplasts. *Planta* **180**, 262–271.

Gutierrez, M., Kanai, R., Huber, S. C., Ku, S. B. and Edwards, G. E. (1974a). Photosynthesis in mesophyll protoplasts and bundle sheath cells of various types of C_4 plants I. Carboxylases and CO_2 fixation studies. *Z. Pflanzenphysiol.* **72**, 305–319.

Gutierrez, M., Gracen, V. E. and Edwards, G. E. (1974b). Biochemical and cytological relationships in C_4 plants. *Planta* **119**, 279–300.

Gutierrez, M., Huber, S. C., Ku, S. B., Kanai, R. and Edwards, G. E. (1975). Intracellular localization of carbon metabolism in mesophyll cells of C_4 plants. *In* "III International Congress on Photosynthesis Research" (M. Avron, ed.), Vol. II, pp. 1219–1230. Elsevier Sci. Publ. Co., Amsterdam.

Harel, E., Lea, P. J. and Miflin, B. J. (1977). The localization of enzymes of nitrogen assimilation in maize leaves and their activities during greening. *Planta* **134**, 195–200.

Hatch, M. D. (1986). Has plant biochemistry finally arrived? *TIBS* **11**, 9–10.

Hatch, M. D. (1987). C_4 photosynthesis: A unique blend of modified biochemistry, anatomy and ultrastructure. *Biochim. Biophys. Acta* **895**, 81–106.

Hatch, M. D. and Burnell, J. N. (1990). Carbonic anhydrase activity in leaves and its role in the first step of C_4 photosynthesis. *Plant Physiol.* **93**, 825–828.

Hatch, M. D. and Kagawa, T. (1974). Activity, location and role of NAD malic enzyme in leaves with C_4-pathway photosynthesis. *Aust. J. Plant Physiol.* **1**, 357–369.

Hatch, M. D and Osmond, C. B. (1976). Compartmentation and transport in C_4 photosynthesis. *In* "Encyclopedia of Plant Physiology New Series" (C. R. Stocking and U. Heber, eds.) Vol. III, pp. 144–184. Springer-Verlag, Berlin.

Hatch, M. D. and Slack, C. R. (1969a). NADP-specific malate dehydrogenase and glycerate kinase in leaves and evidence for their localization in chloroplasts. *Biochem. Biophys. Res. Commun.* **34**, 589–593.

Hatch, M. D. and Slack, C. R. (1969b). Studies on the mechanism of activation and inactivation of pyruvate, phosphate dikinase. A possible regulatory role for the enzyme in the C_4 dicarboxylic acid pathway of photosynthesis. *Biochem. J.* **112**, 549–558.

Hatch, M. D., Mau, S.-L. and Kagawa, T. (1974). Properties of leaf NAD malic enzyme from plants with C_4 pathway photosynthesis. *Arch. Biochem. Biophys.* **165**, 188–200.

Hatch, M. D., Kagawa, T. and Craig, S. (1975). Subdivision of C_4-pathway species based on differing C_4 decarboxylating system and ultrastructural features. *Aust. J. Plant Physiol.* **2**, 111–128.

Hatch, M. D., Dröscher, L., Flügge, U. I. and Heldt, H. W. (1984). A specific translocator for oxaloacetate transport in chloroplasts. *FEBS Lett.* **178**, 15–19.

Hatch, M. D., Agostino, A. and Jenkins, C. L. D. (1995). Measurement of the leakage of CO_2 from bundle-sheath cells of leaves during C_4 photosynthesis. *Plant Physiol.* **108**, 173–181.

Hattersley, P. W., Watson, L. and Osmond, C. B. (1977). *In situ* immunofluorescent labelling of ribulose-1,5-bisphosphate carboxylase in leaves of C_3 and C_4 plants. *Aust. J. Plant Physiol.* **4**, 523–539.

He, D. and Edwards, G. E. (1996). Estimation of diffusive resistance of bundle sheath cells to CO_2 from modeling of C_4 photosynthesis. *Photosynthesis Res.* **49**, 195–208.

Henderson, S. A., von Caemmerer, S. and Farquhar, G. D. (1992). Short-term measurements of carbon isotope discrimination in several C_4 species. *Aust. J. Plant Physiol.* **19**, 263–285.

Hinkle, P. C., Kumar, M. A., Resetar, A., and Harris, D. L. (1991). Mechanistic stoichiometry of mitochondrial oxidative phosphorylation. *Biochemistry* **30**, 3576–3582.

Huber, S. C. and Edwards, G. E. (1977a). Transport in C_4 mesophyll chloroplasts. Characterization of the pyruvate carrier. *Biochim. Biophys. Acta* **462**, 583–602.

Huber, S. C. and Edwards, G. E. (1977b). Transport in C_4 mesophyll chloroplasts. Evidence for an exchange of inorganic phosphate and phosphoenolpyruvate. *Biochim. Biophys. Acta* **462**, 603–612.

Iglesias, A. A. and Andreo, C. S. (1990). Kinetic and structural properties of NADP-malic enzyme from sugarcane leaves. *Plant Physiol.* **92**, 66–72.

Jenkins, C. L. D. and Boag, S. (1985). Isolation of bundle sheath cell chloroplasts from the NADP-ME type C_4 plant *Zea mays*. *Plant Physiol.* **79**, 84–89.

Jenkins, C. L. D. and Russ, V. J. (1984). Large scale, rapid preparation of functional mesophyll chloroplasts from *Zea mays* and other C_4 species. *Plant Sci. Lett.* **35**, 19–24.

Jenkins, C. L. D., Burnell, J. M. and Hatch, M. D. (1987). Form of inorganic carbon involved as a product and as an inhibitor of C_4 acid decarboxylases operating in C_4 photosynthesis. *Plant Physiol.* **85**, 952–957.

Jenkins, C. L. D., Furbank, R. T. and Hatch, M. D. (1989). Mechanism of C_4 photosynthesis. A model describing the inorganic carbon pool in bundle sheath cells. *Plant Physiol.* **91**, 1372–1381.

Johnson, H. S. and Hatch, M. D. (1970). Properties and regulation of leaf nicotinamide-adenine dinucleotide phosphate dehydrogenase and 'malic' enzyme in plants with the C_4-dicarboxylic acid pathway of photosynthesis. *Biochem. J.* **119**, 273–280.

Kagawa, T. and Bruno, P. L. (1988). NADP-malate dehydrogenase from leaves of *Zea mays*: Purification and physical, chemical, and kinetic properties. *Arch. Biochem. Biophys.* **260**, 674–695.

Kanai, R. and Edwards, G. E. (1973a). Separation of mesophyll protoplasts and bundle sheath cells from maize leaves for photosynthetic studies. *Plant Physiol.* **51**, 1133–1137.

Kanai, R. and Edwards, G. E. (1973b). Purification of enzymatically isolated mesophyll protoplasts from C_3, C_4, and Crassulacean acid metabolism plants using an aqueous dextran-polyethylene glycol two-phase system. *Plant Physiol.* **52**, 484–490.

Kane, H. J., Viil, J., Entsch, B., Paul, K., Morell, M. K., and Andrews, J. T. (1994). An improved method for measuring the CO_2/O_2 specificity of ribulosebisphosphate carboxylase-oxygenase. *Aust. J. Plant Physiol.* **21**, 449–461.

Krall, J. P. and Edwards, G. E. (1992). Relationship between photosystem II activity and CO_2 fixation in leaves. *Physiol. Plant.* **86**, 180–187.

Krause, G. H. and Weis E. (1991). Chlorophyll fluorescence and photosynthesis: The basics. *Annu. Rev. Plant Physiol. Plant Mol. Biol.* **42**, 313–349.

Ku, M. S. B. and Edwards, G. E. (1975). Photosynthesis in mesophyll protoplasts and bundle sheath cells of various types of C_4 plants. IV. Enzymes of respiratory metabolism and energy utilizing enzymes of photosynthetic pathways. *Z. Pflanzenphysiol.* **77**, 16–32.

Ku, M. S. B., Schmitt, M. R., and Edwards, G. E. (1979). Quantitative determination of RuBP carboxylase–oxygenase protein in leaves of several C_3 and C_4 plants. *J. Exp. Bot.* **30**, 89–98.

Laetsch, W. M. (1971). Chloroplast structural relationships in leaves of C_4 plants. *In* "Photosynthesis and Photorespiration" (M. D. Hatch, C. B. Osmond and R. O. Slatyer, eds.), pp. 323–349. Wiley-Interscience, New York.

Lal, A. and Edwards, G. E. (1995). Maximum quantum yields of O_2 evolution in C_4 plants under high CO_2. *Plant Cell Physiol.* **36**, 1311–1317.

Leegood, R. C. (1985). The intercellular compartmentation of metabolites in leaves of *Zea mays* L. *Planta* **164**, 163–171.

Leegood, R. C. and von Caemmerer, S. (1988). The relationship between contents of photosynthetic metabolites and the rate of photosynthetic carbon assimilation in leaves of *Amaranthus edulis* L. *Planta* **174**, 253–262.

Leegood, R. C. and Edwards, G. E. (1996). Carbon metabolism and photorespiration: Temperature dependence in relation to other environmental factors. *In* "Photosynthesis and the Environment" (N. R. Baker ed.), Advances in Photosynthesis Series, Kluwer Academic Publ. pp. 191–221.

Leegood, R. C. and Osmond, C. B. (1990). The flux of metabolites in C_4 and CAM plants. *In* "Plant Physiology, Biochemistry and Molecular Biology" (D. T. Dennis and D. H. Turpin eds.), Longman, pp. 274–298.

Long, J. J., Wang, J.-L., and Berry, J. O. (1994). Cloning and analysis of the C_4 photosynthetic NAD-dependent malic enzyme of amaranth mitochondria. *J. Biol. Chem.* **269**, 2827–2833.

Marek, L. F. and Stewart, C. R. (1983). Photorespiratory glycine metabolism in corn leaf discs. *Plant Physiol.* **73**, 118–120.

Monson, R. K., Littlejohn, R. O. and Williams, G. J. (1982). The quantum yield for CO_2 uptake in C_3 and C_4 grasses. *Photosyn. Res.* **3**, 153–159.

Moore, B., Ku, M. S. B. and Edwards, G. E. (1984). Isolation of leaf bundle sheath protoplasts from C_4 dicots and intracellular localization of selected enzymes. *Plant Sci. Lett.* **35**, 127–138.

Murata, T., Ohsugi, R. Matsuoka, M. and Nakamoto, H. (1989). Purification and characterization of NAD malic enzyme from leaves of *Eleusine coracana* and *Panicum dichotomiflorum*. *Plant Physiol.* **89**, 316–324.

Numazawa, T., Yamada, S., Hase, T. and Sugiyama, T. (1989). Aspartate aminotransferase from *Panicum maximum* Jacq. var. *trichoglume* Eyles, a C_4 plant: Purification, molecular properties, and preparation of antibody. *Arch. Biochem. Biophys.* **270**, 313–319.

Oberhuber, W. and Edwards, G. E. (1993). Temperature dependence of the linkage of quantum yield of photosystem II to CO_2 fixation in C_4 and C_3 plants. *Plant Physiol.* **101**, 507–512.

Oberhuber, W., Dai, Z.-Y. and Edwards, G. E. (1993). Light dependence of quantum yields of photosystem II and CO_2 fixation in C_3 and C_4 plants. *Photosyn. Res.* **35**, 265–274.

Ohnishi, J. and Kanai, R. (1987a). Pyruvate uptake by mesophyll and bundle sheath chloroplasts of a C_4 plant, *Panicum miliaceum* L. *Plant Cell Physiol.* **28**, 1–10.

Ohnishi, J. and Kanai, R. (1987b). Light-dependent uptake of pyruvate by mesophyll chloroplasts of a C_4 plant, *Panicum miliaceum* L. *Plant Cell Physiol.* **28**, 243–251.

Ohnishi, J. and Kanai, R. (1987c). Na^+-induced uptake of pyruvate into mesophyll chloroplasts of a C_4 plant, *Panicum miliaceum*. *FEBS Lett.* **219**, 347–350.

Ohnishi, J. and Kanai, R. (1988). Glycerate uptake into mesophyll and bundle sheath chloroplasts of a C_4 plant, *Panicum miliaceum*. *J. Plant Physiol.* **133**, 119–121.

Ohnishi, J. and Kanai, R. (1990). Pyruvate uptake induced by a pH jump in mesophyll chloroplasts of maize and sorghum, NADP-malic enzyme type C_4 species. *FEBS Lett.* **269**, 122–124.

Ohnishi, J., Yamazaki, M. and Kanai, R. (1985). Differentiation of photorespiratory activity between mesophyll and bundle sheath cells of C_4 plants II. Peroxisomes of *Panicum miliaceum* L. *Plant Cell Physiol.* **26,** 797–803.

Ohnishi, J., Flügge, U. I. and Heldt, H. W. (1989). Phosphate translocator of mesophyll and bundle sheath chloroplasts of a C_4 plant, *Panicum miliaceum* L. Identification and kinetic characterization. *Plant Physiol.* **91,** 1507–1511.

Ohnishi, J., Flügge, U.-I., Heldt, H. W. and Kanai, R. (1990). Involvement of Na^+ in active uptake of pyruvate in mesophyll chloroplasts of some C_4 plants. Na^+/pyruvate cotransport. *Plant Physiol.* **94,** 950–959.

Ohsugi, R., Murata, T. and Chonan, N. (1982). C_4 syndrome of the species in the *Dichotomiflora* group of the genus *Panicum* (Gramineae). *Bot. Mag. Tokyo* **95,** 339–347.

Osmond, C. B. (1971). Metabolite transport in C_4 photosynthesis. *Aust. J. Biol. Sci.* **24,** 159–163.

Osmond, C. B. and Grace, S. C. (1995). Perspectives on photoinhibition and photorespiration in the field: Quintessential inefficiencies of the light and dark reactions of photosynthesis? *J. Exp. Bot.* **46,** 1351–1362.

Osmond, C. B. and Smith, F. A. (1976). Symplastic transport of metabolites during C_4 photosynthesis. *In* "Intercellular Communication in Plants: Studies on Plasmodesmata" (B. E. S. Gunning and A. W. Robards, eds.) pp. 229–240. Springer Verlag, New York.

Prendergast, H. D. V., Hattersley, P. W., Stone, N. E. and Lazarides, M. (1986). C_4 acid decarboxylation type in *Eragrostis* (Poaceae): Patterns of variation in chloroplast position, ultrastructure and geographical distribution. *Plant Cell Environ.* **9,** 333–344.

Rathnam, C. K. M. and Edwards, G. E. (1976). Distribution of nitrate-assimilating enzymes between mesophyll protoplasts and bundle sheath cells in leaves of three groups of C_4 plants. *Plant Physiol.* **57,** 881–885.

Rebeille, F. and Hatch, M. D. (1986). Regulation of NADP-malate dehydrogenase in C_4 plants: Relationship among enzyme activity, NADPH to NADP ratios, and thioredoxin redox states in intact maize mesophyll chloroplasts. *Arch. Biochem. Biophys.* **249,** 171–179.

Rumberg, B. and Berry, S. (1995). Refined measurement of the H^+/ATP coupling ratio at the ATP synthase of chloroplasts. *In* "Photosynthesis Research. Xth International Photosynthesis Congress" (P. Mathis, ed.), Vol. III, pp. 139–142, Kluwer Academic Pubishers, Dordrecht.

Sage, R. F., Pearcy R. W. and Seemann, J. R. (1987). The nitrogen use efficiency of C_3 and C_4 plants. *Plant Physiol.* **85,** 355–359.

Samejima, M and Miyachi, S. (1971). Light-enhanced dark carbon dioxide fixation in maize leaves. *In* "Photosynthesis and Photorespiration" (M. D. Hatch, C. B. Osmond and R. O. Slatyer, eds.), pp. 211–217, Wiley-Interscience, New York.

Samejima, M. and Miyachi, S. (1978). Photosynthetic and light-enhanced dark fixation of $^{14}CO_2$ from the ambient atmosphere and ^{14}C-bicarbonate infiltrated through vascular bundles in maize leaves. *Plant Cell Physiol.* **19,** 907–916.

Seemann, J. R., Badger, M. R. and Berry, J. A. (1984). Variations in the specific activity of ribulose-1,5-bisphosphate carboxylase between species utilizing differing photosynthetic pathways. *Plant Physiol.* **74,** 791–794.

Sheen, J.-Y. and Bogorad, L. (1985). Differential expression of the ribulose bisphosphate carboxylase large subunit gene in bundle sheath and mesophyll cells of developing maize leaves is influenced by light. *Plant Physiol.* **79,** 1072–1076.

Slack, C. R. (1969). Localization of photosynthetic enzymes in mesophyll and parenchyma sheath chloroplasts of maize and *Amaranthus palmeri*. *Phytochemistry* **8,** 1387–1391.

Slack, C. R. and Hatch, M. D. (1967). Comparative studies on the activity of carboxylases and other enzymes in relation to the new pathway of photosynthetic carbon dioxide fixation in tropical grasses. *Biochem. J.* **103,** 660–665.

Slack, C. R., Hatch, M. D. and Goodchild, D. J. (1969). Distribution of enzymes in mesophyll and parenchyma-sheath chloroplasts of maize leaves in relation to the C_4-dicarboxylic acid pathway of photosynthesis. *Biochem. J.* **114,** 489–498.

Son, D., Jo, J. and Sugiyama, T. (1991). Purification and characterization of alanine aminotransferase from *Panicum miliaceum* leaves. *Arch. Biochem. Biophys.* **289,** 262–266.

Stitt, M. and Heldt, H. W. (1985). Generation and maintenance of concentration gradients between the mesophyll and bundle sheath in maize leaves. *Biochim. Biophys. Acta* **808,** 400–414.

Sugiyama, T. (1973). Purification, molecular, and catalytic properties of pyruvate phosphate dikinase from maize leaf. *Biochemistry* **12,** 2862–2868.

Taniguchi, M. (1986). $^{14}CO_2$ fixation, transport and decarboxylation of malate in isolated intact chloroplasts from bundle sheath cell of maize. MSc Thesis, Graduate School of Saitama University.

Taniguchi, M. and Sugiyama, T. (1990). Aspartate aminotransferase from *Eleusine coracana*, a C_4 plant: Purification, characterization, and preparation of antibody. *Arch. Biochem. Biophys.* **282,** 427–432.

Taniguchi, M., Kobe, A., Kato, M. and Sugiyama, T. (1996). Aspartate aminotransferase isozymes in *Panicum miliaceum* L, an NAD-malic enzyme-type C_4 plant: Comparison of enzymatic properties, primary structures, and expression patterns. *Arch. Biochem. Biophys.* **318,** 295–306.

Uedan, K. and Sugiyama, T. (1976). Purification and characterization of phosphoenolpyruvate carboxylase from maize leaves. *Plant Physiol.* **57,** 906–910.

Usuda, H. and Edwards, G. E. (1980a). Localization of glycerate kinase and some enzymes for sucrose synthesis in C_3 and C_4 plants. *Plant Physiol.* **65,** 1017–1022.

Usuda, H. and Edwards, G. E. (1980b). Photosynthetic formation of glycerate in isolated bundle sheath cells and its metabolism in mesophyll cells of the C_4 plant *Panicum capillare* L. *Aust. J. Plant Physiol.* **7,** 655–662.

Utsunomiya, E. and Muto, S. (1993). Carbonic anhydrase in the plasma membranes from leaves of C_3 and C_4 plants. *Physiol. Plant.* **88,** 413–419.

Valle, E. M., Craig, S., Hatch, M. D. and Heldt, H. W. (1989). Permeability and ultrastructure of bundle sheath cells isolated from C_4 plants: structure–function studies and the role of plasmodesmata. *Bot. Acta* **102,** 276–282.

Walker, R. P. and Leegood, R. C. (1996). Phosphorylation of phosphoenolpyruvate carboxykinase in plants. Studies in plants with C_4 photosynthesis and Crassulacean acid metabolism and in germinating seeds. *Biochem. J.* **317,** 653–658.

Walker, R. P., Acheson, R. M., Técsi, L. I., and Leegood, R. C. (1997). Phosphoenolpyruvate carboxykinase in C_4 plants: Its role and regulation. *Aust. J. Plant Physiol.* **24,** 459–468.

Wang, J.-L., Klessig, D. F. and Berry, J. O. (1992). Regulation of C_4 gene expression in developing amaranth leaves. *The Plant Cell* **4,** 173–184.

Wang, J.-L., Turgeon, R., Carr, J. P., and Berry, J. O. (1993). Carbon sink-to-source transition is coordinated with establishment of cell-specific gene expression in a C_4 plant. *The Plant Cell* **5,** 289–296.

Watanabe, M., Ohnishi, J. and Kanai, R. (1984). Intracellular localization of phosphoenolpyruvate carboxykinase in bundle sheath cells of C_4 plants. *Plant Cell Physiol.* **25,** 69–76.

Weber, A., Menzlaff, E., Arbinger, B., Gutensohn, M., Eckerskorn, C. and Flügge, U.-I. (1995). The 2-oxoglutarate/malate translocator of chloroplast envelope membrane: molecular cloning of a transporter containing a 12-helix motif and expression of the functional protein in yeast cells. *Biochemistry* **34,** 2621–2627.

Weiner, H., Burnell, J. N., Woodrow, I. E., Heldt, H. W. and Hatch, M. D. (1988). Metabolite diffusion into bundle sheath cells from C_4 plants. Relation to C_4 photosynthesis and plasmodesmatal function. *Plant Physiol.* **88,** 815–822.

Werner-Washburne, M. and Keegstra, K. (1985). L-Aspartate transport into pea chloroplasts: Kinetic and inhibitor evidence for multiple transport systems. *Plant Physiol.* **78,** 221–227.

Yeoh, H.-H., Badger, M. R. and Watson, L. (1980). Variations in $K_m(CO_2)$ of ribulose-1,5-bisphosphate carboxylase among grasses. *Plant Physiol.* **66,** 1110–1112.

Yeoh, H.-H., Badger, M. R. and Watson, L. (1981). Variations in kinetic properties of ribulose-1,5-bisphosphate carboxylases among plants. *Plant Physiol.* **67,** 1151–1155.

4

Regulation of the C$_4$ Pathway

Richard C. Leegood and Robert P. Walker

I. Introduction

The flux of carbon through the C$_4$ pathway must be regulated so that it is coordinated with other components of photosynthetic metabolism under different environmental conditions. For example, the rate of fixation of CO$_2$ in the mesophyll and its subsequent release in the bundle sheath needs to be coordinated with the rate at which it can be assimilated by the Benson–Calvin cycle. The regulation of photosynthetic metabolism in leaves of C$_4$ plants is, therefore, more complex than in C$_3$ plants and is inextricably linked with the transport of metabolites between the two cell types. Most, if not all, of the mechanisms used to regulate photosynthetic metabolism in C$_3$ plants are present in C$_4$ plants, although many of these mechanisms have been modified and additional mechanisms have arisen. Additional complexity may also arise in the variants of the C$_4$ cycle that are found within the acknowledged decarboxylation subtypes (Meister *et al.*, 1996; Walker *et al.*, 1997; see Chapter 3 by Kanai and Edwards, this volume). In this chapter we discuss how the C$_4$ cycle is regulated and how other components of photosynthetic metabolism have been modified in order to coordinate them with the C$_4$ cycle.

II. An Overview of the Regulation of the C$_4$ Cycle

The regulation of the C$_4$ cycle and its integration with other components of photosynthetic metabolism is dependent on several interacting mechanisms. The compartmentation of different parts of the cycle, both in different cells and subcellular compartments, forms the structural basis for this regulation. Superimposed on this is the modulation of both enzymes of the cycle and the proteins that transport the intermediates across membranes. Fluxes and the concentration of intermediates of the cycle, and of other interacting metabolic pathways, are important in coordinating the modulation of these enzymes and membrane transport proteins.

Compartmentation between different cell types is important; for example, CO_2 enters the C_4 cycle in mesophyll cells and is released from it in bundle-sheath cells, thus preventing a futile cycle between carboxylation and decarboxylation. Compartmentation within cells is also important; for example, in PEP carboxykinase (PEP-CK)–type species, decarboxylation of oxaloacetate by PEP-CK occurs in the cytosol. This avoids a high concentration of oxaloacetate in the mitochondria, which would deplete the mitochondria of NADH by the action of mitochondrial malate dehydrogenase.

A. Light-Dependent Changes in the Concentration of Intermediates Regulate the C_4 Cycle

An essential part of the C_4 cycle is the transfer of the products of the carboxylation reaction from the mesophyll to the bundle sheath and the return to the mesophyll of the products of the decarboxylation reaction (Fig. 1). In this example, involving the NADP-malic enzyme species *Zea mays* (maize), up to 50% of the glycerate-3-P produced by the Benson–Calvin cycle in the bundle sheath is transported to the mesophyll, where it is reduced to triose-P, which is then returned to the bundle sheath for regeneration of RuBP. On illumination of leaves of C_4 plants these metabolites increase to a high concentration, and concentration gradients are established between the mesophyll and bundle sheath (Leegood, 1985; Stitt and Heldt, 1985a,b). Metabolite movement between the bundle sheath and the mesophyll is sustained by diffusion via numerous plasmodesmata, driven by gradients in their concentrations. A striking feature of the leaves of C_4 plants is their high content of metabolites when compared with C_3 plants. For example, contents of triose-P may be 20 times higher in maize than in barley (Fig. 1), reflecting the internal metabolite gradients that are established during photosynthesis. Hatch and Osmond (1976) estimated that an intercellular gradient of 10 mM would be needed to sustain observed rates of photosynthesis in maize (which has centrifugally arranged bundle-sheath chloroplasts and a short diffusion path) and a gradient of 30 mM would be needed in *Amaranthus* (which has centripetally arranged bundle-sheath chloroplasts and a longer diffusion path). However, subsequent direct measurement of diffusion constants in isolated bundle sheath strands for a range of small molecular mass compounds (values of *ca.* 3 μmol min^{-1} mg^{-1} chlorophyll mM^{-1}) has resulted in a revision of the required gradient down to 2 mM (Weiner *et al.*, 1988).

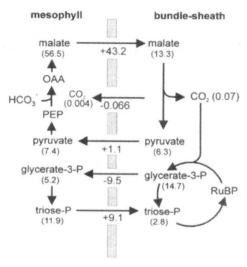

Figure 1 Metabolite concentrations in, and metabolite gradients between, the mesophyll and bundle-sheath cells of maize. Mean values for metabolite concentrations (mM) are taken from Leegood (1985) and Stitt and Heldt (1985a) and for cytosolic CO_2 from Jenkins *et al.* (1989b). The gradient in concentration relative to the bundle-sheath cells is shown across the plasmodesmata. [Redrawn from Leegood, R. C., von Caemmerer, S., and Osmond, C. B. (1997). Metabolite transport and photosynthetic regulation in C_4 and CAM plants. *In* "Plant Metabolism" (D. T. Dennis, D. H. Turpin, D. D. Lefebvre, and D. B. Layzell, eds.), pp. 341–369, Longman, London. Reprinted by permission of Addison Wesley Longman Ltd.]

Direct measurements of metabolite levels in maize leaves have shown the intercellular concentration gradients present during photosynthesis (Fig. 1) are substantial enough to support the shuttling of these compounds at rates greater than required for CO_2 assimilation (Leegood, 1985; Stitt and Heldt, 1985b; Weiner and Heldt, 1992). The most striking feature is that most of the triose-P is in the mesophyll, whereas most of the glycerate-3-P is in the bundle sheath.

Within the C_4 cycle, a gradient of pyruvate undoubtedly exists, but is obscured by the fact that the mesophyll chloroplasts actively take up pyruvate (Flügge *et al.*, 1985), providing a gradient between the mesophyll cytosol and the bundle sheath that is not evident when pyruvate is measured in the mesophyll as a whole. The gradient of malate is difficult to estimate, because much of the malate pool is nonphotosynthetic [99% in *Chloris gayana;* Hatch (1979)] and is present in the vacuole or in nonphotosynthetic leaf cells. No direct measurements have been made of metabolite gradients in leaves of C_4 plants other than maize, although in leaves of *Amaranthus edulis,* an NAD-ME–type species, it has been shown that aspartate and alanine gradients are sufficient to account for diffusion-driven transport of these compounds between the mesophyll and bundle sheath (Leegood and von Caemmerer, 1988). Transport in PEP-CK type plants may be rather more complex, because both malate and aspartate must be transferred from the mesophyll to the bundle sheath, and both PEP and alanine (or pyruvate) return to the mesophyll.

Changes in concentration of metabolites are important in coordinating the C_4 cycle with other processes of photosynthetic metabolism in several ways (Leegood, 1997):

1. In NADP-ME species, such as maize, the C_4 cycle is coupled to the Benson–Calvin cycle because NADPH produced by the decarboxylation of malate in bundle-sheath chloroplasts in used to reduce glycerate-3-P, an intermediate of the Benson–Calvin cycle.

2. In all C_4 plants a large proportion of glycerate-3-P is transported from bundle-sheath chloroplasts to mesophyll chloroplasts, where NADPH produced by noncyclic electron transport reduces it to triose phosphate.

3. Interconversion of glycerate-3-P and PEP by phosphoglycerate mutase and enolase in the mesophyll cytosol provides metabolic communication between the C_4 and Benson–Calvin cycles.

4. A number of intermediates of the C_4 and Benson–Calvin cycles act as effectors of enzyme activity. For example, triose phosphates and hexose phosphates stimulate PEP-C activity whereas malate inhibits it (Gadal *et al.*, 1996).

5. An increase in glycerate-3-P concentration on illumination is thought to increase the pH of the mesophyll cytosol, which brings about an increase in the concentration of cytosolic Ca^{2+} by increasing the permeability of Ca^{2+} channels in the tonoplast. An increase in cytosolic Ca^{2+} is proposed to be a component of a signal transduction pathway that leads to the phosphorylation and therefore activation of PEP-C (Giglioli-Guivarc'h *et al.*, 1996).

We have seen that the C_4 pathway involves metabolites moving between the cytosols of the mesophyll and bundle sheath by diffusion driven by large concentration gradients, and that it may also involve other organelles, such as the mitochondria. This means that competing metabolic processes, such as entry of carbon into respiratory pathways and into sucrose synthesis, are potentially flooded with substrates and require regulation to prevent the dissipation of the metabolite shuttles involved in C_4 photosynthesis.

B. Intracellular Transport of Metabolites Regulates the C_4 Cycle

The transport of metabolites between different subcellular compartments is mediated by transport proteins and is an essential component of the C_4 cycle. The exchange of metabolites occurs at a much greater rate than in

C_3 plants and, in addition, C_4 plants transport a number of metabolites that are not transported to any great extent in C_3 plants (for a more detailed account, see Leegood, 1997). Organelles from leaves of C_4 plants possess the translocators that occur in C_3 plants but also possess translocators with unique, or considerably altered, kinetic properties (Heldt and Flügge, 1992). The rate at which metabolites are transported between subcellular compartments have a pronounced effect on flux through the C_4 cycle. Modulation of the activity of translocators could, therefore, be of great importance in the regulation of the C_4 cycle. Many C_4 translocators are antiports—that is, they exchange one metabolite for another; for example, the P_i translocator in mesophyll cell chloroplast envelopes exchanges P_i for PEP, glycerate-3-P, or triose phosphate. The concentration of P_i in the cytosol can therefore regulate the export of these metabolites from the chloroplast. There are some suggestions that translocators that participate in the C_4 cycle are regulated. For example, the pyruvate translocator from mesophyll chloroplasts is much more active in illuminated leaves (Flügge et al., 1985; Chapter 10 by Sage, Wedin, and Li, this volume). Ohnishi et al. (1989) showed that, in mesophyll chloroplasts prepared from *Panicum miliaceum* leaves, light enhances the transport of PEP compared to glycerate-3-P and triose phosphate by the P_i translocator. This would favor the export of PEP, which is regenerated from pyruvate by pyruvate,P_i dikinase in mesophyll chloroplasts.

C. Modulation of Enzyme Activity Regulates the C_4 Cycle

Photosynthetic carbon metabolism in either C_3 or C_4 plants is a remarkable process in that few, if any, other metabolic pathways are capable of such large and rapid flux changes, even increasing rapidly from zero in the case of a darkened leaf. These changes in flux can occur in a matter of few minutes or seconds under natural conditions, generally in response to fluctuations in light intensity. As Fell (1997) points out, these large flux changes in a pathway cannot be brought about by changes in the activities of one or two enzymes, and multisite coordinated modulation is the only effective way to achieve large changes in metabolic flux. As in the Benson–Calvin cycle, the majority (if not all) of the enzymes involved in the C_4 pathway are subject to control either directly, or indirectly, by light. On illumination, changes in the concentration of many metabolites occur (Fig. 2). These changes can modulate enzyme activity in two ways: (1) they may be effectors of an enzyme and alter its activity by directly interacting with it, or (2) they may cause a change in the activation state of the enzyme by activating, either directly or indirectly, signal transduction pathways that result in covalent modification of the enzyme by reversible protein phosphorylation (Giglioli-Guivarc'h et al., 1996). Often these two mechanisms act together and increase the precision of regulation. For example,

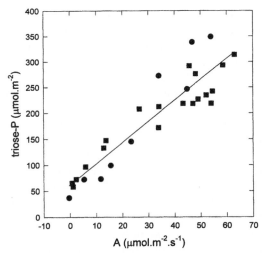

Figure 2 Relationship between the content of triose-P and the CO_2 assimilation rate in leaves of *Amaranthus edulis*. The rate of photosynthesis was varied by changing the intercellular CO_2 concentration (●) or the photon flux density (■). 40 μmol.m^{-2} of triose-P would correspond to a concentration of 1mM if distributed between the cytosol and chloroplasts of the mesophyll and bundle sheath with a collective volume of 100 μl.mg^{-1} chlorophyll (see Leegood and von Caemmerer, 1988). [Reprinted with permission from Leegood, R. C. (1997). The regulation of C_4 photosynthesis. *Adv. Bot. Res.* **26**, 251–316.]

glycerate-3-P activates PEP-C by interacting with it directly, and it is also thought to be an important signal that brings about its phosphorylation and therefore activation. Phosphorylation is also involved in the regulation of pyruvate, P_i dikinase and, in a few C_4 plants, in the regulation of PEP-CK. Light can also activate C_4 enzymes by bringing about a change in conformation as a result of reduction of sulphydryl groups. This is mediated by a coupling of photosynthetic electron transport to the reduction of thioredoxin, a soluble protein that reduces disulphide groups on proteins. NADP-malate dehydrogenase (NADP-MDH) is regulated in this way. Illumination can also modulate enzyme activity by bringing about changes in concentrations of ions and pH. For example, on illumination the pH of the chloroplast stroma rises from 7 to 8 and the concentration of Mg^{2+} rises from 1–3 mM to 2–6 mM (see Leegood *et al.*, 1985). Both these factors bring about an increase the activity of NADP-ME and probably other enzymes, such as pyruvate,P_i dikinase.

Another important factor to consider is the speed at which an enzyme's activity is modulated by covalent modification. For example, NADP-MDH is rapidly activated by the thioredoxin system, whereas for PEP-C it takes an hour or more to be fully phosphorylated on illumination of leaves (Gadal *et al.,* 1996). The slow response of some enzymes, for example PEP-C, may reduce sudden fluctuations in concentration of metabolites that are situated at the junctions of metabolic pathways (Gadal *et al.,* 1996) and will also tend to buffer the system against sudden changes in light intensity.

III. Regulation of the C_4 Cycle

A. Phosphoenolpyruvate Carboxylase

In C_4 plants, PEP-C activity must be regulated in response to light intensity in order to coordinate C_4 and C_3 photosynthetic metabolism, and in the dark the use of PEP by PEP-C must be regulated because PEP lies at an important branchpoint in metabolism. Measurement of the amount of PEP in leaves of C_4 plants provides evidence that PEP-C activity is modulated *in vivo*. For example, although the amount of PEP in leaves of C_4 plants increases on illumination, it soon decreases to the amount found in darkened leaves (Furbank and Leegood, 1984; Usuda, 1986; Roeske and Chollet, 1989). The simplest explanation of this is that, on illumination, PEP-C is activated but that activation is slower than the activation of other photosynthetic processes. It is now known that this activation of PEP-C occurs by reversible protein phosphorylation (see later this chapter). Similarly, during steady-state photosynthesis the amount of PEP is little affected by changes in light intensity (Usuda, 1987; Leegood and von Caemmerer, 1988, 1989), showing that the activity of PEP-C is regulated.

The mechanisms that regulate PEP-C activity in plants have been the subject of much research since the late 1970s and has led to the view that PEP-C activity is regulated both by metabolites, which interact directly with the enzyme to modulate its activity, by changes in cytosolic pH, and also by reversible phosphorylation of a serine residue that alters the sensitivity of the enzyme to these metabolite effectors (Figs. 3 and 4; for reviews, see Chollet *et al.,* 1996; Gadal *et al.,* 1996; Nimmo *et al.,* 1996). Our understanding of these mechanisms has come about by study of both C_4 and CAM plants. Early studies, in which the effects of various metabolites on PEP-C activity *in vitro* were characterized, established that phosphorylated intermediates (triose phosphate and hexose phosphate) were activators of PEP-C, whereas organic acids such as malate, and amino acids such as aspartate and glutamate, were inhibitors (Lowe and Slack, 1971; Nishikido and Takanashi, 1973; Ting and Osmond, 1973; Coombs *et al.,* 1975; Huber and Edwards, 1975; Uedan and Sugiyama, 1976). However, it was not clear how these metabolites could regulate PEP-C activity *in vivo*. For example, the amount of malate, an inhibitor of PEP-C, increases in illuminated maize leaves when the enzyme should be active. A key observation was that PEP-C activity in rapidly prepared extracts of leaves from the CAM plant *Mesembryanthemum*

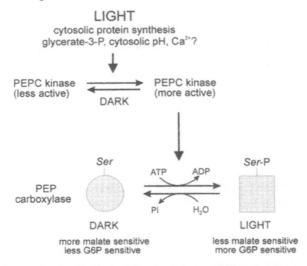

Figure 3 Schematic representation of the regulation of PEP carboxylase in C_4 plants by light/dark signals. Phosphorylation of a single seryl residue (Ser^{15} in maize or Ser^8 in sorghum) induces a change in the properties of PEP carboxylase. (With permission, from the *Annual Review of Plant Physiology and Plant Molecular Biology*, Volume 47, © 1996, by Annual Reviews, Inc.)

crystallinum differed in its sensitivity to inhibition by malate (Winter, 1982) depending on whether the leaves had been illuminated or darkened before extraction. Wu and Wedding proposed that this change in malate sensitivity was brought about by a change in the oligomerization state of PEP-C (Wu and Wedding, 1985, 1987), but this is now thought to be artefactual (McNaughton *et al.*, 1989). The mechanism for the regulation of PEP-C in the CAM plants was elucidated by Nimmo's group (Nimmo *et al.*, 1984, 1986, 1987a; Carter *et al.*, 1990). A combination of feeding *Bryophyllum fedtschenkoi* leaves $^{32}P_i$ and subsequent SDS–PAGE analysis of immunoprecipitates established that PEP-C was phosphorylated at night and dephosphorylated during the day (Nimmo *et al.*, 1984). The phosphorylated night form was purified and dephosphorylated *in vitro* by treatment with alkaline phosphatase, and dephosphorylation correlated with a marked increase in malate sensitivity (Nimmo *et al.*, 1986).

The work on PEP-C in CAM plants stimulated similar studies in C_4 plants, and several groups soon reported that the malate sensitivity of PEP-C changed in response to illumination. PEP-C was found to be more sensitive to inhibition by malate when extracted from darkened leaves (Huber *et al.*, 1986; Doncaster and Leegood, 1987; Nimmo *et al.*, 1987b). The correlation between changes in malate sensitivity and phosphorylation state was soon established. Budde and Chollet (1986) showed that PEP-C was phosphory-

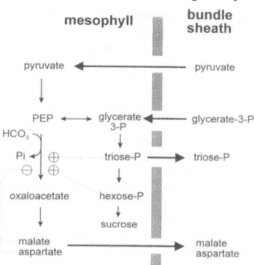

Figure 4 Feedback regulation of PEP carboxylase by metabolites, showing the inhibitory influence of the products (malate and aspartate), activating metabolites deriving from the Benson–Calvin cycle in bundle sheath (triose-P and hexose-P) and the interconversion of PEP and glycerate-3-P, by which the Benson–Calvin cycle and the C_4 cycle can interact directly. Most cofactors and so on are omitted for clarity.

lated *in vitro* exclusively on serine residues by an ATP-dependent soluble protein kinase present in illuminated maize leaves. By feeding maize leaves $^{32}P_i$, it was shown that PEP-C underwent marked changes in the degree of phosphorylation *in vivo* and that changes in phosphorylation state were correlated with changes in malate sensitivity (Nimmo *et al.*, 1987b; Jiao and Chollet, 1988). The enzyme from darkened leaves had a K_i for malate of 0.3 mM, whereas from illuminated leaves the K_i was 0.95 mM (McNaughton *et al.*, 1989). To show unequivocally that changes in the phosphorylation state of PEP-C are responsible for changes in malate sensitivity, Jiao and Chollet (1989) phosphorylated PEP-C purified from darkened maize leaves *in vitro* using a partially purified kinase preparation from maize leaves. They found that changes in maximum catalytic activity and malate sensitivity were directly correlated with changes in the degree of phosphorylation. The importance of the phosphorylation of PEP-C was shown by Bakrim *et al.* (1993), who fed detached leaves a protein synthesis inhibitor. The inhibitor did not perturb the Benson–Calvin cycle, photosynthetic electron flow, photophosphorylation, or stomatal functioning. However, the reduction in the phosphorylation state of PEP-C was correlated with a reduction in photosynthetic rate. A similar result was obtained with mesophyll cell protoplasts (Pierre *et al.*, 1992).

Protein sequencing of PEP-C phosphorylated *in vitro* established that, in maize, the phosphorylation site was located at Ser^{15}, which is within the small N-terminal extension (Jiao and Chollet, 1989). Subsequently it was shown that this serine, or in sorghum, its structural homolog, was the only residue labelled *in vivo* (Jiao *et al.*, 1991b). Other evidence supports this view. McNaughton *et al.* (1989) showed that PEP-C in extracts of darkened maize leaves rapidly lost its malate sensitivity, and this was correlated with loss of a 4 kDa fragment from the protein by proteolysis. Jiao and Chollet (1990) propose that the structural motif Lys/Arg-X-X-Ser is the regulatory phosphorylation site in all plant PEP-Cs. To provide unequivocal proof that phosphorylation of a single serine residue was responsible for modification of the enzyme properties, PEP-C, in which the serine residue had been changed to a different amino acid by site-directed mutagenesis, was overexpressed in *Escherichia coli*. Changing the serine to aspartate (Wang *et al.*, 1992) or cysteine (Duff *et al.*, 1993) clearly established that this was the phosphorylation site and that the effect of phosphorylation could be mimicked by introduction of a negative charge (aspartate) into the N-terminal domain of the protein. Li *et al.* (1997) used a series of synthetic peptides based on the region of the protein containing the phosphorylation site and a complementary set of recombinant mutant target proteins, and found that although the recombinant protein was phosphorylated by a highly purified preparation of PEP-C kinase, the complementary peptide was not. This suggests that a region removed from the N-terminal domain is necessary for its interaction with PEP-C kinase.

Although the kinetic properties of C_4 PEP-C have been characterized by several groups, a problem with many of these studies was the uncertain state of protein phosphorylation and concomitant malate sensitivity (see Echevarria *et al.*, 1994; Duff *et al.*, 1995). To avoid these uncertainties, the kinetic characteristics of sorghum PEP-C, which was overexpressed in *E. coli* and phosphorylated *in vitro*, were studied. PEP-C was much less sensitive to inhibition by malate when assayed at its pH optimum (8.0) rather than at the pH of the cytosol (7.3) (Echevarria *et al.*, 1994; Duff *et al.*, 1995). These studies showed that phosphorylation of the enzyme from sorghum had minor effects on both its maximum catalytic activity and affinity for PEP but a dramatic effect on its activation by glucose-6-phosphate and inhibition by malate.

The discovery that PEP-C is regulated by reversible protein phosphorylation prompted the search for its protein kinase (PEP-C kinase). Measurement of PEP-C kinase activity *in vitro* has relied on incubating purified PEP-C with a source of PEP-C kinase and either measuring the amount of ^{32}P incorporated into PEP-C from [γ-^{32}P]ATP or measuring changes in malate sensitivity of PEP-C. A problem with this approach is that many protein kinases, even cyclic AMP–dependent protein kinase from mammals, phosphorylate PEP-C at the regulatory serine residue and bring about changes in the malate sensitivity. The activity of a physiologically relevant kinase should be higher in illuminated leaves than darkened leaves because in leaves of C_4 plants PEP-C is only phosphorylated in the light. Two types of protein kinase, one Ca^{2+}-dependent and one not, have been shown to phosphorylate PEP-C at the regulatory serine residue. However, the Ca^{2+}-dependent protein kinase described by Ogawa et al. (1992) does not show clear light dependency, whereas the Ca^{2+}-independent protein kinase does (Jiao and Chollet, 1989; McNaughton et al., 1991; Li and Chollet, 1993; Wang and Chollet, 1993). This protein kinase is of low abundance and, although a highly purified preparation has been obtained from maize leaves (Wang and Chollet, 1993), it has not been purified to homogeneity nor has it been cloned. Much less work has been done on the phosphatase that dephosphorylates PEP-C in darkened leaves. Plant tissues contain protein phosphatase activities that are very similar to protein phosphatase type 1 (PP-1) and protein phosphatase type 2A (PP-2A) of other eukaryotes. PP-2A was found to dephosphorylate purified PEP-C *in vitro* with a concomitant increase in malate sensitivity, unlike PP-1 (McNaughton et al., 1991). In maize leaves the activity of PP-2A was not affected by illuminating or darkening leaves (McNaughton et al., 1991).

Considerable attention has been paid to how PEP-C kinase activity is regulated in response to illumination. Maximal phosphorylation of PEP-C is dependent on light intensity and takes up to 100 min after a leaf is illuminated (Nimmo et al., 1987b; Echevarria et al., 1990). In addition, increasing the light intensity in a stepwise fashion is paralleled by a progressive decrease in malate sensitivity (Bakrim et al., 1992). These observations suggest that PEP-C kinase activity is coupled to the functioning of photosynthesis rather than to a receptor such as phytochrome. Evidence for this was provided by feeding inhibitors of photosystems I and II and of the Benson–Calvin cycle to detached leaves. All these inhibitors prevent phosphorylation of PEP-C (Bakrim et al., 1992; Jiao and Chollet, 1992). There are several ways in which the functioning of photosynthesis could modulate PEP-C kinase activity. One possibility is that changes in the concentrations of metabolites in leaves could act as the signal. A good candidate for the signal metabolite in maize is glycerate-3-P, which is transported from the bundle sheath to the mesophyll. In the mesophyll, glycerate-3-P is taken up by the chloroplasts and this transport involves an increase in the pH of the cytosol because the trivalent form of glycerate-3-P predominates at the pH of the cytosol, whereas the divalent form is transported. The observation by Duff et al. (1996) that incubation of mesophyll protoplasts in a solution containing glycerate-3-P brings about an increase in PEP-C kinase activity, whereas glycerate-2-P or other metabolites are ineffective, supports this view. Agents that reduce cytosolic Ca^{2+} concentration, such as chelators or Ca^{2+} channel blockers, prevented phosphorylation of PEP-C (Giglioli-Guivarc'h et al., 1996) suggesting that, although Ca^{2+} is not required by PEP-C kinase, it is required for an event earlier in the signalling pathway. Nimmo's group have recently developed a novel method for measuring the abundance of translatable mRNA encoding PEP-C kinase (Hartwell et al., 1996). RNA isolated from a leaf was translated *in vitro*, and the products of the reaction incubated with purified PEP and [γ-^{32}P]ATP. Phosphorylation of PEP-C was assessed by autoradiography of immunoprecipitated PEP-C after SDS–PAGE. The amount of translatable mRNA encoding PEP-C kinase was measured in leaves that had been fed various inhibitors of transcription and translation. Previously it had been shown that in both C_4 (Jiao et al., 1991a; Pierre et al., 1992) and CAM plants (Carter et al., 1991), protein synthesis was necessary for the appearance of PEP-C kinase activity. This work extended these earlier observations and showed that, at least in *Bryophyllum fedtschenkoi*, protein synthesis is required for an event upstream of the synthesis of PEP-C kinase and also for the *de novo* synthesis of PEP-C kinase itself. Collectively, these results can be used to formulate a hypothesis in which illumination leads to an increase in the pH of the mesophyll cytosol as a result of glycerate-3-P transport into chloroplasts. This increase in pH leads to an increase in the concentration of Ca^{2+} in the cytosol, possibly by the activation of Ca^{2+} channels in the tonoplast. The increase in cytosolic [Ca^{2+}] activates a Ca^{2+}-dependent protein kinase which is involved in the regulation of the synthesis of a protein required for the induction of PEP-C kinase activity.

Thus, reversible phosphorylation of PEP-C, changes in cytosolic pH on illumination of a leaf, and the direct effects of metabolites on the enzyme (Fig. 3) all interact to regulate PEP-C. In illuminated leaves PEP-C has to function in the presence of high concentrations of its inhibitor malate. Phosphorylation of PEP-C in illuminated leaves increases its catalytic activity, reduces the inhibitory effect of malate and increases the effect of its activator glucose-6-phosphate. An increase in the pH of the cytosol leads to an increase in its catalytic rate and also modulates its response to malate and glucose-6-phosphate. Taken together with the changes in metabolites that occur in illuminated leaves, these effects are sufficient to allow PEP-C to operate at appropriate rates *in vivo* (Doncaster and Leegood, 1987; Gao and Woo, 1996). It is therefore apparent that the activity of PEP-C is regulated by an interaction between direct effects of pH and metabolites on the enzyme and a modulation of these effects by phosphorylation. These regulatory mechanisms are intimately connected and are controlled ultimately by factors that control the rate of photosynthesis.

B. Pyruvate,P_i Dikinase

Pyruvate,P_i dikinase (PPDK) catalyses a reversible reaction, converting pyruvate and P_i to PEP and PP_i, and can be considered to be the principal regenerative step of the C_4 cycle. PPDK is light activated by a mechanism involving phosphorylation and its maximum activity is closely related to the prevailing rate of photosynthesis (Usuda et al., 1984). Its activity also shows rapid responses (of the order of a few minutes) to changes in light intensity (Slack, 1968; Hatch and Slack, 1969; Yamamoto et al., 1974). The phosphorylation of PPDK is unusual in that regulation is part of the catalytic mechanism of the enzyme (Edwards et al., 1985; Huber et al., 1994). Interconversion between active and inactive forms is catalysed by a regulatory protein that is bifunctional in that it has two active sites (Roeske and Chollet, 1987). ADP inactivates the enzyme by acting, rather unusually, as a phosphoryl donor. P_i activates the enzyme by phosphorolytic cleavage of the enzyme-P to yield PP_i (Hatch and Slack, 1969; Chapman and Hatch, 1981; Burnell and Hatch, 1983, 1985, 1986; Ashton et al., 1984; Budde et al., 1985).

The catalytic mechanism of PPDK proceeds via three steps (upper portion of Fig. 5), which occur on a histidine residue (His^{458}) (Burnell and Hatch, 1984; Carroll et al., 1990). ADP-dependent inactivation of the enzyme, catalyzed by the regulatory protein, occurs by phosphorylation of a nearby regulatory threonine residue (Thr^{456}) (Burnell, 1984b; Roeske et al., 1988). There is, therefore, competition between the regulatory and catalytic cycles for the catalytically phosphorylated form (E-His-P). The substrate for P_i-dependent activation (by removal of the phosphate attached to Thr^{456})

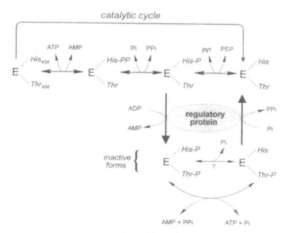

Figure 5 Regulation of pyruvate, P_i dikinase, showing the three-step catalytic reaction cycle and the regulatory cycle, comprising the ADP-dependent phosphorylation (inactivation) and the P_i-dependent dephosphorylation (activation) of Thr^{456} catalysed by the bifunctional regulatory protein. [Redrawn from Huber, S. C., Huber, J. L., and McMichael, R. W. (1994). Control of plant enzyme activity by reversible protein phosphorylation. *Intern. Rev. Cytol.* **149**, 47–98.]

lacks a phosphorylated histidine (Burnell, 1984a). There is, therefore, a regulatory cycle that also involves removal of the phosphate from histidine (lower portion of Fig. 5), but the mechanism by which this occurs remains unclear (Huber et al., 1994). It may either be labile, occurring spontaneously, it may be hydrolyzed by an unidentified phosphatase, or it may be removed by a reversal of the catalytic mechanism, which is unlikely because pyrophosphate would be virtually absent from the chloroplast stroma.

The single subunit regulatory protein (45–48 kDa) was first partially purified by Burnell and Hatch (1983). It is an extremely low abundance protein (<0.04% total leaf protein, <1% of its target protein, PPDK) (Smith et al., 1994). Burnell and Hatch (1985) noted a pH-dependent interconversion between native forms comprising a dimer (90 kDa) and a tetramer (180 kDa). Whether this change in aggregation state is due to contaminants or the regulatory protein itself remains unclear. However, there is no evidence that the activity of the regulatory protein is modified by illumination, by pH, or by covalent modification (Burnell and Hatch, 1985; Budde et al., 1986; Smith et al., 1994).

The pH optimum of PPDK is 8.3 and the enzyme requires free Mg^{2+}, suggesting that it will be maximally active in the illuminated stroma. Its activity is stimulated several-fold by K^+ and by NH_4^+ (Jenkins and Hatch, 1985; Hocking and Anderson, 1986). The latter could be important under conditions that stimulate photorespiratory ammonia release (e.g., water stress, leading to low intercellular CO_2, requiring increased turnover of the C_4 cycle). This may be the reason why PEP-C is also activated by glycine (Nishikido and Takanashi, 1973), which rises dramatically at low CO_2 in *Flaveria bidentis* (Leegood and von Caemmerer, 1994).

Figure 5 shows that, in principle, the mechanism of regulation of PPDK could allow metabolites, such as pyruvate, P_i and adenylates, to influence the amount of catalytically active enzyme. Enhanced pyruvate concentrations prevent inactivation *in vitro* (Budde et al., 1986; Burnell et al., 1986), but studies of changes in PPDK activity in relation to changes in the amounts of adenylates, pyruvate, and PEP in leaves and isolated chloroplasts have not demonstrated any obvious relationships (Roeske and Chollet, 1989; Usuda, 1988; Nakamoto and Young, 1990). One possibility is that there could be compartmentation of metabolites, such as ADP, within organelles. For example, the K_m (ADP) for inactivation of PPDK *in vitro* is 52 μM (Burnell and Hatch, 1985), whereas the concentration of ADP in the mesophyll chloroplast can be 200 μM or higher, yet PPDK can be fully activated *in vivo* (Usuda, 1988).

The C_4 plant *F. bidentis* has been transformed with antisense to PPDK (Furbank et al., 1997). At low light intensities, the rate of CO_2 assimilation was unaffected even in plants with only 15–20% of the wild-type activity, whereas at high light intensities PPDK exerted appreciable control over the rate of photosynthesis (Table I).

C. NADP-malate Dehydrogenase

NADP-malate dehydrogenase is located in the mesophyll chloroplasts, where it converts oxaloacetate, generated by PEP-C, to malate. It is the only enzyme of the C_4 cycle that is regulated by the thioredoxin system, in contrast to the Benson–Calvin cycle, in which four enzymes interact with thioredoxin. Photosynthetically generated electrons are used to reduce, in turn, ferredoxin, thioredoxin, and a disulphide bridge on the target enzyme. NADP-MDH is activated by thioredoxin m (Kagawa and Hatch, 1977; Jacquot et al., 1981; Edwards et al., 1985) and has a midpoint potential of -330 mV, slightly more negative than that of thioredoxin at -300 mV (Rebeille and Hatch, 1986a,b). The light dependence of the activation state closely follows the light dependence of photosynthesis in leaves (Usuda et al., 1984; Rebeille and Hatch, 1986b) and changes in activation state occur rapidly in response to changes in light intensity. The activation of NADP-MDH is dependent on reduction of a disulphide bridge located at the amino terminus of the enzyme between Cys^{10} and Cys^{15} (Kagawa and Bruno, 1988; Decottignies et al., 1988). The oxidized enzyme from maize is essentially inactive.

NADP-MDH from maize has an alkaline pH optimum and a notably low affinity for malate, with a K_m (malate) of 24–32 mM (Ashton and Hatch, 1983a; Kagawa and Bruno, 1988). The activation of NADP-MDH (Edwards et al., 1985) is inhibited by NADP, but there is no evidence that other metabolites play any direct role in the regulation of this enzyme. NADP-MDH is, therefore, sensitive to the redox states of both thioredoxin and

Table I Flux Control Coefficients for CO_2 Assimilation in Wild-type C_4 Plants [*Flaveria bidentis* (NADP-ME Species for NADP-MDH, Rubisco, and PPDK) and *Amaranthus edulis* (NAD-ME Species for PEP-C)] in High Light and Ambient CO_2

Enzyme	Flux control coefficient, C_E^J	Refs.
NADP-malate dehydrogenase	0.0	Trevanion et al., 1997
Rubisco	0.5–0.7	Furbank et al., 1997
Pyruvate,P_i dikinase	0.2–0.3	Furbank et al., 1997
PEP carboxylase	0.26	Dever et al., 1997

The control that any particular enzyme exerts over the flux through a pathway can be measured in mutants and in transgenic plants by asking how much the flux (J) changes when the amount of an enzyme (E) is changed by a known amount. The slope of this relationship is known as a flux control coefficient (C_E^J) and its magnitude can vary from zero (no control) to 1.0 (total control) (or it can even be negative in a branched pathway) (Kacser, 1987). Note that the degree of control may well differ in other subtypes. For example, control by PPDK may be less in PEP-CK type C_4 plants.

of pyridine nucleotides, which tend to change in parallel (Rebeille and Hatch, 1986b). The interconversion of reduced and oxidized enzymes is strongly influenced by the NADPH/NADP ratio. The reduction of oxaloacetate is also strongly inhibited by NADP, which is competitive with respect to NADPH (Ashton and Hatch, 1983a,b). Usuda (1988) has shown that the NADPH/NADP ratio rises in the mesophyll chloroplasts as the light intensity is increased. Modelling of the response of the activity of the enzyme to the ratio NADPH/NADP indicates that, at any particular ratio of oxidized to reduced thioredoxin, high proportions of active NADP-MDH, and hence high rates of oxaloacetate reduction, can only occur at very high NADPH/NADP ratios, a prediction that was confirmed by Rebeille and Hatch (1986a). Perhaps for this reason, the activation state of the enzyme in maize mesophyll chloroplasts is very sensitive to competing electron acceptors that oxidize NADPH (Leegood and Walker, 1983).

The C_4 plant *F. bidentis* has been transformed with the cDNA for NADP-MDH from sorghum (Trevanion *et al.*, 1997). Although constructs were designed to overexpress the sorghum enzyme, all showed reduced activity of the enzyme in both mesophyll and bundle sheath due to cosense suppression. Unlike C_4-monocot grasses, *F. bidentis* has appreciable NADP-MDH activity in the bundle sheath because it transports significant amounts of carbon from the mesophyll as aspartate. This is converted to oxaloacetate and then to malate in the bundle sheath (Meister *et al.*, 1996). However, in plants with a very low activity of the enzyme in the bundle sheath, photosynthesis was only slightly inhibited, showing that the presence of the enzyme in the bundle sheath is not essential for maximum rates of photosynthesis, even though isolated bundle sheath strands had reduced rates of aspartate and 2-oxoglutarate–dependent O_2 evolution. Rates of photosynthesis in plants with less than about 10% of the activity of the wild type were reduced at high light and high CO_2 concentrations, but were unaffected in low light or low CO_2, although the activation state of the enzyme was increased, showing that NADP-MDH has a low control coefficient for photosynthesis in this species (Table I).

D. NADP-malic Enzyme

NADP-malic enzyme (NADP-ME) occurs in the chloroplasts of bundle-sheath cells and, like the other decarboxylases of C_4 photosynthesis, it releases CO_2, which can be fixed directly by Rubisco. Evidence that light can regulate the activity of NADP-ME *in vivo* comes from the observation that the transfer of ^{14}C from the C_4 carboxyl group of C_4 acids ceases when leaves of sugarcane are transferred to darkness (Hatch and Slack, 1966). In addition, the regulatory properties of NADP-ME suggest that light-dependent changes in pH and the concentrations of malate and Mg^{2+} may be important in its regulation (Johnson and Hatch, 1970; Asami *et al.*,

1979). This shows similarities to the regulation of certain enzymes of the Benson–Calvin cycle, such as fructose-1,6-bisphosphatase and ribulose-5-P kinase, that are also regulated by interactions between substrates, pH and Mg^{2+} (see Leegood et al., 1985). Although no measurements of pH and Mg^{2+} have been made on the bundle-sheath chloroplasts of C_4 plants, there is no reason to suppose that they differ substantially from those measured in spinach chloroplasts, in which the pH of the stroma rises in the light from approximately pH 7 to approximately pH 8, and the concentration of Mg^{2+} rises from approximately 1 to 3 mM in the dark to approximately 2 to 6 mM in the light (see Leegood et al., 1985). The concentration of malate in bundle-sheath chloroplasts isolated nonaqueously from illuminated maize leaves is about 1.5 mM (Weiner and Heldt, 1992). For NADP-ME from maize leaves, concentrations of malate higher than 0.4–0.8 mM led to appreciable inhibition at pH 7.5 (a pH close to that of the darkened stroma), but higher concentrations of malate did not inhibit at pH 8.4 (a pH close to that of the illuminated stroma). The inhibitory effect of malate was also more pronounced at 1 mM Mg^{2+} than at 5–10 mM Mg^{2+}. Thus NADP-ME is likely to be activated by the change in stromal conditions occurring on illumination.

The enzyme requires Mg^{2+} or Mn^{2+}. Organic acids also modulate the activity of NADP-ME. Glyoxylate, oxaloacetate, and 2-oxoglutarate inhibit the activity of the maize enzyme, and inhibition is more evident at the lower pH (Asami et al., 1979; see also Iglesias and Andreo, 1989). However, the importance of these compounds as regulators remain unclear until changes in metabolites and their compartmentation in relation to photosynthetic fluxes are better understood.

E. NAD-malic Enzyme

There is evidence for control of NAD-ME by both adenylates and NADH/NAD ratio. Its activity shows a sigmoidal response to malate concentration in some C_4 species (Hatch et al., 1974) and there is an interaction between malate sensitivity and regulation by adenylates. NAD-ME from the NAD-ME–type plants, *Atriplex spongiosa* and *Panicum miliaceum*, is inhibited by physiological concentrations of ATP, ADP, and AMP, with ATP the most inhibitory species at subsaturating concentrations of Mn^{2+} and of the activator, malate. The presence of adenylates increases the $K_{1/2}$ for malate for the enzyme from *A. spongiosa* considerably and accentuates the sigmoidicity of response to malate concentration. Hence, if the mitochondrial malate concentration were to fall from 5 mM to 1.5 mM (as might occur in the dark or in low light), NAD-ME activity would be reduced by approximately 60% in the presence of 2 mM ATP, but by only 20% in the absence of adenylates. However, this type of sensitization may vary because the degree of sigmoidicity with respect to malate varies among species (Murata et al., 1989). The response of NAD-ME to adenylates is unlikely to result in regulation of NAD-ME by energy charge (Furbank et al., 1991). In contrast, the enzyme from *Urochloa panicoides* (PEP-CK type) is activated up to 10-fold by ATP and inhibited by ADP and AMP (Furbank et al., 1991).

The properties of NAD-ME from C_4 plants therefore differ between different C_4 subtypes, and are in complete contrast to the enzyme from CAM plants that is activated by AMP (Wedding and Black, 1983) and that from *Arum* spadix, which is activated by ADP (Wedding and Whatley, 1984). The enzyme requires Mn^{2+}, is stimulated by CoA and acetyl CoA, fructose-1,6-bisphosphate (whether this compound occurs in the mitochondria is questionable) and sulphate (Hatch et al., 1974). It is inhibited by HCO_3^- (to a lesser degree in *U. panicoides* than the enzyme from NAD-ME type C_4 species) and nitrate. It is also inhibited by NADH so that the enzyme could be regulated by the NADH/NAD ratio. All these features are common to NAD-ME isolated from other sources. However, the enzyme from *U. panicoides* is also inhibited by oxaloacetate, 2-oxoglutarate and pyruvate (Burnell, 1987), which suggests that these metabolites may also be important in regulating the enzyme *in vivo*.

F. Phosphoenolpyruvate Carboxykinase

Phosphoenolpyruvate carboxykinase (PEP-CK) catalyzes the adenine nucleotide and Mn^{2+}-dependent interconversion of oxaloacetate and PEP. Although this reaction is freely reversible *in vitro* (Wood et al., 1966), it is likely that the carboxylation reaction is restricted under physiological conditions because of the low affinity of the enzyme for CO_2 (Ray and Black, 1976; Urbina and Avilan, 1989). PEP-CK comprises subunits that, in C_4 plants, range in size from 67 to 74 kDa (Finnegan and Burnell, 1995; Walker and Leegood, 1996; Walker et al., 1997). In C_3 and CAM plants the enzyme undergoes reversible phosphorylation in darkness or at night. The effect of this phosphorylation is unclear, but it seems likely to involve a change in the enzyme's properties (Walker et al., 1997). However, PEP-CK is only phosphorylated in those C_4 plants with the larger (71 kDa) form of the enzyme—for example, *Panicum maximum* (Walker and Leegood, 1996). Here, the enzyme is phosphorylated in darkened leaves and dephosphorylated in illuminated leaves, suggesting that the dephosphorylated enzyme is active. However, in many other C_4 plants it appears that the N-terminal extension has been truncated and that the phosphorylation site has been lost. A comparison of the sequences of the N-terminal regions of PEP-CK from cucumber and the PEP-CK type C_4 grass *U. panicoides* (in which PEP-CK is not phosphorylated) show that the putative phosphorylation motif is absent from the latter (Finnegan and Burnell, 1995). In contrast, a 74 kDa form of PEP-CK, which is present in some NADP-ME species such as maize, is not phosphorylated in either darkened or illuminated maize leaves (Walker et al., 1997), suggesting that either a small change has occurred in the N-terminal extension or that PEP-CK–kinase activity is not present.

The reason PEP-CK is not phosphorylated in most C_4 plants may be related to its role in C_4 photosynthesis. For example, in C_4 plants, PEP-C and PEP-CK are located in different photosynthetic cell types, the mesophyll and bundle sheath, respectively, whereas in C_3 and CAM plants, both enzymes are located in the cytosol of the same cells. There is, therefore, less potential for the operation of a futile cycle with PEP-C. However, it could be that whatever change is brought about by phosphorylation and dephosphorylation is very slow (cf. PEP-C) and that it exerts a constraint on photosynthesis in, for example, fluctuating light environments. It may be significant that neither of the other decarboxylases in C_4 photosynthesis is regulated by covalent modification.

There must be alternative mechanisms in C_4 plants that could activate the enzyme in the light and inactivate the enzyme in darkness [it might otherwise drain the cytosol of OAA (Carnal et al., 1993) or ATP]. The N-terminal portion of PEP-CK is subject to rapid proteolysis, which may affect regulatory properties such as sensitivity to effectors. Burnell (1986) purified the cleaved form of PEP-CK from a range of C_4 plants (*U. panicoides, Chloris gayana,* and *P. maximum*) and showed that the enzyme is inhibited by a number of metabolites, including glycerate-3-P, fructose-6-P, and fructose 1,6-bisphosphate, at physiological concentrations. If these metabolites were similarly effective inhibitors of the native enzyme, they could act to coordinate electron transport and carbon metabolism with the activity of PEP-CK. Thus if ATP became limiting in either the mesophyll or the bundle sheath, for example in low light, this would limit the reduction of glycerate-3-P, which would accumulate and inhibit PEP-CK. However, if the maximum capacity for sucrose synthesis were reached, this would lead to the accumulation of fructose-1,6-bisphosphate and hexose phosphates, and again inhibit PEP-CK. PEP-CK catalyzes a freely reversible reaction and is very sensitive to changes in the ATP/ADP ratio (Walker et al., 1997; see also Kobr and Beevers, 1971). Because the supply of ATP is generated by NAD-ME, there will be close coupling between the two enzymes (see later this chapter). However, these regulatory properties are insufficient to formulate a satisfactory theory of regulation of PEP-CK because they fail to explain how the enzyme could be switched off in the dark (Carnal et al., 1993).

G. Aminotransferases

In leaves of C_4 plants, oxaloacetate formed by the carboxylation of PEP by PEP-C is either converted to malate by malate dehydrogenase (MDH) or to aspartate by AspAT. Whether malate or aspartate is formed from oxaloacetate differs between the C_4 subtypes. The concentration of oxaloacetate in the leaves of NADP-ME–type monocots, which synthesize mainly malate, is low (Furbank and Leegood, 1984) because oxaloacetate is rapidly reduced to malate by NADP-MDH in the chloroplast. In contrast, in aspartate-forming C_4 plants the concentration of oxaloacetate can be several millimolar (Hatch, 1979; Leegood and von Caemmerer, 1988). A high concentration of oxaloacetate is necessary for the equilibrium of the reaction catalyzed by AspAT to be displaced in favor of the formation of aspartate. In order to accumulate large amounts of oxaloacetate in the cytosol of mesophyll cells it is important that little NAD-MDH activity be present. Whether regulation of cytosolic NAD-MDH activity is due to a reduction in NAD-MDH protein or is a result of inhibition of enzyme activity is not clear, although Doncaster and Leegood (1990) showed that, in maize and *P. miliaceum*, high concentrations of malate and oxaloacetate inhibited NAD-MDH.

The activities of AspAT and AlaAT in leaves of aspartate-forming C_4 plants are about 20-fold higher than in C_3 plants and about 10-fold higher than in C_4 plants that transport predominantly malate. The activities are distributed equally between mesophyll and bundle sheath (Hatch and Mau, 1973). Leaves of the C_4 plants, *A. spongiosa* and *P. miliaceum*, contain three forms of AspAT (Hatch and Mau, 1973). In the mesophyll of both NAD-ME and PEP-CK types the predominant form of AspAT is located in the cytosol, whereas in the bundle sheath the predominant form in NAD-ME types is mitochondrial and in PEP-CK types it is cytosolic (Hatch and Mau, 1973; Numazawa *et al.*, 1989; Taniguchi and Sugiyama, 1990; Taniguchi *et al.*, 1995). In all C_4 subtypes, AlaAT is located in the cytosol in both bundle sheath and mesophyll (Hatch and Mau, 1973; Chapman and Hatch, 1983). Meister *et al.* (1996) have found that the NADP-ME type dicot *F. bidentis*, unlike monocot members of this subtype, transports substantial amounts of aspartate between the mesophyll and bundle sheath, and they propose that in the bundle sheath a large proportion of aspartate is metabolised by plastidic AspAT. Three forms of AlaAT (two minor and one major) are present in leaves of the C_4 plants *A. spongiosa* and *P. miliaceum* (Hatch, 1973; Hatch and Mau, 1973; Son *et al.*, 1991) and barley roots (Muench and Good, 1994). The same isoform was induced on greening of *P. miliaceum* leaves and anaerobiosis of barley roots (Son *et al.*, 1991; Muench and Good, 1994). No regulatory properties of the purified enzymes were reported (Son *et al.*, 1991; Muench and Good, 1994). Hatch (1973) reported that 5 mM malate inhibited AlaAT activity from *A. spongiosa* by about 25%, however the physiological importance of this is not clear. In leaves of C_4 plants the flux through aminotransferases appears to be controlled mainly by the concentrations of their substrates, rather than by modulation of enzyme activity by either covalent modification or interaction with metabolite effectors.

IV. Interactions between the C_4 Cycle and Mitochondrial Metabolism

Both NAD-ME and PEP-CK type C_4 plants use NAD-ME to decarboxylate malate in the bundle-sheath mitochondria (Burnell and Hatch, 1988a,b; Hatch et al., 1988; Carnal et al., 1993) (Fig. 6). Because photosynthetic fluxes vastly exceed respiratory fluxes, this requires some uncoupling of C_4 acid decarboxylation from normal mitochondrial metabolism.

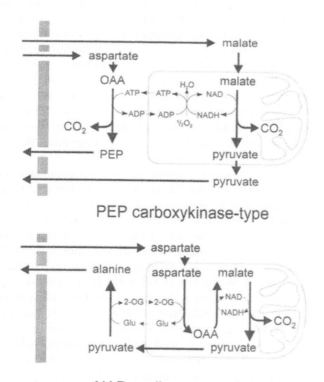

Figure 6 Mechanisms of C_4 acid decarboxylation in the bundle-sheath cells of NAD-malic enzyme–type and PEP carboxykinase–type C_4 plants. The bundle-sheath cell wall is shown on the left, with metabolites transported to and from the mesophyll. The mitochondrion is shown as the enclosed space, the remainder of the reactions occur in the cytosol. In PEP carboxykinase plants, NADH oxidation is linked to ATP generation via the respiratory electron transport chain. Glu, glutamate; 2-OG, 2-oxoglutarate; OAA, oxaloacetate. [Redrawn from Agostino, A., Heldt, H. W., and Hatch, M. D. (1996). Mitochondrial respiration in relation to photosynthetic C_4 acid decarboxylation in C_4 species. *Aust. J. Plant Physiol.* **23,** 1–7.]

In NAD-ME–type species, it has been proposed that aspartate from the mesophyll cells enters the bundle-sheath mitochondria and is converted to OAA (with accompanying transamination of 2-oxoglutarate) and malate. Malate is oxidized using NAD^+ generated by the reduction of OAA (Fig. 6). The alternative route, direct oxidation of malate, may operate as a reserve mechanism if the generation of aspartate becomes limiting, because isolated mitochondria oxidize malate. NAD-ME could potentially be limited by the availability of NAD if NADH oxidation were too slow. In this case, NAD-ME appears to be freed of respiratory control by engagement of the alternative, cyanide-insensitive pathway of respiration. For example, in *P. milaceum* (NAD-ME type), malate oxidation by mesophyll mitochondria is sensitive to cyanide and shows good respiratory control. In contrast, in bundle-sheath mitochondria malate oxidation is largely insensitive to KCN, shows no respiratory control, and is sensitive to an inhibitor of the alternative oxidase, salicylhydroxamate (Gardeström and Edwards, 1983; Agostino et al., 1996). The finding that the alternative oxidase is activated by pyruvate in soybean cotyledons (Day et al., 1995; Hoefnagel et al., 1995) may also be important if it also occurs in mitochondria of C_4 plants. The metabolism of pyruvate also has to be controlled so that the C_4 cycle is not drained of carbon, perhaps by the regulation of pyruvate dehydrogenase (Randall and Rubin, 1977; see Huber et al., 1994, for review)

In PEP-CK–type species, mitochondrial respiration generates the ATP required for the operation of PEP-CK (Hatch et al., 1988; Carnal et al., 1993), which is largely due to oxidation of NADH generated in malate oxidation via NAD-ME. The minimum requirement for NAD-ME is that it should operate at one-third of the rate of PEP-CK (assuming a maximum of 3ATP/NADH generated by coupled electron transport). NAD-ME may make an additional contribution to C_4 acid decarboxylation, because maximum rates of malate decarboxylation substantially exceeded the minimum rates necessary for providing ATP for cytosolic OAA decarboxylation (Carnal et al., 1993). As in NAD-ME–type species, there may be some oxidation of NADH via the alternative oxidase (Agostino et al., 1996). However, it would appear preferable to decarboxylate via PEP-CK, because direct production of PEP and its transport to the mesophyll means that the ATP requirement for PPDK is circumvented. This partitioning of decarboxylation between PEP-CK and NAD-ME could be regulated by several means. An important feature of NAD-ME from *U. panicoides*, unlike that from NAD-ME type C_4 plants, is that it is activated by ATP (up to 10-fold) and inhibited by ADP and AMP (Furbank et al., 1991). As malate oxidation commences in the light, ATP can be expected to act as a feed-forward activator of NAD-ME, increasing the ATP/ADP ratio and thereby promoting decarboxylation of oxaloacetate by PEP-CK. Furbank et al. (1991) suggest that when the maximum capacity of PEP-CK is reached (for example, in high light), then its substrates, ATP and oxaloactetate, will increase in the bundle sheath. The increased oxaloacetate will oxidize NADH in the mitochondria, producing NAD, which, together with the increase in ATP, will promote the activity of NAD-ME to provide additional capacity for the decarboxylation of C_4 acids. Carnal *et al.* (1993) suggest possible mechanisms that might regulate partitioning of C_4 acid decarboxylation between PEP-CK and NAD-ME. They suggest that, because PEP-CK has a low K_m for OAA (12–25 μM) (Burnell, 1986), the reaction requires low cytosolic OAA, which may limit uptake into the mitochondria. However, if the supply of aspartate and OAA increases to saturate PEP-CK, then OAA uptake into the mitochondria increases and accelerates malate decarboxylation by oxidizing NADH, although NAD-MDH activity in the mitochondria is much lower than in NAD-ME species (Carnal *et al.*, 1993). Carnal *et al.* (1993) also suggest that the respiration-induced mitochondrial proton gradient would regulate decarboxylation via a feed-forward effect. Increases in the proton gradient brought about by increased activity of NAD-ME would favor inward transport of ADP, OAA, and P_i and outward transport of ATP (Heldt and Flügge, 1987; Douce and Neuburger, 1989). Cytosolic P_i concentrations may also play a role in that P_i appears to be required for malate uptake into the mitochondria (Carnal *et al.*, 1993).

V. Regulation of the Benson–Calvin Cycle in C_4 Plants

A. Ribulose-1,5-bisphosphate Carboxylase–Oxygenase

All the C_4 acid decarboxylases release CO_2 for fixation by ribulose-1,5-bisphosphate carboxylase–oxygenase (Rubisco) in the bundle sheath. The absence of carbonic anhydrase in bundle-sheath cells is of critical importance in ensuring that this CO_2 (which is the inorganic carbon substrate for Rubisco) is not converted to bicarbonate (Furbank and Hatch, 1987). The regulation of Rubisco in C_4 plants is a neglected area when compared with its C_3 counterparts. The activity of Rubisco is considerably lower in C_4 plants, but its specific activity is higher. Rubisco from C_4 plants has a $K_m(CO_2)$ (28–63 μM), which is appreciably higher than that of C_3 and CAM plants (8–26 μM) but comparable to those of aquatic plants with CO_2 concentrating mechanisms (Yeoh *et al.*, 1980, 1981). However, the specificity factor of Rubisco from C_4 plants is little different from that of Rubisco in C_3 plants. Analysis of other enzymes of the Benson–Calvin cycle in C_4 plants shows that most are rather similar to their C_3 counterparts, so that the fundamental features of regulation are unlikely to be different in the bundle-sheath chloroplasts of C_4 plants (Ashton *et al.*, 1990).

There are two mechanisms of regulating Rubisco in C_3 plants (Portis, 1992). The first mechanism is a change in carbamylation state, catalyzed by Rubisco activase. The second is regulation by the naturally occurring inhibitor of Rubisco, carboxyarabinitol-1-P, which binds tightly to the enzyme in darkened leaves, rendering Rubisco inactive. There is evidence that most C_4 plants regulate Rubisco by changes in carbamylation state between light and dark (Sage and Seemann, 1993), although there are a few exceptions (such as maize) in which such light–dark modulation of total Rubisco activity is very weak (Vu et al., 1984; Usuda, 1985; Sage and Seemann, 1993), possibly because the activity of the CO_2 pump may limit photosynthesis at low light intensities (Sage and Seemann, 1993). There is little evidence for regulation of Rubisco activity by carboxyarabinitol-1-P in most C_4 plants, although it may be present (Sage and Seemann, 1993). However, guinea grass, *P. maximum* (PEP-CK type), showed a 2.5-fold increase in Rubisco activity on illumination (Vu et al., 1984) and had a CA1P content that was 44% of Rubisco active site content (Moore et al., 1991). The parent sugar, 2'-carboxyarabinitol is also present in leaves of maize and *P. maximum* (Moore et al., 1992), suggesting that CA1P is metabolized in the leaves of these species. However, this mode of regulation may operate more as a light–dark switch, rather than as a means of modulating Rubisco activity at different light intensities (Sage and Seemann, 1993).

Flaveria bidentis (NADP-ME type) has been transformed to reduce the amount of Rubisco by antisense (Furbank et al., 1996; von Caemmerer et al., 1997). In ambient CO_2 and saturating light, even modest reductions in Rubisco led to substantial reductions in the photosynthetic rate (see also Edwards et al., 1988), indicating substantial control of the rate of CO_2 assimilation (Table I). Thus, despite the high bundle-sheath CO_2 concentrations attained in C_4 plants, there is only just enough Rubisco in leaves of *Flaveria* to support maximum rates of photosynthesis. Transformants showed a decrease in the CO_2-saturated rate of photosynthesis, but no change in the initial slope of the assimilation rate versus intercellular concentration of CO_2. It has been suggested that, unlike C_3 plants, this initial slope is largely determined by the kinetic characteristics and activity of PEP-C (Edwards and Walker, 1983; Collatz et al., 1992) and, consistent with this, mutants of *Amaranthus edulis* with less PEP-C, or leaves treated with an inhibitor of PEP-C, show a decrease in the initial slope (Jenkins et al., 1989a; Dever et al., 1995). This means that reduced Benson–Calvin cycle turnover in the Rubisco transformants does not affect PEP-C activity, indicating an imbalance in the operation of mechanisms that coordinate the C_3 and C_4 cycles. There was no effect of a decrease in Rubisco on the photosynthetic rate at low light intensities (Furbank et al., 1996), indicating that at low light, C_4 photosynthesis is limited by RuBP regeneration, as in C_3 plants (von Caemmerer and Farquhar, 1981) or by the rate of regeneration of PEP, both of which would be dependent on the provision of ATP and NADPH by electron transport.

B. Why Is Glycerate-3-P Reduced in the Mesophyll?

Although most enzymes of the Benson–Calvin cycle are absent from the mesophyll cells, all C_4 subtypes possess the enzymes for glycerate-3-P reduction in the mesophyll chloroplasts (Hatch and Osmond, 1976). This aspect of the Benson–Calvin cycle is unique to C_4 plants and has consequences for the regulation of the C_4 cycle, of carbohydrate synthesis, and of electron transport. In NADP-ME–type species, such as maize, the low PSII activity in the bundle sheath means that NADPH can only be generated by NADP-ME, which is sufficient to reduce only 50% of the glycerate-3-P generated in the bundle sheath. The remainder is exported to the mesophyll for reduction, but a minimum of two-thirds of the triose-P must be returned to the bundle sheath to maintain pools of Benson–Calvin cycle intermediates. However, even in those C_4 subtypes that have photosystem II in the bundle sheath, glycerate-3-P is reduced in the mesophyll. It can thus be inferred that the glycerate-3-P/triose-P shuttle between the bundle sheath, and mesophyll serves an important function in C_4 metabolism. These functions could be:

1. Coordination of Benson–Calvin cycle and C_4 cycle turnover. Coordination of the rate at which the Benson–Calvin and C_4 cycles fix CO_2 is necessary if photosynthesis is to proceed efficiently under different environmental conditions. The breakdown of such coordination during light flecks, for example, has been shown to result in inefficient CO_2 assimilation (Krall and Pearcy, 1993). Coordination could occur in a variety of ways (see Section II.B), all of which are linked to metabolite fluxes between the bundle sheath and the mesophyll cells. In addition, electron transport in the mesophyll chloroplasts not only powers conversion of pyruvate to malate in the C_4 cycle, but also drives glycerate-3-P reduction.

2. The shuttle is a means of ensuring H^+ transport, and hence charge balance, between the two cell types. The reduction of glycerate-3-P to triose-P in the mesophyll consumes a proton, after which the triose-P is transported to the bundle sheath. The consumption of a proton in this reaction is necessary because a proton is released when CO_2 is hydrated and the resulting HCO_3^- is fixed by PEP-C.

3. The shuttle will result in a decrease in the amount of reductant (NADPH) required for reduction of glycerate-3-P in the bundle sheath and hence a decrease in photosynthetic O_2 evolution in the bundle sheath. Photosynthetically generated O_2 may present a problem in all C_4 subtypes because reduced permeability to CO_2 also reduces permeability to O_2, which could lead to a build up of O_2 in the bundle sheath. Any reduction in the O_2 concentration in the bundle sheath would augment the elevated CO_2 concentration brought about by the operation of the C_4 cycle and further favor the carboxylation of RuBP over its oxygenation. In addition, both

NAD-ME and PEP-CK type species show enhanced respiratory uptake of O_2 in the bundle sheath because of the involvement of the mitochondria in the C_4 cycle.

VI. Regulation of Product Synthesis in Leaves of C_4 Plants

There are many similarities in the mechanisms that regulate carbon partitioning in the leaves of C_3 and C_4 plants. These include the regulation of fructose-1,6-bisphosphatase by the regulatory metabolite, fructose-2,6-bisphosphate, and the covalent modification of sucrose-P synthase (SPS) (Kalt-Torres et al., 1987a; Stitt et al., 1987; Usuda et al., 1987; Huber and Huber, 1991). However, in C_4 plants, the regulation of carbon partitioning must be coordinated not only between chloroplast and cytosol, but also between mesophyll and bundle sheath.

In maize, the weight of evidence suggests that the synthesis of sucrose occurs in the mesophyll cells, and that the synthesis of starch occurs in the bundle-sheath cells. During $^{14}CO_2$ fixation in maize, labelled sucrose appears first in the mesophyll cells (Furbank et al., 1985). The majority of the sucrose–phosphate synthetase, fructose-6-P,2-kinase, fructose–2,6-bisphosphatase, and fructose-2,6-bisphosphate is also present in the mesophyll (Stitt and Heldt, 1985a; Lunn and Furbank, 1997). Although under some conditions in maize, appreciable amounts of fructose-2,6-bisphosphate and PP_i-dependent phosphofructokinase are present in the bundle sheath (Clayton et al., 1993), the fructose-2,6-bisphosphate may be associated solely with the regulation of glycolysis rather than sucrose synthesis. Sucrose–phosphate synthase activity has also been shown in the bundle-sheath cells of maize (Ohsugi and Huber, 1987), but the association of appreciable amounts of SPS with the bundle-sheath chloroplasts (Cheng et al., 1996) is probably artefactual, perhaps due to nonspecific binding of antibodies to Rubisco (Lunn and Furbank, 1997). However, in C_4 plants in general there appears to be some flexibility in the site of sucrose synthesis. Up to one-third of the SPS may occur in the bundle sheath of some C_4 plants, such as Sorghum bicolor or P. miliaceum, although there is no correlation of the distribution with C_4 subtype (Lunn and Furbank, 1997; Lunn et al., 1997). Within the genus Panicum, each C_4 decarboxylation type studied had appreciable SPS activity in the mesophyll and the bundle sheath (Ohsugi and Huber, 1987). The major function of bundle-sheath cell sucrose–phosphate synthase is probably sucrose synthesis following starch degradation (Ohsugi and Huber, 1987).

Lunn and Hatch (1995) have examined differences among a number of C_4 plants in their partitioning of fixed carbon into starch or sucrose within the leaves. The amount of sucrose and starch stored during the day varied widely, from sucrose contents only a few percent of those of starch in starch storers (e.g. *Amaranthus edulis, Atriplex spongiosa,* and *Flaveria trinervia*) to sucrose contents being about equal to starch in sucrose storers (*Eleusine indica, E. coracana*). None accumulated fructan and only one, *A. spongiosa*, accumulated significant amounts of hexoses.

A. The Phosphate Translocator

When triose-P is exported from the chloroplast of a C_3 plant and is converted to sucrose, the P_i released reenters the chloroplast via the phosphate translocator in exchange for more triose-P. This provides a direct link between the provision of triose-P and its use in sucrose synthesis. The situation in maize is rather more complicated. The triose-P that is available for sucrose synthesis in the mesophyll contains P_i that was incorporated into glycerate-3-P in the bundle sheath. Hence, P_i released in sucrose synthesis in the mesophyll must be transported back to the bundle sheath to regenerate RuBP. However, the P_i will, nevertheless, play a part in regulating the transport of glycerate-3-P, triose-P, and PEP across the mesophyll chloroplast envelope by the P_i translocator.

In addition to the exchange of glycerate-3-P, triose-P, and P_i, the mesophyll chloroplasts also catalyze the export of PEP, formed in the chloroplast by the action of pyruvate,P_i dikinase, in exchange for P_i to sustain PEP-C in the cytosol. This exchange of PEP, P_i, glycerate-3-P, and triose-P occurs on a common P_i translocator in the chloroplast envelope of C_4 plants (Chapter 3). Important features that distinguish transport of phosphorylated intermediates in C_4 plants from that in C_3 plants are both the rate and the direction of transport across the chloroplast envelope. The rate of transport must be equivalent to the rate of CO_2 fixation in C_4 plants, whereas in C_3 plants it is maximally one-sixth of the rate of steady-state CO_2 assimilation (Edwards and Walker, 1983). In addition, during photosynthesis in C_4 plants, the mesophyll chloroplasts import glycerate-3-P and export triose-P, and the bundle-sheath chloroplasts export glycerate-3-P and import triose-P that has been reduced in the mesophyll chloroplasts. Although transport processes across the envelope of C_4 mesophyll chloroplasts are now adequately characterized, relatively little is known of transport into the bundle-sheath chloroplasts, largely because of the difficulty of isolating these intact and in appreciable quantities from any C_4 plants (although see previously mentioned). Bundle-sheath chloroplasts are unusual because they must export glycerate-3-P export at high rates *in vivo*. In C_3 plants, glycerate-3-P is not exported by chloroplasts to any great extent because glycerate-3-P is transported by the translocator as the glycerate-3-P^{2-} ion, whereas glycerate-3-P^{3-} is the form that predominates at the pH that occurs in the illuminated stroma (Heldt and Flügge, 1992). It is possible that, in C_4 plants, the stromal pH is lower in bundle-sheath chloroplasts or that glycerate-3-P within the chloroplast reaches such high internal concentrations that transport of glycerate-3-P^{2-} from the chloroplast becomes inevitable.

B. Cytosolic Fructose-1,6-Bisphosphatase

The rate at which triose-P is withdrawn from the Benson–Calvin cycle for the synthesis of sucrose depends partly on the availability of substrate for the cytosolic fructose-1,6-bisphosphatase, and partly on alterations in the concentrations of its effectors, such as fructose-2,6-bisphosphate. In maize, rapid photosynthesis is accompanied by a three- to four-fold decrease in fructose-2,6-bisphosphate and dramatic increases in triose-P, glycerate-3-P, and fructose-1,6-bisphosphate in the mesophyll when the metabolite shuttles are operational (Stitt and Heldt, 1985b; Usuda *et al.*, 1987) (Fig. 1). Thus, the cytosolic fructose-1,6-bisphosphatase is stimulated by increases in the concentration of its substrate and by decreasing concentrations of its inhibitor, fructose-2,6-bisphosphate. These changes in the amount of fructose-2,6-bisphosphate come about by regulation of the enzyme that generates it, fructose-6-P,2-kinase. This enzyme is inhibited by triose-P (although less than in spinach) and by glycerate-3-P and by intermediates in the C_4 cycle, such as PEP and oxaloacetate, which are present at high concentrations in the mesophyll during photosynthesis (Soll *et al.*, 1983; Stitt and Heldt, 1985a,b). Thus, fructose-2,6-bisphosphate will build up and inhibit sucrose synthesis when metabolites of the Benson–Calvin cycle and of the C_4 cycle are low. At the end of the day, fructose-2,6-bisphosphate increases in maize leaves as partitioning to starch increases (Kalt-Torres *et al.*, 1987a; Usuda *et al.*, 1987).

The cytosolic fructose 1,6-bisphosphatase acts as a valve in photosynthetic metabolism in that it not only controls the rate at which carbon is withdrawn for sucrose synthesis, but also, via its substrate affinity, sets a threshold for such withdrawal (Herzog *et al.*, 1984). In maize, the cytosolic fructose 1,6-bisphosphatase does not receive triose-P directly from the bundle-sheath chloroplasts, but instead receives triose-P on its return from the mesophyll to the bundle sheath. Consequently, the regulation of fructose 1,6-bisphosphatase has been modified so that surplus carbon can be siphoned off for sucrose production, but allowing most of the triose-P to be returned to the Benson–Calvin cycle to allow the regeneration of RuBP. As we have seen, triose-P is present in the mesophyll of C_4 plants at far higher concentrations than in C_3 plants (Figs. 1 and 4), hence the amount of fructose-1,6-bisphosphate, formed through the action of aldolase, is also higher in the mesophyll cytosol. In maize, the cytosolic fructose 1,6-bisphosphatase shows a much higher K_m for FBP (20 mM in maize compared with 3 μM in spinach and, in the presence of fructose-2,6-bisphosphate, 3–

5 μM in maize and 20 μM in spinach) (Stitt and Heldt, 1985a). In practice, this means that effective regulation of this enzyme would occur as the concentration of triose-P varies between 0.5 and 2 mM (Stitt and Heldt, 1985a). If any C_4 plants make significant amounts of sucrose in the bundle sheath, then it would be expected that the properties of FBPase in these cells would be different.

C. Sucrose Phosphate Synthase

Light activation of SPS is the result of a dephosphorylation of the enzyme that, *in vivo,* is accompanied either by increases in V_{max} or by a higher affinity for the substrates, UDPglucose and fructose-6-P, and the activator, glucose-6-P, and decreased inhibition by P_i, depending on the species (Huber *et al.,* 1989). Much of what is known about the regulation of SPS and its kinase and phosphatase comes from studies of the enzyme from spinach (Huber *et al.,* 1994; Huber and Huber, 1996) and is summarized in Fig. 7. Until recently, the only SPS from a C_4 plant that has been studied in any depth is that from maize. Sicher and Kremer (1985) first showed that maize leaf SPS was activated by light, and Kalt-Torres *et al.* (1987b)

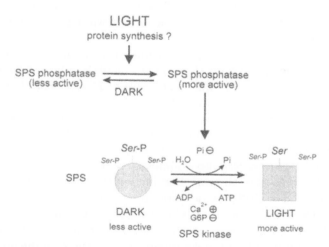

Figure 7 Schematic representation of the regulation of spinach leaf sucrose-P synthase (SPS) by reversible seryl phosphorylation. Multisite phosphorylation and the identification of the major regulatory site (Ser[158]) is indicated. An increase in glucose-6-P and a decrease in P_i, as might occur during a dark-to-light transition, would favor dephosphorylation, and hence activation, of SPS and also increase catalytic activity as a result of allosteric regulation. Other important factors may be the light modulation of SPS phosphatase and changes in cytosolic Ca^{2+}. Most of this scheme remains to be demonstrated in C_4 plants. [Adapted from Huber, S. C., and Huber, J. L. (1996). Role and regulation of sucrose–phosphate synthase in higher plants. *Ann. Rev. Plant Physiol. Plant Mol. Biol.* **47,** 431–444.]

showed pronounced diurnal changes in SPS activity in maize, which has similarly been shown to be regulated by protein phosphorylation (Huber and Huber, 1991). In spinach, activation results from the dephosphorylation of a serine residue (Ser[158]). In the deduced sequence of maize SPS (Worrell et al., 1991; Huber et al., 1994), it appears that the regulatory seryl residue found in spinach has been conserved, indicating that, in maize, Ser[162] is a likely candidate for regulatory phosphorylation. The maize enzyme has been classified in Group I by Huber et al. (1989) because it shows light-dependent changes in V_{max}. This is in contrast to the enzyme from spinach, which shows changes in metabolite modulation, but no change in V_{max} following dephosphorylation (Group II) and that from soybean, which lacks both covalent modification and shows only weak control by metabolites (Group III). It remains to be established whether SPS from different C_4 plants shows such regulatory differences, although a survey of a number of *Panicum* species by Ohsugi and Huber (1987) showed that only one C_4 species, *P. virgatum*, did not show dark inactivation of SPS. The importance of these differences *in vivo* is emphasized by the fact that overexpression of maize SPS in tomato leads to considerable changes in leaf carbohydrate partitioning because the activity of the maize enzyme is not down-regulated in tomato (Worrell et al., 1991; Galtier et al., 1993), whereas spinach SPS expressed in tobacco is down-regulated (Stitt and Sonnewald, 1995).

The limitations of extrapolating too readily from what is known about the regulation of SPS in C_3 plants (Fig. 7) to C_4 plants is indicated by studies of the activation of SPS in bundle-sheath cells of *P. miliaceum*, *F. bidentis*, and *S. bicolor* and mesophyll cells of *D. sanguinalis*. These suggest that a different mechanism of regulation may be operative in both the mesophyll and bundle sheath. ATP activated the enzyme in the dark, largely by decreasing the K_m for UDP-glucose about 10-fold (down to 0.7 mM). A decrease in the K_m for UDP-glucose was also evident when SPS was rapidly extracted from leaves of *P. miliaceum* following illumination (Lunn et al., 1997). These changes in the K_m for UDP-glucose were much more marked than in SPS from spinach (Stitt et al., 1988). It has been suggested that this dependence on ATP could either reflect the operation of a protein kinase that normally maintains a constitutive phosphorylation site on SPS (Fig. 7) or the presence of a novel site that is phosphorylated (Lunn et al., 1997).

Diurnal changes in SPS activity from the two cell types in maize have been studied by Ohsugi and Huber (1987). SPS in maize mesophyll was high and relatively constant in the middle of the light period, dropped rapidly after sunset, and increased again prior to the light period. SPS activity in the bundle sheath was lower and followed the prevailing light intensity more closely than the enzyme from the mesophyll. It was suggested that SPS in the bundle sheath may require higher light intensities to saturate activation.

D. Enzymes of Starch Synthesis

In leaves from a wide range of C_4 plants grown under normal light conditions, starch is only found in the bundle sheath (Downton and Trugunna, 1968; Laetsch, 1971; Lunn and Furbank, 1997). However, growth of maize plants in continuous light or at low temperatures leads to starch accumulation in the mesophyll (Hilliard and West, 1970; Downton and Hawker, 1973) and in another NADP-ME species, *Digitaria pentzii*, starch is synthesised both in the mesophyll and in the bundle-sheath cells (Mbaku et al., 1978). The activities of starch synthase, of the branching enzyme, and of ADP-glucose pyrophosphorylase are present in both cell types, but are higher in the bundle sheath of maize than in the mesophyll (Huber et al., 1969; Downton and Hawker, 1973; Echevarria and Boyer, 1986; Spilatro and Preiss, 1987). However, the enzymes of starch degradation, starch phosphorylase, and amylase, are evenly distributed and may be slightly higher in the mesophyll (Spilatro and Preiss, 1987). The activity of ADP-glucose pyrophosphorylase in the bundle sheath was about 20-fold that in the mesophyll, and ADP-glucose pyrophosphorylase from both cell types was comparable to the enzyme from spinach in its affinity for ATP and glucose-1-P. Like the enzyme from other sources, glycerate-3-P is an activator and P_i an inhibitor, with ratios of glycerate-3-P/P_i for half-maximal activation between 7 and 10 for the bundle-sheath enzyme and 9–16 for the mesophyll enzyme, compared to a ratio of less than 1.5 for half-maximal activation of the spinach enzyme. However, the bundle-sheath enzyme was more sensitive to activation by glycerate-3-P than the mesophyll enzyme (by a factor of 10 in the presence of 0.4 mM P_i) (Spilatro and Preiss, 1987). This lower sensitivity to glycerate-3-P, together with the lower concentration of glycerate-3-P and the relatively low activities of enzymes of starch synthesis in the mesophyll, would appear to favor starch synthesis in the bundle sheath. The sigmoidicity of the response of ADP-glucose pyrophosphorylase to P_i and glycerate-3-P would also ensure that the enzyme acts as a valve, bleeding off hexose-P only when photosynthesis is running at appreciable rates. The leaf content of hexose-P increases as the rate of photosynthesis increases in both maize and *Amaranthus edulis* (Fig. 8; Usuda, 1987; Leegood and von Caemmerer, 1988, 1989; Sage and Seemann, 1993). Jeannette and Prioul (1994) reported that ADP-glucose pyrophosphorylase in maize underwent diurnal changes in activity, although these were not related to changes in the rate of starch accumulation. Because such diurnal changes do not occur in C_3 plants, this potentially important observation requires confirmation.

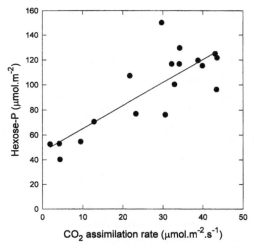

Figure 8 Relationship between the content of hexose-P and the CO_2 assimilation rate in leaves of maize. The rate of photosynthesis was varied by changing photon flux density. [Replotted from data in Leegood, R. C., and von Caemmerer, S. (1989). Some relationships between the contents of photosynthetic intermediates and the rate of photosynthetic carbon assimilation in leaves of *Zea mays* L. *Planta* **178**, 258–266.]

VII. Summary

The regulation of the C_4 cycle and its integration with other components of photosynthetic metabolism is dependent on several interacting mechanisms. The compartmentation of different parts of the cycle in both different cells and subcellular compartments forms the structural basis for this regulation. Fluxes of metabolites and the concentrations of intermediates of the C_4 cycle, and metabolites of other interacting metabolic pathways such as the Benson–Calvin (C_3) cycle, are important regulating the activities of enzymes of the C_4 cycle and in coordinating the activities of the C_3 and C_4 cycles. Superimposed on this is the modulation by light of the enzymes of the C_4 cycle by a variety of mechanisms that include light-dependent changes in pH and ion concentrations (NAD-ME), redox activation via the thioredoxin system (NADP-MDH) and phosphorylation (PEP-C, PPDK, and, in some species, PEP-CK). In addition, enzymes of the Benson–Calvin cycle and of product synthesis are regulated by mechanisms that are similar to their counterparts in C_3 plants. Much still needs to be discovered about the importance of these various control mechanisms, particularly by studying structure–function relationships for the enzymes themselves and by studying the control of photosynthesis in transgenic plants that express either different amounts of the various enzymes or enzymes whose properties have been altered by site-directed mutagenesis.

References

Agostino, A., Heldt, H. W. and Hatch, M. D. (1996). Mitochondrial respiration in relation to photosynthetic C_4 acid decarboxylation in C_4 species. *Aust. J. Plant Physiol.* **23,** 1–7.

Asami, S., Inoue, K. and Akazawa, T. (1979). NADP-malic enzyme from maize leaf: Regulatory properties. *Arch. Biochem. Biophys.* **196,** 581–587.

Ashton, A. R. and Hatch, M. D. (1983a). Regulation of C_4 photosynthesis: Physical and kinetic properties of active (dithiol) and inactive (disulfide) NADP-malate dehydrogenase from *Zea mays. Arch. Biochem. Biophys.* **227,** 406–415.

Ashton, A. R. and Hatch, M. D. (1983b). Regulation of C_4 photosynthesis: Regulation of activation and inactivation of NADP-malate dehydrogenase by NADP and NADPH. *Arch. Biochem. Biophys.* **227,** 416–424.

Ashton, A. R., Burnell, J. N. and Hatch, M. D. (1984). Regulation of C_4 photosynthesis: Inactivation of pyruvate, Pi dikinase by ADP-dependent phosphorylation and activation by phosphorolysis. *Arch. Biochem. Biophys.* **230,** 492–503.

Ashton, A. R., Burnell, J. N., Furbank, R. T., Jenkins, C. L. D. and Hatch, M. D. (1990). Enzymes of C_4 photosynthesis. *In* "Methods in Plant Biochemistry" (P. J. Lea, ed.) Vol. 3, pp. 39–72. Academic Press, London.

Bakrim, N., Echevarria, C., Crétin, C., Arrio-Dupont, M., Pierre, J. N., Vidal, J., Chollet, R. and Gadal, P. (1992). Regulatory phosphorylation of *Sorghum* leaf phosphoenolpyruvate carboxylase. Identification of the protein–serine kinase and some elements of the signal transduction cascade. *Europ. J. Biochem.* **204,** 821–830.

Bakrim, N., Prioul, J. L., Deleens, E., Rocher, J. P., Arrio-Dupont, M., Vidal, J., Gadal, P. and Chollet, R. (1993). Regulatory phosphorylation of C_4 phosphoenolpyruvate carboxylase: A cardinal event influencing the photosynthesis rate in sorghum and maize. *Plant Physiol.* **101,** 891–897.

Budde, R. J. A. and Chollet, R. (1986). *In vitro* phosphorylation of maize leaf phosphoenolpyruvate carboxylase. *Plant Physiol.* **82,** 1107–1114.

Budde, R. J. A., Holbrook, G. P. and Chollet, R. (1985). Studies on the dark/light regulation of maize leaf pyruvate, orthophosphate dikinase by reversible phosphorylation. *Arch. Biochem. Biophys.* **242,** 283–290.

Budde, R. J. A., Ernst, S. M. and Chollet, R. (1986). Substrate specificity and regulation of the maize (*Zea mays*) leaf ADP:protein phosphotransferase catalysing phosphorylation/inactivation of pyruvate, orthophosphate dikinase. *Biochem. J.* **236,** 579–584.

Burnell, J. N. (1984a). Regulation of C_4 photosynthesis. Catalytic dephosphorylation and P_i-mediated activation of pyruvate P_i dikinase. *Biochem. Biophys. Res. Comm.* **120,** 559–565.

Burnell, J. N. (1984b). Regulation of C_4 photosynthesis: Proximity of a histidine involved in catalysis and a threonine involved in regulation of pyruvate, P_i dikinase. *Biochem. Intern.* **9,** 683–689.

Burnell, J. N. (1986) Purification and properties of phosphoenolpyruvate carboxykinase from C_4 plants. *Aust. J. Plant Physiol.* **13,** 577–587.

Burnell, J. N. (1987). Photosynthesis in phosphoenolpyruvate carboxykinase-type C_4 species: Properties of NAD-malic enzyme from *Urochloa panicoides. Aust. J. Plant Physiol.* **14,** 517–525.

Burnell, J. N. and Hatch, M. D. (1983). Dark–light regulation of pyruvate, P_i dikinase in C_4 plants: Evidence that the same protein catalyses activation and inactivation. *Biochem. Biophys. Res. Comm.* **111,** 288–293.

Burnell, J. N. and Hatch, M. D. (1984). Regulation of C_4 photosynthesis: Identification of a catalytically important histidine residue and its role in the regulation of pyruvate, P_i dikinase. *Arch. Biochem. Biophys.* **231**, 175–182.

Burnell, J. N. and Hatch, M. D. (1985). Regulation of C_4 photosynthesis: Purification and properties of the protein catalyzing ADP-mediated inactivation and P_i-mediated activation of pyruvate, P_i dikinase. *Arch. Biochem. Biophys.* **237**, 490–503.

Burnell, J. N. and Hatch, M. D. (1986). Activation and inactivation of an enzyme catalyzed by a single, bifunctional protein: A new example and why. *Arch. Biochem. Biophys.* **245**, 297–304.

Burnell, J. N. and Hatch M. D. (1988a). Photosynthesis in phosphoenolpyruvate carboxykinase–type C_4 plants: Photosynthetic activities of isolated bundle sheath cells from *Urochloa panicoides*. *Arch. Biochem. Biophys.* **260**, 177–186.

Burnell, J. N. and Hatch M. D. (1988b). Photosynthesis in phosphoenolpyruvate carboxykinase–type C_4 plants: Pathways of C_4 acid decarboxylation in bundle sheath cells of *Urochloa panicoides*. *Arch. Biochem. Biophys.* **260**, 187–199.

Burnell, J. N., Jenkins, C. L. D. and Hatch, M. D. (1986). Regulation of C_4 photosynthesis: A role for pyruvate in regulating pyruvate, P_i dikinase activity *in vivo*. *Aust. J. Plant Physiol.* **13**, 203–210.

Carnal, N. W., Agostino, A. and Hatch, M. D. (1993). Photosynthesis in phosphoenolpyruvate carboxykinase–type C_4 plants: Mechanism and regulation of C_4 acid decarboxylation in bundle sheath cells. *Arch. Biochem. Biophys.* **306**, 360–367.

Carroll, L. J., Dunaway-Mariano, D., Smith, C. M. and Chollet, R. (1990). Determination of the catalytic pathway of C_4-leaf pyruvate, orthophosphate dikinase from maize. *FEBS Lett.* **274**, 178–180.

Carter, P. J., Nimmo, H. G., Fewson, C. A. and Wilkins, M. B. (1990). *Bryophyllum fedtschenkoi* protein phosphatase type 2A can dephosphorylate phosphoenolpyruvate carboxylase. *FEBS Lett.* **263**, 233–236.

Carter, P. J., Nimmo, H. G., Fewson, C. A. and Wilkins, M. B. (1991). Circadian rhythms in the activity of a plant protein kinase. *EMBO J.* **10**, 2063–2068.

Chapman, K. S. R. and Hatch, M. D. (1981). Regulation of C_4 photosynthesis: Mechanism of activation and inactivation of extracted pyruvate, Pi dikinase in relation to dark/light regulation. *Arch. Biochem. Biophys.* **210**, 82–89.

Chapman, K. S. R. and Hatch, M. D. (1983). Intracellular location of phosphoenolpyruvate carboxykinase and other C_4 photosynthetic enzymes in mesophyll and bundle sheath protoplasts of *Panicum maximum*. *Plant Sci. Lett.* **29**, 145–154.

Cheng, W.-H., Im, K. H., and Chourey, P. S. (1996). Sucrose phosphate synthase expression at the cell to tissue level is coordinated with sucrose sink-to-source transitions in maize leaf. *Plant Physiol.* **111**, 1021–1029.

Chollet, R., Vidal, J., and O'Leary, M. H. (1996). Phosphoenolpyruvate carboxylase: A ubiquitous, highly regulated enzyme in plants. *Ann. Rev. Plant Physiol. Plant Mol. Biol.* **47**, 273–298.

Clayton, H., Ranson, J., and Rees, T. (1993). Pyrophosphate-fructose-6-phosphate 1-phosphotransferase and fructose-2,6-bisphosphate in the bundle sheath of maize leaves. *Arch. Biochem. Biophys.* **301**, 151–157.

Collatz, G. J., Ribas-Carbo, M. and Berry, J. A. (1992). Coupled photosynthesis–stomatal conductance model for leaves of C_4 plants. *Aust. J. Plant Physiol.* **19**, 519–538.

Coombs, J., Baldry, C. W. and Bucke, C. (1975). The C-4 pathway in *Pennisetum purpureum*. I. The allosteric nature of PEP carboxylase. *Planta* **110**, 95–107.

Day, D. A., Whelan, J., Millar, A. H., Siedow, J. N. and Wiskich, J. T. (1995). Regulation of the alternative oxidase in plants and fungi. *Aust. J. Plant Physiol.* **22**, 497–509.

Decottignies, P., Schmitter, J. M., Miginiac-Maslow, M., Le Maréchal, P., Jacquot, J. P. and Gadal, P. (1988). Primary structure of the light-dependent regulatory site of corn NADP-malate dehydrogenase. *J. Biol. Chem.* **263**, 11780–11785.

Dever, L. V., Blackwell, R. D., Fullwood, N. J., Lacuesta, M., Leegood, R. C., Onek, L. A., Pearson, M. and Lea, P. J. (1995). The isolation and characterization of mutants of the C_4 photosynthetic pathway. *J. Exper. Botany* **46**, 1363–1376.

Dever, L. V., Bailey, K. J., Leegood, R. C. and Lea, P. J. (1997). Control of photosynthesis in *Amaranthus edulis* mutants with reduced amounts of phosphoenolpyruvate carboxylase. *Aust. J. Plant Physiol.* **24**, 469–476.

Doncaster, H. D. and Leegood, R. C. (1987). Regulation of phosphoenolpyruvate carboxylase activity in maize leaves. *Plant Physiol.* **84**, 82–87.

Doncaster, H. D. and Leegood, R. C. (1990). Factors regulating the metabolism of oxaloacetate to malate in leaves of C_4 plants. *J. Plant Physiol.* **136**, 330–335.

Douce, R. and Neuburger, M. (1989). The uniqueness of plant mitochondria. *Ann. Rev. Plant Physiol. Plant Mol. Biol.* **40**, 371–414.

Downton, W. J. S., and Hawker, J. S. (1973). Enzymes of starch and sucrose metabolism in *Zea mays* leaves. *Phytochemistry* **12**, 1551–1556.

Duff, S. M. G., Lepiniec, L., Crétin, C., Andreo, C. S., Condon, S. A., Sarath, G., Vidal, J., Gadal, P. and Chollet, R. (1993). An engineered change in the L-malate sensitivity of a site-directed mutant of sorghum phosphoenolpyruvate carboxylase: The effect of sequential mutagenesis and S-carboxymethylation at position 8. *Arch. Biochem. Biophys.* **306**, 272–276.

Duff, S. M. G., Andreo, C. S., Pacquit, V., Lepiniec, L., Sarath, G., Condon, S. A., Vidal, J., Gadal, P. and Chollet, R. (1995). Kinetic analysis of the non-phosphorylated, *in vitro* phosphorylated, and phosphorylation-site-mutant (Asp^8) forms of intact recombinant C_4 phosphoenolpyruvate carboxylase from sorghum. *Europ. J. Biochem.* **228**, 92–95.

Duff, S. M. G., Giglioli-Guivarc'h, N., Pierre, J.-N., Vidal, J., Condon, S. A., and Chollet, R. (1996). *In-situ* evidence for the involvement of calcium and bundle-sheath–derived photosynthetic metabolites in the C_4 phosphoenolpyruvate–carboxylase kinase signal-transduction chain. *Planta* **199**, 467–474.

Echevarria, E., and Boyer, C. D. (1986). Localization of starch biosynthetic and degradative enzymes in maize leaves. *Am. J. Botany* **73**, 167–171.

Echevarria, C., Vidal, J., Jioa, J.-A. and Chollet, R. (1990). Reversible light-activation of the phosphoenolpyruvate carboxylase protein–serine kinase in maize leaves. *FEBS Lett.* **275**, 25–28.

Echevarria, C., Pacquit, V., Bakrim, N., Osuna, L., Delgado, B., Arrio-Dupont, M. and Vidal, J. (1994). The effect of pH on the covalent and metabolic control of C_4 phosphoenolpyruvate carboxylase from *Sorghum* leaf. *Arch. Biochem. Biophys.* **315**, 425–430.

Edwards, G. E. and Walker, D. A. (1983). C_3, C_4: Mechanisms, and cellular and environmental regulation, of photosynthesis. Blackwell, Oxford.

Edwards, G. E., Nakamoto, H., Burnell, J. N. and Hatch, M. D. (1985). Pyruvate, P_i dikinase and NADP-malate dehydrogenase in C_4 photosynthesis: Properties and mechanism of light/dark regulation. *Ann. Rev. Plant Physiol.* **36**, 255–286.

Edwards, G. E., Jenkins, C. L. D., and Andrews, J. (1988). CO_2 assimilation and activities of photosynthetic enzymes in high chlorophyll fluorescence mutants of maize having low levels of ribulose 1,5-bisphosphate carboxylase. *Plant Physiol.* **86**, 533–539.

Fell, D. (1997). "Understanding the control of metabolism." Portland Press, London.

Finnegan, P. M. and Burnell, J. N. (1995). Isolation and sequence analysis of cDNAs encoding phosphoenolpyruvate carboxykinase from the PCK-type C_4 grass *Urochloa panicoides*. *Plant Mol. Biol.* **27**, 365–376.

Flügge, U. I., Stitt, M. and Heldt, H. W. (1985). Light-driven uptake of pyruvate into mesophyll chloroplasts from maize. *FEBS Lett.* **183**, 335–339.

Furbank, R. T. and Hatch, M. D. (1987). Mechanism of C_4 photosynthesis. The size and composition of the inorganic carbon pool in bundle-sheath cells. *Plant Physiol.* **85**, 958–964.

Furbank, R. T. and Leegood, R. C. (1984). Carbon metabolism and gas exchange in leaves of Zea mays L. Interaction between the C_3 and C_4 pathways during photosynthetic induction. Planta 162, 457–462.

Furbank, R. T., Stitt, M. and Foyer, C. H. (1985). Intercellular compartmentation of sucrose synthesis in leaves of Zea mays. Planta 164, 172–178.

Furbank, R. T., Agostino, A. and Hatch, M. D. (1991). Regulation of C_4 photosynthesis: Modulation of mitochondrial NAD-malic enzyme by adenylates. Arch. Biochem. Biophys. 289, 376–381.

Furbank, R. T., Chitty, J. A., von Caemmerer, S., and Jenkins, C. L. D. (1996). Antisense RNA inhibition of RbcS gene expression reduces Rubisco level and photosynthesis in the C_4 plant Flaveria bidentis. Plant Physiol. 111, 725–734.

Furbank, R. T., Chitty, J. A., Jenkins, C. L. D., Taylor, W. C., Trevanion, S. J., von Caemmerer, S. and Ashton, A. R. (1997). Genetic manipulation of key photosynthetic enzymes in the C_4 plant Flaveria bidentis. Aust. J. Plant Physiol. 24, 487–493.

Gadal, P., Pacquit, V., Giglioli, N., Bui, V.-L., Pierre, J. N., Echevarria, C. and Vidal, J. (1996). The role of PEPC phosphorylation in the regulation of C_4 photosynthesis. In "Protein Phosphorylation in Plants" (P. R. Shewry, N. G. Halford and R. Hooley, eds.), pp. 53–64, Oxford Science Publications, Oxford.

Galtier, N., Foyer, C. H., Huber, J., Voelker, T. A., and Huber, S. C. (1993). Effects of elevated sucrose–phosphate synthase activity on photosynthesis, assimilate patitioning, and growth in tomato (Lycopersicon esculentum var UC82B). Plant Physiol. 101, 535–543.

Gao, Y. and Woo, K. C. (1996). Regulation of phosphoenolpyruvate carboxylase in Zea mays by protein phosphorylation and metabolites and their roles in photosynthesis. Aust. J. Plant Physiol. 23, 25–32.

Gardeström, P., and Edwards, G. E. (1983). Isolation of mitochondria from leaf tissue of Panicum milaceum, a NAD-malic enzyme type C_4 plant. Plant Physiol. 71, 24–29.

Giglioli-Guivarc'h, Pierre, J.-N., Brown, S., Chollet, R., Vidal, J. and Gadal, P. (1996). The light-dependent transduction pathway controlling the regulatory phosphorylation of C_4 phosphoenolpyruvate carboxylase in protoplasts from Digitaria sanguinalis. Plant Cell 8, 573–586

Hartwell, J., Smith, L. H., Wilkins, M. B., Jenkins, G. I. and Nimmo, H. G. (1996). Higher plant phosphoenolpyruvate carboxylase kinase is regulated at the level of translatable mRNA in response to light or a circadian rhythm. Plant J. 10, 1071–1078.

Hatch, M. D. (1973). Separation and properties of leaf aspartate aminotransferase and alanine aminotransferase isoenzymes operative in the C_4 pathway of photosynthesis. Arch. Biochem. Biophys. 156, 207–214.

Hatch, M. D. (1979). Mechanism of C_4 photosynthesis in Chloris gayana: pool sizes and kinetics of $^{14}CO_2$ incorporation into 4-carbon and 3-carbon intermediates. Arch. Biochem. Biophys. 194, 117–127.

Hatch, M. D. and Mau, S.-L. (1973). Activity, location, and role of aspartate aminotransferase isoenzymes in leaves with C_4 pathway photosynthesis. Arch. Biochem. Biophys. 156, 195–206.

Hatch, M. D. and Slack, C. R. (1966). Photosynthesis by sugar-cane leaves. A new carboxylation reaction and the pathway of sugar formation. Biochem. J. 101, 103–111.

Hatch, M. D. and Slack, C. R. (1969). Studies on the mechanism of activation and inactivation of pyruvate, P_i dikinase. A possible regulatory role for the enzyme in the C_4 dicarboxylic acid pathway of photosynthesis. Biochem. J. 112, 549–558.

Hatch, M. D. and Osmond, C. B. (1976). Compartmentation and transport in C_4 photosynthesis. In "Encyclopedia of Plant Physiology" (C. R. Stocking and U. Heber, eds.), New Series Vol. 3, pp. 144–184. Springer-Verlag, Berlin.

Hatch, M. D., Mau, S.-L. and Kagawa, T. (1974). Properties of leaf NAD malic enzyme from plants with C_4 pathway photosynthesis. Arch. Biochem. Biophys. 165, 188–200.

Hatch, M. D., Agostino, A. and Burnell, J. N. (1988). Photosynthesis in phosphoenolpyruvate carboxykinase-type C_4 plants: Activity and role of mitochondria in bundle sheath cells. *Arch. Biochem. Biophys.* **261,** 357–367.

Heldt, H. W. and Flügge, U. I. (1987). Subcellular transport of metabolites in plant cells. *In* "Biochemistry of Plants: Physiology of Metabolism" (D. D. Davies, ed.), Vol. 12, pp. 49–85, Academic Press, New York.

Heldt, H. W. and Flügge, U. I. (1992). Metabolite transport in plant cells. *In* "Plant Organelles" (A. K. Tobin, ed.), pp. 21–47. Cambridge University Press, Cambridge.

Herzog, B., Stitt, M., and Heldt, H. W. (1984). Control of photosynthetic sucrose synthesis by fructose 2,6-bisphosphate. III. Properties of the cytosolic fructose-1,6-bisphosphatase. *Plant Physiol.* **75,** 561–566.

Hilliard, J. H., and West, S. H. (1970). Starch accumulation associated with growth reduction at low temperatures in a tropical plant. *Science* **168,** 494–496.

Hocking, C. G. and Anderson, J. W. (1986). Survey of pyruvate, P_i dikinase activity of plants in relation to the C_3, C_4 and CAM mechanisms of CO_2 assimilation. *Phytochemistry* **25,** 1537–1543.

Hoefnagel, M. H. N., Millar, A. H., Wiskich, J. T. and Day, D. A. (1995). Cytochrome and alternative respiratory pathways compete for electrons in the presence of pyruvate in soybean mitochondria. *Arch. Biochem. Biophys.* **318,** 394–400.

Huber, S. C. and Edwards, G. E. (1975). Inhibition of phosphoenolpyruvate carboxylase from C_4 plants by malate and aspartate. *Can. J. Botany* **53,** 1925–1933.

Huber, S. C., and Huber, J. L. (1991). Regulation of maize leaf sucrose–phosphate synthase by protein phosphorylation. *Plant Cell Physiol.* **32,** 319–326.

Huber, S. C., and Huber, J. L. (1996). Role and regulation of sucrose–phosphate synthase in higher plants. *Ann. Rev. Plant Physiol. Plant Mol. Biol.* **47,** 431–444.

Huber, S. C., Sugiyama, T. and Akazawa, T. (1986). Light modulation of maize leaf phosphenolpyruvate carboxylase. *Plant Physiol.* **82,** 550–554.

Huber, S. C., Nielsen, T. H., Huber, J. L. A., and Pharr, D. M. (1989). Variation among species in light activation of sucrose–phosphate synthase. *Plant Cell Physiol.* **30,** 277–285.

Huber, W., DeFekete, A., and Ziegler, H. (1969). Enzymes of starch metabolism in chloroplasts of the bundle sheath and palisade cells of *Zea mays*. *Planta* **87,** 360–364.

Huber, S. C., Huber, J. L. and McMichael, R. W. (1994). Control of plant enzyme activity by reversible protein phosphorylation. *Intern. Rev. Cytol.* **149,** 47–98.

Iglesias, A. A. and Andreo, C. S. (1989). Purification of NADP-malic enzyme and phosphoenolpyruvate carboxylase from sugar cane leaves. *Plant Cell Physiol.* **30,** 399–405.

Jeannette, E. and Prioul, J.-L. (1994). Variations of ADP-glucose pyrophosphorylase activity from maize leaf during day/night cycle. *Plant Cell Physiol.* **35,** 869–878.

Jacquot, J.-P. P., Buchanan, B. B., Martin, F. and Vidal, J. (1981). Enzyme regulation in C_4 photosynthesis. Purification and properties of thioredoxin-linked NADP-malate dehydrogenase from corn leaves. *Plant Physiol.* **68,** 300–304.

Jenkins, C. L. D. and Hatch, M. D. (1985). Properties and reaction mechanism of C_4 leaf pyruvate, P_i dikinase. *Arch. Biochem. Biophys.* **239,** 53–62.

Jenkins, C. L. D., Furbank, T. R., and Hatch, M. D. (1989a). Inorganic carbon diffusion between C_4 mesophyll and bundle sheath cells: Direct bundle sheath CO_2 assimilation in intact leaves in the presence of an inhibitor of the C_4 pathway. *Plant Physiol.* **91,** 1356–1363.

Jenkins, C. L. D., Furbank, T. R., and Hatch, M. D. (1989b). Mechanism of C_4 photosynthesis. A model describing the inorganic carbon pool in bundle sheath cells. *Plant Physiol.* **91,** 1372–1381.

Jiao, J. A. and Chollet, R. (1988). Light/dark regulation of maize leaf phosphoenolpyruvate carboxylase by *in vivo* phosphorylation. *Arch. Biochem. Biophys.* **269,** 526–535.

Jiao, J. A. and Chollet, R. (1989). Regulatory seryl-phosphorylation of C_4 phosphoenolpyruvate carboxylase by a C_4-leaf protein-serine kinase. *Arch. Biochem. Biophys.* **269**, 526–535.

Jiao, J. and Chollet, R. (1990). Regulatory phosphorylation of serine-15 in maize phosphoenolpyruvate carboxylase by a C_4-leaf protein-serine kinase. *Arch. Biochem. Biophys.* **283**, 300–305.

Jiao, J. and Chollet, R. (1992). Light activation of maize phosphoenolpyruvate carboxylase protein–serine kinase activity is inhibited by mesophyll and bundle sheath–directed photosynthesis inhibitors. *Plant Physiol.* **98**, 152–156.

Jiao, J.-A., Echevarria, C., Vidal, J. and Chollet, R. (1991a). Protein turnover as a component in the light/dark regulation of phosphoenolpyruvate carboxylase protein–serine kinase activity in C_4 plants. *Proc. Natl. Acad. Sci. USA* **88**, 2712–2715.

Jiao, J.-A., Vidal, J., Echevarria, C. and Chollet, R. (1991b). *In vivo* regulatory phosphorylation site in C_4-leaf phosphoenolpyruvate carboxylase from maize and sorghum. *Plant Physiol.* **96**, 297–301.

Johnson, H. S. and Hatch, M. D. (1970). Properties and regulation of leaf nicotinamide-adenine dinucleotide phosphate–malate dehydrogenase and malic enzyme in plants with the C_4-dicarboxylic acid pathway of photosynthesis. *Biochem. J.* **119**, 273–280.

Kacser H. (1987). Control of metabolism. *In* "The Biochemistry of Plants: Biochemistry of Metabolism" (D. D. Davies, ed.), Vol. 11, 39–67. Academic Press, London.

Kagawa, T. and Bruno, P. L. (1988). NADP-malate dehydrogenase from leaves of *Zea mays*: Purification and physical, chemical, and kinetic properties. *Arch. Biochem. Biophys.* **260**, 674–695.

Kagawa, T. and Hatch, M. D. (1977). Regulation of C_4 photosynthesis: Characterization of a protein factor mediating the activation and inactivation of NADP-malate dehydrogenase. *Arch. Biochem. Biophys.* **184**, 290–297.

Kalt-Torres, W., Kerr, P. S., Usuda, H., and Huber, S. C. (1987a). Diurnal changes in maize leaf photosynthesis. I. Carbon exchange rate, assimilate export rate, and enzyme activities. *Plant Physiol.* **83**, 283–288.

Kalt-Torres, W., Kerr, P. S., and Huber, S. C. (1987b). Isolation and characterization of maize leaf sucrose-phosphate synthase. *Physiologia Plantarum* **7**, 653–658.

Kobr, M. J. and Beevers, H. (1971). Gluconeogenesis in the castor bean endosperm. I. Changes in glycolytic intermediates. *Plant Physiol.* **47**, 48–52.

Krall J. K. and Pearcy R. W. (1993). Concurrent measurements of oxygen and carbon dioxide exchange during lightflecks in maize (*Zea mays* L.). *Plant Physiol.* **103**, 823–828.

Leegood, R. C. (1985). The intercellular compartmentation of metabolites in leaves of *Zea mays*. *Planta* **164**, 163–171.

Leegood, R. C. (1997). The regulation of C_4 photosynthesis. *Adv. Bot. Res.* **26**, 251–316.

Leegood, R. C. and Walker, D. A. (1983). Modulation of NADP-malate dehydrogenase activity in maize mesophyll chloroplasts. *Plant Physiol.* **71**, 513–518.

Leegood, R. C. and von Caemmerer, S. (1988). The relationship between contents of photosynthetic intermediates and the rate of photosynthetic carbon assimilation in leaves of *Amaranthus edulis* L. *Planta* **174**, 253–262.

Leegood, R. C. and von Caemmerer, S. (1989). Some relationships between the contents of photosynthetic intermediates and the rate of photosynthetic carbon assimilation in leaves of *Zea mays* L. *Planta* **178**, 258–266.

Leegood, R. C. and von Caemmerer, S. (1994). Regulation of photosynthetic carbon assimilation in leaves of C_3–C_4 intermediate species of *Moricandia* and *Flaveria*. *Planta* **192**, 232–238.

Leegood, R. C., Foyer, C. H. and Walker, D. A. (1985). Regulation of the Benson–Calvin cycle. *In* "Photosynthetic Mechanisms and the Environment" (J. Barber and N. R. Baker, eds.), pp. 189–258. Elsevier, Amsterdam.

Leegood, R. C., von Caemmerer, S. and Osmond, C. B. (1997). Metabolite transport and photosynthetic regulation in C_4 and CAM plants. In "Plant Metabolism" (D. T. Dennis, D. H. Turpin, D. D. Lefebvre, and D. B. Layzell, eds.), pp. 341–369, Longman, London.

Li, B. and Chollet, R. (1993). Resolution and identification of C-4 phosphoenolpyruvate carboxylase protein kinase and their reversble light activation in maize leaves. *Arch. Biochem. Biophys.* **307**, 416–419.

Li, B., Pacquit, V., Jiao, J., Duff, S. M. G., Maralihalli, G. B., Sarath, G., Condon, S. A., Vidal, J. and Chollet, R. (1997). Structural requirements for phosphorylation of C_4-leaf phosphoenolpyruvate carboxylase by its highly regulated protein kinase. A comparative study with synthetic peptide substrates and native, mutant target proteins. *Aust. J. Plant Physiol.* **24**, 443–449.

Lowe, J. and Slack, C. R. (1971). Inhibition of maize leaf phosphoenolpyruvate carboxylase by oxaloacetate. *Biochim. Biophys. Acta* **235**, 207–209.

Lunn, J. E., and Hatch, M. D. (1995). Primary partitioning and storage of photosynthate in sucrose and starch in leaves of C_4 plants. *Planta* **197**, 385–391.

Lunn, J. E., and Furbank, R. T. (1997). Localisation of sucrose–phosphate synthase and starch in leaves of C_4 plants. *Planta* **202**, 106–111.

Lunn, J. E., Furbank, R. T., and Hatch, M. D. (1997). Adenosine 5′-triphosphate–mediated activation of sucrose–phosphate synthase in bundle sheath cells of C_4 plants. *Planta* **202**, 249–256.

Mbaku, S. B., Fritz, G. J. and Bowes, G. (1978). Photosynthetic and carbohydrate metabolism in isolated leaf cells of *Digitaria pentzii*. *Plant Physiol.* **62**, 510–515.

McNaughton, G. A. L., Fewson, C. A., Wilkins, M. B. and Nimmo, H. G. (1989). Purification, oligomerization state and malate sensitivity of maize leaf phosphoenolpyruvate carboxylase. *Biochem. J.* **261**, 349–355.

McNaughton, G. A. L., MacKintosh, C., Fewson, C. A., Wilkins, M. B. and Nimmo, H. G. (1991). Illumination increases the phosphorylation state of maize leaf phosphoenolpyruvate carboxylase by causing an increase in the activity of a protein kinase. *Biochim. Biophys. Acta* **1093**, 189–195.

Meister, M., Agostino, A. and Hatch, M. D. (1996). The roles of malate and aspartate in C_4 photosynthetic metabolism of *Flaveria bidentis* (L.). *Planta* **199**, 262–269.

Moore, B. D., Kobza, J. and Seemann, J. R. (1991). Measurement of 2-carboxyarabinitol 1-phosphate in plant leaves by isotope dilution. *Plant Physiol.* **96**, 208–213.

Moore, B. D., Sharkey, T. D., Kobza, J. and Seemann, J. R. (1992). Identification and levels of 2′-carboxyarabinitol in leaves. *Plant Physiol.* **99**, 1546–1550.

Muench, D. G. and Good, A. G. (1994). Hypoxically inducible barley alanine aminotransferase: cDNA cloning and expression analysis. *Plant Mol. Biol.* **24**, 417–427.

Murata, T., Ikeda, J., Takano, M. and Ohsugi, R. (1989). Comparative studies of NAD-malic enzyme from leaves of various C_4 plants. *Plant Cell Physiol.* **30**, 429–437.

Nakamoto, H. and Young, P. S. (1990). Light activation of pyruvate, orthophosphate dikinase in maize mesophyll chloroplasts: A role for adenylate energy charge. *Plant Cell Physiol.* **31**, 1–6.

Nimmo, G. A., Nimmo, H. G., Fewson, C. A. and Wilkins, M. B. (1984). Diurnal changes in the properties of phosphoenolpyruvate carboxylase in *Bryophyllum* leaves: A possible covalent modification. *FEBS Lett.* **178**, 199–203.

Nimmo, G. A., Nimmo, H. G., Hamilton, I. D., Fewson, C. A. and Wilkins, M. B. (1986). Purification of the phosphorylated night form and dephosphorylated day form of phosphoenolpyruvate carboxylase from *Bryophyllum fedtschenkoi*. *Biochem. J.* **239**, 213–220.

Nimmo, G. A., Wilkins, M. B., Fewson, C. A and Nimmo, H. G. (1987a). Persistent circadian rhythms in the phosphorylation state of phosphoenolpyruvate carboxylase from *Bryophyllum fedtschenkoi* leaves and in its sensitivity to inhibition by malate. *Planta* **170**, 408–415.

Nimmo, G. A., McNaughton, G. A. L., Fewson, C. A., Wilkins, M. B. and Nimmo, H. G. (1987b). Changes in the kinetic properties and phosphorylation state of phosphoenolpyruvate carboxylase in *Zea mays* leaves in response to light and dark. *FEBS Lett.* **213**, 18–22.

Nishikido, T. and Takanashi, H. (1973). Glycine activation of PEP carboxylase from monocotyledonous C_4 plants. *Biochem. Biophys. Res. Comm.* **53**, 126–133.

Numazawa, T., Yamada, S., Hase, T. and Sugiyama, T. (1989). Aspartate aminotransferase from *Panicum maximum* Jacq. var trichoglume Eyles, a C_4 plant: Purification, molecular properties, and preparation of antibody. *Arch. Biochem. Biophys.* **270**, 313–319.

Ogawa, N, Okumura, S. and Izui, K. (1992). A Ca^{2+}-dependent protein kinase phosphorylates phosphoenolpyruvate carboxylase in maize. *FEBS Lett.* **302**, 86–88.

Ohnishi, J., Flügge, U.-I. and Heldt, H. W. (1989). Phosphate translocator of mesophyll and bundle sheath chloroplasts of the C_4 plant, *Panicum miliaceum*. Identification and kinetic characterization. *Plant Physiol.* **91**, 1507–1511.

Ohsugi, R. and Huber, S. C. (1987). Light modulation and localization of sucrose phosphate synthase activity between mesophyll cells and bundle sheath cells in C_4 species. *Plant Physiol.* **84**, 1096–101.

Pierre, J. N., Pacquit, V. and Gadal, P. (1992). Regulatory properties of phosphoenolpyruvate carboxylase in protoplasts from *Sorghum* mesophyll cells and the role of pH and Ca^{2+} as possible components of the light-transduction pathway. *Eur. J. Biochem.* **210**, 531–537.

Portis, A. R., Jr. (1992). Regulation of ribulose 1,5-bisphosphate carboxylase/oxygenase activity. *Ann. Rev. Plant Physiol. Plant Mol. Biol.* **43**, 415–437.

Randall, D. D. and Rubin, P. M. (1977). Plant pyruvate dehydrogenase complex. II. ATP-dependent inactivation and phosphorylation. *Plant Physiol.* **59**, 1–3.

Ray, T. B., Black, C. C. Jr. (1976). Characterization of phosphoenolpyruvate carboxykinase from *Panicum maximum*. *Plant Physiol.* **58**, 603–607.

Rebeille, F. and Hatch, M. D. (1986a). Regulation of NADP-malate dehydrogenase in C_4 plants: Effect of varying NADPH to NADP ratios and thioredoxin redox state on enzyme activity in reconstituted systems. *Arch. Biochem. Biophys.* **249**, 164–170.

Rebeille, F. and Hatch, M. D. (1986b). Regulation of NADP-malate dehydrogenase in C_4 plants: Relationship among enzyme activity, NADP to NADP ratios, and thioredoxin redox states in intact maize mesophyll chloroplasts. *Arch. Biochem. Biophys.* **249**, 171–179.

Roeske, C. A. and Chollet, R. (1987). Chemical modification of the bifunctional regulatory protein of maize leaf pyruvate, orthophosphate dikinase. Evidence for two distinct active sites. *J. Biol. Chem.* **262**, 12575–12582.

Roeske, C. A. and Chollet, R. (1989). Role of metabolites in the reversible light activation of pyruvate, orthophosphate dikinase in *Zea mays* mesophyll cells *in vivo*. *Plant Physiol.* **90**, 330–337.

Roeske, C. A., Kutny, R. M., Budde, R. J. A. and Chollet, R. (1988). Sequence of the phosphothreonyl regulatory site peptide from inactive maize leaf pyruvate, orthophosphate dikinase. *J. Biol. Chem.* **263**, 6683–6687.

Sage, R. F. and Seemann, J. R. (1993). Regulation of ribulose-1,5-bisphosphate carboxylase/oxygenase activity in response to reduced light intensity in C_4 plants. *Plant Physiol.* **102**, 21–28.

Sicher, R. C., Kremer, D. F. (1985). Possible control of maize leaf sucrose–phosphate synthase activity by light. *Plant Physiol.* **79**, 695–698.

Slack, C. R. (1968). The photoactivation of a phosphopyruvate synthase in leaves of *Amaranthus palmeri*. *Biochem. Biophys. Res. Comm.* **30**, 483–488.

Smith, C. M., Duff, S. M. G. and Chollet, R. (1994). Partial purification and characterization of maize-leaf pyruvate, orthophosphate dikinase regulatory protein: A low abundance, mesophyll chloroplast stromal protein. *Arch. Biochem. Biophys.* **308**, 200–206.

Soll, J., Wörzer, C. and Buchanan, B. B. (1983). Fructose-2,6-bisphosphate and C_4 plants. In "Advances in Photosynthesis Research" (C. Sybesma, ed.), Vol. 3, pp. 485–488, Martinus-Nijhoff, The Hague.
Son, D., Jo, J. and Sugiyama, T. (1991). Purification and characterization of alanine aminotransferase from *Panicum miliaceum* leaves. *Arch. Biochem. Biophys.* **289**, 262–266.
Spilatro, S. R. and Preiss, J. (1987). Regulation of starch synthesis in the bundle-sheath and mesophyll of *Zea mays* L. Intercellular compartmentation of enzymes of starch metabolism and the properties of the ADPglucose pyrophosphorylases. *Plant Physiol.* **83**, 621–627.
Stitt, M. and Heldt, H. W. (1985a). Control of photosynthetic sucrose synthesis by fructose-2,6-bisphosphate. Intercellular metabolite distribution and properties of the cytosolic fructose bisphosphatase in leaves of *Zea mays* L. *Planta* **164**, 179–188.
Stitt, M. and Heldt, H. W. (1985b). Generation and maintenance of concentration gradients between the mesophyll and bundle-sheath in maize leaves. *Biochim. Biophys. Acta* **808**, 400–414.
Stitt, M. and Sonnewald, U. (1995). Regulation of metabolism in transgenic plants. *Ann. Rev. Plant Physiol. Plant Mol. Biol.* **46**, 341–368.
Stitt, M., Huber, S. C. and Kerr, P. (1987). Control of photosynthetic sucrose formation. In "The Biochemistry of Plants" (M. D. Hatch and N. K. Boardman, eds.), Vol. 10, pp. 327–409, Academic Press, New York.
Stitt, M., Wilke, I., Feil, R., and Heldt, H. W. (1988). Coarse control of sucrose–phosphate synthase in leaves: Alterations of the kinetic properties in response to the rate of photosynthesis and the accumulation of sucrose. *Planta* **174**, 217–230.
Taniguchi, M. and Sugiyama, T. (1990). Aspartate aminotransferase from *Eleusine coracana*, a C_4 plant—purification, characterisation and preparation of antibody. *Arch. Biochem. Biophys.* **282**, 427–432.
Taniguchi, M., Kobe, A., Kato, M. and Sugiyama, T. (1995). Aspartate aminotransferase isoenzymes in *Panicum miliaceum* L., an NAD-malic enzyme–type C_4 plant: Comparison of enzymatic properties, primary structures, and expression patterns. *Arch. Biochem. Biophys.* **318**, 295–306.
Ting, I. P. and Osmond, C. B. (1973). Activation of plant P-enolpyruvate carboxylases by glucose-6-phosphate: A particular role in Crassulcean acid metabolism. *Plant Sci. Lett.* **1**, 123–128.
Trevanion, S. J., Furbank, R. T. and Ashton, A. R. (1997). NADP-malate dehydrogenase in the C_4 plant *Flaveria bidentis*. Cosense suppression of activity in mesophyll and bundle-sheath cells and consequences for photosynthesis. *Plant Physiol.* **113**, 1153–1165.
Uedan, K. and Sugiyama, T. (1976). Purification and characterization of phosphoenolpyruvate carboxylase from maize leaves. *Plant Physiol.* **57**, 906–910.
Urbina, J. A. and Avilan, L. (1989). The kinetic mechanism of phosphoenolpyruvate carboxykinase from *Panicum maximum*. *Phytochemistry* **28**, 1349–1353.
Usuda, H. (1985). The activation state of ribulose 1,5-bisphosphate carboxylase in maize leaves in dark and light. *Plant Cell Physiol.* **26**, 1455–1463.
Usuda, H. (1986). Non-autocatalytic build-up of ribulose 1,5-bisphosphate during the initial phase of photosynthetic induction in maize leaves. *Plant Cell Physiol.* **27**, 745–749.
Usuda, H. (1987). Changes in levels of intermediates of the C_4 cycle and reductive pentose phosphate pathway under various light intensities in maize leaves. *Plant Physiol.* **84**, 549–554.
Usuda, H. (1988). Adenine nucleotide levels, the redox state of the NADP system, and assimilatory force in nonaqueously purified mesophyll chloroplasts from maize leaves under different light intensities. *Plant Physiol.* **88**, 1461–1468.
Usuda, H., Ku, M. S. B. and Edwards, G. E. (1984). Activation of NADP-malate dehydrogenase, pyruvate, P_i dikinase, and fructose 1,6-bisphosphatase in relation to photosynthetic rate in maize. *Plant Physiol.* **76**, 238–243.

Usuda, H., Kalt-Torres, W., Kerr, P. S., and Huber, S. C. (1987). Diurnal changes in maize leaf photosynthesis. I. Levels of metabolic intermediates of sucrose synthesis and the regulatory metabolite fructose-2,6-bisphosphate. *Plant Physiol.* **83**, 289–293.

von Caemmerer, S. and Farquhar G. D. (1981). Some relationships between the biochemistry of photosynthesis and the gas exchange of leaves. *Planta* **153**, 376–387.

von Caemmerer, S., Millgate, A., Farquhar, G. D. and Furbank, R. T. (1997). Reduction of ribulose-1,5-bisphosphate carboxylase/oxygenase by antisense RNA in the C_4 plant *Flaveria bidentis* leads to reduced assimilation rates and increased carbon isotope discrimination. *Plant Physiol.* **113**, 469–477.

Vu, J. C. V., Allen, L. H. and Bowes, G. H. (1984). Dark/light modulation of ribulose bisphosphate carboxylase activity in plants from different photosynthetic categories. *Plant Physiol.* **76**, 843–845.

Walker, R. P. and Leegood, R. C. (1996). The regulation of phosphoenolpyruvate carboxykinase activity in plants. *In* "Current Research in Photosynthesis" (P. Mathis, ed.), Vol. 5, pp. 29–34. Kluwer, Dordrecht.

Walker, R. P., Trevanion, S. J. and Leegood, R. C. (1995). Phosphoenolpyruvate carboxykinase from higher plants: Purification from cucumber and evidence of rapid proteolytic cleavage in extracts from a range of plant tissues. *Planta* **196**, 58–63.

Walker, R. P., Acheson, R. M., Técsi, L. I. and Leegood, R. C. (1997). Phosphoenolpyruvate carboxykinase in C_4 plants: Its role and regulation. *Aust. J. Plant Physiol.* **24**, 459–468.

Wang, Y.-H. and Chollet, R. (1993). Partial purification and characterization of phosphoenolpyruvate carboxylase protein–serine kinase from illuminated maize leaves. *Arch. Biochem. Biophys.* **304**, 496–502.

Wang, Y.-H., Duff, S. M. G., Lepiniec, L., Crétin, C., Sarath, G., Condon, S. A., Vidal, J., Gadal, P. and Chollet, R. (1992). Site-directed mutagenesis of the phosphorylatable serine (Ser[8]) in C_4 phosphoenolpyruvate carboxylase from sorghum. The effect of negative charge at position 8. *J. Biol. Chem.* **267**, 16759–16762.

Wedding, R. T. and Black, M. T. (1983). Physical and kinetic properties and regulation of the NAD malic enzyme purified from leaves of *Crassula argentea*. *Plant Physiol.* **72**, 1021–1028.

Wedding, R. T. and Whatley, F. R. (1984). Malate oxidation by *Arum spadix* mitochondria: Participation and characterstics of NAD malic enzyme. *New Phytol.* **96**, 505–517.

Weiner, H. and Heldt, H. W. (1992). Inter- and intracellular distribution of amino acids and other metabolites in maize (*Zea mays* L.) leaves. *Planta* **187**, 342–246.

Weiner, H., Burnell, J. N., Woodrow, I. E., Heldt, H. W. and Hatch, M. D. (1988). Metabolite diffusion into bundle sheath cells from C_4 plants. Relation to C_4 photosynthesis and plasmodesmatal function. *Plant Physiol.* **88**, 815–822.

Winter, K. (1982). Properties of phosphoenolpyruvate carboxylase in rapidly prepared, desalted leaf extracts of the Crassulacean acid metabolism plant *Mesembryanthemum crystallinum* L. *Planta* **154**, 298–308.

Wood, H. G., Davis, J. J. and Lochmüller, H. (1966). The equilibria of reactions catalyzed by carboxytransphosphorytase, phosphoenolpyruvate carboxykinase, and pyruvate carboxylase and the synthesis of phosphoenolpyruvate. *J. Biol. Chem.* **241**, 5692–5704.

Worrell, A. C., Bruneau, J.-M., Summerfelt, K., Boersig, M., and Voelker, T. A. (1991). Expression of a maize sucrose phosphate synthase in tomato alters leaf carbohydrate partitioning. *Plant Cell* **3**, 1121–1130.

Wu, M. X. and Wedding, R. T. (1985). Regulation of phosphoenolpyruvate carboxylase from *Crassula* by interconversion of oligomeric forms. *Arch. Biochem. Biophys.* **240**, 655–662.

Wu, M. X. and Wedding, R. T. (1987). Regulation of phosphoenolpyruvate carboxylase from *Crassula argentea*: Further evidence on the dimer–tetramer interconversion. *Plant Physiol.* **84**, 1080–1083.

Yamamoto, E., Sugiyama, T. and Miyachi, S. (1974). Action spectrum for light activation of pyruvate, phosphate dikinase in maize leaves. *Plant Cell Physiol.* **15,** 987–992.

Yeoh, H. H., Badger, M. R. and Watson, L. (1980). Variation in $K_m(CO_2)$ of ribulose-1,5-bisphosphate carboxylase among grasses. *Plant Physiol.* **66,** 1110–1112.

Yeoh, H. H., Badger, M. R. and Watson, L. (1981). Variations in kinetic properties of ribulose-1,5-bisphosphate carboxylases among plants. *Plant Physiol.* **67,** 1151–1155.

5

Leaf Structure and Development in C_4 Plants

Nancy G. Dengler and Timothy Nelson

I. Introduction

Haberlandt (1882, 1914) initially called attention to the presence of two kinds of chlorenchymatous cells in the leaf blades of certain grasses and sedges and suggested that differences in chloroplast size and number between the two cell types might represent a division of labor. Haberlandt used the term *Kranz* to refer to the wreath of radially arranged mesophyll cells that surrounded the conspicuous bundle sheath; however, the term came to be applied to both the enlarged, chloroplast-rich bundle sheath cells (Kranz cells) and to the entire suite of distinctive structural characteristics (Kranz anatomy) (Brown, 1975). The intimate linkage between Kranz anatomy and C_4 photosynthesis was demonstrated almost immediately after discovery of the C_4 pathway, and the functional significance of the division of labor between two chlorenchymatous cell types, the spatial arrangement of tissues, and the role of a diffusion barrier at the mesophyll/bundle sheath interface was recognized within the first decade of research on C_4 photosynthesis (Hatch *et al.*, 1967, 1975; Laetsch, 1971, 1974; Gutierrez *et al.*, 1974).

Kranz anatomy and C_4 photosynthesis are invariably associated in species of two monocotyledonous families, the Poaceae and Cyperaceae, and of 16 dicotyledonous families (Downton, 1975; Ragavendra and Das, 1978; Chapter 17, this volume). Because these families lack a common C_4 ancestor, C_4 photosynthesis is thought to have evolved independently in each family (Ehleringer and Monson, 1993; Monson, 1996; Chapter 12, this volume), as well as multiple times within the Poaceae, where C_4 species occur in three distinct subfamilies (Hattersley and Watson, 1992; Sinha and Kellogg, 1996). Considerable variation in leaf anatomy occurs among the C_3 ancestral types in each of these families, yet the key features of Kranz anatomy have evolved repeatedly from these diverse genetic backgrounds, indicating strong selection for this suite of structural characteristics and their importance for the operation of C_4 biochemistry and physiology.

In this chapter we first describe the key features of Kranz anatomy in relation to their functional significance and review and illustrate some of the structural diversity known for C_4 plants. Second, we describe the ontogenetic development of Kranz anatomy and biochemical compartmentation and what is known about regulation of these developmental events. The distinctive anatomy of C_4 plants has become an important model system for studying plant structure–function relationships, the origin of tissue pattern during leaf development, and the differentiation of specialized cell types (Furbank and Foyer, 1988; Nelson and Langdale, 1989, 1992; Nelson and Dengler, 1992; Furbank and Taylor, 1995; Ku et al., 1996).

II. Kranz Anatomy and Biochemical Compartmentation

The division of labor hypothesized by Haberlandt (1882; 1914) is now known to involve an initial primary carbon assimilation (PCA) step in the leaf mesophyll and a second photosynthetic carbon reduction (PCR) step in the bundle sheath (Chapter 3, this volume). PCA activity invariably occurs within C_4 mesophyll, but, in a few taxa, PCR activity occurs in nonbundle sheath chlorenchymatous tissue (Crookston and Moss, 1973; Raynal, 1973; Brown, 1975; Carolin et al., 1975; Shomer-Ilan et al., 1975). We deal with these variants in this chapter, and refer to cells of the PCA/mesophyll tissue of C_4 species as M cells and to PCR/bundle sheath tissue as BS cells, regardless of whether the PCR tissue is topographically bundle sheath (as in most C_4 plants) or only its functional equivalent (in certain anatomic variants in which PCR activity occurs in nonbundle sheath tissue).

A. Key Features of Kranz Anatomy

Despite considerable variation in the architecture of leaf tissues among the families in which C_4 photosynthesis has evolved, certain anatomical features are invariably associated with the pathway and are regarded as being essential to its operation. These are: (1) specialization of two types of photosynthetic cells; (2) spatial configuration of tissues so that M cells are toward the exterior of the leaf and in contact with the intercellular airspace whereas BS cells are to the interior of the M cells and closer to the vascular tissues of the leaf; (3) a short diffusion pathlength for photosynthetic metabolites between M and BS cells, reflected in high vein density and low ratios of M to BS tissue; and (4) features that limit the rate of CO_2 and HCO_3 leakage from BS cells, including minimal exposure of BS surface area to intercellular space and chemical modification of the BS wall (Hattersley et al., 1977; Hattersley and Watson, 1992).

In plants with typical Kranz anatomy, the bundle sheath at the periphery of the vascular bundles is modified as PCR tissue. BS cells are large in comparison to their C_3 counterparts; they have large, numerous chloroplasts; form starch; and have an asymmetric arrangement of cytoplasmic components (Fig. 1). M cells are similar to their counterparts in C_3 species, but are typically enlarged in a radial direction, an arrangement that permits each cell to be in contact with PCR tissue. The volume of intercellular airspace is lower in C_4 species than in their C_3 relatives, and both the shape of the BS cells and the arrangement of M cells reduces contact between the BS cell surface and intercellular air space (Fig. 1). The cell wall at the PCA/PCR interface is often highly modified: a suberin lamella, thought to reduce apoplastic leakage of CO_2, is deposited in the BS portion of the cell wall in certain C_4 types (Table I), and numerous plasmodesmata connecting M and BS cells provide a pathway for the symplastic diffusion of metabolites.

B. Variation in Kranz Anatomy

Although all known C_4 plants share the features listed previously, comparative anatomic surveys have identified many variations on this theme. Recognition of this diversity has been important in the identification of structural features that are essential for the operation of the C_4 pathway (Hattersley et al., 1977; Hattersley, 1987) and for understanding the developmental mechanisms that form the same suite of structural features in distantly related plant groups (Sinha and Kellogg, 1996).

1. Poaceae Structural variation in the grass family in relation to C_4 photosynthesis is the best characterized of any group (Hattersley and Watson, 1975, 1976; Brown, 1977; Hattersley and Browning, 1981; Hattersley, 1984; Prendergast and Hattersley, 1987; Prendergast et al., 1987; Dengler et al., 1994). In C_3 grasses, the large longitudinal vascular bundles are surrounded by two sheath layers: an inner mestome sheath, consisting of nonchlorenchymatous sclerenchyma cells and an outer parenchymatous bundle sheath with chloroplast-containing cells (Fig. 1A) (Esau, 1965). Smaller vascular bundles may lack a continuous mestome sheath layer. These bundle sheath layers are modified in different ways in C_4 grasses. Each of the three C_4 biochemical types (differentiated on the basis of BS cell decarboxylation reactions; see Chapter 2, this volume) is associated with a "classical" suite of structural characteristics, and a majority of C_4 grass species are typified by one of these three main biochemical–anatomical types (Hattersley and

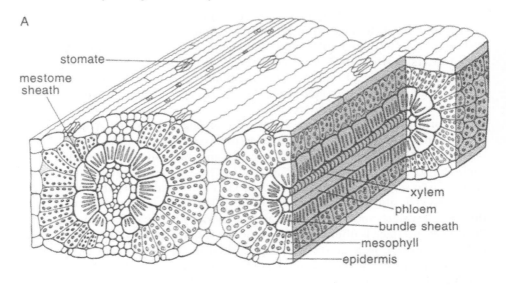

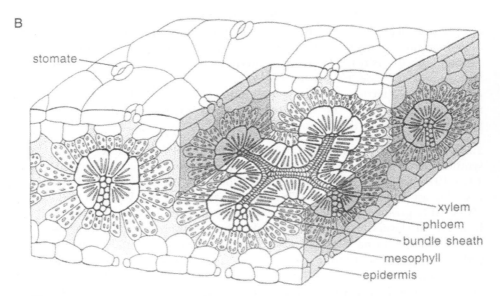

Figure 1 Diagram of Kranz anatomy in the C_4 grass *Panicum capillare* (A) and the C_4 dicot *Atriplex rosea*. (A) The large longitudinal vascular bundles are surrounded by two sheath layers: an inner sclerenchymatous mestome sheath and an outer parenchymatous bundle sheath, the site of PCR activity. Smaller longitudinal bundles may lack an inner mestome sheath. (B) Veins are surrounded by a single parenchymatous bundle sheath layer. Note radial enlargement of M cells in both (A) and (B).

Table 1 Anatomical and Ultrastructural Characteristics of C$_4$ Structural Types in the Grasses (Poaceae)[a]

Structural type	Biochemical type	Bundle sheath number	PCR/BS chloroplast position	PCR/BS chloroplast grana	PCR/BS cell wall suberin lamella	PCR/BS cell outline
"Classical" NADP-ME	NADP-ME	1	Centrifugal	−	+	Uneven
"Classical" NAD-ME	NAD-ME	2	Centripetal	+	−	Even
"Classical" PCK	PCK	2	Centrifugal or even	+	+	Uneven
Aristidoid	NADP-ME	2	Centrifugal	−	−	Even
Neurachneoid	NADP-ME or PCK	2	Centrifugal	+	−	Even
Arundinelloid	NADP-ME	1	Centrifugal	−	+	Uneven
Triodeoid	NAD-ME	2	Centrifugal	+	+	Even
Eriachneoid	NADP-ME	2	Centrifugal or centripetal	+	−	Even or uneven

[a] Sources: Hattersley and Watson, 1976; Hattersley and Browning, 1981; Prendergast and Hattersley, 1987; Hattersley and Watson, 1992.

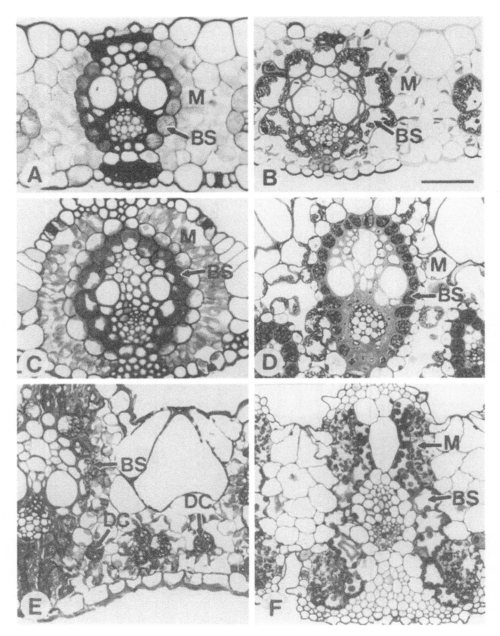

Figure 2 Light micrographs of cross sections of leaf blades of C$_4$ grasses representing some of the variation found in the Poaceae. (A) *Panicum bulbosum* ("classical" NADP-ME type). (B) *Melinus minutiflora* ("classical" PCK type). (C) *Aristida jerichoensis* (aristidoid type, NADP-ME). (D) *Alloteropsis semialata* (neurachneoid type, PCK). (E) *Arundinella nepalensis* (arundinelloid type, NADP-ME). Note distinctive cells that, like the BS cells, carry out

Watson, 1992). The three classical types differ in: (1) number of bundle sheath layers, (2) position of BS chloroplasts, (3) presence or absence of well-developed grana in BS chloroplasts, (4) presence or absence of a suberized lamella in BS cell walls, and (5) evenness of the BS outline in cross section (Table I). For example, in the "classical" NADP-malic enzyme (NADP-ME) type, only a single bundle sheath layer with an uneven outline and suberized walls is present, chloroplasts have reduced grana, and organelles have a centrifugal arrangement (Fig. 2A). In the "classical" NAD-malic enzyme (NAD-ME) type, two bundle sheaths are present. the outer sheath, the site of PCR activity, lacks a suberin lamella and has a smooth outline in cross section, granal chloroplasts, numerous mitochondria, and centripetally arranged organelles (Fig.1A). The "classical" PEP carboxykinase (PCK) type also has two sheaths, and both have a suberin lamella. The outer BS has an uneven outline, granal chloroplasts, and numerous mitochondria; BS cell organelles are either centrifugal or scattered in arrangement (Fig. 2B). Based on positional, structural, and developmental criteria, the mestome sheaths of "classical" NAD-ME and "classical" PCK grasses are regarded as the equivalent of the inner sclerenchymatous mestome sheaths of C_3 grasses and the PCR/BS layers as the equivalent of the outer parenchymatous bundle sheath (Brown, 1975). In the "classical" NADP-ME type with one bundle sheath, the mestome sheath layer is the one modified as PCR tissue, and the outer parenchymatous bundle sheath layer is missing (Brown, 1975; Dengler et al., 1985).

In addition to these three main biochemical–structural types, other well-known variants occur in the grasses (Table I). The aristidoid type is NADP-ME and possesses two chlorenchymatous bundle sheaths; the inner has typical PCR/BS structure and activity, whereas the outer functions to refix CO_2 that has leaked from the PCR sheath (Fig. 2C) (Hattersley, 1987; Ueno, 1992; Sinha and Kellogg, 1996). The neurachneoid type possesses two bundle sheaths and is NADP-ME, but only the inner sheath is chlorenchymatous (Fig. 2D) (Hattersley et al., 1982). The arundinelloid type is also NADP-ME and possesses a single bundle sheath; in addition, longitudinal strands of PCR tissue, not associated with vascular tissue, occur within the mesophyll (Fig. 2E). These strands of "distinctive cells" have typical PCR enzyme activity and structure, including a suberin lamella (Crookston and

---

PCR function. (F) *Triodia scariosa* (trioidioid type, NAD-ME). Scale = 50 μm. BS, bundle sheath cells; DC, distinctive cells; M, mesophyll cells. [Reprinted with permission from Dengler, N. G., Dengler, R. E., Donnelly, P. M., and Hattersley, P. W. (1994). Quantitative leaf anatomy of C_3 and C_4 grasses (Poaceae): Bundle sheath and mesophyll surface area relationships. *Ann. Bot.* **73**, 241–255.]

Moss, 1973; Hattersley et al., 1977; Hattersley and Browning, 1981; Dengler et al., 1990, 1996). The triodioid type (NAD-ME) is an unusual variant with BS extensions that "drape" from the vein to the patches of M tissue (Fig. 2F) (Craig and Goodchild, 1977; Prendergast et al., 1987). More recently, other unexpected variants were recognized based on biochemical typing: the eriachneoid type, with typical PCK-like anatomy but NADP-ME biochemistry (Prendergast and Hattersley, 1987; Prendergast et al., 1987), and certain species of *Eragrostis, Panicum, Enneapogon,* and *Triraphis,* all with PCK-like anatomy, but NAD-ME biochemistry (Ohsugi and Murata, 1980, 1986; Ohsugi et al., 1982; Prendergast et al., 1986, 1987). Although the pattern of variation within the grasses is more complex than originally envisaged, these unusual variants are taxonomically restricted, and most grasses can be typed based on anatomy alone (Hattersley, 1987; Hattersley and Watson, 1992).

2. Cyperaceae Four distinct structural types and two C_4 biochemical types, NADP-ME and NAD-ME, are found in the sedge family (Lerman and Raynal, 1972; Raynal, 1973; Takeda et al., 1980; Ueno et al., 1986; Bruhl et al., 1987; Estelita-Teixeira and Handro, 1987). The longitudinal vascular bundles of sedge leaves and photosynthetic stems (culms) are similar to those of grasses in having both an inner mestome and an outer parenchymatous bundle sheath. In the rhynchosporoid type (NADP-ME), the mestome sheath layer is modified as the PCR tissue and only a partial parenchymatous sheath is present (Fig. 3A) (Takeda et al., 1980). In the other C_4 types, vascular parenchyma that is internal to the mestome sheath is modified as PCR tissue (Brown, 1975). In the eleocharoid type (NAD-ME), PCR tissue forms a continuous layer (Fig. 3B) (Bruhl et al., 1987; Ueno and Samejima, 1989) but, in the fimbristyloid and chlorocyperoid types (largely NADP-ME, although some species of *Eleocharis* are NAD-ME) (Ueno et al., 1988), the layer is interrupted by the large metaxylem vessel elements (Fig. 3C,D). Non-PCR parenchymatous bundle sheath, thought to function as PCA tissue, is present, forming a continuous layer in the fimbristyloid type and a partial layer in chlorocyperoid sedges. Thus, C_4 sedges (except for the rhynchosporoid type) differ strikingly in tissue pattern from C_4 grasses in that the PCR tissue is completely isolated from intercellular air space of the PCA tissue by the suberized cell walls of the mestome sheath layer (Bruhl et al., 1987).

3. Dicotyledons The occurrence of standard Kranz anatomy in 16 dicotyledonous plant families reinforces the view that this constellation of features is essential for the operation of the C_4 pathway. Although the typical pattern of dicot leaf venation is more complex than the predominantly longitudinal venation of grasses and sedges, the arrangement of photosynthetic tissues as seen in cross section is similar (Fig. 1B). A single bundle sheath surrounds

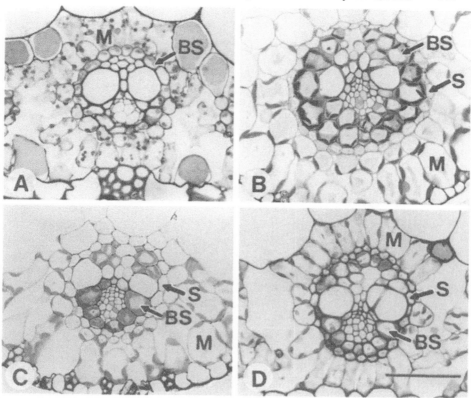

Figure 3 Light micrographs of cross sections of leaf blades and culms of C_4 Cyperaceae illustrating structural types. (A) *Rhyncospora* sp. (rhycosporoid type, NADP-ME). (B) *Eleocharis retroflexa* (eleocharoid type, NAD-ME). (C) *Fimbristylis dichotoma* (fimbristyloid type, NADP-ME). (D) *Cyperus polystachyos* (chlorocyperoid type, NADP-ME). Note PCR tissue in position of mestome sheath (A) or internal to the mestome sheath, in position of vascular parenchyma (B–D). Scale = 50 μm. BS, bundle sheath cells; M, mesophyll cells; S, mestome sheath. (Provided by C. L. Soros, University of Toronto).

the leaf veins in typical dicots, and this tissue is specialized as PCR/BS tissue in C_4 species (Fig. 4A–C). Typically, M cells are radially enlarged and arranged radially in relation to the veins (Fig. 1B, 4A), although species vary in mesophyll cell arrangement, so that either palisade and spongy mesophyll are present (Fig. 4B), or M cells occur toward the exterior of the leaf and interior ground tissue cells have few or no chloroplasts (Fig. 4C–F). A survey of anatomical types in the Chenopodiaceae designated this generalized pattern with BS surrounding the veins for C_4 dicots as the atriplicoid type (Carolin *et al.*, 1975, 1978). Variants of the atriplicoid pattern

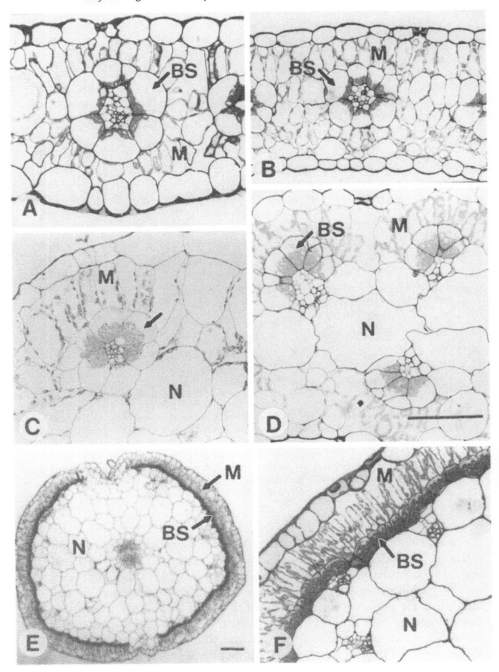

have been recognized based on mesophyll arrangement and continuity of the BS layer (Nyakas and Kalapos, 1996). Three other structural types are found in C_4 species of the Chenopodiaceae and other families with thick fleshy leaves: (1) the kochioid type, in which vascular bundles occur at the periphery of central nonchlorenchymatous tissue and BS tissue is present only at the exterior of the veins (Fig. 4D) (Carolin *et al.*, 1975, 1978; Nyakas and Kalapos, 1996); (2) the salsoloid type, in which the PCR/BS tissue forms a continuous layer to the exterior of the vascular bundles (Fig. 4E,F) (P'yankov *et al.*, 1997); and (3) the kranz-suaedoid type, which is similar to the salsoloid type but vascular bundles have a more central position within the leaf and are not directly associated with BS tissue (not illustrated) (Shomer-Ilan *et al.*, 1975; Fisher *et al.*, 1997).

C. Diffusion Barriers to CO_2 Leakage

An essential function of BS cell structure is to maintain the higher CO_2 concentration that results from pumping activity of the PCA cycle. The CO_2 diffusion pathway from the BS cells back to intercellular space is minimized by the reduced volume of intercellular air space within leaf mesophyll and the low surface to volume relationships of the BS cells themselves (Byott, 1976; Dengler *et al.*, 1994), but the most signficant factor for maintenance of high CO_2 concentration with these cells is thought to be the chemical modification of the cell walls by the deposition of suberin (Laetsch 1971, 1974; Carolin *et al.*, 1973; Hatch and Osmond, 1976; Hattersley and Browning, 1981; Hatch, 1987). In all grasses that have been examined, the cell walls of the mestome sheath possess a suberin lamella, consisting of two dark bands separated by a lighter band (Fig. 5) (O'Brien and Carr, 1970; Hattersley and Browning, 1981; Botha *et al.*, 1982; Eastman *et al.*, 1988a). The dark bands are thought to represent suberin polymer and the light band waxes that form the major diffusion barrier to water and other molecules (Espelie and Kolattukudy, 1979; Soliday *et al.*, 1979; Kolattukudy, 1980). Typically the suberin lamella extends throughout the outer tangential wall of each sheath cell and partway through the radial walls (Fig. 5A)

Figure 4 Light micrograph of cross sections of leaf blades of C_4 dicots illustrating structural types. (A) *Amaranthus lividus* (atriplicoid type) with radiate mesophyll. (B) *Tribulus terrestris* (atriplicoid type) with palisade and spongy type mesophyll. (C) *Portulaca villosa* (atriplicoid type) with enlarged, nonchlorenchymatous cells in center of fleshy leaf. (D) *Bassia hyssopfolia* (kochioid type) with incomplete BS layer and non-chlorenchymatous cells in center of fleshy leaf. (E,F) *Salsola kamarovii* (salsoloid type) with a large central zone of nonchlorenchymatous parenchyma cells and peripheral layers of BS and M cells not directly associated with individual veins. BS, bundle sheath; M, mesophyll cells; N, nonchlorenchymatous cells. Scale = 50 μm. (C,E,F provided by Dr. Insun Kim, Keimyung University, Korea).

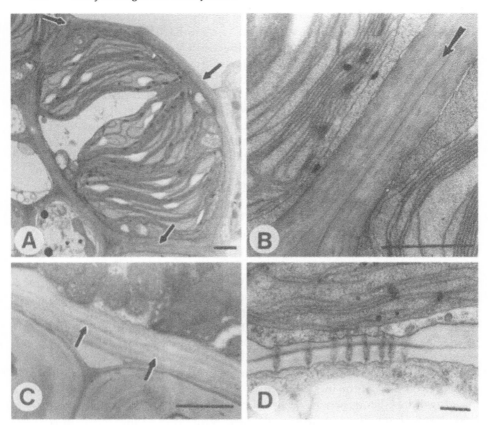

Figure 5 Electron micrographs of suberin lamellae in leaf blades of C_4 grasses. (A) BS cell of *Setaria glauca* ("classical" NADP-ME type) showing suberin lamella in radial and outer tangential wall (arrows). (B) Radial wall of adjacent BS cells of *Setaria glauca* showing lack of fusion between suberin lamellae. Arrow indicates putative pathway for apoplastic movement of water and solutes between suberin lamellae in adjacent radial walls of BS cells. (C) Outer tangential wall of BS cell of *Panicum capillare* ("classical" NAD-ME type) showing light and dark striations (arrows). (D) Primary pit field with clustered plasmodesmata in PCR–PCA interface in *Arundinella hirta* (arundinelloid type, NADP-ME). Scales = 1 μm. [A–C, reprinted with permission from Eastman, P. A. K., Dengler, N. G., and Peterson, C. A. (1988). Suberized bundle sheaths in grasses (Poaceae) of different photosynthetic types. I. Anatomy, ultrastructure and histochemistry. *Protoplasma* **142**, 92–111, with permission. D, reprinted with permission from Dengler, N. G., Donnelly, P. M., and Dengler, R. E. (1996). Differentiation of bundle sheath, mesophyll, and distinctive cells in the, C_4 grass *Arundinella hirta* (Poaceae). *Amer. J. Bot.* **83**, 1391–1405.]

(O'Brien and Carr, 1970; Hattersley and Browning, 1981; Eastman *et al.*, 1988a). Suberin lamellae of adjacent mestome sheath cells do not fuse, leaving a potential apoplastic pathway for radial movement of water and solutes across the sheath layer (Fig. 5B) Hattersley and Browning, 1981; Evert et al., 1985; Eastman et al., 1988b).

Within the grass family, differences occur among the C_4 biochemical–structural types in number and location of suberized layers (Table I) (Hattersley and Browning, 1981). In the "classical" NADP-ME type, PCR/BS tissue develops from the mestome sheath layer and cells have a characteristic suberin lamella. In the "classical" PCK type, both the mestome sheath itself and the outer sheath (PCR/BS tissue) possess suberin lamellae, but in "classical" NAD-ME grasses, a suberin lamella is absent in the BS cells (but present in mestome sheath cells). This variation in bundle sheath suberization indicates that the structural types might vary in the conductance properties of their BS layers, and early comparisons found that "classical" NAD-ME grasses had more negative carbon isotope ratios, presumably reflecting leakage of CO_2 to intercellular space (Hattersley, 1982). More recent measurements of gas exchange properties and carbon isotope discrimination did not detect differences among the structural types, however, suggesting that other traits might compensate for the absence of a suberin lamella in BS cells of NAD-ME grasses (Henderson et al., 1992). Leakiness of the BS is determined in part by the biochemical capacity of the M cells to provide CO_2 and of the BS cells to fix CO_2; thus activities of PCA and PCR enzymes might compensate for higher permeability of the BS cell walls (Henderson et al., 1992). In addition, the centripetal arrangement of chloroplasts and mitochondria might also increase resistance to CO_2 leakage by increasing the pathlength for diffusion and permitting fixation of CO_2 released from the mitochondria before it escapes to the intercellular space (Hattersley and Browning, 1981).

BS cell walls that lack a defined suberin lamella may still be relatively impermeable to diffusion of water and solutes. The BS cell walls of "classical" NAD-ME grasses often have a striated appearance, with diffuse light and dark bands that could represent the equivalent chemical components of suberin and waxes and function in the same way (Fig. 5C) (Hattersley and Browning, 1981; Eastman et al., 1988a). Wilson and Hattersley (1989) observed that BS cell walls of "classical" NAD-ME grasses were more resistant to rumen bacterial digestion than were those of "classical" NADP-ME and "classical" PCK type grasses, a property that might be correlated with apoplastic impermeability. No C_4 dicot species has been observed to possess a suberin lamella, yet BS cell walls are usually described as being thickened and as staining more darkly and/or having light and dark striations when observed by electron microscopy (Carolin et al., 1975, 1978; Liu and Dengler, 1994). Further research is needed to address unresolved questions of the nature of these wall modifications and their significance as CO_2 barriers.

In the Cyperaceae, both outer and inner tangential walls of the mestome sheath possess a suberin lamella that is similar in structure to that of the Poaceae (Carolin et al., 1977; Ueno et al., 1988; Ueno and Samejima, 1989; Bruhl and Perry, 1995). Thus, in the fimbristyloid, chlorocyperoid, eleocharoid types, BS cells not only lack direct contact with intercellular air space, but also are separated from intercellular air space by at least two suberized cell walls. C_4 sedges have carbon isotope ratios that are typical for the C_4 pathway (Ueno et al., 1989), and mean values for the structural types do not differ significantly (Bruhl and Perry, 1995).

D. Intercellular Diffusion of Metabolites

The physiological requirement for rapid flux of metabolites between M and BS tissue limits the volume of PCA tissue that can be associated with a given volume of PCR tissue in all C_4 plants. Because PCR tissue typically develops in the location of the sheath surrounding the vascular bundles, this relationship is expressed in the characteristic close vein spacing of C_4 taxa (Fig. 6) (Chonan, 1972; Crookston and Moss, 1974; Kanai and Kashiwagi, 1975; Morgan and Brown, 1979; Kawamitsu et al., 1985; Oguro et al., 1985; Ohsugi and Murata, 1986; Dengler et al., 1994). For instance, Kawamitsu et al. (1985) found that mean interveinal distance was about 300 μm for C_3 grasses, but only about 100 μm for C_4 grasses. Furthermore, "classic" NADP-ME type grasses tend to have shorter interveinal distances than do "classic" NAD-ME type or "classic" PCK type grasses, which is apparently related to the absence of an outer bundle sheath (Kawamitsu et al., 1985; Dengler et al., 1994). Similar differences among C_4 subtypes in vein spacing have been documented in surveys of smaller numbers of species within the Cyperaceae ($n = 13$) (Li and Jones, 1994) and in a range of dicot species ($n = 10$) (Rao and Rajendrudu, 1989).

Although measurement of interveinal distance continues to be one the most reliable means of identifying Kranz anatomy (Sinha and Kellogg, 1996), it is an oversimplification of the more complex three-dimensional architecture of leaf tissue. M cells tend to be radially elongate, allowing each cell contact with a BS cell (Fig. 1A,B). Ground tissue cells that are more distant from the PCR tissue usually lack PCA/M features such as formation of conspicuous chloroplasts (Figs. 2C, 2F, 4B) (Carolin et al., 1978; Araus et al., 1990; Kim and Fisher, 1990; Li and Jones, 1994) or accumulation of PEPCase (Dengler et al., 1995). In the first broad comparative survey (119 species) of grass leaf anatomy in relation to photosynthetic pathway, Hattersley and Watson (1975) formalized these relationships as the "maximum cells distant count" (the number of chlorenchymatous cell diameters through which metabolites would have to diffuse between the furthest M cell and a BS cell). Despite considerable variation in leaf thickness and interveinal distances, they found that M cells were never more than one cell removed from the BS cells of C_4 species (the "one cell distant"

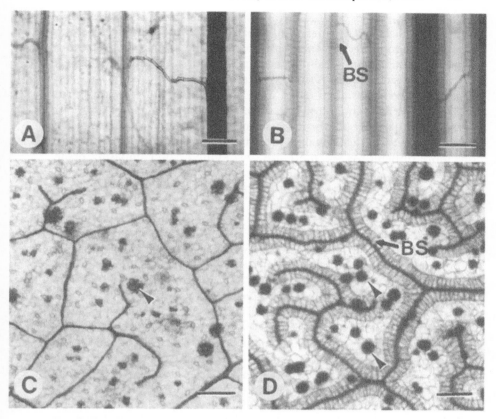

Figure 6 Vein pattern in related C_3 and C_4 species. (A) C_3 grass *Bromus tectorum*. (B) C_4 grass *Zea mays*. (C) C_3 dicot *Atriplex hastata*. (D) C_4 dicot *Atriplex rosea*. Note closer vein spacing of C_4 species. Unlabelled arrowheads indicate druse crystals in mesophyll of *Atriplex* species. Scale = 100 μm.

criterion for Kranz anatomy), whereas in C_3 species mesophyll cells could be up to 10 cell diameters from the veins.

The physiological requirement for short diffusional pathlength between PCA and PCR tissue can also be quantified as the relative amounts of the two tissue types (Hattersley, 1984; Dengler *et al.*, 1994). Expressed as area fractions of leaf cross sections, mean volumes of M tissue are lower for C_4 grasses compared to C_3 grasses. C_4 grass leaves are usually thinner than C_3 grass leaves and have less intercellular airspace, both contributing to reduced M volume (Dengler *et al.*, 1994). Mean volumes for BS tissue are greater in C_4 than C_3 species (Hattersley, 1984; Dengler *et al.*, 1994). Differences in BS volume also occur among the C_4 subtypes: for instance, "classical" NADP-ME species tend to have less BS tissue than "classical" NADP-ME and "classical" PCK types. This broad pattern of differences among the C_4 types within the Poaceae holds up, even when species representing the more unusual anatomical–biochemical types are included (Hattersley, 1984; Dengler et al., 1994). A similar pattern is found for C_3 and C_4 species of the Cyperaceae, although based on a smaller sample of species (Soros and Dengler, 1998). In contrast to the longitudinal arrangement of veins in leaves and culms of both the Poaceae and Cyperaceae, the more complex venation patterns of dicots makes quantitative assessment of tissue proportions more difficult. As yet, there are only limited data comparing tissue proportions of C_4 and C_3 dicots, but where these have been made, patterns similar to those in monocots have been found (Rao and Rajendrudu, 1989; Araus et al., 1990).

The presence of suberin lamellae and other wall modifications that restrict apoplastic movement of water and solutes indicates that metabolite flux between M and BS tissue follows a symplastic pathway. This is supported by the high frequency of plasmodesmata at the PCA/PCR interface (Osmond and Smith, 1976; Evert et al., 1977; Fisher and Evert, 1982; Botha and Evert, 1988; Robinson-Beers and Evert, 1991a; Botha, 1992). Groups of plasmodesmata are typically clustered in thin areas (primary pit fields) in the thickened common wall (Fig. 5D) and are a common feature across all groups of C_4 plants where ultrastructural observations have been made (e.g., Johnson and Brown, 1973; Carolin et al., 1977; Craig and Goodchild, 1977; Evert et al., 1977; Fisher and Evert, 1982; Botha and Evert, 1988; Dengler et al., 1990; Robinson-Beers and Evert, 1991a,b; Botha et al., 1993; Liu and Dengler, 1994; Dengler et al., 1996; Evert et al., 1996). Primary pit fields have an additional structural complexity when a suberin lamella is present in the PCR/BS cell walls. Typically, the suberin lamella is thickened and consists of multiple parallel layers within the pit field, and the plasmodesmata are constricted in diameter where they cross the suberin lamella (Fig. 5D) (Laetsch, 1971; Hattersley and Browning, 1981; Eastman et al., 1988a; Ueno et al., 1988; Robinson-Beers and Evert, 1991b; Botha et al., 1993; Evert et al., 1996). This specialized pitfield structure may function to isolate symplastic movement of water and solutes in the cytoplasmic annulus from the apoplastic flux occurring within the cell wall (Canny, 1986; Evert et al., 1996).

E. Specialized Features of PCR and PCA Cells

1. Biochemical Compartmentation As detailed in Chapters 3 and 4 of this volume, the C_4 pathway relies on the compartmentation of biochemical activities for carbon assimilation (e.g., PEPCase, NADP-MDH) and for carbon reduction (e.g., NADP-ME, Rubisco) in mesophyll cells and bundle sheath cells, respectively. This biochemical compartmentation has been demonstrated most effectively in several species from which it is possible to separate intact M cells from BS strands for direct measurement of enzymatic activities. The activities for assimilatory steps of the pathway (CO_2 to C_4 acid) are exclusively found in M cells; those for CO_2 release and refixation are exclusively found in BS cells (Chapter 3, this volume). Antibodies against individual C_4 enzymes reveal the same patterns of compartmentation of the proteins that correspond to these activities, when used in immunolocalization experiments on histological sections of intact leaf tissues (Fig. 7) (Chapter 3, this volume). It should be noted that such experiments consistently indicate that PPdK is present predominantly but not exclusively in M cells. When examined at high resolution using the immunogold/EM visualization method, the subcellular localization of enzymes in intact tissue confirms the compartmentalization measured by enzymatic assays of subcellular fractions (see Chapter 3, this volume). In C_4 species with anatomical variations on Kranz anatomy, such as *Arundinella hirta* and *Eleocharis vivipara*, immunolocalization studies have been useful in confirming the distribution of C_4 pathway activities between PCA and PCR tissue in unconventional positions (Fig. 7B,C) (Ueno and Samejima, 1990; Ueno, 1995, 1996a).

The mRNAs that encode C_4 pathway enzymes exhibit BS- or M-exclusive accumulation patterns in mature leaves (for review, see Nelson and Dengler, 1992; Nelson and Langdale, 1992). In illuminated leaves of maize, the mRNAs encoding Rubisco large and small subunits and NADP-ME are detected exclusively in BS cells by the *in situ* hybridization method (Fig. 7D), whereas mRNAs encoding PEPCase, NADP-MDH, and PPdK are detected exclusively in M cells (Langdale *et al.*, 1987). Similarly, mRNAs for Rubisco and NAD-ME are found only in BS cells of mature leaves of *Amaranthus hypochondriacus*, whereas mRNAs for PPdK and PEPCase are M-specific (Wang *et al.*, 1992, 1993a; Long and Berry, 1996; Ramsperger *et al.*, 1996). In both of these species, the mRNA patterns observed in developing or dark-grown leaves are less cell-specific than those observed in mature illuminated leaves. As described in Section III.B., experiments with several C_4 species suggest that the mechanisms for achieving the localization of individual mRNAs in BS versus M cells vary from species to species and from gene to gene, with documented examples of cell-specific transcription, posttranscriptional processes, and translation producing the observed patterns of activity in various systems.

In addition to the enzymes of the C_4 pathway, a variety of other activities and/or their corresponding proteins have been localized exclusively or preferentially in the BS or M cells of C_4 plants, including activities related to nitrogen assimilation, respiration, and a variety of other metabolic processes (see Chapter 3, this volume). Electrophoretic comparisons of the total proteins accumulated in BS and M cells (Potter and Black, 1982) suggest that the number of proteins differentially accumulated is large. It is likely

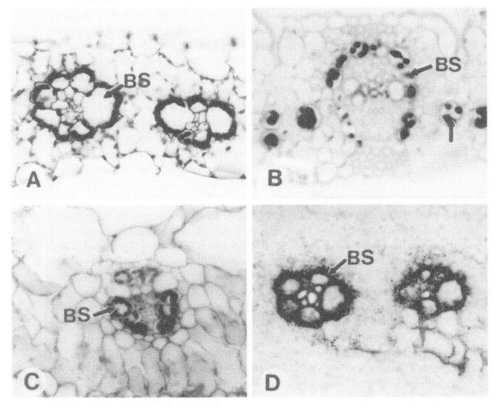

Figure 7 Cell-specific localization of enzymes and their RNAs. (A) Leaf cross section of maize (*Zea mays*) showing immunolocalization of NADP-malic enzyme in BS cells. (B) Leaf cross section of grass *Arundinella hirta* showing immunolocalization of NADP-ME in both BS and distinctive cells. (C) Culm cross section of sedge *Eleocharis vivipara* showing localization of RuBPCase in vascular parenchyma functioning as BS/PCR tissue. (D) *In situ* hybridization of Ssu (small subunit of RuBPCase) RNA in BS cells of *Zea mays*. Scale = 50 μm. BS, bundle sheath cells; unlabeled arrow, distinctive cell. (C, provided by C. L. Soros, University of Toronto).

that this reflects the observed structural differences between BS and M cells, as well their distinct roles in metabolic processes in addition to the C_4 pathway. It will be interesting to learn whether the mechanism used by a given C_4 species to achieve these cell-specific differences (e.g., transcriptional, posttranscriptional, or translational) is the same as it uses for compartmentalization of its C_4 pathway activities. That is, to what extent do BS versus M cell compartmentation differences evolve as a syndrome rather than a gene at a time?

2. Additional Structural Features Chloroplast dimorphism between BS and M cells has long been recognized (Rhoades and Carvalho, 1944; Laetsch and Price, 1969; Laetsch, 1971). BS chloroplasts generally are larger than those of M cells and, where quantitative data have been gathered, are more numerous per cell and occupy a greater fraction of cell cross-sectional area (Liu and Dengler, 1994; Dengler et al., 1996; Ueno, 1996b). Structural dimorphism is most conspicuous in NADP-ME species: BS chloroplasts are not only larger than M chloroplasts, but also have greatly reduced grana (Fig. 8) (Laetsch, 1971, 1974; Laetsch et al., 1965; Laetsch and Price, 1969; Brangeon, 1973; Kirchanski, 1975, P'yankov et al., 1997), which is correlated with reduced Photosystem II activity (Meierhoff and Westhoff, 1993). In NAD-ME and PCK-type grasses, BS chloroplasts have well-developed grana, but are still larger than M cell chloroplasts. In mature leaves of most C_4 species, starch is typically present in the chloroplasts of BS but not M cells (Rhoades and Carvalho, 1944; Laetsch 1968, 1971), although this difference may simply represent source–sink gradients within leaves (Black and Mollenhauer, 1971).

Another conspicuous difference between BS and M chloroplasts is the degree of elaboration of the inner chloroplast envelope into a system of tubules and vesicles called *peripheral reticulum*. In C_4 dicots, peripheral reticulum is better developed in M cells than in BS cells (Laetsch, 1971, 1974; Chapman et al., 1975; Carolin et al., 1978; Sprey and Laetsch, 1978; Liu and Dengler, 1994), but in the C_4 Cyperaceae, particularly the fimbistyloid and chlorocyperoid types, peripheral reticulum is highly developed in the BS cells, forming extensive arrays of vesicles or anastomosing tubules at the periphery of the chloroplasts (Carolin et al., 1977; Ueno et al., 1988; Ueno and Samejima, 1989). The functional significance of chloroplast

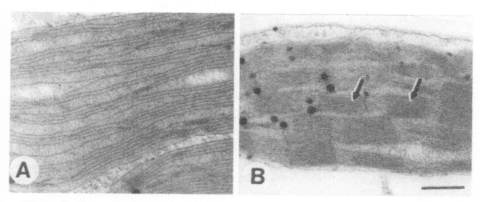

Figure 8 Electron micrographs of chloroplasts in leaves of the C_4 grass *Arundinella hirta* (arundinelloid type, NADP-ME). (A) BS cell showing that few thylakoids are aggregated into grana. (B) M cell showing aggregation of thylakoids into conspicuous grana (arrows). Scale = 1 μm. [Reprinted with permission from Dengler, N. G., Donnelly, P. M., and Dengler, R. E. (1996). Differentiation of bundle sheath, mesophyll, and distinctive cells in the C_4 grass *Arundinella hirta* (Poaceae). *Amer. J. Bot.* **83**, 1391–1405.]

peripheral reticulum is incompletely understood, but is thought to facilitate the transfer of metabolites across the chloroplast envelope (Laetsch, 1968, 1971; Sprey and Laetsch, 1978; Ueno et al., 1988).

Other observations indicate additional differences between BS and M cells in the organization of chloroplast thylakoids. Perhaps the most striking structural dimorphism of chloroplasts occurs in fimbistyloid and chlorocyperoid sedges, where the thylakoids of BS cell chloroplasts form convoluted loops and swirls, whereas those of the M cells have conventional grana and parallel thylakoids (Laetsch, 1971; Carolin et al., 1977; Jones et al., 1981; Estelita-Teixeira and Handro, 1987; Ueno et al., 1988; Ueno and Samejima, 1989). Here, too, functional significance is unknown, and variation in thylakoid ultrastructure is not correlated with C_4 biochemical type.

Dimorphism in mitochondrial size, number, and internal organization was observed early in the exploration of C_4 photosynthesis and the greatest differences were subsequently shown to be correlated with localization of the decarboxylation step in mitochondria of "classical" NAD-ME–type grass species (Laetsch, 1968, 1971; Osmond et al., 1969; Boynton et al., 1970; Black and Mollenhauer, 1971; Downton, 1971; Frederick and Newcomb, 1971; Carolin et al., 1975, 1977, 1978; Chapman et al., 1975; Hatch et al., 1975; Shomer-Ilan et al., 1975). When quantitative measurements have been made, BS cells have been shown to have 5- to 20-fold more mitochondria than M cells (Frederick and Newcomb, 1971; Laetsch 1971; Hatch et al., 1975; Dengler et al., 1986; Liu and Dengler, 1994) and mitochondria that are twice as large (Liu and Dengler, 1994). BS mitochondria in NAD-ME species frequently have well-developed internal membrane systems, resulting in a greater membrane surface area, thought to facilitate large metabolite fluxes required between mitochondria and cytoplasm (Hatch et al., 1975). In "classical" NAD-ME grasses, the spatial relationship between mitochondria and chloroplasts is close, and there is even a suggestion of physical adherence of mitochondrial and chloroplast membranes during experimental manipulation (Miyake et al., 1985).

Photorespiration is suppressed in healthy leaves of C_4 plants under normal atmospheric levels of O_2. Because BS cells have Calvin–Benson cycle activity and thus produce substrate for the photorespiratory glycolate pathway, it is thought that photorespired CO_2 is refixed within the bundle sheath cells (Edwards and Huber, 1981). Peroxisomes, the site of part of the photorespiratory pathway, were histochemically detected in BS cells of four C_4 grasses (Frederick and Newcomb, 1971) and microbodies (the morphological equivalent of biochemically defined peroxisomes) are more numerous in BS than in M cells of many C_4 species (Laetsch, 1968, 1971; Boynton et al., 1970; Frederick and Newcomb, 1971; Black and Mollenhauer, 1971; Huang and Beevers, 1972; Craig and Goodchild, 1977; Liu and Dengler, 1994) and may be larger than in adjacent M cells (Liu and Dengler, 1994).

III. Development of the C_4 Syndrome

Anatomical differences between C_4 taxa and their closest C_3 relatives indicate that several levels of leaf organization must be developmentally modified from the C_3 "default" condition for the efficient operation of the C_4 pathway. These include changes at the levels of (1) overall tissue pattern, (2) cell pattern within tissues, and (3) specialized cell structure. For example, the physiological requirement for rapid intercellular diffusion of metabolites is met by development of a denser vein pattern, thus reducing the volume of M associated with a given volume of BS tissue. Direct contact between individual M cells and BS cells is accomplished by alteration of patterns of cell division and expansion during mesophyll development, resulting in a radial arrangement of cells. In C_4 grasses, the requirement for CO_2 tightness is met by modifying tissues that already possess a suberin lamella or by modifying cell walls that are unsuberized in C_3 relatives. The division of labor between the two types of chlorenchymatous cells requires many further alterations of cell-specific biochemistry and structure.

A. Vascular Pattern

1. Vascular Pattern Ontogeny Precursors of dermal and ground tissues are present from leaf inception, but precursors of the vascular system become established during the early stages of leaf development (Esau, 1965). It is only after the formation of vascular pattern that BS and M cells become delimited in relation to the veins (Langdale and Nelson, 1991; Nelson and Dengler, 1992). Thus, close vein spacing is the first aspect of the C_4 syndrome to be expresssed developmentally and vascular pattern clearly provides the positional framework for the differentiation of BS and M cells. In dicots, leaf venation becomes established in three phases: (1) The midvein provascular strand develops in an acropetal direction from the stem vasculature into a new leaf primordium, (2) secondary vein provascular strands grow progressively from the midvein toward the margin concurrent with the formation of the leaf lamina, and (3) minor vein provascular strands form a network of small veins between the secondary veins, usually in a basipetal direction (Esau, 1965; Nelson and Dengler, 1997). It is this last stage of minor vein formation that must be modified during leaf development in C_4 dicots: The developmental programs that determine vein spacing patterns must be altered to result in the short interveinal distances in the leaves of C_4 species (e.g., Fig. 6D). During the development of the striate venation pattern of grass leaves such as maize, the largest longitudinal veins initially appear without connection to prexisting vascular bundles, suggesting that vascular pattern can be organ-autonomous (Esau, 1965; Evert et al., 1996; Dengler et al., 1997; Nelson and Dengler, 1997). As a grass leaf grows in width, new longitudinal veins are intercalated between adjacent veins. This process is modified in C_4 grasses so that the formation of small longitudinal veins is prolonged or accelerated, resulting in a greater number of closely spaced veins in mature leaf blades (e.g., Fig. 6B).

B. Regulation of Vascular Pattern Formation

The C_4 biochemical system resides in cooperating cell types organized around the leaf vasculature. It is important to understand the ontogeny of the vascular pattern and its role in the ontogeny and function of C_4 leaves. For example, there is evidence in maize that the vascular pattern provides positional landmarks for the specialization of BS, M, and other cell types. In contrast to the excellent descriptions of the ontogeny and form of leaf vascular patterns in a large number of dicot and monocot species, including many C_4 species (for reviews, see Sachs, 1991; Nelson and Dengler, 1997), remarkably little is understood of the regulation of pattern formation in adjacent tissues, including BS and M cells.

What we know of the process of vascular pattern formation is largely the result of observations on the alteration of normal vascular patterns by wounding, hormone application, and mutation. Experiments in dicot species suggest that vascular differentiation is a response to a local shoot-to-root oriented flow of auxin and that vascular systems are the product of the "canalization" of polar auxin flow such that maximally conductive files of cells differentiate (reviewed in Sachs, 1991; Nelson and Dengler, 1997). The regeneration of vascular tissue at wounds, graft junctions, and other interruptions of a preexisting system occurs in an orientation that obeys the original polarity of auxin flow in participating tissues when possible, but in any event reforms a continuous system that reestablishes shoot-to-root auxin flow. Similarly, the effect of exogenous applications of auxin depends on the physical possibility of oriented flow from the application. Generalized application usually does not result in vascular differentiation. The identification of the cellular components responsible for maintaining and responding to polarized auxin flow is the subject of much investigation, but at present it is unclear which if any of several characterized auxin-binding activities are responsible (Napier and Venis, 1995; Venis and Napier, 1997).

It should also be noted that the bulk of the evidence implying a role for polarized auxin flow in vascular patterning and differentiation comes from treatments of differentiated tissues, particularly of stem. In some cases it is difficult to reconcile the proposed key role of auxin flow in vascular patterning with the ontogeny of patterns. This is particularly evident in the case of monocot leaf vasculature in which acropetal vein initiation is followed by adjacent basipetal initiation of veins (Nelson and Dengler, 1997). It remains possible that the differentiation of vascular tissues during normal ontogeny, particularly of leaves, is subject to different or additional controls. A threshold supply of auxin may be a necessary but insufficient feature for establishment of patterned vascular differentiation.

Mathematical models can generate two-dimensional patterns that describe a broad array of monocot and dicot vascular systems in a manner that mimics even the ontogeny of the living systems. Diffusion–reaction models can reproduce a variety of vascular networks by invoking the interaction in spatial patterns of inducers and inhibitors of vascular differentiation without assuming the identity of either (Meinhardt, 1996). Similarly, fractal mathematics can generate vascular network patterns with astonishing similarity to those found in nature, again without assumptions about the identity of factors (Kull and Herbig, 1995). One study incorporated estimates of the physiological demands of the developing leaf on its developing vascular system as a factor in such mathematical patterning models, with some predictive success (Kull and Herbig, 1995). It would be interesting to perform such calculations for C_4 leaves, incorporating the high vein density and local transitions from sink to C_3 to C_4 metabolism as cell differentiation occurs.

To complement the physiological studies of vascular differentiation and pattern formation, genetic screens have been initiated to identify mutations in genes influencing vascular patterns. The natural ontogenetic variation in vascular pattern in the leaves of many species implies that pattern is controlled genetically. For example, the dense vascularization of the maize leaf blade is formed by successive initiation of lateral (major), large intermediate, and small intermediate (minor) veins, yet initiation is limited to laterals and few intermediates in the sheath regions of the same leaves (reviewed in Nelson and Dengler, 1997). Ethyl methanesulfonate (EMS)-induced variants of the C_4 grass *Panicum maximum* have been identified that lack the last-initiated veins in this hierarchy in the blade (Fladung, 1994). A variety of other mutants have been described with alterations in leaf venation patterns in C_3 species. In the *monopteros* mutant of *Arabidopsis thaliana*, leaf marginal veins are missing or interrupted, with little apparent effect on overall leaf morphology (Berleth and Jurgens, 1993; Przemeck *et al.*, 1996). Mutant *monopteros* plants exhibit a reduced capacity for polar transport of auxin, but it is not yet clear whether this is the cause or a consequence of the vascular pattern abnormality. The *lopped1* (and allelic *tornado*) mutant also exhibits a defect in auxin transport, and its leaves are narrowed with a bifurcated and twisted midvein (Carland and McHale, 1996; Cnops *et al.*, 1996). Several narrow leaf mutants have been described in maize, *Antirrhinum*, tobacco, and other species in which leaves approach radial symmetry with an associated reduction in the vascular pattern to a single midvein with zero to a few secondary veins (Miles, 1989; McHale, 1992, 1993; Waites and Hudson, 1995). It is not yet possible to determine whether the vascular pattern defects in any of these mutants is a cause or consequence of the morphological phenotype.

Dominant mutant alleles of the maize homeodomain gene *Knotted* (normally down-regulated at leaf initiation) cause ectopic accumulation of Knotted product along the lateral veins of leaves (Smith *et al.*, 1992). In the blade, such lateral veins exhibit a differentiation pattern characteristic of the sheath, in which the bundle sheath is discontinuous and the immediately surrounding parenchyma lacks chloroplasts (Sinha and Hake, 1994). Midribless mutants have been recovered from several C_4 grasses (Rao *et al.*, 1989; Fladung *et al.*, 1991; Fladung, 1994; Paxson and Nelson, unpublished observations). In grasses, the midrib region of the leaf blade normally exhibits a vascular pattern distinct from the adjacent laminae with few basipetal veins between laterals, a pattern typical of the sheath. In affected leaves of midribless mutants, adaxial thickenings are absent over the midvein, and the vascular pattern in the median region is the same as in the laminae, with the full complement of basipetal veins. Although the currently available mutants offer few clues to the nature of the pattern formation process, the existence of these mutants suggest that leaf vascular pattern is subject to genetic control.

B. Cell Differentiation

1. Molecular Mechanisms for Cell-Specific Enzyme Expression As described previously and in Chapter 3 of this volume, the cell-specific accumulation of the enzymes of the C_4 pathway in BS and M tissues has been documented in several species, as has the cellular distribution of the corresponding proteins and mRNAs in developing and mature leaves. In some cases, the close correspondence between patterns of mRNA accumulation and enzymatic activity suggests that the compartmentation of activities is achieved largely at the RNA level. In other cases, cell-specific activities correspond to mRNAs that are significantly accumulated in both BS and M cells, suggesting that posttranscriptional and/or posttranslational processes are subject to cell-specific controls. Studies of the cell-specific localization of Rubisco during light-dependent development in maize and amaranth suggest that the appearance of this activity in BS cells is subject to transcriptional, translational, and posttranslational cell-specific controls (Berry *et al.*, 1985, 1986, 1988, 1990; Sheen and Bogorad, 1985; Langdale *et al.*, 1988a,b; Boinski *et al.*, 1993). Although cell-specific expression of the other C_4 pathway genes has not yet been subjected to as intensive an analysis, most also appear to exhibit a variety of levels of control, based on comparisons of promoter activity assays, mRNA and protein accumulation patterns, and enzyme activity patterns (e.g., Langdale et al., 1987, 1988a,b; Sheen and Bogorad, 1987a,b; Wang et al., 1993a; Long and Berry, 1996; Ramsperger et al., 1996). The current evidence indicates that several cell-specific transcriptional and posttranscriptional processes have a role in achieving BS- or M-cell localization of the C_4 pathway constituents during development, and that the relative roles of the processes may vary among species (reviewed in Furbank and Taylor, 1995). Recently, transgenic experiments in the C_4 dicot *Flaveria bidentis* showed that the promoter of the PEPCase gene is sufficient for directing the accumulation of the GUS reporter activity to only M cells (Stockhaus et al., 1997). Additional experiments of this type should permit the definition of the molecular basis for BS- and M-cell specificity of each of the pathway genes.

2. Role of Cell–Cell Communication

a. Clonal Relationship of BS and M Cells It is likely that the complementary differentiation of BS and M cells is coordinated during leaf development. However, direct evidence of this developmental interdependence is currently lacking. Genetic and histological analysis of the clonal relationships between BS and M cells has been performed on several C_4 grasses (Bosabalidis et al., 1994; Dengler et al., 1985, 1996; Langdale et al., 1989). Together, these studies suggest that the differentiation of BS and M cells need not depend on a single clonal relationship between the two types. Histological studies of the leaf ontogeny of the NADP-ME grasses maize, *Digitaria brownii, Panicum bulbosum,* and *Cymbopogon procerus* (all of the single-sheath anatomical type), revealed that BS cells are largely or entirely derived from provascular cell divisions, whereas in the double-sheath NAD-ME– and PCK-type C_4 grasses *Panicum effusum, Eleusine coracana,* and *Sporobolus elongatus,* in which vasculature is surrounded by a mestome sheath, BS is derived from ground tissue (Fig. 9A,B) (Dengler et al., 1985). However, clonal studies in maize in which lineages of BS and M cells were tracked with a variety of independent genetic pigmentation markers, suggest that more than one pattern of derivation can coexist surrounding the same vein (Langdale et al., 1989). That is, BS can be derived from provascular or ground cells, although the provascular derivation is by far the most frequent origin. This suggests that if lineage does play a role in distinguishing BS from M cell fates, it can be overridden by cell position or by regulative cell interactions such as the onset of metabolic cooperation. The appearance of BS-like distinctive cells in the NADP-ME–type C_4 grass *Arundinella hirta* at leaf interveinal sites that are assigned to minor veins in other C_4 grasses suggests that positional information rather than provascular lineage may be the fundamental signal for BS differentiation, at least in that species (Figs. 2E, 7B) (Dengler et al., 1997).

An apparent overriding of cell position occurs in both the maize *pigmy* (*py* and allelic *tangled*) mutant and in the *Panicum maximum* aberrant

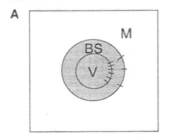 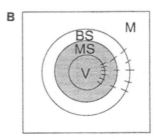

Figure 9 Clonal relationships of BS and M cells. (A) In grasses with a single bundle sheath layer, the BS/PCR layer develops from the provascular tissue. (B) In grasses with a double bundle sheath, the BS/PCR layer develops from ground tissue. Density of plasmodesmata (bars) reflects common clonal relationship—that is BS with vascular tissue in (A) and BS with mesophyll tissue in (B). BS, bundle sheath cells; M, mesophyll cells; V, vascular tissue; stippling, provascular tissue.

bundle sheath (*abs*) mutant (Fladung *et al.*, 1991; Fladung, 1994; Smith *et al.*, 1996). In both cases, BS cells appear in the interveinal mesophyll region. However, these ectopic bundle sheath cells are always linked in files to the true BS surrounding veins. It is possible that these represent aberrant cell division patterns that generate BS cells distal to veins after the normal BS proximal to veins has already been committed to differentiation. That is, they represent a differentiated BS lineage. It will be of interest to learn more of the ontogeny of the various cell types in these cases, and to establish the developmental timing of cell divisions that create ectopic BS.

b. Possible Intercellular Signalling During Differentiation The apparently varied clonal relationships of BS and M cells is indirect evidence of a mechanism that guides the differentiation of the complementary cell types based on their relative positions, possibly through signals originating one or more cells distant and acting within a few plastochrons of the provascular divisions within a given area of the leaf primordium. Additional evidence for such a hypothesis comes from observations of the differentiation of BS and M cells in maize foliar and husk leaves exposed to varying light conditions (Fig. 10) (Langdale *et al.*, 1988b). In foliar leaves grown in darkness, mRNAs encoding C_4 pathway enzymes are not accumulated in BS or M cells, and mRNAs for Rubisco (exclusively accumulated in BS cells in the light) are accumulated in a non–cell-specific pattern (Fig. 10A,B) (Sheen and Bogorad, 1985; Langdale *et al.*, 1988b). Pigment accumulation and plastid differentiation patterns and CO_2 compensation points suggest that both BS and M cells differentiate in a C_3 pattern, although it should be noted that morphological differentiation of BS and M cells occurs apparently normally. In husk leaves, which correspond to enlarged sheath regions of foliar leaves, the sparse venation makes it possible to estimate the influence

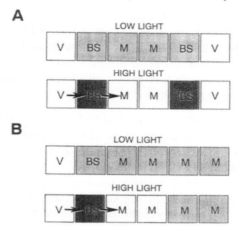

Figure 10 Effect of light and position on accumulation of Rubisco mRNAs in maize. (A) In foliar leaves, low light results in non–cell-specific Rubisco mRNA accumulation. Under high light, Rubisco mRNA accumulates to high levels in BS cells only. (B) In husk leaves, low light results in non–cell-specific Rubisco mRNA accumulation. Under high light, Rubisco mRNA accumulates at high levels in BS cells, but remains at low levels in cell that are more than two cells distant from BS. Hypothetical signal (arrows) from provascular tissues induces C_4 pattern of enzyme expression in presence of high light but does not reach more distant M cells. (Reprinted from *Trends in Genetics,* Vol. 7, J. A. Langdale and T. Nelson, Spatial regulation of photosynthetic development in C_4 plants, pp. 191–196, copyright 1991, with permission from Elsevier Science.)

of light intensity on C_4-type cell differentiation to extend at least several cells distant from the nearest vein, with cells more distant than that remaining in a default C_3 pattern regardless of light intensity (Fig. 10C,D). Again, light appears to influence cell-specific photosynthetic gene expression patterns, but not the morphologic differences between BS and M cells (beyond plastid morphology). Similar observations have been made of amaranth cotyledons and leaves in which cell-specific patterns of C_4 mRNA and protein accumulation develop gradually from an initially nonspecific accumulation in BS and M cells (Wang *et al.*, 1992, 1993a; Long and Berry, 1996; Ramsperger *et al.*, 1996).

In species where BS and M are derived from ground cells, the two cell types are expected to be in symplastic contact over the course of their differentiation (Fig. 9B). However, even in the case of BS and M cells with distinct derivation (the NADP-ME type grasses; for example, Fig. 9A), primary M-BS plasmodesmata are apparent from the earliest stages of development. Whether a phase of secondary plasmodesmata formation between M and BS cells occurs is unclear (Robinson-Beers and Evert, 1991a; Dengler *et al.*, 1996; Evert *et al.*, 1996). Although these BS–M plasmodesmata are not morphologically mature until relatively late in vascular differentiation, it does indicate that symplastic contact may be available as a means of coordination throughout BS and M differentiation. BS and M cells differentiate in synchrony at all locations surrounding veins in the maize leaf, and symplastic continuity among cells of potentially different ontogeny may be a basis for maintaining this synchrony.

3. Mutants with Differential Effects on BS and M Cells Several workers have taken a genetic mutational approach to identify the genes responsible for the differentiation of BS and M cells, screening for mutants with BS- or M-specific defects. Most of these mutations appear to affect photosynthetic differentiation and associated plastid morphology without affecting overall Kranz anatomy or other aspects of the C_4 syndrome. Several mutants of maize have been identified that exhibit BS-specific defects. In the *bundle sheath defective1* (*bsd1*) mutant, BS-specific C_4 mRNAs fail to accumulate, yet C_4 mRNAs accumulate normally in M cells (Langdale and Kidner, 1994). The gene has been cloned, but its sequence does not yet provide clues as to its function. In the *bsd2* mutant, the plastid-encoded *rbcL* mRNA accumulates in both BS and M cells, a C_3-like pattern also observed in wild-type leaf sheaths (Roth *et al.*, 1996). In addition, BS chloroplasts are abnormal, but nucleus-encoded C_4-mRNAs still accumulate in the normal cell-specific pattern. This suggests that the normal role of the *bsd2* gene is to repress *rbcL* mRNA accumulation in M cells and that the aberrant BS chloroplasts are a secondary effect of the resulting metabolic disturbance.

It is likely that the distinct metabolic roles of BS and M cells provide a significant number of targets for mutation that result in BS-specific phenotypes that have little to do with the regulation of cell differentiation. For example, both the maize *leaf permease1* (*lpe1*) mutant and the *Non-Chromosomal Stripe2* (*NCS2*) mutant exhibit BS-specific chloroplast defects, yet the primary mutations are in the genes for a putative membrane nucleobase permease and for mitochondrial Complex I subunits, respectively (Marienfeld and Newton, 1994; Schultes *et al.*, 1996). It is curious that no corresponding collection of M-specific mutations has been reported in C_4 species, with the possible exception of certain nutritional/uptake defects leading to interveinal yellowing. Roth *et al.* (1996) suggest that the lack of M-specific mutants could represent a dependence of BS differentiation on M cell function or that BS cells are more susceptible than M cells to perturbations in photosynthetic metabolism. Additional BS- and M-specific C_4 mutants should become available as gene-directed mutants are recovered from transposon insertional lines of maize using a PCR screening strategy for identifying lines with insertions in specific C_4 genes.

It is of interest to note that the *CAB underexpressed1* (*cue1*) mutant of the C_3 species *Arabidopsis thaliana* exhibits defects in the expression of light-regulated genes in mesophyll cells but not in photosynthetic BS cells (Li et al., 1995). In this case, there is little rationale for a metabolic interdependence of the two cell types beyond the normal channeling of photosynthate. Mutants of the C_4 dicot *Amaranthus edulis* were recovered from a screen for plants capable of growth in high CO_2 but unable to grow in air (Dever et al., 1995; Lacuesta et al., 1997). The defects in these mutants, which included individual lines with severely reduced PEPCase, absent NAD-ME, or high glycine accumulation, were limited to metabolic imbalances and the consequent growth inhibition in air, with no obvious general influence on BS and M development. The metabolic defects in these mutants have been useful in characterizing the photorespiratory pathway in this C_4 species, which is particularly well revealed in the mutant backgrounds (Lacuesta et al., 1997).

Few mutants have been described that affect the development of Kranz anatomy or other aspects of the C_4 syndrome. This may be because the resulting phenotypes are more subtle than the pigmentation deficiencies and stunted growth that usefully mark the failure of photosynthetic differentiation. Alternatively, the genes that guide Kranz differentiation may be so essential that their corresponding mutants have embryo- or seedling-lethal phenotypes. In *P. maximum,* genetic variants have been recovered with abnormal BS morphology *aberrant bundle sheath* (*abs*) and with variegated leaves containing yellow-green sectors with defective BS chloroplasts (*variegated1, var1*) (Fladung et al., 1991; Fladung, 1994). In *abs* variants, the BS and ensheathed veins are irregularly formed, in some cases with extra cells in thickness. The *abs* phenotype resembles that of the maize *pygmy1* (*py1,* allelic to *tangled1*) mutant, in which irregular planes of cell division generate files of BS cells extending into the mesophyll (Smith et al., 1996). Both *abs1* and *py1* mutants are pleiotropic and are therefore unlikely candidates for Kranz-specific mutants. In the maize *leafbladeless1* (*lbl1*) mutant, pinform leaves are formed with a single radially symmetric vascular cylinder surrounded by layers of BS and M cells that accumulate C_4 pathway enzymes in the expected pattern (Miles, 1989, 1992). True Kranz mutants may yet turn up in careful screens of seedling-lethal mutants or in brute-force screens for variation in leaf morphology and anatomy.

C. Role of Environmental Signals

As described in Chapter 3 of this volume, C_4 photosynthesis is a measurable advantage under conditions of high light intensity, high temperature, and high atmospheric oxygen. However, the pathway has an energetic cost that can outstrip the benefit under less extreme conditions. Therefore, it is not surprising that a variety of studies have noted that at least some aspects of commitment to the C_4 syndrome are inducible and that some species appear to be plastic in the degree of C_4 development they carry out. The effect of light on the degree of C_4 gene expression pattern has been noted in several studies, as described previously (e.g., Sheen and Bogorad, 1985; Langdale et al., 1988b; Wang et al., 1993a). In general, higher light levels favor full compartmentalization of C_4 activities and corresponding gene activities, whereas lower light levels or darkness lead in some species to a "default" pattern resembling C_3 patterns of photosynthetic gene expression. For example, in dark-grown seedings, transcripts for the otherwise BS-specific RuBPCase accumulate in both BS and M cells of maize leaves (Langdale et al.,1988b). It is possible that most or all aspects of the pathway itself are facultative, with Kranz anatomy the only "hardwired" component. At continuous temperatures below about 18°C, individual C_4 pathway activities such as PEPCase begin to drop, interrupting pathway function (Slack et al., 1974; Labate et al., 1990; Kleczdowski and Edwards, 1991; Nie et al., 1992; Krall and Edwards, 1993; Robertson et al., 1993). This appears to occur primarily at the level of enzyme inactivation, as no such temperature effects have been observed for expression of the corresponding genes, and no effects have been observed on the differentiation of Kranz anatomy.

In some remarkable species, leaf anatomy and physiology is plastic and fashioned according to environment. *Eleocharis vivipara* is an amphibious sedge that exhibits C_4 traits when growing in a terrestrial environment, but C_3 traits when growing submerged (Ueno et al., 1988). The plastic traits include the differentiation of a photosynthetic bundle sheath layer from boundary parenchyma, the differentiation of vascular bundles, BS/M ratio, organelle population in BS, chloroplast size, initial appearance of $^{14}CO_2$ pulses in C_4 acids, and cell-specific localization of C_4 pathway activities for PEPCase and PPdK (Ueno et al., 1988; Ueno, 1996b). The establishment of the C_4 pathway appears to be particularly plastic in species that lack other aspects of the Kranz syndrome, such as some C_3–C_4 intermediate species and certain aquatic species. In the aquatic monocot *Hydrilla verticillata*, a version of the C_4 pathway is induced under conditions of dense growth, although the plant is C_3 under other conditions (Bowes and Salvucci, 1984; Reiskind et al., 1997). *Hydrilla verticillata* leaves lack Kranz anatomy, and the C_4 pathway appears to act in individual leaf cells to concentrate CO_2 in the chloroplasts without intercellular shuttling. The genera such as *Flaveria* and *Moricandia* that include C_3–C_4 intermediate species exhibit both interspecific variation and intraspecific developmental plasticity in anatomy and biochemistry between the C_3 and C_4 extremes, which has been reviewed elsewhere (Monson et al., 1984; Edwards and Ku, 1987; Chapter 10, this volume).

D. Role of Physiological Signals in C_4 Leaf Development

The development of C_4 leaves, at least at the level of photosynthetic differentiation, is likely to be influenced by a variety of physiological conditions, including developmental age of the plant, internal and external signals for flowering, limitations in nutrients, hormonal balances, short- and long-term responses to pathogens, and, perhaps most importantly, synthetic and catabolic metabolism. However, at present few observations have been reported that correlate C_4 development with physiology. In amaranth, the BS-specific localization of Rubisco mRNA and protein occurs concomitant with the basipetal sink-to-source transition in the leaf (Wang et al., 1993b). Rubisco is found in both BS and M cells prior to this transition. A series of studies of the regulation of transcription of the C_4 genes encoding activities such as PEPCase and PPdK, as well as other photosynthetic activites, suggest that all are subject to regulation by a hexokinase-dependent system that monitors the state of sugar metabolism (Jang and Sheen, 1994; Sheen, 1994; Jang et al., 1997). At present, this regulation has not yet been correlated with cell-specific expression or with possible local metabolic changes during development. Studies in *F. brownii* and *F. trinervia* observed that the onset of C_4 activities is correlated with the degree of leaf expansion, possibly reflecting this same transition (Moore and Edwards, 1988; Cheng et al., 1989).

IV. Summary

Despite considerable variation in leaf anatomy in the 18 flowering plant families in which C_4 photosynthesis has evolved, certain anatomic features (Kranz anatomy) are invariably associated with the C_4 pathway and are regarded as being essential to its operation. These are: (1) specialization of BS and M cells, (2) positioning of these cells in relation to leaf veins, (3) shortening the length of the C_4 metabolite diffusion pathway between M and BS cells, and (4) modification of BS cell walls so that the rate of apoplastic leakage of CO_2 is minimized. The Kranz anatomy of mature leaves in all these C_4 groups is achieved by modifying the developmental processes of the C_3 ancestral type, requiring (1) alteration of leaf venation pattern, (2) interpretation of position in relation to leaf veins by BS and M cells, and (3) cell-specific expression of photosynthetic enzymes, structural features of cell walls, and organelle number, size, and placement.

Important advances are being made in understanding the molecular regulation of cell-specific enzyme expression, although current evidence suggests that transcriptional, posttranscriptional, and translational processes all have a role, and that these roles may vary among groups of C_4 plants. Full expression of C_4 cell-specific enzymes is influenced by light and temperature and appears to be dependent on the physiological age of leaf tissues. In contrast, very little is understood at present about the developmental modification of leaf venation patterns and about the signaling pathways used to interpret positional information by BS, M, and other cell types that differentiate in relation to the leaf veins. The varied clonal relationships of BS and M cells provides indirect evidence for a mechanism that guides the differentiation of complementary cell types based on their relative cell position, possibly through signals originating one or more cells distant and arising early during the formation of provascular tissue. Although a number of described mutations affect aspects of the Kranz syndrome, all have pleiotropic effects, and thus new screens are required to identify true Kranz mutants. Genes identified from such screens may shed light on the developmental regulation of C_4 vein pattern and determination of BS and M cell identities.

References

Araus, J. L., Brown, R. H., Bouton, J. H. and Serret, M.D. (1990). Leaf anatomical characteristics in *Flaveria trinervia* (C_4), *Flaveria brownii* (C_4-like) and their F1 hybrid. *Photosyn Res.* **26**, 49–57.

Berleth, T. and Jurgens, G. (1993). The role of the *monopteros* gene in organizing the basal body region of the *Arabidopsis* embryo. *Development* **118**, 575–587.

Berry, J. O., Breiding, D. E. and Klessig, D. F. (1990). Light-mediated control of translational initiation of ribulose–1,5-bisphosphate carboxylase in amaranth cotyledons. *Plant Cell* **2**, 795–803.

Berry, J. O., Carr, J. P. and Klessig, D. F. (1988). mRNAs encoding ribulose–1,5-bisphosphate carboxylase remain bound to polysomes but are not translated in amaranth seedlings transferred to darkness. *Proc. Natl. Acad. Sci. USA* **85**, 4190–4194.

Berry, J. O., Nikolau, B. J., Carr, J. P. and Klessig, D. F. (1985). Transcriptional and posttranscriptional regulation of ribulose 1,5-bisphosphate carboxylase gene expression in light and dark grown amaranth cotyledons. *Mol. Cell. Biol.* **5**, 2238–2246.

Berry, J. O., Nikolau, B. J., Carr, J. P. and Klessig, D. F. (1986). Translational regulation of light-induced ribulose 1,5-bisphosphate carboxylase gene expression in amaranth. *Mol. Cell. Biol.* **6**, 2347–2353.

Black, C. C. and Mollenhauer, J. H. (1971). Structure and distribution of chloroplasts and other organelles in leaves with various rates of photosynthesis. *Plant Physiol.* **47**, 15–23

Boinski, J. J., Wang, J.-L., Xu, P., Hotchkiss, T. and Berry, J. O. (1993). Posttranscriptional control of cell type-specific gene expression in bundle sheath and mesophyll chloroplasts of *Amaranthus hypochondriacus*. *Plant Mol. Biol.* **22**, 397–410.

Bosabalidis, A. M., Evert, R. F. and Russin, W. A. (1994). Ontogeny of the vascular bundles and contiguous tissues in the maize leaf blade. *Amer. J. Bot.* **81**, 745–752.

Botha, C. E. J. (1992). Plasmodesmatal distribution, structure and frequency in relation to assimilation in C_3 and C_4 grasses in southern Africa. *Planta* **187**, 348–358.

Botha, C. E. J. and Evert, R. F. (1988). Plasmodesmatal distribution and frequency in vascular bundles and contiguous tissues of the leaf of *Themeda triandra*. *Planta* **173**, 433–441.

Botha, C. E. J., Evert, R. F., Cross, R. H. M. and Marshall, D. M. (1982). The suberin lamella, a possible barrier to water movement from the veins to the mesophyll of *Themeda triandra* Forsk. *Protoplasma.* **112**, 1–8.

Botha, C. E. J., Hartley, J. and Cross, R. H. M. (1993) The ultrastructure and computer-enhanced digital image analysis of plasmodesmata at the Kranz mesophyll-bundle sheath

interface of *Themeda triandra* var. imberbis (Retz) A. Camus in conventionally-fixed leaf blades. *Ann. Bot.* **72,** 255–261.
Bowes, G. and Salvucci, M. E. (1984). *Hydrilla:* Inducible C_4-type photosynthesis without Kranz anatomy. *In* "Advances in Photosynthesis Research" (C. Sybesma ed.) p. 829–832, The Hague: Martinus Nijhoff/Junk Publishers.
Boynton, J. E., Nobs, M. A., Bjorkman, O. and Pearcy, R. W. (1970). Hybrids between *Atriplex* species with and without β-carboxylation photosynthesis. Leaf anatomy and ultrastructure. *Carnegie Inst. Wash. Yearbook* **69,** 629–632.
Brangeon, J. (1973). Compared ontogeny of the two types of chloroplasts of *Zea mays*. *J. Micros.* **16,** 233–242.
Brown, W. V. (1975). Variations in anatomy, associations, and origins of Kranz tissue. *Amer. J. Bot.* **62,** 395–402.
Brown, W. V. (1977). The Kranz syndrome and its subtypes in grass systematics. *Mem. Torrey Bot. Club.* **23,**1–97.
Bruhl, J. J., Stone, N. E. and Hattersley, P. W. (1987). C_4 acid decarboxylation enzymes and anatomy in sedges (Cyperaceae): First record of NAD-malic enzyme species. *Aust. J. Plant Physiol.* **14,** 719–728.
Bruhl, J. J. and Perry, S. (1995). Photosynthetic pathway related ultrastructure of C_3, C_4 and C_3-like and C_3–C_4 intermediate sedges (Cyperaceae), with special reference to *Eleocharis*. *Aust. J. Plant Physiol.* **22,** 521–530.
Byott, G. S. (1976). Leaf air space systems in C_3 and C_4 species. *New Phytol.* **76,** 295–299.
Canny, M. J. (1986). Water pathways in wheat leaves. III. The passage of the mestome sheath and the function of the suberised lamellae. *Physiol. Plant.* **66,** 637–647.
Carland, F. M. and McHale, N. A. (1996). *LOP1:* A gene involved in auxin transport and vascular patterning in *Arabidopsis. Development* **122,** 1811–1819.
Carolin, R. C., Jacobs, S. W. L. and Vesk, M. (1973). The structure of the cells of the mesophyll and the parenchymatous bundle sheath of the Gramineae. *Bot. J. Linn. Soc.* **66,** 259–275.
Carolin, R. C., Jacobs, S. W. L. and Vesk, M. (1975). Leaf structure in Chenopodiaceae. *Bot. Jahr. Syst. Pflazengesch. Pflanzengeogr.* **95,** 226–255.
Carolin, R. C., Jacobs, S. W. L. and Vesk, M. (1977). The ultrastructure of Kranz cells in the family Cyperaceae. *Bot. Gaz.* **138,** 413–419.
Carolin, R. C., Jacobs, S. W. L. and Vesk, M. (1978). Kranz cells and mesophyll in the Chenopodiales. *Aust. J. Bot.* **26,** 683–698.
Chapman, E. A., Bain, J. M. and Gove, D. W. (1975). Mitochondria and chloroplast peripheral reticulum in the C_4 plants *Amaranthus edulis* and *Atriplex spongiosa. Aust. J. Plant Physiol.* **2,** 207–223.
Cheng, S.-H., Moore, B. D., Wu, J., Edwards, G. E. and Ku, M. S. B. (1989). Photosynthetic plasticity in *Flaveria brownii*. Growth irradiance and the expression of C_4 photosynthesis. *Plant Physiol.* **89,** 1129–1135.
Chonan, N. (1972). Differences in mesophyll structures between temperate and tropical grasses. *Proc. Crop Sci. Soc. Japan.* **41,** 414–419.
Cnops, G., den Boer, B., Gerats, A., Van Montagu, M. and Van Lijsebettens, M. (1996). Chromosome landing at the *Arabidopsis TORNADO1* locus using an AFLP-based strategy. *Mol. Gen. Genet.* **253,** 32–41.
Craig, S. and Goodchild, D. J. (1977). Leaf ultrastructure of *Triodia irritans:* A C_4 grass possessing an unusual arrangement of photosynthetic tissues. *Aust. J. Bot.* **25,** 277–290.
Crookston, R. K. and Moss, D.N. (1973). A variation of C_4 leaf anatomy in *Arundinella hirta* (Gramineae). *Plant Physiol.* **52,** 397–402.
Crookston, R. K. and Moss, D. N. (1974). Interveinal distance for carbohydrate transport in leaves of C_3 and C_4 grasses. *Crop Sci.* **14,**123–125.
Dengler, N. G., Dengler, R. E. and Hattersley, P. W. (1985). Differing ontogenetic origins of PCR ("Kranz") sheaths in leaf blades of C_4 grasses (Poaceae). *Amer. J. Bot.* **72,** 284–302.

Dengler, N. G., Dengler, R. E. and Hattersley, P. W. (1986). Comparative bundle sheath and mesophyll differentiation in the leaves of the C_4 grasses *Panicum effusum* and *P. bulbosum*. *Amer. J. Bot.* **73,**1431–1442.

Dengler, N. G., Dengler, R. E. and Grenville, D. J. (1990). Comparison of photosynthetic carbon reduction (Kranz) cells having different ontogenetic origins in the C_4 NADP-malic enzyme grass *Arundinella hirta. Can. J. Bot.* **68,**1222–1232.

Dengler, N. G., Dengler, R. E., Donnelly, P. M. and Hattersley, P. W. (1994). Quantitative leaf anatomy of C_3 and C_4 grasses (Poaceae): Bundle sheath and mesophyll surface area relationships. *Ann. Bot.* **73,** 241–255.

Dengler, N. G., Dengler,R. E., Donnelly, P. M. and Filosa, M. (1995). Expression of the C_4 pattern of photosynthetic enzyme accumulation during leaf development in *Atriplex rosea* (Chenopodiaceae). *Amer. J. Bot.* **82,** 318–327.

Dengler, N. G., Donnelly, P. M. and Dengler, R. E. (1996). Differentiation of bundle sheath, mesophyll, and distinctive cells in the C_4 grass *Arundinella hirta* (Poaceae). *Amer. J. Bot.* **83,** 1391–1405.

Dengler, N. G., Woodvine, M. A., Donnelly, P. M. and Dengler, R. E. (1997) Formation of vascular pattern in developing leaves of the C_4 grass *Arundinella hirta. Int. J. Plant Sci.* **158,** 1–12.

Dengler, R. E. and Dengler, N. G. (1990). Leaf vascular architecture in the atypical C_4 NADP-malic enzyme grass *Arundinella hirta. Can. J. Bot.* **68,** 1208–1221.

Dever, L. V., Blackwell, R. D., Fullwood, N. J., Lacuesta, M., Leegood, R. C., Onek, L.A., Pearson, M. and Lea, P. J. (1995). The isolation and characterization of mutants of the C_4 photosynthetic pathway. *J. Exp. Bot.* **46,** 1363–1376.

Downton, W. J. S. (1971). The chloroplasts and mitochondria of bundle sheath cells in relation to C_4 photosynthesis. In "Photosynthesis and Photorespiration" (M. D. Hatch, C. B. Osmond, and R. O. Slatyer, eds.) pp. 419–425. Wiley Interscience, New York.

Downton, W. J. S. (1975). The occurrence of C_4 photosynthesis among plants. *Photosynthetica* **9,** 96–105.

Eastman, P. A. K., Dengler, N. G. and Peterson, C. A. (1988a). Suberized bundle sheaths in grasses (Poaceae) of different photosynthetic types I. Anatomy, ultrastructure and histochemistry. *Protoplasma* **142,** 92–111.

Edwards, G. E. and Huber, S. C. (1981). "The C_4 Pathway." Academic Press, New York.

Edwards, G. E. and Ku, M. S. B. (1987). Biochemistry of C_3–C_4 intermediates. *In* "The Biochemistry of Plants," (M. D. Hatch and N. K. Boardman, eds.) pp. 275–325, *New York: Academic Press.*

Ehleringer, J. R. and Monson, R. K. (1993). Evolutionary and ecological aspects of photosynthetic pathway variation. *Ann. Rev. Ecol. Syst.* **24,** 411–439.

Esau, K. (1965). "Plant Anatomy." John Wiley & Sons, New York.

Espelie, K. E. and Kolattukudy, P. E. (1979). Composition of the aliphatic components of 'suberin' from the bundle sheaths of *Zea mays* leaves. *Plant Sci. Lett.* **15,** 225–230.

Estelita-Teixeira, M. E. and Handro, W. (1987). Kranz pattern in leaf, scape and bract of *Cyperus* and *Fimbristylis* species. *Rev. Brasil. Bot.* **10,**105–111.

Evert, R. F., Eschrich, W. and Heyser, W. (1977). Distribution and structure of the plasmodesmata in mesophyll and bundle-sheath cells of *Zea mays. Planta* **136,** 77–89.

Evert, R. F., Botha, C. E. J. and Mierza, R. J. (1985). Free-space marker studies in the leaf of *Zea mays* L. *Protoplasma* **126,** 62–73.

Evert, R. F., Russin, W. A. and Bosabalidis, A. M. (1996). Anatomical and ultrastructural changes associated with sink-to-source transition in developing maize leaves. *Int. J. Plant Sci.* **157:** 247–261.

Fisher, D. D., Schenk, H. J., Thorsch, J. A., and Ferren, Jr., W. R. (1997). Leaf anatomy and subgeneric affiliations of C_3 and C_4 species of *Suaeda* (Chenopodiaceae) in North America. *Amer. J. Bot.* **89,** 1198–1210.

Fisher, D. G. and Evert, R. F. (1982). Studies on the leaf of *Amaranthus retroflexus* (Amaranthaceae): Chloroplast polymorphism. *Bot. Gaz.* **143,** 146–155.
Fladung, M. (1994). Genetic variants of *Panicum maximum* (Jacq.) in C_4 photosynthetic traits. *J. Plant Physiol.* **143,** 165–172.
Fladung, M., Bossinger, G., Roeb, G. W. and Salamini, F. (1991). Correlated alterations in leaf and flower morphology and rate of leaf photosynthesis in a midribless (*mbl*) mutant of *Panicum maximum* Jacq. *Planta* **184,** 356–361.
Frederick, S. E. and Newcomb, E. H. (1971). Ultrastructure and distribution of microbodies in leaves of grasses with and without CO_2-photorespiration. *Planta* **96,** 152–174.
Furbank, R. T. and Foyer, C. H. (1988). C_4 plants as valuable model experimental systems for the study of photosynthesis. *New Phytol.* **109,** 265–277.
Furbank, R. T. and Taylor, W. C. (1995). Regulation of photosynthesis in C_3 and C_4 plants: A molecular approach. *Plant Cell* **7,** 797–807.
Gutierrez, M., Gracen, V. E. and Edwards, G. E. (1974). Biochemical and cytological relationships in C_4 plants. *Planta* **119,** 279–300.
Haberlandt, G. (1882). Vergleichende Anatomie des Assimilatorischen Gewebesystems der Pflanzen. *Jahrb. Wiss. Bot.* **13,** 74–188.
Haberlandt, G. (1914). "Physiological Plant Anatomy." Macmillan, London.
Hatch, M. D., Slack, C. D. and Johnson, H. S. (1967). Further studies on a new pathway of photosynthetic carbon dioxide fixation in sugar cane and its occurrence in other plant species. *Biochem. J.* **102,** 417–422.
Hatch, M. D., Kagawa, T. and Craig, S. (1975). Subdivision of C_4 pathway species based on differing C_4 acid decarboxylating systems and ultrastructural features. *Aust. J. Plant Physiol.* **2,** 111–128.
Hatch, M. D. and Osmond, C. B. (1976). Compartmentation and transport in C_4 photosynthesis. *In* "Transport in Plants" (C. R. Stocking and U. Heber, eds.) pp. 144–184, Springer Verlag, Berlin.
Hatch, M. D. (1987). C_4 photosynthesis: A unique blend of modified biochemistry, anatomy and ultrastructure. *Biochim. Biophys. Acta* **895,** 81–106.
Hattersley, P. W. (1982). $\delta 13C$ values of C_4 types in grasses. *Aust. J. Plant Physiol.* **9,** 139–154.
Hattersley, P. W. (1984). Characterization of C_4 leaf anatomy in grasses (Poaceae). Mesophyll: bundle sheath area ratios. *Ann. Bot.* **53,** 163–179.
Hattersley, P. W. (1987). Variation in photosynthetic pathway. *In* "Grass Systematics and Evolution" (T. R. Soderstrom, C. S. Campbell, and M. E. Barkworth, eds.), pp. 49–64. Smithsonian Institution Press, Washington.
Hattersley, P. W. and Browning, A. J. (1981). Occurrence of the suberized lamella in leaves of grasses of different photosynthetic types. I. In parenchymatous bundle sheaths and PCR ("Kranz") sheaths. *Protoplasma* **109,** 371–401.
Hattersley, P. W. and Watson, L. (1975). Anatomical parameters for predicting photosynthetic pathways of grass leaves: The 'maximum lateral cell count' and the 'maximum cells distant count.' *Phytomorphology* **25,** 325–333.
Hattersley, P. W. and Watson, L. (1976). C_4 grasses: An anatomical criterion for distinguishing between NADP-malic enzyme species and PCK or NAD-ME enzyme species. *Aust. J. Bot.* **24,** 297–308.
Hattersley, P. W. and Watson, L. (1992). Diversification of photosynthesis. *In* "Grass Evolution and Domestication" (G. P. Chapman, ed.), pp. 38–116, Cambridge University Press, Cambridge.
Hattersley, P. W., Watson, L. and Johnston, C. R. (1982). Remarkable leaf anatomical variations in *Neurachne* and its allies (Poaceae) in relation to C_3 and C_4 photosynthesis. *Bot. J. Linn. Soc. Lond.* **84,** 265–272.
Hattersley, P. W., Watson, L. and Osmond, C. B. (1977). *In situ* immunofluorescent labelling of ribulose-1.5 bisphosphate carboxylase in leaves of C_3 and C_4 plants. *Aust. J. Plant Physiol.* **4,** 523–539.

Henderson, S. A., von Caemmerer, S. and Farquhar, G. D. (1992). Short-term measurements of carbon isotope discrimination in several C_4 species. *Aust. J. Plant Physiol.* **19**, 263–285.

Huang, A. H. C. and Beevers, H. (1972). Microbody enzymes and carboxylases in sequential extracts from C_4 and C_3 leaves. *Plant Physiol.* **50**, 242–248.

Jang, J.-C., Leon, P., Zhou, L. and Sheen, J. (1997). Hexokinase as a sugar sensor in higher plants. *Plant Cell* **9**, 5–19.

Jang, J.-C. and Sheen, J. (1994). Sugar sensing in higher plants. *Plant Cell* **6**, 1665–1679.

Johnson, S. C. and Brown, W. V. (1973). Grass leaf ultrastructural variations. *Amer. J. Bot.* **60**, 727–735.

Jones, M. B., Hannon, G. E. and Coffey, M. D. (1981). C_4 photosynthesis in *Cyperus longus* L., a species occurring in temperate climates. *Plant Cell Environ.* **4**, 161–168.

Kanai, R. and Kashiwagi, M. (1975). *Panicum milioides*, a Gramineae plant having Kranz leaf anatomy without C_4 photosynthesis. *Plant Cell Physiol.* **16**, 669–679.

Kawamitsu, Y., Hakoyama, S. Agata, W. and Takeda, T. (1985). Leaf interveinal distances corresponding to anatomical types in grasses. *Plant Cell Physiol.* **26**, 589–593.

Kim, I. and Fisher, D. G. (1990). Structural aspects of the leaves of seven species of *Portulaca* growing in Hawaii. *Can. J. Bot.* **68**,1803–1811.

Kirchanski, S. J. (1975). The ultrastructural development of the dimorphic plastids of *Zea mays*. *Amer. J. Bot.* **62**, 695–705.

Kleczdowski, L. A. and Edwards, G. E. (1991). A low temperature–induced reversible transition between different kinetic forms of maize leaf phosphoenolpyruvate carboxylase. *Plant Physio. Biochem.* **29**, 9–18.

Kolattukudy, P. E. (1980). Cutin, suberin, and waxes. In "The Biochemistry of Plants" (P. K. Stumpf, ed.), pp. 571–645, Academic Press, New York.

Krall, J. P. and Edwards, G. E. (1993). PEP carboxylases from 2 C_4 species of *Panicum* with markedly different sysceptibilites to cold inactivation. *Plant Cell Physiol.* **34**, 1–11.

Ku, M. S. B., Kano-Murakami, Y. and Matsuoka, M. (1996). Evolution and expression of C_4 photosynthetic genes. *Plant Physiol.* **111**, 949–957.

Kull, U. and Herbig, A. (1995). Das Blattadersystem der Angiospermen: Form und Evolution. *Naturwissenschaften* **82**, 441–451.

Labate, C. A., Adcock, M. D. and Leegood, R. G. (1990). Effects of temperature on the regulation of photosynthetic carbon assimilation in leaves of maize and barley. *Planta* **181**, 547–554.

Lacuesta, M., Dever, L. V., Munoz-Rueda, A. and Lea, P. J. (1997). A study of photorespiratory ammonia production in the C_4 plant *Amaranthus edulis*, using mutants with altered photosynthetic capacities. *Physiol. Plant.* **99**, 447–455.

Laetsch, W. M. (1971). Chloroplast structural relationships in leaves of C_4 plants. In "Photosynthesis and Photorespiration" (M. D. Hatch and R. O. Slayter, eds.) pp. 323–349, Wiley-Interscience, New York.

Laetsch, W. M. (1974). The C_4 syndrome: A structural analysis. *Ann. Rev. Plant Physiol.* **25**, 27–52.

Laetsch, W. M., Stetler, D. A. and Vlitos, A. J. (1965). The ultrastructure of sugar cane chloroplasts. *Z. Pflanzenphysiol.* **54**, 472–474.

Laetsch, W. M. and Price, I. (1969). Development of the dimorphic chloroplasts of sugar cane. *Amer. J. Bot.* **56**, 77–87.

Laetsch, W. M. (1968) Chloroplast specialization in dicotyledons possessing the C_4-dicarboxylic acid pathway of photosynthetic CO_2 fixation. *Amer. J. Bot.* **55**, 875–883.

Langdale, J. A. and Kidner, C. A. (1994). *bundle sheath defective*, a mutation that disrupts cellular differentiation in maize leaves. *Development* **120**, 673–681.

Langdale, J. A., Lane, B., Freeling, M. and Nelson, T. (1989). Cell lineage analysis of maize bundle sheath and mesophyl cells. *Devel. Biol.* **133**, 128–139.

Langdale, J. A., Metzler, M. C. and Nelson, T. (1987). The *argentia* mutation delays normal development of photosynthetic cell-types in *Zea mays*. *Devel. Biol.* **122**, 243–255.

Langdale, J. A. and Nelson, T. (1991). Spatial regulation of photosynthetic development in C$_4$ plants. *Trends Gen.* **7**, 191–196.
Langdale, J. A., Rothermel, B. A. and Nelson, T. (1988a). Cellular pattern of photosynthetic gene expression in developing maize leaves. *Genes Devel.* **2**, 106–115.
Langdale, J. A., Zelitch, I., Miller, E. and Nelson, T. (1988b). Cell position and light influence C$_4$ versus C$_3$ patterns of photosynthetic gene expression in maize. *EMBO J.* **7**, 3643–3651.
Lerman, J. C. and Raynal, J. (1972). Biologie vegetale—La teneur en isotopes stables du carbone chez les Cyperacees: Sa valeur taxonomique. *C.R. Acad. Sci. Paris.* **275**, 1391–1394.
Li, H.-M., Culligan, K., Dixon, R. A. and Chory, J. (1995). *Cue1*: A mesophyll cell-specific positive regulator of light-controlled gene expression in *Arabidopsis. Plant Cell* **7**, 1599–1610.
Li, M. and Jones, M. B. (1994). Kranzkette, a unique C$_4$ anatomy occurring in *Cyperus japonicus* leaves. *Photosynthetica* **30**,117–131.
Liu, Y.-Q. and Dengler, N. G. (1994). Bundle sheath and mesophyll cell differentiation in the C$_4$ dicotyledon *Atriplex rosea*: Quantitative ultrastructure. *Can. J. Bot.* **72**, 644–657.
Long, J. J. and Berry, J. O. (1996). Tissue-specific and light-mediated expression of the C$_4$ photosynthetic NAD-dependent malic enzyme of amaranth mitochondria. *Plant Physiol.* **112**, 473–482.
Marienfeld, J. R. and Newton, K. J. (1994). The maize NCS2 abnormal growth mutant has a chimeric nad4–nad7 mitochondrial gene and is associated with reduced Complex I function. *Genetics* **138**, 855–863.
McHale, N. A. (1992). A nuclear mutation blocking initiation of the lamina in leaves of *Nicotiana sylvestris. Planta* **186**, 355–360.
McHale, N. A. (1993). *LAM1* and *FAT* genes control development of the leaf blade in *Nicotiana sylvestris. Plant Cell* **5**, 1029–1038.
Meierhoff, K. and Westhoff, P. (1993). Differential biogenesis of photosystem II in mesophyll and bundle-sheath cells of monocotyledonous NADP-malic enzyme–type C$_4$ plants: The non-stoichiometric abundance of the subunits of photosystem II in the bundle sheath chloroplasts and the translational activity of the plastome-encoded genes. *Planta* **191**, 23–33.
Meinhardt, H. (1996). Models of biological pattern formation: Common mechanism in plant and animal development. *Int. J. Dev. Biol.* **40**, 123–134.
Miles, D. (1989). An interesting leaf developmental mutant from a Mu active line that causes a loss of leaf blade. *Maize Genetics Coop. Newsletter* **63**, 66.
Miles, D. (1992). Further characterization of the leaf developmental mutant, *lbl. Maize Genetics Coop. Newsletter* **66**, 40.
Miyake, H. Furukawa, A., and Totsuka, T. (1985). Structural associations between mitochondria and chloroplasts in the bundle sheath cells of *Portulaca oleracea. Ann. Bot.* **55**, 815–817.
Monson, R. K. (1996). The use of phylogenetic perspective in comparative plant physiology and developmental biology. *Ann. Miss. Bot. Gard.* **83**, 3–16.
Monson, R. K., Edwards, G. E. and Ku, M. S. B. (1984). C$_3$–C$_4$ intermediate photosynthesis in plants. *Bioscience* **34**, 563–573.
Moore, B. D. and Edwards, G. E. (1988). The influence of leaf age on C$_4$ photosynthesis and the accumulation of inorganic carbon in *Flaveria trinervia*, a C$_4$ dicot. *Plant Physiol.* **88**, 125–130.
Morgan, J. A. and Brown, R. H. (1979). Photosynthesis in grass species differing in carbon dioxide fixation pathways. II. A search for species with intermediate gas exchange and anatomical characteristics. *Plant Physiol.* **64**, 257–262.
Napier, R. M. and Venis, M. A. (1995). Auxin action and auxin-binding proteins. *New Phytol.* **129**, 167–201.
Nelson, T. and Dengler, N. G. (1992). Photosynthetic tissue differentiation in C$_4$ plants. *Int. J. Plant Sci.* **153**, S93–S105.
Nelson, T. and Dengler, N. G. (1997). Leaf vascular pattern formation. *Plant Cell* **9**, 1121–1135.

Nelson, T. and Langdale, J. A. (1989). Patterns of leaf development in C_4 plants. *Plant Cell.* **1,** 3–13.

Nelson, T. and Langdale, J. A. (1992). Developmental genetics of C_4 photosynthesis. *Ann. Rev. Plant Physiology and Plant Mol. Biol.* **43,** 25–47.

Nie, G.-Y., Long, S. P. and Baker, N. R. (1992). The effects of development at suboptimal growth temperatures on photosynthetic capacity and susceptibility to chilling dependent photoinhibition in *Zea mays. Physiol. Plant.* **85,** 554–560.

Nyakas, A. and Kalapos, T. (1996). Variation in C_4 type leaf anatomy in the Hungarian angiosperm flora. *Abstract. Bot.* **20,** 93–104.

O'Brien, T. P. and Carr, D. J. (1970). A suberized layer in the cell walls of the bundle sheath of grasses. *Aust. J. Biol. Sci.* **23,** 275–287.

Oguro, H., Hinata, K. and Tsunoda, S. (1985). Comparative anatomy and morphology of leaves between C_3 and C_4 species of *Panicum. Ann. Bot.* **55,** 859–869.

Ohsugi, R. and Murata, T. (1980). Leaf anatomy, post-illumination CO_2 burst and NAD-ME enzyme activity of *Panicum dichotomiflorum. Plant Cell Physiol.* **21,** 1329–1333.

Ohsugi, R. and Murata, T. (1986). Variations in the leaf anatomy among some C_4 *Panicum* species. *Ann. Bot.* **58,** 443–453.

Ohsugi, R., Murata, T. and Chonan, N. (1982). C_4 syndrome of the species in the Dichotomiflora group of the genus *Panicum* (Gramineae). *Bot. Mag.* **95,** 339–347.

Osmond, C. B. and Smith, F. A. (1976). Symplastic transport of metabolites during C_4-photosynthesis. *In* "Intercellular Communication in Plants: Studies on Plasmodesmata" (B. E. S. Gunning and A. W. Robards, eds.), pp. 229–241, Springer Verlag, Berlin.

Potter, J. W. and Black, C. C. (1982). Differential protein composition and gene expression in leaf mesophyll cells and bundle sheath cells of the C_4 plant *Digitaria sanguinalis* (L.) Scop. *Plant Physiol.* **70,** 590–597.

Prendergast, H. D. V., Hattersley, P. W., Stone, N. E. and Lazarides, M. (1986). C_4 acid decarboxylation type in *Eragrostis* (Poaceae): Patterns of variation in chloroplast position, ultrastructure and geographical distribution. *Plant Cell Environ.* **9,** 333–344.

Prendergast, H. D. V. and Hattersley, P. W. (1987). Australian C_4 grasses (Poaceae): Leaf blade anatomical features in relation to C_4 acid decarboxylation types. *Aust. J. Bot.* **35,** 355–382.

Prendergast, H. D. V., Hattersley, P. W. and Stone, N. E. (1987). New structural/biochemical associations in leaf blades of C_4 grasses (Poaceae). *Aust. J. Plant Physiol.* **14,** 403–420.

Przemeck, G. K. H., Mattsson, J., Hardtke, C. S., Sung, Z. R. and Berleth, T. (1996). Studies on the role of the Arabidopsis gene MONOPTEROS in vascular development and plant cell axialization. *Planta* **200,** 229–237.

P'yankov, V. I., Voznesenskaya, E. V., Kondratschuk, A. V., and Black, Jr., C. C. (1997). A comparative anatomical and biochemical analysis in *Salsola* (Chenopodiaceae) species with and without a Kranz type leaf anatomy: A possible reversion of C_3 to C_4 photosynthesis. *Amer. J. Bot.* **84,** 597–606.

Ragavendra, A. S. and Das, V. S. R. (1978) The occurrence of C_4-photosynthesis: A supplementary list of C_4 plants reported during late 1974–mid 1977. *Photosynthetica* **12,** 200–208.

Ramsperger, V. C., Summers, R. G. and Berry, J. O. (1996). Photosynthetic gene expression in meristems and during initial leaf development in a C_4 dicotyledonous plant. *Plant Physiol.* **111,** 999–1010.

Rao, S. A., Mengesha, M. H., Rao, Y. S. and Reddy, C. R. (1989). Leaf anatomy of midribless mutants in pearl millet. *Curr. Sci.* **58,** 1034–1036.

Rao, A. P. and Rajendrudu, G. (1989). Net photosynthetic rate in relation to leaf anatomical characteristics of C_3, C_3-C_4 and C_4 dicotyledons. *Proc. Indian Acad. Sci.* **99,** 529–537.

Raynal, J. (1973) Notes cyperologiques: 19. Contribution a la classification de la sous-famille des Cyperioidae. *Adansonia* **13,** 145–171.

Reiskind, J. B., Madsen, T. V., Van Ginkel, L. C. and Bowes, G. (1997). Evidence that inducible C_4-type photosynthesis is a chloroplastic CO_2-concentrating mechanism in *Hydrilla*, a submersed monocot. *Plant Cell Environ.* **20,** 211–220.

Rhoades, M. M. and Carvalho, A. (1944). The function and structure of the parenchyma sheath plastids of the maize leaf. *Bull. Torrey Bot. Club* **71**, 335–346.

Robertson, E. J., Baker, N. R. and Leech, R. M. (1993). Chloroplast thylakoid protein changes induced by low growth temperature in maize revealed by immunocytology. *Plant Cell Environ.* **16**, 809–818.

Robinson-Beers, K. and Evert, R. F. (1991a). Ultrastructure of and plasmodesmatal frequency in mature leaves of sugarcane. *Planta* **184**, 291–306.

Robinson-Beers, K. and Evert, R. F. (1991b). Fine structure of plasmodesmata in mature leaves of sugarcane. *Planta* **184**, 307–318.

Roth, R., Hall, L. N., Brutnell, T. P. and Langdale, J. A. (1996). *bundle sheath defective2*, a mutation that disrupts the coordinated development of bundle sheath and mesophyll cells in the maize leaf. *Plant Cell* **8**, 915–927.

Sachs, T. (1991). "Pattern Formation in Plant Tissues." Cambridge: Cambridge University Press.

Schultes, N. P., Brutnell, T. P., Allen, A., Dellaporta, S. L., Nelson, T. and Chen, J. (1996). The *Leaf permease1* gene of maize is required for chloroplast development. *Plant Cell* **8**, 463–475.

Sheen, J. (1994). Feedback control of gene expression. *Photosyn. Res.* **39**, 427–438.

Sheen, J.-Y. and Bogorad, L. (1985). Differential expression of the ribulose bisphosphate carboxylase large subunit gene in bundle sheath and mesophyll cells of developing maize leaves is influenced by light. *Plant Physiol.* **79**, 1072–1076.

Sheen, J.-Y. and Bogorad, L. (1987a). Differential expression of C_4 pathway genes in mesophyll and bundle sheath cells of greening maize leaves. *J. Biol. Chem.* **262**, 11726–11730.

Sheen, J.-Y. and Bogorad, L. (1987b). Regulation of levels of nuclear transcripts for C_4 photosynthesis in bundle sheath and mesophyll cells of maize leaves. *Plant Mol. Biol.* **8**, 227–238.

Shomer-Ilan, A., Beer, S. and Waisel, Y. (1975). *Suaeda monoica*, a C_4 plant without typical bundle sheaths. *Plant Physiol.* **56**, 676–679.

Sinha, N. and Hake, S. (1994). The Knotted leaf blade is a mosaic of blade, sheath, and auricle identities. *Devel. Genet.* **15**, 401–414.

Sinha, N. R. and Kellogg, E. A. (1996). Parallelism and diversity in multiple origins of C_4 photosynthesis in the grass family. *Amer. J. Bot.* **83**, 1458–1470.

Slack, C. R., Roughan, P. G. and Bassett, H. C. M. (1974). Selective inhibition of mesophyll chloroplast development in some C_4-pathway species by low night temperature. *Planta* **118**, 57–73.

Smith, L. G., Greene, B., Veit, B. and Hake, S. (1992). A dominant mutation in the maize homeobox gene, *Knotted1*, causes its ectopic expression in leaf cells with altered fates. *Development* **116**,

Smith, L. G., Hake, S. and Sylvester, A. W. (1996). The *tangled-1* mutation alters cell division orientations throughout maize leaf development without altering leaf shape. *Development* **122**, 481–489.

Soliday, C. L., Kolattukudy, P. E. and Davis, R. W. (1979). Chemical and ultrastructural evidence that waxes associated with the suberin polymer constitute the major diffusion barrier to water vapor in potato tuber (*Solanum tuberosum* L.). *Planta* **146**, 607–614.

Soros, C. L. and Dengler, N. G. (1998). Quantitative leaf anatomy of C_3 and C_4 Cyperaceae and comparisons with the Poaceae. *Int. J. Plant Sci.* **159**, 480–491.

Sprey, B. and Laetsch, W. M. (1978). Structural studies of peripheral reticulum in C_4 plant chloroplasts of *Portulaca oleracea*. *Z. Pflanzenphysiol.* **87**, 37–53.

Stockhaus, J., Schlue, U., Koczor, M., Chitty, J. A., Taylor, W. C. and Westhoff, P. (1997). The promoter of the gene encoding the C_4 form of phosphoenolpyruvate carboxylase directs mesophyll-specific expression in transgenic C_4 *Flaveria* spp. *Plant Cell* **9**, 479–489.

Takeda, T., Ueno, O. and Agata, W. (1980). The occurrence of C_4 species in the genus *Rhynchospora* and its significance in Kranz anatomy of the Cyperaceae. *Bot. Mag.* **93**, 55–65.

Ueno, O. (1992). Immunogold localization of photosynthetic enzymes in leaves of *Aristida latifolia*, a unique C_4 grass with a double chlorenchymatous bundle sheath. *Physiol. Plant.* **85**, 189–196.

Ueno, O. (1995). Occurrence of distinctive cells in leaves of C_4 species in *Arthraxon* and *Microstegium* (Andropogeae–Poaceae) and the structural and immunochemical characterization of these cells. *Int. J. Plant Sci.* **156**, 270–289.

Ueno, O. (1996a). Immunocytochemical localization of enzymes involved in the C_3 and C_4 pathways in the photosynthetic cells of an amphibious sedge, *Eleocharis vivipara*. *Planta* **199**, 394–403.

Ueno, O. (1996b). Structural characterization of photosynthetic cells in an amphibious sedge, *Eleocharis vivipara*, in relation to C_3 and C_4 metabolism. *Planta* **199**, 382–393.

Ueno, O. and Samejima, M. (1989). Structural features of NAD-malic enzyme type C_4 *Eleocharis*: An additional report of C_4 acid decarboxylation types in the Cyperaceae. *Bot. Mag.* **102**, 393–402.

Ueno, O. and Samejima, M. (1990). Immunogold localization of ribulose 1,5-bisphosphate carboxylase in amphibious *Eleocharis* species in relation to C_3 and C_4 photosynthesis. In "Current Research in Photosynthesis" (M. Baltscheffsky, ed.) pp. 867–870, Kluwer Academic Publishers, Dordrecht.

Ueno, O., Samejima, M. and Koyama, T. (1989). Distribution and evolution of C_4 syndrome in *Eleocharis*, a sedge group inhabiting wet and aquatic environments, based on culm anatomy and carbon isotope ratios. *Ann. Bot.* **64**, 425–438.

Ueno, O., Samejima, M., Muto, S. and Miyachi, S. (1988). Photosynthetic characteristics of an amphibious plant, *Eleocharis vivipara*: Expression of C_4 and C_3 modes in contrasting environments. *Proc. Natl. Acad. Sci. USA* **85**, 6733–6737.

Ueno, O., Takeda, T. and Maeda, E. (1988). Leaf ultrastructure of C_4 species possessing different Kranz anatomical types in the Cyperaceae. *Bot. Mag.* **101**,141–152.

Ueno, O., Takeda, T. and Murata, T. (1986). C_4 acid decarboxylating enzyme activities of C_4 species possessing different anatomical types in the Cyperaceae. *Photosynthetica* **20**, 111–116.

Venis, M. A. and Napier, R. M. (1997). Auxin perception and signal transduction. In "Molecular and Cell Biology Updates: Signal Transduction in Plants" (P. Aducci, ed.) pp. 45–63. Birkhaeuser Verlag, Basel, Switzerland.

Waites, R. and Hudson, A. (1995). *phantastica*: a gene required for dorsoventrality of leaves in *Antirrhinum majus*. *Development* **121**, 2143–2154.

Wang, J.-L., Klessig, D. G. and Berry, J. O. (1992). Regulation of C_4 gene expression in developing amaranth leaves. *Plant Cell* **4**, 173–184.

Wang, J.-L., Long, J. J., Hotchkiss, T. and Berry, J. O. (1993a). C_4 photosynthetic gene expression in light- and dark-grown amaranth cotyledons. *Plant Physiol.* **102**, 1085–1093.

Wang, J.-L., Turgeon, R., Carr, J. P. and Berry, J. O. (1993b). Carbon sink-to-source transition is coordinated with establishment of cell-specific gene expression in a C_4 plant. *Plant Cell* **5**, 289–296

Wilson, J. R. and Hattersley, P. W. (1989). Anatomical characteristics and digestibility of leaves of *Panicum* and other grass genera with C_3 and different types of C_4 photosynthetic pathway. *Aust. J. Agr. Res.* **40**,125–136.

6

Modeling C_4 Photosynthesis

Susanne von Caemmerer and Robert T. Furbank

I. Introduction

C_4 photosynthesis requires the integrated functioning of mesophyll and bundle-sheath cells of leaves and is characterized by a CO_2 concentrating mechanism that allows Rubisco, located in the bundle-sheath cells, to function at high CO_2 concentrations. This offsets the low affinity Rubisco has for CO_2 and largely inhibits its oxygenation reaction, reducing photorespiration rates in air. In the mesophyll, CO_2 is initially fixed by phosphoenolpuruvate (PEP) carboxylase into C_4 acids that are then decarboxylated in the bundle sheath to supply CO_2 for Rubisco. The coordinated functioning of C_4 photosynthesis requires a very specialized leaf anatomy, where photosynthetic cells are organized in two concentric cylinders. Thin-walled mesophyll cells adjacent to intercellular airspace radiate from thick-walled bundle-sheath cells, adjacent to the vasculature (Chapter 5, this volume; Hatch 1987). The high CO_2 concentration in the bundle sheath is linked to both the structure of the bundle-sheath wall (which has a low permeability to CO_2) and to the relative biochemical capacities of the C_3 cycle in the bundle sheath and C_4 acid cycle that operates across the mesophyll bundle-sheath interface.

The mathematical modeling of the C_4 pathway is not as frequently used as that of the C_3 photosynthetic pathway. The complexity inherent in the two compartment C_4 photosynthetic mechanism is necessarily reflected in the complexity of accurate C_4 models. Nevertheless, many of the gas exchange characteristics observed with intact leaves have been predicted using models based on biochemical and anatomic characteristics. The two major biochemical models considered in this chapter are those of Berry and Farquhar (1978) and Peisker (1979) (see also Peisker 1986; Peisker and Henderson 1992). The two models are very similar in their basic design and differ only in details. Collatz *et al.* (1992) have modified the model by Berry and Farquhar (1978) so that it could be coupled to a stomatal model. He and Edwards (1996) have used these models to estimate diffusive resistances of the bundle sheath.

Here we build on these previous models and test the resultant model by exploring the relationships between gas exchange characteristics and leaf biochemistry. We examine the predictions of the model and discuss the analysis of plants in which levels of key photosynthetic enzymes have been altered by genetic engineering techniques.

II. Basic Model Equations

Figure 1 shows a schematic representation of the proposed carbon fluxes in C_4 photosynthesis. Rubisco and the complete C_3 photosynthetic pathway are located in the bundle-sheath cells, bounded by a relatively gas-tight cell wall such that the C_3 cycle relies on C_4 acid decarboxylation as the source of CO_2. C_4 acids are, in turn, generated by CO_2 fixation by PEP carboxylase

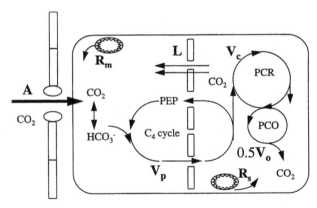

Figure 1 Schematic representation of the main features of the C_4 model. CO_2 diffuses into the mesophyll cell, where it is converted to HCO_3^- and fixed by PEP carboxylase at the rate V_p. It is assumed that C_4 acid decarboxylation in the bundle sheath occurs at the same rate. In the bundle sheath, CO_2 released by C_4 acid decarboxylation either leaks back to the mesophyll (L) or is fixed by Rubisco (V_c) in the photosynthetic carbon reduction (PCR) cycle. In the photosynthetic carbon oxydation (PCO) cycle, CO_2 is released at half the rate of Rubisco oxygenations (V_o). CO_2 is also released in the mesophyll and bundle sheath from mitochondrial respiration (R_m, R_s) not involved in photorespiration.

in the mesophyll cytosol, then diffuse to the bundle-sheath cells, where they are decarboxylated.

The net rate of CO_2 fixation for C_4 photosynthesis can be given by two equations. The first describes Rubisco carboxylation in the bundle sheath. Because all carbon fixed into sugars ultimately must be fixed by Rubisco, overall CO_2 assimilation, A, can be given by

$$A = V_c - 0.5V_o - R_d \quad (1)$$

where V_c and V_o are the rates of Rubisco carboxylation and oxygenation and R_d is the rate of mitochondrial respiration not associated with photorespiration. A list of symbols is given in the appendix.

Mitochondrial respiration may occur in the mesophyll as well as in the bundle sheath and, as CO_2 released in the bundle sheath may be more readily refixed by Rubisco, we also describe R_d by its mesophyll and bundle-sheath components

$$R_d = R_m + R_s \quad (2)$$

Because the bundle sheath compartment is a semiclosed system and is dependent for its supply of CO_2 on the decarboxylation of C_4 acids formed in the mesophyll cells, the CO_2 assimilation rate, A, can also be written in terms of the mesophyll reactions as

$$A = V_p - L - R_m \quad (3)$$

where V_p is the the rate of PEP carboxylation, R_m is the mitochondrial respiration occurring in the mesophyll, and L is the rate of CO_2 leakage from the bundle sheath to the mesophyll. The leakage, L, is given by

$$L = g_s (C_s - C_m) \quad (4)$$

where g_s is the physical conductance to CO_2 leakage and is determined by the properties of the bundle-sheath cell wall; C_s and C_m are the bundle sheath and mesophyll CO_2 partial pressures. We have assumed that there is a negligible amount of HCO_3^- leakage from the bundle sheath because the HCO_3^- pool should be small due to the absence of carbonic anhydrase activity in these cells (Farquhar, 1983; Jenkins et al., 1989).

The C_4 cycle consumes energy during the regeneration of PEP. Thus, leakage of CO_2 from the bundle sheath is an energy cost to the leaf that represents a compromise between retaining CO_2, allowing efflux of O_2, and permitting metabolites to diffuse in and out at rates fast enough to support the rate of CO_2 fixation. The CO_2 leakage depends on the balance between the rates of PEP carboxylation and Rubisco activity and the conductance of the bundle sheath to CO_2. Leakiness (ϕ) defines leakage as a fraction of the rate of PEP carboxylations and thus describes the efficiency of the C_4 cycle

$$\phi = L/V_p \tag{5}$$

A related term, *overcycling*, has also frequently been used (Jenkins et al., 1989; Furbank et al., 1990). Overcycling defines leakage as a fraction of CO_2 assimilation rate and gives the fraction by which the flux through the C_4 acid cycle has to exceed net CO_2 assimilation rate

$$\text{Overcycling} = L/A = (V_p - (A + R_m))/A \tag{6}$$

A. Equations for Enzyme Limited Photosynthesis

Many important features of the C_4 model can be examined with the enzyme-limited rates that are presumed to be appropriate under conditions of high irradiance. We consider these first before discussing light- and electron transport–limited photosynthesis with its added complexity.

1. CO_2 Assimilation Rate in the Bundle Sheath As is the case in C_3 models of photosysnthesis (Farquhar et al., 1980; Farquhar and von Caemmerer, 1982), Rubisco carboxylations at high irradiance can be described by its RuBP saturated rate

$$V_c = \frac{C_s V_{cmax}}{C_s + K_c(1 + O_s/K_o)} \tag{7}$$

where V_{cmax} is the maximum carboxylation rate, K_c and K_o are the Michaelis–Menten constants for CO_2 and O_2, and O_s is the O_2 concentration in the bundle sheath. Following the oxygenation of one mol of RuBP, 0.5 mol of CO_2 is evolved in the photorespiratory pathway and Farquhar et al. (1980) showed that the ratio of oxygenation to carboxylation can be expressed as

$$V_o/V_c = 2\, \Gamma_*/C_s \tag{8}$$

where Γ_* is the CO_2 compensation point in a C_3 plant in the absence of other mitochondrial respiration, and

$$\Gamma_* = 0.5[V_{omax}K_c/(V_{cmax}K_o)]O_s = \gamma_* O_s \tag{9}$$

where V_{omax} is the maximal oxygenase activity and the term in the bracket is the reciprocal of Rubisco specificity, $S_{c/o}$.

Subsituting for V_c and V_o in Eq. 1, it can be shown that the Rubisco limited rate of CO_2 assimilation can be given by

$$A = \frac{(C_s - \gamma_* O_s)V_{cmax}}{C_s + K_c(1 + O_s/K_o)} - R_d \tag{10}$$

In Section II.B.4, we discuss the RuBP regeneration or electron transport-limited rate.

2. Bundle Sheath CO_2 Concentration

To derive an overall expression for CO_2 assimilation rate as a function of mesophyll CO_2 and O_2 partial pressure, C_m and O_m, one needs to derive an expression for C_s and O_s. Equation 10 can be used to derive an expression for C_s:

$$C_s = \frac{\gamma_* O_s + K_c(1 + O_s/K_o)((A + R_d)/V_{cmax})}{1 - (A + R_d)/V_{c\,max}} \qquad (11)$$

This equation is analogous to the equation for the C_3 compensation point (Farquhar and von Caemmerer, 1982). If V_{cmax} could be estimated accurately from biochemical measurements together with A, it provides a means of estimating bundle sheath CO_2 concentration. One can also obtain an expression for C_s from Eqs. 3 and 4:

$$C_s = C_m + \frac{Vp - A - R_m}{g_s} \qquad (12)$$

3. Bundle Sheath O_2 Concentration

Photosystem II (PSII) activity and O_2 evolution in the bundle sheath varies widely among the C_4 species. Some NADP-ME species such as *Zea mays* and *Sorghum bicolor* have little or none (Chapman *et al.*, 1980; Hatch 1987). NADP-ME dicots and NAD and PCK species can have high PSII activity. Because the bundle sheath is a fairly gas-tight compartment, this has implications for the steady-state concentration of bundle sheath O_2 concentration (Raven, 1977; Berry and Farquhar, 1978). Following Berry and Farquhar (1978), we assume that the net O_2 evolution, E_o, in the bundle-sheath cells equals its leakage, L_o, out of the bundle sheath, that is

$$E_o = L_o = g_o(O_s - O_m) \qquad (13)$$

The conductance to leakage of O_2 across the bundle sheath, g_o can be related to the conductance to CO_2 by way of the ratio of diffusivities and solubilities by

$$g_o = g_s(D_{O2}S_{O2}/D_{CO2}S_{CO2}), \qquad (14)$$

where D_{O2} and D_{CO2} are the diffusivites for O_2 and CO_2 in water, respectively, and S_{O2} and S_{CO2} are the respective Henry constants such that

$$g_o = 0.047 g_s \qquad (15)$$

at 25°C (Farquhar, 1983). If we set $E_o = \alpha A$, where α ($0 < \alpha > 1$) denotes the fraction of O_2 evolution occurring in the bundle sheath, this gives the following expression for O_s:

$$O_s = \frac{\alpha A}{0.047 g_s} + O_m \qquad (16)$$

4. The Rate of PEP Carboxylation

Like Berry and Farquhar (1978), we make

the assumption that a steady-state balance exists between the rate of PEP carboxylation and the release of C_4 acids in the bundle sheath. In Eq. 3 it is assumed that PEP carboxylation provides the rate limiting step and not, for example, carbonic anhydrase. As PEP carboxylase uses HCO_3^- rather than CO_2, hydration of CO_2 is really the first step in carbon fixation in C_4 species (Hatch and Burnell, 1990).

When CO_2 is limiting the rate of PEP carboxylation is given by a Michaelis–Menten equation

$$V_p = \frac{C_m V_{pmax}}{C_m + K_p} \tag{17}$$

where V_{pmax} is the maximum PEP carboxylation rate, K_p is the K_m for CO_2. This assumes that the substrate PEP is saturating under these conditions. When the rate of PEP regeneration is limiting, for example, by the capacity of pyruvate orthophosphate dikinase (PPDK), V_p is set constant ($V_p = V_{pr}$) as discussed by Peisker (1986) and Peisker and Henderson (1992).

5. Quadratic Expression for the Enzyme Limited CO_2 Assimilation Rate To obtain an overall rate equation for CO_2 assimilation as a function of the mesophyll CO_2 and O_2 partial pressures (C_m and O_m) one combines Eq. 10, 12, and 16. The resulting expression is a quadratic of the form

$$aA_c^2 + bA_c + c = 0 \tag{18}$$

where

$$A_c = (-b + \sqrt{b^2 - 4ac})/(2a) \tag{19}$$

and

$$a = 1 - \frac{\alpha}{0.047}\frac{K_c}{K_o} \tag{20}$$

$$b = -\left\{(V_p - R_m + g_s C_m) + (V_{cmax} - R_d) + g_s(K_c(1 + O_m/K_o)) \right. $$
$$\left. + \frac{\alpha\gamma}{0.047}(\gamma_* V_{cmax} + R_d K_c/K_o)\right\} \tag{21}$$

$$c = (V_{cmax} - R_d)(V_p - R_m + g_s C_m) - $$
$$(V_{cmax} g_s \gamma_* O_m + R_d g_s K_c(1 + O_m/K_o)) \tag{22}$$

Equation 19 can be approximated by:

$$A_c = \min\{(V_p - R_m + g_s C_m), (V_{cmax} - R_d)\} \tag{23}$$

where min { } stands for minimum of (that is, the smaller of the bracketed terms). At low CO_2 concentration, CO_2 assimilation rate is given by

$$A_c = \frac{C_m V_{pmax}}{C_m + K_p} - R_m - g_s C_m \tag{24}$$

and linerarly related to the maximum PEP carboxylase activity, V_{pmax}. The product $g_s C_m$ is the inward diffusion of CO_2 into the bundle sheath and because g_s is low (0.003 mol m^{-2} sec^{-1}), the flux is only 0.3 μmol m^{-2} sec^{-1} at C_m of a 100 μbar and can thus be ignored. At high CO_2 concentration CO_2 assimilation rate is given by either the maximal Rubisco activity, V_{cmax} or the rate of PEP regeneration.

B. Light and Electron Transport Limited Photosynthesis

1. Rates of ATP and NADPH Consumption The energy requirements for the regeneration of RuBP in the bundle sheath are the same as in a C_3 leaf (Farquhar *et al.*, 1980; Farquhar and von Caemmerer, 1982). There is, however, the additional cost of 2 mol ATP for the regeneration of one mol of PEP from pyruvate in the mesophyll such that:

$$\text{Rate of ATP consumption} = 2V_p + (3 + 7\gamma_* O_s/C_s) V_c \tag{25}$$

where $(7\gamma_* O_s/C_s) V_c$ is the energy requirement due to photorespiration (because $V_o/V_c = 2\gamma_* O_s$, Eqs. 8 and 9). In the PCK type C_4 species some of the ATP for PEP regeneration may come form the mitochondria such that the photosynthetic requirement may be less (see later discussion on differences between C_4 types) (Burnell and Hatch, 1988; Carnal *et al.*, 1993).

There is no net NADPH requirement by the C_4 cycle itself, although in NADP-ME species, NADPH consumed in the production of malate from OAA in the mesophyll is released in the bundle sheath during decarboxylation (Hatch and Osmond, 1976). This may have implications for the response of C_4 photosynthesis to fluctuating light environments (Krall and Pearcy, 1993). The rate of NADP consumption is given by the requirement of the PCR cycle:

$$\text{Rate of NADP consumption} = (2 + 4\gamma_* O_s/C_s) V_c \tag{26}$$

where $(4\gamma_* O_s/C_s) V_c$ is the NADPH requirement of the photorespiratory cycle (Farquhar and von Caemmerer, 1982). It is important to note that, under most situations, C_s is sufficiently large that this term can be ignored, but it does become relevant at very low mesophyll CO_2 concentrations (Siebke *et al.*, 1997).

2. Partitioning of Electron Transport Rate between C_3 and C_4 Cycle NADPH and ATP are produced by the chloroplast's electron transport chain. The reduction of NADPH$^+$ to NADPH + H$^+$ requires the transfer of 2 electrons through the whole chain electron transport, which, in turn, requires 2 photons each at PSII and Photosystem I (PSI) (Table I). The generation of ATP can be coupled to the proton production via whole chain electron transport, or ATP can be generated via cyclic electron transport around PSI. Table I gives the various stoichiometries.

As discussed previously, PSII activity in the bundle sheath varies among C_4 species with different C_4 decarboxylation types. Presumably, when PSII is deficient or absent from the bundle-sheath chloroplasts, some ATP is generated via cyclic photophosphorylation and 50% of the NADPH required for the reduction of PGA is derived from NADPH generated by $NADP^+$ malic enzyme (Chapman et al., 1980). The remainder of the PGA must be exported to the mesophyll chloroplast, where it is reduced and then returned to the bundle sheath (Hatch and Osmond, 1976). Measurements of metabolite pools of *Amaranthus edulis*, a NAD^+ malic enzyme species with PSII activity in the bundle sheath, indicate it also exports some PGA to the mesophyll for reduction (Leegood and von Caemmerer, 1988; Chapter 4, this volume). It appears therefore that energy is shared between mesophyll and bundle-sheath cells. Here we have taken a very simple approach and modeled the electron transport as a whole, allocating a different fraction of it to the C_4 and C_3 cycle rather than compartmenting it to mesophyll and bundle sheath chloroplasts. That is, whole chain electron transport

$$J_t = J_m + J_s \qquad (27)$$

and $J_m = xJ_t$ and $J_s = (1 - x)J_t$ where $0 < x > 1$. Because at most 2 of 5 ATP are required in the mesophyll, $x \approx 0.4$ (Eq. 25). Peisker (1988) has modeled the optimization of x at low light in some detail.

Assuming the ATP requirements shown in Table I, and a stoichiometry of $3H^+$ required per ATP produced, an expression for the whole chain electron transport required for C_4 acid regeneration can be derived. However, because the efficiency of proton partitioning through the cytochrome B_6f complex is uncertain, two cases are considered: one in which photophosphorylation operates without Q-cycle activity, and another in which it operates with Q-cycle activity (Eqs. 28 and 29, respectively). This is discussed in more detail later.

$$J_m = 3V_p \qquad (28)$$

$$J_m = 2V_p \qquad (29)$$

Similarly, the whole chain electron transport rates required for the C_3 cycle are

$$J_s = 4.5(1 + 7\gamma_* O_s/3C_s)V_c \qquad (30)$$

in the absence of the Q-cycle activity. With the Q-cycle activity it is

Table I Stoichiometries of the Chloroplastic Electron Transport Chain

Electron transport	Quanta	e^-	H^+	NADPH	H^+/e^-	H^+/Quanta	ATP	e^-/ATP	Quanta/ATP
Whole chain	4	2	4	1	2	1	1.33[a]	1.5	3
							(1)	(2)	(4)
Whole chain + Q Cycle	4	2	6	1	3	1.5	2	1	2
							(1.5)	(1.33)	(2.66)
Cyclic through PSI	4	4	4	—	1	1	1.33	3	3
							(1)	(4)	(4)
Cyclic through PSI + Q-cycle	4	4	8	—	2	2	2.66	1.5	1.5
							(2)	(2)	(2)

[a] Assuming 3 H^+/ATP, numbers in brackets assume 4 H^+/ATP.

$$J_s = 3((1 + 7\gamma_* O_s/3C_s)V_c \tag{31}$$

3. Light Dependence of Electron Transport Rate The relationship between the electron transport, J, and the absorbed irradiance that we have used is at present empiric and is one that has also been used for C_3 models:

$$\theta J^2 - J(I_2 + J_{max}) + I_2 J_{max} = 0 \tag{32}$$

where I_2 is the photosynthetically useful light absorbed by PSII and J_{max} is the maximum electron transport. The curvature factor, θ is empirical and 0.7 is a good average value. I_2 is related to incident irradiance I by

$$I_2 = I(\text{absorptance})(1 - f)/2 \tag{33}$$

where f is to correct for spectral quality of the light (~0.15) (Evans, 1987). The 2 is in the denominator because the light absorbed is used by two photosystems. The absorptance of leaves is commonly about 0.85. Ögren and Evans (1993) give a detailed discussion on the parameters of Eq. 32. The equation can be solved for J as follows

$$J = \frac{I_2 + J_{max} - \sqrt{(I_2 + J_{max})^2 - 4\theta I_2 J_{max}}}{2\theta} \tag{34}$$

4. Quadratic Expression for Electron Transport Limited CO_2 Assimilation Rate Here, we assume that an obligatory Q-cycle operates (i.e., we use Eq. 29 and 31), and from Eqs. 3 and 4 one can derive two equations for an electron transport limited CO_2 assimilation rate

$$A_j = \frac{xJ_t}{2} + g_s(C_s - C_m) - R_m \tag{35}$$

and

$$A_j = \frac{(1 - \gamma_* O_s/C_s)(1 - x)J_t}{3(1 + 7\gamma_* O_s/C_s)} - R_d \tag{36}$$

Equation 36 is similar to the equation describing the electron transport limited rate of C_3 photosynthesis (Farquhar and von Caemmerer, 1982). The bundle-sheath CO_2 concentration under these conditions is given by

$$C_s = \frac{(\gamma_* O_s)(7/3(A_j + R_d) + (1 - x)J_t/3)}{(1 - x)J_t/3 - (A_j + R_d)} \tag{37}$$

Combining Eq. 16, 35, and 37 then yields a quadaratic expression of the form

$$aA_j^2 + bA_j + c = 0 \tag{38}$$

where

$$A_j = (-b + \sqrt{b^2 - 4ac})/(2a) \qquad (39)$$

and

$$a = 1 - \frac{7\gamma_*\alpha}{3*0.047} \qquad (40)$$

and

$$b = -\left\{\left(\frac{xJ_t}{2} - R_m + g_sC_m\right) + \left(\frac{(1-x)J_t}{3} - R_d\right) + g_s\left(\frac{7\gamma_*O_m}{3}\right) \right.$$
$$\left. + \frac{\alpha\gamma_*}{0.047}\left(\frac{(1-x)J_t}{3} + R_d\right)\right] \qquad (41)$$

$$c = \left(\left(\frac{xJ_t}{2} - R_m + g_sC_m\right)\left(\frac{(1-x)J_t}{3} - R_d\right)\right) - g_s\gamma_*O_m$$
$$\left(\frac{(1-x)J_t}{3} + \frac{7R_d}{3}\right) \qquad (42)$$

Equation 39 can be approximated by

$$A_j = \min\left[\left(\frac{xJ_t}{2} - R_m + g_sC_m\right), \left(\frac{(1-x)J_t}{3} - R_d\right)\right\} \qquad (43)$$

where min { } stands for minimum of.

C. Summary of Equations

Equations 19 and 39 are the two basic equations of the C_4 model and

$$A = \min\{A_c, A_j\} \qquad (44)$$

As pointed out by Peisker and Henderson (1992), both Rubisco and PEP carboxylase reactions can be limited by either the enzyme activity or the substrate regeneration rate, and, in theory, four types of combinations of rate limitations are possible. In the way we have presented the equations here, we have assumed that light or the electron transport capacity limit both PEP and RuBP regeneration rates simultaneously. Furthermore, in the model of C_3 photosynthesis by Farquhar et al. (1980) and von Caemmerer and Farquhar (1981), it was assumed that the limitation of RuBP regeneration could be adequately modeled by an electron transport limitation without consideration of other Calvin cycle enzymes. In the case of C_4, we have included the possiblity that PEP regeneration may also be limited by the enzyme activity of enzymes such as PPDK.

III. Analysis of the Model

A. Parameterization of the Model

Many of the model's parameters can be assigned *a priori* (Table II), leaving only key variables like V_{cmax}, V_{pmax}, V_{pr}, g_s, and J_{max} to be assigned.

It is important to note that the kinetic constants of Rubisco from C_4 species differ from those of C_3 species (Badger *et al.*, 1974; Yeoh *et al.*, 1980, 1981; Jordan and Ogren, 1981, 1983; Seemann *et al.*, 1984; Badger and Andrews, 1987; Wessinger *et al.*, 1989; Hudson *et al.*, 1990). The K_m for CO_2 can be between 1.5 and 3 times the C_3 value. The K_m for O_2 has been measured less frequently, but appears to also be greater (Badger *et al.*, 1974; Jordan and Ogren, 1981; Badger and Andrews, 1987). There seems, however, to be little difference in the relative specificity of Rubisco between C_3 and C_4 species (Jordan and Ogren, 1981). The kinetic constants used here are derived from values measured for tobacco *in vivo* (von Caemmerer *et al.*, 1994) by multiplying both the K_m CO_2 and O_2 for tobacco by 2.5 and using the same relative specificity as for tobacco (Table II). Seemann *et al.* (1984) estimated the catalytic turnover rate, k_{cat}, to be 1.2 times that of C_3 Rubiscos. We have used a $k_{cat} = 4$ sec^{-1} as a conversion factor from Rubisco catalytic site concentration to maximal activity. Figure 2 illustrates the difference in CO_2 assimilation rate as a function of bundle sheath CO_2 concentration (Eq. 10) between this hypothetical C_4 Rubisco and the to-

Table II Parameters (at 25°C) used in the Model

V_{cmax}	60 μmol m^{-2} sec^{-1} or variable	Maximum Rubisco activity
K_c	650 μbar[a]	Michaelis constant of Rubisco for CO_2
K_o	450 mbar[a]	Michaelis constant of Rubisco for O_2
γ_*	0.000193 (0.5/2590)[b]	0.5/($S_{c/o}$) half the reciprocal of Rubisco specificity
V_{pmax}	120 μmol m^{-2} sec^{-1} or variable	Maximum PEP carboxylase activity
V_{pr}	80 μmol m^{-2} sec^{-1} or variable	PEP regeneration
K_p	80 μbar[c]	Michaelis constant of PEP carboxylase for CO_2
g_s	3 mmol m^{-2} sec^{-1}	Bundle-sheath conductance to CO_2
g_o	0.047g_s[d]	Bundle-sheath conductance to CO_2
R_d	0.01 V_{cmax}	Leaf mitochondrial respiration
R_m	0.5 R_d	Mesophyll mitochondrial respiration
α	$0 < \alpha > 1$	Fraction of PSII activity in the bundle sheath
x	0.4	Partitioning factor of electron transport rate
J_{max}	400 μmol electrons m^{-2} sec^{-1} or variable	Maximal electron transport rate

[a] 2.5 times the *in vivo* values of tobacco (von Caemmerer *et al.*, 1994) given in the legend to Figure 2.
[b] von Caemmerer *et al.* (1994).
[c] Bauwe 1986.
[d] Farquhar 1983.

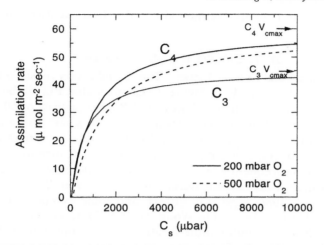

Figure 2 Comparison of CO_2 assimilation rates, A, as a function of bundle sheath CO_2, C_s, with kinetic parameters for a C_4 and C_3 Rubisco, respectively, at a bundle-sheath O_2 concentration of 200 mbar. Constants for the C_4 Rubisco are given in Table II. For the C_3 Rubisco, kcat = 3 sec^{-1} such that V_{cmax} = 45 μmol m^{-2} sec^{-1}, K_c = 260 μbar, K_o = 179 mbar, γ^* = 0.000193 = 0.5/2590 (von Caemmerer et al., 1994). A is also shown for the C_4 Rubisco at a bundle-sheath O_2 concentration of 500 mbar.

bacco C_3 Rubisco. In the case of the C_4 Rubisco, the V_{cmax} is greater per Rubisco protein because of the higher k_{cat}, but saturation occurs at higher CO_2 concentrations due to the larger K_m (Table II). The difference between C_3 and C_4 at low CO_2 concentrations is exacerbated when the higher bundle sheath O_2 concentrations for C_4 species that do evolve O_2 in the bundle sheath are taken into account (Fig. 2).

For PEP carboxylase, a K_m CO_2 of 80 μbar has been chosen (Bauwe, 1986). The maximal activities of PEP carboxylase, Rubisco, and bundle-sheath conductance vary with leaf age and environmental growth conditions; however, the relative scaling of these capacities of the mesophyll and bundle sheath is also of vital importance, as it greatly affects bundle-sheath CO_2 and O_2 concentrations and leakiness. We have used a ratio of 2 for V_{pmax}/V_{cmax} for standard outputs, but see further discussion (Table II). We chose a bundle sheath conductance of 3 mmol m^{-2} s^{-1}, which is greater than estimates by Furbank and Hatch (1989), Jenkins et al. (1989), and Brown and Boyd (1993), but less than estimates by He and Edwards (1996), (see Table 4 in He and Edwards, 1996).

It is difficult to know how to scale the leaf mitochondrial respiration with photosynthetic capacity. In C_3 models it has usually been scaled with Rubisco, and that means that the CO_2 compensation point remains unchanged. Justification for this was derived from a loose correlation between

R_d and leaf protein content. We have also scaled R_d to Rubisco here (Table II).

B. The Model at High Irradiance

1. PEP Carboxylase and Rubisco Activity

The model is very sensitive to the balance between V_{cmax} and V_{pmax}. Figure 3 examines what PEP carboxylase activities are required for the efficient use of Rubisco activity of 60 μmol m^{-2} sec^{-1} at a mesophyll CO_2 concentration of 100 μbar. Above a PEP carboxylase activity of 120 μmol m^{-2} sec^{-1} CO_2 assimilation rate saturates and bundle-sheath CO_2 concentration and leakiness, ϕ, increase rapidly. V_{pmax} and V_{cmax} have frequently been measured in response to leaf age and nitrogen nutrition. Generally measured ratios range from 2 to 8 (Usuda, 1984; Hunt et al., 1985; Sage et al., 1987), and some exceptionally high ratios of 14 to 20 were observed by Wong et al. (1985). It is clear that ratios beyond 4 result in very high bundle-sheath CO_2 concentrations and high values of leakiness (Fig. 3). There are several reasons one can suggest why the ratios of V_{pmax} to V_{cmax} measured in leaf extracts may be high. PEP carboxylase is a highly regulated enzyme with many of the metabolites such as malate acting as negative effectors (Doncaster and Leegood, 1987), and thus it is plausible that the *in vivo* V_{pmax} is down-regulated relative to the *in vitro* measurements. It is also possible that PEP is not saturating at low CO_2 concentration because PEP pools have been shown to increase with

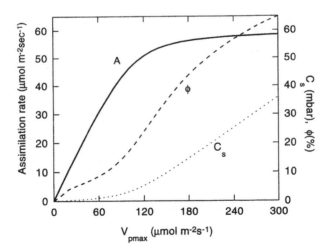

Figure 3 CO_2 assimilation rate as a function of maximum PEP carboxylase activity, V_{pmax}, at a mesophyll CO_2 concentration, C_m, = 100 μbar. Also shown are bundle-sheath CO_2 concentration, C_s, and leakiness, ϕ. Maximal Rubisco activity, V_{cmax} = 60 μmol m^{-2} sec^{-1}, mitochondrial respiration, R_d, and α where assumed zero. Other constants are as in Table II.

increasing CO_2 concentration (Leegood and von Caemmerer, 1988, 1989). As a result, it is important to bear in mind that the rate equation for PEP carboxylase (Eq. 17) may be an oversimplification that may need modification in future. Another possibility discussed in a later section is the effect a diffusion limitation on the passage of CO_2 from the intercellular airspace to the mesophyll cytoplasm has on the requirement for PEP carboxylase activity.

2. Bundle Sheath Conductance A low conductance to CO_2 diffusion across the bundle sheath is an essential feature of the C_4 pathway. This effectively eliminates CO_2 exchange with the normal atmosphere such that C_4 acid decarboxylation is the major source of CO_2 in that compartment. For example, inhibition of PEP carboxylase with an inhibitor reduced CO_2 assimilation rate almost completely at ambient CO_2 partial pressures (Jenkins *et al.*, 1989). The conductance of CO_2 diffusion across the mesophyll–bundle sheath interface has been measured for different C_4 species by various techniques. Resulting estimates range from 0.6 to 2.5 mmol m^{-2} sec^{-1} on a leaf area basis, which correspond to resistances between 400 and 1600 m^2 sec^1 mol^{-1} (Jenkins *et al.*, 1989; Brown and Byrd, 1993). Figure 4 shows CO_2 assimilation rate under standard conditions as a function of bundle sheath resistance ($1/g_s$). The low oxygen sensitivity, a characteristic of the C_4 photosynthetic pathway, is a good test for what bundle sheath

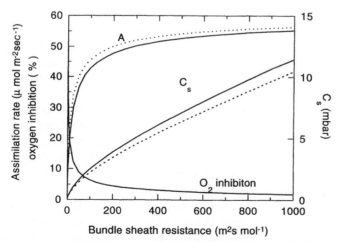

Figure 4 CO_2 assimilation rate as a function of bundle sheath resistance, r_s, at a mesophyll CO_2 concentration, C_m, = 100 μbar, and mesophyll oxygen concentrations of 200 mbar (solid line) and 20 mbar (dotted line). Also shown is the bundle-sheath CO_2 concentration, C_s, at both oxygen concentrations, and the percent oxygen inhibition calculated from $100*((A(20) - A(200))/A(200))$.

resistance is required to adequately reflect the C_4 syndrome. However, He and Edwards (1996), who used this modeling approach, have estimated very low bundle sheath resistances between 15 and 90 m^2 sec^1 mol^{-1} for *Zea mays* from measurements made by Dai *et al.* (1993, 1995) on leaves of different ages. For the standard outputs here, a bundle sheath conductance of 3 mmol m^{-2} sec^{-1} (resistance of 333 m^2 sec mol^{-1}) was chosen.

It is well known that PEP carboxylase activity (V_{pmax}) and Rubisco activity (V_{cmax}) vary with leaf age, nitrogen nutrition, and light environment during the growth of plants. At present we know little about variations of morphologic characters such as bundle sheath conductance with leaf age and growth conditions. Bundle sheath conductance is the product of the diffusion conductance across the bundle sheath cell wall interface times the bundle sheath surface area. Only a few measurements of bundle sheath surface area per leaf area have been made, ranging between 0.6 to 2.16 m^2 m^{-2} (Apel and Peisker, 1978) and 1.13 to 3.1 m^2 m^{-2} (Brown and Byrd, 1993). It is likely that bundle sheath surface area per leaf area may be larger in young expanding leaves relative to mature leaves. Measurements of bundle sheath surface area and conductance are required for leaves grown under different environmental conditions.

3. CO_2 Response Curves In C_3 species, CO_2 response curves have been useful in the analysis of photosynthetic capacities (von Caemmerer and Farquhar, 1981). In Figs. 5, 6, and 7 we analyze how V_{pmax} and V_{cmax}, and g_s affect the shape of the CO_2 response curves of C_4 photosynthesis. C_4 photosynthesis is characterised by low CO_2 compensation points and CO_2 response curves that saturate abruptly. The model (Eqs. 19 and 23) mirrors these unique features (Figs. 5 and 6). The initial slope of the curve is proportional to the PEP carboxylase activity, whereas the saturated rate is proportional to Rubisco activity, the rate of PEP regeneration, or an electron transport limitation.

Different maximal PEP carboxylase activities primarily affect the initial slope of the CO_2 response curves. When PEP carboxylase activity is very low, however, there is also a reduction in the saturated rate of CO_2 assimilation because Rubisco in the bundle sheath is not completely saturated with CO_2 (See Fig. 3). Variations in the initial slope with variations in PEP carboxylase activity have frequently been demonstrated; however, they always occur with concurrent changes in Rubisco activity (Hunt *et al.*, 1985; Sage *et al.*, 1987). Dever *et al.* (1995) and Dever (1997) have generated mutants of *Amaranthus edulis* with reduced PEP carboxylase activity, but with no changes in Rubisco activity. These plants show similar gas exchange characteristics to those predicted in Fig. 6A.

The model predicts that maximal Rubisco activity affects only the saturated part of the A versus C_m response curve. Recent gas exchange measurements on transgenic *Flaveria bidentis* with reduced amounts of Rubisco

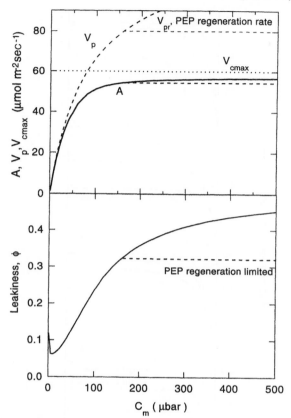

Figure 5 (A) CO_2 assimilation rate, A (Eq. 19, solid or dashed line), as a function of mesophyll CO_2 concentration, C_m. Also shown are the rates of PEP carboxylation, V_p, and and maximal Rubisco activity, V_{cmax} (see Eq. 23). At high C_m, A is limited by V_{cmax} (solid line) when PEP carboxylase is PEP saturated, or by a PEP regeneration rate of 80 μmol m^{-2} sec^{-1} (dashed line). In this simulation, the fraction of O_2 evolution in the bundle sheath, $\alpha = 0$, and mitochondrial respiration, $R_d = 0$. Other parameters are as given in Table II. (B) Leakiness, ϕ, as a function of mesophyll CO_2 concentration, C_m. When PEP carboxylase is PEP saturated (solid line), and when PEP regeneration is limiting at a rate of 80 μmol m^{-2} sec^{-1} (dashed line).

confirm this role of Rubisco (Fig. 7) (Furbank *et al.*, 1996; von Caemmerer *et al.*, 1997). These plants have unaltered PEP carboxylase activities *in vitro* and the fact that the initial slope is not affected by the drastic reductions in Rubisco activity indicates that activity of PEP carboxylase *in vivo* is unaffected by low CO_2.

A factor that affects the curvature of the CO_2 response curve is bundle sheath conductance (g_s) as it is a key determinant of the bundle sheath CO_2 concentration. Although variations in the saturation characteristics

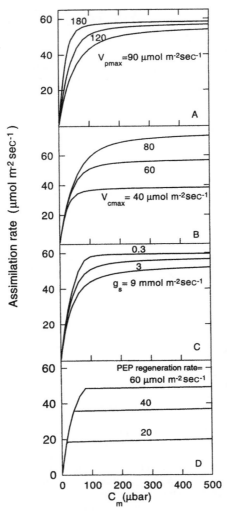

Figure 6 CO_2 assimilation rate as a function of mesophyll CO_2 concentration, C_m, at (A) various maximum PEP carboxylase activities, V_{pmax}; at (B) various maximal Rubisco activities, V_{cmax}; at (C) different bundle sheath conductances; and (D) at different rates of PEP regenerations. Other constants are taken from Table II. R_d and $\alpha = 0$.

have been noted between young and old leaves and between species (Pearcy *et al.*, 1982), no systematic studies have yet correlated the shape of the CO_2 response curves with variations in g_s. Ludwig *et al.* (1998) analyzed transgenic *Flaveria bidentis* that ectopically expressed tobacco carbonic anhydrase in the bundle-sheath cells. They found that this led to increased leakage of inorganic carbon out of the bundle sheath. This was because carbonic anhydrase, normally absent from bundle-sheath cells, would increase the conversion of CO_2 to HCO_3^-, and hence the steady state concentration of HCO_3^- in the bundle sheath of the transgenics, which would in turn increase the total leak of inorganic carbon. Figure 7B shows that the change in CO_2 response curve observed in these plants is similar to that predicted by an increase in conductance by the model (Fig. 6C).

Variations in the rate of PEP regeneration also affects the CO_2 saturated rate of CO_2 assimilation in a similar way to changes in Rubisco activity, and it is often impossible to distinguish between these two limitations. PEP regeneration at high irradiance can be limited by the capacity of enzymes of the C_4 cycle such as pyruvate orthophosphate dikinase (PPDK), NADP-malate dehydrogenase, or by the capacity of the chloroplastic electron transport chains. Measurements of the CO_2 response curves of *Zea mays* at two irradiances (Leegood and von Caemmerer, 1989) are consistent with a limitation to PEP regeneration, because the initial slopes remain independent of irradiance. Figure 7C shows a CO_2 response curve for transgenic *Flaveria bidentis* with reduced activities of NADP-malate dehydrogenase (Trevanion *et al.*, 1998), which is also consistent with this hypothesis.

4. Oxygen Sensitivity of C_4 Photosynthesis The low O_2 sensitivity and low CO_2 compensation points are key characteristics of the C_4 pathway. PEP carboxylase itself is not sensitive to O_2 concentrations, and the lack of O_2 sensitivity is attributed to the high bundle-sheath CO_2 concentrations and favorable CO_2/O_2 ratios that inhibit photorespiration (Hatch and Osmond, 1976; Hatch, 1987). Figure 4 supports this, showing a decrease in O_2 sensitivity with increasing bundle-sheath resistance and bundle-sheath CO_2 concentration. However, there is another important reason for the lack of O_2 sensitivity when bundle sheath resistance is high that was first pointed out by Berry and Farquhar (1978) and is illustrated in Fig. 8. Figure 8 examines in more detail CO_2 assimilation rate at low mesophyll CO_2 concentrations (0–40 μbar) and two O_2 concentrations. Under these conditions, CO_2 assimilation rate in the model is limited by PEP carboxylation rate. At the CO_2 compensation point, the predicted bundle-sheath CO_2 concentration is typical of C_3 compensation points increasing rapidly with increasing mesophyll CO_2 concentration. There is an appreciable amount of photorespiration occuring at 200 mbar O_2 but not at 20 mbar O_2 under these conditions that is not reflected in differences in CO_2 assimilation rates. Instead, because the bundle-sheath compartment is almost gas tight, bundle-sheath CO_2 concentrations change in response to changes in O_2 concentration rather than the CO_2 assimilation rate itself (Fig. 8). Although the ratio of bundle-sheath O_2 and CO_2 uniquely define the ratio of oxygen-

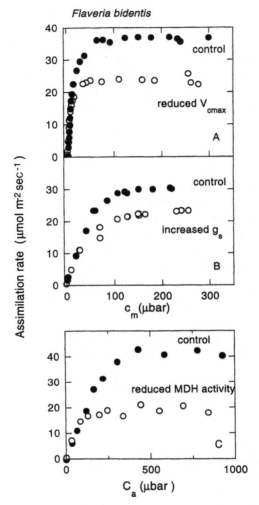

Figure 7 CO_2 assimilation rate as a function of mesophyll CO_2 concentration, C_m in *Flaveria bidentis*. (A) In a control plant and transgenic plant with reduced amount of Rubisco due to an antisense construct targeted to the small subunit of Rubisco. [Data are redrawn from von Caemmerer, S., Millgate, A., Farquhar, G. D., and Furbank, R. T. (1997). Reduction of Rubisco by antisene RNA in the C_4 plant *Flaveria bidentis*, leads to reduced assimilation rates and increased carbon isotope discrimination. *Plant Physiol.* **113**, 469–477.] Measurements were made at an irradiance of 2000 μmol quanta m^{-2} sec^{-1}, and a leaf temperature of 25°C. The control leaf had a Rubisco site concentration of 10.5 μmol m^{-2} and the transgenic leaf, 5.2 μmol m^{-2}. (B) In a control plant and transgenic plant in which tobacco carbonic anyhdrase is expressed in the bundle sheath. In these plants, bundle-sheath conductance was effectively increased through an increased leak rate of bundle sheath HCO_3^- (Ludwig *et al.*, 1998). Measurements were made under same conditions as in (A). (C) In a control plant and transgenic plant with reduced amounts of NADP-linked malate dehydrogenase activity due

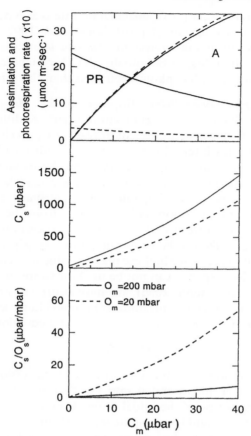

Figure 8 Modeled CO_2 assimilation rate, A, and photorespiration rate at low mesophyll CO_2 concentrations, C_m, and a mesophyll O_2 concentration of 20 (dashed lines) and 200 mbar (solid lines). Also shown are bundle-sheath CO_2 concentrations, C_s, and the ratio of bundle sheath CO_2 and O_2 partial pressures. Parameters are the same as in Fig. 5.

to transformation with a cosense DNA contruct. [Data are redrawn from Trevanion, S. J., Furbank, R. T., and Ashton, A. R. (1998). NADP malate dehydrogenase in the C_4 plant *F. bidentis:* Cosense suppression of activity in mesophyll and bundle sheath cells and consequences for photosynthesis. *Plant Physiol.* **113,** 1153–1165. Measurements were made at a temperature of 25°C and an irradiance of 1150 μmol quanta m^{-2} sec^{-1}. The NADP-MDH activity in the control leaf was 110 μmol m^{-2} sec^{-1} and in the transgenic leaf, 17 μmol m^{-2} sec^{-1}.

ase to carboxylase rate (Eq. 8), CO_2 assimilation rate is also dependent on bundle sheath CO_2 concentration (Eq. 10). Siebke et al. (1997) and Maroco et al. (1998) have shown that transgenic *Flaveria bidentis* with reduced amounts of Rubisco have higher bundle-sheath CO_2 concentrations at low CO_2 and, as a consequnce, less photorespiration than the wild type.

5. CO_2 Diffusion from Intercellular Airspaces to the Mesophyll Cytosol

Studies have shown that there is a significant resistance to diffusion of CO_2 from the intercellular airspace to the chloroplast in C_3 species where it is possible to quantify this internal conductance, g_i, with measurements of carbon isotope discrimination (Evans et al., 1986; Evans and von Caemmerer, 1996). In tobacco it was shown that g_i was correlated with the chloroplast surface area appressing intercellular airspace (Evans et al., 1994). In C_4 species the diffusion path for CO_2 is from the intercellular airspace to the cytoplasm were PEP carboxylase and carbonic anhydrase reside. Therefore, g_i is likely to be proportional to mesophyll surface area exposed to intercellular airspace in these plants. Unfortunately, techniques used to measure g_i in C_3 species can not be used for C_4 species (for review, see Evans and von Caemmerer, 1996). We have at present no estimate of g_i nor for the drop in CO_2 concentration from intercellular airspace, C_i to mesophyll, C_m. C_i and C_m are related in the following equation:

$$A = g_i(C_i - C_m) \tag{45}$$

Incorporating Eq. 45 into Eq. 19 results in a cubic expression that is not easily solved. It can be incorporated into Eq. 39 giving a slightly more complex quadratic. In case of the initial slope of the CO_2 response curve, one can again use Eq. 24 and ignoring the term $g_i C_m$ and combining it with Eq. 45, a quadratic similar to the one given for C_3 leaves is obtained (von Caemmerer and Evans, 1991).

$$A^2 - A(g_i(C_i + K_p) + V_{pmax} - R_m) + g_i(V_{pmax}C_i - R_m(C_i + K_p)) = 0 \tag{46}$$

the first derivative with respect to C_i at $C_i = 0$ is given by

$$\frac{dA}{dC_i} = \frac{g_i V_{pmax}}{g_i K_p + V_{pmax}} \tag{47}$$

Peffer and Peisker (1995) used this equation together with measurements of PEP carboxylase activity and initial slope (dA/dC_i) to estimate g_i in plants grown under different light intensities. However, they conceded that their estimate of $g_i = 0.87$ mol m^{-2} sec^{-1} is based on the assumption that g_i does not change with growth conditions. In C_3 plants, g_i has been found to be positively correlated with photosynthetic rate by several authors (Evans and von Caemmerer, 1996), and it may also hold for C_4 leaves.

Figure 9 illustrates the effect g_i has on the initial slope of the CO_2 response curve. The curve was generated from Eq. 19 deriving C_i from A and C_m with the help of Eq. 45. The presence of a substantial resistance to CO_2 diffusion across the mesophyll–cell wall provides one reason why measured ratios of V_{pmax} and V_{cmax} need to be greater than 2.

C. CO_2 Fixation at Limiting Light

1. Optimal Partition of Electron Transport Figures 10 and 11 show typical modeled light response curves of CO_2 assimilation. The shapes of the curves are determined by Eq. 32. Figure 10 shows the light response curve of CO_2 assimilation rate when electron transport is partitioned optimally between C_4 and C_3 cycles. The solutions were found by an optimization procedure but can also be calculated from the first and second derivatives of Eq. 39 (Peisker, 1988). It is noteworthy that the fraction of electron transport allocated to the C_4 cycle, x, equals 0.404 over a wide range of irradiances but drops at very low irradiance. Under low light, the bundle-sheath CO_2 concentration is close to the mesophyll CO_2 concentration and electron transport is required for recycling of photorespiratory CO_2. The optimal partitioning increases from 0.404 to 0.417 if oxygen is evolved in the bundle

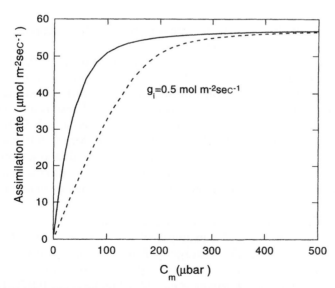

Figure 9 The effect of a limited conductance to CO_2 diffusion between intercellular airspace and mesophyll cytoplasm, g_i, on the shape of the CO_2 response curve under enzyme limited conditions. Parameters are as in Fig. 5 and $R_d = 0$ and $\alpha = 0$. $g_i = 0.5$ mol m^{-2} sec^{-1} (dashed line) or $g_i = \infty$ (solid line).

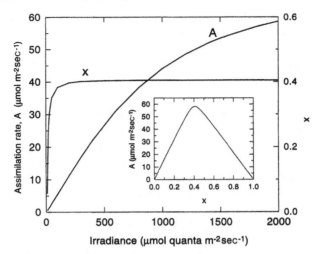

Figure 10 Light reponse of CO_2 assimilation rate, A, with optimal partitioning, x, of electron transport between C_4 and C_3 cycle at at a mesophyll CO_2 concentration $C_m = 100$ μbar. Parameters are given in Table II except that $R_d = 0$ and $\alpha = 0$. Inset: CO_2 assimilation rate, A, at 2000 μmol quanta m^{-2} sec^{-1} as a function of the fraction x of electron transport partioned to the C_4 cycle.

sheath ($\alpha = 1$). It also increases with increasing bundle-sheath conductance, reaching 0.415 at $g_s = 0.01$ mol m^{-2} sec^{-1}.

For all further modeling we have used a constant value of $x = 0.4$. Figure 11 shows a typical light reponse curve for CO_2 assimilation rate under these conditions. Bundle-sheath CO_2 concentration increases with increasing irradiance, whereas leakiness, ϕ, is predicted to decrease. These results are in accordance with experimental measurements of bundle-sheath CO_2 concentration (Furbank and Hatch, 1987). Concerning leakiness, it is interesting to note that an optimal partitioning would have predicted a decrease in leakiness at very low irradiance. The rise in leakiness at low irradiance with a constant x is, however, consistent with experimental evidence (Henderson et al., 1992).

C_4 photosynthesis is characterized by light reponse curves that saturate only at very high irradiance. An example of this is shown for a *Flaveria bidentis* wild type plant (Fig. 12). Transgenic *Flaveria bidentis* plants in which amount of Rubisco has been reduced through antisense saturate at much lower irradiances (Fig. 12) (Furbank et al., 1996; Siebke et al., 1997). These limitations are easily reproduced by the model. Saturation at lower irradiance would also occur at lower CO_2 concentration where A may limited by PEP carboxylase activity, or because of limitations to the regeneration of PEP (Eq. 44).

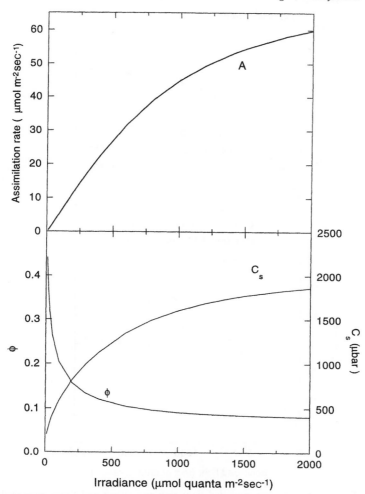

Figure 11 (A) Light reponse of CO_2 assimilation rate, A, with a constant partitioning, $x = 0.4$, of electron transport between C_4 and C_3 cycle at a mesophyll CO_2 concentration $C_m = 100$ μbar, $J_{max} = 400$ μmol electrons m^{-2} sec^{-1}, $R_d = 0$ and $\alpha = 0$. (B) The bundle sheath CO_2 concentration and leakiness under the previously mentioned conditions.

2. Leakiness Leakiness, ϕ, was defined by Eq. 5 as the ratio of the rate of CO_2 leakage out of the bundle sheath to the rate of PEP carboxylation. Leakiness defines the efficiency of the CO_2 concentrating function of the C_4 pathway and has become and important parameter in the description of C_4 photosynthesis. It can only be measured indirectly and estimates range from 10% to 40% (Farquhar, 1983; Krall and Edwards, 1990; Henderson *et al.*, 1992; Hatch *et al.*, 1995). Measurements of carbon isotope discrimina-

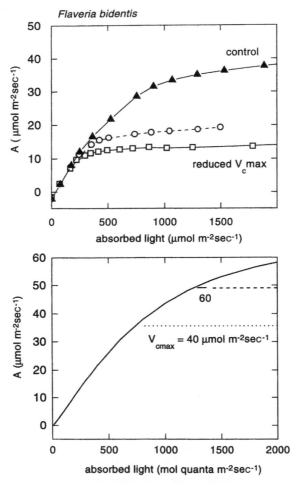

Figure 12 Comparison between modelled light response curves and those measured in control and transgenic *Flaveria bidentis* with reduced amounts of Rubisco. [Data are redrawn form Siebke, K., von Caemmerer, S., Badger, M. R., and Furbank, R. T. (1997). Expressing an RbcS antisense gene in transgenic *Flaveria bidentis* leads to an increased quantum requirement per CO_2, fixed in photosystem I and II. *Plant Physiol.* **115,** 1163–1174.] Measurements were made at an ambient CO_2 concentration of 350 μbar and a a leaf temperature of 25°C. Rubisco site concentrations were 10.5, 5, and 3 μmol m^{-2} for control and transgenic leaves repectively. Modeled curves show the effect of a Rubisco limitation at high irradiance. Parameters are given in Table II except that $R_d = 0$ and $\alpha = 0$.

tion during CO_2 exchange (Henderson et al., 1992) and measurements of chlorophyll fluorescence and CO_2 exchange (Krall and Edwards, 1990; Edwards, and Baker, 1993) indicate that the C_3 and the C_4 cycle are regulated such that ϕ remains constant over a range of CO_2 concentrations, irradiances, and temperatures. In the model, ϕ is determined by the bundle-sheath conductance to CO_2 and the CO_2 gradient between bundle sheath and mesophyll (Eq. 4). This gradient in turn is dependent on the relative capacities of PEP carboxylase or Rubisco at high irradiance (Figs. 3 and 5) or the electron transport capacities at low and moderate irradiances (Fig. 11). The model has no feedback regulation from the C_3 cycle to the C_4 cycle, so that when Rubisco becomes limiting at high CO_2 concentrations (Fig. 5), the model predicts a rapid increase in ϕ unless PEP regeneration is also curtailed. Although there appears to be little variation in ϕ with short-term environmental pertubation in many species, carbon isotope and chlorophyll fluorescence studies on transgenic *F. bidentis* with artificially reduced amounts of Rubisco have shown that ϕ does increase in these circumstances (von Caemmerer et al., 1997; Siebke et al., 1997). These results show that the relative amounts of Rubisco and C_4 cycle enzymes are important in affecting leakiness but that this ratio is normally under strict genetic control.

3. Quantum Yield Traditionally, quantum yield has been measured as the initial slope of the light response curve of CO_2 fixation. The quantum yield of PSII can also be measured at other irradiances with the use of chlorophyll fluorescence (Genty et al., 1989). Quantum yield among unstressed C_3 plants varies little across a wide range of species when measured under standard CO_2, O_2, and temperatures (see Ehleringer and Pearcy, 1983; Hatch, 1987). C_4 plants, however, show considerable interspecific variation, ranging from 0.05 to 0.069 across 32 C_4 species (Ehleringer and Pearcy, 1983). It has been suggested that this range of quantum yields might reflect interspecific variation in leakiness (Ehleringer and Pearcy, 1983; Farquhar, 1983; Furbank et al., 1990). Considering that there are three subpathways responsible for the C_4 photosynthetic mechanism, one might expect some trends in the energy requirements and, hence, quantum yields between decarboxylation types and there is some evidence for such a correlation. However, the largest differences observed are between NAD-ME dicot types and the remainder of the C_4 monocots and dicots, in which the former have average quantum yields of around 0.053 compared to 0.0625 for the latter (Ehleringer and Pearcy, 1983). It is interesting to note that there appears to be no consistent correlation between measured quantum yields and leak rates estimated either by carbon isotope discrimination or radioisotope methods (see Hatch et al., 1995).

Despite the reservations outlined previously, it is possible to model the effect of leakiness on the quantum requirement of CO_2 fixation in C_4 plants, and this has been done on at least two occasions (Farquhar, 1983; Furbank et al., 1990). Furbank et al. (1990) constructed a model that predicts the quantum requirement of photosynthesis from a given leakiness, based on the ATP and NADPH requirements of the mesophyll reactions. Because the increased energy demands result in an increased ATP requirement per net CO_2 fixed, the following relationship can be derived from Eq. 25:

$$\text{ATP requirement} = (2/(1 - \phi) + 3)(A + R_d) \qquad (48)$$

With certain assumptions, this ATP requirement can be related to quantum requirement per net CO_2 fixed (see Table I and Fig. 13). In these calcula-

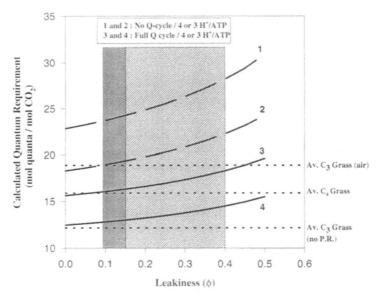

Figure 13 Quantum requirement of CO_2 assimilation calculated from the ATP requirement of CO_2 fixation with different leakage values (Eq. 48). The quanta required to produce 1 mol ATP was varied according to the following assumptions: 1. No Q-cycle operation in proton partitioning and 4 H^+ per ATP produced; 2. No Q-cycle operation but 3 H^+ per ATP; 3. Electron transport operating with full Q-cycle efficiency and 4H^+ per ATP and 4. Electron transport operating with full Q-cycle efficiency and 3H^+ per ATP. The shaded area represents the range of leakiness values recently obtained by Hatch et al. (1995) for C_4 grasses (dense shading) or other values from the literature (light shading). The average measured quantum requirements per CO_2 fixed for C_4 grasses and for C_3 plants at 30°C with and without photorespiration are also shown (from Ehleringer and Pearcey, 1983). Calculated quantum requirement was corrected for light quality and energy dissipation to nonphotosynthetic processes as described in Furbank et al. (1990). It was assumed that the ATP over and above the basal requirement of 5 mol per mol CO_2 fixed was provided by cyclic photophosphorylation, which functioned with or without Q-cycle efficiency as specified for linear electron flow (see Table I).

tions, it became apparent that a key assumption was the number of protons partitioned per electron transported in the chloroplast electron transport chain, affecting the number of ATP produced per quanta absorbed (see Table I). Opinion is, at present, divided as to the mechanism of proton partitioning during electron transport through the thylakoid cytochrome b_6f complex. One mechanism, which now has widespread support, involves a cycle around cytochrome b_6f (the "Q-cycle") and essentially doubles the efficiency of proton pumping through this complex per electron transported. A second key assumption in this model is the number of protons that must be partitioned per ATP produced by the thylakoid ATPase. The two most common values reported in the literature are 3 and 4 (Table I). Figure 13 describes the model of Furbank et al. (1990) by plotting the relationship between calculated quantum requirement and leakiness and comparing this with measured quantum yields for C_3 and C_4 plants (adapted from Hatch, 1992). Four calculated relationships are shown using the following assumptions: (1) No Q-cycle operation in proton partitioning and 4 H^+ per ATP produced; (2) no Q-cycle operation but 3 H^+ per ATP; (3) electron transport operating with full Q-cycle efficiency and $4H^+$ per ATP; and (4) electron transport operating with full Q-cycle efficiency and $3H^+$ per ATP. The shaded areas represent the range of leakiness values obtained by Hatch et al. (1995) for C_4 grasses and other measured values. The average measured quantum requirements per CO_2 fixed for C_4 grasses and for C_3 plants at 30°C with and without photorespiration are also shown (from Ehleringer and Pearcey, 1983). In the absence of a Q-cycle, C_4 plants would be no more efficient than C_3 plants, even at 30°C and assuming 3 H^+ per ATP produced. Assuming even low to moderate leakage rates [e.g., the range shown here, estimated for C_4 dicots by the radioisotope method (Hatch et al., 1995)], there would be no energetic advantage in the operation of the C_4 CO_2 concentrating mechanism. The only way we can account for the low measured values for quantum requirement in C_4 monocots is to assume the operation of the Q-cycle, and we have proposed that the C_4 mechanism could not have evolved in its absence (see Furbank et al., 1990; Hatch, 1992).

The treatment described previously is, of course, a slight oversimplification as it does not include the possibility that at low light intensities some photorespiratory activity might occur although even then photorespiration is unlikely to be a large component. However, Berry and Farquhar (1978) noted in their model that the light response is concave upwards at low irradiance so that there is no unique slope as there is in C_3 models. This is because in the model, ϕ varies at low light and bundle-sheath CO_2 concentrations may be sufficiently low for some photorespiration to occur (Figs. 10 and 11). Figure 14 shows the quantum requirement calculated at 50 μmol quanta m^{-2} s^{-1} of incident light as a function of bundle-sheath

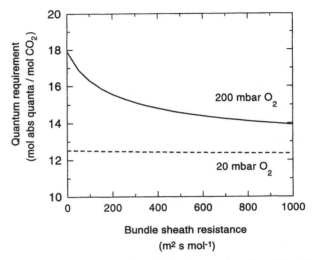

Figure 14 Quantum requirement of CO_2 assimilation as a function of bundle-sheath resistance at a mesophyll concentration of 20 and 200 mbar O_2 and a mesophyll CO_2 concentration of 300 μbar. Calculations assume the full operation of a Q-cycle and $3H^+$ per ATP. Parameters are given in Table II, except that $R_d = 0$ and $\alpha = 0$.

resistance. The quantum requirement drops as bundle-sheath resistance increases, but above 250 m^2 s mol^{-1} there would be little experimentally detectable variation in quantum yield. This would support the notion that even with the wide variations reported for CO_2 leak rates from bundle-sheath cells, they are unlikely to account for the observed variations in quantum yield between C_4 species. However, the model also predicts low O_2 and CO_2 sensitivity at low irradiance as low light limits the accumulation of bundle-sheath CO_2. Low bundle-sheath inorganic carbon concentrations have been observed experimentally under such conditions (Furbank and Hatch, 1987). Peisker, who modeled the optimization of energy balance between the C_3 and the C_4 cycle in detail at low irradiance, also pointed out this oxygen dependence (Peisker, 1988).

D. Modeling Different Decarboxylation Types

1. Bundle Sheath Conductance C_4 species have been subdivided into three biochemical types based on the enzymes catalyzing the decarboxylation of C_4 acids in the bundle sheath: NAD-malic enzyme (NAD-ME), NADP-malic enzyme (NADP-ME), and phosphoenolpyruvate carboxykinase (PCK) (see Chapter 3). These biochemical variations are accompanied by a suite of anatomical features such as the presence or absence of a suberized lamella in the cell wall between bundle-sheath and mesophyll cells, and the centripetal or centrifugal location of chloroplasts in bundle-sheath cells (Hatch, 1987). There is inferential evidence that this suberized lamella is responsible for a lower bundle-sheath conductance (Hatch and Osmond, 1976), but at present there is no clear experimental confirmation of this. Species lacking suberization almost invariably have centripetally located chloroplasts creating a longer liquid diffusion path, as Hattersley and Browning (1981) suggested, that may overcome this shortcoming (Furbank et al., 1989; Jenkins et al., 1989). However, variations in carbon isotope discrimination of dry matter suggest that systematic differences in leakiness may exist between the different biochemical subtypes (Hattersley, 1982). Either bundle-sheath conductance or the ratio PEP carboxylase to Rubisco activity could be responsible for these differences (Henderson et al., 1992). With the current level of knowledge it is not easy to refine the model to account for such decarboxylation-type specific characteristics.

2. O_2 Evolution in the Bundle Sheath One factor that can be readily tailored to an individual species or subtype is α, which aportions the amount of oxygen evolution that occurs in the bundle sheath relative to the mesophyll chloroplasts. Some NADP-ME monocots such as sorghum and maize have very little PSII activity in the bundle sheath, whereas higher levels occur in other grass and NADP-ME dicots. The NAD-ME and PCK types have bundle sheath PSII activities similar to C_3 species (see Chapter 3, this volume). For maize and sorghum, α will be zero (Eq. 16), whereas it will approach or even exceed 0.5 in many other cases.

Apart from the consequences on movement of PGA to the mesophyll for reduction (which is difficult to model), the major effect of changes in α will be bundle-sheath O_2 concentration. This is shown in Fig. 15, which plots the effect of a varying α between 0 and 1 on CO_2 assimilation rate and the bundle-sheath CO_2 and O_2 concentrations. The model predicts that bundle-sheath O_2 concentrations can be several-fold greater than ambient O_2 concentration because bundle-sheath conductance to O_2 is modeled to be 20-fold less than that to CO_2 due to the low solubility of O_2 (Eq. 15) (Berry and Farquhar, 1978; Farquhar, 1983).

3. Energy Requirements Because of the different biochemical mechanisms used for decarboxylation in the three C_4 subtypes, one might predict differences in the energy balance and quantum requirements between these types. For example, a unique feature of the PCK types is the involvement of mitochondrial oxidative phosphorylation in the provision of ATP for the PEP carboxykinase reaction in the bundle sheath (Hatch, 1987; Carnal et al., 1993). In these species, some malate is transported directly to the bundle sheath, some is decarboxylated in mitochondria to give CO_2 and NADH with ATP being generated during the subsequent oxidation of NADH (see Chapter 3). Minimally, assuming 3 ATP are produced per

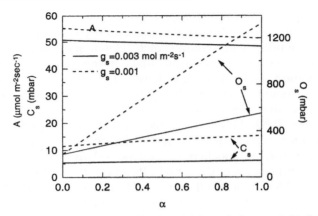

Figure 15 The effect of varying the proportion of O_2 evolution occurring in the bundle-sheath chloroplast (α) on CO_2 assimilation rate and bundle sheath CO_2, C_s, and O_2 concentration, O_s. Other parameters are as in Fig. 5.

NADH oxidized in mitochondrial electron transport, one-quarter of the C_4 acids decarboxylated would have to follow this route to fully support the ATP requirements of oxaloacetate decarboxylation. Such a partitioning between the two pathways is supported by measurements of relative amounts of PEP carboxykinase and NAD-malic enzyme in PCK types and from *in vitro* studies that suggest, however, that there may be some flexibility in this process (Hatch, 1987; Carnal *et al.*, 1993). For the PCK types the ATP requirement is therefore given by

$$\text{ATP requirement} = \left\{ \frac{2}{(1 + \beta)(1 - \phi)} + 3 \right\}(A + R_d) \quad (49)$$

and the NADPH requiremnent by

$$\text{NADPH requiremnent} = \left(\frac{1}{(1 + \beta)(1 - \phi)} + 2 \right)(A + R_d) \quad (50)$$

where β is the mol of ATP produced in the mitochondria per mol of NADPH. If $\beta = 3$, 3.5 ATP and 2.25 NADPH are required from thylakoid electron transport per CO_2 fixed (assuming no CO_2 leak). Converting this ATP requirement to quanta required as in Fig. 13, and assuming no Q-cycle, the basal energy requirement for PCK types is around 14 quanta per CO_2 fixed, compared to 18.3 for NAD-ME types using the same assumptions (Fig. 13). This large difference is dependent on the value taken for β, which in turn depends on the P/O ratio for mitochondrial respiration (i.e., its "efficiency") in an analogous fashion to the ATP/2e ratio for thylakoid electron transport assumed in Table I. Differences of 4 quanta per CO_2 fixed would be easily observed *in vivo;* however, PCK-type quantum requirements do lie at the lower end of values observed for C_4 plants by Ehleringer and Pearcy (1983): 15.6 for PCK, compared to 18.9 for NAD-ME dicots and 16.6 for NAD-ME grasses, but similar to values obtained for NADP-ME grasses (15.4). The reason for the lack of large differences in practice may be interspecific variability in the other factors discussed previously that affect the efficiency of the C_4 process.

IV. Summary

A biochemical model of C_4 photosynthesis is reviewed. The model is based on the assumption that CO_2 fixation into C_4 acids by phosphoenolpyruvate (PEP) carboxylase is a mechanism for concentrating CO_2 in the bundle sheath, where Rubisco is located and CO_2 is refixed via the C_3 pathway. A steady-state balance is assumed between the transport of CO_2 into the bundle sheath, its refixation by the C_3 pathway, and the leakage of CO_2 from the bundle sheath. Equations are developed that allow predictions of bundle-sheath CO_2 and O_2 concentrations. We show that model is able to explain key characteristics of the C_4 photosynthetic pathway and use it to analyze the influence of PEP carboxylase activity, Rubisco activity, and bundle-sheath conductance on the CO_2 assimilation rate. The model integrates current biochemical knowledge of C_4 photosynthesis in a small number of equations. We test the output of the model using the response of both transgenic and wild-type plants to environmental parameters and address some of the common "What if?" questions posed about C_4 plants. Hopefully, this approach will be valuable for integrating our current knowledge into a thesis of how C_4 plants respond to their environment and the regulatory processes involved.

References

Apel, P. and Peisker, M. (1978). Einfluss hoher Sauerstoffkonzentrationen auf den CO_2-Kompensationspunkt von C_4-Pflanzen. *Kulturplanze* **XXVI,** 99–103.

Badger, M. R., Andrews, T. J. and Osmond, C. B. (1974). Detection in C_3, C_4 and CAM plants of a low-K_m (CO_2) form of RuDP carboxylase, having high RuDP oxygenase activity at physiological pH. *In* "Proceedings of the Third International Congress on Photosynthesis" (M. Avron, ed.) Elsevier Scientific Publishing, Amsterdam, Netherlands, pp. 1421–1429.

Badger, M. R. and Andrews T. J. (1987). "Co-evolution of Rubisco and CO_2 concentrating mechanisms" *In* "Progress in Photosynthesis Research" (J. Biggens, ed.) Vol. III, Martinus Nijhoff Publishers, Dordrecht, Netherlands, pp. 601–609.

Bauwe, H. (1986). An efficient method for the determination of K_m values for HCO_3 of phosphoenolpyruvate carboxylase. *Planta* **169,** 356–360.

Berry, J. A. and Farquhar, G. D. (1978). The CO_2 concentration function of C_4 photosynthesis: a biochemical model. *In* "Proceedings of the 4th International Congress on Photosynthesis" (D. Hall, J. Coombs and T. Goodwin, eds.) Biochemical Society, London, pp. 119–131.

Brown, R. H. and Byrd, G. T. (1993). Estimation of bundle sheath cell conductance in C_4 species and O_2 insensitivity of photosynthesis. *Plant Physiol* **103**, 1183–1188

Burnell, J. N. and Hatch, M. D. (1988). Photosynthesis in phosphoenolpyruvate carboxykinase-type C_4 plants: Pathways of C_4 acid decarboxylation in bundle sheath cells of *Urochloa panicoides*. *Arch. Biochem. Biophys.* **260**, 187–199.

Carnal, N. W., Agostino, A. and Hatch, M. D. (1993). Photosynthesis in phosphoenolpyruvate carboxykinase-type C_4 plants: Mechanism and regulation of C_4 acid decarboxylation in bundle sheath cells. *Arch. Biochem. Biophys.* **306**, 360–367.

Chapman, K. S. R., Berry, J. A. and Hatch, M. D. (1980). Photosynthetic metabolism in bundle sheath cells of the C_4 species *Zea mays*: Sources of ATP and NADPH and the contribution of Photosystem II. *Arch. Biochem. Biophys.* **202**, 330–341.

Collatz, G. J., Ribas-Carbo, M. and Berry, J. A. (1992). Coupled photosynthesis–stomatal model for leaves of C_4 plants. *Aust. J. Plant. Physiol.* **19**, 519–538.

Dai, Z., Edwards, G. E. and Ku, M. S. B. (1993). C_4 photosynthesis. The CO_2 concentrating mechanism and photorespiration. *Plant Physiol.* **103**, 83–90.

Dai, Z., Ku, M. S. B. and Edwards, G. E. (1995). C_4 photosynthesis: The effect of leaf development on the CO_2-concentrating mechanism and photorespiration in maize. *Plant Physiol.* **107**, 815–825.

Dever, L. V., Blackwell, R. D., Fullwood, N. J., Lacuesta, M., Leegood, R. C., Onek, L. A., Pearson, M. and Lea, P. J. (1995). The isolation and characterisation of mutants of the C_4 phtosynthetic pathway. *J. Exp. Bot.* **46**, 1363–1376.

Dever, L. V. (1997). Control of photosynthesis in *Amaranthus edulis* mutants with reduced amounts of PEP carboxylase. *Aust. J. Plant Physiol.* **24**, 469–476.

Doncaster, H. D. and Leegood, R. C. (1987). Regulation of phosphoenolpyruvate carboxylase activity in maize leaves. *Plant Physiol.* **84**, 82–87.

Edwards, G. E. and Baker, N. R. (1993). Can assimilation in maize leaves be predicted accurately from chlorophyll fluorescence analysis? *Photosyn. Res.* **37**, 89–102.

Ehleringer, J. and Pearcy, R. W. (1983). Variation in quantum yields for CO_2 uptake among C_3 and C_4 plants. *Plant Physiol.* **73**, 555–559.

Evans, J. R. (1987). The dependence of of quantum yield on wavelength and growth irradiance. *Aust. J. Plant Physiol.* **14**, 69–79.

Evans, J. R., Sharkey, T. D., Berry, J. A. and Farquhar, G. D. (1986). Carbon isotope discrimination measured concurrently with gas exchange to investigate CO_2 diffusion in leaves of higher plants. *Aust. J. Plant Physiol.* **13**, 281–292.

Evans, J. R., von Caemmerer, S., Setchell, B. A. and Hudson, G. S. (1994). The relationship between CO_2 transfer conductance and leaf anatomy in transgenic tobacco with reduced content of Rubisco. *Aust. J. Plant Physiol.* **21**, 475–495.

Evans, J.R. and von Caemmerer, S. (1996). Carbon dioxide diffusion inside leaves. *Plant Physiol.* **110**, 339–346.

Farquhar, G. D. (1983). On the nature of carbon isotope discrimination in C_4 species. *Aust. J. Plant Physiol.* **10**, 205–226.

Farquhar, G. D., von Caemmerer, S. and Berry, J. A. (1980). A biochemical model of photosynthetic CO_2 assimilation in leaves of C_3 species. *Planta* **149**, 78–90.

Farquhar, G. D. and von Caemmerer, S. (1982). Modelling of photosynthetic responses to environmental conditions. *In* "Physiological Plant Ecology II. Water Relations and Carbon Assimilation" (O. L. Lange, P. S. Nobel, C. B. Osmon and H. Ziegler, eds.) *Encyclopedia of Plant Physiology* (New Ser.), Vol. 12B, Springer Verlag, Berlin, pp. 549–587.

Furbank, R. T. and Hatch, M. D. (1987). Mechanism of C_4 photosynthesis. The size and composition of the inorganic carbon pool in bundle-sheath cells. *Plant Physiol.* **85**, 958–964.

Furbank, R. T. and Hatch, M. D. (1989). CO_2 concentrating mechanism of C_4 photosynthesis. Permeability of isolated bundle-sheath cells to inorganic carbon. *Plant Physiol.* **91,** 1364–1371.
Furbank, R. T., Jenkins, C. L. D. and Hatch, M. D. (1990). C_4 photosynthesis: Quantum requirements, C_4 acid overcycling and Q-cycle involvement. *Aust. J. Plant Physiol.* **17,** 1–7.
Furbank, R. T., Chitty, J. A., von Caemmerer, S. and Jenkins, C. L. D. (1996). Antisense RNA inhibition of RbcS gene expression in the C_4 plant *Flaveria bidentis. Plant Physiol.* **111,** 725–734.
Genty, B., Briantais, J. M. and Baker, N. (1989). The relationship between the quantum yield of photosynthetic electron transport and quenching of chlorophyll fluorescence. *Biochem. Biophys. Acta* **990,** 87–92.
Hatch, M. D. (1987). C_4 photosynthesis a unique blend of modified biochemistry, anatomy and ultra structure. *Biochim. Biophys. Acta* **895,** 81–106.
Hatch, M. D. (1992). The making of the C_4 pathway. *In* "Research in Photosynthesis" Vol. III, (N. Murata, ed.), pp. 747–756.
Hatch, M. D. and Burnell, J. N. (1990). Carbonic anhydrase activity in leaves and its role in the first step of C_4 photosynthesis. *Plant Physiol.* **93,** 825–828.
Hatch, M. D. and Osmond, C. B. (1976). Compartmentation and transport in C_4 photosynthesis. *In* "Transport in Plants III. Intracellular Interactions and Transport Processes.(C. R. Stocking and U. Heber, eds.) *Encyclopedia of Plant Physiology New Series,* Vol 3, Springer-Verlag, Berlin, pp. 144–184.
Hatch, M. D., Agostino, A. and Jenkins, C. L. D. (1995). Measurements of leakage of CO_2 from bundle-sheath cells of leaves during C_4 photosynthesis. *Plant Physiol.* **108,** 173–181.
Hattersley, P. W. and Browning, A. J. (1981). Occurrence of the suberized lamella in leaves of grasses of different photosynthetic types. I. In parenchymatous bundle sheath and PCR ('Kranz') sheaths. *Protoplasma* **109,** 333–371.
Hattersley, P. W. (1982). δ^{13} C values of C_4 types in grasses. *Aust. J. Plant Physiol.* **9,** 139–154.
He, D. and Edwards, G. E. (1996). Estimation of diffusive resistance of bundle sheath cells to CO_2 from modelling of C_4 photosynthesis. *Photosyn. Res.* **49,** 195–208.
Henderson, S. A., von Caemmerer, S. and Farquhar, G. D. (1992). Short-term measurements of carbon isotope discrimination in several C_4 species. *Aust. J. Plant Physiol.* **19,** 263–285.
Hudson, G. S., Mahon, J. D., Anderson, P. A., Gibbs, M. J., Badger M. R., Andrews, T. J. and Whitefeld, P. R. (1990). Comparisons of rbcL genes for the large subunit of ribulose bisphosphate carboxylase from closeley related C_3 and C_4 plant species. *J. Biol. Chem.* **265,** 808–814.
Hunt, E. R., Weber, J. A. and Gates, D. M. (1985). Effects of nitrate application on *Amaranthus powellii* wats. III. Optimal allocation of leaf nitrogen for photosynthesis and stomatal conductance. *Plant Physiol.* **79,** 619–624.
Jenkins, C. L. D. (1989). Effects of the phosphoenolpyruvate carboxylase inhibitor 3,3-dichloro-2-(dihydroxyphosphinoylmethyl) propenoate on photosynthesis. C_4 selectivity and studies on C_4 photosynthesis. *Plant Physiol.* **89,** 1231–1237.
Jenkins, C. L. D., Furbank, R. T. and Hatch, M. D. (1989). Inorganic carbon diffusion between C_4 mesophyll and bundle sheath cells. *Plant Physiol.* **91,** 1356–1363.
Jordan, D. B. and Ogren, W. L. (1981). Species variation in the specificity of ribulose carboxylase/oxygenase. *Nature (London)* **291,** 513–515.
Jordan, D. B. and Ogren, W. L. (1983). Species variation in kinetic properties of ribulose 1,5,-bisphosphate carboxylase/oxygenase. *Arch. Biochem. Biophys.* **227,** 425–433.
Krall, J. K. and Edwards, G. E. (1990). Quantum yields of photosytem II electron transport and carbon fixation in C_4 plants. *Aust. J. Plant Physiol.* **17,** 579–588.

Krall, J. K. and Pearcy, R. W., (1993). Concurrent measurements of oxygen and carbon dioxide exchange during lightflecks in maize (*Zea mays* L.). *Plant Physiol.* **103**, 823–828.

Leegood, R. C. and von Caemmerer, S. (1988). The relationship between contents of photosynthetic metabolites and the rate of photosynthetic carbon assimilation in the leaves of *Amaranthus edulis* L. *Planta* **174**, 253–265.

Leegood, R. C. and von Caemmerer, S. (1989). Some relationships between contents of photosynthetic intermediates and the rate of photosynthetic carbon assimilation in leaves of *Zea mays* L. *Planta* **178**, 258–266.

Ludwig, M., von Caemmerer, S., Price G. D., Badger, M. R. and Furbank, R. T. (1998). Expression of tobacco carbonic anhydrase in the C_4 dicot *Flaveria bidentis*. *Plant Physiol.* **117**, 1071–1082.

Maroco, J. P., Ku, M. S. B., Lea, P., Furbank R. T. and Edwards, G. E. (1998). Oxygen requirement and inhibition of C_4 photosynthesis. An analysis on C_4 and C_3 cycle deficient C_4 plants. *Plant Physiol.* **116**, 823–832.

Ögren, E. and Evans, J. R. (1993). Photosynthetic light response curves. I. The influence of CO_2 partial pressure and leaf inversion. *Planta* **189**, 182–190.

Peisker, M. (1979). Conditions for low and oxygen-independent CO_2 compensations concentrations in C_4 plants as derived from a simple model. *Photosynthetica* **13**, 198–207.

Peisker, M. (1986). Models of carbon metabolism in C_3–C_4 intermediate plants as applied to the evolution of C_4 photosynthesis. *Plant Cell Env.* **9**, 627–635.

Peisker, M. (1988). Modelling of C_4 photosynthesis at low quantum flux densities. *In* "International Symposium on Plant Mineral Nutrition and Photosynthesis '87. Vol. II. Photosynthesis" (S. Vaklinova, V. Stanev and M. Dilova, eds.), pp. 226–234. Bulgarian Academy of Sciences, Sofia.

Peisker, M. and Henderson S. A. (1992), Carbon: Terrestrial C_4 plants. (Commissioned review). *Plant Cell Environ.* **15**, 987–1004.

Pfeffer, M. and Peisker, M. (1995). In vivo K_m for CO_2 (K_p) of phospho*enol*pyruvate carboxylase (PEPC) and mesophyll CO_2 resistance (rm) in leaves of *Zea mays* L. *In* "Photosynthesis: From Light to Biosphere," (P. Mathis, ed.) Kluwer Academic Publishers, Netherlands, Vol. V, pp. 547–550.

Raven, J. A. (1977). Ribulose bisphosphate carboxylase activity in terrestrial plants: Significance of O_2 and CO_2 diffusion. *Curr. Adv. Plant. Sci.* **9**, 579–794.

Sage, F. R., Pearcy, W. R. and Seemann, J. R. (1987). The nitrogen use efficiency of C_3 and C_4 plants. III. Leaf nitrogen effects on the activity of carboxylating enzymes in *Chenopodium album* (L.) and *Amaranthus retroflexus* (L.). *Plant Physiol.* **85**, 355–359.

Seemann, J. R., Badger, M. R. and Berry, J. A. (1984). Variations in the specific activity of ribulose-1,5,bisphopshate carboxylase between species utilizing differering photosynthetic pathways. *Plant Physiol.* **74**, 791–794.

Siebke, K., von Caemmerer, S., Badger, M. R. and Furbank, R. T. (1997). Expressing an RbcS antisense gene in transgenic *Flaveria bidentis* leads to an increased quantum requirement per CO_2 fixed in Photosystem I and II. *Plant Physiol.* **115**, 1163–1174.

Trevanion, S. J., Furbank, R. T. and Ashton, A. R. (1998). NADP malate dehydrogenase in the C_4 plant *F. bidentis*: Cosense suppression of activity in mesophyll and bundle sheath cells and consequences for photosynthesis. *Plant Physiol.* **113**, 1153–1165.

Usuda, H. (1984). Variations in the photosynthesis rate and activity of photosynthetic enzymes in maize leaf tissue of different ages. *Plant Cell Physiol.* **25**, 1297–1301.

von Caemmerer, S. and Farquhar, G. D. (1981). Some relationships between the biochemistry of photosynthesis and the gas exchange of leaves. *Planta* **153**, 376–387.

von Caemmerer, S. and Evans, J. R. (1991). Determination of the average partial pressure of CO_2 in chloroplasts from leaves of several C_3 plants. *Aust. J. Plant Physiol.* **18**, 287–305.

von Caemmerer, S., Evans, J. R., Hudson G. S. and Andrews, T. J. (1994). The kinetics of ribulose-1,5-bisphosphate carboxylase/oxygenase *in vivo* inferred from measurements of photosynthesis in leaves of transgenic tobacco. *Planta* **195**, 88–97.

von Caemmerer, S., Millgate, A., Farquhar, G. D. and Furbank, R. T. (1997). Reduction of Rubisco by antisense RNA in the C_4 plant *Flaveria bidentis*, leads to reduced assimilation rates and increased carbon isotope discrimination. *Plant Physiol.* **113**, 469–477.

Wessinger, M. E., Edwards G. E. and Ku, M. S. B. (1989). Quantity and kinetic properties of ribulose-1,5-bisphosphate carboxylase in C_3, C_4, and C_3–C_4 intermediate species of *Flaveria* (Asteraceae). *Plant Cell Physiol.* **30**, 665–671.

Wong, S.-C., Cowan, I. R. and Farquhar, G. D. (1985). Leaf conductance in relation to rate of CO_2 assimilation. I. Influence of nitrogen nutrition, phosphorus nutrition, photon flux densitiy, and ambient partial pressure of CO_2 ontogeny. *Plant Physiol.* **78**, 821–825.

Yeoh, H.-H., Badger, M. R. and Watson, L. (1980). Variations in $k_m(CO_2)$ of ribulose-1,5-bisphosphate carboxylase among grasses. *Plant Physiol.* **66**, 1110–1112.

Yeoh, H.-H., Badger, M. R. and Watson, L. (1981). Variations in kinetic properties of ribulose-1,5-bisphosphate carboxylases among plants. *Plant Physiol.* **67**, 1151–1155.

Appendix 1

List of symbols	Definition
A (μmol m^{-2} sec^{-1})	Rate of CO_2 assimilation
A_c (μmol m^{-2} sec^{-1})	Enzyme limited rate of CO_2 assimilation
A_j (μmol m^{-2} sec^{-1})	Electron transport limited rate of CO_2 assimilation
α	Fraction of PSII activity in the bundle sheath
β	mol of ATP produced in mitochondria per mol of NADPH
C_i (μbar)	Intercellular airspace CO_2 partial pressure
C_m (μbar)	Mesophyll CO_2 partial pressure
C_s (μbar)	Bundle-sheath CO_2 partial pressure
D_{O2}	Diffusivity of O_2 in air
D_{CO2}	Diffusivity of CO_2 in air
E_o (μmol m^{-2} sec^{-1})	Net O_2 evolution
ϕ	Leakiness of the bundle sheath
f	Spectral light quality correction factor
g_i (mol m^{-2} sec^{-1})	Conductance for CO_2 diffusion from intercellular airspace to mesophyll cytosol
g_s (mol m^{-2} sec^{-1})	Bundle-sheath conductance to CO_2
g_o (mol m^{-2} sec^{-1})	Bundle-sheath conductance to O_2
Γ_* (μbar)	C_3 compensation point in the abscence of mitochondrial respiration

γ_*	Half of the reciprocal of Rubisco specificity $(0.5/S_{c/o})$
I (μmol quanta m^{-2} sec^{-1})	Incident irradiance
I_2 (μmol quanta m^{-2} sec^{-1})	Photosynthetically active irradiance absorbed by PSII
J (μmol electrons m^{-2} sec^{-1})	Electron transport rate
J_{max} (μmol electrons m^{-2} sec^{-1})	Maximal electron transport rate
J_t (μmol electrons m^{-2} sec^{-1})	Total electron transport rate (to support C_3 and C_4 cycle activity)
J_m (μmol electrons m^{-2} sec^{-1})	Electron transport rate to support C_4 cycle activity
J_s (μmol electrons m^{-2} sec^{-1})	Electron transport rate to support C_3 cycle activity
K_c (μbar)	Michaelis–Menten constant of Rubisco for CO_2
K_o (μbar)	Michaelis–Menten constant of Rubisco for O_2
K_p (μbar)	Michaelis–Menten constant of PEP carboxylase for CO_2
L (μmol m^{-2} sec^{-1})	Leakrate of CO_2 out of the bundle sheath
L_o (μmol m^{-2} sec^{-1})	Leakrate of O_2 out of the bundle sheath
O_m (μbar)	O_2 partial pressure in the mesophyll
O_s (μbar)	O_2 partial pressure in the bundle sheath
R_d (μmol m^{-2} sec^{-1})	Mitochondrial respiration in the light
P/O	Number of P_i molecules used in ADP phosphorylation per oxygen consumed (in effect, the ATP/O ratio)
R_m (μmol m^{-2} sec^{-1})	Mitochondrial respiration in the mesophyll
R_s (μmol m^{-2} sec^{-1})	Mitochondrial respiration in the bundle sheath
r_s (m^{-2} sec^{-1} mol^{-1})	Bundle-sheath resistance $(1/g_s)$
S_{O2}	Solubility coefficient of O_2 in water
S_{CO2}	Solubility coefficient of CO_2 in water
V_c (μmol m^{-2} sec^{-1})	Rubisco carboxylation rate
V_{cmax} (μmol m^{-2} sec^{-1})	Maximal Rubisco carboxylation rate
V_o (μmol m^{-2} sec^{-1})	Rubisco oxygenation rate
V_{omax} (μmol m^{-2} sec^{-1})	Maximal Rubisco oxygenation rate

V_p (μmol m^{-2} sec^{-1})	PEP carboxylation rate
$V_{p\max}$ (μmol m^{-2} sec^{-1})	Maximal PEP carboxylation rate
V_{pr} (μmol m^{-2} sec^{-1})	PEP regeneration rate
x	Partitioning factor of electron transport rate between the C$_4$ and C$_3$ cycle

III

Ecology of C_4 Photosynthesis

7

Environmental Responses

Steve P. Long

I. Introduction

The primary physiological effect of the unique combination of metabolism and leaf anatomy that characterize the C_4 syndrome is the elevation of the CO_2 concentration at the site of Rubisco in the bundle sheath (Hatch, 1987; Kanai and Edwards, Chapter 3; see appendix for list of abbreviations). This elevated concentration has two effects. First, it causes competitive inhibition of the oxygenase reaction of Rubisco, eliminating most of the PCO pathway activity and associated energy expenditure that underlies photorespiration. Second, it allows Rubisco to approach its maximum rate of catalysis despite its low affinity for CO_2. This chapter explains why these two effects underlie the potentially greater efficiencies of light, water, and nitrogen use at the leaf level of C_4 species relative to C_3 species. In the following sections, the theoretical implications of the C_4 mechanism for the responses of photosynthesis to the environmental variables of light, water, nitrogen, and temperature are explored. These theoretical expectations are then compared to observed plant performance and geographic distributions. The first three sections consider the process in the warm environments where C_4 species are most abundant, but they also explore some of the apparent anomalies between C_4 distributions and theoretical expectations, in particular their dominance of some tropical wetland communities. The final section examines their rarity in cool temperate and cold climates, and the question of whether this reflects an inherent disadvantage of C_4 photosynthesis at low temperatures.

II. Light

A. Theory

One of the first physiological features noted of C_4 plants, following the discovery of their unique biochemistry in 1965, was their high rate of photosynthesis at full sunlight under tropical conditions (Black, 1971; Hatch, 1992a). Photosynthesis in C_3 species at light saturation is often colimited by the quantity of Rubisco and by the capacity for regeneration of the acceptor RuBP molecule (Evans and Farquhar, 1991). The maximum rate of CO_2 assimilation that both of these limitations can support is lowered by oxygenation of RuBP and the resulting photorespiratory evolution of CO_2 (Farquhar et al., 1980; Long, 1994). Photorespiration is estimated to decrease net carbon gain in C_3 species by 20% to 50% (Zelitch, 1973; Bainbridge et al., 1995). C_4 species avoid most of this loss by concentrating CO_2 at the site of Rubisco in the bundle sheath (Hatch, 1987). This allows C_4 species to attain potentially higher rates of photosynthesis in full sunlight. Although additional energy is required to assimilate CO_2 via the C_4 dicarboxylate cycle, this can become irrelevant at light-saturation, because, by definition, absorbed light energy is in excess of requirements.

In dim light, when photosynthesis is linearly dependent on the photon flux, the rate of CO_2 assimilation depends entirely on the energy requirements of carbon assimilation (Long et al., 1993). The additional ATP required for assimilation of one CO_2 in C_4 photosynthesis, compared to C_3 photosynthesis, increases the photon requirement in C_4 plants (Hatch, 1987; von Caemmerer and Furbank, Chapter 6). However, when the temperature of a C_3 leaf exceeds approximately 25°C, the amount of light energy diverted into photorespiratory metabolism in C_3 photosynthesis exceeds the additional energy required for CO_2 assimilation in C_4 photosynthesis (Hatch, 1992a).

C_4 species usually show no or minimal light-dependent CO_2 loss. The low CO_2 compensation point (Γ) of photosynthesis (<1 Pa) is clear evidence of this. Despite an apparent absence of photorespiration, C_4 species possess the mechanism for PCO metabolism and, as a result, show some oxygen inhibition of photosynthesis; thus some energy is still lost in this pathway (Dai et al., 1993; von Caemmerer and Furbank, Chapter 6). Leakage of CO_2 from the bundle sheath to the mesophyll also decreases the efficiency of light use (Hatch et al., 1995).

The theoretical quantum efficiency of C_4 photosynthesis, regardless of photosynthesis type, in the absence of leakage is calculated to be 0.070 when the operation of a partial Q cycle during photosynthetic electron transport is assumed (Furbank et al., 1990). For nine C_4 grasses, three of each C_4 subtype, the average leakage rate was 11%. There was little variability in leakage rate between the different C_4 subtypes (Hatch et al., 1995). An 11% leakage rate would lower the maximum quantum efficiency from the theoretical 0.070 to 0.063. This is in close agreement to the measured range of 0.060 to 0.065 for C_4 grasses (Table I) (Pearcy and Ehleringer, 1984). This compares to an average maximum of 0.093 for C_3 species, measured in a low pO_2 to eliminate photorespiration (Long et al., 1993). In normal air, photorespiration lowers the maximum quantum yield of C_3 photosynthesis to about 0.054 at 30°C (Table I) (Ehleringer and Björkman, 1977). Thus, leaf photosynthesis in C_4 species at warm temperatures is more efficient in both light-limiting and light-saturating conditions. Does this greater efficiency at the leaf level translate into higher efficiency of light use at the plant and stand level?

B. Higher Efficiency of Light Use at the Stand Level?

Assuming a similar partitioning of absorbed light energy between photochemical and nonphotochemical processes in C_4 species by comparison to C_3, the potential efficiency of light use in photosynthesis at the leaf level will be higher in C_4 species at all light levels when leaf temperature is 25°C or higher (Ehleringer and Björkman, 1977; Hatch, 1992b). It therefore follows that the efficiency of light use of a C_4 leaf canopy has to be higher than that for a C_3 canopy of identical size and architecture at these temperatures. However, translation of this higher canopy photosynthetic efficiency into a higher productivity and a higher efficiency of biomass accumulation depends on both the rate of respiration per unit mass and the resources invested in photosynthetic tissue versus heterotrophic stem and root tissue.

C_4 species are often suggested to partition a lower fraction of total dry mass into leaves than in C_3 species of similar growth form (Slatyer, 1970; Tilman and Wedin, 1991). When related C_3 and C_4 species were compared, leaf longevity was shorter in the C_4 partner (Slatyer, 1970; Kalapos et al.,

Table I Maximum Quantum Yields for CO_2 Uptake in Different Photosynthetic Types of Dicots and Grass[a]

Photosynthetic type	Dicots	Grasses
C_3	0.052 ± 0.001 (14)	0.053 ± 0.001 (9)
C_4 NAD-ME	0.053 ± 0.001 (9)	0.060 ± 0.001 (3)
C_4 NADP-ME	0.061 ± 0.001 (6)	0.065 ± 0.001 (8)
C_4 PCK	—	0.064 ± 0.001 (5)

[a] Quantum yields are calculated on the basis of the quanta absorbed by the leaves being measured. Values are means (± 1 sd) for different species (number in parenthesis) measured at 30°C and in normal air. [Data from: Pearcy, R. W., and Ehleringer, J. (1984). Comparative ecophysiology of C_3 and C_4 plants. Plant Cell Environ. **7**, 1–13.]

1996). These factors all tend to offset the higher energy use efficiency at the leaf level, with the result that growth rates are not automatically higher when C_4 species or stands are compared with C_3 species or stands in the same warm environment. Indeed, Gifford (1974) and Evans (1975) both noted a poor correlation between the high leaf photosynthetic rates of C_4 species and high rates of plant production. However, Monteith (1978) suggested that the key point of comparison of the performance of C_4 versus C_3 canopies in the field is the efficiency with which each stand converts its intercepted radiation into biomass. This approach separates variation in canopy size, architecture, and duration from the effects of differences in photosynthetic capacity. On the basis of dry matter formed per unit of total solar radiation intercepted for a range of healthy crops growing under optimum conditions, the C_3 species attained a maximum of 1.4 g J^{-1} for C_3 compared to 2.0 g J^{-1} for the C_4 species (Monteith, 1978). Assuming that photosynthetically active radiation is 50% of the total and that the average energy content in plant dry mass is 17 MJ kg^{-1}, this equates to efficiencies of conversion of photosynthetically active radiation into biomass (ε_c) of 0.070 (C_4) and 0.049 (C_3). Subsequently, Snaydon (1991) again highlighted the poor correlation between the high maximum leaf photosynthetic rates of C_4 species and annual biomass production. In criticizing the assumption that C_4 species are more efficient, Snaydon (1991) observed that much of a plant canopy would never receive full sunlight. However, this ignores the fact that at temperatures of approximately 25°C and higher, C_4 photosynthesis increases photosynthetic rate and efficiency regardless of light level (Hatch, 1992a). This may explain why the only seven species in the survey of Snaydon (1991) to show annual productivities above 60 t (dry matter) ha^{-1} were all C_4.

C. Does the Higher Production Potential Have Ecological Significance?

C_4 species are most abundant in the semiarid tropics and subtropics, in particular savanna and subtropical grasslands (Cerling et al., 1993; Sage, Wedin, and Li, Chapter 10). Whereas many C_4 savanna grasses achieve exceptional light conversion efficiencies when irrigated and fertilized (Monteith, 1978), in natural ecosystems, moisture and nutrient limitations typically prevent the native C_4 species from achieving such high productivity (Fig. 1) (Jones et al., 1992). Codominance of these habitats by C_4 species often results more from their high water-use efficiency (WUE) and nitrogen-use efficiency (NUE) rather than high production potential (Tieszen et al., 1979; Jones et al., 1992). C_4 species are also common early in succession following forest clearance and in areas where the nutrient status of the soil is so poor that secondary succession is arrested (Kamnalrut and Evenson, 1992). Tropical sites that are moist and fertile enough to allow C_4 plants to achieve high efficiency of light conversion and thus approach

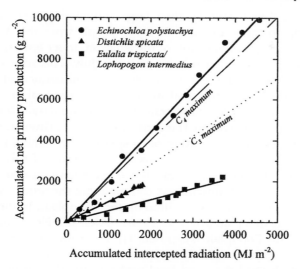

Figure 1 Accumulated monthly mean values of net primary production plotted against the accumulated quantity of total solar radiation (all wavelengths) intercepted by the same communities. Solid lines are the best-fitting straight lines to the illustrated data points. Data are for: (●) monotypic stands of *Echinochloa polystachya* on the Amazon floodplain near Manaus, Brazil (Piedade *et al.*, 1991); (▲) monotypic stands of *Distichlis spicata* on saline grassland close to Mexico City (Jones *et al.*, 1992); and (■) mixed C_4 grass stands codominated by *Eulalia trispicata* and *Lophopogon intermedius* in moist savanna near Hat Yai, Thailand (Kamnalrut and Evenson, 1992). The broken lines indicate the maxima suggested for the two photosynthetic types (after Monteith, 1978). Data redrawn from Piedade *et al.* (1991) and Jones *et al.* (1992).

their theoretical productive potentials are generally dominated by forests—so long as humans have not intervened. Whereas some C_4 species can survive in forest shade (Robichaux and Pearcy, 1980; Winter *et al.*, 1982; Ohwi, 1984), the dominant vegetation of tropical forest ecosystems is exclusively C_3, with one important exception—nutrient-rich tropical wetlands. This appears one case where high light conversion efficiency, rather than inherently high WUE and NUE of C_4 species, plays a significant role in their dominance of the habitat.

1. Emergent Vegetation on Floodplains of Equatorial River Systems The most productive emergent macrophytes of tropical rivers and lake margins are C_4 species such as *Cyperus papyrus* and *Echinochloa polystachya*. Both form extensive monotypic stands occupying many thousands of square kilometers of tropical swamps and the margins of lakes and rivers (Jones, 1986; Junk, 1993). *Echinochloa polystachya* together with the C_4 grasses *Paspalum dilatum* and *Paspalum repens* may occupy much of the 200,000 km² of the nutrient rich Várzea floodplains of the Amazon region (Junk, 1993).

The Amazon and its tributaries show large annual amplitudes of water level responding to the seasonality of water supply from the Andes. At Manaus, near the geographic center of the Amazon, water level rises 16 m from a minimum in November to a maximum in July. This annual "tide" results in a large floodplain along the many rivers and connected lakes of the Amazon; about 10% of the total surface area of the region (Piedade *et al.*, 1994). The higher areas are occupied by C_3 inundation forest, but the lower areas are C_4 grassland (Piedade *et al.*, 1991). Given the strong correlations between the distribution of C_4 species with semiarid habitats and their high NUE, their almost exclusive domination of this wet and nutrient rich habitat appears an anomaly. Of the emergent macrophytes (species that maintain a photosynthetic canopy above the water's surface), *E. polystachya* occupies the lowest position on the floodplain. In the central Amazon region, the new plants of *E. polystachya* develop in November, when this lower level of the floodplain is exposed. The rising waters of the Amazon cover the floodplain in December or January, and rise by approximately 2–3 m each month until the water level peaks in July. The sediment rich waters are too turbid to support any photosynthesis of submerged leaves. To survive, the shoots of *E. polystachya* grow upward at a rate sufficient to maintain a LAI of two to three above the water surface (Piedade *et al.*, 1991). As older leaves become submerged, new leaves are formed above, so that the canopy size remains sufficient to intercept some 90% of the available light. Over the 10 months of flooding, these stands accumulate a remarkable 100 t ha^{-1} of dry matter, the highest annual productivity recorded for any vegetation (Piedade *et al.*, 1991). Why is this area vegetated exclusively by C_4 species?

One explanation, suggested by Piedade *et al.* (1991), is that the remarkable rate of growth required for the plant to keep pace with the rising water level is only possible with the higher maximum ε_c conferred by C_4 photosynthesis. This suggestion is supported by the observation that the average dry matter gain per unit of intercepted light averaged over the year is 2.3 g J^{-1} (Fig. 1). This is comparable to the highest values for conversion reported for other C_4 species and more than 40% higher than the recorded maximum for C_3 species (Piedade *et al.*, 1991). The high conversion of intercepted radiation into dry matter in *E. polystachya* is also reflected in high rates of leaf photosynthesis (35 μmol m^{-2} sec^{-1}) (Piedade *et al.*, 1994). The stem elongation necessary to maintain a canopy above the rising water surface could of course be achieved without additional biomass; however, the stems must also be sufficiently robust, and therefore expensive to make, to resist the flow of the floodwater that would otherwise drag the shoot and its leaves into the turbid water. *Echinochloa polystachya* is an important example where the higher potential ε_c and potential productivity of C_4 photosynthesis has a direct and essential influence on the distribution of the species and its ability to colonize this niche.

2. Temperate Coastal Saltmarshes in Northwestern Europe Another example where high ε_c may allow a species to colonize a difficult habitat may be coastal mudflats. The genus *Spartina* is exclusively C_4 and include the most frequent primary colonizers of coastal mudflats, leading to the development of saltmarsh communities (Long and Woolhouse, 1978; Long and Mason, 1983). *Spartina anglica* arose late in the nineteenth century from the hybridization of the European species *S. maritima* with *S. alterniflora* introduced from North America to the south coast of England (Chapman, 1996). Since that time, with the aid of deliberate introductions, this new C_4 species has invaded thousands of hectares of coastal mudflats in estuaries around the British Isles and Western Europe from Spain to Denmark (Long and Woolhouse, 1978). Although, the mature coastal saltmarshes of Western Europe are dominated by C_3 species, *S. anglica* is now the common primary colonizer, and its rapid spread throughout the twentieth century has allowed considerable expansion of saltmarsh area (Long *et al.*, 1975; Long and Woolhouse, 1978). Why is this species able to colonize mudflats that other saltmarsh species do not? Colonization of mudflat by seedlings requires a rapid rate of growth and development of a root and/or rhizome system in the summer that is sufficient to prevent uprooting by winter storms (Long and Mason, 1983). Relative to C_3 species of similar form, C_4 photosynthesis allows young plants, in a limited period of time, to accumulate more mass or to achieve the same mass but with higher root to shoot ratios. For example, on eastern English mudflats, *S. anglica* shows similar net primary productivity as the C_3 grass *Puccinellia maritima* from the adjacent saltmarsh. However, *S. anglica* invests a significantly greater portion of its biomass belowground, thereby establishing a more robust root system that effectively anchors the plant into the sediment (Long and Mason, 1983).

III. Nitrogen

In the context of this chapter, two terms are used to describe productivity per unit of nitrogen resource. Photosynthetic nitrogen use efficiency (PNUE) is the net rate of leaf CO_2 uptake in full sunlight per unit leaf nitrogen content. Nitrogen use efficiency (NUE) is the ratio of increase in plant biomass to increase in plant nitrogen, which in the context of annual plants and crops approximates the ratio of biomass to N at completion of growth.

A. Theory of Higher Nitrogen Use Efficiency

As a direct result of CO_2 concentration at the site of Rubisco in C_4 species, the theoretical requirement for nitrogen in photosynthesis is lower than in C_3 species. Rubisco can account for up to 30% of total leaf nitrogen in C_3 species (Evans, 1989; Bainbridge et al., 1995). At the current atmospheric pCO_2, the efficiency of carboxylation via Rubisco is depressed in C_3 photosynthesis because the enzyme is not CO_2 saturated and because O_2 competes with CO_2 for RuBP (Long, 1991). Furbank and Hatch (1987) calculated a CO_2 concentration at Rubisco of C_4 species of 10–100 times that found in C_3 species. At these CO_2 concentrations and 30°C, following the calculations of Long (1991), a C_4 leaf would require 13.4% to 19.8% of the amount of Rubisco in a C_3 leaf to achieve the same A_{sat}. This is in close agreement with the 3- to 6-fold lower Rubisco concentrations found in leaves of C_4 species compared to C_3 (Ku et al., 1979; Sage et al., 1987). However, Dai et al. (1993) calculated a smaller elevation of CO_2 concentration in the bundle sheath to about three times the concentration in the mesophyll of C_3 species. At this concentration, the leaf of a C_4 species would require 42% of the Rubisco of a C_3 leaf to achieve the same photosynthetic rate, suggesting a greater requirement for Rubisco in C_4 leaves than is the amount usually observed. Because the specificity of Rubisco for CO_2 increases as temperature decreases, this "benefit" of C_4 photosynthesis declines with falling temperature. Assuming a concentration within the bundle sheath of C_4 species of 10 times the concentration within the mesophyll of C_3 species, then at 30°C the same A_{sat} could be supported by 19.8% of the amount of Rubisco in a C_3 leaf. Following the calculations of Long (1991), this requirement would rise to 39.7% at 20°C and 53% at 15°C. The theoretical temperature dependence of decreased Rubisco requirement may explain variation in Rubisco contents between C_4 species. Rubisco constitutes only 4% to 8% of soluble leaf protein in the summer active C_4 species *Tidestromia oblongifolia*, which grows at approximately 45°C in Death Valley, but constitutes 19% to 21% of soluble leaf protein in *Atriplex* spp. from cool coastal sites (Osmond et al., 1982). Transgenic reduction in the amount of Rubisco within C_4 leaves by 60% results in decreased photosynthetic capacity of 33% (von Caemmerer et al., 1997). Sage et al., (1987) found that the CO_2-saturated activities of Rubisco extracted from the dicotyledonous weeds *Chenopodium album* (C_3) and *Amaranthus retroflexus* (C_4) were just sufficient to support observed in vivo rates of photosynthesis in these species. The C_4 leaf, however, required 60% to 75% less Rubisco to match the photosynthetic capacity of the C_3 leaf.

By the reasoning discussed previously, it could be expected that lower Rubisco content would incur a penalty at low temperatures. As temperature declines, the amount of Rubisco required to support C_4 photosynthesis approaches that of C_3 photosynthesis. If the requirement for more Rubisco in C_4 plants at low temperature were not met, increased leakage from the bundle sheath would result. This could be detected in plants grown at low temperatures by changes in the carbon isotope ratio (Hattersley, 1983; von

Caemmerer et al., 1997). However, Jackson et al. (1986) found no evidence of any increase in ^{13}C discrimination in *S. anglica* growing on a saltmarsh in eastern England with average temperatures of $\leq 16°C$. Troughton and Card (1975) similarly found no difference in ^{13}C discrimination in five C_4 species regardless of whether they were grown at 14° or 40°C, and Kebede et al. (1989) observed a small decrease in ^{13}C discrimination in *Eragrostis tef* with decrease in temperature. These results indicate that the increased requirement for Rubisco at low temperature is met in C_4 species able to acclimate to low temperature. This may also explain in part the observed decline of NUE of C_4 species, relative to C_3, with decrease in growth temperature (Wilson and Ford, 1971; Long, 1983).

The large quantities of Rubisco in C_3 leaves reflect the low catalytic capacity of this enzyme and its low affinity for CO_2 in the presence of O_2 (Bainbridge et al., 1995). Rubisco from C_3 plants will catalyze 2 to 4 carboxylations per second per active site ($k_{cat} = 2-4\ sec^{-1}$) (Bainbridge et al., 1995). In comparing Rubisco specificity (τ) and catalytic turnover rate (k_{cat}) from different photosynthetic organisms, a strong negative correlation is observed, suggesting that high values of τ in terrestrial C_3 plants and red algae have been achieved at the expense of k_{cat} (Bainbridge et al., 1995). Because Rubisco in C_4 species functions in a higher CO_2 concentration, where specificity is less important, it might be expected that evolution could have favored the emergence of C_4 forms of Rubisco with lower specificity and higher k_{cat}, so that less enzyme would be required. Rubico from C_4 plants generally exhibits higher k_{cat} than C_3 Rubisco, and this contributes to higher C_4 PNUE (Seemann et al., 1984; Sage and Seemann, 1993). Unequivocal evidence of lower specificities of Rubisco from C_4 species is lacking, however. For example, τ for *Zea mays* appears about 10% less than the average for C_3 angiosperms (Kane et al., 1994; Kent and Tomany, 1995), but this difference is close to the apparent resolution of measurement of τ (Uemura et al., 1996).

The benefit of a lower requirement for Rubisco in C_4 leaves is partially offset by the N requirement for the enzymes of the photosynthetic C_4 dicarboxylate cycle, primarily PEP carboxylase (PEPCase). However, because of its 10-fold higher k_{cat}, much smaller concentrations of PEPCase are required. For *C. album* (C_3) and *A. retroflexus* (C_4) growing in the same field, Rubisco accounted for 29% of total leaf N in the C_3 species, but only 9% in the C_4 species, with PEPc accounting for a further 5% of total N. Thus, PNUE was still significantly higher in C_4 species despite increased investment in PEPCase (Sage et al., 1987).

B. Leaf and Plant Nitrogen Use Efficiencies

Maximum leaf nitrogen concentrations in well-fertilized plants are 120–180 mmol m^{-2} for C_4 species, which is, on average, 65% less than the range of 200–260 mmol m^{-2} for C_3 species (Ehleringer and Monson, 1993). When combined with the higher leaf photosynthetic rates of C_4 species, this results in a PNUE that is about twice that of C_3 species (Brown, 1978; Wong, 1979; Osmond *et al.*, 1982; Hatch, 1987; Sage and Pearcy, 1987b; Wong and Osmond, 1991; Li, 1993). In well-fertilized crops this appears to result in substantial increases in NUE. The NUE from five studies of C_4 crops at harvest ranged from 66 to 130 kg dry mass kg^{-1} N, with an average of 106. This compares to a range of 53–81 with an average of 63 for C_3 crops (calculated from: Hanway, 1962; Lidgate, 1984; Karlen *et al.*, 1987a,b; Wild and Jones, 1988; Squire, 1990). However, end-of-season NUE for C_4 perennial grasses in natural habitats or grown without fertilizers can be considerably higher. *Miscanthus sinensis* in a grassland system in Japan attained an NUE of 110, *E. polystachya* on the Amazon floodplain 200, and the temperate grassland species *Spartina pectinata* 250 (calculated from: Shoji *et al.*, 1990; Piedade *et al.*, 1991, 1997; Beale and Long, 1997).

Hocking and Meyer (1991) analyzed the growth of *Z. mays* (C_4) and *Triticum aestivum* (C_3) at five N-supply levels, and at current ambient and an elevated pCO_2. At ambient pCO_2, NUE was always higher in *Z. mays*, ranging from 28% higher at the lowest rate of N-supply, where both crops were nitrogen deficient, to 75% at the highest N-supply, where N was assumed to be nonlimiting. Relative nitrogen accumulation rates in the two crops were similar, but the C_4 crop invested a significantly larger proportion of its N into the roots relative to the shoots. The difference in NUE between the photosynthetic types was eliminated when *T. aestivum* was grown at a pCO_2 of 150 Pa. This could be explained by the elimination of photorespiration in the C_3 species at this high pCO_2 resulting in increased PNUE (Polley *et al.*, 1995; Drake *et al.*, 1997). These results agree well with the predicted higher NUE of C_4 species at the plant level, excepting that the gain in NUE was least at the lowest N-supply. Other comparisons of pairs of C_3 and C_4 species of similar growth form have also found a higher NUE in the C_4 species at high rates of N-supply, but a lower NUE when N-supply is deficient (Sage and Pearcy, 1987a; Wong and Osmond, 1991).

In natural environments, nitrogen is commonly a limiting resource, and it has been suggested that the higher PNUE of C_4 plants may confer an ecological advantage in situations of N limitation (Brown, 1978). The results that showed reduced or reversed NUE differences between C_3 and C_4 crops and weeds in N-deficient conditions do not appear to support this hypothesis. When species are considered in their natural habitats, however, it becomes clear that C_4 plants exploit their PNUE advantage to enhance ecological performance relative to C_3 plants.

Exploitation of higher NUE does not simply mean improved performance under N deficiency, because a higher NUE may enable a plant to allocate more N to tissues involved in acquisition of the resources that are most limiting to plant growth (Ehleringer and Monson, 1993). For example, two contrasting strategies can be hypothesized for how C_4 species might exploit a 60% higher PNUE than C_3 species. First, they may produce two-thirds more leaf area with the same amount of N resource and thereby have higher whole-plant photosynthetic rates and a greater ability to compete for light. Second, they may produce the same leaf area as a C_3 plant, and partition the 40% of the N saved into roots, favoring the C_4 species in competition for limiting soil resources, including the remaining soil N.

In plants adapted to fertile soils (crops, weeds, and tropical wetland plants such as *E. polystachya*), the former strategy appears to predominate (Piedade *et al.*, 1991). In such environments, N is commonly limiting the rate of leaf area development, and when N becomes deficient, production is limited more by the loss of leaf area than decline in photosynthetic rate per unit leaf area (Scott *et al.*, 1994). For example, the weedy C_4 dicot *A. retroflexus* appears to exploit its higher PNUE by investing the saved N into a more rapid rate of leaf production than the ecologically similar C_3 annual *C. album* (Sage and Pearcy, 1987b). This enhances competitive performance of *A. retroflexus* relative to *C. album* in mixed stands.

Under strong N deficiency, however, the PNUE advantage of C_4 relative to C_3 species adapted to eutrophied habitats can decline and may even be reversed. In *A. retroflexus* (C_4) relative to *C. album* (C_3) (Sage and Pearcy, 1987b) and *E. frumentacea* (C_4) relative to *T. aestivum* (C_3) (Wong and Osmond, 1991), PNUE in the C_4 plant was inferior at the lowest N-supply. How can C_4 photosynthesis cause a lower PNUE at low N in these species? In both studies, photosynthesis in the N-deficient C_4 plants became increasingly responsive to elevation of pCO_2 above ambient, suggesting a partial breakdown of the CO_2 concentrating mechanism (Wong, 1979; Sage and Pearcy, 1987b). Apparently, leaf area production continued at inadequate N, leading to an insufficient N supply for the development of a competent C_4 photosynthetic apparatus. Peisker and Henderson (1992) observed similar results in N-deficient *Panicum* spp. and suggested that partial failure of the C_4 mechanism at low N may result from partitioning of the limited resource to the bundle sheath so that the development of mesophyll was impaired. In an ecological setting, however, this loss of photosynthetic integrity and with it, any advantage in nitrogen use that occurs at low N, is of little consequence because these species are not competitive on infertile soils for a wide range of reasons. At high N, where these species are most competitive, the superior PNUE of the C_4 species is effectively exploited to enhance capture of light, which is typically a highly limiting resource. To address whether PNUE differences are exploited to enhance C_4 performance on N-deficient soils, it is best to compare species adapted to soil infertility. In such species, decreased partitioning of biomass into leaves in C_4 species in comparison to C_3 species of similar growth form (e.g., Slatyer,

1970; Wedin and Tilman, 1993; Kalapos *et al.*, 1996) indicates that evolution has favored the strategy where the N savings are allocated to belowground tissues.

Tilman (1990) examined the characters that would theoretically favor the competitive ability of a species on nitrogen-poor soils. It was concluded that when nitrogen was the limiting nutrient, the most important determinant of competitive ability for nitrogen was the concentration (R^*), to which nitrogen could be decreased by a species in monoculture. This theory predicts that nitrogen deficient habitats should be dominated by the species with the lowest R^* (Tilman and Wedin, 1991; Wedin and Tilman,1993). In a comparison of five perennial prairie grass species in monoculture, the two C_4 grasses *Schizachyrium scoparium* and *Andropogon gerardii* had a significantly lower R^* than the three C_3 species *Agropyron repens, Poa pratensis*, and *Agrostis scabra* (Tilman and Wedin, 1991). This was attributed both to the lower tissue N contents and their ability to partition a higher proportion of biomass into roots, both features that could be explained by C_4 photosynthesis. In pairwise competition experiments between *S. scoparium* and each of the C_3 species, *S. scoparium* either displaced or greatly reduced the biomass of the C_3 competitor. The converse occurred in plots to which nitrogen fertilizer was added. The results are indicative of C_4 photosynthesis conferring an advantage in an N-limiting situation and demonstrate that the C_4 mechanism does not inherently exclude C_4 species from low N environments, as might have been deduced from studies of C_4 crops and weeds at low N.

In summary, C_4 plants show a higher efficiency of N-use at the whole plant and crop level. This difference between C_4 and C_3 species diminishes with decreasing temperature. The higher intrinsic PNUE allows both increased investment of N into roots and development of additional leaf area. In plants adapted to high N, the higher NUE apparently allows for greater leaf area production and light acquisition. However, when N is severely deficient, the NUE of C_4 crops and their weeds may become inferior, possibly because of impairment of the CO_2 concentrating mechanism. For wild perennial grasses of low nitrogen habitats, the C_4 species may tolerate the greater depletion of soil N, allowing them to outcompete C_3 species of the same habitat.

IV. Water

The worldwide abundance of C_4 grasses in a regional flora is more closely correlated with growing season temperature than any other variable (Teeri, 1988). However, their high efficiency of water use is a key determinant of their fitness and distribution. C_4 species form a high proportion of floras in warm and hot semiarid climates with summer or warm season rainfall (Ehleringer and Monson, 1993; Amundson et al., 1994). They constitute almost the entire grass flora of tropical savanna, the major component of herbaceous vegetation in monsoonal climates, and are codominants in much of the natural prairie and steppe vegetation of warm temperate zones (Cerling et al., 1993; Sage, Wedin, and Li, Chapter 10). Many desert therophytes (herbaceous annuals) are C_4. Here the efficiency of C_4 photosynthesis may allow these species to grow rapidly and complete their life cycles in the brief period of good soil moisture availability following heavy rainfall (Mulroy and Rundell, 1977). Shrub floras of cool semideserts also include many C_4 species (Winter, 1981; Cerling et al., 1993; Ehleringer and Monson, 1993). In communities of short-grass prairie, niche separation is apparent with the C_3 grasses active during the spring and the C_4 grasses active during the summer when temperatures are higher and soil moisture availability lower (Kemp and Williams, 1980). Analyses of the water relations and performance of these species indicate that this temporal separation results from higher summer temperatures rather than lower soil moisture (Kemp and Williams, 1980; Martin et al., 1991). However, high temperature greatly enhances evaporative demand and the probability of water stress, even in generally wet environments. Under these conditions, the greater WUE may be most relevant because it allows for a conservative water use strategy without greatly compromising photosynthesis, as would be the case in C_3 plants exhibiting conservative water use. Indeed, the higher WUE of C_4 species may enable primary productivity on sites to hot and dry for C_3 plants to survive (Schulze et al., 1996).

A. Theory

Water vapor efflux from the surface of the mesophyll cell walls to the bulk atmosphere occurs through the same gaseous diffusive pathway as CO_2 assimilation. Both gases are therefore subject to the same diffusive controls. In C_4 photosynthesis, PEPCase is saturated at a much lower CO_2 concentration than Rubisco, primarily because there is no oxygen inhibition of PEPCase activity. As a result, A_{sat} is saturated at a p_i of about 10–15 Pa (Pearcy and Ehleringer, 1984; Ehleringer and Monson, 1993). By contrast, A will not approach saturation in C_3 species until p_i exceeds 60 to 100 Pa at approximately 25°C (Harley and Tenhunen, 1991) (Fig. 2). Any decrease in stomatal or boundary layer conductance in a C_3 species will therefore decrease A, but in a C_4 species it will only affect A if p_i is lowered below 10–15 Pa. The effect of this difference on WUE is illustrated for "typical" responses of A to p_i for C_3 and C_4 leaves in Fig. 2. For any given leaf conductance (g_l), p_i will equal p_a when $A = 0$. As A increases, p_i will decline as a linear function (the supply function) of g_l, as illustrated by the dotted lines in Fig. 2. For a given g_l, the point at which the supply function intersects

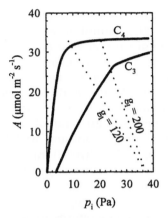

Figure 2 Solid lines indicate the responses of CO_2 uptake per unit leaf area (A) to intercellular CO_2 partial pressure (p_i) for "typical" C_3 and C_4 leaves. Curves are generated from the biochemical models of leaf photosynthesis of Farquhar et al. (1980) and Collatz et al. (1992), respectively, for light saturation and a leaf temperature of 25°C (Humphries and Long, 1995). The dotted lines show how p_i will decline with increasing A at two leaf conductances (g_l, μmol m^{-2} sec^{-1}).

the A/p_i response indicates the assimilation rate that will be achieved at the current ambient pCO_2. If C_3 and C_4 leaves have equal g_l and also the same leaf temperature and ambient humidity, then transpiration will be the same in both. The higher A_{sat} of the C_4 leaf at a common g_l will thus result in a higher leaf WUE; in this specific example, 32% higher. On average, however, leaf conductance is about 40% lower in C_4 leaves relative to C_3 leaves under the same atmospheric conditions, and this leads to substantially higher WUE than when g_l is equal (Long, 1985). For example, in Fig. 2, the C_4 leaf with a 40% lower leaf conductance would have a WUE 2.1 times that of the C_3 leaf, because reducing g_l from 200 to 120 mmol m^{-2} sec^{-1} reduces transpiration by 40%, but has a negligible effect on net CO_2 assimilation in the the C_4 leaf. Similar reductions in g_l and transpiration in the C_3 leaf would result in a considerable loss of carbon gain, as illustrated in Fig. 2.

B. Water Use Efficiency in Whole Plants

Is the theoretical and observed higher WUE at the leaf level translated into improved WUE at the whole plant level? For grasses grown at 30° in the same environment, the ratio of dry weight gained per unit water transpired (WUE) was approximately 1.2×10^{-3} for the C_3 species, but approximately 3.6×10^{-3} for the C_4 species (Downes, 1969). Similar differences have been reported from other studies (reviewed: Osmond et al., 1982; Ehleringer and Monson, 1993). Functionally, the advantage of increased efficiency of water use seems most apparent in conserving soil moisture and this may explain their predominance in hot semiarid climates with warm season rainfall. For example, Kalapos *et al.,* (1996) compared growth and stomatal conductance of C_3 *T. aestivum* with its C_4 grass weed *Tragus racemosus.* Under adequate soil moisture conditions, growth rates were similar, but the stomatal conductance of the C_4 species was much lower than that of the C_3 species. This difference disappeared in severe water stress. By using its soil moisture reserves more sparingly, the C_4 species was able to assimilate for a much longer period after a rainfall event than was the C_3 species. However, where the C_4 species shares the same root zone as a competing C_3 species, this advantage might be lost because the moisture not used by the C_4 species could instead be used by its competitor.

Is this higher WUE affected by other factors? The relevance of the higher WUE of C_4 relative to C_3 species is limited to comparisons of species in the same environment. Furthermore, this leaf difference will only be translated into an equivalent whole plant difference if the plants have the same degree of coupling to the bulk atmosphere. If, for example, because of differences in stature or canopy structure, the C_4 species is more closely coupled to the bulk atmosphere than the C_3 species to which it is compared, the difference in WUE observed at the leaf level will be decreased at the whole plant level. The advantage of a decreased stomatal conductance for a given rate of CO_2 assimilation will be proportional to the humidity gradient between the mesophyll surface and the bulk atmosphere. Because the humidity in the saturated air adjacent to the mesophyll wall surface rises exponentially with temperature, this gradient is likely to be greatest in hot environments. Exceptions are where high rainfall or large water bodies result in the bulk air being close to water vapor saturation. The difference in WUE should therefore increase with temperature, explaining in part the closer correlation of C_4 distributions with warm-season as opposed to cold-season precipitation. Downes (1969) showed that WUE averaged across seven grass species was 10 times higher for the C_4 species at 35°C, but only 1.7 times higher at 20°C.

The inherently higher WUE depends on the efficient operation of the C_4 process. Any factor that could raise the point at which p_i saturates photosynthesis would lower WUE. Leakage of CO_2 from the bundle sheath or loss of the PEPc activity could both lower WUE. Increased leakage from the bundle sheath would be apparent in increased isotopic discrimination. If all carbon is assimilated through the C_4 pathway without leakage, a constant and low discrimination against ^{13}C should be obtained. However, C_4 species can show considerable variation in ^{13}C discrimination in response to drought. Buchmann *et al.* (1996) showed that drought in a high-light treatment caused a significant increase in the ^{13}C discrimination in 7 of 14

C_4 grass species. This raises the possibility that although the C_4 syndrome may allow a plant to conserve soil moisture and avoid drought (Kalapos et al., 1996), the C_4 advantage in severe drought may be compromised.

C. Salinity

C_4 species form a particularly high proportion of the herbaceous and shrub flora of saline environments, even in cool temperate regions (Long and Mason, 1983). Although C_4 species, like CAM species, differ from C_3 species in their requirement for Na^+ as a micronutrient (Brownell and Bielig, 1996), this requirement is so minute that it can only be coincidental with the predominance of C_4 species on saline sites. The presence of C_4 species does not prove an advantage of the C_4 mechanism because other independent characters are of selective advantage. The strong rhizome system and "anchor" roots of *Spartina* spp. are of obvious advantage in gaining a foothold on mudflats subjected to tidal currents (Long and Woolhouse, 1978). However, C_3 halophytes such as *Scirpus* spp. have a similar capability. Salt glands, which can actively excrete Na^+ or Cl^- from the shoot, occur in some taxa that include C_4 species; for example, the grass tribe Chlorideae and the dicotyledonous family Chenopodiaceae. However, salt glands are also found in families that lack C_4 species [e.g., Plumbaginaceae (Long and Mason, 1983)].

The inherently higher WUE of C_4 species would have two theoretical advantages in saline environments. First, soils infiltrated with seawater have a soil water potential of around -2.5 MPa. To extract water, the halophyte must generate a lower water potential, even though this exceeds limits that can apparently be tolerated by many mesophytic vascular plants (Long and Mason, 1983). Transpiration, therefore, must be minimal, and the higher WUE of C_4 species would confer the advantage of maximizing carbon gain per unit of water lost. Second, plant mineral content is inversely correlated to WUE as an assumed result of increased passive uptake with increased transpiration (Masle et al., 1992). For a halophyte, increased transpiration increases the energy needed to exclude Na^+ and Cl^- (Long and Mason, 1983).

Halophytes are commonly xeromorphic (Long and Mason, 1983). This serves to protect the water reserves of the plant in periods of drought or high potential evapotranspiration when soil water potential falls to levels where only minimal water extraction is possible. A cost of xeromorphy is increased resistance to diffusion of CO_2 to the mesophyll (Long, 1985). Again, because of the low p_i necessary to saturate C_4 photosynthesis (Fig. 2), this cost is minimized in C_4 species. As in the case of semiarid habitats, C_4 species become a less prominent component of the herbaceous flora of saline habitats with latitude, and are absent from sites where the mean minimum temperature of the warmest month does not exceed 9°C (Long,

1983). This again suggests a loss of advantage, or a disadvantage, of the C_4 syndrome in cooler climates, the issue considered in the following section.

V. Temperature

As noted earlier, on a regional basis the proportion of C_4 species within a herbaceous or shrub flora is more closely correlated with temperature than with any other variable (Teeri, 1988). This is particularly apparent in the grasses, both across continental and along altitude transects (Teeri and Stowe, 1976; Tieszen et al., 1979; Sage, Wedin, and Li, Chapter 10). Not only does the proportion of C_4 species within a flora decline with decrease in growing season temperature, but on latitudinal gradients they also are generally absent from sites where the average daily minimum for the warmest month of the year is less than about 8°C (Long, 1983).

The temperature optimum of light-saturated CO_2 uptake (A_{sat}) in C_4 species is typically about 10°C higher than that of C_3 species from the same environment. For example, the temperate perennial C_4 grass *S. anglica* shows a temperature optimum of 30°C, whereas C_3 grasses from the same region have a temperature optimum of approximately 20°C (Long et al., 1975). This difference may be attributed to the elimination of photorespiration (Long, 1991). If photosynthesis of the C_3 species is measured at low pO_2 or high pCO_2 to eliminate photorespiration, then the temperature optimum in C_3 species can be shifted upwards by about 10°C (Long, 1983; Long, 1991). Depressed temperature optima in C_3 species result because both the affinity of Rubisco for, and the aqueous solubility of, CO_2 relative to O_2 declines with increase in temperature (Jordan and Ogren, 1984). This favors oxygenation and photorespiration relative to carboxylation and photosynthesis as temperature rises (Sage, Chapter 1). As a result, the potential increase in photosynthesis resulting from the C_4 syndrome relative to C_3 will rise with temperature (Long, 1991).

One of the most obvious environmental features relating to the distribution of C_4 species, however, is their rarity in the cool temperate zones and their absence from the cold climate zones of the globe; zones D and $C_{b/c}$ respectively, following Köppen's system of climate classification (Money, 1989). Comparisons have frequently shown that A_{sat} in C_4 species declines rapidly with decrease in temperature below 20°C, and that few species are capable of any net CO_2 uptake below 10°C (reviewed in Long, 1983). *Zea mays* is the only C_4 crop of significance in temperate climate zones. However, exposure of mature *Z. mays* leaves to temperatures of 17°C and below in the light depresses subsequent photosynthetic rates of these leaves through slow or irreversible photoinhibition of photosynthesis (Long et al., 1983; 1994). Although young leaf tissue is less vulnerable to this damage (Long *et al.*, 1989), leaves that develop at suboptimal temperatures also show impaired development of the photosynthetic apparatus and a depressed photosynthetic capacity (Nie *et al.*, 1992, 1995).

Observations of the poor performance of C_4 species in cool climates and their absence from cold climates has led to the suggestion that C_4 species are inherently limited to warm and hot climates (Black, 1971). It is likely that the C_4 syndrome evolved in the tropics and subtropics within C_3 taxa adapted to warm conditions (Ehleringer *et al.*, 1991; Cerling *et al.*, 1997), and thus, absence of C_4 species from colder environments could simply reflect the general lack of fitness of these taxa for colder climates. Radiation of the C_4 syndrome into colder climates would be further slowed by the fact that the resource use efficiency advantages over C_3 species decline with temperature (Long, 1983; Hatch, 1992a). This, however, does not necessarily mean that C_4 species are inherently limited from, or disadvantaged in, cool and cold climates. Is there any theoretical disadvantage of C_4 photosynthesis at low temperature?

A. Theory

Because photorespiration relative to photosynthesis rises with increased temperature in C_3 plants, the net energy expended per molecule of CO_2 assimilated will also rise. C_4 species largely avoid this cost, but do use additional energy in the C_4 dicarboxylate cycle, as discussed previously. By avoiding photorespiration, their energy requirement for CO_2 assimilation is largely independent of temperature. Figure 3 illustrates the modeled dependency of the maximum absorbed quantum yield (ϕ_{abs}) in C_3 photosynthesis on temperature. The curve illustrated for ϕ_{abs} at the current atmospheric pCO_2 (36 Pa) shows close agreement to the measured temperature dependence (e.g., Ehleringer and Björkman, 1977). This predicts that below about 25°C, more light energy is required to assimilate a single mole of CO_2 in C_4 species, whereas C_3 species require more light above 25°C (Fig. 3). The predicted temperature at which C_4 species achieve a higher ϕ_{abs} would have been about 7°C lower than today at the postglacial pCO_2 minimum of about 18 Pa and 3°C lower at the preindustrial pCO_2 of 28 Pa. Conversely, it is predicted that the temperature at which ϕ_{abs} of the two photosynthetic types intersects will be 4°C higher when the atmospheric pCO_2 reaches 55 Pa during the next century. These predictions indicate that the areas of the globe where C_4 photosynthesis is more efficient under light-limiting conditions have declined since the nineteenth century, and will show a further and potentially dramatic decline with rising atmospheric pCO_2 over the twenty-first century.

Ehleringer (1978) noted the close correlation between the distribution of C_4 species over the Great Plains and the current transition temperature, and from this suggested a key role for ϕ_{abs} in determining species distribu-

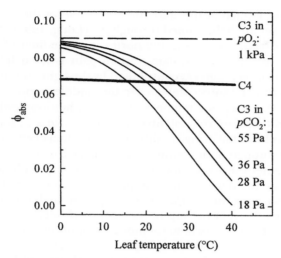

Figure 3 Predicted response of the maximum absorbed light quantum yield of CO_2 uptake (ϕ_{abs}) versus leaf temperature. Plots illustrated are for a C_3 leaf (broken line) in the absence of photorespiration ($pO_2 = 1$ kPa), a C_4 leaf in normal air (thick solid line), and for C_3 leaves at normal partial pressures of oxygen, in the atmospheric pCO_2 minimum (18 Pa), the preindustrial pCO_2 (28 Pa), the current pCO_2 (36 Pa) and the pCO_2 anticipated for the middle of the next century (55 Pa). Rates are calculated from the equations and parameters of Long (1991), Humphries and Long (1995), and adapted from Collatz *et al.* (1992).

tion. Maximum quantum yield determines the maximum rate of CO_2 assimilation in deep shade environments and has a diminishing influence on photosynthetic rate as light saturation is approached (e.g., Long *et al.*, 1994). A lower maximum quantum yield is therefore likely to depress fitness in a shaded environment. Any individual that begins life in a shaded environment could therefore be at a disadvantage. The occurrence of some species well north of the transition temperature would be consistent with the proposal of Ehleringer (1978), providing that these were species of open habitats or primary colonizers where interspecific competition for light is unlikely. *Spartina anglica*, which extends in distribution to 58°N in Western Europe, colonizes bare mudflats, whereas *Salsola kali*, which occurs at 61°N, colonizes the bare sand of coastal fore-dunes (Long, 1983). Should the lower maximum quantum yield preclude C_4 species from cool climates when competition is present?

The occurrence of C_4 species such as *Cyperus longus* in wetland communities in Southern England and *Miscanthus sinensis* in subalpine grasslands in Japan indicates that some species, at least, are not precluded (Long, 1983; Ohwi, 1984). At the canopy level, net photosynthetic gain is a function of both light-saturated and light-limited photosynthesis. Within a canopy, all leaves are light limited at the start and at the end of the day, and the lower canopy leaves are light limited throughout (Long, 1993). Analysis of carbon gain by *Z. mays* canopies at the northern range limit of this crop, suggests that about 50% of the daily carbon gain is from light-limited photosynthesis. The remainder is contributed by the light-saturated photosynthesis of the upper canopy leaves over the middle period of the day (Baker *et al.*, 1988). Even at low temperature, some photorespiration occurs in C_3 species and therefore the potential A_{sat} is higher in a C_4 species at all temperatures, as illustrated in Fig. 4. Although additional energy is required to achieve light-saturated photosynthesis, this can have no effect when light is in excess. Whether there is a theoretical advantage of C_4 photosynthesis at low temperature will depend on a complex balance between lower rates of light-limited photosynthesis within the canopy and higher rates of light-saturated photosynthesis at the top of the canopy (Long, 1993).

Figure 5 integrates a C_3 biochemical model of photosynthesis (modified from: Farquhar *et al.*, 1980) with a canopy microclimate model (Norman, 1980), as described previously (Long, 1991; Humphries and Long, 1995). The assumed benefit of C_4 photosynthesis is elimination of photorespiration at the cost of additional energy. This is simulated by eliminating oxygenation, assuming an 11% leakage of CO_2 from the bundle sheath and a

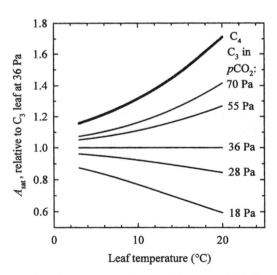

Figure 4 Predicted light saturated net photosynthesis (A_{sat}) as a function of leaf temperature and plotted as a fraction of the rate calculated for a C_3 leaf in the current atmospheric pCO_2 (36 Pa). The upper (thick) line indicates the predicted rate for a C_4 leaf in the same atmosphere. Other lines indicate C_3 rates at past and anticipated future atmospheric pCO_2; see Fig. 3 for details.

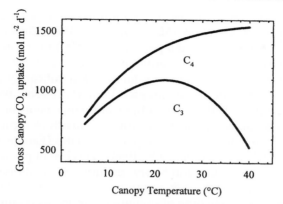

Figure 5 Predicted rates of gross canopy CO_2 uptake integrated over a diurnal course for a range of canopy temperatures. The simulation is for a leaf area index of three assuming a spherical distribution of foliar elements, on Julian day 190 and with clear sky conditions (atmospheric transmittance = 0.75) at a latitude of 52°N. Equations and parameter from Long (1991) and Humphries and Long (1995). Details of the modifications made to simulate the C_4 canopy are given in the text.

minimum of 14.2 absorbed photons per CO_2 assimilated (Hatch, 1987; Hatch *et al.*, 1995). For a summer day at latitude 52°N and assuming clear sky conditions, the diurnal course of light and temperature at each canopy layer is predicted, and from this the leaf photosynthetic rates are calculated and summed to give the daily integral for the canopy. Dark respiration is excluded, and an average leaf area index of three was chosen with an assumed random, (i.e., spherical, distribution of foliar elements). The simulation shows a clear theoretical advantage of C_4 photosynthesis at all temperatures. This advantage would be lower for overcast sky days and times of year with lower maximum photon fluxes, when light-limited photosynthesis would constitute a larger proportion of total canopy CO_2 gain. A leaf area index of three, however, is high for natural and seminatural grasslands, and many natural herbaceous communities (Jones *et al.*, 1992). To simplify the simulation, a uniform canopy temperature has been imposed. The result illustrated in Fig. 5 suggests that there is a theoretical advantage of eliminating photorespiration on canopy carbon gain, despite a higher minimum photon requirement per CO_2 assimilated, even at temperatures of 10°C and below. However, achieving this theoretical advantage assumes that the C_4 photosynthetic apparatus can function at low temperatures.

B. Low Temperature Impairment of C_4 Photosynthesis

As noted earlier, *Z. mays* is the only C_4 crop of significance that is grown in temperate climates. Interest in extending its range as a grain and forage crop into colder climates has focused attention on the aspects of chilling damage that depress production (Miedema et al., 1987; Greaves, 1996). Exposure of seedlings to chilling temperatures ($\leq 12°C$) in the light results in slowly reversible damage or, after prolonged exposure, irreversible damage to photosynthetic capacity. This is evident as a significantly decreased maximum quantum yield of photosynthesis (Long et al., 1987; Baker et al., 1989). Measurements of leaf photosynthesis and conversion of absorbed light energy into biomass (ε_c), in field grown plants show significant decreases during early season growth in temperate climates (Farage and Long, 1987; Baker et al., 1989; Long et al., 1990, 1994). This loss of photosynthetic efficiency results from a photoinhibitory impairment of photosystem II. This has been associated with both reaction center damage (Baker et al., 1989) and xanthophyll quenching of excitation energy decreasing the efficiency of energy transfer to the reaction center (Fryer et al., 1995). The primary cause of this damage appears to be decreased capacity of photosynthetic carbon assimilation at low temperatures. This in turn decreases the rate of photochemical utilization of absorbed light energy, increasing "excitation pressure" within the thylakoid membrane and the probability of photoinhibition (Long et al., 1994; Huner et al., 1995). In addition to this damage to mature leaves, leaves developed at 14°C show a ϕ_{abs} and an A_{sat} of about 25% of values for leaves developed at 20°C, when both are measured at 20° (Nie et al., 1992). This developmental impairment of photosynthesis is related to suppressed or incomplete processing of specific polypeptides, in particular those coded by the chloroplast DNA (Hayden and Baker, 1990; Ortiz-Lopez et al., 1990; Bredenkamp et al., 1992, Bredenkamp and Baker, 1992, 1994; Fryer et al., 1995, 1998; Nie et al., 1995). However, even within *Zea* there is considerable variation in low temperature tolerance. Comparison of high altitude land races of *Z. mays* and the perennial *Zea perennis* with modern cultivars of *Z. mays* indicates that significant variation in ability to resist photoinhibition at low temperatures exists (Long et al., 1987). Similar wide variation has been identified and analyzed in the C_4 weed *Echinochloa crosgalli* that extends its distribution into Ontario and Quebec, Canada (Potvin and Simon, 1989). Furthermore, the C_3 crop *Phaseolus vulgaris*, which shares the same center of domestication around the Tehuacan valley of Mexico as *Z. mays* (Heiser, 1965), appears equally vulnerable to photoinhibition at chilling temperatures (Powles, 1984). This suggests that low temperature impairment may be a function of the tropical origins of the crop, rather than an inherent feature of C_4 photosynthesis *per se*.

Assimilation of CO_2 in C_4 species involves the additional steps of the C_4 cycle, transport between the mesophyll and bundle sheath, and translocation of C_4 cycle intermediates between cytoplasm and chloroplast. Are any of these steps inherently unable to function efficiently at low temperatures?

Considerable attention has focussed on PPDK because the enzyme shows the lowest extractable activity of the C_4 cycle enzymes, with activity only sufficient to support *in vivo* rates of photosynthesis. It is also subject to light/dark regulation, is activated by inorganic phosphate and deactivated by ADP (reviewed: Edwards *et al.*, 1985). Further support for its role as the key regulating step in the C_4 cycle comes from studies of *Flaveria bidentis* transformed to produce specific decreases in photosynthetic enzymes. These analyses suggest that control of light-saturated photosynthesis is shared between PPDK (0.2–0.3) and Rubisco (0.7) (Furbank *et al.*, 1997). *In vitro*, the enzyme from *Z. mays* is low temperature labile, with a transition temperature of about 10°C (Sugiyama, 1973).

Bypassing the C_4 cycle in CO_2 assimilation is one theoretical mechanism by which C_4 species could avoid any disadvantage at low temperature. However, there is no evidence that CO_2 assimilation bypasses the C_4 cycle at low temperatures. In the cool temperate C_4 species *S. anglica* a low Γ of less than 1Pa was maintained at 5°C (Long *et al.*, 1975) and the kinetics of ^{14}C labeling of intermediates at 10°C also appeared typical of C_4 species (Thomas and Long, 1978). In PEP-CK type C_4 species, PEP is formed in the bundle sheath by the decarboxylation reaction. Normally, this PEP is then converted to pyruvate or alanine and translocated back to the mesophyll, where pyruvate is converted to PEP via PPDK (Hatch, 1987; Kanai and Edwards, Chapter 3). One solution to the low temperature lability of PPDK, would be to bypass this step in the cycle and use the PEP formed in the bundle sheath as the source for subsequent carboxylation in the mesophyll. Smith and Woolhouse (1983) showed that the activities of PPDK in the mesophyll of *S. anglica* were far too low to support observed rates of photosynthesis *in vivo*. They hypothesized that PEP must be supplied to the mesophyll by translocation of the PEP formed by decarboxylation in the bundle sheath. Matsuba *et al.*, (1997), using further enzyme assays and flux analysis, have confirmed this hypothesis for *S. anglica*, and have also shown that this is not a feature of PEP-CK species of warmer climates. Other C_4 types do not form PEP in the bundle sheath and would not have this option. However, whereas PPDK may be labile at $\leq 10°C$ *in vitro*, compatible solutes and intermediates can stabilize PPDK down to 0°C (Edwards *et al.*, 1985; Krall *et al.*, 1989). Analysis of the *in vitro* low temperature stability of PPDK among *Z. mays* genotypes has revealed considerable variation which can be related to sites of origin (Sugiyama *et al.*, 1979; Sugiyama and Hirayma, 1983). In summary, there is no evidence that the low temperature lability of PPDK is an insurmountable barrier to the evolution of species capable of efficient C_4 photosynthesis at low temperature. Assuming a tropical or subtropical origin of C_4 species and assuming no inherent barrier to C_4 photosynthesis in cold climates, can any C_4 species be found that are efficient at low temperatures? Given the much shorter dispersal distances, the most likely sites of evolution of low temperature tolerance in C_4 species might be along altitude gradients in the tropics and subtropics. Here, low temperature habitats can occur close to warmer climates that support a range of C_4 species. It may be hypothesized therefore, that there will be a greater likelihood of finding chilling-tolerant C_4 taxa on altitudinal gradients in the tropics than along latitudinal gradients. Some evidence to support this reasoning comes from analyses of latitudinal and altitudinal gradients. Threshold temperatures for C_4 occurrence are generally lower along altitude gradients in the tropics than along altitude or latitude gradients in temperature zones (Rundel, 1980; Cabido *et al.*, 1997; Sage, Wedin, and Li, Chapter 10), and the number of C_4 taxa occurring at high altitude in the tropics is much greater than occurs at high latitudes and similar temperature during the growing season. For example, near the northern margin of C_4 occurrence in Canada (north of the 16°C mean July isotherm) there are estimated to be 10 native C_4 grass species in the genera *Distichilis*, *Muhlenbergia*, and *Spartina*, and three introduced weed species in *Echinochloa* and *Setaria* (estimates by R. F. Sage, personal communication, using plant distribution descriptions in Scoggan, 1978, and Schwarz and Redmann, 1988, and climate data in Hare and Hay, 1974). By contrast, above 2500 m in Peru (which corresponds to a mean annual isotherm of 16°C in the central Andes, Johnson, 1976), 75 C_4 species are recorded (Table II). In Canada, C_4 species are absent north of about the 13°C mean July isotherm (R.F. Sage, personal communication), whereas in the Peruvian Andes, 48 species occur above 3000 m, where the mean annual temperature is 13°C. As well, many additional C_4 genera are found at high altitude in the tropics that do not occur at high latitudes. In Peru, for example, 23 C_4 genera occur above 2500 m, compared to the 5 C_4 genera in northern Canada. Among the 23 Peruvian genera are a number such as *Axonopus*, *Paspalum*, and *Pennisetum*, which are generally absent from colder regions of the temperate zones (Cabido *et al.*, 1997).

Miscanthus, a C_4 genus that includes a number of species from high altitude locations in the Asian subtropics, provides clear evidence of efficient C_4 photosynthesis at low temperatures.

C. Miscanthus

Miscanthus is a genus of about 20 species found in India and eastwards to Japan. A number of the species are found at high altitude (Ohwi, 1984). *Miscanthus transmorrisonensis* dominates subalpine grasslands above 2400 m in Taiwan (Kao, 1997; Kao *et al.*, 1998). *Miscanthus sinensis* is found on mountains of Kyushu and Honshu, where it can form dense and persistent stands (Ohwi, 1984; Shoji *et al.*, 1990). All species are herbaceous perennials, forming reedlike and often highly productive stands (Ohwi, 1984). *Miscanthus* spp. can form fertile hybrids with *Saccharum* (Chen *et al.*, 1993; Sobral

Table II C₄ Grass Genera and Species Number Found Above Various Altitudes in the Peruvian Andes

Genus	Number of C₄ species occurring above indicated altitude			
	>2500 m	>3000 m	>3500 m	4000–4500 m
Mean annual temperature (°C)	16°	13°	9°	6° to 3°
Andropogon	2	0	0	0
Aristida	5	4	2	0
Axonopus	2	2	0	0
Bothriochloa	3	3	1	0
Bouteloua	1	1	1	0
Cenchrus	1	0	0	0
Chloris	2	1	1	0
Chondrosum (= *Bouteloua*)	1	1	1	0
Distichilis	1	1	1	0
Eragrostis	9	5	3	1
Gouinia	1	0	0	0
Heteropogon	1	0	0	0
Lycurus	3	2	1	0
Microchloa	1	0	0	0
Muhlenbergia	9	7	5	3
Paspalum	16	12	6	4
Pennisetum	7	4	2	0
Rhynchelytrum	1	0	0	0
Schizachyrium	2	1	0	0
Setaria	3	2	0	0
Sorghastrum	1	0	0	0
Sporobolus	2	2	2	0
Trachypogon	1	0	0	0
Total	75	48	26	8

Climate data from Johnson, A. M. (1976). "Climates of Central and South America" (World Survey of Climatology Vol. 12), (W. Schwerdtfeger, ed), pp. 147–218. Elsevier, Amsterdam. Species data from Brako, L. and Zarucchi, J. L. (eds.). (1993). "Catalogue of the Flowering Plants and Gymnosperms of Peru," Missouri Botanic Garden, St. Louis, Missouri. No C₄ grasses are reported above 4500 m.

et al., 1994). Like *Saccharum*, *Miscanthus* is classified as a genus of the grass tribe Andropogoneae, which is composed exclusively of C₄ NADP-ME species (Chapman, 1996).

Miscanthus x giganteus, a hybrid of *M. sinensis* and *M. sacchariflorus*, has been grown in agricultural trials as a potential biomass energy crop in Western Europe (Greef and Deuter, 1993). When grown in the cool temperate climate of southern England (lat. 52°N), the crop achieved an efficiency of conversion of intercepted photosynthetically active radiation into biomass energy (ε_c) of 0.077 to 0.078. This is equivalent to the maximum ε_c recorded for C_4 species in warm climates. This was also significantly in excess of the ε_c achieved by temperate C_4 *Spartina* species at the same location (Beale and Long, 1995). The crop canopy developed in early April and ε_i reached 80% by the end of the month and 95% by the end of May, approximately 2 months in advance of equivalent canopy growth in *Z. mays* crops at this latitude (Beale and Long, 1995). Total aboveground dry matter production by the end of the growing season in October was 27 t ha^{-1}, exceeding the dry matter yields achieved by C_3 grasses and *Spartina* spp. at this latitude (Beale and Long, 1995). The result provides the first evidence that the theoretical potential of C_4 photosynthesis can be realized in a cool temperate climate (Beale and Long, 1995). Analysis of photosynthesis in the leaves of this crop showed, in sharp contrast to *Z. mays*, no depression of quantum yield by development in the cool conditions of early spring and showed values of A_{sat} in excess of those of C_3 species over the growing season (Beale *et al.*, 1996). Kao *et al.* (1998) has similarly shown that *M. transmorriensis* can develop leaves at 10°C capable of light saturated rates of photosynthesis in excess of those of temperate C_3 species. This contrasts to the cool temperate C_4 species *S. anglica*, which showed a loss of assimilatory capacity in leaves grown at 10°C (Dunn *et al.*, 1987). These findings with *Miscanthus* provide proof that the theoretical gains in efficiency of C_4 photosynthesis (Fig. 5) can be realized in cool climates, and that there is no inherent flaw in C_4 photosynthesis that prevents efficient operation at low temperatures. The relatively few C_4 species found at high altitudes and latitudes indicate the combination of characters required to achieve low temperature tolerance is not easily acquired. Furthermore, any improvements in fitness with respect to resource use efficiency in C_4 photosynthesis are minimal at low temperature.

VI. Summary

The greater productivity of C_4 plants per unit of nitrogen, water, and intercepted light can all be explained as a consequence of active CO_2 accumulation at the site of Rubisco. Because this fundamental difference between C_3 and C_4 species is at the level of photosynthesis, translation of these differences into altered environmental responses of the whole plant will be modulated by patterns of biomass allocation, phenology, growth form, and interactions with other physiological processes. The higher light conversion efficiency of C_4 plants results in higher potential productivity, but these are often not realized because of water and nutrient limitations. Managed agricultural settings where limiting resources are provided, and naturally resource rich sites such as the floodplains along tropical rivers, do allow C_4 species to meet their potentially high conversion efficiencies and production rates. In water and nutrient limited environments, the higher WUE and NUE allow C_4 plants to be highly competitive for limiting resources, but often through indirect means. C_4 plants could achieve the same rates of carbon gain as C_3 competitors while investing fewer resources into individual leaves, thereby allowing larger investments into total leaf area, root mass, storage, or reproduction. On nutrient- or water-limited sites, increased allocation to roots enhances the ability of the C_4 species to capture proportionally more of the limiting resource. Globally, distributions of C_4 species are most closely related to growing season temperatures. C_4 plants are uncommon to absent in cool temperate and colder climates leading to suggestions that C_4 plants may have an inherent flaw in their photosynthetic pathway. No consistent flaw has been found, however, and lesions caused by cold are often attributable to the tropical nature of the species being examined. The example of *Miscanthus* shows there is no obvious constraint inherent in the design of the C_4 mechanism that prevents efficient operation in colder climates.

Appendix I List of Abbreviations

A, net rate of leaf CO_2 uptake per unit leaf area (μmol m^{-2} sec^{-1})
A_{sat}, A at light saturation
g_l leaf conductance to water vapor (mmol m^{-2} sec^{-1})
LAI, leaf area index (dimensionless)
NAD-ME, NAD-dependent "malic enzyme"
NADP-ME, NADP-dependent "malic enzyme"
NUE, nitrogen use efficiency given as the ratio of increase in plant biomass to increase in plant nitrogen (dimensionless)
pCO_2, partial pressure of CO_2 (PA)
p_a, pCO_2 of the air around a leaf
p_i leaf intercellular pCO_2
PCO, C_2-photosynthetic oxidative or photorespiratory pathway
PCK, phosphoenolpyruvate carboxykinase
PCR, C_3-photosynthetic carbon reduction or Calvin cycle
PEP, phosphoenolpyruvate
PEPCase, PEP carboxylase
PNUE, photosynthetic nitrogen use efficiency given as A_{sat} per unit of leaf nitrogen (μmol^{-2} sec^{-1} kg [N]$^{-1}$)
PPDK, pyruvate inorganic phosphate dikinase
RuBP, ribulose-1,5-bisphosphate
Rubisco, RuBP carboxylase/oxygenase
WUE, efficiency of water use given as the amount of dry matter formed per unit mass of water used (dimensionless)
ε_c, efficiency of conversion of the visible radiation intercepted by a leaf canopy into biomass energy (0–1, dimensionless)
ε_i, efficiency with which a canopy intercepts incident visible radiation (0–1, dimensionless)

(continues)

Appendix I *(Continued)*

ϕ, maximum quantum efficiency of CO_2 per incident photon (dimensionless)
ϕ_{abs}, ϕ on the basis of absorbed photons
Γ, CO_2 compensation point of photosynthesis (Pa)
τ, the specificity of Rubisco for CO_2 relative to O_2 (dimensionless)

Acknowledgments

I wish to thank R. F. Sage for supplying Table II and data on Canadian C_4 grasses, and the reviewers for their valuable comments. Also thanks to Dr. Meirong Li for valuable editorial assistance.

References

Amundson, R., Francovizcaino, E., Graham, R. C., and Deniro, M. (1994). The relationship of precipitation seasonality to the flora and stable-isotope chemistry of soils in the Vizcaino Desert, Baja-California, Mexico. *J. Arid Environ.* **28,** 265–279.

Bainbridge, G., Madgwick, P., Parmar, S., Mitchell, R., Paul, M., Pitts, J., Keys, A. J., and Parry, M. A. J. (1995). Engineering Rubisco to change its catalytic properties. *J. Exp. Bot.* **46,** 1269–1276.

Baker, N. R., Bradbury, M., Farage, P. K., Ireland, C. R., and Long, S. P. (1989). Measurements of the quantum yield of carbon assimilation and chlorophyll fluorescence for assessment of photosynthetic performance of crops in the field. *Phil. Trans. Roy. Soc. London B Biol. Sci.* **323,** 295–308.

Baker, N. R., Long, S. P., and Ort, D. R. (1988). Photosynthesis and temperature, with particular reference to effects on quantum yield. *In* "Plants and Temperature. Society for Experimental Biology Symposium No. XXXXII" (S. P. Long and F. I. Woodward, eds.), pp. 347–375. Company of Biologists Ltd., Cambridge, England.

Beale, C. V., and Long, S. P. (1995). Can perennial C_4 grasses attain high efficiencies of radiant energy conversion in cool climates. *Plant Cell Environ.* **18,** 641–650.

Beale, C. V., and Long, S. P. (1997). Seasonal dynamics of nutrient accumulation and partitioning in the perennial C_4 grasses *Miscanthus* x *giganteus* and *Spartina cynosuroides*. *Biomass Bioenergy* **12,** 419–428.

Beale, C. V., Bint, D. A., and Long, S. P. (1996). Leaf photosynthesis in the C_4 grass *Miscanthus* x *giganteus*, growing in the cool temperate climate of southern England. *J. Exp. Bot.* **47,** 267–273.

Black, C. C. (1971). Ecological implications of dividing plants into groups with distinct photosynthetic production capacities. *Adv. Ecol. Res.* **7,** 87–114.

Brako, L. and Zarucchi, J. L. (eds.). (1993). "Catalogue of the Flowering Plants and Gymnosperms of Peru," Missouri Botanic Garden, St. Louis, Missouri.

Bredenkamp, G. J., and Baker, N. R. (1992). Temperature effects on thylakoid protein-metabolism in maize. *Photosynth. Res.* **34,** 215.

Bredenkamp, G. J., and Baker, N. R. (1994). Temperature-sensitivity of D1 protein–metabolism in isolated *Zea mays* chloroplasts. *Plant Cell Environ.* **17,** 205–210.

Bredenkamp, G. J., Nie, G. Y., and Baker, N. R. (1992). Perturbation of chloroplast development in maize by low growth temperature. *Photosynthetica* **27,** 401–411.

Brown, R. H. (1978). A difference in N use efficiency in C_3 and C_4 plants and its implications for adaptation and evolution. *Crop Sci.* **18,** 93–98.

Brownell, P. F., and Bielig, L. M. (1996). The role of sodium in the conversion of pyruvate to phospho*enol*pyruvate in mesophyll chloroplasts of C_4 plants. *Aust. J. Plant Physiol.* **23,** 171–177.

Buchmann, N., Brooks, J. R., Rapp, K. D., and Ehleringer, J. R. (1996). Carbon-isotope composition of C_4 grasses is influenced by light and water-supply. *Plant Cell Environ.* **19,** 392–402.

Cabido, M., Ateca, N., Astegiano, M and Anton, A. M. (1997). Distribution of C_3 and C_4 grasses along an altitudinal gradient in central Argentina. *J. Biogeog.* **24,** 197–204.

Cerling, T. E., Wang, Y., and Quade, J. (1993). Expansion of C_4 ecosystems as an indicator of global ecological change in the Late Miocene. *Nature* **361,** 344–345.

Cerling, T. E., Harris, J. M., MacFadden, B. J., Leakey, M. G., Quade, J., Eisenmann, V., and Ehleringer, J. R. (1997). Global vegetation change through the Miocene and Pliocene. *Nature* **389,** 153–158.

Chapman, G. P. (1996). "The Biology of Grasses," CAB International, Wallingford, United Kingdom.

Chen, Y. H., Chen, C., and Lo, C. C. (1993). Studies on anatomy and morphology in *Saccharum–Miscanthus* nobilized hybrids. 1. Transmission of tillering, ratooning, adaptation and disease resistance from *Miscanthus* spp. *J. Agric. Assoc. China* **5,** 31–45.

Collatz, G. J., Ribascarbo, M., and Berry, J. A. (1992). Coupled photosynthesis–stomatal conductance model for leaves of C_4 plants. *Aust. J. Plant Physiol.* **19,** 519–538.

Dai, Z. Y., Ku, M. S. B., and Edwards, G. E. (1993). C_4 photosynthesis—the CO_2 concentrating mechanism and photorespiration. *Plant Physiol.* **103,** 83–90.

Downes, R. W. (1969). Differences in transpiration rate between tropical and temperate grasses under controlled conditions. *Planta* **88,** 261–273.

Drake, B. G., Gonzalez-Meler, M., and Long, S. P. (1997). More efficient plants: A consequence of rising atmospheric CO_2? *Annu. Rev. Plant Physiol. Plant Mol. Biol.* **48,** 609–639.

Dunn, R., Thomas, S. M., Keys, A. J., and Long, S. P. (1987). A comparison of the growth of the C_4 grass *Spartina anglica* with the C_3 grass *Lolium perenne* at different temperatures. *J. Exp. Bot.* **38,** 433–441.

Edwards, G. E., Nakamoto, H., Burnell, J. N., and Hatch, M. D. (1985). Pyruvate, P_i dikinase and NADP-Malate dehydrogenase in C_4 photosynthesis—properties and mechanism of light–dark regulation. *Annu. Rev. Plant Physiol.* **36,** 255–286.

Ehleringer, J. R. (1978). Implications of quantum yield differences on the distributions of C_3 and C_4 grasses. *Oecologia* **31,** 255–267.

Ehleringer, J. R. and Björkman, O. (1977). Quantum yield for CO_2 uptake in C_3 and C_4 plants. *Plant Physiol.* **59,** 86–90.

Ehleringer, J. R., Sage, R. F., Flanagan, L. B. and Pearcy, R. W. (1991). Climate change and the evolution of C4 photosynthesis. *Trends Ecol. Evolution.* **6,** 95–99.

Ehleringer, J. R., and Monson, R. K. (1993). Evolutionary and ecological aspects of photosynthetic pathway variation. *Annu. Rev. Ecol. Syst.* **24,** 411–439.

Evans, J. R. (1989). Photosynthesis and the nitrogen relationship in leaves of C_3 plants. *Oecologia* **78,** 9–19.

Evans, J. R., and Farquhar, G. D. (1991). Modeling canopy photosynthesis from the biochemistry of the C_3 chloroplast. *In* "Modeling Crop Photosynthesis—from Biochemistry to Canopy" (K. J. Boote and R. S. Loomis, eds.), pp. 1–16. Crop Science Society of America, Inc., Madison, WI.

Evans, L. T. (1975). The physiological basis of yield. *In* "Crop Physiology" (L. T. Evans, ed.), pp. 327–355. Cambridge University Press, Cambridge, England.

Farage, P. K., and Long, S. P. (1987). Damage to maize photosynthesis in the field during periods when chilling is combined with high photon fluxes. *In* "Progress in Photosynthesis Research" Vol. 4. (J. Biggins, ed.), pp. 139–142. M. Nijhoff Publ., Dordrecht.

Farquhar, G. D., von Caemmerer, S., and Berry, J. A. (1980). A biochemical model of photosynthetic CO_2 assimilation in leaves of C_3 species. *Planta* **149**, 78–90.

Fryer, M. J., Andrews, J. R., Oxborough, K., Blowers, D. A., and Baker, N. R. (1998). Relationship between CO_2 assimilation, photosynthetic electron transport, and active O_2 metabolism in leaves of maize in the field during periods of low temperature. *Plant Physiol.* **116**, 571–580.

Fryer, M. J., Oxborough, K., Martin, B., Ort, D. R., and Baker, N. R. (1995). Factors associated with depression of photosynthetic quantum efficiency in maize at low growth temperature. *Plant Physiol.* **108**, 761–767.

Furbank, R. T., Chitty, J. A., Jenkins, C. L. D., Taylor, W. C., Trevanion, S. J., von Caemmerer, S., and Ashton, A. R. (1997). Genetic manipulation of key photosynthetic enzymes in the C_4 plant *Flaveria bidentis*. *Aust. J. Plant Physiol.* **24**, 477–485.

Furbank, R. T. and Hatch, M. D. (1987). Mechanism of C_4 photosynthesis: The size and composition of the inorganic carbon pool in bundle sheath cells. *Plant Physiol.* **85**, 958–964.

Furbank, R. T., Jenkins, C. L. D., and Hatch, M. D. (1990). C_4 Photosynthesis—quantum requirement, C_4 acid overcycling and Q-cycle involvement. *Aust. J. Plant Physiol.* **17**, 1–7.

Gifford, R., M. (1974). A comparison of potential photosynthesis and yield of plant species with differing photosynthetic metabolism. *Aust. J. Plant Physiol.* **1**, 107–117.

Greaves, J. A. (1996). Improving suboptimal temperature tolerance in maize—the search for variation. *J. Exp. Bot.* **47**, 307–323.

Greef, J. M., and Deuter, M. (1993). Syntaxonomy *of Miscanthus x giganteus* Greef-Et-Deu. *Angewandte Botanik* **67**, 87–90.

Hanway, J. J., (1962). Corn growth and composition in relation to soil fertility: III. Percentages of N, P and K in different plant parts in relation to stage of growth. *Agron. J.* **54**, 222–226.

Hare, F. K. and Hay, J. E. (1974). "Climates of North America" (World Survey of Climatology Vol. 11), (R. A. Bryson, and F. K. Hare, eds.), pp. 49–192. Elsevier, Amsterdam.

Harley, P. C., and Tenhunen, J. D. (1991). Modeling the photosynthetic response of C_3 leaves to environmental factors. *In* "Modeling Crop Photosynthesis—from Biochemistry to Canopy" (K. J. Boote and R. S. Loomis, eds.), pp. 17–39. Crop Sci. Soc. Amer., Inc., Madison, WI.

Hatch, M. D. (1987). C_4 photosynthesis—a unique blend of modified biochemistry, anatomy and ultrastructure. *Biochim. Biophys. Acta* **895**, 81–106.

Hatch, M. D. (1992a). C4 photosynthesis: An unlikely process full of surprises. *Plant Cell Physiol.* **33**, 333–342.

Hatch, M. D. (1992b). C_4 photosynthesis—evolution, key features, function, advantages. *Photosynth. Res.* **34**, 81.

Hatch, M. D., Agostino, A., and Jenkins, C. L. D. (1995). Measurement of the leakage of CO_2 from bundle-sheath cells of leaves during C_4 Photosynthesis. *Plant Physiol.* **108**, 173–181.

Hattersley, P. W. (1983). The distribution of C_3 and C_4 grasses in Australia in relation to climate. *Oecologia* **57**, 255–267.

Hayden, D. B., and Baker, N. R. (1990). Damage to photosynthetic membranes in chilling-sensitive plants—maize, a case-study. *Crit. Rev. Biotechenol.* **9**, 321–341.

Heiser, C. B. (1965). Cultivated plants and cultural diffusion in nuclear America. *Amer. Anthropol.* **67**, 930–949.

Hocking, P. J., and Meyer, C. P. (1991). Effects of CO_2 enrichment and nitrogen stress on growth, and partitioning of dry-matter and nitrogen in wheat and maize. *Aust. J. Plant Physiol.* **18**, 339–356.

Humphries, S. W., and Long, S. P. (1995). Wimovac—a software package for modeling the dynamics of plant leaf and canopy photosynthesis. *Comp. Appl. Biosci.* **11**, 361–371.

Huner, N. P. A., Maxwell, D. P., Gray, G. R., Savitch, L. V., Laudenbach, D. E., and Falk, S. (1995). Photosynthetic response do light and temperature—PSII excitation pressure and redox signaling. *Acta Physiol. Plant.* **17**, 167–176.

Jackson, D., Harkness, D. D., Mason, C. F., and Long, S. P. (1986). *Spartina anglica* as a carbon source for salt-marsh invertebrates—a study using Delta ^{13}C values. *Oikos* **46**, 163–170.

Johnson, A. M. (1976). "Climates of Central and South America" (World Survey of Climatology Vol. 12), (W. Schwerdtfeger, ed.), pp. 147–218. Elsevier, Amsterdam.

Jones, M. B. (1986). Wetlands. *In* "Photosynthesis in Contrasting Environments" (N. R. Baker and S. P. Long, eds.), pp. 103–138. Elsevier, Amsterdam.

Jones, M. B., Long, S. P., and Roberts, M. J. (1992). Synthesis and conclusions. *In* "Primary Productivity of Grass Ecosystems of the Tropics and Sub-tropics" (S. P. Long, M. B. Jones, and M. J. Roberts, eds.), pp. 212–255. Chapman & Hall, London.

Jordan, D. B., and Ogren, W. L. (1984). The CO_2/O_2 specificity of ribulose-1,5-bisphosphate carboxylase/oxygenase. Dependence on ribulose-bisphosphate concentration, pH and temperature. *Planta* **161**, 308–313.

Junk, W. J. (1993). Wetlands of tropical South America. *In* "Wetlands of the World" (D. Whigham, S. Hejny, and D. Dykjova, eds.), pp. 679–739. Kluwer, Dordrecht.

Kalapos, T., van den Boogaard, R., and Lambers, H. (1996). Effect of soil drying on growth, biomass allocation and leaf gas exchange of two annual grass species. *Plant Soil* **185**, 137–149.

Kamnalrut, A., and Evenson, J. P. (1992). Monsoonal grassland in Thailand. *In* "Primary Productivity of Grass Ecosystems of the Tropicas and Sub-tropics" (S. P. Long, M. B. Jones, and M. J. Roberts, eds.), pp. 100–126. Chapman & Hall, London.

Kane, H. J., Viil, J., Entsch, B., Paul, K., Morell, M. K., and Andrews, T. J. (1994). An improved method for measuring the CO_2/O_2 specificity of ribulosebisphosphate carboxylase-oxygenase. *Aust. J. Plant Physiol.* **21**, 449–461.

Kao, W. Y. (1997). Contribution of *Miscanthus transmorrisonensis* to soil organic carbon in a mountain grassland: Estimated from stable carbon isotope ratio. *Bot. Bull. Acad. Sinica* **38**, 45–48.

Kao, W. Y., Tsai, T. T., and Chen, W. H. (1998). A comparative study of *Miscanthus floridulus* (Labill) Warb and *M. transmorrisonensis* Hayata: Photosynthetic gas exchange, leaf characteristics and growth in controlled environments. *Ann. Bot.* **81**, 295–299.

Karlen, D. L., Flannery, R. L. and Sadler, E. J. (1987a). Nutrient and dry matter accumulation rates for high yielding maize. *J. Plant Nutr.* **10**, 1409–1417.

Karlen, D. L., Sadler E. J. and Camp, C. R. (1987b). Dry matter, nitrogen, phosphorus and potassium accumulation rates by corn on Norfolk Loamy Sand. *Agron. J.* **79**, 649–656.

Kebede, H., Johnson, R. C., and Ferris, D. M. (1989). Photosynthetic response of Eragrostis tef to temperature. *Physiol. Plant.* **77**, 262–266.

Kemp, P. R., and Williams, G. J. (1980). A physiological basis for niche separation between *Agropyron smithii* (C_3) and *Bouteloua gracilis* (C_4). *Ecology* **61**, 846–858.

Kent, S. S., and Tomany, M. J. (1995). The differential of the ribulose 1,5-bisphosphate carboxylase oxygenase specificity factor among higher plants and the potential for biomass enhancement. *Plant Physiol. Biochem.* **33**, 71–80.

Krall, J. P., Edwards, G. E., and Andreo, C. S. (1989). Protection of pyruvate, P_i dikinase from maize against cold lability by compatible solutes. *Plant Physiol.* **89**, 280–285.

Ku, M. S. B., Schmitt, M. R., and Edwards, G. E. (1979). Quntitative determination of RuBP carboxylase–oxygenase protein in leaves of several C_3 and C_4 plants. *J. Exp. Bot.* **114**, 89–98.

Li, M. R. (1993). Leaf photosynthetic nitrogen-use efficiency of C_3 and C_4 species of *Cyperus*. *Photosynthetica* **29**, 117–130.

Lidgate, H. J. (1984). Nitrogen uptake of winter wheat. *In* "The Nitrogen Requirement of Cereals, MAFF Reference Book 385," pp. 177–182. MAFF/ADAS, London.

Long, S. P. (1983). C_4 Photosynthesis at low temperatures. *Plant Cell Environ.* **6**, 345–363.

Long, S. P. (1985). Leaf gas exchange. *In* "Photosynthetic Mechanisms and the Environment" (J. Barber and N. R. Baker, eds.), pp. 453–500. Elsevier, Amsterdam.

Long, S. P. (1991). Modification of the response of photosynthetic productivity to rising temperature by atmospheric CO_2 concentrations—has its importance been underestimated? *Plant Cell Environ.* **14**, 729–739.

Long, S. P. (1993). The significance of light-limited photosynthesis to crop canopy carbon gain and productivity—a theoretical analysis. *In* "Photosynthesis: Photoreactions to Plant Productivity" (Y. P. Abrol, P. Mohanty, and Govindjee, eds.), pp. 547–560. Oxford & IBH Publishing, New Delhi.

Long, S. P. (1994). Resource capture by single leaves. *In* "Resource Capture by Crops: 52nd Easter School." (J. L. Monteith, R. K. Scott, and M. H. Unsworth, eds.), pp. 17–34. Nottingham University Press, Nottingham.

Long, S. P., and Jones, M. B. (1992). Introduction, aims, goals and general methods. *In* "Primary Productivity of Grass Ecosystems of the Tropics and Sub-tropics" (S. P. Long, M. B. Jones, and M. J. Roberts, eds.), pp. 1–24. Chapman & Hall, London.

Long, S. P., and Mason, C. F. (1983). "Saltmarsh Ecology," Blackie, Glasgow.

Long, S. P., and Woolhouse, H. W. (1978). Primary production in *Spartina* marshes. *In* "Ecological Processes in Coastal Environments" (R. L. Jefferies and A. J. Davy, eds.), pp. 333–352. Blackwell Science, Oxford.

Long, S. P., Bolharnordenkampf, H. R., Croft, S. L., Farage, P. K., Lechner, E., and Nugawela, A. (1989). Analysis of spatial variation in CO_2 uptake within the intact leaf and its significance in interpreting the effects of environmental stress on photosynthesis. *Phil. Trans. Roy. Soc. London B Biol. Sci.* **323**, 385–395.

Long, S. P., East, T. M., and Baker, N. R. (1983). Chilling damage to photosynthesis in young *Zea mays.* 1. Effects of light and temperature-variation on photosynthetic CO_2 assimilation. *J. Exp. Bot.* **34**, 177–188.

Long, S. P., Farage, P. K., Aguilera, C., and Macharia, J. M. N. (1990). Damage to photosynthesis during chilling and freezing, and its significance to the photosynthetic productivity of field crops. *In* "Trends in Photosynthesis Research" (J. Barber, M. G. Guerrero, and H. Medrano, eds.), pp. 344–356. Intercept, Andover.

Long, S. P., Humphries, S., and Falkowski, P. G. (1994). Photoinhibition of photosynthesis in nature. *Annu. Rev. Plant Physiol. Plant Mol. Biol.* **45**, 633–662.

Long, S. P., Incoll, L. D., and Woolhouse, H. W. (1975). C_4 photosynthesis in plants from cool temperate regions, with particular reference to *Spartina townsendii. Nature* **257**, 622–624.

Long, S. P., Nugawela, A., Bongi, G., and Farage, P. K. (1987). Chilling dependent photoinhibition of photosynthetic CO_2 uptake. *In* "Progress in Photosynthesis Research" Vol. 4. (J. Biggins, ed.), pp. 131–138, M. Nijhoff Publ., Dordrecht.

Long, S. P., Postl, W. F., and Bolharnordenkampf, H. R. (1993). Quantum yields for uptake of carbon-dioxide in C_3 vascular plants of contrasting habitats and taxonomic groupings. *Planta* **189**, 226–234.

Martin, C. E., Harris, F. S., and Norman, F. J. (1991). Ecophysiological responses of C_3 forbs and C_4 grasses to drought and rain on a tallgrass prairie in northeastern Kansas. *Bot. Gaz.* **152**, 257–262.

Masle, J., Farquhar, G. D., and Wong, S. C. (1992). Transpiration ratio and plant mineral content are related among genotypes of a range of species. *Aust. J. Plant Physiol.* **19**, 709–721.

Matsuba, K., Imaizumi, N., Kaneko, S., Samejima, M., and Ohsugi, R. (1997). Photosynthetic responses to temperature of phosphoenolpyruvate carboxykinase type C_4 species differing in cold sensitivity. *Plant Cell Environ.* **20**, 268–274.

Miedema, P., Post, J., and Groot, P. (1987). "The Effects of Low Temperature on Seedling Growth of Maize Genotypes," Pudoc, Wageningen.

Money, D. C. (1989). "Climate and Environmental Systems," Unwin Hyman, London.

Monteith, J. L. (1978). A reassessment of maximum growth rates for C_3 and C_4 crops. *Exp. Agric.* **14**, 1–5.

Mulroy, T. W., and Rundell, P. W. (1977). Annual plants: Adaptations to desert environments. *BioScience* **27**, 109–114.

Nie, G. Y., Long, S. P., and Baker, N. R. (1992). The effects of development at suboptimal growth temperatures on photosynthetic capacity and susceptibility to chilling-dependent photoinhibition in *Zea mays. Physiol. Plant.* **85**, 554–560.

Nie, G. Y., Robertson, E. J., Fryer, M. J., Leech, R. M., and Baker, N. R. (1995). Response of the photosynthetic apparatus in maize leaves grown at low temperature on transfer to normal growth temperature. *Plant Cell Environ.* **18,** 1–12.

Norman, J. M. (1980). Interfacing leaf and canopy light interception models. In "Predicting Photosynthesis for Ecosystem Models" (J. D. Hesketh and J. W. Jones, eds.), Vol. 2, pp. 49–67. CRC Press, Boca Raton.

Ohwi, J. (1984). "Flora of Japan," Smithsonian Institution, Washington, D.C.

Ortiz-Lopez, A., Nie, G. Y., Ort, D. R., and Baker, N. R. (1990). The involvement of the photoinhibition of photosystem-II and impaired membrane energization in the reduced quantum yield of carbon assimilation in chilled maize. *Planta* **181,** 78–84.

Osmond, C. B., Winter, K., and Ziegler, H. (1982). Functional significance of different pathways of CO2 fixation in photosynthesis. In "Physiological Plant Ecology II. Encyclopedia of Plant Physiol. New Series." (O. L. Lange, P. S. Nobel, C. B. Osmond, and H. Ziegler, eds.), Vol. 12B, pp. 479–547. Springer, Berlin.

Pearcy, R. W., and Ehleringer, J. (1984). Comparative ecophysiology of C_3 and C_4 plants. *Plant Cell Environ.* **7,** 1–13.

Peisker, M., and Henderson, S. A. (1992). Carbon—terrestrial C_4 plants. *Plant Cell Environ.* **15,** 987–1004.

Piedade, M. T. F., Junk, W. J., and Long, S. P. (1991). The productivity of the C_4 grass *Echinochloa polystachya* on the Amazon floodplain. *Ecology* **72,** 1456–1463.

Piedade, M. T. F., Junk, W. J., and Long, S. P. (1997). Nutrient dynamics of the highly productive C_4 macrophyte *Echinochloa polystachya* on the Amazon floodplain. *Funct. Ecol.* **11,** 60–65.

Piedade, M. T. F., Long, S. P., and Junk, W. J. (1994). Leaf and canopy photosynthetic CO_2 uptake of a stand of *Echinochloa polystachya* on the Central Amazon floodplain—are the high-potential rates associated with the C_4 syndrome realized under the near-optimal conditions provided by this exceptional natural habitat. *Oecologia* **97,** 193–201.

Polley, H. W., Johnson, H. B., and Mayeux, H. S. (1995). Nitrogen and water requirements of C_3 plants grown at glacial to present carbon dioxide concentrations. *Funct. Ecol.* **9,** 86–96.

Potvin, C., and Simon, J. P. (1989). The evolution of cold temperature adaptation among populations of a widely distributed C_4 weed barnyard grass. *Evol. Trends Plants* **3,** 98–105.

Powles, S. B. (1984). Photoinhibition of photosynthesis induced by visible-light. *Ann. Rev. Plant Physiol.* **35,** 15–44.

Robichaux, R. H., and Pearcy, R. W. (1980). Photosynthetic responses of C_3 and C_4 species from cool shaded habitats in Hawaii. *Oecologia* **47,** 106–109.

Rundel, P. W. (1980). The ecological distribution of C_4 and C_3 grasses in the Hawaiian Islands. *Oecologia* **45,** 354–359.

Sage, R. F. and Seemann, J. R. (1993). Regulation of ribulose-1,5-bisphosphate carboxylase/oxygenase activity in response to reduced light intensity in C_4 plants. *Plant Physiol.* **102,** 21–28.

Sage, R. F., and Pearcy, R. W. (1987a). The nitrogen use efficiency of C_3 and C_4 plants. I. Leaf nitrogen, growth, and biomass partitioning in *Chenopodium album* (L) and *Amaranthus retroflexus* (L). *Plant Physiol.* **84,** 954–958.

Sage, R. F., and Pearcy, R. W. (1987b). The nitrogen use efficiency of C_3 and C_4 plants. II. Leaf nitrogen effects on the gas exchange characteristics of *Chenopodium album* (L) and *Amaranthus retroflexus* (L). *Plant Physiol.* **84,** 959–963.

Sage, R. F., Pearcy, R. W., and Seemann, J. R. (1987). The nitrogen use efficiency of C_3 and C_4 plants. III. Leaf nitrogen effects on the activity of carboxylating enzymes in *Chenopodium album* (L) and *Amaranthus retroflexus* (L). *Plant Physiol.* **85,** 355–359.

Schulze, E.-D., Ellis, R., Schulze, W., and Trimborn, P. (1996). Diversity, metabolic types and Delta ^{13}C carbon isotope ratios in the grass flora of Namibia in relation to growth form, precipitation and habitat conditions. *Oecologia* **106,** 352–369.

Schwarz, A. G. and Redmann, R. E. (1988). C_4 grasses from the boreal forest region of northern Canada. *Can. J. Bot.* **66**, 2424–2430.

Scoggan, H. J. (1978). "The Flora of Canada, Part 2," National Museum of Natural Sciences and National Museums of Canada, Ottawa.

Scott, R. K., Jaggard, K. W., and Sylvester-Bradley, R. (1994). Resource capture by arable crops. In "Resource Capture by Crops: 52nd Easter School" (J. L. Monteith, R. K. Scott, and M. H. Unsworth, eds.), pp. 279–302. Nottingham University Press, Nottingham.

Seemann, J. F., Badger, M. R. and Berry, J. A. (1984). Variations in specific activity of ribulose-1,5-bisphsphate carboxylase between species utlizing differing photosynthetic pathways. *Plant Physiol.* **74**, 791–794

Shoji, S., Kurebayashi, T., and Yamada, I. (1990). Growth and chemical composition of Japanese pampas grass (*Miscanthus sinensis*) with special reference to the formation of dark-colored Andisols in Northeastern Japan. *Soil Sci. Plant Nutrit.* **36**, 105–120.

Slatyer, R. O. (1970). Comparative photosynthesis, growth and transpiration of two speices of *Atriplex. Planta* **93**, 175–189.

Smith, A. M., and Woolhouse, H. W. (1983). Metabolism of phosphoenolpyruvate in the C_4 cycle during photosynthesis in the phosphoenolpyruvate-carboxykinase C_4 grass *Spartina anglica* Hubb. *Planta* **159**, 570–578.

Snaydon, R. W. (1991). The productivity of C_3 and C_4 plants—a reassessment. *Funct. Ecol.* **5**, 321–330.

Sobral, B. W. S., Braga, D. P. V., Lahood, E. S., and Keim, P. (1994). Phylogenetic analysis of chloroplast restriction enzyme site mutations in the *Saccharinae* Griseb. Subtribe of the *Andropogoneae* Dumort. Tribe. *Theor. Appl. Genetics* **87**, 843–853.

Squire G. R. (1990). "The Physiology of Tropical Crop Production," CAB International: Wallingford.

Sugiyama, T. (1973). Purification, molecular, and catalytic properties of pyruvate phosphate dikinase from the maize leaf. *Biochemistry* **12**, 2862–2868.

Sugiyama, T., and Hirayma, Y. (1983). Correlation of the activities of phosphoenolpyruvate carboxylase and pyruvate, orthophosphate dikinase with biomass in maize seedlings. *Plant Cell Physiol.* **24**, 783–787.

Sugiyama, T., Schmitt, M. R., Ku, S. B., and Edwards, G. E. (1979). Differences in cold lability of pyruvate, orthophosphate dikinase among C_4 species. *Plant Cell Physiol.* **20**, 965–971.

Teeri, J. A. (1988). Interaction of temperature and other environmental variables influencing plant distribution. In "Plants and Temperature. Society for Experimental Biology Symposium No. XXXXII" (S. P. Long and F. I. Woodward, eds.), pp. 77–89. Company of Biologists Ltd., Cambridge.

Teeri, J., and Stowe, L. G. (1976). Climatic patterns and the distribution of C_4 grasses in N. America. *Oecologia* **23**, 1–12.

Thomas, S. M., and Long, S. P. (1978). C_4 photosynthesis in *Spartina townsendii* (sensu lato) at low and high temperatures. *Planta* **142**, 171–174.

Tieszen, L. L., Senyimba, M. M., Imbamba, S. K., and Troughton, J. H. (1979). The distribution of C_3 and C_4 grasses and carbon isotope discrimination along an altitudinal and moisture gradient in Kenya. *Oecologia* **37**, 337–350.

Tilman, D. (1990). Mechanisms of plant competition for nutrients: The elements of a predictive theory of competition. In "Perspectives on Plant Competition" (J. Grace and D. Tilman, eds.), pp. 117–141. Academic Press, New York.

Tilman, D., and Wedin, D. (1991). Plant traits and resource reduction for five grasses growing on a nitrogen gradient. *Ecology* **72**, 685–700.

Troughton, J. H., and Card, K. A. (1975). Temperature effects on the carbon-isotope ratio of C_3, C_4 and Crassulacean acid metabolism (CAM) plants. *Planta* **123**, 185–190.

Uemura, K., Suzuki, Y., Shikanai, T., Wadano, A., Jensen, R. G., Chmara, W., and Yokota, A. (1996). A rapid and sensitive method for determination of relative specificity of

Rubisco from various species by anion-exchange chromatography. *Plant Cell Physiol.* **37**, 325–331.

von Caemmerer, S., Millgate, A., Farquhar, G. D., and Furbank, R. T. (1997). Reduction of ribulose-1,5-bisphosphate carboxylase/oxygenase by antisense RNA in the C_4 plant *Flaveria bidentis* leads to reduced assimilation rates and increased carbon isotope discrimination. *Plant Physiol.* **113**, 469–477.

Wedin, D., and Tilman, D. (1993). Competition among grasses along a nitrogen gradient: initial conditions and mechanisms of competition. *Ecol. Monogr.* **63**, 199–229.

Wild, A. and Jones, L. H. P. (1988). Mineral nutrition of crop plants. *In* "Russell's Soil Conditions and Plant Growth" (A. Wild, ed.), 11th ed., pp. 69–112. Longmans, Harlow.

Wilson, J. R., and Ford, C. W. (1971). Temperature influences of the growth, digestibility, and carbohydrate composition of two tropical grasses, *Panicum maximum* and *Setaria sphacelata* and two cultivars of temperate grass Lolium. *Aust. J. Agric. Res.* **22**, 563–571.

Winter, K. (1981). C_4 plants of high biomass in arid regions of Asia—occurrence of C_4 photosynthesis in Chenopodiaceae and Polygonaceae from the Middle East and USSR. *Oecologia* **48**, 100–106.

Winter, K., Schmitt, M. R., and Edwards, G. E. (1982). *Microstegium vimineum*, a shade adapted C_4 grass. *Plant Sci. Lett.* **24**, 311–318.

Wong, S. C. (1979). Elevated atmospheric partial pressure of CO_2 and plant growth. I. Interactions of nitrogen nutrition and photosynthetic capacity in C_3 and C_4 plants. *Oecologia* **44**, 68–74.

Wong, S. C., and Osmond, C. B. (1991). Elevated atmospheric partial pressure of CO_2 and plant growth. III. Interactions between *Triticum aestivum* (C_3) and *Echinochloa frumentacea* (C_4) during growth in mixed culture under different CO_2, N-nutrition and irradiance treatments, with emphasis on belowground responses estimated using the Delta ^{13}C value of root biomass. *Aust. J. Plant Physiol.* **18**, 137–152.

Zelitch, I. (1973). Plant productivity and the control of photorespiration. *Proc. Nat. Acad. Sci. U.S.A.* **70**, 579–584.

8

Success of C_4 Photosynthesis in the Field: Lessons from Communities Dominated by C_4 Plants

Alan K. Knapp and Ernesto Medina

I. Introduction

Plants with the C_4 photosynthetic pathway occur in a broad array of biomes as both minor and major components of plant communities (Ehleringer and Monson, 1993). However, in a few community types, notably some temperate grasslands and tropical or subtropical savannas, C_4 plants dominate in terms of biomass, productivity, and cover. It is in these communities that studies of C_4 photosynthesis and its relationship to environmental stress can help identify the abiotic and biotic factors that ultimately control the ecological success of C_4 plants. Similarly, a comparative assessment of functional traits in the dominant C_4 and subdominant C_3 species can provide insight into a variety of adaptive characteristics associated with the C_4 photosynthetic pathway.

The goal of this chapter is to synthesize research from field studies conducted in two plant communities dominated by C_4 grasses: the North American tallgrass prairie, a temperate, subhumid grassland type in the central United States, and the neotropical savannas of South America. In both of these communities, multiple abiotic and biotic factors shape overall plant community composition and interact to determine which species are dominant. Clearly, use of the C_4 photosynthetic pathway by the dominant grasses is only one of several morphological and physiological characteristics responsible for their success. Yet this pathway provides one of the sharpest distinctions between the dominant and subdominant species in these grasslands and savannas. We do not limit our consideration of field responses to those directly related to C_4 photosynthesis; rather, we attempt to highlight an array of adaptive characteristics that, in combination with the C_4 photosynthetic pathway, enables these plants to dominate the flow of energy into these grassland and savanna ecosystems.

II. The C_4-Dominated Tallgrass Prairies of North America

The tallgrass (true or bluestem) prairie was once one of the largest grassland types in North America (Samson and Knopf, 1994), extending from Canada to Texas, and eastern Kansas to Ohio as part of the "prairie peninsula" (Transeau, 1935; Risser et al., 1981). Across this broad geographic area (>680,000 km^2) there were a variety of prairie "types" (e.g., wet prairies, sand prairies) due to specific edaphic or topographic features. Today, most of this once extensive grassland has been plowed and is used for row crop agricultural production (Samson and Knopf, 1994). Throughout most of its original range, tallgrass prairie was, and is still is, dominated by the C_4 grasses of the Andropogoneae and Paniceae tribes (Waller and Lewis, 1979): *Andropogon gerardii* Vitman., *Sorghastrum nutans* (L.) Nash, *Schizachyrium scoparium* Michx., and *Panicum virgatum* L. At its northern extent, however, the proportion of C_3 grasses increases (Terri and Stowe, 1976; Tieszen et al., 1997; Sage and Wedin, Chapter 10, this volume). The C_4 grasses are tropical to subtropical in origin and are thought to have been restricted to the southeastern U.S. flora prior to their expansion into the Great Plains (Risser et al., 1981). Tallgrass prairie can be very productive, although spatial and interannual variability is high. For example, annual aboveground production averaged 417 g m^{-2} over a 20-year period in northeastern Kansas (Knapp et al., 1998), but ranged from 179 to 756 g m^{-2} (Briggs and Knapp, 1995). Belowground biomass is estimated to be 2 to 4 times higher than aboveground biomass (Rice et al., 1998). Today, the largest unplowed tract of tallgrass prairie lies in the Flint Hills of eastern Kansas (Knapp et al., 1998). The previously mentioned C_4 grasses are clearly dominant in the Flint Hills as well as at the Konza Prairie Research Natural Area, a long-term ecological research site (Callahan, 1984) where most of the tallgrass prairie field studies reviewed in this chapter were completed.

There are three factors, climate, fire and grazing, recognized as critical for the origin and persistence of the tallgrass prairie. None of these have been constant in time or space, but their independent and interactive roles are widely accepted (Borchert, 1950; Wells, 1970; Risser et al., 1981; Axelrod, 1985; Anderson, 1990; Seastedt and Knapp, 1993). The continental climate of the central Great Plains, although extremely variable, has been characterized as having warm summers and cold winters with most precipitation occurring in the spring and early summer months (Borchert, 1950). On average, annual precipitation is sufficient to support dominance by C_3

woody vegetation even at the drier, western extent of the tallgrass prairie (830 mm/yr at Konza Prairie; Knapp et al., 1998). Periodic summer droughts and less frequent, but severe, multiple-year droughts are probably more important than average precipitation in controlling the floristic nature of this grassland (Weaver, 1954; Weakly, 1962).

Nonclimatic factors such as fire and grazing also are critical factors controlling the extent of the tallgrass prairie ecosystem (e.g., Wells, 1970; Vogl, 1974; Axelrod, 1985; Anderson, 1990). In the absence of fire, expansion of forest vegetation into grasslands is well documented (Bragg and Hulbert, 1976; Abrams et al., 1986; Knight et al., 1994; Loehle et al., 1996). Fire affects community composition, primary production, and a variety of organismic through landscape-level patterns and processes (Collins and Wallace, 1990). The most notable native grazer in the tallgrass ecosystem is the North America bison (*Bos bison*). Axelrod (1985) noted that the presence of vast herds of these ungulates paralleled the postglacial expansion of North America grasslands. The impact of bison on plant community composition, nutrient cycling, and energy flow is well documented (Coppock et al., 1983; Vinton and Hartnett, 1992; Vinton et al., 1993).

A. The C_4 Grasses and Their C_3 Competitors in Tallgrass Prairie

At Konza Prairie there are more than 500 species of vascular plants, with the vast majority of these being C_3 forbs, grasses, and sedges (Freeman, 1998). However, it is the C_4 grasses that compose most of the canopy cover and biomass constituting as much as 80% of aboveground net primary productivity (Briggs and Knapp, 1995). The dominant grasses differ from the forbs in a variety of ways besides photosynthetic pathway (Table I). Many of these characteristics, such as leaf size and root habit, affect the overall ecophysiological performance of these species and thus, interact with photosynthetic carbon uptake. A review of these interactions follows.

1. Plant Water Relations Several studies have compared the water relations of C_4 grasses in the tallgrass ecosystem with co-occurring C_3 forb and woody species (Hake et al., 1984; Knapp, 1986a; Knapp and Fahnestock, 1990; Martin et al., 1991; Axmann and Knapp, 1993; Fahnestock and Knapp 1993, 1994; Turner et al., 1995; Turner and Knapp, 1996). In general, the C_4 grasses maintain leaf water status at levels similar to the forbs and other growth forms during the wetter portions of the growing season, but lower leaf water potentials during the drier periods (Fig. 1; Knapp, 1986a). These differences may reflect the different rooting habits of the forbs compared to the grasses (Weaver, 1958). Although the grasses are noted for having deep roots (>2 m in deeply developed soils; Weaver 1968), the majority of root biomass occurs in the upper 25 cm (Jackson et al., 1996; Blair, Kansas State University, unpublished data, 1997). Deep roots may be important for

Table I Characteristics of the Dominant Grasses Compared to Characteristics of the Subdominant Forbs (Numerous Herbaceous, Nongraminoid Species) in a Temperate, Subhumid Grassland

	Dominant grasses	Subdominant forbs
Photosynthetic pathway	C_4	C_3
Taxonomic affiliation	Monocot One family	Dicot Numerous families
Leaf shape size	Linear, narrow	Many ovate, broader
Leaf orientation	Mostly vertical	More horizontal
Root habit	Fibrous, most shallow, some very deep roots	Taproot, many deeper roots
Pollination mode	Exclusively wind	Mostly insect
Flowering phenology	Late season	Throughout the season
Examples	*Andropogon gerardii* *Sorghastrum nutans* *Panicum virgatum*	*Artemisia ludoviciana* *Salvia pitcherii* *Psoralea tenuiflora* *Solidago canadensis* *Vernonia baldwinii*

survival during severe drought, but apparently these roots are not capable of meeting demands of transpiration during the drier portions of the growing season. Thus, during the late summer, water potential declines more in grasses than in forbs (Fig. 1). Forbs may have as deep or deeper roots than the grasses, and studies of their depth distribution indicates that a greater proportion of forb root biomass occurs deeper in the soil profile (Weaver 1958).

2. Leaf-level Gas Exchange Maximum rates of leaf-level photosynthesis (A_{max}) can exceed 40 μmol m^{-2} sec^{-1} in *A. gerardii* and usually occur in the spring (Knapp, 1985). Comparative studies of the dominant C_4 grasses and C_3 forbs indicate that most forbs have lower A_{max} (Turner *et al.*, 1995; Turner and Knapp, 1996). Photosynthetic rates in *A. gerardii* may not reach light saturation even at full sunlight levels, whereas rates of photosynthesis in the C_3 forbs increase only marginally at light levels above half full sunlight (Fig. 2). Although there are some C_3 forbs in the tallgrass ecosystem that behave as spring ephemerals (Knapp, 1986b), the majority are active throughout the growing season and are thus similar in phenological pattern to the C_4 grasses. This contrasts with other grasslands such as the semi-arid, shortgrass steppe in Colorado, where strong seasonal partitioning in photosynthetic activity between C_4 and C_3 species has been documented (Kemp and Williams, 1980; Monson *et al.*, 1986). Throughout the tallgrass

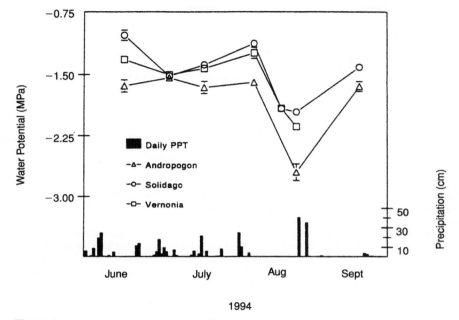

Figure 1 Seasonal course of midday leaf water potential (mean ± SE) in the dominant C_4 tallgrass prairie grass at Konza Prairie, *Andropogon gerardii*, and two C_3 forbs, *Solidago canadensis* and *Vernonia baldwinii*. Also shown is the seasonal pattern of precipitation. *Vernonia* leaves senesced prior to the last sampling period. Plants measured were growing in a lowland topographic position in annually burned tallgrass prairie. [From Davy, A. L. (1996). "Responses of two forbs, *Solidago canadensis* and *Vernonia baldwinii* to variation in nitrogen availability and light environments in lowland, annually burned tallgrass prairie." Masters thesis, Division of Biology, Kansas State University, Manhattan, Kansas.]

prairie growing season, photosynthetic rates in forbs are lower than in grasses when water is not limiting (although see Turner and Knapp, 1996, for exceptions). However, differences between the two are marginal during dry periods (Fig. 3).

Stomatal conductance in the C_4 grasses is usually lower than in the forbs, which is typical of comparisons of C_4 and C_3 species (Martin *et al.*, 1991; Fahnestock and Knapp, 1993, 1994; Turner *et al.*, 1995; Turner and Knapp, 1996; Fig. 3). When these gas exchange differences are coupled with differences in leaf size and temperature (both of which are higher in forbs), leaf-level photosynthetic water use efficiency (moles CO_2/moles H_2O) is higher in the C_4 grasses (Turner *et al.*, 1995; Turner and Knapp, 1996). The higher water use efficiency may enable these species to use water more conservatively between rain events and delay drought stress in the summer. However, the C_4 grasses still experience lower water potentials late in the

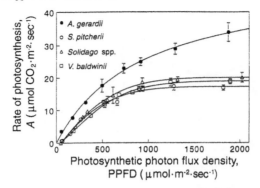

Figure 2 Response of net photosynthesis (mean ± SE) to increasing light (PPFD) in the dominant C_4 grass in tallgrass prairie, *Andropogon gerardii*, and three C_3 forbs, *Salvia pitcherii*, *Solidago canadensis*, and *Vernonia baldwinii*. Note that photosynthesis in the forbs saturates at approximately half full-sunlight whereas photosynthesis in the C_4 grass continues to increase with increasing light. Plants measured were under no apparent water stress in an annually burned watershed at Konza Prairie. [Reprinted with permission from Turner, C. T., and Knapp, A. K. (1996). Responses of a C_4 grass and three C_3 forbs to variation in nitrogen and light in tallgrass prairie. *Ecology* **77**, 1738–1749.]

season relative to the forbs, perhaps reflecting the aforementioned shallow root distribution of the grasses.

3. Leaf Nitrogen Several field studies have shown that leaf N concentrations are consistently lower in C_4 grasses than in co-occurring C_3 species (Wedin and Tilman, 1990; Fig. 4). Fertilization studies indicate that leaf N concentrations and aboveground productivity will increase in response to additional N in both forbs and grasses (Owensby *et al.*, 1970; Turner *et al.*, 1995; Turner and Knapp, 1996). In contrast, additions of P have no effect (Seastedt and Ramundo, 1990). In general, C_4 grasses are thought to have lower N requirements and greater photosynthetic nitrogen-use efficiency than co-occurring C_3 species. Moreover, in sites dominated by C_4 grasses, the C_4 species themselves may help maintain low levels of N. In such sites, N cycling is slowed, leading to reduced levels of available N in the soil (Wedin and Tilman, 1990, 1993). This is due to the poor quality (low N) of senescent C_4 grass tissue that immobilizes N and reduces the rate of mineralization. Long-term fertilization with N decreases the dominance of C_4 grasses in the tallgrass prairie and other grasslands (Gibson *et al.*, 1993; Wedin and Tilman, 1993).

B. Factors That Enhance Dominance by C_4 Grasses in Tallgrass Prairie

1. Fire Annual or frequent spring (April) fire in tallgrass prairie enhances the dominance of C_4 grasses (Gibson and Hulbert, 1987; Collins and Gib-

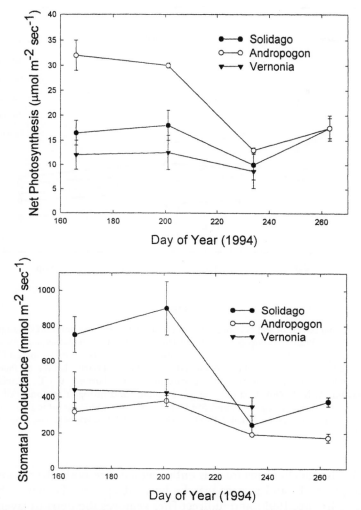

Figure 3 Seasonal course of net photosynthesis and stomatal conductance (mean ± SE) in the dominant C_4 tallgrass prairie grass at Konza Prairie, *Andropogon gerardii*, and two C_3 forbs, *Solidago canadensis* and *Vernonia baldwinii*. *Vernonia* leaves senesced prior to the last sampling period. Plants measured were growing in a lowland topographic position in annually burned tallgrass prairie. [From Davy, A. L. (1996). "Responses of two forbs, *Solidago canodensis* and *Vernonia baldwinii* to variation in nitrogen availability and light environments in lowland, annually burned tallgrass prairie." Masters thesis, Division of Biology, Kansas State University, Manhattan, Kansas.]

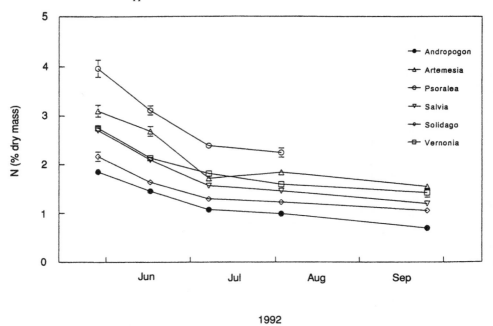

Figure 4 Seasonal course of leaf N concentration (mean ± SE) in the dominant C_4 tallgrass prairie grass at Konza Prairie, *Andropogon gerardii*, and five C_3 forbs, *Artemisia ludoviciana, Psoralea tennuiflora, Salvia pitcherii, Solidago canadensis,* and *Vernonia baldwinii*. Data are combined from plants in annually burned and unburned watersheds at Konza Prairie. Leaves of *Psoralea* senesced at midseason. [Reprinted with permission from Turner, C. T., Kneisler, J. R., and Knapp, A. K. (1995). Comparative gas exchange and nitrogen responses of the dominant C_4 grass, *Andropogon gerardii*, and five C_3 forbs to fire and topographic position in tallgrass prairie during a wet year. *Intern. J. Plant Sci.* **156,** 216–226. Copyright © The University of Chicago Press.]

son, 1990). The mechanisms for this enhancement are both direct (some early spring C_3 forbs and grasses may be harmed by fire; Hartnett, 1991; Johnson and Knapp, 1995) and indirect [fire removes the detritus layer in tallgrass prarie (Knapp and Seastedt, 1986), and alters surface radiation and temperature regimes that favors C_4 grasses]. Detritus accumulation occurs because above-ground primary production in this grassland is substantial (averaging 350–500 g m^{-2} year^{-1}; Briggs and Knapp, 1995) but decomposition of aboveground detritus is relatively slow (Kucera *et al.*, 1967). Thus, detritus may accumulate to as much as 1000 g m^{-2} and to 1 m in depth (Weaver and Rowland, 1952; Knapp, 1984a). The detritus layer normally functions as an energy, water, and nutrient filter and fire serves a vital function in this grassland by removing the litter layer (Knapp and Seastedt 1986).

Immediately after fire, the spring microclimate for plant growth can be characterized as one of high light and warm soil temperatures relative to unburned sites (Knapp, 1984a). Such a microclimate is favorable to CO_2 assimilation by plants with the C_4 pathway. Later in the season, soil and plant water stress may be greater in burned sites (due to a greater amount of transpiring leaf area), compromising the advantage of a warmer microclimate (Knapp, 1985; Briggs and Knapp, 1995). Although leaf N may be higher in some C_4 grasses immediately after fire (Knapp, 1985; Svejcar and Browning, 1988), in general, fire (particularly annual fires) reduces N availability (Ojima *et al.*, 1994; Blair, 1997). The ability of the C_4 grasses to dominate sites with low N availability is key to their successful postfire response.

Leaves of *A. gerardii* that develop in the high light, postfire environment are thicker and wider, have higher specific leaf mass, greater chlorophyll a/b ratios, higher N concentration, and greater stomatal density (Fig. 5; Knapp and Gilliam, 1985; Knapp *et al.*, 1998). As a result of these attributes, leaf-level photosynthetic rates as well as photosynthetic water- and nitrogen-use efficiencies are greater in burned sites (Fig. 5; Knapp 1985; Svejcar and Browning, 1988). These differences are consistent with those documented from comparative studies of sun versus shade plants (Boardman, 1977), attesting to the importance of the spring light environment for these C_4 species. Moreover, DeLucia *et al.* (1992) found that soil temperatures

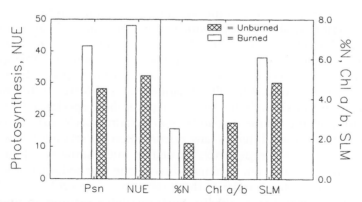

Figure 5 Comparison of maximum net photosynthetic rate (μmol m^{-2} sec^{-1}), nitrogen-use efficiency (μmol CO_2 g N^{-1} sec^{-1}), leaf N (%), chlorophyll a/b ratio, and specific leaf mass (SLM, mg leaf tissue/cm^2) in *Andropogon gerardii* in annually burned vs. unburned tallgrass prairie. [Data are from Knapp, A. K. (1985). Effect of fire and drought on the ecophysiology of *Andropogon gerardii* and *Panicum virgatum* in a tallgrass prairie. *Ecology* **66**, 1309–1320; and Knapp, A. K., and Gilliam, F. S. (1985). Response of *Andropogon gerardii* (Poaceae) to fire-induced high vs. low irradiance environments in tallgrass prairie: Leaf structure and photosynthetic pigments. *Amer. J. Bot.* **72**, 1668–1671.]

less than 20°C, such as those found in unburned sites, reduce N transport, foilar N concentration, and leaf-level photosynthesis in *A. gerardii*. Thus, increased light and warmer soil temperatures in the spring have measurable positive effects on the ecophysiology of this dominant C_4 grass in tallgrass prairie.

In some years, factors other than microclimate may offset the positive aspects of fire and decrease production. Primary among these is the degree of water stress the grasses experience during the growing season. As noted earlier, leaf area is greater after fire, as is tiller density and stomatal conductance (Knapp, 1985). These responses can lead to reduced soil moisture and increased water stress in burned tallgrass prairie (Knapp, 1985; Knapp, et al., 1993a; Briggs and Knapp, 1995). In an analysis of 19 years of production data from Konza prairie, Briggs and Knapp (1995) found that spring fire significantly increased aboveground production in lowland sites with deep soils, but not in upland sites where water stress is more severe (Knapp, et al., 1993a). Moreover, in drought years, aboveground primary production in unburned sites may be equal to, or greater than, that in burned sites (Knapp et al., 1998). Thus, soil and plant water relations can mediate production responses to fire. Given the prevalence of drought in this region (Borchert, 1950; Weaver, 1954; Weakly, 1962), selective forces are expected to strongly favor species with high photosynthetic water-use efficiency and the ability to tolerate and survive severe water stress. Such characteristics are found in the dominant C_4 grasses.

2. Drought A unique feature of the tallgrass prairie climate is that it favors plants that must be able to cope with extreme interannual variation in water availability (Knapp et al., 1998). In years with "normal" precipitation (which is sufficient to support forest), the C_4 grasses must be able grow rapidly in this highly competitive environment. In other years, droughts that may be severe enough to kill established trees (Weaver, 1968) must be tolerated and survived by the grasses. Hence, these species must have characteristics of both competitive and stress tolerating species to be successful (Grime, 1979). Several studies have documented the very low leaf water potentials these grasses experience, as well as interannual variability in water stress that ranges from minimal stress in one year followed by years with leaf water potentials similar to those recorded in desert biomes (Fig. 6) (Hake et al., 1984; Knapp, 1984b). Such variability requires an unique set of drought tolerance and recovery mechanisms.

The C_4 dominant *A. gerardii* tolerates drought periods through a variety of mechanisms. This species and other C_4 codominants are capable of significant osmotic adjustment (0.8–1.3 MPa) in response to drought (Knapp, 1984b). Leaves also fold or roll as turgor is reduced (Redmann, 1985; Heckathorn and DeLucia, 1991) and this may reduce damage to the photosynthetic system by lowering leaf temperature and transpiration and reducing damage from photoinhibition. Surprisingly, stomatal closure in *A. gerardii* is not complete in leaves that fold at −2.5 to −3.0 MPa and CO_2 uptake still occurs (Fig. 6; Knapp, 1985). Leaf N is translocated belowground to rhizomes during drought (Hayes, 1985; Heckathorn and DeLucia, 1994), which protects N from loss, but may limit photosynthetic recovery once water stress is relieved. Nonetheless, Knapp (1985) found that even when leaves of *A. gerardii* were subjected to extremely low water potentials (−6.6 MPa), leaf senescence and chlorophyll degradation did not occur and photosynthesis in *A. gerardii* recovered to approximately 28% to 48% of predrought levels after soil water stress was alleviated.

In summary, relative to their C_3 competitors, C_4 grasses have lower stomatal conductance, higher photosynthetic water-use efficiency, and potentially higher photosynthetic and growth rates, as long as soil water is readily available. However, their relatively shallow root systems subject these grasses to greater water stress than experienced by the forbs or invading woody species during periods of drought (Axmann and Knapp, 1993). Although C_4 grasses of the tallgrass prairie can recover from extreme droughts, the recovery capabilities of are probably not as great as those documented for dominant C_4 species in the more xeric shortgrass steppe of North America (Sala and Lauenroth, 1982). It is likely that the trade-off between maximum growth rate and drought tolerance is shifted more towards growth in the more mesic tallgrass prairie.

C. Factors That Reduce the Tendency Toward Dominance by C_4 Grasses in Tallgrass Prairie

1. Grazing Native ungulates have grazed the tallgrass prairie since its formation (Axelrod, 1985). The grazing activities of ungulates such as bison generally decrease the dominance of *A. gerardii* (Collins, 1987). This C_4 dominant is preferred forage by bison, and ungulates select grazing sites based on its abundance (Vinton *et al.*, 1993). Because the competitively superior tall grasses are selectively grazed, whereas many of the co-occurring forbs are not, species richness increases in sites grazed by bison (Collins *et al.*, 1998). Similar responses have been documented in other grasslands (Belsky, 1992). Despite its reduction in dominance, *A. gerardii* does not disappear from tallgrass sites unless grazing pressures are heavy and chronic. Indeed, the long-term coexistence of the dominant C_4 grasses with the reported large herds of bison in the American Great Plains suggest that historic grazing pressures were not chronically severe, and that these species have mechanisms for coping with herbivory.

When *A. gerardii* is grazed, it, like most grasses, uses both intrinsic and extrinsic mechanisms to compensate for tissue lost (McNaughton, 1983). For example, leaf-level photosynthesis (Wallace, 1990) and leaf production

(Archer and Detling 1984) may increase after defoliation whereas the biomass of belowground storage organs may decrease (Vinton and Hartnett, 1992). This indicates that aboveground compensation occurs at the expense of belowground reserves. The water relations of this grass may also be improved in grazed sites (Fahnestock and Knapp, 1993) leading to increased photosynthesis (Fig. 6). Overall, rapid regrowth after defoliation may compensate for tissue lost, but this compensation is not sustainable in the face of chronic grazing (Vinton and Hartnett, 1992; Turner *et al.*, 1993), and the timing and intensity of defoliation strongly affect regrowth potential (Tate *et al.*, 1994).

D. The Ideal Environment for C_4 Grass Dominance in Temperate Grasslands

Although fire, climate variability (drought), and grazing are all intrinsic features of the tallgrass prairie environment, only fire strongly favors dominance by C_4 grasses. In particular, it is the high-light, warm, postfire microenvironment that benefits plants with the C_4 pathway. Severe drought may cause mortality of invading tree species, but drought also affects C_4 grasses more than the deep-rooted C_3 forbs (Fig. 1). Thus, frequent fire is the critical factor allowing C_4 grasses to remain dominant. Increased moisture availability might be expected to increase the invasion of woody species, or less drought-tolerant C_3 species, but annual fire has been coupled with high water availability since 1991 at Konza Prairie as part of a long-term irrigation study (Knapp *et al.*, 1994). Even with this coupling, the C_4 grasses have maintained their dominance in this study site (Knapp, unpublished data, 1997).

Grazing tolerance mechanisms are also well documented in *A. gerardii*, but they are not sufficient to offset the negative effects of herbivory on C_4 grass dominance. In addition to differentially impacting the grasses directly, grazing may enhance the fertility of grasslands by rapidly recycling N (Detling, 1988; Frank *et al.*, 1994; Holland and Detling, 1994). Higher N availability favors C_3 species (Wedin and Tilman, 1993; Fig. 4) and again, fire is critical for the success of the C_4 grasses even in grazed sites (Collins, 1987). Bison graze in relatively discrete patches throughout the growing season (Day and Detling, 1990; Fahnestock and Knapp 1993, 1994; Vinton *et al.*, 1993) and in unburned sites they may return to these established patches year after year. After fire has swept through a grassland, these patches are less apparent and bison may be more likely to establish patches in new areas, allowing the previous year's patches to recover (Catchpole, 1996). Moreover, frequent fire reduces soil N availability (Blair, 1997), and thus may partially offset the increase in soil N caused by grazing.

It appears that the ideal environment for C_4 dominance in tallgrass prairie is one with annual spring fire, no ungulate grazing, and abundant growing season soil moisture. Such an environment provides high light and warm

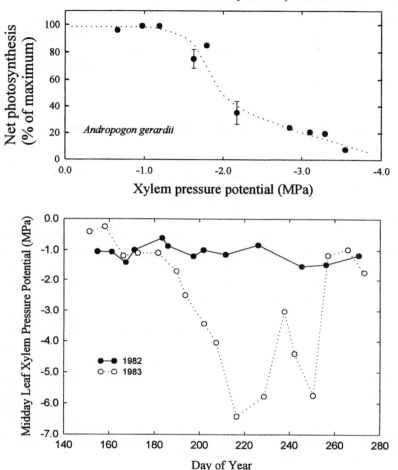

Figure 6 Top: Response of net photosynthesis in *Andropogon gerardii* to decreasing leaf xylem pressure potential. Reprinted with permission from Knapp *et al.*, 1993a. Bottom: Interannual variation in midday leaf xylem pressure potential in *A. gerardii* in unburned tallgrass prairie. In 1982, there was a relatively normal amount of precipitation, whereas 1983 was significantly drier. Leaves from the same site were measured in both years. [Reprinted with permission from *Oecologia*. Water relations and growth of three grasses during wet and drought years in a tallgrass prairie. A. K. Knapp, **65**, 37, Fig. 1, 1984. Copyright © 1984, Springer-Verlag GmbH & Co. KG.]

temperatures in the spring and sufficient soil moisture in the summer to allow the grasses to grow tall and outcompete the shorter C_3 species. Competition for light aboveground and the immobilization and preemption of N belowground due to low quality litter and the relatively shallow fibrous roots of the grasses appear to be the primary mechanisms by which the C_4 species outcompete and dominate C_3 species in this grassland. Clearly, the C_4 photosynthetic pathway is critical to this success with traits such as high photosynthetic rates, low N requirements, and high water-use efficiency.

III. The C_4-Dominated Neotropical Savannas

Tropical savannas are ecosystems characterized by a continuous herbaceous cover of bunch grasses and sedges, almost exclusively with the C_4 photosynthetic pathway. Growth in these species exhibits clear seasonality, which is related to water availability (Sarmiento, 1984; Cole, 1986; Frost et al., 1986). Woody species (shrubs, trees, and palms) may occur in varying densities depending on soil depth and water availability (Cole, 1986; Medina and Silva, 1990). Legumes are commonly found in savanna vegetation as herbaceous, shrub, and tree species. In South American savannas, legume diversity can be high, although their biomass seldom surpasses 1% of the total aboveground biomass (Medina and Bilbao, 1991).

Tropical savanna ecosystems cover a large expanse of South America, and although one can identify physiognomic variation among ecosystems from different locations, they have generally converged towards C_4 dominance (Sarmiento, 1983). C_4-dominated tropical savannas are similar to the temperate tallgrass prairies described previously in that they are characterized by strong seasonality in water availability, mostly low soil fertility, and the frequent occurrence of fires. In modern times, fires in these ecosystems are most commonly associated with human activities. However, as is the case with tallgrass prairie, fires were an important ecological factor in South America even before the arrival of humans about 15,000 years ago.

A. Distribution of Savannas in Tropical South America

The largest savanna ecosystems in South America are located in Brazil (Cerrado, Gran Pantanal, Rio Branco–Rupununi, Amazon campos), Venezuela and Colombia (Llanos, Gran Sabana), Bolivia (Llanos de Mojos), and the coastal campos of the Guayanas (for detailed descriptions see Sarmiento, 1983, and detailed literature listings in Huber, 1974). The surface covered by savanna vegetation in tropical South America is a bit larger than 2×10^6 km². Within this area, considerable variation in savanna physiognomy can be expected, ranging from (1) dry savannas in areas with low rainfall and medium to light soils, to (2) seasonal savannas (including woodlands of variable density), and (3) pure grasslands and wet savannas that are flooded during the rainy season and possess medium to heavy soils (Johnson and Tothill, 1985).

Savannas have been differentiated physiognomically according to the proportion, height, and cover of woody perennials (Sarmiento, 1984). The greatest diversity of these physiognomic groups can be found in the Brazilian Cerrados (Eiten, 1972; Sarmiento, 1984). These include: (1) grasslands ("campo limpo") without woody species taller than the herbaceous stratum; (2) tree or shrub savannas ("campo sujo") with woody components less than 2 m in height and less than 2% in cover; (3) wooded savanna ("campo cerrado") with woody components up to 10 m tall, and cover between 2% and 15%; and (4) savanna woodland ("cerrado") with a woody cover between 15% and 40%. As would be expected, there is almost continuous variation as one of these groups geographically merges with another.

A classification for ecophysiological purposes based on the seasonality of soil water regime has been proposed by Sarmiento (1983, 1984). This classification includes the semi-seasonal savannas with a moderate dry season that interrupts a mostly wet climate. These savannas are found within the Amazon basin (Amazon campos). The seasonal savannas, with a well defined rainless period that lasts between 2-6 months. In areas covered by these savannas, fire is common with average frequencies between 0.5–1 per year. The hyperseasonal savannas, characterized by the frequent alternation of periods of drought and soil saturation. These seasonally flooded savannas are common in the Orinoco llanos (Ramia, 1967), in the Gran Pantanal region of Brazil, and in Bolivia (Haase and Beck, 1989). Finally, there are the flooded savannas (Esteros), characterized by a period of excess water that lasts for more than 8 months (Medina *et al.*, 1976; Escobar and González-Jiménez, 1979).

B. The C_4 Grasses and Their C_3 Competitors in Tropical Savannas

The grass cover in South American savannas is dominated by the same tribes of the subfamily Panicoideae (tribes Andropogoneae and Paniceae) and, in some cases, the same genera that dominate the temperate tallgrass prairies. A third group, however, that is also important in South American savannas belong to the subfamily Chloridoideae *sensu latu* (including tribes Eragrostideae and Chloridideae *s.s.*). A clear example of the taxonomic and ecological differentiation of grass species in a representative savanna region of Central Brazil has been produced by Filgueiras (1991), who listed all the grass species of the Federal District of Brazil, representing an area of 5814 km². Species were ranked in importance in several physiognomic types of savannas chosen to reflect broad variation in soil depth and soil water availability, with probable additional variation in soil fertility. These types are characterized by the presence of a continuous grass layer and are classified as wet grasslands, nonflooded grasslands ("Campo Limpo"), shrublands ("Campo Sujo"), and woodland ("Campo Cerrado"). Most of the native species can be assigned to a well-defined photosynthetic metabolic type using the descriptions of Hattersley and Watson (1992) and Zuloaga (1987). Details of the distribution of species within the Brazilian savannas, as derived from the Filgueiras (1991) study are presented in Fig. 7 and discussed next.

In all physiognomic types both C_3 and C_4 grass species are present; however, C_4 species always dominate. C_3 species are most frequent in the wet grasslands where they represent more than 25% of all species (Fig. 7). Malate-forming C_4 species belonging to the NADP-malic enzyme (NADPme) group constitute more than 70% of the species in nonflooded savannas. The importance of this group decreases to about 50% in wet savannas, but they remain as the most frequent group in this environment. Aspartate formers that include the NAD malic enzyme (NADme) and PEP carboxykinase (PEPck) C_4 types are evenly distributed within all physiognomic types, where they represent approximately 20% of all species.

The widespread distribution of certain genera and species along broad environmental gradients within these savanna types is remarkable. *Arthropogon villosus,* a C_4 grass, occurs in all savanna types although it is more frequent in the savanna shrublands. Two C_3 grasses, *Echinolaena inflexa* and *Ichnanthus camporum,* appear in all three types of nonflooded savannas,

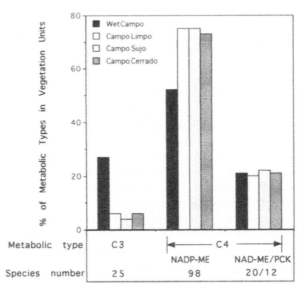

Figure 7 Relative distribution of photosynthetic types in savannas of Central Brazil, Federal District. Compiled from Filgueiras (1991) using photosynthetic pathway and C_4 subtypes listed in Hattersley and Watson (1992).

often in shade (see also Klink and Joly, 1989). However, of the 25 C_3 species recorded, 17 are restricted to the wet campos. The two recorded C_3-C_4 intermediate species of *Panicum* [*P. decipiens* and *P. hians* (=*P. milioides*), Zuloaga, 1987; Chapter 16] are also restricted to the wet campos.

Within the NADPme C_4 type, several species occur in all savannas (*Agenium goyazense, Andropogon selloanus, Schyzachirium condensatum*), but a significant number are restricted to the wet campos (e.g., *Aristida capillacea, Andropogon bicornis, Hypoginium virgatum, Axonopus comans, Paspalum decumbens*). Within this biochemical type the genus *Paspalum* has 32 recorded species of which 11 are restricted to the wet campos. This pattern is generally consistent with past studies that have observed higher frequencies of NADPme C_4 species in moist environments, especially in Australia, South Africa, and the United States (Hattersley, 1983; Ehleringer and Monson, 1993). The genus *Panicum* contains C_3 and C_4 species (Chapter 16). From the 21 recorded native species, all 6 of the C_3 species are restricted to the wet campos, together with 2 of the C_4 species (*P. aquaticum* and *P. dichotomiflorum*, both NADme). The remaining 13 C_4 species are native to drier habitats, and are predominately NADme as well.

Based on the patterns of geographic distribution described, it can be concluded that C_4 species dominate both flooded and nonflooded physiognomic types. Nevertheless, there are conspicuous C_3 species that are distributed through all savanna types. A second conclusion is that C_4 metabolism does not reduce the competitive ability of C_4 grasses in wet environments. In fact, within the photosynthetically diversified genus *Panicum* both aquatic and semi-aquatic C_3 and C_4 species coexist. For example, within the wettest physiognomic types, C_3 species such as *P. laxum, P. schwackeanum*, and *P. subtiramulosum* (belonging to the subgenus *Phanopyrum*) grow together with C_4 species such as *P. dichotomiflorum* and *P. aquaticum*.

C. Ecophysiological Properties of Tropical C_4 Grasses That Enhance Competitive Ability

1. Photosynthetic Characteristics Under optimum growth conditions, tropical C_4 grasses have photosynthetic and respiration rates per unit leaf area that are about twice as high as those of co-occurring C_3 grasses and legumes (Ludlow and Wilson, 1971; Baruch *et al.*, 1985). Optimum leaf temperature for C_4 photosynthesis in tropical grasses is 35° to 40°C (Ludlow and Wilson, 1971; San José and García-Miragaya, 1981; Baruch *et al.*, 1985), leading to the suggestion that C_4 plants have an inherent photosynthetic advantage in warm tropical climates. In the field, however, photosynthetic capacity can be strongly reduced due to nutrient and water limitations (San José and García-Miragaya, 1981). Thus, caution should be used in extrapolating gas-exchange patterns from controlled-growth studies to patterns of performance in the field. Even with this cautionary note, however, there are cases in which photosynthetic performance measured under controlled-growth conditions appears to correlate with observed ecological patterns.

In one intensive study, Baruch et al. (1985) examined three species, two African C_4 species that have been introduced into Venezuelan savannas (*Hyparrhenia rufa* and *Melinis minutiflora*) and one native Venezuelan species (*Trachypogon plumosus*). Following disturbance by fire, the introduced African species can successfully compete with *T. plumosus*, often replacing it in large areas. One of the introduced species, *H. rufa*, displaces *T. plumosus* in lowland regions, where the climate is warmer and drier, whereas the other *M. minutiflora* displaces *T. plumosus* at higher elevations (>600 m), where the climate is cooler and wetter. In the driest and least fertile sites, *T. plumosus* effectively resists displacement. When grown in controlled environments and observed under optimal environmental conditions, the introduced species exhibit higher growth rates, higher photosynthetic capacities, and higher photosynthetic temperature optima (in the case of *H. rufa* only), compared to the native species (Baruch et al., 1985). The higher rates of carbon assimilation may explain the ability of the introduced species to successfully invade recently disturbed sites. The native species exhibits higher photosynthetic rates at lower leaf water potentials, and low leaf nutrient concentrations, which may explain its persistence in the driest, least fertile sites. One of the introduced species, *H. rufa*, exhibits a higher photosynthetic temperature optimum and higher rates of photosynthesis at low water potentials compared to the other introduced species, *M. minutiflora* (Baruch and Fernandez, 1993). This may explain the differential invasion success of these species at high and low elevations. When taken together, these results demonstrate that like the case for C_3 species, C_4 species have the potential to evolve differential photosynthetic responses to the environment and exhibit photosynthetic responses to the environment that appear to reflect ecological performance in the field.

2. Rates of Organic Matter Production When grown in a greenhouse with a tropical climate, C_4 savanna plants tend to have higher growth rates than C_3 species, in part because of their higher photosynthetic capacity and higher photosynthetic temperature optima (Ludlow, 1985). However, it is now clear that productivity of plant populations under natural conditions is not always directly related to photosynthetic capacity per unit leaf area (Snaydon, 1991). The amount of leaf area produced per unit soil surface is often the parameter most closely related to productivity because it determines the degree of interception of photosynthetic active radiation (Walter, 1960). In this regard, C_4 savanna grasses also produce greater leaf area indexes than herbaceous C_3 counterparts (Sarmiento, 1992).

Measurements of the annual course of biomass accumulation in above- and belowground plant tissues show a clear pattern associated with the seasonality of rains (water availability in the soil) and the occurrence of fire. As in temperate tallgrass prairies, root biomass of tropical savanna grasses is generally higher than that of the aboveground organs (Sarmiento, 1992). In studies of two dominant tropical savanna grasses, Susachs (1984) found significantly higher amounts of carbon allocation to belowground tissues (Table II), and an inverse relationship between total root biomass and the production of green aboveground biomass (Fig. 8). In the two species studied, root biomass decreased at the beginning of the growing season as new culms and leaves were produced. Toward the end of the growing season, aboveground biomass stabilized and root biomass increased markedly. The aboveground/belowground biomass ratios were less than unity throughout the growing season (Fig. 8), though values for *Axonopus canescens* were generally higher (0.70) than those for *T. plumosus* (0.46; Table II). Burning late in the growing season caused an increase in the production of organic matter in *A. canescens*, whereas this effect is observed in *T. plumosus* only 1 year after fire.

The majority of production values available in the literature for tropical grasslands correspond to aboveground biomass, and in most cases they do not take into account mortality and standing dead tissue. As a result, current values in the literature are likely underestimates of true production in tropical savannas. Bulla *et al.* (1981) developed a simple model to determine productivity of grasslands that included mortality and the amount of organic matter decomposed during the evaluation period. The method was originally developed for aboveground biomass but has been extended subsequently to evaluate the production of belowground organs (e.g., the studies of Susachs, 1984). Long *et al.* (1989) used a similar methodology to measure primary productivity in a range of contrasting grasslands in the tropics and the subtropics. They were able to show that productivity values of tropical

Table II Productivity of C_4 (NADP-ME) Grasses in Central Llanos of Venezuela during Two Years after a Fire Late in the Growing Season

Period after burning	Productivity (g m^{-2}year^{-1})		A/B	Total
	Aboveground (A)	Belowground (B)		
Axonopus canescens				
April '82/March '83	556	1071	0.56	1627
March '83/March '84	551	784	0.70	1135
Trachypogon plumosus				
April '82/March '83	258	663	0.39	921
March '83/March '84	375	822	0.46	1197

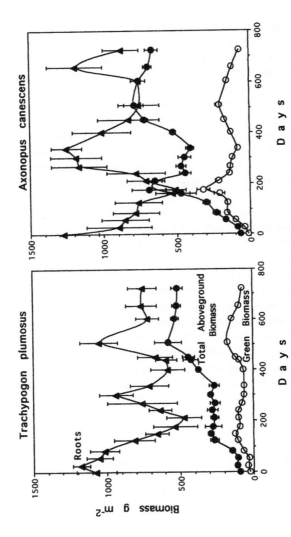

Figure 8 Annual course of belowground and aboveground biomass in seasonal savannas dominated by the C_4 grasses *Trachypogon plumosus* and *Axonopus canescens* in Central Llanos of Venezuela. [From Susachs, F. (1984). "Caracterización ecológica de las sabanas de un sector de las Llanos Bajos Centrales de Venezuela." Tesis Doctoral. Universidad Central de Venezuela. Facultad de Ciencias, Caracas.]

grasslands are considerably higher than previously thought, and in some cases reached levels reported for tropical forests. It should be mentioned that, in contrast to the patterns reported for South American tropical savannas, aboveground production in the African savannas studied by Long et al. (1989) was higher than belowground production.

3. Nitrogen Requirements and Growth Tropical C_4 grasses and sedges have been found to be more efficient than their C_3 counterparts in the use of N in photosynthesis (Wilson, 1975; Baruch et al., 1985; Field and Mooney, 1986; Li, 1993). It has been suggested that this efficiency may be a critical factor in their dominance of tropical savanna environments (Brown, 1978). Photosynthetic nitrogen–use efficiency (PNUE) is better expressed as a potential (Field and Mooney, 1986; Potential Photosynthetic NUE (PPNUE) = A_{max}/N_{leaf} [μmol CO_2 mmol^{-1}N sec^{-1}]). Other authors define photosynthetic nitrogen–use efficiency as the slope of the relationship between A_{max} and leaf N.

Li (1993) compared the PPNUE of species of Cyperaceae with different photosynthetic pathways. In this study, C_4 species of *Cyperus* exhibited PP-NUEs of 0.28–0.39, whereas those for the C_3 species were 0.18–0.25 μmol CO_2 mmol^{-1} N sec^{-1}. When the entire group of species is considered, it is clear that C_3 Cyperaceae have a larger range of variation in PPNUE than C_4 Cyperaceae. Maximum PPNUEs for C_4 grasses under optimal growth conditions are approximately 0.5 μmol CO_2 mmol^{-1} N sec^{-1} (Baruch et al., 1985).

The photosynthetic capacity of tropical grass species under greenhouse and natural conditions is linearly correlated with total leaf nitrogen content (Baruch et al., 1985; Simoes and Baruch, 1991; LeRoux and Mordelet, 1995; Anten et al., 1998). The slope of the relationship indicates a high PPNUE in all C_4 species measured. The analysis of Anten et al. (submitted) provides a comparison of the PPNUE in terrestrial and aquatic grasses. The PPNUE's values at average leaf N content in the field ranged from 0.16 to 0.23 μmol CO_2 mmol^{-1} N sec^{-1} in the C_3 aquatic grasses *Hymenachne amplexicaulis* and *Leersia hexandra*. These values ranged from 0.23 mmol CO_2 mol^{-1} N sec^{-1} in an aquatic C_4 grass (*Paspalum fasciculatum*) to 0.30 μmol CO_2 mmol^{-1} N sec^{-1} in a terrestrial tuft grass *H. rufa*.

4. Changes in Species Composition and Growth Induced by Fertilization The consequences of interspecific differences in nutrient requirements, particularly those involving N and P, has been shown with fertilization experiments. Native South American C_4 grasses can persist and function normally with low soil P availability (Medina et al., 1978; San José and Garcia-Miragaya, 1981). Dominant C_4 species in northern South American savannas, such as *T. plumosus* and *A. canescens*, respond vigorously to added N and P. In addition, retranslocation of N from leaves to belowground storage sites at the end of the growing season has been shown to be more efficient per unit leaf weight, than retranslocation of P (Medina, 1987). Interspecific differences have been observed in the nutritional requirement of C_4 savanna grasses from different ecological regions. Bilbao and Medina (1990) compared the use of N by *Andropogon gayanus* and *Paspalum plicatulum* under fertilization with N and P. *A. gayanus* always responded more markedly to P addition compared to *P. plicatulum*. In both species, however, N use-efficiency for organic matter production was improved by the simultaneous addition of P. Similar responses have been observed in fertilization experiments conducted in Australian and South African savannas (Medina, 1987). Thus, C_4 tropical grasses are similar to C_4 temperate grasses in their strong response to fertilization with N. However, P limitation does appear to play a stronger role in determining the ecological performance of C_4 tropical grasses in that P modulates the degree to which these grasses can respond to N acquisition.

Long-term fertilization experiments conducted in South Africa (O'Connor, 1985) revealed that grasses of different C_4 photosynthetic subtypes exhibit contrasting responses to an additional supply of N. Fertilization resulted in an increase of the frequency of C_4 grasses of the NADme and PEPck photosynthetic types, but reduced the frequency of the NADPme type in grass swards with highly diversified species composition (Fig. 9). The latter type is dominant in most South American savannas (Johnson and Tothill, 1985; Filgueiras, 1991). It can be hypothesized that the aspartate-formers (NADme and PEPck types) have a higher N requirement than malate-formers (NADPme type), a characteristic that may be associated with the greater abundance of malate-formers in C_4 South American savannas. As discussed previously, the aspartate-forming species tend to predominate in drier and subtropical savannas, whereas the malate-formers dominate in wetter tropical climates on dystrophic soils (Huntley, 1982).

D. Tropical Savannas: The Cradle of C_4 Metabolism in Grasses?

As noted in our review of tallgrass prairie, temperate C_4 grasses are believed to have evolved in tropical regions of the world (Brown and Smith, 1972; Smith and Robbins, 1974; Hattersley, 1983). It is often assumed that the environmental characteristics of lowland tropical savannas [which have been described in detail by several authors (Frost *et al.*, 1986; Sarmiento, 1992; Table III), provide insight into the factors that contributed to the evolution of the C_4 pathway. Thus, the seasonal droughts, frequent fires, high radiation loads, warm temperatures, and nutrient limitations of many lowland tropical savannas would appear to have caused selection favoring certain traits of the C_4 photosynthetic pathway. These traits include:

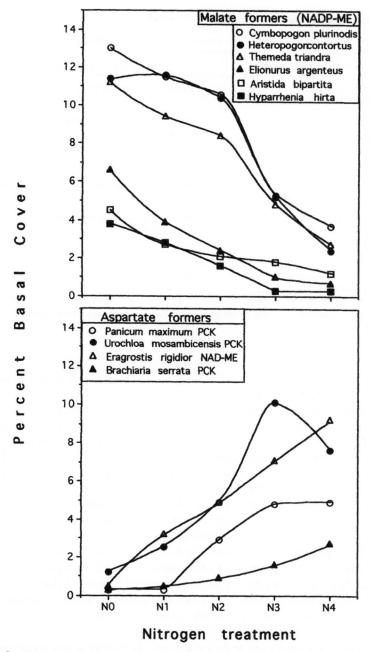

Figure 9 Changes in frequency of grass species after 13 years of nitrogen fertilization in Towoomba, South Africa. [From O'Connor, T. G. (1985). A synthesis of field experiments concerning the grass layer in the savanna regions of southern Africa. South African National Scientific Programmes Report No. 114. FRD, CSIR, Pretoria.]

Table III Ecological Constraints Typical of Lowland Savanna Environments

Environmental variable	Physiological stress
Constant high irradiation (midday 1500–2000 μmol m^{-2} sec^{-1})	Photoinhibition, water stress
Constant high air temperature (midday 25–32°C)	Water stress, increased respiration and photorespiration
Constant high evaporative demand (midday leaf/air VPD = 2–3 kPa)	Water stress, reduced CO_2 uptake
Seasonal drought (2–7 months soil ψ < 1.5 Mpa)	Water stress, leaf area reduction, nutrient deficiency
Low nutrient availability (Σ bases = 0.1–5 cmol kg^{-1}; N = 0.1–0.15% pH = 4.5–5.5)	Nutrient deficiency, reduced protein synthesis
Fire (annual frequency: 0.3, 0.5, 1)	Reduction of leaf area, nutrient losses from system (N_2)

Modified from Sarmiento, G. (1992). Adaptive strategies of perennial grasses of South American Savannahs. *J. Veget. Sci.* **3**, 291–417.

1. *The C_4 photosynthetic pathway is more efficient in the use of water to assimilate CO_2 than the C_3 pathway. This results from the capability of C_4 plants to have a positive carbon balance at lower internal CO_2 concentrations.* Under conditions of high water availability, this trait does not constitute a competitive advantage for C_4 plants. Other factors such as the maximum rate of CO_2 assimilation, plant growth form, and biomass allocation patterns are more critical for successful establishment and reproduction at high water availability. Nevertheless, C_4 grasses dominate tropical and temperate savannas with seasonal dry periods and tend to extend their range locally during extended droughts. It is possible that the higher water-use efficiency of the C_4 photosynthetic pathway (1) allows C_4 grasses to achieve superior growth rates during those seasonal dry periods when water is not as available, and (2) is coupled to the evolution of other morphological and physiological traits that directly contribute to the tolerance of extended drought. For example, high photosynthetic water-use efficiency of many C_4 grasses is associated with substantial osmotic adjustment, leaf rolling, and metabolic tolerance of tissue dehydration.

2. *Savanna C_4 grasses have to be fire tolerant.* This trait is not necessarily unique to C_4 plants, but seems to depend most strongly on a perennial, tufted growth habit, with large biomass investment in underground organs capable of storing nutrient resources. Accumulation of carbon and nutrient

reserves allows for quick recovery of photosynthetic biomass after fires. However, the C_4 pathway probably does contribute to high growth rates following fire.

3. *The CO_2 concentrating mechanism of the C_4 pathway allows for a higher capacity to use high solar radiation loads and to reduce the temperature-dependent effects of photorespiration, thus increasing the photosynthetic temperature optimum.* The existence of these traits in C_4 grasses would appear to render them uniquely adapted to the warm, high-light environment of lowland tropical savannas.

4. *The CO_2 concentrating mechanism of the C_4 pathway allows for a more efficient use of nitrogen to assimilate atmospheric CO_2 compared to C_3 grasses.* C_3 plants depend on large nitrogen investments in RuBP-carboxylase for their photosynthetic productivity.

It is tempting to interpret the existence of these aspects of C_4 photosynthesis as evidence that lowland tropical savannas, because of their unique environmental character, represent the evolutionary cradle of C_4 metabolism. However, all of these traits do not seem to be strong enough, alone or in combination, to explain the dominance of C_4 grasses in savannas. This is principally because ecophysiological studies since the late 1970s have revealed that plants possessing the C_3 photosynthetic pathway are capable of inhabiting, and in fact dominating, ecosystems with frequent fire regimes, warm, dry desert ecosystems with high solar radiation loads, and ecosystems characterized with extremely low nutrient availability. Thus, although the C_4 photosynthetic pathway may facilitate certain aspects of growth and persistence in environments such as those that characterize the lowland tropics, there is no reason to expect that ancestral plants with the C_3 photosynthetic pathway could not also have evolved to dominate these ecosystems.

A new hypothesis that may help to explain the dominance of tropical and temperate savanna ecosystems by C_4 grasses proposes that the high capacity for C_4 plants to assimilate carbon under low atmospheric CO_2 partial pressures (<300 μbar) provided them with a competitive advantage over C_3 plants during past episodes of diminished CO_2 availability (Ehleringer *et al.*, 1991). During the recent past (60,000 to 18,000 years before present) the atmospheric CO_2 partial pressure was relatively low (< 250 μbar). Only most recently has it increased to around 280 μbar (prior to the industrial revolution), and then to around 360 μbar today (Lorius and Oeschger, 1994). Partial pressures of CO_2 below 1000 μbar or so reduce the capacity of C_3 plants to produce organic matter and may have favored the evolution of more efficient mechanisms for fixing CO_2 (Ehleringer and Monson 1993). In Mount Kenya, changes in vegetation dominated by contrasting photosynthetic types during the last 25,000 years have been associated with arid climates and lower atmospheric CO_2 concentrations (Street-Perrott 1994). Combined analyses of carbon/nitrogen ratios and $\delta^{13}C$ in Sacred Lake of Mount Kenya indicate that past changes in the dominant vegetation toward a C_4 type preceded reductions in water level, suggesting that the main driving factor was reductions in atmospheric CO_2 concentration and not aridity (Street-Perrott, 1994).

It is possible that the tropical savanna environment is indeed the cradle of C_4 metabolism, and that the C_4 pathway arose in response to selection by the unique environment found in this ecosystem. However, in an atmosphere characterized by high CO_2 partial pressure, C_3 grasses should have been competitively equal, or superior, to C_4 grasses (Ehleringer and Monson, 1993). It is only in combination with an atmosphere of relatively low CO_2 partial pressure, that the adaptive potential of C_4 photosynthesis reaches its highest level in tropical savannas. It is probably due to historic reductions in the atmospheric CO_2 partial pressure that C_4 grasses were able to achieve broader geographic distribution and dominate many of the tropical, subtropical, and temperate grassland ecosystems. This is an aspect of the relationship between C_4 photosynthesis and its success in natural environments that has not been fully realized until recently (Cerling, Chapter 13).

IV. Summary

1. Similarities in Temperate and Tropical Grasslands It is clear that the C_4 grasses that dominate the North American temperate tallgrass prairies and the South American tropical savannas share many ecophysiological characteristics. These include high photosynthetic capacity, a requirement for high light levels, relatively low stomatal conductance, high photosynthetic water-use efficiency, and low N requirements. These may contribute to additional traits that are critical to the success of these grasses, including: rapid growth rates; high productivity; substantial allocation of carbon belowground; tolerance of, and perhaps dependence on, frequent fire; and the ability to withstand drought. Although C_3 species may share many of these traits, others are exclusive to plants with the C_4 photosynthetic pathway. The presence of frequent fire, low soil fertility, and strong seasonal variation in water availability in many tropical savannas mirrors the environment of the temperate tallgrass prairie. Thus, it should not be surprising that these two grasslands are dominated by C_4 grasses and share many genera.

2. C_4 Grasslands and Increasing CO_2 The prediction that increasing CO_2 concentration in the atmosphere as a result of human activities will favor the expansion of the C_3 flora, particularly forest trees, is supported by models differentiating the effects of CO_2 concentration and aridity on the changes in the relative proportion of grasslands and coniferous forests during the last 18,000 years (Robinson, 1994). Actual changes that will occur in the grass flora of the savannas and tallgrass prairies during the process of atmospheric CO_2 enrichment are uncertain. Floristic studies show that C_4 grasses can share wet and nutrient rich environments with C_3 grasses and compete sucessfully with them. C_4 grasses are certainly not limited to seasonally dry environments, and studies on growth and community responses of tallgrass prairie species to elevated CO_2 have not shown any decreases in abundance (Owensby et al., 1996). On the contrary, many studies have shown that positive responses, although not as strong as in C_3 plants, may be expected. These positive responses may be related to CO_2-induced reductions in stomatal conductance and enhanced water relations of C_4 grasses at elevated CO_2 (Knapp et al., 1993b). However, most C_4 grasses in savannas and tallgrass prairie are restricted to high light environments, and if tree density increases as a result of CO_2 enrichment, it may lead to a reduction of the area occupied by C_4 grasses. As a result, a high fire frequency may become even more critical for maintaining the dominance of C_4 grasses in these ecosystems in a high CO_2 world.

References

Abrams, M. D., Knapp, A. K. and Hulbert, L. C. (1986). A ten-year record of aboveground biomass in a Kansas tallgrass prairie: Effect of fire and topographic position. *Amer. J. Bot.* **73**, 1509–1515.

Anderson, R. C. (1990). The historic role of fire in the North American Grassland. In "Fire in North American Tallgrass Prairies" (S. L. Collins and L. L. Wallace, eds.), pp. 8–18. University of Oklahoma Press, Norman, Oklahoma.

Anten, N. P. R., Werger, M. J. A. and Medina, E. (1998). Nitrogen distribution and leaf area indices in relation to photosynthetic nitrogen use efficiency in savanna grasses. *Plant Ecology* (in press).

Archer, S. and Detling, J. K. (1984). The effects of defoliation and competition on regrowth of tillers of two North American mixed-grass prairie graminoids. *Oikos* **43**, 351–357.

Axelrod, D. I. (1985). Rise of the grassland biome, central North America. *Bot. Rev.* **51**, 163–201.

Axmann, B. D. and Knapp, A. K. (1993). Water relations of *Juniperus virginiana* and *Andropogon gerardii* in an unburned tallgrass prairie watershed. *Southwest. Natur.* **38**, 325–330.

Baruch, Z. and Fernández, D. (1993). Water relations of native and introduced grasses in a neotropical savanna. *Oecologia* **96**, 179–185

Baruch, Z., Ludlow, M. M. and Davis, R. (1985). Photosynthetic responses of native and introduced C_4 grasses from Venezuelan savannas. *Oecologia* **67**, 388–393.

Belsky, J. (1992). Effects of grazing, competition, disturbance, and fire on species composition and diversity in grassland communities. *J. Veg. Sci.* **3**, 187–200.

Bilbao, M. and Medina, E. (1990). Nitrogen use efficiency for growth in a cultivated African grass and a native South American pasture grass. *J. Biogeog.* **17**, 421–425.

Blair, J. M. (1997). Fire, N availability, and plant responses in grasslands: A test of the transient maxima hypothesis. *Ecology* **78**, 2359–2368.

Boardman, N. K. (1977). Comparative photosynthetic rates of sun and shade plants. *Ann. Rev. Plant Physiol.* **28**, 355–377.

Borchert, J. R. (1950). The climate of the central North American grassland. *Ann. Assoc. Amer. Geog.* **40**, 1–39.

Bragg, T. B. and Hulbert, L. C. (1976). Woody plant invasion of unburned Kansas bluestem prairie. *J. Range Manag.* **29**, 19–23.

Briggs, J. M. and Knapp, A. K. (1995). Interannual variability in primary production in tallgrass prairie: Climate, soil moisture, topographic position and fire as determinants of aboveground biomass. *Amer. J. Bot.* **82**, 1024–1030.

Brown, R. H. (1978). A difference in N use efficiency in C_3 and C_4 plants and its implications in adaptation and evolution. *Crop Sci.* **18**, 93–98.

Brown, W. V. and Smith, B. N. (1972). Grass evolution, the Kranz syndrome, $^{13}C/^{12}C$ ratios, and continental drift. *Nature* **239**, 345–346.

Bulla, L., Pacheco, J. M. and Miranda, R. (1981). A simple model for the measurement of primary production in grasslands. *Boletin Socie. Venez. Cienc. Naturale* **35**, 281–304.

Callahan, J. T. (1984). Long-term ecological research. *BioScience* **34**, 363–367.

Catchpole, F. B. (1996). "The dynamics of bison (*Bos bison*) grazing patches in tallgrass prairie." Masters Thesis, Division of Biology, Kansas State University, Manhattan.

Cole, M. (1986). "The Savannas, Biogeography and Geobotany." Academic Press. London.

Collins, S. L. (1987). Interaction of disturbances in tallgrass prairie: A field experiment. *Ecology* **68**, 1243–1250.

Collins, S. L. and Gibson, D. J. (1990). Effect of fire on community structure in tallgrass and mixed-grass prairie. *In*: "Fire in North American Tallgrass Prairies" (S. L. Collins and L. L. Wallace, eds.), pp. 81–98. University of Oklahoma Press, Norman, Oklahoma.

Collins, S. L. and Wallace, L. L. (1990). "Fire in North American Tallgrass Prairies." University of Oklahoma Press, Norman, Oklahoma.

Collins, S. L., Knapp, A. K., Briggs, J. M., Blair, J. M., and Steinauer, E. M. (1998). Modulation of diversity by grazing and mowing in native tallgrass prairie. *Science* **280**, 745–747.

Coppock, D. L., Detling, J. K., Ellis, J. E. and Dyer, M. L. (1983). Plant herbivore interactions in a North American mixed-grass prairie. I. Interactions of black-tailed prairie dogs on intraseasonal aboveground plant biomass and nutrient dynamics and plant species diversity. *Oecologia* **56**, 1–9.

Davy, A. L. (1996). "Responses of two forbs, *Solidago canadensis* and *Vernonia baldwinii* to variation in nitrogen availability and light environments in lowland, annually burned tallgrass prairie." Masters thesis, Division of Biology, Kansas State University, Manhattan, Kansas.

Day, T. A. and Detling, J. K. (1990). Grassland patch dynamics and herbivore grazing preference following urine deposition. *Ecology* **63**,180–188.

DeLucia, E. H., Heckathorn, S. A. and Day T. A. (1992). Effects of soil temperature on growth, biomass allocation and resource acquisition of *Andropogon gerardii*. *New Phytol.* **120**, 543–549.

Detling, J. K. (1988). Grasslands and savannas: Regulation of energy flow and nutrient cycling by herbivores. *In* "Concepts of Ecosystems Ecology" (Ecological Studies 67), (L. R. Pomeroy and J. J. Alberts, eds.), pp. 131–148. Springer-Verlag, New York.

Ehleringer, J. R., Sage, R. F. and Flanagan, L. R. (1991). Climate change and the evolution of C_4 photosynthesis. *Trends Ecol. Evol.* **6**, 95–99.

Ehleringer, J. R. and Monson, R. K. (1993). Evolutionary and ecological aspects of photosynthetic pathway variation. *Ann. Rev. Ecol. System.* **24**, 411–439.

Eiten, G. (1972). The Cerrado vegetation of Brazil. *Botan. Rev.* **38**, 201–241.

Escobar, A. and González-Jiménez, E. (1979). La production primaire de la sabane inondable d'Apure (Venezuela). *Geo-Eco-Trop.* **3**, 53–70.

Fahnestock, J. T. and Knapp, A. K. (1993). Water relations and growth of tallgrass prairie forbs in response to selective herbivory by bison. *Internat. J. Plant Sci.* **154**, 432–440.

Fahnestock, J. T. and Knapp, A. K. (1994). Responses of forbs and grasses to selective grazing by bison: interactions between herbivory and water stress. *Vegetatio* **115,** 123–131.
Field C. and Mooney H. A. (1986). The photosynthesis–nitrogen relationships in wild plants. *In* "The Economy of Plant Form and Function" (T. J. Givnish, ed.), pp. 25–55. Cambridge University Press, Cambridge, UK.
Filgueiras, T. S. D. (1991). A floristic analysis of the Graminae of Brazil's Federal District and a list of the species occurring in the area. *Edin. J. Bot.* **48,** 73–80.
Frank, D. A., Inouye, R. S., Huntly, N., Minshall, G. W. and Anderson, J. E. (1994). The biogeochemistry of a north-temperate grassland with native ungulates: Nitrogen dynamics in Yellowstone National Park. *Biogeochemistry* **26,** 163–188.
Freeman, C. C. (1998). The flora of Konza Prairie: A historical review and contemporary patterns. *In* "Grassland Dynamics: Long-term Ecological Research in Tallgrass Prairie" (A. K. Knapp, J. M. Briggs, D. C. Hartnett and S. L. Collins, eds.), pp. 69–80. Oxford University Press, New York.
Frost, P., Medina, E., Menaut, J. -C., Solbrig, O., Swift, M., Walker, B. (1986). "Responses of Savannas to Stress and Disturbance" (Biology International Special Issue-10), 78 pp. International Union of Biological Sciences, Paris.
Gibson, D. J. and Hulbert, L. C. (1987). Effects of fire, topography and year-to-year climatic variation on species composition in tallgrass prairie. *Vegetatio* **72,** 175–185.
Gibson, D. J., Seastedt T. R. and Briggs, J. M. (1993). Management practices in tallgrass prairie: Large- and small-scale experimental effects on species composition. *J. Appl. Ecol.* **30,** 247–255.
Grime, J. P. (1979). "Plant Strategies and Vegetation Processes." John Wiley and Sons, New York.
Haase, R. and Beck, S. G. (1989). Structure and composition of savanna vegetation in northern Bolivia: A preliminary report. *Brittonia* **41,** 80–100.
Hake, D. R., Powell, J., McPherson, J. K., Claypool, P. L. and Dunn, G. L. (1984). Water stress of tallgrass prairie plants in central Oklahoma. *J. Range Manag.* **37,** 147–151.
Hartnett, D. C. (1991). Effects of fire in tallgrass prairie on growth and reproduction of prairie coneflower (*Ratibida columnifera*: Asteraceae). *Amer. J. Bot.* **78,** 429–435.
Hattersley, P. W. (1983). The distribution of C_3 and C_4 grasses in Australia in relation to climate. *Oecologia* **57,** 113–128.
Hattersley P. W. and Watson L. (1992). Diversification of photosynthesis. *In* "Grass Evolution and Domestication" (E. G. Chapman, ed.), pp. 38–116. Cambridge University Press, Cambridge, UK.
Hayes, D. C. (1985). Seasonal nitrogen translocation in big bluestem during drought conditions. *J. Range Manag.* **38,** 406–410.
Heckathorn, S. A. and DeLucia, E. H. (1991). Effect of leaf rolling on gas exchange and leaf temperature of *Andropogon gerardii* and *Spartina pectinata*. *Botan. Gaz.* **152,** 263–268.
Heckathorn, S. A. and DeLucia, E. H. (1994). Drought-induced nitrogen retranslocation in perennial C_4 grasses of tallgrass prairie. *Ecology* **75,** 1877–1886.
Holland, E. A. and Detling J. K. (1990). Plant response to herbivory and belowground nitrogen cycling. *Ecology* **71,** 1040–1049.
Huber, O. (1974). "The Neotropical Savannas." Istituto Latinoamericano, Roma.
Huntley, B. J. (1982). Southern African savannas. *In* "Ecology of Tropical Savannas" (Ecological Studies 42), (B. J. Huntley and B. H. Walker, eds.), pp. 101–119. Springer-Verlag, Berlin.
Jackson, R. B., Canadell, J., Ehleringer, J. R., Mooney, H. A., Sala, O. E. and Schulze, E. D. (1996). A global analysis of root distributions for terrestrial biomes. *Oecologia* **108,** 389–411.
Johnson, R. W. and Tothill, J. C. (1985). Ecology and management of world savannahs. Definition and broiad geographic outline of savannah lands. *In* "Ecology and Management of World Savannahs" (J. C. Tothill and J. J. Mott, eds.), pp. 1–13. Australian Academy of Sciences, Canberra.

Johnson, S. R. and Knapp, A. K. (1995). The influence of fire on *Spartina pectinata* wetland communities in a northeastern Kansas tallgrass prairie. *Can. J. Bot.* **73**, 84-90.

Kemp, P. R. and Williams, G. J. (1980). A physiological basis for niche separation between *Agropyron smithii* (C_3) and *Bouteloua gracilis* (C_4). *Ecology* **61**, 846-858.

Knapp, A. K. (1984a). Post-burn differences in solar radiation, leaf temperature and water stress influencing production in a lowland tallgrass prairie. *Amer. J. Bot.* **71**, 220-227.

Knapp, A. K. (1984b). Water relations and growth of three grasses during wet and drought years in a tallgrass prairie. *Oecologia* **65**, 35-43.

Knapp, A. K. (1985). Effect of fire and drought on the ecophysiology of *Andropogon gerardii* and *Panicum virgatum* in a tallgrass prairie. *Ecology* **66**, 1309-1320.

Knapp, A. K. (1986a). Post-fire water relations, production and biomass allocation in the shrub, *Rhus glabra*, in tallgrass prairie. *Botan. Gaz.* **147**, 90-97.

Knapp, A. K. (1986b). Ecophysiology of *Zigadenus nuttallii*, a toxic spring ephemeral in a warm season grassland: Effect of defoliation and fire. *Oecologia* **71**, 69-74.

Knapp, A. K. and Gilliam, F. S. (1985). Response of *Andropogon gerardii* (Poaceae) to fire-induced high vs. low irradiance environments in tallgrass prairie: Leaf structure and photosynthetic pigments. *Amer. J. Bot.* **72**, 1668-1671.

Knapp, A. K. and Fahnestock, J. T. (1990). Influence of plant size on the carbon and water relations of *Cucurbita foetidissima*. *Functional Ecology* **4**, 789-797.

Knapp, A. K. and Seastedt, T. R. (1986). Detritus accumulation limits productivity in tallgrass prairie. *BioScience* **36**, 662-668.

Knapp, A. K., Koelliker, J. K., Fahnestock, J. T. and Briggs, J. M. (1994). Water relations and biomass responses to irrigation across a topographic gradient in tallgrass prairie. *In* "Proceedings of the 13th North American Prairie Conference" (R. G. Wickett, P. D. Lewis, A. Woodliffe, and P. Pratt, eds.), pp. 215-220. Windsor, Ontario, Canada.

Knapp, A. K., Briggs, J. M., Hartnett, D. C. and Collins, S. L. (1998). "Grassland Dynamics: Long-Term Ecological Research in Tallgrass Prairie." Oxford University Press, NY.

Knapp, A. K., Fahnestock, J. T., Hamburg, S. J., Statland, L. B., Seastedt, T. R. and Schimel D. S. (1993a). Landscape patterns in soil-plant water relations and primary production in tallgrass prairie. *Ecology* **74**, 549-560.

Knapp, A. K., Hamerlynck, E. P. and Owensby, C. E. (1993b). Photosynthetic and water relations responses to elevated CO_2 in the C_4 grass *Andropogon gerardii*. *Intern. J. Plant Sci.* **154**, 459-466.

Knight, C. L., Briggs, J. M. and Nellis, M. D. (1994). Expansion of gallery forest on Konza Prairie Research Natural Area, Kansas, USA. *Land. Ecol.* **9**, 117-125.

Kucera, C. L., Dahlman, R. C. and Koelling, M. (1967). Total net productivity and turnover on an energy basis for tallgrass prairie. *Ecology* **48**, 536-541.

LeRoux, X. and Mordelet, P. (1995). Leaf and canopy assimilation in a West African humid savanna during the early growing season. *J. Trop. Ecol.* **11**, 529-545.

Li, M. (1993). Leaf photosynthetic nitrogen-use efficiency of C_3 and C_4 Cyperus species. *Photosyn.* **29**, 117-130.

Loehle, C., Li, B. and Sundell, R. C. (1996). Forest spread and phase transitions at forest-prairie ecotones in Kansas, USA. *Land. Ecol.* **11**, 225-235.

Long, S., García-Moya, E., Imbamba, S., Kamnalrut, A., Piedade, M., Scurlock, J., Shen, Y. and Hall, D. (1989). Primary productivity of natural grass ecosystems of the tropics: A reappraisal. *Plant Soil* **115**, 155-166.

Lorius, C. and Oeschger, H. (1994). Palaeo-perspectives: Reducing uncertainties in global change? *Ambio* **23**, 3036.

Ludlow, M. M. and Wilson, G. L. (1971). Photosynthesis of tropical pastures. I. Illuminance, CO_2 concentration, leaf temperature and leaf-air vapour pressure difference. *Aust. J. Biol. Sci.* **24**, 449-470.

Martin, C. E., Harris, F. S. and Norman, F. J. (1991). Ecophysiological responses of C_3 forbs and C_4 grasses to drought and rain on a tallgrass prairie in northeastern Kansas. *Botan. Gaz.* **152**, 257–262.

McNaughton, S. J. (1983). Compensatory plant growth as a response to herbivory. *Oikos* **40**, 329–336.

Medina, E. (1987). Requirements, conservation, and cycles of nutrients in the herbaceous layer. *In* "Determinants of Tropical Savannas" (B. H.Walker, ed.), pp. 39–65. IUBS Monographies, Series No. 3. IRL Press Ltd., Oxford, UK.

Medina, E. and Bilbao, B. (1991). Significance of nutrient relations and symbiosis for the competitive interaction between grasses and legumes in tropical savannas. *In* "Modern Ecology, Basic and Applied Aspects" (G. Esser and D. Overdieck, eds.), pp. 295–319. Elsevier Science Publ., Amsterdam.

Medina, E. and Silva J. F. (1990). Savannas of northern South America: A steady state regulated by water–fire interactions on a background of low nutrient availability. *J. Biogeog.* **17**, 403–413.

Medina, E., Mendoza, A. and Montes, R. (1978). Nutrient balance and organic matter production in the Trachypogon savannas of Venezuela. *Trop. Agricult.* (*Trinidad*) **55**, 243–253.

Medina, E., de Bifano, T. and Delgado, M. (1976). Diferenciación fotosintética en plantas superiores. *Interciencia* **1**, 96–104.

Monson, R. K., Sackschewsky, M. R. and Williams, G. J. III. (1986). Field measurements of photosynthesis, water-use efficiency, and growth in *Agropyron smithii* (C_3) and *Bouteloua gracilis* (C_4) in the Colorado shortgrass steppe. *Oecologia* **68**, 400–409.

O'Connor, T. G. (1985). A synthesis of field experiments concerning the grass layer in the savanna regions of southern Africa. South African National Scientific Programmes Report No. 114. FRD, CSIR, Pretoria.

Ojima, D. S., Schimel, D. S., Parton, W. J. and Owensby, C. E. (1994). Long- and short-term effects of fire on nitrogen cycling in tallgrass prairie. *Biogeochemistry* **24**, 67–84.

Owensby, C. E., Ham, J. M., Knapp, A. K., Rice, C. W., Coyne, P. I. and Auen, L. M. (1996). Ecosystem level responses of tallgrass prairie to elevated CO_2. *In* "Carbon Dioxide and Terrestrial Ecosystems" (H. A. Mooney and G. W. Koch, eds.), pp. 175–193. Academic Press, New York.

Owensby, C. E., Hyde, R. M. and Anderson, K. L. (1970). Effects of clipping and supplemental nitrogen and water on loamy upland bluestem range. *J. Range Manag.* **23**, 341–346.

Ramia, M. (1967). Tipos de sabanas en los llanos de Venezuela. *Boletín de la Sociedad Venez. Ciencias Natur.* **27**, 264–288.

Redmann, R. E. (1985). Adaptation of grasses to water stress: Leaf rolling and stomate distribution. *Ann. Miss. Botan. Gard.* **72**, 833–845.

Rice, C. W., Todd, T. C., Blair, J. M., Seastedt, T. R., Ramundo, R. A. and Wilson, G. W. T. (1998). Belowground Biology and Processes. *In* "Grassland Dynamics: Long-term Ecological Research in Tallgrass Prairie" (A. K. Knapp, J. M. Briggs, D. C. Hartnett and S. L. Collins, eds.), pp. 244–264. Oxford University Press, New York.

Risser, P. G., Birney, C. E., Blocker, J. D., May, S. W., Parton, W. J. and Wiens, J. A. (1981). "The True Prairie Ecosystem." US/IBP Synthesis Series 16. Hutchinson Ross Publishing Company, Stroudsburg, Pennsylvania.

Robinson, J. M. (1994). Speculations on carbon dioxide starvation, late Tertiary evolution of stomatal regulation and floristic modernization. *Plant, Cell Environ.* **17**, 345–354.

Sala, O. E. and Lauenroth, W. K. (1982). Small rainfall events: An ecological role in semiarid regions. *Oecologia* **53**, 301–304.

Samson, F. and Knopf, F. (1994). Prairie conservation in North America. *BioScience* **44**, 418–421.

San José, J. J. and Garcia-Miragaya, J. (1981). Factores ecológicos operacionales en la producción de materia orgánica de las sabanas de Trachypogon. *Boletín de la Sociedad Venez. Ciencias Natur.* **36**, 347–374.

Sarmiento, G. (1983). The savannas of tropical America. In "Tropical Savannas" (F. Bourliere, ed.), pp. 245–288. Ecosystems of the World, Elsevier Publ. Co., Amsterdam.
Sarmiento, G. (1984). "The Ecology of Neotropical Savannas." Harvard University Press. Cambridge.
Sarmiento, G. (1992). Adaptive strategies of perennial grasses of South American savannas. *J. Veg. Sci.* **3**, 291–417.
Seastedt, T. R. and Knapp, A. K. (1993). Consequences of non-equilibrium resource availability across multiple time scales: The transient maximum hypothesis. *Amer. Natur.* **141**, 621–633.
Seastedt, T. R. and Ramundo, R. A. (1990). The influence of fire on belowground processes of tallgrass prairies. In "Fire in North American Tallgrass Prairies" (S. L. Collins and L. L. Wallace, eds.), pp. 99–117. University of Oklahoma Press, Norman, Oklahoma.
Simoes, M. and Baruch, Z. (1991). Responses to simulated herbivory and water stress in two tropical grasses. *Oecologia* **88**, 173–180.
Smith, B. N. and Robbins, M. J. (1974). Evolution of C_4 photosynthesis: An assessment based on 13C/12C ratios and Kranz anatomy. In "Proceedings of the Third International Congress on Photosynthesis" (M. Avron, ed.), pp. 1579–1587. Elsevier Scientific Publ., Amsterdam.
Snaydon, R. W. (1991). The productivity of C_3 and C_4 grasses: A reassessment. *Func. Ecol.* **5**, 321–330.
Street-Perrot, F. A. (1994). Palaeo-perspectives: Changes in terrestrial ecosystems. *Ambio* **23**, 37–43.
Susachs, F. (1984). "Caracterización ecológica de las sabanas de un sector de los Llanos Bajos Centrales de Venezuela." Tesis Doctoral. Universidad Central de Venezuela. Facultad de Ciencias, Caracas.
Svejcar, T. J. and Browning, J. A. (1988). Growth and gas exchange of *Andropogon gerardii* as influenced by burning. *J. Range Manag.* **41**, 239–244.
Tate, K. W., Gillen, R. L., Mitchell, R. L. and Stevens, R. L. (1994). Effect of defoliation intensity on regrowth of tallgrass prairie. *J. Range Manag.* **47**, 38–42.
Terri, J. A. and Stowe, L. G. (1976). Climatic patterns and the distribution of C_4 grasses in North America. *Oecologia* **23**, 1–12.
Tieszen, L. L., Reed, B. C., Bliss, N. B., Wylie, B. K. and Dejong, D. D. (1997). NDVI, C_3 and C_4 production, and distribution in Great Plains grassland cover classes. *Ecol. Appl.* **7**, 59–78
Transeau, E. N. (1935). The prairie peninsula. *Ecology* **16**, 423–437.
Turner, C. L. and Knapp, A. K. (1996). Responses of a C_4 grass and three C_3 forbs to variation in nitrogen and light in tallgrass prairie. *Ecology* **77**, 1738–1749.
Turner, C. T., Kneisler, J. R. and Knapp, A. K. (1995). Comparative gas exchange and nitrogen responses of the dominant C_4 grass, *Andropogon gerardii*, and five C_3 forbs to fire and topographic position in tallgrass prairie during a wet year. *Intern. J. Plant Sci.* **156**, 216–226.
Turner, C. L., Seastedt, T. R. and Dyer, M. I. (1993). Maximization of aboveground grassland production: The role of defoliation frequency, intensity, and history. *Ecol. Appl.* **3**, 175–186.
Vinton, M. A. and Hartnett, D. C. (1992). Effects of bison grazing on *Andropogon gerardii* and *Panicum virgatum* in burned and unburned tallgrass prairie. *Oecologia* **90**, 374–382.
Vinton, M. A., Hartnett, D. C., Finck, E. J. and Briggs, J. M. (1993). Interactive effects of fire, bison (*Bison bison*) grazing and plant community composition in tallgrass prairie. *Amer. Mid. Natur.* **129**, 10–18.
Vogl, R. J. (1974). Effects of fire on grasslands. In "Fire and Ecosystems" (T. T. Kozlowski and C. E. Ahlgren, eds.). Academic Press, New York.
Wallace, L. L. (1990). Comparative photosynthetic responses of big bluestem to clipping versus grazing. *J. Range Manag.* **43**, 58–61.
Waller, S. S. and Lewis, J. K. (1979). Occurrence of C_3 and C_4 photosynthetic pathways in North American grasses. *J. Range Manag.* **32**, 12–28.
Walter, H. (1960). "Grundlagen der Pflanzenverbreitung. I. Standortslehre." Eugen Ulmer Verlag, Stuttgart.

Weakly, H. E. (1962). History of drought in Nebraska. *J. Soil Water Conserv.* **17,** 271–273.
Weaver, J. E. (1954). "North American Prairie." Johnsen Publishing Company, Lincoln, Nebraska.
Weaver, J. E. (1958). Classification of root systems of forbs of grassland and a consideration of their significance. *Ecology* **39,** 393–401.
Weaver, J. E. (1968). "Prairie Plants and Their Environment." University of Nebraska Press, Lincoln, Nebraska
Weaver, J. E. and Rowland, N. W. (1952). Effects of excessive natural mulch on development, yield, and structure of native grassland. *Botan. Gaz.* **114,** 1–19.
Wedin D. A. and Tilman, D. (1990). Species effects on nitrogen cycling: A test with perennial grasses. *Oecologia* **84,** 433–441.
Wedin, D. A. and Tilman, D. (1993). Competition among grasses along a nitrogen gradient: Initial conditions and mechanisms of competition. *Ecol. Mono.* **63,** 199–229.
Wells, P. V. (1970). Postglacial vegetational history of the Great Plains. *Science* **67,** 1574–1582.
Wilson, J. R. (1975). Comparative response to nitrogen deficiency of a tropical and a temperate grass in the interrelation between photosynthesis, growth and the accumulation of non-structural carbohydrate. *Nether. J. Agri. Sci.* **23,** 104–112.
Zuloaga, F. O. (1987). Systematics of the New World species of *Panicum* (Poaceae: Paniceae). *In* "Grass Systematics and Evolution" (T. R. Soderstrom, K. W. Hilu, C. S. Campbell, and M. E. Barkworth, eds.), pp. 287–306. Smithonian Institution Press, Washington D. C.

9

C_4 Plants and Herbivory

Scott A. Heckathorn, Samuel J. McNaughton, and James S. Coleman

I. Introduction

Most C_4 species are either grasses of tropical and subtropical distribution, are the "warm-season" grasses of temperate habitats, or occur in osmotically stressful habitats (Teeri and Stowe, 1976; Stowe and Teeri, 1978; Osmond *et al.*, 1981; Monson, 1989; Ehleringer and Monson, 1993). How the distinctive habitats, anatomy, physiology, and growth forms of C_4 plants interact with herbivory is incompletely understood; however, certain characteristics unique to C_4 species can influence the likelihood of a plant being eaten (resistance) or its ability to recover from being eaten (tolerance). Consequently, C_4 traits probably affected the natural selection and evolution of herbivores, which in turn could have influenced the evolution of C_4 plants (i.e., coevolution) (e.g., McNaughton, 1985).

In this chapter, we first discuss C_4 traits that are thought to, or that we predict may affect plant resistance or tolerance to herbivory. We then review the available evidence pertinent to these traits and their role in plant–herbivore interactions. As we illustrate, the effect of C_4 traits on plant resistance to herbivory is well studied, particularly the relationship between forage quality of C_4 leaves and herbivore selectivity. It has been argued that the generally lower nutrient and higher fiber content of C_4 foliage decreases herbivory on C_4 plants (Caswell *et al.*, 1973), although we suggest that the evidence that herbivores select against C_4 foliage when both C_3 and C_4 plants are available is inconsistent. Paradoxically, we show that on a global basis, herbivores currently consume a greater proportion of foliage in C_4-

dominated ecosystems than in systems dominated by C_3 species, indicating that C_4 traits do not prevent herbivory and that herbivores of C_4 plants are apparently well adapted for the task of eating C_4 leaves. In contrast to C_4 traits and resistance to herbivory, little is known about how C_4 characteristics affect plant tolerance to herbivory (with the exception of the importance of the grass growth form, which is common to most C_4 species). We suggest that certain C_4 traits influence how plants respond to herbivory, and argue that more research effort in this area is needed.

II. Unique Features of C_4 Species That May Affect Herbivores and Plant Responses to Herbivory

Not all of the features unique to C_4 plants directly influence plant–herbivore interactions. However, certain characteristics of C_4 plants are likely to have direct, or near direct, bearing on plant resistance or tolerance to herbivory. We compiled a list of C_4 features already recognized to, or that we predict might also, affect plant–herbivore interactions (Table I). In this section, we discuss why these characteristics are, or may be, important to the resistance or tolerance of herbivory.

A. Nutrient and Fiber Content of C_4 Foliage

It is now well established that leaves of C_4 plants often have lower nitrogen and phosphorus concentrations (N_L and P_L), and hence higher carbon-to-nitrogen or phosphorus ratios (C:N and C:P), than C_3 plants (e.g., Wilson and Haydock, 1971; Caswell et al., 1973; McNaughton et al., 1982; Field and Mooney, 1986; Wilsey et al., 1997). Consequently, C_4 leaves are nutritionally inferior to C_3 leaves, and this has direct impact on herbivore foraging behavior, growth, and reproduction (Van Dyne et al., 1980; Caswell et al., 1973; McNaughton et al., 1982; Jones and Coleman, 1991). Caswell et al. (1973) first proposed that when given a choice, herbivores should select C_3 over similar C_4 foliage, and when forced to rely solely on C_4 foliage, herbivore growth or reproduction may decline. In addition, N_L is usually lower in C_4 than C_3 foliage because of the CO_2-concentrating mechanism and resultant higher photosynthetic nitrogen-use efficiency (PNUE) in C_4 plants (Osmond et al., 1981; Field and Mooney, 1986; Ehleringer and Monson, 1993). Extremely low levels of N_L constrain the type of herbivores that can use C_4 leaves as a food source; for example, C_4 folivory is often limited to slow-growing generalist herbivores with microbial gut symbionts, such as ruminant mammals (Van Dyne et al., 1980; Abe, 1991; Phelan and Stinner, 1992; Wedin, 1996). As a further consequence of high C:N, the carbon:nutrient balance theory of plant defense (e.g., Bryant et al., 1983) predicts that C_4 plants would be less likely to rely on nitrogen-rich secondary

Table I Unique Characteristics of C_4 Species that Potentially Influence Plant–Herbivore Interactions

Low N_L and high C:N	Nutritionally poor foliage, so may decrease herbivore selectivity and performance. Suitable only for slower-growing herbivores and efficient digestive systems with microbial symbionts. Also C_4 plants may not use N-based defensive compounds as often, but instead rely on C-based defenses like toughness and phenolics, which should select for generalist folivores.
High fiber content	Herbivores would need strong mandibles or jaws and teeth to consume C_4 leaves. Must have digestive systems adapted to high-fiber diets. This should select for larger generalist folivores.
Kranz anatomy	Bundle-sheath (BS) cells would tend to discourage feeding by xylem and phloem feeders and hinder access to resources in BS cells and veins for generalist folivores. Less mesophyll would provide less reward to interveinal feeders.
Xylem and phloem nutrients	Lower transpiration rates (E) and higher rates of photosynthesis and translocation of photosynthate from leaves may increase the concentration of vascular tissue resources available to herbivores that can access veins.
Light response	Higher P_{max} and lack of photosynthetic saturation at high-light levels may speed recovery from herbivory because (1) uneaten leaves that were unshaded have higher P_{max}, and (2) uneaten leaves that were self-shaded exhibit greater photosynthetic responses to canopy opening by herbivores.
Temperature response	Because of higher temperature optima, C_4 plants have greater photosynthetic and growth rates at high temperatures and may benefit more from any increase in leaf temperature after canopy opening by herbivores, which may increase herbivory tolerance when temperatures are warm. However, at low temperatures, C_3 plants have higher photosynthetic and growth rates, and thus may have greater herbivore tolerance.
Higher P_{max} and PNUE	A given loss of leaf tissue or N should have a larger impact on whole-plant photosynthesis, which might limit recovery from herbivory. However, less N might be needed to photosynthetically compensate for, or replace, lost tissue, and this may speed recovery.
Stomatal conductance	Lower g_s and E and higher WUE contribute to water conservation and drought tolerance. Mechanisms limiting water use may increase herbivory tolerance in water-limited situations.

(continues)

Table I (*Continued*)

Biomass allocation	Higher P_{max}, PNUE, and WUE may permit C_4 species to allocate more resources to either new leaves or nonphotosynthetic tissue, resulting in different biomass allocation patterns relative to C_3 species. This has many implications for herbivores; for example, increased allocation to shoots could provide more resources for aboveground herbivores, whereas increased root allocation could benefit belowground herbivores.
Mutualistic symbioses	High photosynthetic rates and WUE may allow C_4 species to more readily pay the carbon costs of such symbioses in high-light or water-limited environments, which may increase herbivory tolerance. For example, C_4 grasses are usually mycorrhizal and often associated with diazotrophic N_2-fixing bacteria, which may provide better access to nutrients before and after herbivory. However, it is also possible that such symbioses are a severe carbon drain after leaf removal, and hence slow recovery.
Warm, dry, high-light habitats	The habitats that C_4 species are typically found in may not be as suitable for small aboveground invertebrate herbivores. Such habitats might have less diverse invertebrate communities with large aboveground herbivores or a predominance of belowground grazers due to inhospitable aboveground climates. Vertebrate herbivores may be the major consumers aboveground. Seasonal growth of C_4 species in temperate habitats causes seasonal patterns in herbivore behavior.
Mostly grasses	Usually have both intercalary and protected apical meristems, thus, can easily recover from shoot removal as meristems are not lost to aboveground herbivory.

compounds, such as alkaloids and cyanogens, and instead, more likely to rely on carbon-based chemical defenses, such as phenolics and lignified fibers. If it is true that C_4 species tend to use carbon-rich defensive compounds more than closely related C_3 species, then C_4 plants should tend to be eaten by generalist herbivores, and specialist herbivores should be more common on C_3 plants.

Several aspects of Kranz anatomy should also have a direct influence on C_4 plant resistance to folivory. The characteristic large bundle-sheath cells of C_4 plants, small leaf interveinal distances, and thinner leaves, relative to C_3 plants (Hattersley, 1984; Dengler and Nelson, Chapter 5), mean that C_4 leaves usually have proportionally more cellulose and lignin (Minson, 1971) and less mesophyll than C_3 leaves. In addition, lignified fibers, cork, and silica-containing cells are often concentrated over the vascular bundles of C_4 leaves (Gould, 1968; Esau, 1977), structurally protecting the bundle sheath from herbivory and digestive processes, thereby further reducing food quality. As a result of these features, generalist herbivores feeding on C_4 leaves probably need strong jaws, teeth, or mandibles, as well as digestive systems adapted to high fiber diets, in order to access food resources contained within the bundle sheath and vascular tissue. Also, because of Kranz anatomy, invertebrate folivores that feed strictly on mesophyll tissue should prefer C_3 to C_4 foliage, as should herbivores that feed on phloem and xylem tissue.

It is generally agreed that there is a gradient of cell-wall digestibility, and therefore energy yield, from foliage cells in the decreasing order of phloem, mesophyll, and undifferentiated parenchyma > epidermis > bundle-sheath parenchyma > sclerenchyma and xylem. Sclerenchyma, xylem, and bundle-sheath cells usually contribute more to tissue strength and their greater representation in C_4 foliage increases leaf resistance to shearing and tearing relative to C_3 species. This influences both harvesting effectiveness and digestibility of leaf tissue (Wilson, 1990, 1991). Therefore, as a result of Kranz anatomy, the digestibility of C_4 foliage by animals is generally lower than the digestibility of C_3 foliage.

For example, *in vitro* digestibility (output/input) of leaf blades of *Panicum* spp. differing in photosynthetic pathway varied (mean + 1 s.e.) as follows (Akin *et al.*, 1983; Wilson and Hattersley, 1983):

$$C_3 = 0.88 + 0.07$$
$$C_3/C_4 = 0.93 + 0.07$$
$$C_4 \text{ PCK} = 0.74 + 0.03$$
$$C_4 \text{ NADPE-ME} = 0.70$$
$$C_4 \text{ NAD-ME} = 0.27 + 0.07$$

This digestibility gradient is in general agreement with a concomitant gradient of increasing bundle sheath in proportion to mesophyll from C_3 to C_4 NAD-ME type grasses (Hattersley, 1984; Dengler *et al.*, 1994). In addition to inherent differences in digestibility, differences in habitat growth temperatures may contribute to differences in digestibility of C_3 and C_4 foliage. Forage plants grown at higher temperatures consistently have lower digestibilities than the same species grown at lower temperatures (Dirven and Deinum, 1977). This is mainly due to decreasing digestibility with increasing growth temperature of the cell walls of bundle sheath, xylem, and sclerenchyma, and the effect is more pronounced in stem than in leaf tissue (Wilson *et al.*, 1991). The principal mechanism for this is an increase in cell wall lignification at higher growth temperatures.

The low digestibility of C_4 foliage can limit ungulate performance. Typically, an ungulate eats to rumen-fill, then goes through a period of regurgitating and chewing the accumulated food until particle size is small enough to sink in rumen fluid. The time required to break down food to a small particle size increases with decreasing digestibility of food. Thus, reduction in food digestibility increases passage time through the digestive tract, and thereby acts as a negative feedback on voluntary intake. For this reason, it is common agricultural practice to pelletize high-fiber diets, which circumvents some of their limitation on ungulate performance, and to supplement high-fiber diets with food concentrates (Wilson, 1990, 1991).

B. Food Resources in the Bundle Sheath and Vascular Tissue

Any mechanical difficulties and physiological costs associated with gaining access to resources within the bundle sheath and vascular tissue may be offset to some degree by the large fraction of leaf protein and nutrients contained therein in C_4 species. Because many enzymes involved in CO_2 fixation are localized in the bundle-sheath cells of C_4 leaves (Kanai and Edwards, Chapter 3), a large proportion of total leaf protein in C_4 plants is contained in this tissue (e.g., ca. 50% of total soluble protein; Ku et al., 1979). The rewards for folivores that can gain access to vascular tissue may also be different for C_4 compared to C_3 plants (Raven, 1983). Because transpiration rates (E) are typically lower for C_4, in contrast to C_3, plants (e.g., Osmond et al., 1981), the concentration of nutrients in the xylem sap is higher in C_4 plants when the flux of nutrients in the xylem is not limited by nutrient availability nor the capacity to load nutrients into the xylem (xylem loading $> E$; Kramer and Boyer, 1995; Marschner, 1995). However, when the concentration of nutrients in the xylem is limited by transpiration-dependent mass flow of nutrients in the soil, then xylem nutrient concentrations should be lower in C_4 plants. Any differences in xylem nutrient concentration between C_3 and C_4 plants could be of particular importance to xylem feeders (Raven, 1983), which are important herbivores in many grassland communities (Scott et al., 1979; Andrzejewska and Gyllenberg, 1980; Brown and Gange, 1990). We point out, however, there has not been enough work done to date to confirm these predictions.

The concentration of photosynthate in the phloem may also vary between C_3 and C_4 plants. Translocation of recently fixed carbon from the leaf is generally twice as rapid in C_4 than C_3 plants, even on a phloem-area basis, and a greater proportion of recently fixed carbon is exported in C_4 plants (Lush and Evans, 1974; Osmond et al., 1981). In general, C_4 plants have greater light-saturated photosynthetic rates and PNUE than C_3 plants (Osmond et al., 1981; Edwards and Walker, 1985; Field and Mooney, 1986; Ehleringer and Monson, 1993). This may also contribute to differences in carbon-based resources in the phloem simply through greater production of photosynthate by C_4 plants in warm sunny habitats. Again, however, there is not enough evidence to confirm that the amount of resources in the phloem of C_4 (vs. C_3) plants is greater.

C. Photosynthetic Characteristics

Several aspects of the photosynthetic response to light that are unique to C_4 plants have a bearing on their response to herbivory. In general, C_4 plants have greater maximum rates of photosynthesis (P_{max}) than C_3 plants, are less likely to saturate photosynthetically at full sunlight than C_3 species, and have lower quantum yields than C_3 species at low temperatures and limiting light (Ehleringer and Björkman, 1977; Ehleringer, 1978; Osmond et al., 1981; Edwards and Walker, 1985; Ludlow et al., 1988; Ehleringer and Monson, 1993). As a result of these traits, C_4 species should be better able to capitalize photosynthetically on increases in light intensity that surviving foliage may encounter following severe herbivory; thus, they should benefit more from release of self-shading than C_3 plants.

The photosynthetic response to temperature also differs between C_4 and C_3 plants, and this may have ramifications for plant responses to herbivory. It is well established that C_4 species typically have higher temperature optima for photosynthesis than C_3 plants, in large part due to the lack of temperature-sensitive increases in photorespiration (Osmond et al., 1981; Edwards and Walker, 1985; Ehleringer and Monson, 1993). If leaf temperatures of remaining foliage increase after herbivory due to less self-shading, then the response of photosynthesis of uneaten leaves should usually benefit C_4 more than C_3 plants, as illustrated in Fig. 1. For idealized C_3 and C_4

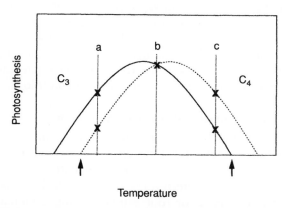

Figure 1 Idealized response of net photosynthesis to temperature for a C_3 (solid curve) and C_4 plant (dashed curve). Within the range of temperatures bounded by the arrows, any increase in temperature results in a larger increase or smaller decrease in photosynthesis for the C_4, compared to the C_3, plant (e.g., from point a to point b, point b to point c, or point a to point c).

plants with the same P_{max}, any positive shift along the X-axis over the range indicated by arrows (this does not necessarily hold outside this range), results in either relatively greater increases, or smaller decreases, in photosynthesis for the C_4 species. For example, temperature increases from point a to point b, point b to c, or point a to point c benefit the C_4 species more than the C_3 species. Presumably, any increases in photosynthesis related to increases in leaf temperature should improve a plant's ability to recover from herbivory. Despite the greater relative benefits that C_4 plants may derive from temperature increases following herbivory, at low temperatures C_3 plants usually have greater absolute rates of photosynthesis than C_4 plants, and the inverse is true at high temperatures. Therefore, at low temperatures C_3 plants may be more tolerant of herbivory than C_4 plants, whereas at high temperatures the situation may be reversed.

An additional consequence of photosynthetic differences between C_4 and C_3 plants for plant–herbivore interactions relates to the impact that folivory has on whole-plant photosynthetic potential. Because C_4 plants have higher P_{max} and PNUE, a given loss of leaf tissue or nitrogen should have a larger impact on whole-plant photosynthesis than in C_3 plants, all other things being equal. As a result, C_4 species might recover more slowly from folivory. However, less nitrogen might be needed by C_4 plants to photosynthetically compensate for or replace lost tissue, and this may speed recovery from folivory.

The difference between C_3 and C_4 species in stomatal behavior may potentially play a role in plant responses to herbivory as well. C_4 plants usually have lower stomatal conductance (g_s), and thus lower rates of transpiration (E) at a given leaf-to-air vapor pressure deficit (LAVPD), and greater water-use efficiency (WUE) than C_3 plants (Osmond *et al.*, 1981; Edwards and Walker, 1983; Ehleringer and Monson, 1993). These traits confer a potential advantage to C_4 plants in terms of water conservation and, thus, it is reasonable to speculate that these characteristics might increase tolerance to herbivory under water-limited conditions.

D. Biomass and Nutrient Allocation

Higher P_{max}, PNUE, and WUE could, in theory, permit C_4 species to allocate carbon or nutrient resources differently than in C_3 species, resulting in different whole-plant biomass or N allocation patterns, which would have important implications for herbivores. For example, increased carbon or nutrient allocation to roots or root exudation might increase belowground herbivory (Seastedt, 1985; Yeates, 1987), and belowground herbivory is usually substantial in grassland ecosystems (Scott *et al.*, 1979; Stanton, 1988; Brown and Gange, 1990). However, increased root allocation may increase nutrient acquisition after herbivory. At present, however, the few studies that provide comparative information on biomass allocation of similar C_3 and C_4 species indicate that, whereas C_3 and C_4 species often differ in biomass allocation, no consistent difference is apparent. For example, Wilson and Haydock (1971) and Wilsey et al. (1997) both found that the fraction of total plant mass allocated to shoots was greater in several tropical C_4 grasses than several temperate C_3 grasses. Similarly, Gebauer et al. (1987) found that shoot:root mass was greater in *Amaranthus retroflexus* (C_4 forb) than in *Atriplex hortensis* (C_3 forb); however, the proportion of total plant N allocated to shoots was less in the C_4 species. In contrast, Sage and Pearcy (1987) found that *A. retroflexus* (C_4 forb) allocated less biomass to shoots at high and low N than did *Chenopodium album* (C_3 forb). Redmann and Reekie (1982) reported that *Bouteloua gracilis* (C_4 grass) allocated a greater fraction of fixed carbon to roots than did *Hordeum distichum* and *Lolium multiflorum* (both C_3 grasses). Interestingly, both Osmond et al. (1981), working with two *Atriplex* species, and Davidson (1969), studying several grass species, observed increases in shoot:root mass at high versus moderate temperatures (e.g., 25° vs. 35°C) in C_4 species but decreases in C_3 plants. Consequently, whereas the C_3 and C_4 species exhibited similar shoot:root mass at moderate temperatures, shoot:root mass was greater in the C_4 species at high temperatures.

E. Additional Factors

Symbioses between mycorrhizal fungi and C_4 species are nearly ubiquitous (Ingham and Molina, 1991; Rabatin and Stinner, 1991), and symbiosis between the roots of C_4 plants and diazotrophic N_2-fixing bacteria is very common, particularly in tropical and subtropical ecosystems (Marschner, 1995). These two mutualistic associations are important for plant acquisition of nutrients, especially P and N, and mycorrhizae may also improve plant water uptake (Ingham and Molina, 1991; Rabatin and Stinner, 1991; Marschner, 1995). Mycorrhizae and N_2-fixing bacteria may, therefore, affect plant–herbivore relations by influencing plant nutrient and water status, which are known to play an important role in plant–herbivore interactions (Jones and Coleman, 1991; McNaughton, 1991b). The effects of these symbionts on plant–herbivore relations may be either positive or negative. For example, improved plant nutrient and water status resulting from these symbioses will most likely increase herbivory tolerance. However, these symbioses are known to be carbon-costly to the host plant and can decrease plant competitive ability when carbon resources are limiting (e.g., in low-light growth conditions, during drought, and following folivory) (Ingham and Molina, 1991). The high photosynthetic rates and WUE of C_4 species may allow them to more readily pay the carbon costs of these and other mutualistic symbioses (e.g., endophytic fungi) in high-light, warm, and water-limited environments. However, recovery from folivory may be slowed if such symbioses are a severe carbon drain and if C_4 species support a greater mass of symbiotic microbes than C_3 species in a given habitat (e.g., mycorrhizae infection in C_4 grasses of tallgrass prairie is greater than in C_3 grasses; Hetrick et al., 1988).

As mentioned before, C_4 species occur primarily in warm, high-light, and often osmotically stressful habitats of tropical-to-warm temperate latitudes. This necessarily means that the herbivores that can potentially feed on C_4 plants are a subset of the herbivores that potentially inhabit the environments in which C_4 plants are found. Clearly, this defines and limits the potential range of plant–herbivore relationships in these ecosystems. For example, these habitats may not be well-suited for small invertebrate herbivores with high surface/volume ratios, and hence high rates of water loss. Thus, the habitats in which C_4 species occur might have less diverse invertebrate communities with larger-sized aboveground herbivores (e.g., grasshoppers), or a predominance of belowground grazers due to inhospitable aboveground climates. Also, vertebrate folivores may predominate in these C_4 habitats.

In many temperate ecosystems, the growth of C_4 species is largely restricted to the warmest part of the growing season. In contrast, competing C_3 species are often dormant during this time, and are instead active during the cool seasons (Teeri and Stowe, 1976; Stowe and Teeri, 1978; Osmond et al., 1981; Monson, 1989; Ehleringer and Monson, 1993). The success of C_4 relative to C_3 species during the warm season is related not only to the higher temperature optima of photosynthesis common in C_4 plants, but also to higher temperature optima for other nonphotosynthetic traits as well (e.g., belowground processes), which likely reflects the tropical and subtropical origins of C_4 plants (Jones, 1985; Marschner, 1995). In temperate habitats containing a mix of C_3 and C_4 species, certain herbivores (e.g., large mammalian grazers) may have to rely on C_4 foliage during the warm season, but have C_3 foliage available to them during the remaining period of the year. If herbivores cannot use C_4 leaves, then they have to either migrate to habitats wherein C_3 species are active, restrict their foraging to those C_3 species that are active during the warm season, or become dormant during the warm season.

Lastly, as pointed out earlier in this chapter and in other chapters of this volume, most C_4 species are grasses and this fact alone has important implications for plant–herbivore interactions. Most grasses have both intercalary meristems and apical and axillary meristems that are not exposed to aboveground grazers; thus, C_4 species are usually relatively tolerant of folivory because shoot meristems are not lost to grazing (Hyder, 1972).

III. How Important Are C_4 Characteristics to Herbivory Tolerance and Resistance?

A. C_4 Photosynthesis and Herbivory in an Ecosystem Context

The forage quality of plants is usually highest during the early period of plant tissue growth, and this period is typically short relative to foliage lifetime; therefore, animals are frequently forced to eat plant material of marginal food quality, or must bear the cost of searching for young tissues. As mentioned previously, Caswell *et al.* (1973) first proposed that the traits comprising the C_4 photosynthetic pathway exacerbate the limitations in quality of C_4 foliage, thereby reducing herbivore growth, survival, and reproduction. This led the authors to predict that herbivore preference for C_4 plants would be lower than for C_3 plants. They also predicted that detritus food webs will be relatively more important than grazing food webs in ecosystems dominated by C_4 producers compared with systems where C_3 producers predominate. As we discuss later, tests of the first prediction have been equivocal, and to our knowledge, the second prediction concerning the greater importance of detritus food webs in C_4 dominated ecosystems has not been tested.

In gross energetic terms, herbivory is an inconsequential pathway of energy flow in most terrestrial ecosystems because less than one-tenth of annual net primary production is generally consumed by herbivores (McNaughton *et al.*, 1991). Therefore, more than nine-tenths of terrestrial energy flow in most ecosystems enters detritus food webs as dead plant material, allowing plants to internally recycle, and thereby conserve, mobile nonstructural nutrients prior to senescence. However, variation between ecosystems in the proportion of primary production consumed by herbivores is substantial. To evaluate global patterns of production and consumption of foliage in C_3- versus C_4-dominated ecosystems, we reanalyzed a previously published data set on energy flow through terrestrial ecosystems ranging from tundra to wet tropical forest (McNaughton *et al.*, 1989, 1991). This new analysis documents higher levels of foliage production and higher proportional consumption of foliage in ecosystems dominated by C_4 plants.

Mean annual foliage production of ecosystems dominated by C_4 vegetation (principally tropical grassland and savanna) was 2.5 times greater than systems dominated by C_3 vegetation. Foliage production of systems dominated by a mix of C_3 and C_4 species (principally temperate to subtropical grassland) was 1.65 times greater than C_3 systems (Fig. 2). These data indicate that C_4-dominated ecosystems can potentially sustain greater levels of aboveground herbivory than C_3 systems. To examine this possibility, we compared aboveground consumption among these same vegetation types.

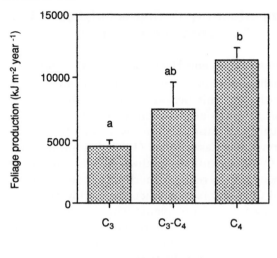

Figure 2 Mean aboveground net foliage production in terrestrial ecosystems dominated by either C_3, a mix of C_3 and C_4 (C_{3-4}), or C_4 species [see McNaughton, S. J., Oesterheld, M., Frank, D. A., and Williams, K. J. (1989). Ecosystem-level patterns of primary productivity and herbivory in terrestrial habitats. *Nature* **341**, 142–144; and McNaughton, S. J., Oesterheld, M., Frank, D. A., and Williams, K. J. (1991). Primary and secondary production in terrestrial ecosystems. *In* "Comparative Analyses of Ecosystems: Patterns, Mechanisms, and Theories" (J. Cole, G. Lovett, and S. Findlay, eds.), pp. 120–139. Springer-Verlag, New York, for detailed explanation of methods and description of study sites]. Foliage production was affected by photosynthetic pathway (ANOVA; $F_{2,71} = 18.07$; $P < 0.000001$) and significant differences (Sheffe's test; $P < .05$) among vegetation types are indicated by different letters. Error bars = 1 se.

Herbivores currently consume nearly 60% of annual foliage produced in C_4-dominated systems, but only 10–11% in C_3 and C_{3-4} systems (Fig. 3). Thus, ecosystems with C_4 species as the major producers currently sustain more than five times the level of proportional folivory as C_3 and C_{3-4} vegetation types. These data indicate, contrary to earlier hypotheses (Caswell *et al.*, 1973), that tropical ecosystems in which C_4 species predominate experience much greater levels of aboveground herbivory than do those ecosystems in which C_3 species are the principal primary producers. In addition, because plant tissue not consumed by herbivores eventually enters detritus pathways, total ecosystem detritus production may be proportionally greater in C_3 and C_{3-4}-mixed vegetation types than in tropical C_4 grasslands.

There are some caveats to this analysis, however. First, the relatively high consumption rate in C_4 systems may be due, in part, to the fact that native mammalian herbivores are still present and dominant in many of these habitats, whereas they have been eliminated from many C_3 systems. For

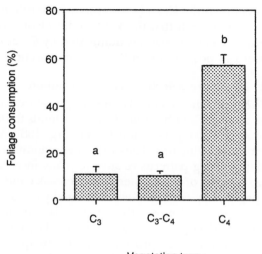

Figure 3 The proportion of annual foliage consumed by herbivores in terrestrial ecosystems dominated by C_3, a mix of C_3 and C_4 (C_{3-4}), or C_4 species [see McNaughton, S. J., Oesterheld, M., Frank, D. A., and Williams, K. J. (1989). Ecosystem-level patterns of primary productivity and herbivory in terrestrial habitats. *Nature* **341,** 142–144; and McNaughton, S. J., Oesterheld, M., Frank, D. A., and Williams, K. J. (1991). Primary and secondary production in terrestrial ecosystems. *In* "Comparative Analyses of Ecosystems: Patterns, Mechanisms, and Theories" (J. Cole, G. Lovett, and S. Findlay, eds.), pp. 120–139. Springer-Verlag, New York, for detailed explanation of methods and description of study sites]. Consumption was affected by photosynthetic pathway (ANOVA; $F_{2,71} = 53.68$; $P < 0.000001$) and significant differences (Sheffe's test; $P < .05$) among vegetation types are indicated by different letters. Error bars = 1 se.

example, the consumption rate of herbivores in North American C_3 or C_{3-4} grasslands may have been quite high when buffalo herds roamed the prairies in great numbers, but we have no good estimates on past consumption rates in these grasslands. Our analysis also does not provide insight as to the underlying mechanisms accounting for differences among vegetation types in primary foliage production and herbivore consumption, and certainly does not indicate that the observed patterns are the direct result of differences in photosynthetic pathway.

Nonetheless, this broad-scale analysis demonstrates that (1) the C_4 syndrome does not, by itself, prevent folivory—relative to other present-day ecosystems, C_4 ecosystems sustain greater rates of folivory; and (2) the herbivores that feed on C_4 foliage in C_4-dominated systems are apparently well adapted to this task. These results do not preclude the possibility that in ecosystems containing a mix of C_3 and C_4 species, herbivore-feeding patterns vary between C_3 and C_4 pathways. Nor do they preclude the possibility that differences between the C_3 and C_4 pathways have influenced herbivore success and evolution such that the kinds of herbivores that feed on foliage are different between ecosystems dominated by C_3 and C_4 species. We discuss the latter two possibilities in the following section.

B. Herbivore Selectivity and Distribution

1. Mammalian Herbivores In mammals, feeding preference for forbs rather than grasses and for C_3 rather than C_4 foliage tends to increase with decreasing body mass (Andrzejewska and Gyllenberg, 1980; Van Dyne *et al.*, 1980). Furthermore, grazing ungulates and other mammals are typically less selective in their feeding patterns relative to many insects due to the comparatively large mouths of the former (Andrzejewska and Gyllenberg, 1980). Whereas insects may feed between sclerenchyma- and phytolith-overlaid tissues, ungulates and most other mammals cannot. Therefore, mammals in general, and large ungulates in particular, must cope with and are well-adapted for physically processing and subsequently digesting heterogeneous and often fiberous forage. For example, within the prairie ecosystems of the central United States, preference for C_3 versus C_4 grasses and for forbs versus grasses increases roughly with decreasing body mass from bison, to cattle, to sheep, to pronghorn antelope and white-tailed deer (Andrzejewska and Gyllenberg, 1980; Van Dyne *et al.*, 1980; Plumb and Dodd, 1993). Thus, in ecosystems where vegetation is comprised of a mix of C_3 and C_4 species, photosynthetic pathway probably plays a decreasing role in determining resistance to mammalian herbivory as herbivore size increases.

The tendency for avoidance of grasses as a food source as herbivore mass decreases may partly explain why the most conspicuous herbivores in extensive grassland ecosystems prior to their destruction in the mid-to-late 19th century were large, grazing ungulates. These animals are adapted to traveling over large distances to find localized bursts of productivity in a stochastic climate. They evolved hypsodont dentition capable of shredding fiber and silica-rich tissues, and they have a multicompartmented digestive system supporting a symbiotic fauna that efficiently extracts energy from poor quality forage (McNaughton *et al.*, 1985; McNaughton, 1991a). The Miocene expansion of grassland ecosystems accompanying increasing terrestrial aridity coincides with the origin and convergent evolution of the C_4 pathway in several plant families (Ehleringer *et al.*, 1991). Although less than 5% of all plant species have this pathway (Ehleringer *et al.*, 1991), C_4 grasses are often exclusively represented in tropical grasslands and are dominant members of the warm-season flora of temperate grasslands. In the grasslands dominated by C_4 species, a greater proportion of aboveground net primary production that is consumed by herbivores is eaten by large mammalian grazers, compared to that fraction consumed by small mammalian and invertebrate herbivores, relative to grasslands containing C_3 or C_{3-4}-mixed vegetation (Detling, 1988).

There are clear morphological, physiological, and behavioral patterns of evolution in herbivores that can be related to feeding on grasses versus dicots (McNaughton *et al.*, 1984; McNaughton, 1991). However, these patterns seem unlikely to result from herbivore evolution in response to photosynthetic pathways but, instead, to reflect the evolution of abrasive defenses in grasses, such as silica accumulation, and chemical defenses in dicots. For example, African herbivores are subdivided into three classes according to their feeding patterns (Lamprey, 1963; McNaughton and Georgiadis, 1986), and those patterns also reflect the importance in their diets of forages with C_4 and C_3 photosynthesis. Browsers feed virtually exclusively on the arborescent component of savanna ecosystems and, therefore, are specialists on plants with C_3 photosynthesis. Typical examples are giraffe (*Girrafa camelopardalis*) and dik-dik (*Madoqua kirkii*). At the other extreme of specialization are such pure grazers as wildebeest (*Connochaetes taurinus*) and buffalo (*Syncerus caffer*), which feed only on the C_4 grasses of the herb layer. Intermediate- or mixed-feeders eat principally C_4 grasses during the wet season, when those grasses are actively growing, then switch to C_3 browse when the dry season arrives. Among the mixed feeders are elephant (*Loxodonta africana*) and impala (*Aepyceros melampus*).

Interactions between these herbivore classes and their forages have major effects in organizing the structure of African vegetation (Laws, 1970; McNaughton and Sabuni, 1988). Elephants are a key element in this organizing fauna (Laws, 1970). They are capable of killing mature trees, most commonly by stripping the bark during the dry season and pushing them over to feed on otherwise inaccessible canopy foliage. Browsers and mixed-feeders, in turn, forage heavily on the saplings and seedlings of the C_3 canopy plants, which keeps those plants in fire-susceptible size classes. Finally, grazers have an important role in determining the dry season fuel load. In areas or during periods with minor grazing, substantial fuel loads accumulate, which, inevitably, result in dry season fires. Those fires kill some seedlings and saplings, but more commonly burn them back to the soil surface, arresting them in small size classes. Browsing, grazing, or fire, alone, cannot prevent regeneration of the woody canopy—all three acting in concert are required to generate the typical open savanna of African landscapes (Norton-Griffiths, 1979; Pellew, 1983). Although these patterns are clearly related to the partitioning of photosynthetic pathways among vegetation canopy classes in African ecosystems, it seems unlikely that those pathways are necessary contributors to the dynamics involved. Instead, photosynthetic pathways are partitioned among the growth forms by the constraints of climate and water availability, whereas the relative importance of various canopy components is regulated proximally by grazing, browsing, and fire.

2. Arthropod Herbivores As mentioned previously, Caswell et al. (1973) hypothesized that C_4 plants may be poorer host plants for herbivores than C_3 plants and, therefore, should be avoided by insect herbivores. A number of studies support the notion that C_4 plants may be nutritionally poorer than C_3 plants because of low N_L and high fiber content (e.g., Caswell and Reed, 1976; Heidorn and Joern, 1984; Barbehenn and Bernays, 1992). There is some evidence for greater consumption of C_3 versus C_4 plants by insect herbivores in communities containing mixtures of both types of plants (Boutton et al., 1978; Pinder and Kroh, 1987). This is despite our previous observation that folivory levels are much greater in C_4 versus C_3 systems.

Clearly, there are still a number of unanswered questions regarding insect use of C_3 versus C_4 plants. First, when both C_3 and C_4 foliage is available, do studies showing greater insect consumption of C_3 relative to C_4 plants in the field necessarily support the hypothesis of Caswell et al. (1973)? The answer to this question is complicated because feeding preferences do not always correlate with insect performance. For example, it is well known that insects may increase consumption rates when fed nutritionally poorer tissue (Scriber and Slansky, 1981; Jones and Coleman, 1991). Thus, it could be possible to observe increased consumption of C_4, relative to C_3, plants if C_4 foliage was of poor nutritional quality compared to C_3 foliage.

More importantly, rates of consumption in the field on C_3 versus C_4 plants have rarely been normalized to the relative distribution of leaf material available to insects from these types of plants (see Pinder and Kroh, 1987). Thus, it is not always clear whether greater consumption rates are due to greater availability of leaf material or to feeding preference. Even when this factor has been taken into account, the results are equivocal. For example, Pinder and Kroh (1987) analyzed the stable carbon isotope ratio of insects in order to determine the relative amount of C_3 versus C_4 plant material that had been consumed and compared that to the relative abundance of C_3 and C_4 plants. They found no difference in the relative consumption of C_3 versus C_4 plants in two of the plant communities examined, but they did observe greater consumption of C_3 plants in a third community. Even in the latter case, in which C_3 plants were consumed in larger quantities, it is not known for certain whether this occurred because C_3 plants were truly preferred by insects because no preference or choice assays were conducted in the field. Determination of feeding preference in the field is important, because Heidorn and Joern (1984) found that selection by grasshoppers of C_3 versus C_4 grasses in the field did not follow what would be predicted from food preference tests in the lab. This led them to conclude that photosynthetic pathway was at best a minor determinant of food selection by insects in the field. The major point here is that testing the Caswell et al. (1973) hypothesis requires that insect feeding preferences, consumption patterns, and overall performance all be explicitly examined in the field, along with the relative availability of C_3 and C_4 plant material, in order to draw any concrete conclusions.

Although the hypothesis that insects prefer and feed more on C_3 plants versus C_4 plants is at best marginally supported by the data, might there be other clear relationships between photosynthetic pathway and insect herbivores? For example, we would predict that the low nitrogen and high fiber content of C_4 plants would select for insect herbivores with a larger digestive tract and more of a generalist feeding strategy. Thus, we would predict larger body masses of insects preferentially feeding on C_4 over C_3 plants, and a trend toward larger herbivores in the insect communities of C_4-dominated ecosystems. Anecdotal evidence supports this notion; for example, the observation that relatively large-bodied grasshoppers seem to be the dominant insect foragers in C_4-dominated systems. However, Pinder and Jackson (1988) found that carbon isotope composition of a number of species in the Orthoptera indicated some preferentially fed on C_3 plants, others on C_4, and some on C_3–C_4 mixtures; however, the authors did not determine if preference was correlated with body mass. Thus, the idea that body mass of insects differs as a function of the photosynthetic pathway of food plants has not yet been tested. When conducted, it will have to be in a phylogenetically rigorous manner.

It is also unclear whether the low nitrogen, high-fiber content of C_4 plants has selected for an insect community that has a greater proportion of generalists than would be found on C_3 plants, as this analysis has not yet been done. It, too, would need to be phylogenetically rigorous. It seems likely to us, however, that differences in the nutritional quality of C_3 versus C_4 plants will explain certain aspects of insect community composition to a greater degree than it appears to explain insect feeding preferences and consumption rates in the field.

C. Importance of Factors Other Than N_L, $C:N$, and Fiber

As is clear from the preceding sections of this chapter, the influence of variation between C_3 and C_4 plants in N_L, $C:N$, and foliage fiber content on herbivore feeding behavior, performance, and herbivore community composition has been reasonably well studied, although is not completely understood. In contrast, with the exception of the importance of the grass growth form, virtually no direct evidence is available concerning how the remaining factors proposed in Table I affect resistance or tolerance to herbivory in C_3, compared to C_4, species. However, there is some indirect evidence available for some of the remaining characteristics in Table I to indicate that these distinguishing features of C_4 plants probably play a role in plant–herbivore relations.

1. Food Resources in the Bundle Sheath and Vascular Tissue Regarding our prediction that Kranz anatomy should discourage xylem and phloem feeders, we could find no studies that addressed this issue. Raven (1983) discusses the energetic costs and benefits of feeding on vascular tissue of C_3 and C_4 species. Both Raven (1983) and Dixon (1985) allude to potential differences in the concentration of xylem and phloem resources between C_3 and C_4 plants, pointing out that plant nutrient status and sap nutrient concentration are important determinants of host plant selection by sap feeders. However, in neither of these extensive reviews do the authors provide any indication that the influence of photosynthetic pathway on sap feeding has been directly studied. Whereas most phloem-feeding aphids are found in temperate habitats and feed on trees (Dixon, 1985), C_3 and C_4 grasses are known to experience infestation by aphids as well as other phloem feeders. Furthermore, xylem feeders are relatively common herbivores on C_4 grasses (Scott *et al.*, 1979; Andrzejewska and Gyllenberg, 1980; Brown and Gange, 1990).

Interestingly, most phloem and xylem feeders gain access to the vascular tissue by mechanical means (i.e., involving pressure-dependent insertion of mouth parts), and reaching the vascular tissue is difficult and time consuming (e.g., requiring 15–60 min for aphids) (Raven, 1983; Dixon, 1985). Once the vascular tissue is reached, both xylem and phloem feeding can have negative consequences on plant growth (Raven, 1983). Although little is known about interspecific differences in mechanical impedence involved in reaching leaf veins, it is reasonable to expect that the difficulty in reaching the vascular tissue for sap feeders is greater for C_4 than for comparable C_3 leaves, and that this plays a role in host plant selection.

2. Light and Temperature Responses There is also indirect evidence to indicate that the unique responses of photosynthesis to light and temperature of C_4 species influence plant–herbivore interactions. Öztürk *et al.* (1981) showed that under well-watered conditions, *Amaranthus edulis* (C_4 forb) outcompeted *Atriplex hortensis* (C_3 forb) and *Panicum maximum* (C_4 grass) outcompeted *Avena sativa* (C_3 grass) when plants were grown under natural sunlight. By contrast, when plants were grown under shaded conditions, (*ca.* one-third to one-half of natural light), the C_3 species out-competed the C_4. Because competitive ability and herbivory tolerance are related to plant vigor and physiological status, these results indicate that under high-light conditions, C_4 plants may be relatively more tolerant of herbivory than C_3 species, all other things being equal. Under low light, C_3 species might display relatively greater herbivory tolerance.

A similar argument can be made to allow extrapolation from competitive ability to herbivory tolerance with regards to the impact of temperature on plant–herbivore relations. Pearcy *et al.* (1981) used a "de Wit" replacement-series experimental design to compare the competitive ability of *A. retroflexus* (C_4 forb) and *C. album* (C_3 forb) under daytime temperatures of 17°, 25°, and 34°C. The C_4 species out-competed the C_3 species at 25° and 34°C, but the C_3 species was the superior competitor at 17°C. Using a similar experimental design, Christie and Detling (1982) found that *Bouteloua curtipendula* (C_4 grass) out-competed *Agropyron smithii* (C_3 grass) at 30°C daytime temperature under two levels of soil-N availability; however, at 20°C, the C_3 species was the better competitor at both N levels. Both of these competition studies (Pearcy *et al.*, 1981; Christie and Detling, 1992) used cooccurring, ecologically similar species, thereby bolstering our confidence that these patterns may be typical.

Additional evidence for an important role of temperature in determining the relative herbivory tolerance (not resistance) of C_3 and C_4 plants comes from comparison of studies examining the response of North American prairie grasses to clipping (Table II). Included are all studies that we could find in which cooccurring C_3 and C_4 grasses (at least one species of each) from the central plains of North America were subjected to controlled clipping and monitored for growth afterward. If not already done so by the authors, we evaluated and ranked the relative clipping tolerance of each species by comparing the change in clipped plant mass relative to

Table II A Comparison of Studies That Examined Responses to Controlled Clipping in Cooccurring C_3 and C_4 Grasses from the Great Plains of North America

Authors	Location	Setting	Species examined and relative tolerance to herbivory
Holscher (1945)	Montana	Field site	*Agropyron smithii* (C_3) > *Bouteloua gracilis* (C_4)
Day and Detling (1990)	South Dakota	Field site	*Poa pratensis* (C_3) > *Schizachyrium scoparium* (C_4)
Robertson (1933)	Nebraska	Greenhouse winter	*Koeleria cristata* (C_3) > *P. pratensis* (C_3) > *B. gracilis* (C_4), *Bromis inermis* (C_3), *Sorghum sudanensis* (C_4), and *Stipa spartea* (C_3)
Biswell and Weaver (1933)	Nebraska	Garden	*Bouteloua curtipendula*, *Buchloe dactyloides*, and *Schizachyrium scoparium* (all C_4) > *P. pratensis* (C_3) > *Andropogon gerardii*, *B. gracilis*, and *Panicum virgatum* (all C_4)
Anderson and Briske (1995)	Texas	Garden	*S. scoparium* (C_4) > *Bothriochloa saccharoides* (C_4), and *Stipa leucotricha* (C_3)

unclipped controls. [Note: for the Holscher (1945) study, we based our ranking strictly on how clipped plants compared to conspecific controls over the 4 year period, although the author also compares regrowth between species and reaches a partly different conclusion]. Data for total plant biomass was available in Robertson (1933) and Biswell and Weaver (1933), but only aboveground mass was measured in the remaining studies. With one exception, data were collected during the normal growing season from plants grown outside; in Robertson (1933), plants were grown in a greenhouse during winter under somewhat cooler than summertime temperatures.

In those studies conducted at more northern latitudes (Holscher, 1945; Day and Detling, 1990) or under cooler conditions (Robertson, 1933), the greatest tolerance to herbivory was exhibited by C_3 species, whereas the reverse was true at more southern latitudes (Biswell and Weaver, 1933; Anderson and Briske, 1995). Because each of these studies used different species, one must be cautious in intepreting these results as anything more than supportive of the potential importance of growth temperature to the relative response of C_3 and C_4 plants to herbivory. Furthermore, this analysis does not take into consideration differences in resistance to herbivory. At the community level, vegetation responses to herbivory will be determined by how plants both resist and tolerate herbivory [see MacMahon (1988) and Simms (1988) for a review of the response of North American desert and grassland C_3–C_4 (mixed) vegetation to grazing]. For example, heavy grazing favors *Bouteloua gracilis*, a C_4 grass, in mixed-grass prairie in North Dakota, because C_3 species are preferentially eaten (Dormmar *et al.*, 1994). Still, given the importance of temperature in determining the distribution and competitive ability of C_4 species, it is likely that temperature also plays an influential role in the herbivory tolerance of C_3 versus C_4 plants.

In temperate habitats dominated by both C_3 and C_4 species, the higher temperature optima for photosynthesis and growth that underlies the success of C_4 species during the warm season also results in seasonal patterns of herbivore behavior. To illustrate, in the prairies of the Great Plains of the United States, both bison and cattle preferentially graze C_4 grasses during the warm summer months, but consume mostly C_3 grasses during other times of the year (Plumb and Dodd, 1993; Vinton *et al.*, 1993; Steuter *et al.*, 1995). Preferential grazing by large mammals in these prairie systems can drive shifts in plant community composition and increase plant community heterogeneity, and thus can influence the behavior and success of other herbivores, including those that consume primarily C_3 species (e.g., Collins, 1987; Detling, 1988; Simms, 1988; Steuter *et al.*, 1995).

In addition, the more positive response, in general, of C_4 (versus C_3) species to increases in temperature and light availability may largely explain why C_4 species tend to benefit more from fire than co-occurring C_3 species in grassland ecosystems of the central U.S. (Bazzaz and Parrish, 1982). Soil temperatures and light levels increase following fire in grasslands, and this usually stimulates plant growth (Boerner, 1982; McNaughton, 1991b). Because large mammalian herbivores often prefer to graze on recently burned areas (Vinton and Harnett, 1992; Vinton et al., 1993) and C_4 species are more responsive to fire than C_3 species, photosynthetic pathway, therefore, can have consequences for plant–herbivore relations after fire.

3. Patterns of Water Use As with the response of photosynthesis to temperature and light, there is some indirect data available relating to our prediction that lower transpiration rates and greater WUE in C_4 plants may affect plant–herbivore relations during drought. In contrast to temperature and light, however, the available evidence is inconsistent in its support of differences between C_3 and C_4 species. For example, in the study by Öztürk et al. (1981) mentioned previously, the two C_4 species were generally competitively superior to the two C_3 species under conditions of decreased water availability. Interestingly, shoot:root mass increased with decreasing water availability in the C_4 species, but not in the C_3 species; yet, allocation of total plant N to roots increased, and was greater, at low water availability in the C_4 species. However, in the Pearcy et al. (1981) study discussed previously, water availability did not influence the competitive outcome between the C_3 and the C_4 species. At the community level, similar inconsistencies in the response of C_3 and C_4 vegetation to drought are observed. For example, perennial C_4 grasses increase in abundance during drought relative to cooccurring C_3 grasses in Minnesota prairie (Tilman and El-Haddi, 1992), whereas changes in species abundance during severe drought were unrelated to photosynthetic pathway in short-, mixed-, and tall-grass prairies of Kansas (Weaver and Albertson, 1936; Albertson and Weaver, 1942). In a study of the effects of grazing during drought in Great Basin desert species, Chambers and Norton (1993) found that grazing decreased plant survival to a greater extent in *Atriplex confertifolia* (C_4 shrub) than in *Artemisia spinescens* (C_3 shrub), even though *A. confertifolia* is more grazing resistant in years of normal precipitation. Despite the fact that C_4 species typically use water more efficiently than C_3 species under the same environmental conditions, it is thought that the C_4 pathway is not inherently more drought-tolerant than the C_3 pathway (Osmond et al., 1981). For this reason, it may be that differences in water economy between C_3 and C_4 plants have no consistent bearing on plant–herbivore interactions, as the previously mentioned studies collectively indicate.

4. Herbivory Defenses There is abundant evidence in support of our point that because most C_4 species are grasses and the grass growth-form is, by itself, an effective mechanism to increase tolerance to aboveground herbivory (e.g., Hyder, 1972) that, on average, C_4 species may be more tolerant of folivory than C_3 species. This characteristic may perhaps explain, to a large degree, the global patterns of aboveground primary consumption discussed earlier. We stress that the grass growth-form does not necessarily increase resistance to herbivory; in fact, the greater consumption in C_4-dominated systems (versus C_3 systems) suggests that the opposite is true. Whether the general lack of production and accumulation of chemical defenses in grasses (Siegler, 1981) and potential differences between C_3 and C_4 species in chemical defenses contribute to the global patterns of primary consumption is unknown. Most, if not all, C_4 dicots are members of families known to produce various C- and N-based antiherbivore compounds (Siegler, 1981), and there are a few documented examples of C_4 grasses that accumulate toxic levels of either C- or N-based compounds; for example, cyanogenic compounds (Siegler, 1981); nitrate, oxalate, and prussic acid (Jones, 1985); phytohemagglutinins (Liener, 1979); cinnamic acid and thionins (Swain, 1979; Carpita, 1996); and tannins (McMillian et al., 1972). Furthermore, variation in phenolic compounds in the C_4 cereal *Sorghum bicolor* has been shown to be related to herbivory (McMillian et al., 1972; Swain, 1979; Cherney et al., 1991); thus, differences in the production of chemical defenses among C_4 species is likely to play a role in herbivory resistance. However, no systematic study or comparison has been conducted to establish whether closely related C_3 and C_4 species vary with respect to the types and relative amounts of chemical defenses that they produce and whether this relates to differences in resistance to herbivory.

IV. Summary

C_3 and C_4 plants have a great number of differences in their physiological, morphological, and ecological attributes that should affect their relative susceptibility and tolerance to herbivory. To date, most emphasis has been placed on how aspects of C_4 anatomy and physiology might make them a nutritionally poorer food source for herbivores, and therefore experience less herbivory, as herbivores should choose to feed on higher quality C_3 plants. Our analysis suggests, however, that this hypothesis is does not necessarily hold. On a global scale, ecosystems dominated by C_4 plants currently experience far greater levels of folivory than ecosystems dominated by C_3 plants. Within a given ecosystem containing both C_3 and C_4 species, the evidence that herbivores prefer and eat more of C_3 versus C_4 plants is equivocal at best. Clearly, the relationship between photosynthetic pathway and herbivory is more complex than previously thought.

However, it is likely that differences between C_3 and C_4 plants may explain a number of other important aspects of plant–herbivore interactions that have not yet been investigated. For example, the significantly greater magnitude of folivory occurring in ecosystems dominated by C_4 plants indicates that something about the C_4 syndrome, be it physiology (e.g., water-use efficiency, nitrogen-use efficiency, photosynthetic response to release from self-shading), morphology (e.g., protected apical meristems, biomass allocation), or the habitat occupied by C_4 plants (e.g., warmer and drier) allows these plants to tolerate greater levels of herbivory. This needs to be investigated more fully, and comparisons of the response to herbivory between closely related C_3 and C_4 species is likely to yield much valuable information in this regard. In addition, morphological and physiological differences between C_3 and C_4 plants may explain a great deal of variation in the structure and dynamics of herbivore communities. For example, the poor nutritional quality of C_4 plants should lead to an herbivore community composed of larger animals that tend to feed in a more generalist way in comparison to the herbivore communities of C_3 plants. Also, the C_4 syndrome might select against phloem- and xylem-feeding insects, perhaps decreasing their relative abundance in the insect community of C_4 plants in comparison to those with the C_3 photosynthetic pathway. Clearly, there is much work to be done in order to fully understand the contribution of photosynthetic pathway to the wide global variation in herbivore community composition.

Acknowledgments

We thank Russell Monson, Rowan Sage, and David Wedin for helpful and insightful comments on preliminary drafts of this chapter.

References

Abe, T. (1991). Cellulose centered perspective on terrestrial community structure. *Oikos* **60,** 127–133.
Akin, D. E., Wilson, J. R., and Windham, W. R. (1983). Site and rate of tissue digestion in leaves of C_4, C_3, and intermediate *Panicum* species. *Crop. Sci.* **23,** 155–157.
Albertson, F. W., and Weaver, J. E. (1942). History of the native vegetation of western Kansas during seven years of continuous drought. *Ecol. Mono.* **12,** 23–51.
Anderson, V. A., and Briske, D. D. (1995). Herbivore-induced species replacement in grasslands: is it driven by herbivory tolerance or avoidance? *Ecol. Applic.* **5,** 1014–1024.
Andrzejewska, L., and Gyllenberg, G. (1980). Small herbivore subsystem. *In* "Grasslands, Systems Analysis and Man" (A. I. Breymeyer, and G. M. Van Dyne, eds.), pp. 201–267. Cambridge University Press, Cambridge.
Barbehenn, R. V., and Bernays, E. A. (1992). Relative nutritional quality of C_3 vs. C_4 grasses for a graminivorous lepidopteran, *Paratrytone melane* (Hesperiidae). *Oecologia* **92,** 97–103.
Bazzaz, F. A., and Parrish, J. A. D. (1982). Organization of grassland communities. *In* "Grasses and Grasslands" (J. R. Estes, R. J. Tyrl, and J. N. Brunken, eds.), pp. 233–254. University of Oklahoma Press, Norman.

Biswell, H. H., and Weaver, J. E. (1933). Effect of frequent clipping on the development of roots and tops of grasses in prairie sod. *Ecology* **14**, 368–390.

Boerner, R. E. (1982). Fire and nutrient cycling in temperate ecosystems. *Bioscience* **32**, 187–192.

Boutton, T. W., Cameron, G. N., and Smith, B. N. (1978). Insect herbivory and C_3 and C_4 grasses. *Oecologia* **36**, 21–32.

Brown, V. K., and Gange, A. C. (1990). Insect herbivory below ground. In "Advances in Ecological Research" (M. Begon, A. H. Fitter, and A. MacFadyen, eds.), pp. 1–58. Academic Press, London.

Bryant, J. P., Chapin, F. S., and Klein, D. R. (1983). Carbon/nutrient balance of boreal plants in relation to vertebrate herbivory. *Oikos* **40**, 357–368.

Carpita, N. C. (1996). Structure and biogenesis of the cell walls of grasses. *Annu. Rev. Plant Physiol. Plant Mol. Biol.* **47**, 445–476.

Caswell, H., and Reed, F. C. (1976). Plant-herbivore interactions: The indigestibility of C_4 bundle sheath cells by grasshoppers. *Oecologia* **26**, 151–156.

Caswell, H., Reed, F., Stephenson, S. N., and Werner, P. A. (1973). Photosynthetic pathways and selective herbivory: A hypothesis. *Amer. Nat.* **107**, 465–480.

Chambers, J. C., and Norton, B. E. (1993). Effects of grazing and drought on population dynamics of salt desert shrub species on the Desert Experimental Range, Utah. *J. Arid Environ.* **24**, 261–275.

Cherney, J. H., Cherney, D. J. R., Akin, D. E., and Axtell, J. D. (1991). Potential of brown-midrib, low-lignin mutants for improving forage quality. *Advances Agronomy* **46**, 157–198.

Christie, E. K., and Detling, J. K. (1982). Analysis of interference between C_3 and C_4 grasses in relation to temperature and soil nitrogen supply. *Ecology* **63**, 1277–1284.

Collins, S. L. (1987). Interaction of disturbances in tallgrass prairie: A field experiment. *Ecology* **68**, 1243–1250.

Day, T. A., and Detling, J. K. (1990). Grassland patch dynamics and herbivore grazing preference following urine deposition. *Ecology* **71**, 180–188.

Davidson, R. L. (1969). Effect of root/leaf temperature differentials on root/shoot ratios in some pasture grasses and clover. *Ann. Bot.* **33**, 561–569.

Dengler, N. G., Dengler, R. E., Donnelly, P. M., and Hattersley, P. W. (1994). Quantitative leaf anatomy of C_3 and C_4 grasses (Poaceae): Bundle sheath and mesophyll surface area relationships. *Annals Bot.* **73**, 241–255.

Detling, J. K. (1988). Grasslands and savannas: regulation of energy flow and nutrient cycling by herbivores. In "Concepts of Ecosystem Ecology" (L. R. Pomeroy, and J. J. Alberts, eds.), pp. 131–148. Springer-Verlag, New York.

Dirven, J. G. P., and Deinum, B. (1977). The effect of temperature on the digestibility of grasses: An analysis. *Forage Res.* **3**, 1–17.

Dixon, A. F. G. (1985). "Aphid Ecology." Blackie, Glasgow.

Dormaar, J. F., Adams, B. W., and Willms, W. D. (1994). Effect of grazing and abandoned cultivation on a *Stipa–Bouteloua* community. *J. Range Manage.* **47**, 28–32.

Edwards, G., and Walker, D. (1983). "C_3, C_4: Mechanisms, and Cellular and Environmental Regulation, of Photosynthesis." University of California Press, Berkeley.

Ehleringer, J. R. (1978). Implications of quantum yield differences on the distribution of C_3 and C_4 grasses. *Oecologia* **31**, 255–267.

Ehleringer, J. R., and Björkman, O. (1977). Quantum yields for CO_2 uptake in C_3 and C_4 plants: Dependence on temperature, CO_2, and O_2 concentration. *Plant Physiol.* **59**, 86–90.

Ehleringer, J. R., and Monson, R. K. (1993). Evolutionary and ecological aspects of photosynthetic pathway variation. *Annu. Rev. Ecol. Syst.* **24**, 411–439.

Ehleringer, J. R., Sage, R. F., Flanagan, L. B., and Pearcy, R. W. (1991). Climate change and the evolution of C_4 photosynthesis. *Trends Ecol. Evol.* **6**, 95–99.

Esau, K. (1977). "Anatomy of Seed Plants," 2nd ed. John Wiley & Sons, New York.
Field, C., and Mooney, H. A. (1986). The photosynthesis-nitrogen relationship in wild plants. In "On the Economy of Plant Form and Function" (T. J. Givnish, ed.), pp. 25–55. Cambridge University Press, Cambridge.
Gebauer, G., Schuhmacher, M. I., Krstic, B., Rehder, H., and Ziegler, H. (1987). Biomass production and nitrate metabolism of *Atriplex hortensis* L. (C_3 plant) and *Amaranthus retroflexus* L. (C_4 plant) in cultures at different levels of nitrogen supply. *Oecologia* **72,** 303–314.
Gould, F. W. (1968). "Grass Systematics." McGraw-Hill, New York.
Hattersley, P. W. (1984). Characterization of C_4 type leaf anatomy in grasses (Poaceae). Mesophyll: bundle sheath area ratios. *Annals Bot.* **53,** 163–179.
Heidorn, T., and Joern, A. (1984). Differential herbivory on C_3 versus C_4 grasses by the grasshopper *Ageneotettix deorum* (Orthoptera: acrididae). *Oecologia* **65,** 19–25.
Hetrick, B. A. D., Kitt, D. G., and Wilson, G. T. (1988). Mycorrhizal dependence and growth habit of warm-season and cool-season prairie plants. *Can. J. Bot.* **66,** 1376–1380.
Holscher, C. E. (1945). The effects of clipping bluestem wheatgrass and bluegrama at different heights and frequencies. *Ecology* **26,** 148–156.
Hyder, D. N. (1972). Defoliation in relation to vegetative growth. In "The Biology and Utilization of Grasses" (V. B. Younger, and C. M. McKell, eds.), pp. 304–317. Academic Press, San Diego.
Ingham, E. R., and Molina, R. (1991). Interactions among mycorrhizal fungi, rhizosphere organisms, and plants. In "Microbial Mediation of Plant–Herbivore Interactions" (P. Barbosa, V. A. Krischik, and C. G. Jones, eds.), pp. 169–197. John Wiley & Sons, New York.
Jones, C. A. (1985). "C_4 Grasses and Cereals: Growth, Development, and Stress Response." John Wiley & Sons, New York.
Jones, C. G., and Coleman, J. S. (1991). Plant stress and insect herbivory: Toward an integrated perspective. In "Response of Plants to Multiple Stresses" (H. A. Mooney, W. E. Winner, and E. J. Pell, eds.), pp. 249–280. Academic Press, San Diego.
Kramer, P. J., and Boyer, J. S. (1995). "Water Relations of Plants and Soils." Academic Press, San Diego.
Ku, M. S. B., Schmitt, M. R., Edwards, G. E. (1979). Quantitative determination of RuBP carboxylase–oxygenase protein in leaves of several C_3 and C_4 plants. *J. Exp. Bot.* **30,** 89–98.
Lamprey, H. F. (1963). Ecological separation of the large mammal species in the Tarangire Game Reserve, Tanganyika. *E. Afr. Wildl. J.* **1,** 63–92.
Laws, R. J. (1970). Elephants as agents of habitat and landscape change in East Africa. *Oikos* **21,** 1–15.
Liener, I. E. (1979). Phytohemagglutinins. In "Herbivores: Their Interaction with Secondary Plant Metabolites" (G. A. Rosenthal, and D. H. Janzen, eds.), pp. 567–597. Academic Press, New York.
Ludlow, M. M., Samarakoon, S. P., and Wilson, J. R. (1988). Influence of light regime and leaf nitrogen concentration on 77K fluorescence in leaves of four tropical grasses: No evidence for photoinhibition. *Aust. J. Plant Physiol.* **15,** 669–676.
Lush, W. M., and Evans, L. T. (1974). Translocation of photosynthetic assimilate from grass leaves, as influenced by environment and species. *Aust. J. Plant Physiol.* **1,** 417–431.
MacMahon, J. A. (1988). Warm deserts. In "North American Terrestrial Vegetation" (M. G. Barbour, and W. D. Billings, eds.), pp. 231–264. Cambridge University Press, New York.
Marschner, H. (1995). "Mineral Nutrition of Higher Plants." Academic Press, San Diego.
McMillian, W. W., Wiseman, B. R., Burns, R. E., Harris, H. B., and Greene, G. L. (1972). Bird resistance in diverse germplasm of sorghum. *Ag. J.* **64,** 821–822.
McNaughton, S. J. (1985). Ecology of a grazing ecosystem: The Serengeti. *Ecol. Monog.* **55,** 259–294.
McNaughton, S. J. (1991a). Evolutionary ecology of large tropical herbivores. In "Plant–Animal Interactions: Evolutionary Ecology in Tropical and Temperate Regions (P. W. Price, T. M.

Lewinsohn, G. W. Fernandes, and W. W. Benson, eds.), pp. 509–522. Wiley-Interscience, New York.
McNaughton, S. J. (1991b). Dryland herbaceous perennials. *In* "Response of Plants to Multiple Stresses" (H. A. Mooney, W. E. Winner, and E. J. Pell, eds.), pp. 307–328. Academic Press, San Diego.
McNaughton, S. J., Coughenour, M. B., and Wallace, L. L. (1982). Interactive processes in grassland ecosystems. *In* "Grasses and Grasslands" (J. R. Estes, R. J. Tyrl, and J. N. Brunken, eds.), pp. 167–193. University of Oklahoma Press, Norman.
McNaughton, S. J., and Georgiadis, N. J. (1986). Ecology of African grazing and browsing mammals. *Ann. Rev. Ecol. Syst.* **17,** 39–65.
McNaughton, S. J., Oesterheld, M., Frank, D. A., and Williams, K. J. (1989). Ecosystem-level patterns of primary productivity and herbivory in terrestrial habitats. *Nature* **341,** 142–144.
McNaughton, S. J., Oesterheld, M., Frank, D. A., and Williams, K. J. (1991). Primary and secondary production in terrestrial ecosystems. *In* "Comparative Analyses of Ecosystems: Patterns, Mechanisms, and Theories" (J. Cole, G. Lovett, and S. Findlay, eds.), pp. 120–139. Springer-Verlag, New York.
McNaughton, S. J., and Sabuni, G. A. (1988). Large African mammals as regulators of vegetation structure. *In* "Plant Form and Vegetation Structure" (M. J. A. Werger, P. J. M. van der Aart, H. J. During, and J. T. A. Verhoeven, eds.), pp. 339–354. SPB Academic Publishing, The Hague.
McNaughton, S. J., Tarrants, J. L., McNaughton, M. M., and Davis, R. H. (1985). Silica as a defense against herbivory and growth promotor in African grasses. *Ecology* **66,** 528–535.
Minson, D. J. (1971). Influence of lignin and silicon on a summative system for assessing the organic matter digestibility of *Panicum. Aust. J. Agric. Res.* **22,** 589–598.
Monson, R. K. (1989). On the evolutionary pathways resulting in C_4 photosynthesis and crassulacean acid metabolism (CAM). *In* "Advances in Ecological Research" (M. Begon, A. H. Fitter, E. D. Ford, and A. MacFadyen, eds.), Vol. 19, pp. 57–110. Academic Press, London.
Osmond, C. B., Winter, K., and Ziegler, H. (1981). Functional significance of different pathways of CO_2 fixation in photosynthesis. *In* "Encyclopedia of Plant Physiology" (O. L. Lange, P. S. Nobel, C. B. Osmond, and H. Ziegler, eds.), Vol. 12B, Physiological Plant Ecology II: Water Relations and Carbon Assimilation, pp. 480–547. Springer-Verlag, Berlin.
Öztürk, M., Rehder, H., and Ziegler, H. (1981). Biomass production of C_3- and C_4-plant species in pure and mixed culture with different water supply. *Oecologia* **50,** 73–81.
Pearcy, R. W., Tumosa, N., and Williams, K. (1981). Relationships between growth, photosynthesis and competitive interactions for a C_3 and a C_4 plant. *Oecologia* **48,** 371–376.
Pellew, R. A. P. (1983). The impacts of elephant, giraffe and fire upon the *Acacia tortilis* woodlands of the Serengeti. *Afr. J. Ecol.* **21,** 41–74.
Phelan, P. L., and Stinner, B. R. (1992). Microbial mediation of plant-herbivore ecology. *In* "Herbivores—Their Interactions with Secondary Plant Metabolites" (G. A. Rosenthal, and M. R. Berenbaum, eds.), pp. 279–315. Academic Press, San Diego.
Pinder, III, J. E., and Kroh, G. C. (1987). Insect herbivory and photosynthetic pathways in old-field ecosystems. *Ecology* **68,** 254–259.
Pinder, III, J. E., and Jackson, P. R. (1988). Plant photosynthetic pathways and grazing by phytophagous orthopterans. *Amer. Mid. Nat.* **120,** 201–211.
Plumb, G. E., and Dodd, J. L. (1993). Foraging ecology of bison and cattle on a mixed prairie: Implications for natural area management. *Ecol. Applicat.* **3,** 631–643.
Rabatin, S. C., and Stinner, B. R. (1991). Vesicular–arbuscular mycorrhizae, plant, and invertebrate interactions in soil. *In* "Microbial Mediation of Plant–Herbivore Interactions" (P. Barbosa, V. A. Krischik, and C. G. Jones, Eds.), pp. 141–168. John Wiley & Sons, New York.
Raven, J. A. (1983). Phytophages of xylem and phloem: a comparison of animal and plant sap-feeders. *In* "Advances in Ecological Research" (A. MacFadyen, and E. D. Ford, eds.), Vol. 13, pp. 135–234. Academic Press, London.

Redmann, R. E., and Reekie, E. G. (1982). Carbon balance in grasses. *In* "Grasses and Grasslands" (J. R. Estes, R. J. Tyrl, and J. N. Brunken, eds.), pp. 195–231. University of Oklahoma Press, Norman.
Robertson, J. H. (1933). Effect of frequent clipping on the development of certain grass seedlings. *Plant Physiol.* **8,** 425–447.
Sage, R. F., and Pearcy, R. W. (1987). The nitrogen use efficiency of C_3 and C_4 plants. I. Leaf nitrogen, growth, and biomass partitioning in *Chenopodium album* (L.) and *Amaranthus retroflexus* (L.). *Plant Physiol.* **84,** 954–958.
Scott, J. A., French, N. R., and Leetham, J. W. (1979). Patterns of consumption in grasslands. *In* "Perspectives in Grassland Ecology" (N. R. French, ed.), pp. 89–105. Springer-Verlag, New York.
Scriber, J. M., and Slansky, F. (1981). The nutritional ecology of immature insects. *Annu. Rev. Entomol.* **26,** 183–211.
Seastedt, T. R. (1985). Maximization of primary and secondary productivity by grazers. *Amer. Nat.* **126,** 559–564.
Siegler, D. S. (1981). Secondary metabolites and plant systematics. *In* "The Biochemistry of Plants" (E. E. Conn, ed.), Vol. 7, pp. 139–176. Academic Press, New York.
Simms, P. L. (1988). Grasslands. *In* "North American Terrestrial Vegetation" (M. G. Barbour and W. D. Billings, eds.), pp. 265–286. Cambridge University Press, New York.
Stanton, N. L. (1988). The underground in grasslands. *Annu. Rev. Ecol. Syst.* **19,** 573–589.
Steuter, A. A., Steinauer, E. M., Hill, G. L., Bowers, P. A., and Tieszen, L. L. (1995). Distribution and diet of bison and pocket gophers in a sandhill prairie. *Ecol. Applic.* **5,** 756–766.
Stowe, L. G., and Teeri, J. A. (1978). The geographic distribution of C_4 species of the dicotyledonae in relation to climate. *Amer. Nat.* **112,** 609–623.
Swain, T. (1979). Tannins and lignins. *In* "Herbivores: Their Interaction with Secondary Plant Metabolites" (G. A. Rosenthal, and D. H. Janzen, eds.), pp. 657–682. Academic Press, New York.
Teeri, J. A., and Stowe, L. G. (1976). Climatic patterns and the distribution of C_4 grasses in North America. *Oecologia* **23,** 1–12.
Tilman, D., and El-Haddi, A. (1992). Drought and biodiversity in grasslands. *Oecologia* **89,** 257–264.
Van Dyne, G. M., Brockington, N. R., Szocs, Z., Duek, J., and Ribic, C. A. (1980). Large herbivore subsystem. *In* "Grasslands, Systems Analysis and Man" (A. I. Breymeyer, and G. M. Van Dyne, eds.), pp. 269–537. Cambridge University Press, Cambridge.
Vinton, M. A., and Hartnett, D. C. (1992). Effects of bison grazing on *Andropogon gerardii* and *Panicum virgatum* in burned and unburned tallgrass prairie. *Oecologia* **90,** 374–382.
Vinton, M. A., Hartnett, D. C., Finck, E. J., and Briggs, J. M. (1993). Interactive effects of fire, bison (*Bison bison*) grazing and plant community composition in tallgrass prairie. *Amer. Midl. Nat.* **129,** 10–18.
Weaver, J. E., and Albertson, F. W. (1936). Effects of the great drought on the prairies of Iowa, Nebraska, and Kansas. *Ecology* **17,** 567–639.
Wedin, D. A. (1996). Species, nitrogen, and grassland dynamics: The constraints of stuff. *In* "Linking Species and Ecosystems" (C. G. Jones, and J. H. Lawton, eds.), pp. 253–262. Chapman and Hall, New York.
Wilsey, B. J., Coleman, J. S., and McNaughton, S. J. (1997). Effects of elevated CO_2 and defoliation on grasses: A comparative ecosystem approach. *Ecol. Applic.* **7,** 844–853.
Wilson, J. R. (1990). Influence of plant anatomy on digestion and fibre breakdown. *In* "Microbial and Plant Opportunities to Improve Ligno-Cellulose Utilization by Ruminants" (D. E. Akin, L. G. Ljungdahl, J. R. Wilson, and P. J. Harris, eds.), pp. 99–117. Elsevier, New York.
Wilson, J. R. (1991). Plant structures: Their digestive and physical breakdown. *Rec. Adv. Nutr. Herbivores* **10,** 207–215.

Wilson, J. R., Deinum, B., and Engles, F. M. (1991). Temperature effects on anatomy and digestibility of leaf and stem of tropical and temperate forage species. *Neth. J. Agri. Sci.* **39,** 31–48.

Wilson, J. R., and Hattersley, P. W. (1983). *In vitro* digestion of bundle sheath cells in rumen fluid and its relation to the suberized lamella and C_4 photosynthetic type in *Panicum* species. *Grass Forage Sci.* **38,** 219–223.

Wilson, J. R., and Haydock, K. P. (1971). The comparative response of tropical and temperate grasses to varying levels of nitrogen and phosphorus nutrition. *Aust. J. Agric. Res.* **22,** 573–587.

Yeates, G. W. (1987). How plants affect nematodes. *In* "Advances in Ecological Research" (A. MacFadyen, and E. D. Ford, eds.), Vol. 17, pp. 61–113. Academic Press, San Diego.

10
The Biogeography of C_4 Photosynthesis: Patterns and Controlling Factors

Rowan F. Sage, David A. Wedin, and Meirong Li

I. Introduction

C_4 photosynthesis is a CO_2-concentrating mechanism that increases the carboxylation rate of Rubisco while simultaneously minimizing oxygenase activity and the inhibitory effects of photorespiration (Chollet and Ogren, 1975). From a photosynthetic standpoint, the greatest advantage of C_4 photosynthesis is in situations that promote photorespiration in C_3 plants, most notably high temperature and low intercellular CO_2 concentrations that result from restricted stomatal aperture and low atmospheric CO_2. Simple predictions based on relative performance of C_3 and C_4 photosynthesis are that C_4 species should outperform C_3 species in warm habitats and conditions promoting stomatal closure (drought, high salinity, and low humidity). Thus, the textbook generalization has become that C_4 photosynthesis is an adaptation for warm, high-light habitats with limited moisture availability (Barbour *et al.*, 1987; Taiz and Zeiger, 1991; Raven *et al.*, 1992).

As with most generalizations, this view of the adaptive significance of C_4 photosynthesis is an oversimplification that can lead to misunderstanding. C_4 plants often dominate sites that are hot and dry, yet they can also dominate warm, wet habitats while being absent in certain hot, dry locations. Frequently, the distribution of C_4 species is more a reflection of interactions between multiple ecological agents, rather than superior performance in heat or drought. To explain effectively why a C_4 species is successful in a particular location, it is therefore necessary to expand the discussion beyond photosynthetic attributes and consider other key ecological controls. In this review, we synthesize more than 30 years of research on C_4 distributions to present a current perspective of C_4 biogeography. We begin by summarizing the global pattern of C_4 distribution (Section II) and then discuss the major ecological factors affecting those distributions (Section III). Finally, because C_4 plants react differently than C_3 species to global change agents such as rising CO_2 and associated climate warming, we discuss how future change may affect C_4 vegetation (Section IV). Our emphasis is on C_4 grasses, largely because they predominate in terms of species numbers, ecological importance and availability of data relative to C_4 sedges and dicots.

II. The Global Distribution of C_4 Photosynthesis

Although it has long been recognized that Kranz anatomy is common in grasses of warm environments, its adaptive significance was not realized until the discovery of the C_4 syndrome (Brown, 1958; Downton and Tregunna, 1968; Laetsch, 1974; Hatch, Chapter 2). With the C_4 discovery, the known geographic distribution of Kranz plants contributed to the rapid realization that C_4 photosynthesis is an adaptation for warm environments, a view substantiated with the elucidation of Rubisco oxygenation (Chollet and Ogren, 1975; Bjorkman, 1976). Within a few years of the C_4 discovery, cursory surveys noted higher C_4 occurrence at the warm ends of latitude (Downton and Tregunna, 1968), altitude (Hofstra et al., 1972), and seasonal gradients (Black et al., 1969).

A. Latitudinal Distribution

1. Floristic Assessments Between 1975 and 1985, the geographical distribution of C_4 photosynthesis was elucidated, largely as a result of detailed surveys of C_4 species representation in regional grass, sedge and dicot floras in North America (Teeri and Stowe, 1976; Mulroy and Rundel, 1977; Stowe and Teeri, 1978; Teeri et al., 1980), South America (Ruthsatz and Hofmann, 1984); Australia (Hattersley, 1983; Takeda et al., 1985), western Asia (Winter and Troughton, 1978; Shomer-Ilan et al., 1981; Winter, 1981; Ziegler et al., 1981; Borchers et al., 1982), Europe (Collins and Jones, 1985), North Africa (Winter et al., 1976), central Africa (Livingstone and Clayton, 1980; Hesla et al., 1982) and southern Africa (Vogel et al., 1978; Ellis et al., 1980; Werger and Ellis, 1981). Subsequent surveys presented C_4 representation in grass and dicot floras of Egypt (Vogel et al., 1986; Batanouny et al., 1988, 1991), the former Soviet Union (Pyankov and Mokronosov, 1993), China (Kuoh and Chiang-Tsai, 1991, Yin and Zhu, 1990; Yin and Wang, 1997), Japan (Takeda et al., 1985a; Okuda, 1987; Okuda and Furakawa, 1990; Ueno and Takeda, 1992), the Indian subcontinent (Takeda, 1985), Namibia (Schulze et al., 1996), and South America (Cavagnaro, 1988; Filgueiras, 1991). In

Fig. 1, we combine the results of published surveys of C_4 representation in grass floras to provide a global image of C_4 grass distribution. To complement published surveys, we have compiled additional estimates from regional floras not previously evaluated for C_4 representation. In Fig. 2, we show the relationship between latitude and C_4 percentage in grass floras from oceanic islands of less than 2000 m in elevation. Climates tend to be more homogeneous across individual island systems because of their limited size and the moderating effects of marine waters. Island comparisons therefore show latitudinal influences with less complication than might be encountered on continental sites where interior heating or extreme seasonality can occur.

In all regions, C_4 representation in grass floras is dependent upon latitude (Figs. 1 and 2). In the tropics and subtropics, more than two-thirds of all grasses are C_4. Grasses from arid regions at low latitudes are almost exclusively C_4 (for example, Somalia, Namibia, northwest India, northwest Australia), and human-modified monsoon regions (central India) have very high C_4 occurrence (>85%). Tropical savannas consistently show very high C_4 representation—more than 90% of the principal savanna grasses of low latitudes are C_4 (Solbrig, 1996) (Table I). Similarly, grasses from small islands and atolls below 25° latitude are almost always C_4 (Fig. 2). In forested regions of the tropics (for example, Columbia, eastern Peru, southern China, the eastern Brazilian state of Bahia, and larger islands such as Java), representation of C_4 photosynthesis in grasses can fall to 45% to 70% as a result of an abundance of shade-adapted C_3 species and high numbers of C_3 bamboo species (Renvoize, 1984).

Outside of shaded habitats, the only factors consistently favoring C_3 grass occurrence in a low latitude flora are an abundance of water (typically in marshlands or swamps), high altitude, or human agriculture. In Java and the Peruvian Amazon region, for example, few C_3 grasses occur below 1000 m except in marshes and shaded areas (Backer and van den Brink 1968; Brako and Zarucchi, 1993). Common C_3 grasses of tropical marshes are in the genera *Leersia, Oryza, Phragmites,* and *Phalaris.* The most common C_3 grass in the lowland tropics is now cultivated rice (*Oryza sativa*), which has been domesticated from ancestors adapted to flooded soils, and is largely grown in flooded paddies (Pande *et al.,* 1994).

In contrast to low latitudes, regions above 60°N have no C_4 grasses with the exception of three to five weedy *Setaria* and *Digitaria* species from Scandinavia, Alaska, and Russia, and two *Spartina* species (*S. gracilis* and *S. pectinata*), plus three species of *Muhlenbergia* (*M. mexicana, M. richardsonis,* and *M. glomerata*) that occur in southern Alaska and northwestern Canada (Schwarz and Redmann, 1988, 1989). The *Spartina* and *Muhlenbergia* species are the only C_4 species known to occur naturally above 60°N and are largely restricted to continental interiors where warmer microsites favorable to C_4

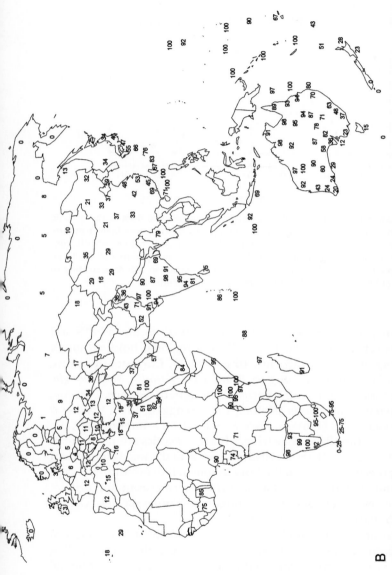

Figure 1 Percentage representation of C_4 photosynthesis in regional grass floras of (A) the western hemisphere and (B) the eastern hemisphere. Data compiled from previously published surveys or by estimating C_4 grass occurrence in regional floras using the C_4 species list in Sage, Li, and Monson (Chapter 16) to type photosynthetic pathways, except for *Panicum*, which relied on Brown (1977) and Zuloaga (1987). See Appendix I for a list of sources. Data generally include species occurring below 2000 m. All grass species reported in an area were included, and no attempt was made to exclude introduced species.

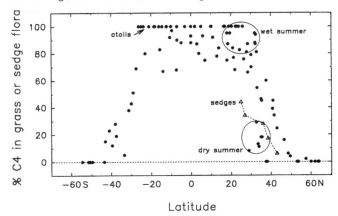

Figure 2 Percentage representation of C_4 photosynthesis in the grass flora of the oceanic islands of the world. Data compiled as for Fig. 1. The upper oval encloses islands from the western Atlantic/Pacific regions having wet summers. The lower oval encloses islands from the eastern Atlantic/Pacific having dry summers. The dashed line and open triangles indicate percentage C_4 representation in the Cyperaceae flora of the Japanese islands according to Ueno and Takeda (1992). Islands and sources presented in Appendix I.

establishment are present (Schwarz and Redman, 1988). Between 50°N and 60°N, C_4 grasses represent as much as 24% of a regional flora, although the general range is between 5% and 16% (Fig. 1). Consistently, native C_4 species at these latitudes belong to *Bouteloua, Distichilis, Muhlenbergia, Spartina,* and *Sporobolus,* and weedy species in the genera *Cynodon, Digitaria, Echinochloa, Eragrostis,* and *Setaria* (Halten, 1971; Scoggan, 1978; Erskine *et al.*, 1985; Vestergaard and Hansens, 1989; Skelton, 1991). Interestingly, the native species are in the Chlorideae tribe, while the exotics often belong to the Paniceae (Watson and Dallwitz, 1998). Removal of the weed component reduces the C_4 grass proportion in the flora by one-half to two-thirds, except in the heavily human-modified habitats of northwestern Europe and Britain, where all C_4 species are exotics except saltmarsh *Spartina* species (Collins and Jones, 1985; Parkhurst and Mullen, 1991).

In contrast to northern latitudes, C_4 grasses (and dicots) in southern latitudes are absent above 46°S (Figs. 1 and 2). This reflects the chilling effect of Antarctica on high southern latitudes and the lack of continental area where interior heating can occur. The southern one-fifth of South America is the only continental area extending above 45°S.

The transition from C_3 to C_4 dominance in the grass floras of the world occurs in the temperate latitudes between 45° and 30° in both hemispheres (Figs. 1 and 2). At these latitudes, the relationship between latitude and C_4 representation varies greatly. The western coasts of North America and

Europe–Africa have low C_4 representation until near 32°N, at which point C_4 representation in the grass flora rises markedly. In contrast, eastern Asia and eastern North America exhibit more gradual relationships between latitude and C_4 occurrence. These differences are well represented in the island floras off the respective coasts of these continents (compare ovals, Fig. 2). At approximately 33°N latitude, the Channel islands of California and Madeira in the eastern Atlantic exhibit less than 30% C_4 species, even when C_4 exotics are included. By contrast, in the southern Japanese islands at similar latitudes the grasses are 55% to 70% C_4, and 75% to 95% C_4 in the Bahamas and the barrier islands of the western Atlantic. Differences in C_4 representation between these locations relate largely to precipitation during the growing season. In the western Atlantic and Pacific regions, summers are wet, providing the moisture required for C_4 grasses to exploit summer heat. The eastern Atlantic and Pacific at latitudes between 25°N and 40°N have Mediterranean climates with winter precipitation and extreme summer drought (Fig. 3). C_3 species exploit the mild winter and abundant moisture in these settings, whereas the lack of significant summer precipitation prevents C_4 productivity except near perennial springs, creeks, and seeps. In central California, for example, a specialized suite of C_4 grasses (*Orcuttia*, *Neostapfia* spp.) has evolved to exploit seasonal pools that remain moist during the summer (Keeley, 1998). These species are the major native C_4 grasses found on non-salinized soils in the region (Crampton, 1974).

C_4 representation in sedges (Cyperaceae) and dicots shows a similar pattern of increasing proportion with decreasing latitude. C_4 sedges and dicots are rare above 45°N except for locally abundant weedy species in Europe, and on saline soils (Collins and Jones, 1985; Kalapos *et al.*, 1997). They become common below 35°N latitude, although the maximum representation is well below that of grasses. In the southern Ryuku islands of Japan (24°N) for example, C_4 representation in sedge genera reaches 44%, compared to 83% in grasses (Ueno and Takeda, 1992) (dashed line, Fig. 2). In the Nagasaki region of the Japanese main islands, 24% to 30% of the sedges are C_4, compared to 55% of grasses. Similar patterns prevail in North America, Africa, and Australia, where the C_4 representation in a given sedge flora is one-third to two-thirds the representation in the grass flora. In northern Australia, for example, 50% to 70% of the sedges are C_4, whereas more than 90% of the grasses are C_4 (Hattersley, 1983, Takeda *et al.*, 1985b). In Kenya, 65% of the Cyperaceae are C_4, whereas more than 90% of the low elevation grasses are C_4 (Livingstone and Clayton, 1980; Hesla *et al.*, 1982). In south Florida, 40% of the sedges are C_4, compared to 80% of the grasses (Teeri and Stowe 1976; Teeri *et al.*, 1980). Reduced C_4 representation in sedge floras is probably because sedges commonly occur on wet, marshy soils where C_4 plants tend to be less abundant.

Table I Common Grass Species from Tropical Savannas of the World

American savannas	African savannas
Andropogon bicornis	*Aristida* spp.
Andropogon hirtiflorus	*Andropogon gayanus*
Andropogon leucostachys	*Brachiaria obtusiflora*
Andropogon selloanus	*Chloris gayana*
Andropogon semiberbis	*Ctenium newtonii*
Aristida capillacea	*Cymbopogon afronardus*
Aristida pallens	*Cymbopogon plurinodis*
Aristida tincta	*Cynodon dactylon*
Axonopus aureus	*Diplachne fusca*
Axonopus canescens	*Echinochloa pyramidalis*
Axonopus capillaris	*Eragrostis pallens*
Axonopus purpusii	*Eragrostis superba*
Ctenium chapadense	*Heteropogon contortus*
Diectomis fastigiata	*Hyparrhenia cymbaria*
Echinolaena inflexa	*Hyparrhenia diplandra*
Eleusine tristachya	*Hyparrhenia dissoluta*
Elyonurus adustus	*Hyparrhenia filipendula*
Elyonorus latiflorus	*Hyparrhenia rufa*
Hymenachne amplexicaule[a]	*Imperata cylindrica*
Leersia hexandra	*Loudetia demeusei*
Leptocoryphium lanatum	*Loudetia kagerensis*
Mesosetum loliforme	*Loudetia simplex*
Panicum cayense	*Monocymbium ceresiiforme*
Panicum laxum	**Oryza barthii**
Panicum olyroides	*Pennisetum clandestinum*
Panicum spectabile	*Pennisetum purpureum*
Parathenia prostata	*Schizachyrium semiberbe*
Paspalum acuminatus	*Schmidta bulbosa*
Paspalum densum	*Setaria* spp.
Paspalum carinatum	*Sporobulus robustus*
Paspalum gardnerianum	*Sporobulus spicatus*
Paspalum fasciculatum	*Stipagrostis unipulumis*
Paspalum pectinatum	*Themeda triandra*
Paspalum plicatulum	*Trachypogon capensis*
Paspalum pulchelum	*Vossia cuspidata*
Paspalum repens	**Asian savannas**
Paspalum virgatum	*Aristida setacea*
Setaria geniculata	*Aristida adscensiones*
Setaria gracilis	*Bothriochloa intermedia*
Sporobulus cubensis	*Cynodon dactylon*
Thrasya paspaloides	*Eleusine indica*
Thrasya petrosa	*Heteropogon contortus*
Trachypogon canescens	*Imperata cylindrica*
Trachypogon montufari	*Miscanthus sinensis*
Trachypogon plumosus	*Panicum repens*
Trachypogon vestitus	*Saccharum arundinaceum*
Tristachya chrysothrix	
Tristachya leiostachya	

Table I   (Continued)

Australian savannas	
Alloteropis semialata	Eriachne trisecta
Aristida contorta	Eulalia fulva
Aristida latifolius	Heteropogon contortus
Aristida leptopda	Heteropogon triticeus
Arundinella nepalensis	Imperata cylindrica
Astrebia elymoides	Iseilema vaginiflorum
Astrebia lappacea	**Oryza spp.**
Astrebia pectinata	Panicum mindanenese
Astrebia squarrosa	Paspalidium spp.
Bothriochloa intermedia	Plectrachne pungens
Chinachne cyathopoda	Pseudopogonatherum irritans
Chloris gayana	Schizachyrium fragile
Chrysopogon fallax	Sehima nervosum
Chrysopogon pallidus	Setaria surgens
Dichanthium fecundum	Sorghum intrans
Dichanthium sericeum	Sorghum plumosus
Elyonurus citreus	Spinifex hirsutus
Eragrostis spp.	Sporobulus virginicus
Eriachne arenacea	Themeda australis
Eriachne obtusa	Themeda aveacea
Eriachne stipacea	Triodia irritans
	Triodia inutilis

^{a} C_3 species are presented in boldface type.
From: Solbrig, O. T. (1996). The diversity of the savanna ecosystems. In "The Biodiversity of the Savanna Ecosystem Processes—A Global Perspective" (O. T. Solbrig, E. Medina, and J. F. Silva, eds.), pp. 1–27. Springer-Verlag, Berlin; and Hayashi, I. (1979). Ecology, phytosociology and productivity of grasses and grassland. I. The autecology of some grassland species. In "Ecology of Grasslands and Bamboolands of the World" (M. Numata, ed.), pp. 141–152. VEB Gustav Fischer Verlag Jena.

C_4 dicots exhibit the same latitudinal trends as grasses and sedges, but because they are a minor fraction of the total dicot flora, their floristic representation is always low (<5% of dicots; Stowe and Teeri, 1978). Describing their representation in specific families or functional groups is thus a more useful way of describing latitude effects. Akhani et al., (1997) showed increasing representation of the C_4 pathway in the Chenopodiaceae with decreasing latitude which reflects responses in the Poaceae. In Europe, about 20% of the Chenopod flora is C_4, whereas in central Africa, more than 90% is C_4. In Egypt, C_4 representation in Euphorbia increases from less than 20% on the Mediterranean coast to 50% in southern Egypt (Batanouny et al., 1991). Functional groups enriched with C_4 dicots include summer annuals in monsoon-affected lowlands of the Sonoron and Indian deserts (Shreve and Wiggens, 1964; Sankhla et al., 1975; Mulroy and Rundel, 1977); halophytes of western North America and central Asia (Walter and

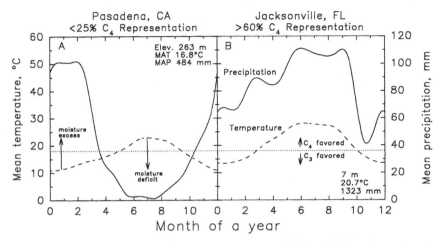

Figure 3 Walter-type climate diagrams for (A) Pasadena, California, a representative of a Mediterranean climate; and (B) Jacksonville, Florida, representative of warm temperate humid zone with wet summers. Adapted from Walter *et al.* (1975). Precipitation represented by solid lines, temperature represented by dashed lines, and the approximate transition from temperatures favoring C_3 versus C_4 grasses shown as a dotted line. Elevation (Elev.), mean annual temperature (MAT), and mean annual precipitation (MAP) also shown for each climate station. C_4 representation estimates approximated from Fig. 1A for regional grass floras. In Walter diagrams, when the precipitation curve is above the temperature curve, moisture excess occurs, indicative of a wet season. When the temperature curve is above the precipitation curve, a moisture deficit occurs, indicative of a dry season.

Box, 1983a; Dzurec *et al.*, 1985; Pyankov and Mokronosov, 1993; Akhani *et al.*, 1997); herbaceous plants of low latitude beach dunes (Barbour, 1992; Seeliger, 1992); and summer broadleaf weeds of cultivated land (Black *et al.*, 1969; Paul and Elmore, 1984).

2. Limitations of the Floristic Approach Although valuable, floristic assessments are limited in that they provide no ecological information such as cover, biomass, or relative dominance. Floristic checklists also tend to underestimate ecological ranges. Failure to recognize these limitations has led to misinterpretations and subsequent misconceptions concerning C_4 distribution. For example, in North America, Stowe and Teeri (1978) noted that 2.5% of the dicot flora in Florida was C_4, whereas 4.4% of Arizona dicots were C_4, and that the percentage of C_4 dicots in a flora was highly correlated with summer pan evaporation. Subsequently, numerous authors cited this evidence in support of the notion that C_4 was an adaptation for aridity (Arizona being drier than Florida), without considering the ecological ranges for the various species. In Arizona and adjacent regions, for example, the higher proportion of C_4 dicots reflects a rich flora of summer annuals which respond to episodic monsoon rains (Kemp, 1983; Smith and Nobel, 1986). These annuals are not necessarily adapted to drought, because they grow when the soils are moist and complete much of their life cycle before severe drought returns. Of greater significance is that the C_4 annuals grow in response to warm-season precipitation, and form a C_4 equivalent to the diverse bloom of C_3 annuals of the same region that appear in response to winter rains (Mulroy and Rundel, 1977; Kemp, 1983; Smith and Nobel, 1986).

3. Vegetation Assessments Detailed assessments of the contribution of C_3 and C_4 vegetation to biomass and cover within a site followed floristic surveys, as did reinterpretation of vegetation research from eras preceding the C_4 discovery (see Pearcy *et al.,* 1981, comment on Clements *et al.,* 1929). Vegetation assessments provide ecological information such as cover, biomass representation, and phenological patterns, but tend to be labor intensive and reflect conditions at discrete times. In addition, identifying a species as C_3 or C_4 can be difficult if it is nonreproductive and listings of C_4 taxa are unavailable. The discovery that C_4 species discriminate less against ^{13}carbon (^{13}C) relative to ^{12}carbon (^{12}C) greatly aided assessments of the relative contribution of C_4 biomass to vegetation in a given locality. Assessments of the ratio of ^{13}C relative to ^{12}C (δ^{13}C) overcome many of the limitations associated with direct quantification of C_3 and C_4 vegetation because the stable isotope procedure involves relatively little labor in the field, integrates photosynthetic contribution over time and space, and can be used for assessing C_4 plant contribution to belowground biomass and organic matter (Tieszen and Archer, 1990; Wong and Osmond, 1991). Analysis of δ^{13}C is currently the preferred method for evaluating C_3 and C_4 contributions to site productivity (Tieszen and Archer, 1990). To integrate over wide areas, standing biomass can be pooled from sampled material; to integrate over time, plant residues in the soil are evaluated. The soil organic matter in the A horizon reflects contributions to primary productivity on time scales of decades to centuries (Tieszen and Boutton, 1989); whereas deeper layers of organic matter and fossilized soils and plant remains allow for longer term evaluations (Cerling, Chapter 13).

At the regional level, vegetation and isotopic assessments of C_3 and C_4 biomass reflect patterns in the floristic studies, but often with an important difference. At tropical to warm temperate latitudes, C_4 species dominate grassland productivity, even though they may represent less than 60% of the grass flora of the region. In North America, for example, the C_4 contribution to grassland productivity in the central Great Plains exceeds 80% south of 42°N latitude, and then declines to near zero above 50°N (Fig. 4). Conversely, floristic assessments show only 40% to 70% C_4 representation in the North American grass flora below 42°N (Fig. 1).

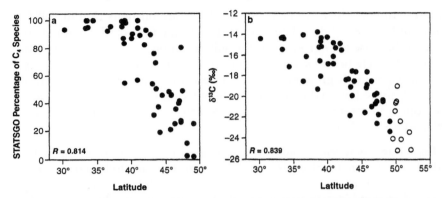

Figure 4 (A) The percentage contribution to primary production of C_4 species in the central plains of the United States, as a function of latitude. (B) Carbon isotope ratios of soil organic matter collected along a latitudinal gradient from United States (filled symbols) or Canadian (open symbols) field sites. STATSGO refers to the State Soil Geographic Database, from which productivity and species dominance data were compiled. [Reprinted from Tieszen, L. L., Reed, B. C., Bliss, N. B., Wylie, B. K. and DeJong, D. D. (1997) NDVI, C_3 and C_4 production, and distributions in Great Plains grassland land cover classes. *Ecol. Appl.* **7**, 59–78, with permission.]

4. Temperature Correlations Commonly, floristic or vegetation assessments along a latitude gradient are accompanied by correlative analyses between C_4 representation and various environmental parameters. In all cases, the contribution of C_4 species to the flora or biomass shows a high correlation with some index of temperature. In North America, C_4 grass, sedge, and dicot representation is well correlated ($r^2 > 0.87$) with July average temperature, July minimum temperature, number of days above 32°C, mean annual degree days, summer pan evaporation, and potential evapotranspiration (Teeri and Stowe, 1976; Stowe and Teeri, 1978; Teeri *et al.*, 1980). In Australia, increases in mean *annual* temperature between 14° and 23°C is well correlated with increasing C_4 representation as indicated by $\delta^{13}C$ ratio (Fig. 5). Below 15°C, C_4 contribution to productivity in Australian grasslands is negligible, whereas above 23°C mean annual temperature, C_3 contribution is negligible. When expressed as a function of *growing season* temperature, C_4 species are infrequent where the mean maximum of the warmest month is below 16°C and the mean minimum of the warmest month is between 6° and 12°C (Hattersley, 1983; Long, 1983). *Spartina, Setaria, Miscanthus,* and *Muhlenbergia* species, however, often persist in climates with lower growth season temperature (Ehleringer and Monson, 1993).

5. Transitions between C_4- and C_3-Dominated Grasslands—the Great Plains Example In North America, analysis of C_4 and C_3 cover in remnant prairies

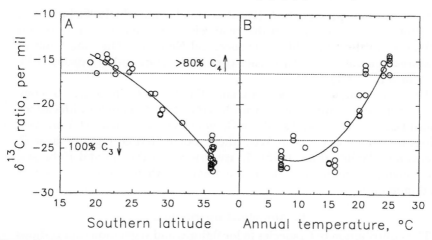

Figure 5 The carbon isotope ratio in soil organic matter along an (A) latitudinal and (B) temperature gradients in Australian grasslands and savannas. Compiled from data presented in Bird and Pousai (1997).

has facilitated extensive analysis of potential productivity and species dominance as a function of photosynthetic type (Tieszen et al., 1997). When coupled with remotely sensed reflectance images of red and near infrared radiation (the normalized difference vegetation index, NDVI) and carbon isotopic assessments, these data allow for the mapping of the contribution of C_3 and C_4 species to potential primary production of the Great Plains biome in the central United States (Fig. 6; see color insert).

The contribution of C_4 biomass to potential productivity illustrated in Fig. 6 generally confirms the C_4 distribution documented at a much coarser scale by the florstic analysis of Terri and Stowe (1976) for C_4 grasses (Fig. 1). The detail shown in Fig. 6 presents a clearer understanding of the patterns of dominance and climatic controls on C_4 distribution. Rather than a systematic decline in C_4 cover as latitude increases, there is a peak of C_4 dominance along a 300 km band stretching from the Texas panhandle to eastern Kansas. To the south of this band in central Texas, and to the north of the band in Nebraska, C_4 production drops below 70% of total production. North of Nebraska, C_4 production drops off further, reflecting the cooler climate associated with higher latitude. North of Oklahoma, a general decline in C_4 dominance occurs along an east–west gradient. This reflects increasing elevation and associated cooler summer temperature, and declining summer precipitation, as one moves west (Epstein et al., 1997; Tieszen et al., 1997). Winter is the driest season on the Great Plains, and the increase in precipitation from west to east largely occurs as increased summer precipitation (Paruelo and Lauenroth, 1996; Epstein *et al.*, 1997). Of special note is the relatively high level of C_4 production (70–80%) in the Sand Hills of west-central Nebraska. The sandy soils of this area generally favor C_4 grasses over C_3 (Barnes and Harrison, 1983; Epstein *et al.*, 1997).

The combination of temperature and climate can be illustrated by comparing central Wyoming regions with less than 20% C_4 biomass with eastern South Dakota and Nebraska, where C_4 biomass is more than 60%. In Wyoming, precipitation peaks in May, when cooler thermal regimes favor C_3 vegetation (Walter *et al.*, 1975). In eastern South Dakota and Nebraska, precipitation is high throughout the summer, when temperature favors C_4 dominance.

B. Altitudinal Distribution

The contribution of C_4 species to local floras and vegetation stands shows a strong decline with increasing altitude (Figs. 7 and 8) (see also Meinzer, 1978; Teeri, 1979; Wentworth, 1983; Pyankov and Mokronosov, 1993). Whereas grasslands are heavily dominated by C_4 vegetation at low altitude, C_3 plants dominate at high altitude, with the shift generally occurring between 1500 and 3000 m (Fig. 7). Surprisingly, no obvious latitude modification of the altitude pattern is present. In Hawaii, at 19.5°N, the transition from complete C_4 to C_3 dominance occurs between 1000 and 2000 m. In Wyoming, at 42°N, the transition in biomass contribution is between 1800 and 2600 m. In Kenya (0°), the transition in isotopic ratios from pure C_4 to pure C_3 occurs between 2500 and 3000 m (Fig. 7), whereas in New Guinea (5–9°S), this transition occurs between 1600 and 3000 m (Fig. 8). Some evidence for a latitudinal interaction is evident in floras from Argentina, where C_4 plants dominate the grass flora to higher elevations in northern Argentina than in central Argentina (compare the "A" curve with the "a" curve in Fig. 7A).

As with latitudinal trends, temperature indices are closely correlated with elevation trends and reveal a decrease in abundance of C_4 species at similar low temperatures. In tropical montane studies, C_4 species generally disappear from a local flora at sites where the mean minimum temperature of the warmest month is between 8° and 10°C. In temperate zones, the transition to C_3 dominance is generally reported to occur where the mean minimum temperature of the warmest month reaches 12° to 14°C (Chazdon, 1978; Tieszen *et al.*, 1979; Rundel, 1980; Wentworth, 1983; Cavagnaro, 1988; Earnshaw *et al.*, 1990). In arid regions of central Asia, C_4 species are generally absent where the growth season temperature is less than 6° to 8°C, although a few dicot species (*Atriplex pamirica, Climacoptera lanata, Halogeton glomeratus*) persist to growth season minimum temperatures averaging as low as 2°C (Pyankov and Mokronsov, 1993). In Hawaii, Rundel (1980) noted

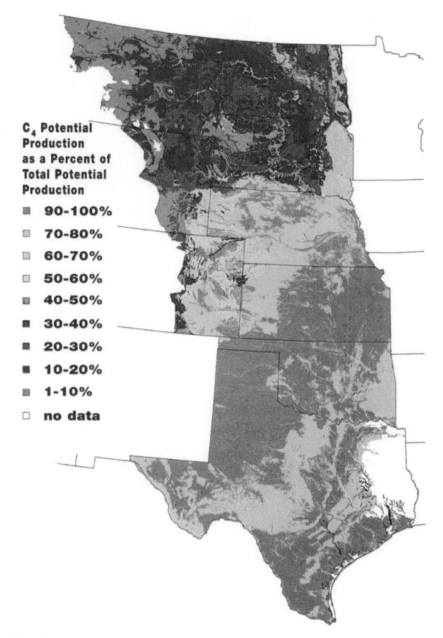

Figure 6 The proportional contribution to primary productivity of C_4 grass biomass in the central Plains of the U.S., developed from the STATSGO data base according to Tieszen *et al.* (1997). [Reprinted from Tieszen, L. L., Reed, B. C., Bliss, N. B., Wylie, B. K., and DeJong, D. D. (1997). NDVI; C_3 and C_4 production, and distributions in Great Plains grassland land cover classes. *Ecol. Appl.* **7**, 59–78, with permission.]

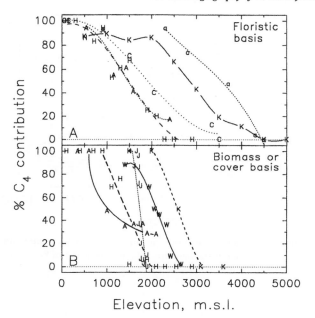

Figure 7 The representation of C_4 photosynthesis along an altitudinal gradient based on (A) the percentage C_4 representation in regional grass floras, or (B) on the basis of C_4 biomass or cover. Symbol codes: A, central Argentina at 31°S (Cavagnero, 1988, for floristic; Cabido et al., 1997 for percentage cover); a, northern Argentina, 23°S (Ruthsatz and Hoffmann, 1984); C, Costa Rica, 10°N (Chazdon, 1978); H, Hawaii, 20°N (Rundel, 1980); J, Japan, Kirigamine grassland, 37°N (Nishimura et al., 1997); K, Kenya 0°; (Livingstone and Clayton, 1980, for floristics, and Tieszen et al., 1979, for biomass); W, Wyoming, 41°N (Boutton et al., 1980).

that the temperature for transition to C_4 dominance occurred at a mean maximum temperature of 21° to 23°C, and a daily mean of about 15°C, similar to the estimated transition temperatures on Mt. Kenya and in Costa Rica (Chazdon, 1978; Teizen et al., 1979). Aridity modifies altitude trends, with C_4 plants reaching higher altitudes on drier sites, for example, Mt. Kenya (Young and Young, 1983) and the Pamirs of central Asia (Pyankov, 1994).

C. The C_4-Dominated Biomes

Floristically, C_4 species comprise a minority of the terrestrial plant species—about 8000 of the estimated 250,000 to 300,000 land plants species, or less than 4% (Sage, Li, and Monson, Chapter 16). On a land surface basis, C_4 species assume an importance far exceeding their floristic contribution in that they are major components of biomes that cover more than 35% of the earth's land surface area (Table II). C_4 species have the potential

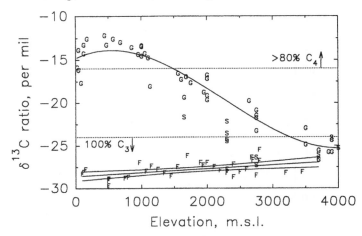

Figure 8 The relationship between carbon isotope ratio in soil organic matter and elevation (meters above sea level) in grasslands (G), swamps (S) and forests (F) of Papua New Guinea. Compiled from data presented in Bird *et al.* (1994).

to dominate any low latitude, low elevation landscape where the competition for light from trees and shrubs has been reduced. Where C_4 species occur, conditions often favor establishment of woody vegetation. Thus the distribution of C_4 vegetation in the tropics and subtropics largely reflects the distribution of woody- versus graminoid-dominated ecotypes.

The most extensive natural C_4 biomes are the various savannas of the tropics, subtropics, and warm temperate zones that range in C_4 grass cover from near 20% (for example, the cerrado of Brazil) to nearly pure C_4 grasslands such as the Serengeti in east Africa. One-eighth of the Earth's land surface is covered by savanna, with Africa and Australia approaching two-thirds savanna coverage, South America, 45%, and India and southeast Asia about 10% each (Cole, 1986; Scholes and Archer, 1997). In temperate North America, the C_4-dominated Great Plains grassland once covered the middle one-third of the continent below 50°N (Coupland, 1992). Much of this grassland system has been converted to row crop agriculture, and in the eastern portion (the area once dominated by tall grass prairie), the native C_4 grasses have been largely replaced by a few C_4 annuals, notably crops such as *Zea maize* and their associated weeds (Fig. 9). Because conversion of natural vegetation to agriculture has been intensive during this century, the most extensive C_4 vegetation type in the world may now consist of C_4 crops, pasture grasses, and weeds (Pearson, 1992; Brown, Chapter 14).

At lower latitudes, C_4 plants are common in coastal habitats. Along the coasts of the Americas, *Spartina* and *Distichilis* species are common domi-

Table II Categorization of the Major Biomes of the World[a]

Biome	Major locations
High C₄ representation	
Tropical and subtropical grassland and savanna	South and Central America, Africa, India, southeast Asia, Australia
Warm temperate grassland and savanna	Central and southeastern North America, northern Argentina, Australia
Arid steppe (low to midlatitude)	Southwestern North America, central Asia, Australia, Africa
Beach dunes, warm temperate to tropics	Global
Tropical, subtropical wetlands (nonarborescent)	Global, especially South America, central Africa, southeast Asia
Saltmarsh (warm temperate to tropical)	Global, but more temperate due to mangrove dominance in tropics
Salt desert (<45° latitude)	Western North America, central Asia, central Australia
Hot deserts and semideserts	Southwest North America, Africa, Australia, southeast Asia
Disturbed ground (low latitudes, low elevation)	Global, more in arid regions
Always C₃ dominated	
Forests (all, including arborescent wetlands)	Global
Tundra (all)	Polar latitudes, high mountains (all latitudes)
Heathlands	Northern Europe, Canada, Russia
Cool temperate grasslands and savanna	Canada, northwest United States, southeastern Europe, southern Russia, Australia, southern Argentina, Tasmania
Montane grasslands	Global, elevations between 2000 to 4000 m
Mediterranean grasslands	California, Chile, southern Europe, Asia Minor, northern and southern Africa, southwestern Australia
Mediterranean-type shrublands (chapparal)	California, Chile, southern Europe, Asia Minor, North Africa, South Africa, southwest Australia
Temperate to boreal wetlands (nonarborecent)	Canada, northern Europe, Russia, New Zealand, Patagonia
Cold deserts and semideserts	Western North America, north-central Asia, Tibet
Saltmarshes (>50° latitude)	Global
Mangles (mangrove swamps)	Tropical, subtropical latitudes
Disturbed ground (high latitude and altitude)	Global

[a] Based on whether there is high C₄ representation (loosely defined as >25% of the vegetative cover, and a potential for C₄ dominance), or whether there is never C₄ dominance and C₄ cover is negligible (<1%).
Developed from D. W. Goodall, (ed.) *"Ecosystems of the World* (1977–1990)," Vol. 1–16. Elsevier, Amsterdam.

Figure 9 Photographs illustrating that fate of most of the tallgrass prairie biome in central North America. (A) Remnant C_4-dominated prairie grassland at the Konza Prairie in eastern Kansas (39°05' × 96°40'W), and (B) the new C_4 grassland dominated by *Zea mays* (the tall grass prominent in the center of the photo) with weedy C_4 species growing along the margins of cultivated fields and highways (as shown in the foreground just beyond the child). Photo from eastern Nebraska, along Interstate 80 (40°50' × 97°W). The child is Katherine Sage. Photographs by R. F. Sage.

nants that often form monospecific stands in hypersaline marsh zones (MacDonald, 1977; West, 1977; Pomeroy and Wiegert, 1983; Costa and Davy, 1992). In Australia, *Sporobolus* and *Distichilis* spp. are the common saltmarsh grasses, whereas in Japan, *Zoysia sinica* is a distinctive C_4 saltmarsh grass (Saenger *et al.*, 1977; Archibold, 1995). In Africa and southwest Asia, numerous C_4 dicots (*Salsola, Sueda,* and *Heliotropium* spp.) become important saltmarsh elements along with C_4 grasses in *Aeleropus* and *Sporobolus* (Blasco, 1977; Zahran, 1977). Coastlines from warm temperate to tropical regions contain a suite of C_4 grasses, sedges, and dicots that often dominate beach dunes, except in Mediterranean climates (Barbour, 1992; Moreno-Casasola and Castillo, 1992; Seeliger, 1992; van der Maarel, 1993).

C_4 grasses, and dicots in the Chenopodiaceae, often dominate salinized soils of arid to semiarid continental interiors (Walter and Box, 1983a,b; West, 1983; Lewis *et al.*, 1985; Guy and Krouse, 1986; Akhani *et al.*, 1997). In these regions, C_4 shrubs and small trees are common, with more than a dozen species able to form dense scrub. In North America, the C_4 shrub *Atriplex confertifolia* is particularly important in saline soils of the Great Basin steppe (West, 1983; Dzurec *et al.*, 1985), whereas in southwestern and central Asia, *Anabasis, Calligonum, Haloxylon,* and *Salsola* spp. are important C_4 shrubs and short-stature trees in saline and sandy deserts (Walter and Box, 1983a; Akhani *et al.*, 1997). Between the Caspian Sea and Alma-Ata, for example, *Haloxylon aphyllum* (= *H. ammodendron*) becomes arborescent, with trunk diameters of 1 m and height up to 9 m (Walter and Box, 1983b). This is probably the tallest C_4 species in the world, and forms forests with individual stands covering up to 600,000 ha. These forests are heavily exploited for fuel wood and are often destroyed by overcutting (Walter and Box, 1983b).

III. Factors Controlling the Distribution of C_4 Species

Over the years, much discussion has addressed the ecological controls affecting the distribution of C_4 species, and the paradigm emerged that C_4 photosynthesis is an adaption to hot, high light, and typically arid environments (Osmond *et al.*, 1982). In reviewing material for this chapter, we arrived at a slightly modified conclusion that reduces the importance of aridity:

> C_4 plants have two primary requirements for success—warm growing seasons and access to moderate to high light intensity. All other factors such as aridity are secondary in that they influence patterns of C_4 distribution and dominance, but are inconsequential in the absence of both moderate light and warm growth conditions.

In the following section, we defend this view from a biogeographical perspective.

A. Temperature

As discussed previously, virtually every index of growing season temperature is highly correlated with C_4 abundance along both latitude and elevation gradients. C_4 plants are rare at altitudes and latitudes where growth season temperatures are less than an average of approximately 16°C and minimum midsummer temperatures average less than 8° to 12°C (Long, 1983). Winter temperature is not an obvious factor because C_4 species dominate many temperate locations where severe winter freezing occurs, and, if dormant, survive freezing as well as their C_3 associates (Long, 1983; Schwarz and Reaney, 1989).

In addition to indices of growing season temperature, the other temperature component of note is the length of the warm growing season relative to the cool growing season (Doliner and Joliffe, 1979). Where the warm season is wet, but the cool season dry, open landscape will be C_4 dominated, as is observed in the monsoon-climates of India and Africa (Fig. 10A). Where the warm season is dry, but the cool season mild and wet (as occurs in the Mediterranean climates of southern Europe, California, western Australia, South Africa and Chile), open landscapes will have few C_4 species, except as weeds in irrigated fields (Fig. 10B) (Beetle, 1947; Collins and Jones, 1985; Baker, 1989). Where C_3 and C_4 grasses cooccur, much of the C_3 element will break dormancy earlier in spring than the C_4 element. In the northern Plains grasslands of North America, the C_3 grasses become active in March to mid-April, while the C_4 species appear in late April to May (Dickinson and Dodd, 1976; Baskin and Baskin, 1977; Ode et al., 1980; Monson et al., 1983). Throughout much of the Plains grasslands, the presence of a mild spring and a moist, hot summer allow for coexistence between C_3 and C_4 grass floras, although a dynamic ebb and flow is apparent from year to year. During moist, mild springs and dry summers, C_3 grasses do well at the expense of C_4; by contrast, in years with dry springs and wet summers, the C_4 species become more competitive (Monson et al., 1983).

If climate conditions consistently favor one photosynthetic type over the other, then it can capture the necessary soil resources and dominate the site. In the plains of Texas, for example, long, hot summers with an abundance of summer precipitation (Fig. 10C) consistently allow for aggressive summer growth of the C_4 dominants. Whereas winter temperatures are mild enough to allow C_3 species to continue activity, the C_4 grasses produce a dense turf that captures space and nutrients to such an extent that cool season grasses are often excluded from the prairie (Tieszen et al., 1996). This leads to an interesting phenomenon in North America where southern, C_4-dominated grasslands green-up at the same time of year as northern, C_3 dominated grasslands (Tieszen et al., 1997). Alternatively, if climates do not allow for complete space and resource capture by one photosynthetic type, then

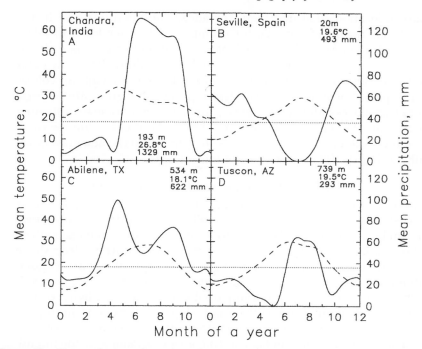

Figure 10 Climate diagrams for regions discussed in the text. (A) Chandra, central India, representative of a monsoonal climate with a hot, wet summer and cool dry season. C_4 grasses dominate (>90%) the grass flora and production in open areas. (B) Seville, Spain, representative of a Mediterranean climate with a moist cool season and dry warm season. C_3 grasses dominate the grass flora and primary production. (C) Abilene, Texas, a warm temperate grassland where the warm season is long enough (6 months over 18°C) and wet enough to allow for C_4 dominance. (D) Tuscon, Arizona, a hot desert where distinct cool and hot seasons of precipitation allow for a temporally segregated C_3 and C_4 flora to coexist in the region. Precipitation (solid line), temperature (dashed line). Climate diagrams adapted from Walter et al. (1975). See legend for Fig. 3 for further climate diagram description.

temporally segregated C_4 and C_3 communities can coexist. This is observed in the Sonoron and Chihuahuan deserts of North America, where winter rains favor a C_3 flora of annual plants, whereas summer rains favor a rich C_4 annual flora (Fig.10D) (Kemp, 1983). Because of severe drought between the winter and summer seasons, moisture is insufficient to enable one community to persist for very long, and thus dominance is prevented.

An underappreciated phenomenon in cool climates at high latitudes and altitudes is growing seasons can be truncated by late-season snowpack or high water levels. This delays the start of the growing season until days are warm, and can often prevent C_3 species from producing a leaf canopy earlier than C_4 plants. Because of this, numerous C_4 species may be able to exist in locations that may seem climatically unsuited to them. In the boreal forest region of Canada, for example, high water levels of lakes, rivers, and fens prevent plant growth along the shoreline until late May to June (R. Sage and D. Kubien, unpublished observations, 1997). As water levels recede, exposed beaches and hummocks become available for colonization by plants including C_4 *Muhlenbergia* species. One grass, *Muhlenbergia mexicana*, produces bulblike vegetative propagules that disperse by floating on water and ice, eventually washing onto beaches as they become exposed. The bulbules then germinate in mid-to-late spring, and the new grass grows rapidly during the long, warm days of midsummer.

B. Light Availability and Disturbance

Because of the higher ATP requirement to run the C_4 cycle, and possibly a reduced capacity to exploit sunflecks (Krall and Pearcy 1993), C_4 photosynthesis can be considered maladapted to most shade environments. Reflecting this, the large majority of C_4 plants are uncompetitive in deep shade (Pearcy and Ehleringer, 1984; Long, Chapter 7). Despite this general understanding, however, the minimum light intensity required for C_4 success is not obvious. C_4 species can occur in partial shade, and their light requirement is dependent on various climate and edaphic interactions. In tropical savannas, C_4 grasses of various species dominate understories of savanna trees such as *Acacia tortilis* and baobab (*Adansonia digitata*) at shade intensities that reduce full sunlight intensity by up to 70% (Belsky *et al.*, 1989; Weltzin and Coughenour, 1990). C_3 representation often increases in response to partial shading, particularly in temperate climates. In the semiarid savannas of West Texas, numerous C_3 grasses increase in importance in the shade of mesquite (*Prosposis* spp.), oaks (*Quercus* spp.), and *Acacia* at the expense of the dominant C_4 grasses such as *Buchloe dactyloides* (Heitschmidt *et al.*, 1986). In the north temperate savannas of Minnesota, C_4 grass cover declines from maximum values (40% of groundcover) at light levels above 50% of canopy openness, to negligible cover at light levels associated with less than 25% of canopy openness (Peterson and Reich, 1996). Similar shifts are reported in Wisconsin savannas and coastal plain pine stands of the southeastern United States (Bray, 1958; Ko and Reich, 1993; Means, 1997). Only with near complete canopy closure, where light levels typically fall below 10% to 20% of the incident radiation (Pearcy, 1990), do C_4 plants become absent, regardless of climate. Because of this, forests are largely free of C_4 species around the world (Medina and Klinge, 1983). (However, see Brown, 1977; Smith and Wu, 1994; and Horton and Neufeld, 1998, for discussions on shade-adapted C_4 grasses). In most instances, therefore, the factors that control the distribution of C_4 vegetation in warmer climates are those that regulate the occurrence of woody vegetation at a high enough density to close the overstory canopy. Because C_3

woody species can potentially occupy any habitat where C_4 grasses occur, the regulation of grass versus woodland dominance is relevant throughout the range of C_4 occurrence.

1. Key Factors Modifying C_4 Grass—C_3 Woodland Dominance   Four primary factors have been identified as critical modulators of grass–woodland dynamics. These are: (1) the moisture regime; (2) edaphic properties and soil nutrition; (3) abiotic disturbance, most notably fire, but also including flooding, wind, and frost; and (4) biotic disturbance from herbivores and disease (Scholes and Archer, 1997; Knapp and Medina, Chapter 8). The relative importance of these four factors varies geographically, and attempts to derive a universally valid ranking have been controversial. Rather than evaluate the relative importance of these factors here, we present a synthetic view that reflects the light requirement of most C_4 species.

Moisture is important because episodic droughts can kill woody vegetation, whereas abundant precipitation promotes rapid tree growth and establishment in grasslands (Goldstein and Sarmiento, 1987; Faber-Langendoin and Tester, 1993). Most grasslands and savannas are characterized by high variability in precipitation, with severe drought a common factor, either seasonally or during dry years of climatic cycles (Goldstein and Sarmiento, 1987; Nichols, 1991; Knapp and Medina, Chapter 8). Regarding edaphic factors, moist, well-drained, nutrient-enriched soils favor woodland over grassland, whereas sandy, saline, poorly drained and/or shallow soils are more likely to have C_4 grasslands (Cole, 1986; Archibold, 1995). Sandy soils have low water and nutrient holding capacity, and are more drought prone in the high evaporative conditions of the tropics. Shallow soils restrict woody species by preventing deep rooting, such that the woody species must directly compete with grasses for belowground resources (Goldstein and Sarmiento, 1987). Shallow soils also drain poorly and are often waterlogged, yet dry frequently because of low capacity for water storage. C_4 grasses are highly competitive on these soils because of superior water use efficiency (WUE) and nutrient use efficiency (NUE) and a high capacity to tolerate both flooding and drought, which are restrictive for woodland species (Medina, 1987; Archibold, 1995).

Abiotic disturbance generally favors herbaceous species over woody species, with fire in particular being critical for maintenance of many C_4 grasslands (Knapp and Seastedt, 1986; Goldammer, 1993; Bond and van Wilgen, 1996). C_4 grasses adapted to dryland soils tend to be pyrophytic, producing highly flammable litter and belowground perennating organs that are protected from fire (Sarmiento, 1992). In contrast, immature woody species tend to be harmed by fire, and frequent burning either kills them outright or prevents them from overtopping and escaping competition from the grass sward (Goldammer, 1993; Bond and van Wilgen, 1996).

In most savannas, protection from fire leads to a gradual increase in woody species, with the eventual transition of much of the landscape to scrub or forest (Bragg and Hurlber, 1976; Hopkins, 1983; San Jose and Farina, 1991; Goldammer, 1993; Scholes and Archer, 1997).

Flooding and wave action also promotes C_4 vegetation because of its negative effects on woody species. Along temperate to tropical coastlines, disturbance from wave action allows a C_4-rich flora of grasses and dicots to dominate beach dunes (Barbour, 1992). Where periodic flooding prevents woodland establishment, fast-growing C_4 grasses and sedges commonly dominate lakeshores and river floodplains (Jones, 1986; Piedade *et al.*, 1994). On each tropical continent, for example, a distinctive suite of fast growing, functionally similar C_4 species has evolved to dominate the resource-rich conditions present on periodically inundated soils. Papyrus (*Cyperus papyrus*) and *Miscanthidium* spp. are common C_4 dominants in flooded wetlands of tropical Africa, whereas in Amazonia, *Echinochloa polystachya* and various *Paspalum* spp. are the functional equivalents (Junk, 1983; Thompson and Hamilton, 1983; Long, Chapter 7).

Large animals are also critical modifiers of C_4 grass/C_3 woodland relationships. In Africa, elephants and giraffes tend to favor C_4 species by selectively killing or suppressing woodland trees (McNaughton, 1994). Grazers tend to favor woody species by opening up sites for their establishment, and by reducing intensity of grass competition and grass fuel accumulation. Overgrazing by cattle and sheep has been a common cause of grassland conversion to shrub or woodland in many semi-arid regions of the world (Lamprey, 1983; Brown and Archer, 1989; Grover and Musick, 1990; Nichols, 1991). Simple rules do not always apply, however, because grazers can also restrict woody establishment if they eat or trample young tree seedlings, and can encourage grass growth by cycling nutrients and preventing inhibitory levels of litter build-up (Knapp and Seastedt, 1986; Scholes and Archer, 1997). In any case, if the animal action in some way prevents woody species establishment and persistence, C_4 grasses will be favored.

Extreme low temperature events can be important in localized situations. In the southeastern United States, ice storms topple trees and remove branches, thereby creating canopy gaps and woody fuel for fire. In pine woodlands, this helps maintain the canopy openness necessary for persistence of a C_4 grass understory (R. Sage, personal observation, 1993). A particularly important consequence of extreme cold events is found in coastal saltmarshes. In the tropics and subtropics, tidal estuaries are dominated by mangrove swamps (all mangroves are woody C_3 trees). Mangroves are intolerant of chilling, and winter temperature below $\sim 10°C$ exclude mangroves swamps from temperate latitudes (Duke *et al.*, 1998). Thus, areas that would have been occupied by mangroves in the absence of lethal cold, are instead dominated by herbaceous C_4 genera such as *Spartina* and *Distichilis* (Chapman, 1977; Archibold, 1995). Along the western Atlantic coast, for example, mangrove swamps extend from the tropics to central Florida, whereas *Spartina*-dominated marshes stretch from north Florida to Newfoundland (Chapman, 1977). Thus, in contrast to the inland latitudinal pattern of C_4 giving way to C_3 vegetation with increasing latitude, the combined effect of salinity and cold yield the opposite pattern of C_3-dominated mangroves giving way to C_4-dominated saltmarshes as one moves to higher latitude.

Rarely do single ecological factors act in isolation. For example, grass–woodland ecotones reflect the balance between factors promoting establishment of woody species (such as moisture and nutrients) versus factors determining intensity and frequency of lethal disturbances. Drought and soil infertility slow woody establishment, prolonging the period when woody species are vulnerable to fire (Kellman, 1988). Seasonal drought enhances fire probability, so that conditions that slow woody establishment also shorten fire intervals. Often, multiple levels of interaction create complex dynamics that are difficult to predict and control, as demonstrated by the large scale grassland–woodland perturbation that followed the lethal epidemic of the ungulate pathogen rinderpest in East Africa beginning in the late nineteenth century (McNaughton, 1992; Dublin, 1995). Following its introduction in the 1890s, rinderpest decimated cattle, giraffe, and wildebeest populations, reducing the negative impact these animals had on woody plants. Brush levels increased, and with it, tsetse fly populations that transmitted trypanosomiasis (sleeping sickness) to humans. As a result of cattle loss and increased threat of disease, most humans left the Serengeti region and human-managed fire decreased in frequency. Much of the C_4-dominated landscape succeeded to C_3 woodland dominated by *Acacia* spp. between 1900 and 1950. With the introduction of disease control in ungulates and humans in the 1950s, and compression of elephant populations into the Serengeti region by human exploitation of surrounding areas, the trends were reversed and expansive grassland returned as woodlands were destroyed. Elephants are highly destructive of the woody species, both directly as a result of browsing and shredding of canopies, and indirectly because the woody debris left by elephants fuels hot fires during the dry season, causing high tree mortality (Dublin *et al.*, 1990). Where grassland has returned, elephants and giraffe continue to select against the woody element because they browse on woody seedlings, which, in combination with fire and wildebeest trampling, restricts woodland regeneration (Dublin *et al.*, 1990; Dublin, 1995). Where elephants have been removed (often by poaching), yet wildebeest populations remain high, fire cycles are disrupted because wildebeest reduce grassy fuel loads. In such areas, *Acacia* thickets are again forming (Sinclair, 1995).

2. Human Disturbance C_4 habitats worldwide have been universally and severely impacted by human activity. Prehistorically, humans affected C_4

distribution by altering fire and grazing regimes to such an extent that much of the world's "natural" savanna and grassland is derived as a result of human activity (Schüle, 1990). Whereas considerable debate surrounds the extent of human-derived savanna, it is generally recognized that savanna in moist areas capable of supporting rainforest, and savanna on Pacific and Caribbean islands is human-derived (Gillison, 1983; Sarmiento, 1983; Cole, 1986; Aide and Cavelier, 1994). In the past few centuries, shifting cultivation in the tropics has promoted the spread of C_4-dominated savanna into forested regions, a process accelerated by rapid population growth of more recent decades (Hopkins, 1983; Mueller-Dombois and Goldammer, 1990). In western Africa, for example, forests have receded by as much as 500 km in response to human activity (Hopkins, 1983). Throughout the tropics and subtropics, large areas once dominated by C_3 forests have been converted to C_4 grasslands and pastures by fire, shifting agriculture, logging, and fuelwood cutting. Dry tropical forests are particularly threatened (Maass, 1995), but increasingly, rainforest is also being cleared. In Madagascar, northeastern Brazil, Southeast Asia, India, Java, and Central America, more than 80% of the natural forests have been lost, in many cases replaced by C_4-dominated grassland (Mueller-Dombois and Goldammer, 1990; Aide and Cavelier, 1994). A common pattern following forest clearing, particularly when the land is initially cultivated, is that soils degrade, losing fertility and structure. Pyrophytic C_4 grasses such as *Imperata cylindrica*, *Hyparrhenia rufa*, and *Melinis minutiflora* often invade recently cleared land and abandoned fields. Once established, these grasses restrict woody colonization by accelerating fire cycles and maintaining low soil fertility. As a result, regeneration of the forest is prevented (Hopkins, 1983; Mueller-Dombois and Goldammer, 1990; D'Antonio and Vitousek, 1992; Aide and Cavelier, 1994).

Although the recent expansion of tropical grassland and pasture at the expense of forest represents the largest shift in the distribution of C_3- versus C_4-dominated systems since the early Holocene, human action enhances the existence of C_3 systems over C_4 in other regions. Overgrazing of C_4-dominated semiarid rangelands has resulted in C_3-dominated shrublands in southwestern North America, central Australia, and southern Africa (Lamprey, 1983; Grover and Musick, 1990; Nicholls, 1991). Fire suppression by European colonists has also contributed to woodland expansion into C_4 biomes, notably in the Americas and Australia (Bragg and Hurlber, 1976; Pyne, 1990). More recently, much of the natural savanna of the tropics and warm temperate zones has become threatened with destruction, in part from agricultural clearing, but increasingly because of management for timber production (Soares, 1990). In the coastal plain of the southeastern United States, large-scale destruction of the *Pinus palustris/Aristida* savanna occurred in the late nineteenth and early twentieth centuries because of logging. This ecosystem originally covered 60% of the southern coastal plain (Drew et al., 1998), and much of it is now timber plantations where trees are maintained at high enough density to shade out the C_4 *Aristida* grasses (Means, 1997).

C. The Role of Water and Salinity

1. Water Because of their higher WUE, C_4 plants are generally thought to be superior competitors in arid sites. Regional and global vegetation models often use algorithms of water use efficiency to help predict occurrence of C_4 versus C_3 vegetation (for example, Haxeltine and Prentice, 1996), and textbooks duly note the adaptive significance of C_4 photosynthesis in arid environments. Although we acknowledge that the C_4 syndrome enhances survival and competitive ability in arid situations, we argue that aridity *per se* is not a prerequisite for C_4 dominance. Water availability has a complex role in promoting C_4 dominance, and no simple trend holds. Excess water, in addition to water deficiency, can favor C_4 species, whereas in Mediterranean zones, summer drought is an important factor in the success of C_3 vegetation. This difference is illustrated by comparing two South African locations of similar latitude. The Capetown–Danger Point region has the wet winters and dry summers of a Mediterranean climate and less than 10% C_4 species in the grass flora, whereas the Durban region, which lies 1000 km east on the Indian Ocean coast, has dry winters and wet summers with more than 75% C_4 representation in the local grass flora (Walter et al., 1975; Vogel et al., 1978). Total precipitation does not appear to matter, as regions in South Africa receiving 100 mm or 1000 mm of rain can be exclusively C_3 or C_4 depending on the timing of the precipitation (Vogel et al., 1978).

Further evidence against a direct role of drought as a primary requirement for C_4 dominance is the observation that C_4 vegetation is absent in arid regions that are chronically cold, such as polar deserts and high-elevation tundra. In contrast, C_4 grasses can dominate both wet and arid zones in hot climates as long as there are other factors limiting the success of woody plants. Along moisture gradients, an inverse relationship exists between precipitation and C_4 cover, but this largely reflects the dominance of woody vegetation at the mesic end of precipitation gradients. If woody vegetation is factored out and only grasslands and savannas of varying moisture status are compared, then the correlation between C_4 dominance and precipitation regime can be reversed, with greater C_4 representation and cover on wetter sites (Paruelo and Lauenroth,1996; Schulze et al., 1996; Epstein et al., 1997). Precipitation patterns also interact strongly with fire frequency. In regions with annual or periodic drought, fire becomes an important agent in favor of C_4 vegetation, but only if the growth season is warm to hot, with some precipitation. If a Mediterranean or low temperature location experiences severe fire, the herbaceous vegetation that follows is C_3 (Kozlowski and Ahlgren, 1974).

Although drought does not override temperature and light requirements for C_4 occurrence, it does have a secondary role in that it allows C_4 species to persist in cooler climates than might otherwise be the case. On Mount Kenya, for example, C_4 grasses at high elevation are found on drier sites (Young and Young, 1983). In the Pamir mountains of central Asia and the Andes of South America, the world's highest stands of C_4 plants (4000+ m) are found on dry, desert slopes (Pyankov, 1994; Cabido et al., 1997). *Muhlenbergia richardsonis,* North America's highest and most northerly C_4 plant, typically occurs on dry soils at its extreme locations (R. Sage, personal observation, 1998; Schwartz and Redman, 1988). Similarly, in central Asia, C_4 plants at high latitudes (55°N) are typically found in arid microsites (Walter and Box, 1993; Pyankov and Mokronosov, 1993).

Relative to the C_3 syndrome, the C_4 syndrome is perhaps most adaptive for aridity in the hot deserts of the world, where conditions are often are too harsh to support herbaceous C_3 vegetation. In Namibia, Schulze et al., (1996) identified 30 to 70 species of C_4 grasses that occur in hot locations that receive between 200 and 400 mm annual precipitation. Without these C_4 species, much of this landscape would be barren. In Death Valley, California, C_4 plants such as *Tidestromia oblongifolia* are active during very hot, arid periods when C_3 annuals and herbaceous perennials are inactive. Competitive exclusion does not appear likely in these hot desert situations because density is low. Instead, C_3 plants may be unable to maintain positive carbon gain at the high stomatal resistances required to prevent excessively high rates of transpiration. *Tidestromia,* for example, exhibits high CO_2 assimilation rates at 50°C, a temperature at which C_3 species exhibit negligible carbon gain (Berry and Björkman, 1980).

2. Salinity Salinity is another important modifier of the control that light and temperature have on C_4 distribution. The C_4 syndrome may be favored under saline conditions because the higher WUE allows C_4 species to have lower water demands than C_3 species, reducing the amount of salt that plants must excrete, store, or cope with metabolically (Osmond et al., 1982; Adam, 1990; Long, Chapter 7). In addition to dominating many salinized soils at low latitude, C_4 halophytes persist on saline soils in climates that otherwise favor C_3 vegetation (Long, 1983; Lewis et al., 1985; Guy et al., 1986). C_4 grasses in the genus *Spartina,* for example, are able to extend to latitudes exceeding 60°N along Atlantic seashores in North America and Europe. These coastal climates generally lack summer heat, and often experience foggy, low light conditions. Similarly, the southernmost C_4 species is *Spartina anglica,* which grows in salt marshes at the southern tip of New Zealand at 46°S (King et al., 1990). In interior basins of western North America and central Asia, precipitation predominately occurs during the winter and early spring, such that the growing season corresponds to cooler months of the year. The vegetation on nonsalinized soils in these basins is largely C_3 (Walter *et al.*, 1975; Walter and Box, 1983a,b). Interior drainage coupled with high summer evaporation creates extensive salinized basins that favor C_4 vegetation, mainly halophytic grasses (for example, *Distichilis* spp.) and Chenopods. In the cold arid steppe of northern Nevada, the C_4 species not considered agricultural weeds are generally C_4 halophytes. Saltgrass (*Distichilis spicata*), alkali sacaton (*Sporobolus aeroides*), and the C_4 shrub *Atriplex confertifolia*, for example, are common on poorly drained soils along marshes and saltflats in the Great Basin region between the Sierra Nevada and Rocky mountains of western North America (West, 1983). In these areas, well-drained, upland soils support C_3 communities (Young and Evans, 1980; West, 1983).

3. Aridity, Salinity, and Photosynthetic Subtypes Although C_4 photosynthesis may not require aridity for ecological success, there is a clear trend for aridity and salinity to affect the distribution of C_4 photosynthetic subtypes. Along precipitation and salinity gradients, species of the NADP-malic enzyme (NADP-ME) subtype are more abundant at the mesic and nonsaline end of the gradient, whereas species of the NAD-malic enzyme (NAD-ME) subtype are the dominant C_4 subtype at the arid and saline ends. These patterns have been noted for the Americas (R. Sage, unpublished, 1998; Knapp and Medina, Chapter 8), Australia (Prendergast, 1989; Henderson *et al.*, 1994), Namibia (Schulze *et al.*, 1996) and central Asia (Pyankov and Mokronosov, 1993). In Namibia, NAD-ME grasses compose more than 70% of the grass flora at 150 mm rainfall, but less than 20% at 600 mm. In Australia, NAD-ME species constitute 45% of the C_4 grass flora at 200 mm, but less than 20% at 850 mm. In both places, NADP-ME species compose 60% or more of the C_4 grasses at the wet end of the gradient. The reasons for these trends are unclear (Hattersley, 1992).

C. Nitrogen

Because C_4 plants have an inherently greater NUE of photosynthesis than C_3 plants (Long, Chapter 7), they often are more nitrogen use efficient at the whole plant level, resulting in lower N requirements and a competitive advantage relative to C_3 plants (Brown, 1978; Wedin and Tilman, 1993; Long, Chapter 7). Whole plant NUE and the ability of a plant to compete for N depend on much more than leaf-level tissue N concentrations, however. Biomass allocation patterns, nutrient retranslocation during senescence, and tissue longevity all contribute to a plant's nitrogen economy (Berendse and Aerts, 1987). For example, an annual C_4 grass may have relatively high N requirements and fare poorly under low N conditions because it cannot retain N from year to year. Thus, low soil nutrient availability, like low water availability, does not appear to be a prime requirement for C_4 dominance. If a warm growing season or high light availability are lacking, C_4 plants generally will be absent regardless of soil nitrogen status. Conversely, C_4 grasses are frequently dominant in tropical and subtropical climates under both high N (eutrophic) and low N (oligotrophic) conditions in which light availability is continuously high because of disturbance, grazing, or management.

Despite these caveats, N appears to play an important role in the competitive balance of C_3 and C_4 vegetation, especially in temperate regions. Experimental N additions in humid temperate grasslands have generally favored C_3 grasses and forbs at the expense of C_4 grasses (Fig. 11) (Wedin, 1995; Wedin and Tilman, 1996). Several mechanisms may underlie this competitive shift. As mentioned previously, the annual N demand of perennial C_4

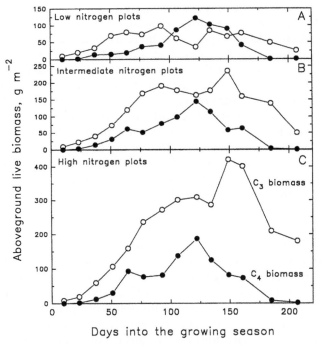

Figure 11 Seasonal patterns of aboveground live biomass for C_3 (open circles) and C_4 (closed circles) vegetation in an east–central Minnesota sand prairie. Low nitrogen plots were unfertilized. Medium nitrogen plots received 5.6 g N m^{-2} yr^{-1}, whereas high nitrogen plots received 17 g N m^{-2} yr^{-1}. Samples were collected in the first year of N addition. Day 0 of the growing season is April 15. (D. Wedin, 1985, unpublished data; see Tilman, 1988, for details of the study site and methods).

grasses is generally lower than that of C_3 grasses. When N availability increases, most of the supplemental N may go to C_3 production. Increased productivity also leads to decreased light availability in the plant canopy and increased light competition (Tilman, 1988), a situation that favors C_3 over C_4 vegetation at moderate temperatures (Knapp and Seastedt, 1986). If N availability and C_3 productivity are high early in the growing season and early season production is not removed (for example, by herbivores, fire, or management), C_4 plants may be at a competitive disadvantage in midgrowing season, when they would otherwise have a competitive advantage in a seasonally hot climate. Increased N availability also appears to extend the C_3 growing season, perhaps by enhancing water use efficiency and net carbon balance under warm temperatures. In Minnesota (United States) sand prairies, C_3 plants normally dominate aboveground live biomass in the spring and fall, whereas C_4 grasses dominate production in June through August, when average daily high temperatures exceed 25°C (Fig. 11). With experimental N addition, the midseason drop in C_3 biomass disappears, and C_3 productivity increases sharply. C_4 production increases less than C_3 production with increased N availability, and as a result, C_4 plants exhibited less midsummer production than the C_3 grasses.

These experimental results are qualitatively consistent with Tilman's (1982) resource-ratio model of plant competition. Because of their lower tissue N requirements and higher light saturation points, C_4 plants should have a competitive advantage when the ratio of light availability to soil N availability is high. In contrast, C_3 plants may, in general, have a competitive advantage when the ratio of light availability to soil N availability is low. Wedin and Tilman (1993) grew the C_4 bunchgrass *Schizachyrium scoparium* and the C_3 grasses *Agropyron repens* and *Poa pratensis* together over a range of soil fertilities and a range of light availabilities. As predicted under the resource-ratio model, decreasing light on a low fertility soil and increasing N supply under full light conditions both gave the C_3 species competitive advantage over the C_4.

IV. C_4 Plants in the Future

In the past 5 million years, C_4 photosynthesis has become a successful adaptive strategy, most likely because of a drop in atmospheric CO_2 in recent geological time (Ehleringer *et al.*, 1997; Cerling, Chapter 13). In turn, the future rise in atmospheric CO_2 from anthropogenic activities may remove the adaptive value of the C_4 syndrome, potentially leading to a shift to C_3 dominance. However, global change is more than simply CO_2 enrichment, and four of the major global change components—climate warming, terrestrial eutrophication, bioinvasions, and human alteration of landscapes (Vitousek, 1994)—are significant modifiers of C_3/C_4 dynamics. Understanding how C_4 vegetation will respond in the future must involve integrating the impacts of the various global change agents. This is difficult in its own right; however the task is especially problematic because any outcome will depend on human demands and social trends, which are difficult to predict in the best of circumstances. In this section, we highlight important observations concerning the major global change agents and C_4 plants.

A. CO_2 Enrichment

C_3 photosynthesis is more responsive than C_4 photosynthesis to atmospheric CO_2 enrichment. In C_4 plants, a doubling of atmospheric CO_2 typically stimulates net CO_2 assimilation by 0% to 25%, whereas for C_3 plants, CO_2 assimilation is usually stimulated by 10% to 75%, depending on temperature (Long and Drake, 1992; Sage, 1994; Tissue *et al.*, 1995; Kirschbaum *et al.*, 1996). Warmer temperatures stimulate photosynthetic responses to CO_2 enrichment, both in C_3 and C_4 plants (Berry and Björkman, 1980; Sage and Sharkey, 1987; Drake *et al.*, 1996). Relative growth responses reflect photosynthetic responses. In a survey of growth in twice ambient atmospheric CO_2, Poorter (1996) observed median response of 44% for wild C_3 herbaceous plants, 40% for C_3 trees, and 58% for C_3 crop plants. In contrast, C_4 plants exhibited a growth stimulation of 14%. Based on these results, one might conclude that elevated CO_2 will favor C_3 species over C_4 species. However, the control carbon has over fitness in natural systems is attenuated by other environmental factors, and competitive outcomes are dependent on many nonphotosynthetic events, notably herbivory, disturbance, phenology, and allocation differences between C_3 and C_4 individuals (Bazzaz and McConnaughay, 1992; Polley , 1997; Wilsey *et al.*, 1997). Here, we address several plausible CO_2 enrichment responses that could enhance either C_3 or C_4 success.

Direct responses to CO_2 enrichment are most apparent on fertile soils where the plants are able to convert increased levels of photosynthate into growth and reproduction (Bazzaz, 1990; Patterson and Flint, 1990; Poorter *et al.*, 1996). Consequently, in mixed stands having species of similar life forms, C_3 species would likely become more competitive relative to C_4 species on nutrient-enriched soils. This has been documented in the ecotone between *Scirpus olneyi* (C_3) and *Spartina patens* (C_4) stands in Chesapeake Bay (Maryland) saltmarshes, and in numerous agricultural or disturbance situations (Patterson and Flint, 1990; Drake *et al.*, 1996). In Chesapeake Bay, *S. olneyi* in mixed C_3 and C_4 communities responded to a doubling of atmospheric CO_2 over 7 years with a 37% increase in net carbon exchange and a 50% to 225% increase in the number of shoots. Root biomass increased in *S. olneyi* along with tolerance of heat, salt, and drought stress (Arp *et al.*, 1992; Drake *et al.*, 1996). Insect herbivory and fungal pathogenesis was also reduced in *Scirpus* (Thompson and Drake, 1994). In contrast, elevated CO_2 had no major direct effect on *S. patens* growing in monoculture; however, *S. patens* produced less biomass in the presence of *S. olneyi*, indicating competitive suppression (Arp et al., 1992; Drake et al., 1996). As a result, invasion of *S. olneyi* into the marsh zone dominated by C_4 grasses is predicted (Arp et al., 1993).

When soil nutrients or rooting volume become limiting, the stimulation of photosynthesis and growth in plants by higher CO_2 is reduced, if not altogether eliminated, such that potential benefits of high CO_2 on photosynthetic carbon gain and growth are less effectively exploited (Thomas and Strain, 1992; Sage, 1994; Poorter et al., 1996). CO_2 enrichment often aggravates nutrient deficiency, both within C_3 plants by aggravating imbalances between nutrient supply and carbohydrate pools, and at the soil level, by increasing nutrient sequestration in microbial pools, and potentially slowing decomposition of litter enriched in carbon (Diaz et al., 1993; Owensby et al., 1996; Polley et al., 1996; Ball, 1997). Where this occurs, the superior ability of C_4 grasses to exploit limited N pools on infertile soils (Wedin and Tilman,1996; Knapp and Medina, Chapter 8) could promote superior C_4 performance following CO_2 enrichment. Alternatively, N fixation by legumes and *Frankia*-nodulated shrubs is stimulated by rising CO_2 (Tissue et al., 1997; Vogel et al., 1997) and should this input of N into ecosystems rise significantly as a result of CO_2 enrichment, C_3 species might be favored.

Increasing atmospheric CO_2 reduces stomatal conductance of both C_4 and C_3 plants by 25% to 50% (Eamus and Jarvis, 1989; Morrison, 1993; Polley, 1997). Whether this will favor C_3 or C_4 vegetation is likely to be system specific; however, because C_4 vegetation is active during warm seasons when potential evapotranspiration is highest, and is generally dominant in regions experiencing periodic drought, the relative significance of reduced stomatal conductance is potentially greatest for C_4 vegetation. Reductions in stomatal conductance improve tissue water status, and despite minor direct effects of high CO_2 on C_4 photosynthesis, the improved water relations lead to greater leaf production, delay senescence, delay soil drying during drought, and reduce drought inhibition of C_4 vegetation (Samarakoon and Gifford, 1995; Polley et al., 1996b). These responses enable C_4 vegetation to spread at the expense of codominant C_3 vegetation. In the Tallgrass Prairie in Kansas, for example, doubling of atmospheric CO_2 over a 5-year period favored the dominant C_4 grass *Andropogon gerardii* over *Poa pratensis*, an important C_3 competitor (Owensby et al., 1996, 1997). This occurred largely because the enhanced water use efficiency and associated improvements in water status allowed *A. gerardii* to delay leaf senescence and maintain growth during drought events that otherwise induce dormancy (Owensby et al., 1993, 1996).

Much of the potential shift in C_4 distribution may occur as a result of woodland expansion into grasslands (Polley, 1997). In particular, a lessening of water use by the C_4 community might favor woody seedlings, allowing for faster shrub and tree establishment and more rapid shading of the grass community. This may partially explain the expansion of C_3 shrubs into arid grasslands of the American southwest observed in recent decades (Polley *et al.*, 1997). Alternatively, if high CO_2 slows growth of woody vegetation by promoting nutrient sequestration, then grasslands may expand. Critical to this outcome will be effects on the fire cycle, which could be altered if CO_2 enrichment accelerates the rate of fuel accumulation. Higher rates of fuel accumulation shorten the burn interval, as well as increasing the fuel load and fire intensity, and thus the probability that a burn will be severe enough to kill the woody element (Sage, 1996).

B. Climate Warming

Consistently, C_4 species become more competitive against C_3 species of similar life form and ecological habitat when grown in higher temperature, demonstrating that warming of a shared growing season should lead to increased C_4 dominance. Pearcy *et al.* (1981) and Christie and Detling (1982) found warmer growth conditions strongly favored C_4 over C_3 vegetation, with an increase from cool (17° to 20°C) to warm (30° to 34°C) daytime growth temperature completely reversing competitive outcomes. Although the C_3 species outperformed the C_4 at the cooler temperature, the C_4 species outperformed the C_3 at warmer temperature, regardless of water or nitrogen availability.

Because activities of C_3 and C_4 species are often segregated in time, the timing of warming may be more critical than the degree of warming. Climate models typically predict that future warming will mainly occur during winter and at higher latitudes (Kattenberg *et al.*, 1996). C_3 species will be favored where cool growing seasons are lengthened, as may occur in temperate regions where a harsh winter inhibits growth. In recent years, C_3 grasses have become more common in the native prairies of southern Saskatchewan (Canada) at the expense of the C_4 species (Peat, 1997). Although this region has experienced warmer winters and springs in recent decades, midsummer temperatures have not increased (Skinner *et al.*, 1998). In other words, the northern Great Plains have become warmer, but not hotter. This longer and warmer spring may be responsible for the improved C_3 performance (Peat, 1997). Even at lower latitudes, moderation of winter temperatures could allow C_3 vegetation to avoid a midwinter inhibition and enhance their ability to capture site resources and compete against summer-dominant C_4 grasses. For example, mangrove vegetation could move to higher latitudes and replace *Spartina* spp. that currently dominate temperate-zone saltmarshes.

In temperate zones, C_4 species should be favored when hot conditions occur earlier in the year, and by any intensification of midsummer heat. However, the C_4 response to climate warming will largely depend on precipitation regimes. Where summer heating is accompanied by increased precipitation, C_4 grasses should do well, unless the increased summer moisture disrupts the fire regime and allows for greater tree establishment. Where summer precipitation does not change, or is reduced, then summer heating and increased evapotranspiration may be detrimental to C_4 vegetation, particularly if milder winters enable C_3 vegetation to capture the limited water supply. Much of the future shift in C_4 vegetation zones will depend on effects of climate change on monsoon precipitation. One location predicted to be affected by monsoon intensification is the arid steppes of southern Nevada and Utah (Lin et al., 1996), where a sparse flora of summer C_4 annuals may be enhanced if summer thunderstorms become more common.

CO_2 enrichment complicates warming scenarios because the photosynthetic responsiveness to high CO_2 generally rises more in C_3 than C_4 species as temperature increases (Berry and Björkman, 1980; Sage et al., 1987; Kimball et al., 1993; Sage et al., 1995). Thus, although elevated temperature alone favors C_4 species, elevated temperature at high CO_2 might allow C_3 species to overcome enhancement of C_4 species by high temperature. Which stimulatory response predominates is not clear, although results to date indicate that elevated CO_2 at high temperature reduces, but generally does not overcome performance advantages of C_4 plants relative to C_3 competitors (Carter and Peterson, 1983). In short grass prairie vegetation of Colorado, Hunt et al., (1996) found that shoot biomass of the C_4 dominant *Bouteloua gracilis* is stimulated approximately 40% by a 4° rise in growth temperature at 350 μmol CO_2 mol^{-1} air, whereas the C_3 dominant *Pascopyrum* (*Agropyron*) *smithii* was unaffected. At double the current CO_2 level, the relative effect of temperature on growth was reduced because the C_3 species had a 10% growth response to a 4° temperature rise, whereas the C_4 dominant species showed only about a 20% thermal stimulation. Similarly, Grise (1997) observed that tripling CO_2 reduced the difference in growth at high temperature (34°C) between C_3 (*Chenopodium album*) and C_4 (*Amaranthus retroflexus*) annual weeds. In mixed stands, the enhancement of growth in *C. album* did not inhibit *A. retroflexus*, indicating little competitive suppression of the C_4 by the C_3 as a result of CO_2 enrichment. Simple competition studies may be limited, however, in that they do not often consider many of the other temperature-sensitive factors that affect field performance. For example, leaf area production, seedling survivorship, allocation patterns, and phenology can alter outcomes of C_3 and C_4 competition relative to predictions based only on photosynthetic responses (Coleman and Bazzaz, 1992; Morse and Bazzaz, 1994).

C. Terrestrial Eutrophication

Anthropogenic fixation of N has more than doubled natural fixation processes, leading to substantial eutrophication of terrestrial ecosystems in Europe and eastern North America, and increasingly in tropical regions as agricultural fertilizers become more readily available (Vitousek *et al.*, 1997). These increases in N availability pose a threat to C_4-dominated native ecosystems, particularly in temperate regions (Wedin and Tilman, 1996). Widespread use of agricultural fertilizers has provided us with an unplanned experiment on the effects of supplemental N on the relative success of C_3 and C_4 vegetation. Grassland "improvement" (primarily N fertilizer additions and legume plantings) has led to C_3-dominated pastures and hayfields in most of eastern and central North America north of roughly 38°N latitude (Burns and Bagley *et al.*, 1995). Grasslands in much of this area, particularly in the former Tallgrass Prairie regions of the Midwest, were C_4-dominated prior to management. In warm climates, N fertilization does not necessarily lead to C_4 displacement because numerous C_4 grasses (for example, *Digitaria* spp.) are aggressive eutrophiles that can dominate nitrogen-enriched soils. In warm temperate regions such as the southeastern United States, grassland management and N fertilizer inputs have resulted in the displacement of native C_4 grasses by Old World C_4 grasses (Kretschmer and Pitman, 1985). However, increased N may also favor woody vegetation in warm climates, although high nitrogen use is often associated with intensive agriculture or urbanization, factors that generally contribute to expansion of C_4 habitat in the tropics. Thus, although increased N availability may be expected to alter the C_3–C_4 balance and favor C_3 vegetation in temperate regions, predictions are more difficult in tropical and subtropical regions.

Elevated CO_2 combined with increasing N deposition might have profound effects on C_3–C_4 dynamics. Based on current understanding, eutrophication increases responsiveness of C_3 species to CO_2 enrichment. For both C_3 grasses in temperate zones, and tropical woody species, the combined effects might strongly favor C_3 invasion of C_4 biomes. This hypothesis is consistent with high CO_2 promotion of C_3 vegetation at the expense of C_4 in the Chesapeake Bay saltmarsh (Drake *et al.*, 1996)

D. Invasive Species and Disturbance Regimes

In natural ecosystems, multiple biotic and abiotic factors generally prevent native species from completely dominating their habitats. By introducing alien plant species into novel habitats, humans often create situations in which the alien escapes natural controls and can explode across landscapes to the detriment of the native flora. C_4 grasses, particularly from Africa, are common invasive species capable of explosive eruptions in novel habitats (Table III) (Chapman, 1996). Originally introduced as pasture plants or weeds, invasive C_4 grasses aggressively spread into sites cleared by logging, fire, and agriculture. Australia, the Americas, and island systems have been particularly susceptible, and losses of both plant and animal diversity have occurred as a result (D'Antonio and Vitousek, 1992). Native vegetation in savannas and forests is often displaced by the invaders, which can form dense monocultures (Backer and Bakhuizen, 1968; Baruch et al., 1985; Mueller-Dombois and Goldhammer, 1990). Once established, invasive grasses accelerate fire cycles, suppressing recolonization by woody species (D'Antonio and Vitousek, 1982; Goldhammer, 1993). Soil fertility may be reduced because regular burning volatilizes N and the high C:N ratios of the weedy C_4 grasses slow nutrient turnover in the soil, reducing recolonization potential of woody species (Frost and Robertson, 1989; Aide and Cavelier, 1994; Johnson and Wedin, 1997). Conversion of forest to exotic grassland often alters regional climate by reducing evapotranspira-

Table III Important Invasive C_4 Grasses of Natural Landscapes of the World

Invaded region	Invaders	Region of origin
North America	*Microstegium vimineum*	Asia
	Eragrostis curvula, Eragrostis lehmanniana	Africa
	Pennisetum spp., *Digitaria* spp.,	
	Echinochloa spp. *Cynodon dactylon,*	
	Sorghum halopense	
Tropical America	*Hyparrhenia rufa, Melinis minuitiflora*	Africa
	Panicum maximum, Brachiaria spp.	
Southeast Asia	*Imperata cylindrica*	Africa
Australia	*Cenchrus ciliaris, Pennisetum polystachyon*	Africa
	M. minutiflora, E. curvula	
Oceania (Hawaii)	*Schizachyrium condensatum, Andropogon virginicus*	North America
	Hyparrhenia rufa, M. minutiflora	Africa
	Cenchrus ciliarus, Digitaria decumbens	
	Pennisetum clandestinum, Pennisetum purpureum	
	Imperata cylindrica	
Europe, New Zealand	*Spartina anglica* (= *S. townsendii*) in coastal saltmarshes	Europe (as a new hybrid from *S. alterniflora* from the Americas and *S. maritima* from Europe)

Bock, J. H. and Bock, C. E. (1992). Vegetation responses to wildfire in native versus exotic Arizona grassland. *J. Veg. Sci.* **3**, 439–446; and D'Antonio, C. M. and Vitousek, P. M. (1992). Biological invasions by exotic grasses, the grass/fire cycle, and global change. *Annu. Rev. Ecol. Syst.* **23**, 63–87.

tion, leading to lower humidity and precipitation, and greater dessication of adjoining woodlands (Kaufman and Uhl, 1990; Salati and Nobre, 1991).

E. Human Land Use Management

Humans are great modifiers of vegetation patterns and have been the agents most responsible for the expansion of C_4 species in recent geological time. With increasing population and economic growth, human exploitation and modification of landscapes will intensify with continued consequences for C_3–C_4 distributions. To a large extent, vegetation on a particular landscape will occur at the prerogative of the people managing the area. Major exceptions are arid and cold deserts where extreme conditions restrict primary productivity. C_4 vegetation will do well where it fits into the management scheme or, as weeds, exploit various aspects of human land use practices.

Looking into the future, arguments could be made for the following trends (all of which are already underway). Slash-and-burn agriculture will continue to convert forest to grasslands (Myers, 1991), allowing a few dozen Old World C_4 pasture and weed species to increase in distribution. As natural forest cover disappears in the beginning of the twenty-first century, efforts to reforest economically marginal C_4 grasslands will intensify. These forests will be low in biodiversity, consisting of a few timber species, and occasionally an understory of C_4 forage grasses (Brown, Chapter 14). Eventually, land use conversion will stabilize with a mix of managed forest, pasture, and cropland, with C_4 species being important components in the latter two as crops and weeds. What appears certain is that natural landscapes will be marginalized, except in the deserts. Natural grasslands and savannas will be heavily converted to managed landscapes to the detriment of much of the C_4 flora. Although some C_4 species will fare very well under human management, many of the 8000 or so C_4 species may, like much of the rest of the biota, find that there is little space left for them in a human-dominated world.

V. Summary

The distribution of C_4 plants is highly dependent on latitude and altitude. C_4 plants dominate most low elevation landscapes in the tropics and subtropics that are not dominated by C_3 trees and shrubs. On a floristic basis, C_4 species generally represent more than 75% of the grass species in hot, open habitats of the tropics. At temperate latitudes, C_4 representation declines with increasing latitude, although high variability occurs because of interactions with other environmental factors. In all regions, C_4 species become uncommon above latitudes of 45° to 50°. C_4 species also become less common with increasing altitude, typically dropping out of the flora above 2000 to 3000 m.

These patterns largely reflect temperature and light regimes, with precipitation, N, and salinity playing secondary roles. C_4 plants are rare where the minimum temperatures during the growth season are below 6° to 12°C, yet dominate high light environments where the minimum temperature during the growth season is higher than 16° to 18°C. Similarly, they are rarely found at light levels below 20% of full sun. Although C_4 plants can exist in hot, dry locations that are too harsh for C_3 plants, aridity is not a prerequisite for C_4 dominance, because C_4 species are common in any high light, low elevation terrestrial habitat in the tropics. However, aridity, salinity, and low N availability do allow C_4 vegetation to remain competitive in some environments that would otherwise be too cold. Seasonality of precipitation plays an important role. Where precipitation is concentrated in the winter and summers are dry, C_4 plants are uncommon. Where precipitation is concentrated in the warm season, C_4 species are common, if not dominant, across open landscapes.

Interactions between C_3 and C_4 vegetation are often between woody C_3 and graminoid C_4 plants. Ecological factors such as fire and large animal herbivory that play an important role in modifying the distribution of grass–woodland ecotones are thus key controllers of the distribution of C_4 dominated ecosystems. Humans also interact with these important ecological controls, and in so doing have greatly modified the distribution of C_4 vegetation. In particular, human-caused deforestation has allowed for widespread advance of C_4-dominated grasslands in the tropics and subtropics. Because of the increasing exploitation of landscapes by humans, the future of C_4 plants largely depends on human management decisions. A few economically significant C_4 species will likely do very well, but the majority of C_4 species may be marginalized in a human-dominated world.

Appendix I Sources for Data Presented in Figs. 1 and 2

Location	Source
North and Central America	
General	Teeri and Stowe, 1976
Alaska	Welsh, 1974
Arizona (Organ Pipe)	E. H. Jordan, 1975, unpublished list
California (Marin)	Howell, 1970
California (Mojave desert)	McLaughlin *et al.*, 1987
California (Santa Barbara)	Smith, 1976
Clay Belt (Ontario/Quebec)	Baldwin, 1958
Florida (north)	Huffmann and Judd, 1998
Florida (south)	Easley and Judd, 1990

(*continues*)

Appendix I (*Continued*)

Location	Source
Georgia (coast)	Jones and Coile, 1988
Georgia	Drew et al., 1998
Honduras (Lancetilla Valley)	Standley, 1931
Kentucky	Conrad, 1990
Louisiana	Anonymous, 1998a
Manitoba	Scoggan, 1957
Mexico (Baja California sur)	Leon and Dominguez, 1989
Mexico (Jalisco)	Lott, 1993
Mexico (Rio Mayo)	Gentry, 1942
Mexico (Tijuana)	Mulroy et al., 1979
Michigan	Voss, 1972
Nevada (central)	Young and Evans, 1980
Northwest Territories	Porsild and Cody, 1980
Nova Scotia	Roland and Smith, 1966
Ontario (Bruce Peninsula)	Anonymous, 1969
Ontario (Haliburton)	Skelton and Skelton, 1991
Ontario (Ottawa)	Dore, 1959
Ontario (Thunder Bay)	Crowe, 1993
Panama	D'Arcy, 1987
Quebec (Gaspe)	Scoggan, 1950
Quebec (north)	Dutilly et al., 1953
Saskatchewan	Looman and Best, 1987
Texas (Trans Pecos)	Powell, 1994
Yukon	Cody, 1996
South America	
Argentina (Central)	Cavagnaro, 1988
Argentina (north)	Ruthsatz and Hofmann, 1984
Beru	Brako and Zarucchi, 1993
Brazil (Bahia)	Renvoize, 1984
Brazil (Distrito Federal)	Filgueiras, 1991
Brazil (Parana)	Renvoize, 1988
Columbia (coast)	Forero and Gentry, 1989
Guianas	Boggan et al., 1997
Surinam	Teunissen and Wildschut, 1970
Tierra del Fuego	Moore, 1983
Africa	
Egypt	Batanouney et al., 1988
Gabon (Lope–Okanda Reserve)	McPherson and Freeland, 1995
Ivory Coast (Nimba)	Schnell, 1952
Kenya	Livingstone and Clayton, 1980
Namibia	Schulze et al., 1996
Somalia	Cope, 1995
South Africa	Vogel et al., 1978
Western tropical Africa	Clayton, 1966
West-central Africa	Anonymous, 1998b

Appendix I (Continued)

Location	Source
Australia	Hattersley, 1983
Asia	
Burma	Takeda, 1985
China (Anhui)	Anonymous, 1992
China (Beijing)	Anonymous, 1984
China (Dalian)	Anonymous, 1982
China (Deserts)	Anonymous, 1985
China (Fujian)	Lin and Zhang, 1995
China (Guangzhou)	Hou, 1956
China (Hebei)	He et al., 1991
China (Intramogolia)	Ma, 1983
China (Manchuria)	Kitagawa, 1939
China (Qinling)	Anonymous, 1976
China (Shandong)	Chen et al., 1990
China (Sichuan)	Anonymous, 1988b
China (South Jiangsu)	Anonymous, 1959
China (Xinjiang)	Anonymous, 1985
India (general)	Takeda, 1985
India (Agra)	Sharma and Dhakre, 1995
India (Cannanore)	Ramachandran and Nair, 1988
India (Jowai)	Balakrishnam, 1983
India (Karnataka)	Sharma et al., 1984
India (northwest desert)	Bhandari, 1990
India (Pachmarhi and Bori)	Mukherjee, 1984
India (Saurastra)	Bole and Pathak, 1988
Iraq	Rechinger, 1964
Israel	Vogel et al., 1986
Korea	Takeda and Hakoyama, 1985
Mongolia	Pyankov, 1997, unpublished observation
Pakistan (general)	Takeda, 1985
Saudi Arabia	Marclaville, 1990
Sinai	Vogel et al., 1986
Turkey	Davis, 1985
Yemen	Wood, 1997
Former USSR (Asian part)	Czerepanov, 1995; Pyankov and Mokronosov, 1993
Europe	
Britain	Clapham et al., 1987
France	Tutin et al., 1964–1980
Germany	Tutin et al., 1964–1980
Hungary	Tutin et al., 1964–1980
Italy	Tutin et al., 1964–1980
Latvia	Fatare, 1989
Northwest Europe	Eric, 1971
Poland	Tutin et al., 1964–1980
Romania	Anonymous, 1979

(*continues*)

Appendix I (Continued)

Location		Source
Scandinavian		Tutin et al., 1964–1980
Spain and Portugal		Tutin et al., 1964–1980
Turkey (European part)		Tutin et al., 1964–1980
Former USSR (European part)		Tutin et al., 1964–1980, Pyankov and Mokronosov 1993; Czerepanov 1995
Greenland		Böcher, Homen, and Jacobsen, 1968
Madagascar		Morat, 1973
Oceanic Islands (with approximate ocean location)		
Macquarie	Antarctic	Hnatiuk, 1993
Marion and Price Edward	Antarctic	Gremmen, 1982
Queen Elizabeth Islands	Arctic	McLachlan et al., 1989
St. Lawrence	Arctic	Young, 1971
Tristan de Cunha	Central Atlantic	Wace and Dickson, 1965
Diego Garcia	Central Indian	Topp, 1988
South Maldives	Central Indian	Spicer and Neuberg, 1979
Sri Lanka	Central Indian	Dassanayake et al., 1994
Bikini	Central Pacific	Taylor, 1950
Christmas	Central Pacific	Chock and Hamilton, 1962
Gilberts	Central Pacific	Fosberg and Sachet, 1987a
Hawaii	Central Pacific	Rundel, 1980
Hawaiian Islands	Central Pacific	B. P. Bishop Museum, 1997
Henderson	Central Pacific	Paulay and Spencer, 1989
Horne and Wallis	Central Pacific	St. John and Smith, 1971
Jaluit Atoll	Central Pacific	Fosberg and Sachet, 1962
Maupiti	Central Pacific	Fosberg and Sachet, 1987b
Nauru	Central Pacific	Thaman et al., 1994
Northern Line	Central Pacific	Wester, 1985
Nui Atoll, Tuvalu	Central Pacific	Woodroffe, 1985
Pheonix	Central Pacific	Fosberg and Stoddard, 1994
Rongiroa	Central Pacific	Sachet, 1969
Society	Central Pacific	Sachet, 1983a
Somoa–Aleipata	Central Pacific	Whistler, 1984
Takapoto	Central Pacific	Sachet, 1983b
Tongatapu	Central Pacific	Ellison, 1990
Wake	Central Pacific	Fosberg and Sachet, 1961–1963
Walpole	Central Pacific	Renevier and Cherrier, 1991
Bahamas	Caribbean	Correl and Correl, 1982
Barbados	Caribbean	Gooding et al., 1965
Jamacian Keys	Caribbean	Stoddart and Fosberg, 1991
Jamaica	Caribbean	Adams, 1972
Lesser Antilles	Caribbean	Gould, 1979
Faroe Islands	Eastern Atlantic	Collins and Jones, 1985
Madeira	Eastern Atlantic	Cope, 1994
Ashmore reef	Eastern Indian	Kenneally, 1993
Christmas	Eastern Indian	Du Puy, 1993

Appendix I (*Continued*)

Location	Source	
Cocos	Eastern Indian	Telford, 1993a
King	Eastern Indian	Hattersley, 1983
Clipperton	Eastern Pacific	Sachet, 1962
Easter	Eastern Pacific	Skottsberg, 1953a
Farallons	Eastern Pacific	Coulter, 1971
Galapagos	Eastern Pacific	Wiggens and Porter, 1971
Guadalupe	Eastern Pacific	Moran, 1996
Juan Fernandez	Eastern Pacific	Skottsberg, 1953b
Desventuradas	Eastern Pacific	Mueller-Dombois and Fosberg, 1998
Santa Catalina	Eastern Pacific	Millspaugh and Nuttall, 1923
Santa Cruz	Eastern Pacific	Junak et al., 1995
Soccoro	Eastern Pacific	Levin and Moran, 1989
Tres Marias	Eastern Pacific	Lenz, 1995
Vancouver	Eastern Pacific	Szczawinski and Harrison, 1973
Perdido Key (USA)	Gulf of Mexico	Looney et al., 1993
Ship (USA)	Gulf of Mexico	Miller and Jones, 1967
Crete	Mediterranean	Turland et al., 1993
Cyprus	Mediterranean	Meilke, 1985
Enousses	Mediterranean	Panitsa et al., 1994
Chisik	North Pacific	Talbot et al., 1995
Ireland	Northeast Atlantic	Collins and Jones, 1985
Outer Hebrides	Northeast Atlantic	Parkhurst and Mullen, 1991
Scilly	Northeast Atlantic	Lousley and Everhand, 1971
Svalbard	Northeast Atlantic	Ronning, 1964
Queen Charlotte Island	Northeast Atlantic	Calder and Taylor, 1968
Falklands	South Atlantic	Moore, 1968
St. Helena	South Atlantic	Roxburgh, 1996
Antipodes	Southwest Pacific	Godley, 1989
Aukland	Southwest Pacific	Johnson and Campbell, 1975
New Zealand–Horowhenua	Southwest Pacific	Duguid, 1990
New Zealand–Opotiki	Southwest Pacific	Heginbotham and Esler, 1985
New Zealand–Westland	Southwest Pacific	Wardle, 1975
Norfolk	Southwest Pacific	Green, 1994
Tasmania	Tasman Sea	Hattersley, 1983
Anegada	Western Atlantic	D'Arcy, 1975
Assateague (Virginia)	Western Atlantic	Stalter and Lamont, 1990
Bermuda	Western Atlantic	Britton, 1918
Monomoy (USA)	Western Atlantic	Lortie et al., 1991
Orient Beach (New York)	Western Atlantic	Lamont and Stalter, 1991
Prince Edward Island	Western Atlantic	Erskine et al., 1985
Staten	Western Atlantic	Buegler and Parisio, 1982
Turtle	Western Atlantic	Stalter, 1973
Aldabra	Western Indian	Hnatiuk, 1980
Seychelles	Western Indian	Robertson, 1989
Abaitu (Taipingtao)	Western Pacific	Huang et al., 1994b
Amami	Western Pacific	Takeda et al., 1985a

(*continues*)

Appendix I (Continued)

Location	Source	
Coral Sea Islands	Western Pacific	Telford, 1993b
Hainan	Western Pacific	Anonymous, 1977
Heron	Western Pacific	Rogers, 1996
Hokaido	Western Pacific	Takeda et al., 1985a
Honshu, Chiba	Western Pacific	Takeda et al., 1985a
Ishigaki	Western Pacific	Takeda et al., 1985a
Java	Western Pacific	Backer and van den Brink, 1968
Kermadec	Western Pacific	Sykes, 1977
Lord Howe	Western Pacific	Hattersley, 1983
N. Barrier Reef	Western Pacific	Fosberg and Stoddart, 1991
Nagasaki	Western Pacific	Takeda et al., 1985a
Okinawa	Western Pacific	Takeda et al., 1985a
Taiwan	Western Pacific	Hus, 1971; Chen and Lai, 1976
Pratas (Tungshatao)	Western Pacific	Huang et al., 1994a

Acknowledgments

The authors wish to thank Alan Knapp, David Kubian, Heather Peat, Robert Pearcy, and David Tissue for valuable reviews, and Larry Tieszen for permission to use Fig. 6. Preparation of this work was supported by National Science and Research Council (Canada) grant #91-37100-6619 to R.F.S.

References

Adam, P. (1990). "Saltmarsh Ecology," Cambridge University Press, Cambridge.
Adams, C. D. (1972). "Flowering Plants of Jamaica," Jamaica University of the West Indies, Mona, Jamaica.
Aide, T. M. and Cavelier, J. C. (1994). Barriers to lowland tropical forest restoration in the Sierra Nevada de Santa Marta, Colombia. *Restoration Ecol.* **2**, 219–229.
Akhani, H., Trimborn, P. and Ziegler, H. (1997). Photosynthetic pathways in Chenopodiaceae from Africa, Asia and Europe with their ecological, phytogeographical and taxonomical importance. *Plant Syst. Evol.* **206**, 187–221.
Anonymous. (1959). "Handbook of Seed Plants of Southern Jiangsu," Science Press, Beijing.
Anonymous. (1969). "Check-list of Vascular Plants of the Bruce Peninsula," Federation of Ontario Naturalists, Don Mills, Ontario.
Anonymous. (1976). "Flora Tsinlingensis," Vol. 1(1). Science Press, Beijing.
Anonymous. (1977). "Flora Hainanica," Vol. 4. Guandong Bot. Inst., Academia Sinica, Science Press, Beijing.
Anonymous. (1979). "The Flora of Romania: Illustrated Determination of Vascular Plants," Editura Academiei, Socialist Republic of Romania.
Anonymous. (1982). "Flora of Dalian Areas," Vol. 2., Dalian Press, Dalian, China.
Anonymous. (1984). "Flora of Beijing," Biology Department, Beijing Normal University, Beiing Press, Beijing.

Anonymous. (1985). "Flora in Desertis Republicae Populorum Sinarum," Vol. 1, Lanzhou Desert Res. Inst., Academia Sinica, Science Press, Beijing.
Anonymous. (1988a). "A Checklist of Plants in Xinjiang," Biological, Soil and Desert Res. Inst., Academia Sinica, Urumqi. Science Press, Beijing.
Anonymous. (1988b). "Flora Sichuanica," Vol. 5(2). Sichuan Sci. and Tech. Press, Chengdu, China.
Anonymous. (1992). "Flora of Anhui," Vol. 5. Anhui Sci. and Tech. Press, Hefei, China.
Anonymous. (1998a). "Louisiana Grasses: Checklist of All Grasses." [http://www.csdl.tamu.edu/FLORA/cgi/].
Anonymous. (1998b). "West Central African Vascular Plant Dataset," [gopher://cossis.mobot.org:70/00/FL_CL/DT_Africa/index/].
Archibold, O. W. (1995). "Ecology of World Vegetation," Chapman and Hall, London.
Arp, W. J., Drake, B. G., Pockmann, W. T., Curtis, P. S. and Whigham, D. F. (1993). Interactions between C_3 and C_4 salt marsh plant species during four years of exposure to elevated atmospheric CO_2. *Vegetatio* **104/105,** 133–143.
B.P. Bishop Museum (1997). "Hawaiian Flowering Plant Checklist Database" The State Museum of Natural and Cultural History, Honolulu, Hawaii. URL *http://www.bishop.hawaii.org/bishop/HBS/*.
Backer, C. A. and van den Brink, R. C. B. (1968). "Flora of Java," Vol. 3. Walters-Noordhoof N. V., Groningen, The Netherlands.
Baker, H. G. (1989). Sources of the naturalized grasses and herbs in California grasslands. *In* "Grassland Structure and Function: California Annual Grassland" (L. F. Huenneke and H. A. Mooney, Eds.), pp. 29–38. Kluwer Acad. Publ., Dordrecht.
Balakrishnam, N. P. (1983). "Flora of Jowai and Vicinity, Meghalaya," Bot. Survey of India, Howrawh.
Baldwin, W. K. W. (1958). "Plants of the Clay Belt of Northern Ontario and Quebec," pp. 84–101. National Museums of Canada, Ottawa.
Ball, A. S. (1997). Microbial decomposition at elevated CO_2 levels: Effect of litter quality. *Global Change Biol.* **3,** 379–386.
Barbour, M. G. (1992). Life at the leading edge: The beach plant syndrome. *In* "Coastal Plant Communities of Latin America" (U. Seeliger, ed.), pp. 291–307. Academic Press, New York.
Barbour, M. G, Burk, J. H. and Pitts, W. D. (1987). "Terrestrial Plant Ecology," 2nd ed., Benjamin/Cummings Publ. Co., Inc., Menlo, California.
Barnes, P. W. and Harrison, A. T. (1982). Species distribution and community organization in a Nebraska sandhills mixed prairie as influenced by plant/soil-water relationships. *Oecologia* **52,** 192–201.
Baruch, Z., Ludlow, M. M. and Davis, R. (1985). Photosynthetic responses of native and introduced C_4 grasses from Venezuelan savannas. *Oecologia* **67,** 388–393.
Baskin, J. M. and Baskin, C. C. (1977). Role of temperature in the germination ecology of three summer annual weeds. *Oecologia* **30,** 377–382.
Batanouny, K. H., Stichler, W. and Ziegler, H. (1988). Photosynthetic pathways and ecological distribution of grass species in Egypt. *Oecologia* **75,** 539–548.
Batanouny, K. H., Stichler, W. and Ziegler, H. (1991). Photosynthetic pathways and ecological distribution of *Euphobia* species in Egypt. *Oecologia* **87,** 565–569.
Bazzaz, F. A. (1990). The response of natural ecosystems to the rising global CO_2 levels. *Annu. Rev. Ecol. Syst.* **21,** 167–196.
Bazzaz, F. A. and McConnaughay, D. M. (1992). Plant–plant interactions in elevated CO_2 environments. *Aust. J. Bot.* **40,** 547–563.
Beetle, A. A. (1947). Distribution of the native grasses of California. *Hilgardia* **17,** 309–354.
Belsky, A. J., Amundson, R. G., Duxbury, J. M., Riha, S. J., Ali, A. R. and Mwonga, S. M. (1989). The effects of trees on their physical, chemical, and biological environments in a semi-arid savanna in Kenya. *J. Appl. Ecol.* **26,** 1005–1024.

Berendse, F. and Aerts, R. (1987). Nitrogen-use-efficiency: A biologically meaningful definition? *Funct. Ecol.* **1**, 293–296.
Berry, J. and Bjorkman, O. (1980). Photosynthetic response and adaptation to temperature in higher plants. *Ann. Rev. Plant Physiol.* **31**, 491–543.
Bhandari, M. M. (1990). "Flora of the Indian Desert," MPS Repros, Jodhpur, India.
Bird, M. I. and Pousai, P. (1997). Variations of Delta ^{13}C in the surface soil organic carbon pool. *Global Biogeochem. Cycles* **11**, 313–322.
Bird, M. I., Haberle, S. G. and Chivas, A. R. (1994). Effect of altitude on the carbon-isotope composition of forest and grassland soils from Papua New Guinea. *Global Biogeochem. Cycles* **8**, 13–22.
Björkman, O. (1976). Adaptive and genetic aspects of C_4 photosynthesis. *In* "CO_2 Metabolism and Plant Productivity" (R. H. Burris and C. C. Black, eds.). University Park Press, Baltimore.
Black, C. C., Chen, T. M. and Brown, R. H. (1969). Biochemical basis for plant competition. *Weed Sci.* **17**, 338–344.
Blasco, F. (1977). Outlines of ecology, botany and forestry of the mangals of the Indian subcontinent. *In* "Ecosystems of the World 1: Wet Coastal Ecosystems" (V. J. Chapman, ed.), pp. 241–260. Elsevier, Amsterdam.
Böcher, T. W., Homen, K. and Jakobsen, K. (1968). "The Flora of Greenland," P. Haase & Sons Publ., Copenhagen.
Bock, J. H. and Bock, C. E. (1992). Vegetation responses to wildfire in native versus exotic Arizona grassland. *J. Veg. Sci.* **3**, 439–446.
Boggan, J. Funk, V. and Kelloff, C. (eds.) (1997). "Flora of the Guianas," Smithsonian Inst. Biol. Div. Guianas Program, Publ. Ser. no. 30., Washington DC.
Bole, P. V. and Pathak, J. M. (1988). "Flora of Saurastra," Part III. Bot. Survey of India, New Delhi.
Bond, W. J. and van Wilgen, B. W. (1996). "Fire and Plants," Chapman & Hall, London.
Boutton, T. W., Harrison, A. T. and Smith B. N. (1980). Distribution of biomass of species differing in photosynthetic pathway along an altitudinal transect in southeastern Wyoming grassland. *Oecologia* **45**, 287–298.
Bragg, T. B. and Hurlber, L. C. (1976). Woody plant invasion of unburned Kansas bluestem prairie. *J. Range Manag.* **29**, 19–24.
Brako, L. and Zarucchi, J. L. (eds.). (1993). "Catalogue of the Flowering Plants and Gymnosperms of Peru," Missouri Botanic Garden, St. Louis, Missouri.
Bray, J. R. (1958). The distribution of savanna species in relation to light intensity. *Can. J. Bot.* **36**, 671–681.
Britton, N. L. (1918). "Flora of Bermuda," Charles Scribner's Sons, New York.
Brown, R. H. (1978). A difference in N use efficiency in C_3 and C_4 plants and its implications in adaptation and evolution. *Crop Sci.* **18**, 93–98.
Brown, W. V. (1958). Leaf anatomy in grass systematics. *Bot. Gaz.* **119**, 170–178.
Brown, W. V. (1977). The Kranz syndrome and its subtypes in grass systematics. *Mem. Torrey Bot. Club* **23**, 1–97.
Brown, J. R. and Archer, S. (1989). Woody plant invasion of grassland: establishment of honey mesquite (*Prosopis glandulosa* var. *glandulosa*) on sites differing in herbaceous biomass and grazing history. *Oecologia* **90**, 18–26.
Buegler, R. and Parisio, S. (1982). "A Comparative Flora of Staten Island," Staten Island Inst. of Arts and Sci., Staten Island, N. Y.
Burns, J. C. and Bagler, C. P. (1996). Cool season grasses for pasture. *In* "Cool Season Forage Grasses" (L. E. Moser, B. R. Buxton, and M. D. Casler, eds.), pp. 321–355. Amer. Soc. Agron., Madison, WI.
Cabido, M., Ateca, N., Astegiano, M. E. and Anton, A. M. (1997). Distribution of C_3 and C_4 grasses along an altitudinal gradient in central Argentina. *J. Biogeogr.* **24**, 197–204.

Calder, J. S. and Taylor, R. L. (1968). "The Flora of the Queen Charlotte Island," Canada Dept. of Agriculture, Ottawa.
Carter, D. R. and Peterson, K. M. (1983). Effects of a CO_2-enriched atmosphere on the growth and competitive interaction of a C_3 and a C_4 grass. *Oecologia* **58**, 188–193.
Cavagnaro, J. B. (1988). Distribution of C_3 and C_4 grasses at different altitudes in a temperate arid region of Argentina. *Oecologia* **76**, 273–277.
Cerling, T. E., Harris, J. M., MacFadden, B. J., Leacey, M. G., Quade, J., Eisenmann, V. and Ehleringer, J. R. (1997). Global vegetation change through the Miocene/Pliocene boundary. *Nature* **389**, 153–158.
Chapman, G. P. (1996). "The Biology of Grasses," CAB International, Willingford, Oxon, UK.
Chapman, V. J. (1977). Introduction. *In* "Ecosystems of the World 1: Wet Coastal Ecosystems" (V. J. Chapman, ed.), pp. 1–29. Elsevier, Amsterdam.
Chazdon, R. L. (1978). Ecological aspects of the distribution of C_4 grasses in selected habitats of Costa Rica. *Biotropica* **10**, 265–269.
Chen, S. S. C. and Lai M.-J. (1976). A synopsis of the Formosan plants. *Taiwania* **21**, 87–115.
Chen, H., Zheng, Y. and Li, F. (1990). "Flora of Shandong," Vol. 1. Qingdao Press, Qingdao, China.
Chock, A. K. and Hamilton, D. C. (1962). Plants of Christmas Island. *Atoll Res. Bull.* **90**, 1–7.
Chollet, R. and Ogren, W. L. (1975). Regulation of photorespiration in C_3 and C_4 species. *Bot. Rev.* **41**, 137–180.
Christie, E. K. and Detling, J. K. (1982). Analysis of interference between C_3 and C_4 grasses in relation to temperature and soil nitrogen supply. *Ecology* **63**, 1277–1284.
Clapham, A. R., Tutin, T. G. and Moore, D. M. (1987). "Flora of the British Isles" 3rd ed. Cambridge University Press, Cambridge.
Clayton, W. D. (1966). Gramineae (Poaceae). *In* "Flora of West Tropical Africa" 2nd ed. (F. N. Hepper, ed.) Vol. 3. pp. 349–512. The Whitefriars Press Ltd., London.
Clements, F. E., Weaver, J. E. and Hanson, H. C. (1929). "Plant Competition," Carnegie Inst. Wash., Washington.
Cody, W. J. (1996). "Flora of the Yukon Territory," NRC Research Press, Ottawa.
Cole, M. M. (1986). "The Savannas—Biogeography and Geobotany," Academic Press, San Diago.
Coleman, J. S. and Bazzaz, F. A. (1992). Effects of CO_2 and temperature on growth and resource use of co-occurring C_3 and C_4 annuals. *Ecology* **73**, 1244–1259.
Collins, R. P. and Jones, M. B. (1985). The influence of climatic factors on the distribution of C_4 species in Europe. *Vegetatio* **64**, 121–129.
Conrad, J. (1990). Annotated checklist of the vascular plants at McClean County, KT, Version 1.0, *Flora Online* **23**.
Cope, T. A. (1994). Gramineace (Poaceae). *In* "Flora of Madeira" (J. R. Press and M. J. Short, eds.), pp. 406–453. The Natural Museum, London.
Cope, T. A. (1995). Poaceae (Gramineae). *In* "Flora of Somalia Vol. 4. Angiospermae (Hydrocharitaceae-Pandanaceae)" (M. Thulin, ed.), pp. 148–270. Royal Botanic Gardens, Kew.
Correl, D. S. and Correl, H. B. (1982). "Flora of the Bahamas Archipelago," J. Cramer, A. R. Gantner Verlag KG., Valdez, FL.
Costa, C. S. B. and Davy, A. J. (1992). Coastal Salt Marsh Communities of Latin America. *In* "Coastal Plant Communities of Latin America" (U. Seeliger, ed.), pp. 179–199. Academic Press, New York.
Coulter, M. (1971). Flora of the Farallon Islands. *Madrono* **21**, 131–137.
Coupland, R. T. (1992). Approach and generalizations. *In* "Ecosystems of the World 8A. Natural Grasslands: Introduction and Western Hemisphere" (R. T. Coupland, ed.), pp. 1–6. Elsevier, Amsterdam.

Crampton, B. (1974). "Grasses in California," University of California Press, Berkeley.
Crowe, J. (1993). "Checklist of Vascular Plants of Thunder Bay District," Thunder Bay Field Naturalists, Thunder Bay, Canada.
Czerepanov, S. K. (1995). "Vascular Plants of Russia and Adjacent States (the Former USSR)." Cambridge University Press, Cambridge.
D'Antonio, C. M. and Vitousek, P. M. (1992). Biological invasions by exotic grasses, the grass/fire cycle, and global change. *Annu. Rev. Ecol. Syst.* **23**, 63–87.
D'Arcy, W. G. (1975). Anegada Island: Vegetation and flora. *Atoll Res. Bull.* **188**, 1–40.
D'Arcy, W. G. (1987). "Flora of Panama—Checklist and Index," Missouri Bot. Gard. St. Louis, Missouri.
Dassanayake, M. D., Fosberg, F. R. and Clayton, W. D. (eds.). (1994). "A Revised Handbook to the Flora of Ceylon," Vol. 8. A. A. Balkema, Rotterdam.
Davis, P. H. (ed.) (1985). "Flora of Turkey and the East Aegean Islands," Vol. 9. University of Edinburgh Press, Edinburgh.
Diaz, S., Grime, J. P., Harris, J. and McPherson, E. (1993). Evidence of a feedback mechanism limiting plant response to elevated carbon dioxide. *Nature* **364**, 616–617.
Dickinson, C. E. and Dodd, J. L. (1976). Phenological Pattern in the Shortgrass Prairie. *Amer. Midl. Nat.* **96**, 367–378.
Doliner, L. H. and Jolliffe, P. A. (1979). Ecological evidence concerning the adaptive significance of C_4 dicarboxylic acid pathway of photosynthesis. *Oecologia* **38**, 23–34.
Dore, W. G. (1959). "Grasses of the Ottawa District," Publ. #1049, Canada Dept. of Agriculture, Ottawa.
Downton, W. J. S., and Tregunna, E. B. (1968). Carbon Dioxide compensation—its relation to photosynthesis carboxylation reactions, systematics of the Gramineae, and leaf anatomy. *Can. J. Bot.* **46**, 207–215.
Drake, B. G., Peresta, G., Beugeling, E. and Matamala, R. (1996). Long-term elevated CO_2 exposure in a Chesapeake Bay wetland: Ecosystem gas exchange, primary production, and tissue nitrogen. *In* "Carbon Dioxide and Terrestrial Ecosystems" (G. W. Koch and H. A. Mooney, eds.) pp. 197–214. Academic Press, New York.
Drew, M. B., Kirkman, L. K. and Gholson, Jr., A. K. (1998). The vascular flora of Ichauway, Baker Country, Georgia: A remnant longleaf pine/wiregrass ecosystem. *Castanea* **63**, 1–24.
Dublin, H. T. (1995). Vegetation dynamics in the Serengeti—Mara ecosystem: The role of elephants, fire and other factors. *In* "Serengeti II: Dynamics, Management, and Conservation of an Ecosystem" (A. R. E. Sinclair and P. Arcese, eds.), pp. 71–90. The University of Chicago Press, Chicago.
Dublin, H. T., Sinclair, A. R. E. and MacGlade, J. (1990). Elephants and fire as causes of multiple stable states in the Serengeti—Mara woodlands. *J. Anim. Ecol.* **59**, 1147–1164.
Duguid, F. C. (1990). Botany of northern Horowhenua lowlands, North Island, New Zealand. *N Z J. Bot.* **28**, 381–437.
Duke, N. C., Ball, M. C. and Ellison, J. C. (1998). Factors influencing biodiversity and distributional gradients in mangroves. *Global Ecol. Biogeogr. Lett.* **7**, 27–47.
Du Puy, D. J. (1993). Christmas Island. *Flora Austrail.* **50**, 1–30.
Dutilly, A., Lepage, E. III and Duman, M. (1953). "Contribution à la flore Du Bassin de la Baie D'ungava." Catholic University of America. Washington, D.C.
Dzurec, R. S., Boutton, T. W., Caldwell, M. M. and Smith, B. N. (1985). Carbon isotope ratios of soil organic matter and their use in assessing community composition changes in Curlew Valley, Utah. *Oecologia* **66**, 17–24.
Eamus, D. and Jarvis, P. G. (1989). The direct effect of increase in the global atmospheric CO_2 concentration on natural and commercial temperate trees and forests. *Adv. Ecol. Res.* **19**, 1–55.
Earnshaw, M. J., Carver, K. A., Gunn, T. C., Kerenga, K., Harvey, V., Griffiths, H. and Broadmeadow, M. S. J. (1990). Photosynthetic pathway, chilling tolerance and cell sap osmotic

potential values of grasses along an altitudinal gradient in Papua New Guinea. *Oecologia* **84**, 280–288.
Easley, M. C. and Judd, W. S. (1990). Vascular flora of the southern upland property of Paynes Prairie State Preserve, Alachua County, Florida. *Castanea* **55**, 142–186.
Ehleringer, J. R. and Monson, R. K. (1993). Evolutionary and ecological aspects of photosynthetic pathway variation. *Annu. Rev. Ecol. Sys.* **24**, 411–439.
Ehleringer, J. R., Cerling, T. E. and Helliker, B. R. (1997). C_4 photosynthesis, atmospheric CO_2, and climate. *Oecologia* **112**, 285–299.
Ellis, R. P., Vogel, J.C. and Fuls, A. (1980). Photosynthetic pathways and the geographical distribution of grasses in Southwest Africa/Namibia. *S. Afri. J. Sci.* **76**, 307–314.
Ellison, J. C. (1990). Vegetation and floristics of the Tongatapu Outliers. *Atoll Res. Bull.* **332**, 1–36.
Epstein, H. E., Lauenroth, W. K., Burke, I. C. and Coffin, D. P. (1997). Productivity patterns of C_3 and C_4 functional types in the U.S. Great Plains. *Ecology* **78**, 722–731.
Erskine, D. S., Catling, P. M., Erskine, D. S. and MacLaren, R. B. (1985). "The Plants of Prince Edward Island," 2nd Ed. Publication 1798, Agriculture Canada, Ottawa.
Faber-Langendoen, D. and Tester, J. R. (1993). Oak mortality in sand savannas following drought in east-central Minnesota. *Bull. Torr. Bot. Club* **120**, 248–256.
Fatare, I. (1989). "Flora Doliny Reki Daugavy (Latvia)," Institute of Biology, CCP Lativian Academy, RIGA, EINATNE.
Filgueiras, T. S. (1991). A floristic analysis of the Gramineae of Brazil's districto federal and a list of the species occurring in the area. *Edinb. J. Bot.* **48**, 73–80.
Forero, E. and Gentry, A. H. (1989). Lista Anotada de las Plantas de Departmento del Choco, Columbia. Biblioteca Jose Sepramino Triana #10. Universada Nacional de Columbia, Bogata.
Fosberg, F. R. (1962). A brief survey of the Cays of Arrecife Alacran, a Mexican Atoll. *Atoll Res. Bull.* **93**, 1–25.
Fosberg, F. R. and Sachet, M.-H. (1961–1963). Wake Island vegetation and flora. *Atoll Res. Bull.* **123**, 1–15.
Fosberg, F. R. and Sachet, M.-H. (1987a). Flora of the Gilbert Islands. *Atoll Res. Bull.* **295**, 1–33.
Fosberg, F. R. and Sachet, M.-H. (1987b). Flora of Maupiti, Society Islands. *Atoll Res. Bull.* **294**, 1–70.
Fosberg, F. R. and Sachet, M.-H. (1962). Vascular plants recorded from Jaluit Atoll. *Atoll Res. Bull.* **92**, 1–39.
Fosberg, F. R. and Stoddart, D. R. (1991). Plants of the Reef Islands of the Northern Great Barrier Reef. *Atoll Res. Bull.* **348**, 1–77.
Fosberg, F. R. and Stoddart, D. R. (1994). Flora of the Pheonix Islands, Central Pacific. *Atoll Res. Bull.* **392**, 1–55.
Frost, P. G. H. and Robertson, F. (1989). Fire, The Ecological Effects of Fire in Savannas. *In* "*Determinants of Tropical Savannas*" (B. H. Walker, ed.), pp. 93–140. IRL Press.
Gentry, H. S. (1942). "Rio Mayo Plants: a Study of the Flora and Vegetation of the Valley of the Rio Mayo, Sonora," Carnegie Inst. Wash., Publ. #527, Washington, D.C.
Gillison, A. N. (1983). Tropical savannas of Australia and the southwest Pacific. *In* "Ecosystems of the World 13: Tropical Savannas" (F. Bourliere, ed.), pp. 183–243. Elsevier, Amsterdam.
Godley, E. J. (1989). The Flora of Antipodes Island. *New Zealand J. Bot.* **27**, 531–563.
Goldammer, J. G. (1993). Historical biogeography of fire: Tropical and subtropical. *In* "The Ecological, Atmospheric, and Climatic Importance of Vegetation Fire" (P. J. Crutzen and J. G. Goldammer, eds.), pp. 297–314. John Wiley and Sons, New York.
Goldstein, G. and Sarmiento, G. (1987). Water relations of trees and grasses and their consequences for the structure of savanna vegetation. *In* "Determinations of Tropical Savannas" (B. H. Walker, Ed.), pp. 13–38. IRL Press.

Goodall, D. W. (ed.). (1977–1990). "Ecosystems of the World." Vol. 1–16. Elsevier, Amsterdam.
Gooding, E. G. B., Loveless, A. R. and Proctor, G. R. (1965). "Flora of Barbados," Ministry of Oversea Development, Publ. #7, Her Majesty's Stationary Office, London.
Gould, F. W. (1979). "Flora of the Lesser Antilles," Vol. 3. (B. Thompson-Mills, ed.). Arnold Arboretum Harvard Univ., Jamaica Plain, Mass.
Green, P. (1994). Norfolk Island Species List. *In* "Flora of Australia," Vol. 49. Aust. Biol. Resources Study & Aust. Gov. Publ. Serv., Canberra.
[http://osprey.anbg.gov.au/norfolk.gardens/norfolk.plant.list.html]
Gremmem, J. (1982). "The vegetation of the Subantarctic Islands Marion and Prince Edward." Dr. W. Junk Publishers, The Hague, Netherlands.
Grise, D. J. (1997). "Effects of Elevated CO_2 and High Temperature on the Relative Growth Rates and Competitive Interactions between a C_3 (*Chenopodium album*) and a C_4 (*Amaranthus hybridus*) Annual" PhD thesis, University of Georgia, Athens, GA.
Grover, H. D. and Musick, H. B. (1990). Shrubland encroachment in southern New Mexico, U.S.A.: An analysis of desertification processes in the American Southeast. *Climatic Change* **17,** 305–330.
Guy, R. D., Reid, D. M. and Krouse, H. R. (1986). Factors affecting $^{13}C/^{12}C$ ratios of inland halophytes. II. Ecophysiological interpretations of patterns in the field. *Can. J. Bot.* **64,** 2700–2707.
Hattersley P. W. (1983). The distribution of C_3 and C_4 grasses in Australia in relation to climate. *Oecologia* **57,** 113–128.
Hattersley, P. W. (1992). C_4 photosynthetic pathway variation in grasses (Poaceae): Its significance for arid and semi-arid lands. *In* "Desertified Grasslands: Their Biology and Management" The Linnean Society of London, pp. 181–212. Academic Press, London.
Haxeltine, A. and Prentice I. C. (1996). BIOME3: An equilibrium terrestrial biosphere model based on ecophysiological constraints, resource availability, and competition among plant functional types. *Global Biogeochem. Cycles* **10,** 693–709.
Hayashi, I. (1979). Ecology, phytosociology and productivity of grasses and grasslands I. The autecology of some grassland species. In "Ecology of Grasslands and Bamboolands in the World" (M. Numata, ed.), pp. 141–152. VEB Gustav Fischer Verlag Jena.
He, S. Y., ed. (1991). "Flora Hebeiensis," Vol. 3. Hebei Sci. and Tech. Press, Shijiazhuang, China.
Heginbotham, M. and Esler, A. E. (1985). Wild vascular plants of the Opotiki–East Cape region, North Islands, New Zealand. *New Zeal. J. Bot.* **23,** 379–406.
Heitschmidt, R. K., Schultz, R. D. and Scifres, C. J. (1986). Herbaceous biomass dynamics and net primary production following chemical control of honey mesquite. *J. Range Manage.* **39,** 67–71.
Henderson, S., Hattersley, P., von Caemmerer, S. and Osmond, C. B. (1994). Are C_4 pathway plants threatened by global climatic change? *In* "Ecophysiology of Photosynthesis" (E. Schulze and M. M. Caldwell, Eds.), pp. 529–549. Springer-Verlag, New York.
Hesla, A. B. I., Tieszen, L. L. and Imbamba, S. K. (1982). A systematic survey of C_3 and C_4 photosynthesis in the Cyperaceae of Kenya, East Africa. *Photosynthetica* **16,** 196–205.
Hnatiuk, R. J. (1993). Subantarctic islands. *Flora of Australia* **50,** 53–62.
Hnatiuk, U. (1980). C_4 photosynthesis in the vegetation of Aldabra Atoll. *Oecologia* **44,** 327–334.
Hopkins, B. (1983). Successional processes. *In* "Ecosystems of the World 13: Tropical Savannas" (F. Bourliere, ed.), pp. 605–616. Elsevier, Amsterdam.
Horton, J. L. and Neufeld, H. S. (1998). Photosynthetic responses of *Microstegium vimineum* (Trin.) A. Camus, a shade-tolerant, C_4 grass, to variable light environments. *Oecologia* **114,** 11–19.
Hou, K. Z. (ed.). (1956). "Flora of Guangzhou," South China Bot. Res. Inst., Academic Sinica, Science Press, Beijing.

Howell, J. T. (1970). "Marin Flora: Manual of the Flowering Plants and Ferns of Marin County, California," 2nd Ed. University of California Press, Berkeley.
Hsu, C.-C. (1971). A guide to the Taiwan grasses, with keys to subfamilies, tribes, genera and species. *Taiwania* **16**, 199–341.
Huang, T.-C, Huang, S.-F. and Hsieh, T.-H. (1994a). The flora of Tungshatao (Pratas Island). *Taiwania* **39**, 27–53.
Huang, T.-C., Huang, S.-F. and Yang, K.-C. (1994b). The flora of Taipingtao (Abaitu Island). *Taiwania* **39**, 1–26.
Huffmann, J. M. and Judd, W. S. (1998). Vascular flora of Myakka River State Park, Sarasota and Manatee Counties, Florida. *Castanea* **63**, 25–50.
Hultén, E. (1971). "Atlas of the Distribution of Vascular Plants in Northwestern Europe," Generalstubens Lirografiska Anstalts Forlag, Stockholm.
Hunt, H. W., Elliott, E. T., Detling, J. K., Morgan, J. A. and Chen, D.-X. (1996). Responses of a C_3 and C_4 perennial grass to elevated CO_2 and temperature under different water regimes. *Global Change Biol.* **2**, 35–47.
Johnson P. N. and Campbell, D. J. (1975). Vascular plants of the Auckland Islands. *New Zeal. J. Bot.* **13**, 665–720.
Johnson, N. C. and Wedin, D. A. (1997). Soil carbon, nutrients, and mycorrhizae during conversion of a dry tropical forest to grassland. *Ecol. Appl.* **7**, 171–182.
Jones, M. B. (1986). Wetlands. *In* "Photosynthesis in Contrasting Environments" (N. R. Baker and S. P. Long, eds.), pp. 103–138. Elsevier, London.
Jones, S. B. and Coile, N. C. (1988). "The Distribution of the Vascular Flora of Georgia," Dept. of Botany, University Georgia, Athens, GA.
Junak, S., Ayers, T., Scott, R., Wilken, D. and Young, D. (1995). "A Flora of Santa Cruz Island," The Santa Barbara Bot. Soc. and the Calif. Native Plant Soc., Santa Barbara.
Junk, W. J. (1983). Ecology of swamps on the middle Amazon. *In* "Ecosystems of the World 4B: MIRES: Swamp, Bog, Fen and Moor" (A. J. P. Gore, ed.), pp. 269–294. Elsevier, Amsterdam.
Kalapos, T., Baloghne-Nyakas, A. and Csontos, P. (1997). Occurrence and ecological characteristics of C_4 dicot and Cyperaceae species in the Hungarian flora. *Photosynthetica* **33**, 227–240.
Kattenberg, A., Giorgi, F., Grassl, H., Meehl, G. A., Mitchell, J. F. B., Stouffer, R. J., Tokioka, T., Weaver, A. J. and Wigley, T. M. L. (1996). Climate models—projections of future climate. *In* "Climate Change 1995: The Science of Climate Change" (J. T. Houghton, L. G. M. Filho, B. A. Callander, N. Harris, A. Kattenberg, and K. Maskell, eds.), pp. 285–357. Cambridge University Press, Cambridge.
Kaufman, J. B. and Uhl, C. (1990). Interactions of anthropogenic activities, fire, and rain forests in the Amazon Basin. *In* "Fire in the Tropical Biota—Ecosystem Processes and Global Challenges" (J. G. Goldammer, ed.), pp. 117–134. Springer-Verlag, Berlin.
Keeley, J. E. (1998). C_4 photosynthesis modification in the evolutionary transition from land to water in aquatic grasses. *Oecologia* **116**, 85–97.
Kellman, M. C. (1988). Disturbance x habitat interactions: fire regimes and the savanna/forest balance in the neotropics. *Can. Geog.* **32**, 80–82.
Kemp, P. R. (1983). Phenological patterns of Chihuahuan Desert plants in relation to the timing of water availability. *J. Ecol.* **71**, 427–436.
Kenneally, K. F. (1993). Ashmore reef and Cartier island. *Flora of Australia* **50**, 43–47.
Kimball, B. A., Mauney, J. R., Nakayama, F. S. and Idso, S. B. (1993). Effects of increasing atmospheric CO_2 on vegetation. *Vegetatio* **104/105**, 67–75.
King, W. M., Wilson, J. B. and Sykes, M. T. (1990). A vegetation zonation from saltmarsh to riverbank in New Zealand. *J. Veg. Sci.* **1**, 411–418.
Kirschbaum, M. U. F., Bullock, P., Evans, J. R., Goulding, K., Jarvis, P. G., Noble, I. R., Rounsevell, M. and Shankey, T. D. (1996). Ecophysiological, ecological, and soil processes in terrestrial ecosystems: A primer on general concepts and relationships. *In* "Climate

Change 1995" (R. T. Watson, M. C. Zinyowera and R. H. Moss, eds.), pp.57–74. Cambridge University Press, Cambridge.

Kitagawa, M. L. (1939). "Florae Manchuricae" (Report of the Inst. of Scientific Research Manchoukuo). Vol. III. (Appendix 1), Hsinking.

Knapp, A. K. and Seastedt, T. R. (1986). Detritus accumulation limits productivity of tallgrass prairie. *BioScience* **36**, 662–668.

Ko, L. J. and Reich, P. B. (1993). Oak tree effects on soil and herbaceous vegetation in savannas and pastures in Wisconsin. *Am. Midl. Nat.* **130**, 31–42.

Kozlowski, T. T. and Ahlgren, C. E. (1974). "Fire and Ecosystems," Academic Press, New York.

Krall, J. P. and Pearcy, R. W. (1993). Concurrent measurements of oxygen and carbon dioxide exchange during lightflecks in maize (*Zea mays* L.). *Plant Physiol.* **103**, 823–828.

Kretschmer, Jr. A. E. and Pitman, W. D. (1995). Tropical and subtropical forages. *In* "Forages" Vol. 1. (R. F. Barnes, D. A. Miller and C. J. Nelson, eds.), pp. 283–304. Iowa University Press, Ames, Iowa.

Kuoh, C-S and Chiang-Tsai, S-H. (1991). A checklist of C_3 and C_4 grasses in Taiwan. *Taiwania* **39**,159–167.

Laetsch, W. M. (1974). The C_4 syndrome: A structural analysis. *Annu. Rev. Plant Physiol.* **25**, 27–52.

Lamont, E. E. and Stalter, R. (1991). The vascular flora of Orient Beach State Park, Long Island, New York. *Bull. Torr. Bot. Club* **118**, 459–468.

Lamprey, H. F. (1983). Pastoralism yesterday and today: The over-grazing problem. *In* "Ecosystems of the World 13: Tropical Savannas" (F. Bourliere, ed.), pp. 643–666. Elsevier, Amsterdam.

Lenz, L. W. (1995). Plants of the Tres Marias Islands, Mayarit, Mexico. *Aliso* **14**, 19–34.

Leon, J. L. L. and Dominguez, R. (1989). Flora of the Sierra De Laguna, Baja California Sur. Mexico. *Madrono* **36**, 61–83.

Levin, G. A. and Moran, R. (1989). "The Vascular Flora of Isla Socorro, Mexico." San Diego Soc. Nat. Hist. Memoir #16, San Diego Nat. Hist. Mus., San Diego, CA.

Lewis, J. P., Collantes, M. B., Pire, E. F., Carnevale, N. J., Boccanelli, S. I., Stofella, S. L. and Prado, D. E. (1985). Floristic groups and plant communities of southeastern Santa Fe, Argentina. *Vegetatio* **60**, 67–90.

Lin, G., Philips, S. L. and Ehleringer, J. R. (1996). Monsoonal precipitation responses of shrubs in a cold desert community on the Colorado Plateau. *Oecologia* **106**, 8–17.

Lin, L. G. and Zhang, Y. (eds.). (1995). "Flora Fujianica," Vol. 6. Fujian Sci. and Tech. Publishing House, Fuzhou, China.

Livingstone, D. A. and Clayton, W. D. (1980). An altitudinal cline in tropical African grass floras and its paleoecological significance. *Quaternary Res.* **13**, 392–402.

Long, S. P. (1983). C_4 photosynthesis at low temperatures. *Plant Cell Environ.* **6**, 345–363.

Long, S. P. and Drake, B. G. (1992). Photosynthetic CO_2 assimilation and rising atmospheric CO_2 concentrations. *In* "Crop Photosynthesis: Spatial and Temporal Determinants" (N. R. Baker and H. Thomas, eds.), pp. 69–95. Elsevier Sci. Publ., Amsterdam.

Looman, J. and Best, K. F. (1987). "Budd's Flora of the Canadian Prairie Provinces," Agriculture Canada, Ottawa.

Looney, P. B., Gibson, P. J., Blyth, A. and Cousens, M. (1993). Flora of the Gulf Islands National Seashore Perdido Key, Florida. *Bull. Torr. Bot. Club* **120**, 327–341.

Lortie, J. P., Sorric, B. A. and Hot, D. W. (1991). Flora of the Monomoy Islands, Chatham, Massachusetts. *Rhodora* **93**, 361–389.

Lott, E. J. (1993). Annotated checklist of the vascular flora of the Chamela Bay region, Jalisco, Mexico. *Occasional Papers Calif. Acad. Sci.* **148**, 1–60.

Lousley, J. E. and Everhand, B. (1971). "Flora of the Isles of Scilly," Pedwood Press Ltd., London.

Ma, Y. (ed.). (1983). "Flora of Intramongolica," Vol. 7. People's Publishing House of Intramongolia, Hohhot.
Maass, J. M. (1995). Conversion of tropical dry forest to pasture and agriculture. In "Seasonal Dry Tropical Forests" (S. H. Bullock, H. A. Mooney and E. Medina, eds.), pp. 399–422. Cambridge University Press, Cambridge.
MacDonald, K. B. (1977). Plant and animal communities of Pacific North American salt marshes. In "Ecosystems of the World 1: Wet Coastal Ecosystems" (V. J. Chapman, ed.), pp. 167–191. Elsevier, Amsterdam.
Marclaville, J. P. (1990). "Flora of Eastern Saudi Arabia," Regan Paul Int., London.
McLachlan, K. I., Aiken, S. G., Lefkovitch, L. P. and Edlund, S. A. (1989). Grasses of the Queen Elizabeth Islands. Can. J. Bot. **67,** 2088–2105.
McLaughlin, S. P., Bowers, J. E. and Hall, K. R. F. (1987). Vascular plants of eastern Imperial County, California. Madrono **34,** 359–378.
McNaughton, S. J. (1994). The propagation of disturbance in savannas through food webs. J. Veg. Sci. **5,** 301–314.
McPherson, G. and Freeland, C. (1995). Checklist of Lope–Okanda Reserve in central Gabon —Explanation of fields. The Museum National d'Histoire Naturelle, Paris. [http://www.mobot.org/MOBOT/research/lope_int.html].
Means, D. B. (1997). Wiregrass restoration: probable shading effects in a slash pine plantation. Restora. Manage. Notes **15,** 52–55.
Medina E. and Klinge, H. (1983). Productivity of tropical forests and tropical woodlands. In "Physiological Plant Ecology IV. Ecosystem Processes: Mineral Cycling, Productivity and Man's Influence" (O. L. Lange, P. S. Nobel, C. B. Osmond and H. Ziegler, eds.), pp. 281–304. Springer-Verlag, Berlin.
Medina, E. (1987). Nutrient requirements, conservation and cycles of nutrients in the herbaceous layer. In "Determinants of Tropical Savannas" (B. H. Walker, Ed.), pp. 39–65. IRL Press.
Meilke, R. D. (1985). "Flora of Cyprus," Vol. 2. Bentham Moxon Trust, Royal Botanic Gardens, Kew.
Meinzer, F. C. (1978). Observaciones sobre la distribucion taxonomica y ecologica de la fatosintesis C_4 en la vegetacion del noroneste de centroamerica. Rev. Biol. Trop. **16,** 359–369.
Miller, G. J. and Jones, Jr. S. B. (1967). The vascular flora of Ship Island, Mississippi. Castanea **32,** 84–99.
Millspaugh, C. F. and Nuttall, L. W. (1923). "Flora of Santa Catalina Island," Publ. 212, Bot. Series Vol. 5. Field Mus. Nat. Hist., Chicago.
Monson, R. K., Littlejohn, R. O., Jr. and Williams, G. J., III. (1983). Photosynthetic adaptation to temperature in four species from the Colorado shortgrass steppe: A physiological model for coexistence. Oecologia **58,** 43–51.
Moore, D. (1983). "Flora of Tierra del Fuego," Missouri Bot. Gard. and Anthony Nelson, Ltd., St. Louis, Missouri.
Moore, D. M. (1968). "The Vascular Flora of the Falkland Islands," British Antarctic Survey, London.
Moran, R. (1996). The flora of Guadalupe Island, Mexico. Mem. Calif. Acad. Sci. **19,** 1–190.
Morat, P. (1973). "Les Savanes du Sud-ouest de Madagascar." O. R. S. T. O. M., Paris.
Moreno-Casasola, P. and Castillo, S. (1992). Dune ecology on the eastern coast of Mexico. In "Coastal Plant Communities in Latin America" (C. S. B. Costa and A. J. Davy, eds.), pp. 309–321. Academic Press, New York.
Morrison, J. I. L. (1993). Response of plants to CO_2 under water limited conditions. Vegetatio **104/105,** 193–209.
Morse, S. R. and Bazzaz, F. A. (1994). Elevated CO_2 and temperature alter recruitment and size hierarchies in C_3 and C_4 annuals. Ecology **75,** 966–975.

Mueller-Dombois, D. and Fosberg, F. R. (1998). "Vegetation of the Tropical Pacific Islands." Springer-Verlag, New York.

Mueller-Dombois, D. and Goldammer, J. B. (1990). Fire in tropical ecosystems and global environmental change: An introduction. In "Fire in the Tropical Biota—Ecosystem Processes and Global Challenges" (J. G. Goldammer, ed.), pp. 1–10. Springer-Verlag, Berlin.

Mukherjee, A. K. (1984). "Flora of Pachmarhi and Bori Reserve," Bot. Survey of India, New Delhi.

Mulroy, T. W. and Rundel, P. W. (1977). Annual plants: Adaptation to desert environments. *BioScience* **27**, 110–114.

Mulroy, T. W., Rundel, P. W. and Bowler, P. A. (1979). Flora of Punta Banda, Baja Californian Norte Mexico. *Madrono* **26**, 69–90.

Myers, N. (1991). Tropical forests: Present status and future outlook. *Climatic Change* **19**, 3–32.

Nicholls, N. (1991). The El Nino/southern oscillation and Australian vegetation. *Vegetatio* **91**, 23–36.

Ode, D. J., Tieszen, L. L. and Lerman, J. C. (1980). The seasonal contribution of C_3 and C_4 plant species to primary production in a mixed prairie. *Ecology* **61**, 1304–1311.

Okuda, T. (1987). The distribution of C_3 and C_4 graminoids on the semi-natural grassland of Southwestern Japan. *J. Japan Grassl. Sci.* **33**, 175–184.

Okuda, T. and Furukawa, A. (1990). Occurrence and distribution of C_4 plants in Japan. *Jpn. J. Ecol.* **40**, 91–121.

Osmond, C. B., Winter, K. and Ziegler, H. (1982). Functional significance of different pathways of CO_2 fixation in photosynthesis. In "Encyclopedia of Plant Physiology, New Series Vol. 12B. Physiological Plant Ecology II. Water Relations and Carbon Assimilation" (O. L. Lange, P. S. Nobel, C. B. Osmond and H. Ziegler, eds.), pp. 479–547. Springer-Verlag, Berlin.

Owensby, C. E., Coyne, P. I., Ham, J. M., Auen, L. M. and Knapp, A. (1993). Biomass production in a tallgrass prairie ecosystem exposed to ambient and elevated CO_2. *Ecol. Appl.* **3**, 644–653.

Owensby, C. E., Ham, J. M., Knapp, A., Bremer, D. and Auen, L. M. (1997). Water vapour fluxes and their impact under elevated CO_2 in a C_4-tallgrass prairie. *Global Change Biol.* **3**, 189–195.

Owensby, C. E., Ham, J. M., Knapp, A., Rice, C. W., Coyne, P. I. and Auen, L. M. (1996). Ecosystem-level responses of tallgrass prairie to elevated CO_2. In "Carbon Dioxide and Terrestrial Ecosystems" (G. W. Koch and H. A. Mooney, eds.), pp. 147–162. Academic Press, New York.

Pande, H. K., Tran, D. V. and That, T. T. (1994). "Improved Upland Rice Farming Systems," FAO, Rome.

Panitsa, M., Dimopoulis, P., Iatrous, G. and Tzanouclakis, D. (1994). Contribution to the study of the Greek flora: Flora and vegetation of the Enousses (Oinousses) Islands (E. Aegean area). *Flora* **189**, 69–78.

Parkhurst, R. J. and Mullen, J. M. (1991). "Flora of the Outer Hebrides," The Natural History Museum, London.

Patterson, D. T. and Flint, E. P. (1990). Implications of increasing carbon dioxide and climate change for plant communities and competition in natural and managed ecosystems. In "Impact of Carbon Dioxide, Trace Gases, and Climate Change on Global Agriculture" (ASA special publication no. 53), pp. 83–109. Amer. Soc. Agron., Crop Sci. Soc. Amer., and Soil Sci. Soc., Madison. WI.

Paul, R. and Elmore, C. D. (1984). Weeds and the C_4 syndrome. *Weeds Today* **15**, 3–4.

Paulay, G. and Spencer, T. (1989). Vegetation of Henderson Island. *Atoll Res. Bull.* **328**, 1–13.

Paruelo, J. M. and Lauenroth, W. K. (1996). Climatic controls of the distribution of plant functional types in a grasslands and shrublands of North America. *Ecol. Appl.* **6**, 1212–1224.

Pearcy, R. W. (1990). Sunflecks and photosynthesis in plant canopies. *Annu. Rev. Plant Physiol Plant Mol. Biol.* **41**, 421–453.

Pearcy, R. W. and Ehleringer, J. (1984). Comparative ecophysiology of C_3 and C_4 plants. *Plant Cell Environ.* **7**, 1–13.
Pearcy, R.W., Tumosa, N. and Williams, K. (1981). Relationships between growth, photosynthesis and competitive interactions for a C_3 and a C_4 plant. *Oecologia* **48**, 371–376.
Pearson, C. J., ed. (1992). "Ecosystems of the World, vol. 18. Field Crop Ecosystems." Elsevier, Amsterdam.
Peat, H. C. L. (1997). "Dynamics of C_3 and C_4 Productivity in Northern Mixed Grass Prairie," M.Sc. thesis, University of Toronto.
Peterson, D. W. and Reich, P. B. (1996). Vegetation structure, composition and diversity in mid-west oak savanna after 32 years of prescribed burning experiment. *Bull. Ecol. Sci. Amer.* **77**, 351 (abstract only).
Piedade, M. T. F., Long, S. P. and Junk, W. J. (1994). Leaf and canopy photosynthetic CO_2 uptake of a stand of *Echinochloa polystachya* on the Central Amazon floodplain. Are the high rates associated with the C_4 syndrome realized under the near-optimal conditions provided by this exceptional natural habitat? *Oecologia* **97**, 193–201.
Polley, H. W. (1997). Implications of rising atmospheric carbon dioxide concentration for rangelands. *J. Range Manage.* **50**, 561–577.
Polley, H. W., Johnson, H. B., Mayeux, H. S. and Tischler, C. R. (1996). Are some of the recent changes in grassland communities a response to rising CO_2 concentrations? *In* "Carbon Dioxide and Terrestrial Ecosystems" (G. W. Koch and H. A. Mooney, eds.), pp. 177–195. Academic Press, New York.
Pomeroy, L. R. and Wiegert, R. G. (eds.). (1983). "The Ecology of a Salt Marsh." Springer-Verlag, Berlin.
Poorter, H., Roumet, C. and Campbell, B. D. (1996). Interspecific variation in the growth response of plants to elevated CO_2: A search for functional types. *In* "Carbon Dioxide and Terrestrial Ecosystems" (G. W. Koch and H. A. Mooney, eds.), pp. 375–412. Academic Press, New York.
Porsild, A. E. and Cody, W. J. (1980). "Vascular Plants of Continental Northwest Territories, Canada," National Museum of Natural Sciences/National Museum of Canada, Ottawa.
Powell, A. M. (1994). "Grasses of the Trans-Pecos and Adjacent Areas," University of Texas Press, Austin.
Prendergast, H.D.V. (1989). Geographical distribution of C_4 acid decarboxylation types and associated structural variants in native Australian C_4 grasses (Poaceae). *Aust. J. Bot.* **37**, 253–273.
Pyankov, V. I. (1994). C_4-species of high-mountain deserts of eastern Pamir. *Russian J. Ecol.* **24**, 156–160.
Pyankov, V. I., Mokronosov, A. T. (1993). General trends in changes of the earth's vegetation related to global warming. *Russ. J. Plant Physiol.* **40**, 443–458.
Pyne, S. J. (1990). Fire conservancy: The origins of wildland fire protection in British India, America and Australia. *In* "Fire in the Tropical Biota—Ecosystem Processes and Global Challenges" (J. G. Goldammer, ed.), pp. 319–336. Springer-Verlag, Berlin.
Ramachandran, V. S. and Nair, V. J. (1988). "Flora of Cannanore," Bot. Survey of India, New Delhi.
Raven, P. H., Evert, R. F. and Eichhorn, S. E. (eds.). (1992). "Biology of Plants," 5th Ed., Worth Publishers, Inc., New York.
Rechinger, K. H. (1964). "Flora of Lowland Iraq," Von J. Cramer, Weinheim, Vienna.
Renevier, A. and Cherrier, J. (1991). Flore et Vegetation De L'ile De Walpole. *Atoll Res. Bull.* **351**, 1–21.
Renvoize, S. A. (1984). "The Grasses of Bahia," Royal Botanic Garden, Kew.
Renvoize, S. A. (1988). "Hatschbach's Parana Grasses," Royal Botanic Gardens, Kew.
Robertson, S. A. (1989). "Flowering Plants of the Seychelles," Royal Botanic Gardens, Kew.

Rogers, R. W. (1996). Flowering and fruiting in the flora of Heron Island, Great Barrier Reef, Australia. *Atoll Res. Bull.* **440,** 1–9.
Roland, A. E. and Smith, E. C. (1966). The Flora of Nova Scotia, Part I. The Pteridophytes, Gymonosperms, and Monocotyledons. *Proc. Nova Scotia Inst. Sci.* **26,** 129–131.
Rønning, O. (1964). "Svalbards Flora," Norsk Polorinstitutt, Oslo.
Roxburgh, Dr. (1996). "St. Helena: Plants seen by Dr. Roxburgh, 1813–4, and commentary 1996," The Geraniaceae Group, London.
Rundel, P. W. (1980). The ecological distribution of C_4 and C_3 grasses in the Hawaiian Islands. *Oecologia* **45,** 354–359.
Ruthsatz, B. and Hofmann, U. (1984). Die Verbreitung von C_4-Pflanzen in den semiariden Anden NW-Argentiniens mit einem Beitrag zur Blattanatomie ausgewahlter Beispiele. *Phytocoenologia* **12,** 219–249.
Sachet, M.-H. (1962). Flora and vegetation of Clipperton Island. *Proc. Calif. Acad. Sci.* **31,** 249–307.
Sachet, M. H. (1969). List of vascular flora of Rongiroa. *Atoll Res. Bull.* **125,** 33–44.
Sachet, M.-H. (1983a). Botanique De L'Ile De Tupai, Iles De La Societe. *Atoll Res. Bull.* **276,** 1–26.
Sachet, M.-H. (1983b). Takapoto Atoll, Tuamutu Archipelago: Terrestrial vegetation and flora. *Atoll Res. Bull.* **277,** 1–41.
Saenger, P., Specht, M. M., Specht, R. L. and Chapman, V. J. (1977). Mangal and coastal saltmarsh communities in Australia. *In* "Ecosystems of the World 1: Wet Coastal Ecosystems" (V. J. Chapman, ed.), pp. 293–345. Elsevier, Amsterdam.
Sage, R. F. (1994). Acclimation of photosynthesis to increasing atmospheric CO_2: The gas exchange perspectives. *Photosyn. Res.* **39,** 351–368.
Sage, R. F. (1996). Modification of fire disturbance by elevated CO_2. *In* "Carbon Dioxide and Terrestrial Ecosystems" (Körner, Ch. and Bazzaz, F.A., eds.), pp. 231–249. Academic Press, New York.
Sage, R. F. and Pearcy, R. W. (1987). The nitrogen use efficiency of C_3 and C_4 plants. I. Leaf nitrogen, growth, and biomass partitioning in *Chenopodium album* L. and *Amaranthus retroflexus* L. *Plant Physiol.* **84,** 954–958.
Sage, R. F., Pearcy, R. W. and Seemann, J. R. (1987). The nitrogen use efficiency of C_3 and C_4 plants. III. Leaf nitrogen effects on the activity of carboxylation enzymes in *Chenopodium album* L. and *Amaranthus retroflexus* L. *Plant Physiol.* **84,** 355–359.
Sage, R. F., Santrucek, J. and Grise, D. J. (1995). Temperature effects on the photosynthetic response of C_3 plants to long-term CO_2 enrichment. *Vegetatio* **121,** 67–77.
Sage, R. F. and Sharkey, T. D. (1987). The effect of temperature on the occurrence of O_2 and CO_2 insensitive photosynthesis in field grown plants. *Plant Physiol.* **84,** 658–664.
Salati, E. and Nobre, C. A. (1991). Possible climatic impacts of tropical deforestation. *Climate Change* **19,** 177–196.
Samarakoon, A. B. and Gifford, R. M. (1995). Soil water content under plants at high CO_2 concentration and interactions with the direct CO_2 effects: A species comparison. *J. Biogeogr.* **22,** 193–202.
San Jose, J. J. and Farinas, M. R. (1991). Temporal changes in the structure of a *Trachypogon* savanna protected for 25 years. *Acta Oecologica* **12,** 237–247.
Sankhla, N., Ziegler, H., Vyas, O.P., Stichler, W. and Trimborn, P. (1975). Eco-physiological studies on Indian arid zone plants: V. screening of some species for the C_4-pathway of photosynthetic CO_2-fixation. *Oecologia* **21,** 123–129.
Sarmiento, G. (1983). The savannas of tropical America. *In* "Tropical Savannas" (Ecosystems of the World 13) (F. Fourliere, ed.), pp. 245–288. Elsevier Sci. Publ., Amsterdam.
Sarmiento, G. (1992). Adaptive strategies of perennial grasses in South American savannas. *J. Veg. Sci.* **3,** 325–336.

Schnell, R. (1952). "Vegetation et Flore de la Region Montugneuse da Nimba (Afrique Occidentale Fruncaise)," Memoires de Iinstitut Francais D'Afrique, Nore.
Scholes, R. I. and Archer, S. R. (1997). Tree–grass interactions in savannas. *Annu. Rev. Ecol. Syst.* **28**, 517–544.
Schüle, W. (1990). Landscapes and climate in prehistory: Interactions of wildlife, man and fire. *In* "Fire in the Tropical Biota—Ecosystem Processes and Global Challenges" (J. G. Goldammer, ed.), pp. 373–318. Springer-Verlag, Berlin.
Schulze, E.-D., Ellis, R., Schulze, W. and Trimborn, P. (1996). Diversity, metabolic types and Delta ^{13}C carbon isotope ratios in the grass flora of Namibia in relation to growth form, precipitation and habitat conditions. *Oecologia* **106**, 352–369.
Schwarz, A. G. and Redmann, R. E. (1988). C_4 grasses from the boreal forest region of northern Canada. *Can. J. Bot.* **66**, 2424–2430.
Schwarz, A. G. and Reaney, J. T. (1989). Perennating structures and freezing tolerance of northern and southern populations of C_4 grasses. *Bot. Gaz.* **150**, 239–246.
Schwarz, A. G. and Redman, R. E. (1989). Photosynthetic properties of C_4 grass (*Spartina gracilis* Trin.) from northern environment. *Photosynthetica* **23**, 449–459.
Scoggan, H. J. (1950). "The Flora of Bic and the Gaspe Peninsula, Quebec," Canada Dept. of Resources and Development, Ottawa.
Scoggan, H. J. (1957). "Flora of Manitoba," pp. 83–139. National Museums of Canada, Ottawa.
Scoggan, H. J. (1978). "The Flora of Canada, Part 2," National Museum of Natural Sciences and National Museums of Canada, Ottawa.
Seeliger, U. (1992). Coastal foredunes of southern Brazil: Physiography, habitats, and vegetation. *In* "Coastal Plant Communities in Latin America" (U. Seeliger, ed.), pp. 367–381. Academic Press, New York.
Sharma A. K. and Dhakre, J. S. (1995). "Flora of Agra District," Bot. Survey of India, New Delhi.
Sharma, B. D., Singh, N. P., Raghaven, R. S. and Peshpauda, U. R. (1984). "Flora of Karnataka," Bot. Survey of India, New Delhi.
Shomer-Ilan, A., Nissenbaum, A. and Waisel, Y. (1981). Photosynthetic pathways and the ecological distribution of the *Chenopodiaceae* in Israel. *Oecologia* **48**, 244–248.
Shreve, F. and Wiggins, I. L. (1964). "Vegetation and Flora of the Sonoran Desert," Vol. 1. Stanford University Press, Stanford, California.
Sinclair, A. R. E. (1995). Equilibria in plant–herbivore interactions. *In* "Serengeti II: Dynamics, Management, and Conservation of an Ecosystem" (P. Arcese and A. R. E. Sinclair Eds.), pp. 91–113. University of Chicago Press, Chicago.
Skelton, E. G. and Skelton, E. (1991). "Haliburton Flora," Royal Ontario Museum, Toronto.
Skinner, W. R., Jefferies, R. L., Carleton, T. J. and Rockwell, R. F. (1998). Prediction of reproductive success and failure in lesser snow geese based on early climatic variables. *Global Change Biol.* **4**, 3–16.
Skottsberg, C. (1953a). The vegetation of Easter Island. *In* "The Natural History of Juan Fernandez and Easter Island Vol. 2. Botany" (C. Skottsberg, ed.), pp. 487–503. Almquist & Wiksells Boktryckeri AB, Uppsala.
Skottsberg, C. (1953b). The vegetation of Juan Fernandez Islands. *In* "The Natural History of Juan Fernandez and Easter Island Vol. 2. Botany" (C. Skottsberg, ed.), pp. 793–959. Almqvist & Wiksells Boktryckeri AB, Uppsala.
Smith, C. F. (1976). "A Flora of the Santa Barbara Region, California," Santa Barbara Mus. Nat. History, Santa Barbara.
Smith, M. and Wu, Y. (1994). Photosynthetic characteristics of the shade-adapted C_4 grass *Muhlenbergia sobolifera* (Muhl.) Trin: Control of development of photorespiration by growth temperature. *Plant Cell Environ.* **17**, 763–769.
Smith, S. D. and Nobel, P. S. (1986). Deserts. *In* "Photosynthesis in Contrasting Environments" (N. R. Baker and S. P. Long, eds.), pp. 13–62. Elsevier, London.

Soares, R. V. (1990). Fire in some tropical and subtropical South American vegetation types: An overview. In "Fire in the Tropical Biota—Ecosystem Processes and Global Challenges" (J. G. Goldammer, ed.), pp. 63–81. Springer-Verlag, Berlin.

Solbrig, O. T. (1996). The diversity of the savanna ecosystems. In "The Biodiversity of the Savanna Ecosystem Processes—A Global Perspective" (O. T. Solbrig, E. Medina and J. F. Silva, eds.), pp. 1–27. Springer-Verlag, Berlin.

Spicer, R. A. and Neuberg, D. M. (1979). The terrestrial vegetation of an Indian Ocean Coral Island: Wilingili- Addu Atoll. Maldive Islands. I. Transect Analysis of vegetation. *Atoll Res. Bull.* **231**, 1–25.

St. John, H. and Smith, A. C. (1971). The vascular plants of Horne and Wallis Islands. *Pacific Sci.* **25**, 313–348.

Stalter, R. (1973). The flora of Turtle Island, Jasper Co., South Carolina. *Castanea* **38**, 35–37.

Stalter, R. and Lamont, E. E. (1990). The vascular flora of Assateague Island, Virginia. *Bull. Torr. Bot. Club* **117**, 48–56.

Standley, P. C. (1931). "Flora of the Lancetilla Valley, Honduras," Publ. #283, Botanical Series Vol. 10. Field Museum of Natural History, Chicago.

Stoddart, D. R. and Fosberg, F. R. (1991). Plants of the Jamaican Keys. *Atoll Res. Bull.* **352**, 1–24.

Stowe, L. G. and Teeri, J. A. (1978). The geographic distribution of C_4 species of the dicotyledonae in relation to climate. *Amer. Nat.* **112**, 609–623.

Sykes, W. R. (1977). "Kermadec Island Flora—An Annotated Checklist," New Zealand Dept. Sci. and Res., DSIR.

Szczawinski, A. F. and Harrison, A. S. (1973). "Flora of Saanich Peninsula," Occasional Paper #16, British Columbia Prov. Museum, Victoria.

Taiz, L. and Zeiger, E. (eds.). (1991). "Plant Physiology" Benjamin/Cummings Publ. Co., Inc., Redwood City, CA.

Takeda, T. (1985). Studies on the ecology and geographical distribution of C_3 and C_4 grasses. III. Geographical distribution of C_3 and C_4 grasses in relation to climate conditions in Indian-subcontinent. *Jpn. J. Crop Sci.* **54**, 365–372.

Takeda, T. and Hakoyama, S. (1985). Studies on the ecology and geographical distribution of C_3 and C_4 grasses. II. Geographical distribution of C_3 and C_4 grasses in Far East and south East Asia. *Jpn. J. Crop Sci.* **54**, 65–71.

Takeda, T., Tanikawa, T, Agata, W. and Hakoyama, S. (1985a). Studies on the ecology and geographical distribution of C_3 and C_4 grasses. I. Taxonomic and geographical distribution of C_3 and C_4 grasses in Japan with special reference to climate conditions. *Jpn. J. Crop Sci.* **54**, 54–64.

Takeda, T., Ueno, O., Samejima, M. and Ohtani, T. (1985b). An investigation for the occurrence of C_4 photosynthesis in the Cyperaceae from Australia. *Bot. Mag. Tokyo* **98**, 393–411.

Talbot, S. S., Talbot, S. C. and Welsh, S. L. (1995). Botanical Reconnaissance of the Tuxedni wilderness area, Alaska. U. S. Department Interior, Nat. Biol. Service, Biol. Sci. Report #6.

Taylor, W. R. (1950). "Plants of Bikini and other Northern Marshall Islands." Mich. Univ. Sci. Ser. #18., University Mich. Press, Ann Arbor.

Teeri, J. A. (1979). The climatology of the C_4 photosynthetic pathway. In "Topics in Plant Population Biology" (O. T. Solbrig, S. Jain, G. B. Johnson and P. H. Raven, eds.), pp. 356–374. Columbia University Press, New York.

Teeri, J. A. and Stowe, L. G. (1976). Climatic patterns and the distribution of C_4 grasses in North America. *Oecologia* **23**, 1–12.

Teeri, J. A., Stowe, L. G. and Livingston, D. A. (1980). The distribution of C_4 species of the Cyperaceae in North America in relation to climate. *Oecologia* **47**, 307–310.

Telford, I. R. H. (1993a). Cocos (Keeling) islands. *Flora of Australia* **50**, 30–42.

Telford, I. R. H. (1993b). Coral sea islands territory. *Flora of Australia* **50**, 47–53.

Teunissen, P. A. and Wildschut, J. T. (1970). "Vegetation and Flora of the Savannas in the Brinkheual Nature Reserve, Northern Surinam." North-Holland Publ., Amsterdam.

Thaman, R. R., Fosberg, F. R., Manner, H. I. and Hassall, D. C. (1994). The flora of Nauru. *Atoll Res. Bull.* **392**, 1–223.

Thomas, R. B. and Strain, B. R. (1991). Root restriction as a factor in photosynthetic acclimation of cotton seedling grown in elevated carbon dioxide. *Plant Physiol.* **96**, 627–634.

Thompson, K. and Hamilton, A. C. (1983). Peatlands and swamps of the African continent. *In* "Ecosystems of the World 4B: MIRES: Swamp, Bog, Fen and Moor" (A. J. P. Gore, ed.), pp. 331–373. Elsevier, Amsterdam.

Tieszen, L. L. and Archer, S. (1990). Isotopic assessment of vegetation changes in grassland and woodland systems. *In* "Plant Biology of the Basin and Range" (C. B. Osmond, L. F. Pitelka and G. M. Hidy, eds.), pp. 293–321. Springer-Verlag, Berlin.

Tieszen, L. L. and Boutton, T. W. (1989). Stable carbon isotopes in terrestrial ecological research. *In* "Stable Isotopes in Ecological Research" (Ecological Studies 68), (P. W. Rundel, J. R. Ehleringer and K. A. Nagy, eds.), pp. 167–195. Springer-Verlag, Berlin.

Tieszen, L. L., Reed, B. C., Bliss, N. B., Wylie, B. K. and DeJong, D. D. (1997). NDVI, C_3 and C_4 production, and distributions in Great Plains grassland land cover classes. *Ecol. Appl.* **7**, 59–78.

Tilman, D. (1982). "Resource Competition and Community Structure," Princeton University Press, Princeton.

Tilman, D. (1988). "Plant Strategies and the Dynamics and Structure of Plant Communities," Princeton University Press, Princeton.

Tissue, D. T., Griffin, K. L., Thomas, R. B. and Strain, B. R. (1995). Effects of low and elevated CO_2 on C_3 and C_4 annuals. II. Photosynthesis and leaf biochemistry. *Oecologia* **101**, 21–28.

Tissue, D. T., Megonigal, J. P. and Thomas, R. B. (1997). Nitrogenase activity and N_2 fixation are stimulated by elevated CO_2 in a tropical N_2-fixing tree. *Oecologia* **109**, 28–33.

Topp, J. M. W. (1988). An annotated check list of the flora of Diego Garcia, British Indian Ocean Territory. *Atoll Res. Bull.* **313**, 1–19.

Turland, N. J., Chilton, L. and Press, J. R. (1993). "Flora of the Cretan Area: Annotated Checklist and Atlas," HMSO, London.

Tutin, T. G., Heywood, U. H., Burges, N. A., Moore, D. M., Valentine, D. H., Waters, S. M. and Webb, D. A. (eds.) (1964–1980). "Flora Europaea," Vol. 5. Cambridge University Press, Cambridge

Ueno, O. and Takeda, T. (1992). Photosynthetic pathways, ecological characteristics, and the geographical distribution of the Cyperaceae in Japan. *Oecologia* **89**, 195–203.

van der Maarel, E. (1993). Dry coastal ecosystems of Africa, America, Asia and Oceania in retrospect. *In* "Ecosystems of the World 2B: Dry Coastal Ecosystems—Africa, America, Asia and Oceania" (E. van der Maarel, ed.), pp. 501–510. Elsevier, Amsterdam.

Vestergaard, P. and Hansens, K. (1989). Distribution of vascular plants of Denmark. *Opeara Botanica* **96**, 137.

Vitousek, P. M. (1994). Beyond global warming: Ecology and global change. *Ecology* **75**, 1861–1876.

Vitousek, P. M., Aber, J. D., Howarth, R. W., Linkens, G. E., Matson, P. A., Schindler, D. W., Schlesinger, W.H. and Tilman, D. G. (1997). Human alteration of the global nitrogen cycle: Sources and consequences. *Ecol. Appl.* **7**, 737–750.

Vogel, C. S., Curtis, P. S. and Thomas, R. B. (1997). Growth and nitrogen accretion of dinitrogen-fixing *Alnus glutinosa* (L.) Gaertn. under elevated carbon dioxide. *Plant Ecol.* **130**, 63–70.

Vogel, J. C., Fuls, A. and Danin, A. (1986). Geographical and environmental distribution of C_3 and C_4 grasses in the Sinai, Negev, and Judean deserts. *Oecologia* **70**, 258–265.

Vogel, J. C., Fuls, A. and Ellis, R. P. (1978). The geographical distribution of Kranz grasses in South Africa. *S. Afr. J. Sci.* **74**, 209–215.

Voss, E. G. (1972). "Michigan Flora," Kingsport Press, Bloomfield Hills, Michigan.
Wace, N. M. and Dickson, J. H. (1965). Part II: The terrestrial botany of the Tristan de Cunha Islands. *Phil. Trans. Roy. Soc. London (B)* **759,** 273–360.
Walter, H. and Box, E. O. (1983a). Middle Asian deserts. In "Ecosystems of the World 5: Temperate Deserts and Semi-deserts" (N. E. West, ed.), pp. 79–104. Elsevier, Amsterdam.
Walter, H. and Box, E. O. (1983b). The Karakum Desert, an example of a well-studied EU-BIOME. In "Ecosystems of the World 5: Temperate Deserts and Semi-deserts" (N. E. West, ed.), pp. 105–159. Elsevier, Amsterdam.
Walter, H., Harnickell, E. and Mueller-Dombois, D. (1975). "Climate-Diagram Maps of the Individual Continents and the Ecological Climatic Regions of the Earth—Supplement to the Vegetation Monographs," Springer-Verlag, Berlin.
Wardle, P. (1975). Vascular plants of Westland National Park (New Zealand) and neighboring lowland and coastal areas. *New Zeal. J. Bot.* **13,** 497–545.
Watson, L. and Dallwitz, M. J. (1998). "Grass Genera of the World: Descriptions, Illustrations, Identification, and Information Retrieval; Including Synonyms, Morphology, Anatomy, Physiology, Cytology, Classification, Pathogens, World and Local Distribution, and References," URL *http://biodiversity.uno.edu/delta/*.
Wedin, D. A. (1995). Species, nitrogen and grassland dynamics: The constraints of stuff. In "Linking Species and Ecosystems" (C. Jones and J. H. Lawton, eds.), pp. 253–262. Chapman & Hall, London.
Wedin, D. A. and Tilman, D. (1993). Competition among grasses along a nitrogen gradient: Initial conditions and mechanisms of competition. *Ecol. Monogr.* **63,** 199–229.
Wedin, D. A. and Tilman, D. (1996). Influence of nitrogen loading and species composition on the carbon balance of grasslands. *Science* **274,** 1720–1723.
Welsh, S. L. (1974). "Anderson's Flora of Alaska and Adjacent Parts of Canada," Brigham Young University Press, Provo, Utah.
Weltzin, J. F. and Coughenour, M. B. (1990). Savanna tree influence on understory vegetation and soil nutrients in northwestern Kenya. *J. Veg. Sci.* **1,** 325–334.
Wentworth, T. R. (1983). Distribution of C_4 plants along environmental and compositional gradients in southeastern Arizona. *Vegetatio* **52,** 21–34.
Werger, M. J. A. and Ellis, R.P. (1981). Photosynthetic pathways in the arid regions of South Africa. *Flora* **171,** 64–75.
West, N. E. (1983). Intermountain salt–desert shrubland. In "Ecosystems of the World 5: Temperate Deserts and Semi-deserts" (N. E. West, ed.), pp. 375–397. Elsevier, Amsterdam.
West, R. C. (1977). Tidal salt-marsh and mangal formations of Middle and South America. In "Ecosystems of the World 1: Wet Coastal Ecosystems" (V. J. Chapman, ed.), pp. 193–213. Elsevier, Amsterdam.
Wester, L. (1985). Checklist of the vascular plants of the Northern Line Islands. *Atoll Res. Bull.* **287,** 1–38.
Whistler, W. A. (1984). Vegetation and flora of Aleipata Islands, western Somoa. *Pacific Sci.* **37,** 227.
Wiggens, I. R. and Porter, D. M. (1971). "Flora of the Galápagos Islands," Stanford University Press, Stanford, California.
Wilsey, B. J., Coleman, J. S. and McNaughton, S. J. (1997). Effects of elevated CO_2 and defoliation on grasses: A comparative ecosystem approach. *Ecol. Appl.* **7,** 844–853.
Winter, K. (1981). C_4 plants of high biomass in arid regions of Asia—occurrence of C_4 photosynthesis in Chenopodiaceae and Polygonaceae from the Middle East and USSR. *Oecologia* **48,** 100–106.
Winter, K. and Troughton, J. H. (1978). Photosynthetic pathways in plants of coastal and inland habitats of Israel and the Sinai. *Flora* **167,** 1–34.
Winter, K., Troughton, J. H. and Card, K. A. (1976). Delta ^{13}C values of grass species collected in the northern Sahara Desert. *Oecologia* **25,** 115–123.

Wong, S. C. and Osmond, C. B. (1991). Elevated atmospheric partial pressure of CO_2 and plant growth. III. Interactions between *Triticum aestivum* (C_3) and *Echinochloa frumentacea* (C_4) during growth in the mixed culture under different CO_2, N nutrition and irradiance treatments, with emphasis on below-ground responses estimated using the Delta ^{13}C value of root biomass. *Aust. J. Plant Physiol.* **18,** 137–152.

Wood, J. R. I. (1997). "A Handbook of the Yemen Flora," Royal Botanic Gardens, Kew.

Woodroffe, C. D. (1985). Vegetation and flora of Nui Atoll, Tuvalu. *Atoll Res. Bull.* **283,** 1–18.

Yin, L. and Wang, P. (1997). Distribution of C_3 and C_4 photosynthetic pathways of plants on the steppe of northeastern China. *Acta Ecologica Sinica* **17,** 113–123.

Yin, L. and Zhu, L. (1990). A preliminary study on C_3 and C_4 plants in the Northeast steppes and their ecological distribution. *J. Appl. Ecol. (Shenyang).* **1,** 237–242.

Young, H. J. and Young, T. P. (1983). Local distribution of C_3 and C_4 grasses in sites of overlap on Mount Kenya. *Oecologia* **58,** 373–377.

Young, J. A. and Evans, R. A. (eds.). (1980). "Physical, Biological, and Cultural Resources of the Gund Research and Demonstration Ranch, Nevada," U. S. Department of Agriculture.

Young, S. B. (1971). The vascular flora of St. Lawrence Island with special reference to floristic condition in the Arctic region. **201,** 11–115. Contrib. Gray Herbarium, Harvard University, Cambridge, MASS.

Zahran, M. A. (1977). Africa A. Wet formations of the African Red Sea coast. *In* "Ecosystems of the World 1: Wet Coastal Ecosystems" (V. J. Chapman, ed.), pp. 215–231. Elsevier, Amsterdam.

Ziegler, H., Batanouny, K. H., Sankhla, N., Vyas, O. P. and Stichler, W. (1981). The photosynthetic pathway types of some desert plants from India, Saudi Arabia, Egypt and Iraq. *Oecologia* **48,** 93–99.

Zuloaga, F. O. (1987). Systematics of New World species of *Panicum* (Poaceae: Paniceae). *In* "Grass Systematics and Evolution" (T. R. Soderstrom, K. W. Hilu, C. S. Campbell and M. E. Barkworth, eds.), pp. 287–306. Smithsonian Institute Press, Washington D. C.

IV

The Evolution of C$_4$ Photosynthesis

11

The Origins of C_4 Genes and Evolutionary Pattern in the C_4 Metabolic Phenotype

Russell K. Monson

I. Introduction

One of the more striking patterns in the evolutionary history of plants is that, with only modest variation, the evolutionary product that we identify as C_4 photosynthesis is fundamentally the same in many species of independent origin. This observation brings to mind several questions concerning the nature of C_4 components and why there is such consistency in the way they are organized. At the molecular level, what is the nature of C_4 genes, and why have they arisen so many times independently of one another? Does evolution at the molecular level reflect (1) new C_4 genes arising through modification of expressed sequences without prior duplication of preexisting C_3 genes (anagenesis of C_4 genes), (2) duplication of preexisting C_3 genes followed by C_3 versus C_4 divergence of expressed sequences (clado-genesis of C_4 genes), and/or (3) modification of regulatory sequences in C_3 genes to produce a C_4 type of expression? Some answers to these questions are now available from studies of the structure and regulation of C_4 genes (Ku et al., 1996).

At the level of plant physiology, questions arise as to which processes were altered during the progression from C_3 to C_4, the order of changes during the evolutionary progression, and whether the path from C_3 to C_4 was similar for various taxa in which C_4 evolved. With the discovery of C_3–C_4 intermediate species, new insight has been gained into specific steps taken during C_3 to C_4 transformation. However, questions about the adaptive context of such steps remain unanswered. For example, were the advantages of C_4 photosynthesis expressed and consistent from the incipient stages of evolutionary transition to the final product? At the level of plant growth forms, why are there so few C_4 trees? Is there an interaction between growth form and the mode of CO_2 assimilation that poses adaptive barriers to the evolution of C_4 photosynthesis?

Evolutionary pattern in the C_4 pathway has been investigated with two main approaches. In the first, diversity is studied in existing C_4 taxa. This approach involves comparison of existing traits and is the one most commonly found in the literature. In the second approach, taxa are studied that appear to represent intermediary evolutionary stages (so-called C_3-C_4 intermediates). This type of study is more recent than the other, and rests on the still debatable assumption that these taxa are indeed evolutionary intermediates (alternative explanations have been offered). In this chapter, I discuss both of these approaches, using them to tackle some of these issues. I specifically address the question of why the independent evolution of C_4 photosynthesis has resulted in such a consistent set of traits.

II. The Evolution of C_4 Genes

A. The PEP Carboxylase Gene

Phospho*enol*pyruvate carboxylase (PEPc) catalyzes the irreversible β-carboxylation of PEP in a broad range of metabolic schemes, including the assimilation of atmospheric CO_2 in plants possessing the C_4 pathway and Crassulacean acid metabolism (CAM), the production of carbon skeletons for amino acid biosynthesis in all plants, the balance of pH and ionic concentrations of guard cells in the leaf and cortical cells in the root, and host cell dicarboxylic acid formation in the root nodules of N_2-fixing plants (for reviews see O'Leary, 1982; Chollet et al., 1996). The various functions of PEPc are carried out by different isoforms that are encoded by a multigene family found in the nuclear genome. In evolutionary terms, the role of PEPc in C_4 photosynthesis is a relatively recent innovation.

The taxonomic pattern of divergence between C_3- and C_4-type PEPc is best understood within the context of the independent origins of C_4 monocots and dicots. In the monocots maize and sorghum, the C_4 PEPc is encoded by a single gene that, at least in maize, is homologous ($\cong 70\%$) with the gene for a C_3 isoform (Kawamura et al., 1992). The C_4 PEPc genes of maize and sorghum are more similar to each other than to the C_4-dicot gene found in *Flaveria trinervia*, or to the C_3-monocot gene found in maize and sorghum (Toh et al., 1994). In contrast to the pattern for monocots, the C_4 isoform in *F. trinervia* is encoded by a small subgroup of PEPc genes (Poetsch et al., 1991). Sequence comparisons for the coding region of one of the C_4 PEPc genes in *F. trinervia* revealed greater similarity to C_3 and C_4 isoforms of the CAM dicot *Mesembryanthum crystallinum* than to the C_4

isoform for sorghum and maize. When amino acid sequences are mapped onto a phylogenetic tree, C_4 PEPc from maize and sorghum are distinct from C_3 PEPc (Fig. 1). The phylogenetic affinity of C_4 PEPc from the dicot *F. trinervia* cannot be unequivocally determined, but it appears distinct from those for C_4 PEPc in maize and sorghum (Toh *et al.*, 1994; Chollet et al., 1996). From this evidence, it appears that the C_4 PEPc genes in maize and sorghum share a common C_3 ancestor gene, and the evolution of C_4-type PEPc has occurred independently in monocots and dicots.

In evolving to serve the C_4 function, few modifications have occurred in the transcribed PEPc coding sequences, but several modifications have occurred within the regulatory elements of promoter sequences. Comparison of the C_3 and C_4 PEPc protein in the dicots *F. trinervia* (C_4) and *F. pringlei* (C_3) have revealed only five C_4-specific amino acid residues (Hermans and Westhoff, 1992). In contrast, the evolution of C_4 expression patterns appears to require more extensive modification, including (1) a higher level of leaf-specific expression, (2) "up-regulation" in mesophyll cells relative to

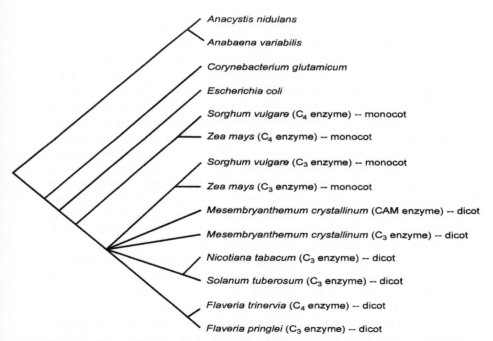

Figure 1 A phylogenetic model depicting relationships among different PEP carboxylase isoforms. Note the apparent divergence of C_4 monocot PEP carboxylase from other isoforms. [Adapted and redrawn from Toh H., Kawamura, T., and Izui, K. (1994). Molecular evolution of phosphoenolpyruvate carboxylate. *Plant Cell Environ.* **17,** 31–43, with permission].

bundle-sheath cells, and (3) light inducibility that increases expression during greening. In the C_4 monocot maize, these traits evolved through acquisition of new *cis*-acting elements in the promoter sequence, including a critical TATTT sequence upstream from the C_4 coding region (Schäffner and Sheen, 1992; Matsuoka *et al.*, 1994). [*Cis*-acting elements are sequences within a gene, or along different domains of the same chromosome, that interact with one another in regulating gene expression. *Trans*-acting elements are gene sequences that interact with expression modulators that originate from outside the chromosome on which the promoter resides (e.g., regulatory proteins).] In comparing the promoter regions of the dicots *F. trinervia* (C_4) and *F. pringlei* (C_3), Hermans and Westhoff (1992) identified critical C_4 regulatory elements including changes in the sequence environment surrounding TATA motifs, the appearance of a unique sequence that is known to interact with the *trans*-regulatory factor GT-1 in the light-induced promoters from other genes (Gilmartin *et al.*, 1990), and the appearance of an adenine–thymine rich attachment region that is characteristic of genes with high expression levels (Hermans and Westhoff, 1992).

B. The Pyruvate, Orthophosphate Dikinase Gene

Pyruvate,P_i dikinase (PPdk) catalyzes the reversible phosphorylation of pyruvate and inorganic phosphate, using ATP as a substrate (see Chapter 3). PPdk exists as cytosolic and chloroplastic isoforms in both C_3 and C_4 plants (Glackin and Grula, 1990; Matsuoka, 1995). In C_3 leaves, stems, and roots, both the cytosolic and plastidic forms are involved in carbohydrate metabolism, presumably in an anapleurotic role providing carbon skeletons (through the production of PEP) for amino acid biosynthesis (Aoyagi and Bassham, 1984). In C_4 leaves, the chloroplastic form has been coopted for the purpose of generating PEP in the mesophyll cell as the HCO_3^- acceptor in the reaction catalyzed by PEPc. In the C_4 monocot, maize, the C_4 chloroplast enzyme coexists with two cytosolic isoforms (Sheen, 1991). In the C_4 dicot *F. trinervia* the C_4 chloroplastic enzyme coexists with a single cytosolic isoform. PPdk is encoded by sequences in the nuclear genome.

As stated previously, the chloroplastic and cytosolic forms of PPdk are found in both C_3 and C_4 plants. Thus, most of the evolution in this gene occurred in response to the chloroplastic versus cytosolic roles for this enzyme, not the C_3 versus C_4 roles. Evolutionary modification has resulted in the chloroplastic gene overlapping the cytosolic gene, with two separate promoters controlling transcription (Fig. 2). The cytosolic and chloroplastic genes share the same code for the expressed PPdk polypeptide, but differ in that expression is controlled by different promoters. The chloroplastic promoter is upstream from the cytosolic promoter, as is a

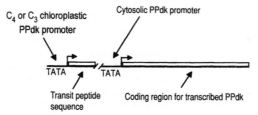

Figure 2 Schematic representation of the overlapped genes of C_3 and C_4 pyruvate,P_i dikinase with distinct promoters for the chloroplastic and cytosolic enzymes. Promoter sequences are marked by characteristic TATA sequences. The right-angled arrows represent transcription initiation sites. The upstream gene contains the coding sequence for a chloroplast transit peptide. [Adapted and redrawn from Sheen, I. (1991). Molecular mechanisms underlying the differential expression of maize pyruvate, orthophosphate dikinase, genes. *Plant Cell* **3**, 225–245, with permission].

coding sequence for the transit peptide (Imaizumi *et al.*, 1992). The chloroplastic promoter includes light-induced regulatory sequences that restrict expression of chloroplast PPdk to greening tissues. This light dependency is characteristic of the chloroplastic PPdk in both C_3 and C_4 leaves (Hata and Matsuoka, 1987; Matsuoka and Yamamoto, 1989), suggesting that this mode of regulation evolved before the divergence of C_3 and C_4 photosynthesis. The unique overlapped structure to the chloroplastic and cytosolic genes has been found in both monocots [rice (C_3) and maize] and dicots (C_3 and C_4 species in *Flaveria*) (Matsuoka, 1995).

Based on limited knowledge about the taxonomic distribution of different PPdk genes, some insight can be gained into phylogenetic pattern. In maize, two different cytosolic PPdk genes have been identified, only one of which is overlapped by the C_4 gene (Sheen, 1991). This is consistent with evolutionary duplication of an ancestral cytosolic gene, followed by addition of a second promoter upstream from the first, and eventual modification to produce C_4-specific expression (Fig. 3). In C_3 and C_4 species of *Flaveria*, only one copy of the overlapped PPdk gene is present (Rosche *et al.*, 1994; Rosche and Westhoff, 1995), leading to the conclusion that modification occurred without duplication (Fig. 3). Analysis of amino acid sequences has revealed greater similarity between the PPdk of C_3 and C_4 monocots, than between C_4 monocots and C_4 dicots (Fig. 4). It is likely that the C_4 PPdk genes in monocots and dicots diverged from a common C_3 ancestral chloroplast gene after divergence of monocots and dicots.

It was only after the evolution of separate cytosolic and chloroplastic promoters that the chloroplastic form was coopted for use in C_4 photosynthesis. In evolving to serve the C_4 system, the PPdk gene has obtained *cis*-acting elements that provide enhanced levels of expression in C_4 mesophyll

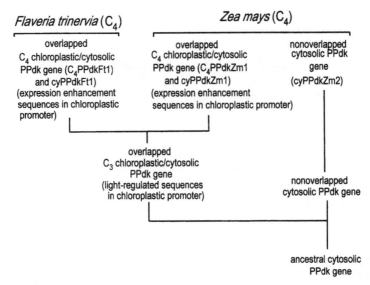

Figure 3 A phylogenetic representation of the divergence in pyruvate,P_i dikinase to produce distinct cytosolic and chloroplastic enzymes in a representative monocot (*Zea mays*) and dicot (*Flaveria trinervia*). Note the evolution of a single-copy gene in the dicot *Flaveria* and the two gene copies in the monocot maize. Nomenclature for the genes follows that of Sheen (1991).

chloroplasts (Sheen, 1991; Matsuoka *et al.*, 1993). By constructing chimeric genes using various regions from the C_4 PPdk promoter of maize and the gene for β-glucuronidase, Matsuoka and Numazawa (1991) demonstrated that the nucleotide sequence between -308 and -289 (relative to the transcription initiation point, ATG) is vital to enhanced expression in maize mesophyll cells. Using a slightly different approach in which the C_4 PPdk maize promoter was spliced to a chloramphenicol acetyltransferase reporter gene, Sheen (1991) demonstrated that two distinct *cis*-acting regions are required for C_4-like expression in maize mesophyll protoplasts. One involves the domain from -347 to -109 and includes elements that control the light-dependency of PPdk expression. The second is found at -108 to -52, and includes elements that regulate PPdk expression within the context of chloroplast development.

When taken together, three observations suggest that few genetic changes are required for adaptation of PPdk to C_4 photosynthesis. First, within the same leaf, the mature PPdk protein is identical with respect to a non-C_4 cytosolic enzyme and a C_4 chloroplastic enzyme. Thus, there are no known essential C_4-specific characteristics required of the active PPdk. Second, the chloroplastic promoter and transit peptide sequence is present in C_3 plants, demonstrating that these features, which are so important

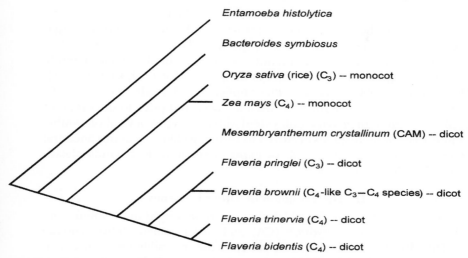

Figure 4 A phylogenetic scheme for the evolution of pyruvate,P_i dikinase (PPdk). Note the apparent divergence of monocot PPdk from dicot PPdk prior to the evolution of C_4 photosynthesis in both groups. [Adapted and redrawn from Matsuoka, M. (1995). The gene for pyruvate, orthophosphate dikinase in C_4 plants: Structure, regulation and evolution. *Plant Cell Physiol.* **36,** 937–943, with permission].

for chloroplast-specific expression in C_4 mesophyll cells, arose in response to selective pressures other than those associated the evolution of C_4 photosynthesis. Finally, the sequence in the promoter region that controls enhanced expression of PPdk in C_4 mesophyll cells is short (only 19 nucleotides; Matsuoka and Numazawa, 1991), indicating few modifications were required to up-regulate C_4 PPdk.

C. The NADP-malic Enzyme Gene

In certain C_4 species, NADP-malic enzyme (NADPme) catalyzes the decarboxylation of malate in the chloroplasts of bundle sheath cells, producing NADPH in the process. Of the three decarboxylating enzymes found in C_4 plants, only the gene for NADPme has been characterized. As with PEPc and PPdk, both C_3 and C_4 isoforms have been identified. With respect to C_3 isoforms, evidence for both cytosolic and chloroplastic enzymes has been reported (Wedding, 1989; Edwards and Andreo, 1992; Marshall *et al.*, 1996). The C_4 isoform is strictly chloroplastic. All isoforms are encoded from the nuclear genome. Sequence comparisons between the C_4 NADPme gene from *F. trinervia* and maize have revealed no sequence identity within the region that encodes the transit peptide, and only moderate identity (75%) at the protein level (Börsch and Westhoff, 1990; Rajeevan *et al.*, 1991). As is the case for PEPc and PPdk, these observations reinforce the conclusion of independent C_4 origins in the monocots and dicots. Within the *Flaveria* dicots, genes for both the C_3 and C_4 forms of NADPme have been isolated from C_3, C_3–C_4 intermediate, and C_4 species (Marshall *et al.*, 1996). The C_3 and C_4 genes have nearly identical sequences in the transit peptide region and high homology (>90%) at the protein level (Lipka *et al.*, 1994; Marshall *et al.*, 1996). It is likely that duplication of an already existing C_3 gene for the chloroplast form of NADPme occurred prior to the origin of C_4 photosynthesis in *Flaveria*. Coincident with the origins of C_4 photosynthesis in *Flaveria*, modifications occurred to *cis*-acting elements of this ancestral gene, resulting in higher levels of expression within leaf bundle-sheath tissue (Marshall *et al.*, 1996).

D. The Evolution of Other C_4 Genes

Two other C_4 proteins for which there is some insight into evolutionary origins are carbonic anhydrase (CA) and NADP malate dehydrogenase (NADPmdh). Carbonic anhydrase catalyzes the reversible interconversion between carbon dioxide and bicarbonate. CA is encoded from the nuclear genome, and is found in the chloroplasts and cytosol of mesophyll cells in both C_3 and C_4 leaves. In C_3 leaves, cytosolic activity is extremely low, whereas chloroplastic activity is high. The role of chloroplastic CA in C_3 leaves has not been completely resolved, but it may enhance the movement of CO_2 to RuBP carboxylase/oxygenase (Rubisco) (Reed and Graham, 1981). In C_4 leaves, CA activity is extremely low in mesophyll cell chloroplasts, but high in the cytoplasm of mesophyll cells where it ensures the availability of bicarbonate to PEP carboxylase (Hatch and Burnell, 1990). PEP carboxylase uses bicarbonate rather than CO_2 as its inorganic carbon substrate, (see Chapter 3). It is apparent that evolution of the C_4 pathway required modifications to the pattern of intracellular expression of CA such that its activity was up-regulated in the mesophyll cytosol and down-regulated in the mesophyll chloroplast.

Phylogenetic comparisons involving three monocot species and eight dicot species have revealed greater similarity between the chloroplastic and cytosolic forms of CA within a group (i.e., within monocots), than between groups (i.e., between monocots and dicots) (Ludwig and Burnell, 1995). Among four *Flaveria* species (representing the C_3, C_3–C_4 intermediate, and C_4 pathways), sequence identity was 99–100% for the 221 amino acids nearest the C-terminal end of the chloroplastic protein, and 90–100% for the transit peptide sequence. The *Flaveria* chloroplastic CA shared 80% sequence identity with the other dicots used in the comparison (including the C_3 species *Arabidopsis thaliana*, *Spinacea oleracea*, *Pisum sativum*, and *Nicotiana tabacum*), and only 60% identity with the monocot species. These results suggest independent divergence of the C_3 (chloroplastic) and C_4 (cytosolic) enzymes after the divergence of monocots and dicots (Fig. 5).

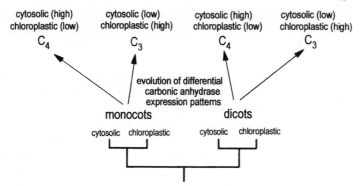

Figure 5 A scheme to account for the divergence of carbonic anhydrase into separate cytosolic and chloroplastic isoforms. The evolution of a cytosolic form with the same length as a chloroplastic form is proposed through mutation to produce an ineffective chloroplast transit sequence. The cytosolic form is most highly expressed in C_4 leaves, whereas the chloroplast form is most highly expressed in C_3 leaves.

Ludwig and Burnell (1995) studied CA cDNA sequences from C_3, C_3–C_4 intermediate, and C_4 representatives of *Flaveria* and concluded that the chloroplastic and cytosolic isoforms from all species share similar polypeptide lengths. It was hypothesized that the C_4 enzyme evolved to reside in the cytoplasm of mesophyll cells through modification to the transit peptide sequence of the chloroplastic isoform (Fig. 6). This would have rendered chloroplast targeting ineffective, and might explain the existence of proteins with similar lengths in the cytosol and chloroplast of C_4 and C_3 species.

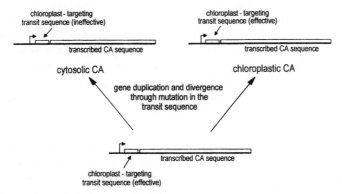

Figure 6 A phylogenetic model depicting the independent divergence of cytosolic and chloroplastic carbonic anhydrases in monocots and dicots, and the evolution of differential expression patterns in C_3 and C_4 species.

Following divergence of the chloroplastic and cytosolic forms, evolutionary modification of the promoter region would have enhanced expression of the cytosolic form in C_4 species and the chloroplastic form in C_3 species (Fig. 5).

Chloroplastic NADP malate dehydrogenase (NADPmdh) is encoded from the nuclear genome and is found in both C_3 and C_4 leaves. In C_3 plants, chloroplastic NADPmdh functions to balance reducing equivalents between the cytosol and chloroplast (Scheibe, 1990). In NADPme-type C_4 plants, chloroplastic NADPmdh is isolated to mesophyll cells, where it converts oxaloacetate, the first product of carboxylation, to malate prior to transport to bundle-sheath cells. The C_4 role requires up-regulation of NADPmdh activity in mesophyll cell chloroplasts and down-regulation of activity in the chloroplasts of bundle-sheath cells. Studies of maize and sorghum have revealed two genes for chloroplastic NADPmdh. One is expressed at low levels and probably encodes the enzyme that functions in balancing reducing equivalents. The other is light induced, expressed at high levels, and probably encodes the C_4-specific enzyme (Luchetta *et al.*, 1991; McGonigle and Nelson, 1995). As with some of the other C_4 genes described, the monocot C_4 NADPmdh probably arose through modification of promoter elements in a duplicated C_3 gene. In *F. trinervia*, NADPmdh is encoded by a single gene (McGonigle and Nelson, 1995). Thus, in this species the C_4 isoform appears to have evolved without prior duplication of a C_3 gene.

E. The Evolution of Differential Expression of Ribulose 1,5-Bisphosphate Carboxylase/Oxygenase (Rubisco) in C_4 Leaves

In C_4 photosynthesis, expression of Rubisco is up-regulated in bundle-sheath cells and suppressed in mesophyll cells. Rubisco is a large, multimeric protein consisting of eight small subunits and eight large subunits, with the large subunits carrying the active sites for catalysis. The gene for the large subunit (typically represented as rbcL) is present as a single copy per chloroplast genome (Gutteridge and Gatenby, 1995). rbcL is actively transcribed in C_4 mesophyll cell chloroplasts, despite the absence of mature Rubisco protein (Schäffner and Sheen, 1991; Boinski *et al.*, 1993). The suppression of Rubisco in C_4 mesophyll cells is due to posttranscriptional processes. Roth *et al.* (1996) have demonstrated a probable role for the nuclear gene (bundle sheath defective 2 Bsd2) in the posttranscriptional control of rbcL transcript accumulation and/or translation in mesophyll and bundle-sheath cells. Although the involvement of nuclear regulation in control over chloroplast gene expression is known from several plant systems, the nature of the control mechanisms is unclear (Mayfield *et al.*, 1995).

More is known about differential expression of the Rubisco small subunit. The small subunit (typically represented as rbcS) is encoded from a nuclear, multigene family (2–12 gene members) (Gutteridge and Gatenby, 1995). The differential expression of rbcS is controlled by a combination of (1) a promoter sequence that flanks the transcription start site and regulates the light-stimulated increase in rbcS expression in bundle-sheath cells (Bansal et al., 1992; Viret et al., 1994), (2) a sequence within the 3′ transcribed region of the gene that is photoactivated and interacts with the promoter sequence upstream from the transcription start site and suppresses transcription in mesophyll cells (Viret *et al.*, 1994), and (3) probable posttranscription suppression of rbcS mRNA in mesophyll cells (Schäffner and Sheen, 1991). The suppresser activity of the 3′ transcribed sequence most likely involves an interaction with *trans*-acting elements unique to C_4 mesophyll cells (Fig. 7).

Insertion of the rbcS promoter from C_4 maize bundle sheath cells into C_3 rice mesophyll cells results in normal C_3-like expression (Matsuoka *et al.*, 1994). In addition, insertion of an rbcS promoter from C_3 wheat mesophyll cells into C_4 maize mesophyll protoplasts results in normal C_3-like expression (Schäffner and Sheen, 1991). Thus, the promoter of the rbcS gene, including those sequences that affect the light dependency of expression, is similar in both C_3 and C_4 monocots. It may be that evolution of the differential expression of Rubisco occurred on the appearance of the 3' suppresser sequence concomitant with *trans*-acting elements unique to C_4 mesophyll cells. Alternatively, the 3′ sequence may have always been present in the C_3 rbcS gene, but only became active as a suppresser in C_4 mesophyll on the appearance of the unique *trans*-acting factor(s).

There are obvious differences in the regulatory systems for rbcS between C_4 dicots and monocots (Schäffner and Sheen, 1991). The rbcS promoter

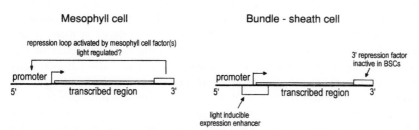

Figure 7 Regulatory interactions among different sequences of the RuBP carboxylase/oxygenase (Rubisco) gene in C_4 plants. The absence of Rubisco expression in mesophyll cells is accomplished through interactions between a 3′ repression sequence and the promoter, an interaction that appears to be regulated by light availability. In bundle-sheath cells, the 3′ repression sequence is inactive, and a light-inducible sequence that overlaps the promoter and transcribed region is active.

from dicot C_3 or C_4 species is not active when fused to a reporter gene and inserted into maize mesophyll protoplasts. The chimeric gene that is constructed through this process lacks the 3' suppresser sequence that would normally inhibit expression in mesophyll cells. However, the rbcS promoter is active when inserted into dicot mesophyll cells, including those from the C_4 dicot *F. trinervia* (Martineau *et al.*, 1989; Schäffner and Sheen, 1991). Conversely, the rbcS promoter from monocot C_3 and C_4 species is only active in monocot cells, but not dicot cells (Schäffner and Sheen, 1991). In dicots, GT-box and G-box motifs upstream from the TATA box have been shown to exert strong regulation in the expression of dicot rbcS (Dean *et al.*, 1989; Ueda *et al.*, 1989) but are inactive in monocot cells (Schäffner and Sheen, 1991). Different sequences upstream from the TATA box appear to be active in monocot rbcS promoters, including a highly conserved GAACGGT constitutive element and unique light-sensitive elements that are absent from the dicot promoters that have been examined (though there is good reason to believe that dicot rbcS promoters contain light-sensitive elements of a different nature; Schäffner and Sheen, 1991). On the basis of these observations, it appears that monocot- and dicot-specific promoter sequences evolved prior to the evolution of C_4 photosynthesis in these groups, and have been retained in the C_4 systems of modern taxa.

Evolutionary modification has also occurred in the kinetic affinity and turnover capacity of the rbcL active site of Rubisco with respect to CO_2 (Seemann *et al.*, 1984). In a survey of eight monocot C_4 species, three dicot C_4 species, four C_3 dicot species, and one C_3 monocot species, it was discovered that C_4 species from both groups exhibited CO_2-saturated specific activities (CO_2 assimilation rate per unit mass of enzyme) that were twice those of C_3 species from both groups. The improvement in turnover capacity of the C_4 Rubisco comes with a potential cost, however, in that the $K_m(CO_2)$ for C_4 species is almost double that of C_3 species (Yeoh *et al.*, 1981). Thus, affinity of the active site for CO_2 has decreased because of increased turnover capacity. It is likely that this tradeoff is due to the coupled dependence of substrate/product binding on properties of the active site. Increases in the turnover capacity of an active site (the capacity to release product) is probably dependent on weakening the binding interaction between product molecules and certain active site moieties (Gutteridge *et al.*, 1995). Such weakening may also decrease the binding affinity of substrate for the active site, resulting in a higher apparent K_m. In the case of Rubisco, selection for increases in turnover number in C_4 leaves probably arose as a result of its exposure to relatively high bundle-sheath CO_2 concentrations, a situation that would favor faster turnover capacity while avoiding the need for tight substrate binding such as might be required at the lower CO_2 concentrations characteristic of C_3 leaves.

F. A Summary of General Evolutionary Patterns in C_4 Genes

There are clear homologies between C_3 and C_4 genes, supporting past hypotheses that the components of C_4 photosynthesis evolved from preexisting metabolic components of C_3 plants (Cockburn, 1983). The evolution of C_4 metabolism is primarily a result of changes in gene regulatory sequences and their influence on expression patterns. Probably because of their modular design, *cis*-acting elements are the frequent target of evolutionary modification to patterns of enzyme expression in a variety of organisms (Dickinson, 1980, 1988; Goldstein *et al.*, 1982; Dobson *et al.*, 1984; Kettler *et al.*, 1986; Thorpe *et al.*, 1993). *Cis*-acting elements can be thought of as independent "circuits" that can be changed in a manner that "rewires" a gene's regulatory system.

Although some C_4 components appear to have arisen through modification of single-copy genes (e.g., NADPmdh and PPdk in *F. trinervia*), the more common pattern is modification of genes that have been duplicated during past evolution. Gene duplication provides the opportunity for evolutionary modification and cooption of one gene copy to produce a novel function while retaining the original function dictated by the other gene copy. Gene duplication occurs through polyploidy or unequal crossing over, and is a fundamental mechanism underlying the divergent evolution of enzymes and the origin of new metabolic pathways (Li, 1983). Two models have been put forth to explain the divergence of genes following duplication. In the first, selectively neutral mutations accumulate in a duplicated gene that is redundant and thus not needed for normal metabolic expression (Ohno, 1973). As mutations accumulate through genetic drift, some genes may evolve to take on a novel role in metabolism. In the second model, duplicated genes diverge through positive selection on their encoded products (Hughes, 1994). Under the influence of selection, two enzymatic isoforms that initially share the same generalized function evolve specialized roles in different metabolic pathways. Studies support the second model for most cases of metabolic evolution (Clark, 1994). Whatever the mechanism, it is clear that metabolic divergence through modification of duplicated genes and regulatory elements is common. When presented in this context, the fact that the C_4 pathway has arisen so many times independently is not surprising.

III. The Evolution of C_4 Metabolism as Evidenced in C_3-C_4 Intermediates

A. General Characteristics of C_3-C_4 Intermediates

C_3-C_4 intermediates have been identified in 27 species of eight genera representing five different families, including monocots and dicots (Chapter 16). The evolutionary intermediacy of C_3–C_4 plants has been supported with evidence from systematics and biogeography (Powell, 1978), morphology (Monson, 1996), life history and ecological traits (Monson, 1989a; Monson and Moore, 1989), and most recently the phylogenetic distribution of DNA sequences (Kopriva *et al.*, 1996). Those C_3–C_4 intermediates that have been identified to date are native to warm climates (Monson, 1989a). Many of the C_3–C_4 intermediate species are weedy in their ecological habit, although this is more obvious in the case of dicots than monocots. C_3–C_4 intermediate gas-exchange patterns can be traced to photorespiration rates that are lower than those of C_3 plants (Monson *et al.*, 1984; Monson, 1989b; Monson and Moore, 1989). This can be seen as a lower apparent CO_2 compensation point for C_3–C_4 intermediate species, compared to C_3 species (Fig. 8).

B. Anatomical Traits of C_3–C_4 Intermediates

If indeed C_3–C_4 intermediates represent transitional stages in the evolution of fully expressed C_4 photosynthesis, then novel anatomical traits are

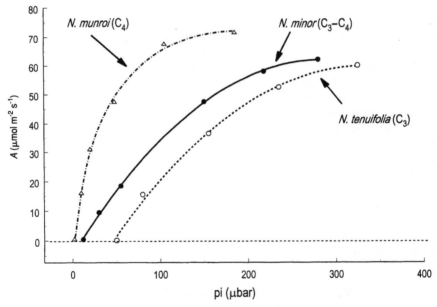

Figure 8 The relationship between CO_2 assimilation rate (A) and intercellular CO_2 partial pressure (pi) for leaves of a C_3, C_4, and C_3–C_4 intermediate *Neurachne* species. Note the shift toward a lower CO_2 compensation point (taken as the X-intercept) in the C_3–C_4 species compared to the C_3 species. [Redrawn from Hattersley, P. W., Wong, S.-C., Perry, S., and Roksandic, Z. (1986). Comparative ultrastructure and gas exchange characteristics of the C_3–C_4 intermediate *Neurachne minor* S. T. Blake (Poaceae). *Plant Cell Environ.* **9**, 217–233, with permission].

among the first evolutionary modifications. In leaves of the C_3–C_4 intermediate *Panicum milioides*, 40% of the mitochondria and 32% of the peroxisomes are concentrated in bundle-sheath cells, compared to only 14% of the mitochondria and 6% of the peroxisomes in leaves of the C_3 congener, *P. rivulare* (Brown et al., 1983a). There is less tendency for the differential concentration of chloroplasts, with 17% of the leaf's total chloroplasts localized in bundle-sheath cells in the C_3–C_4 intermediate species, versus 9% in the C_3 species. Similar anatomical characteristics have been found in C_3–C_4 intermediates within the genera *Flaveria* (Holaday et al., 1984), *Moricandia* (Holaday et al., 1981), *Neurachne* (Hattersley et al., 1986), and *Mollugo* (Laetsch, 1971). In a comprehensive anatomical review of C_3–C_4 intermediates in the genera *Panicum, Flaveria, Moricandia,* and *Neurachne* and their C_3 congeners, Brown and Hattersley (1989) found that on average, the percentages for mitochondria and peroxisomes allocated to bundle-sheath cells were 29–47% for C_3–C_4 intermediate species versus 8–20% for C_3 species.

Not only are organelle concentrations higher in bundle-sheath cells of C_3–C_4 intermediates, but they are also arranged in a novel fashion compared to the leaves of C_3 plants, with mitochondria appressed to the innermost (centripetal) cell wall, surrounded by a layer of chloroplasts (Fig. 9). Thus, there is a very intimate relationship between mitochondria, the organelle responsible for the release of CO_2 during photorespiration, and chloroplasts, the organelle responsible for the assimilation of CO_2 during photosynthesis. In fact in some C_3–C_4 intermediate species of *Panicum*, mitochondria are completely engulfed by chloroplasts (Brown et al., 1983b). These observations have led to the hypothesis that reduced photorespiration rates in C_3–C_4 intermediates are due to the recycling of photorespired CO_2 caused by the abundance and novel arrangement of mitochondria and chloroplasts in bundle-sheath cells (Brown et al., 1983a; Brown and Hattersley, 1989).

C. Differential Expression of Glycine Decarboxylase

The glycine decarboxylase activity of C_3–C_4 intermediate leaves is isolated to bundle-sheath cells (Hylton et al., 1988; Moore et al., 1988; Rawsthorne et al., 1988). Glycine decarboxylase is a mitochondrial enzyme composed of four subunits (P, H, T, and L), that together with serine hydroxymethyltransferase, is responsible for the condensation of two glycine molecules and concomitant release of CO_2 and NH_3 during photorespiratory metabolism (Oliver et al., 1990). Morgen et al. (1993) have demonstrated that the P subunit, which is responsible for decarboxylation, is absent from the mesophyll cells, but abundant in the bundle-sheath cells, of all C_3–C_4 intermediates so far examined. (In C_3–C_4 species from *Flaveria* and *Panicum*, the other three subunits were also missing from mesophyll cells, but they were present in *Moricandia arvensis*.) When taken together, modifications

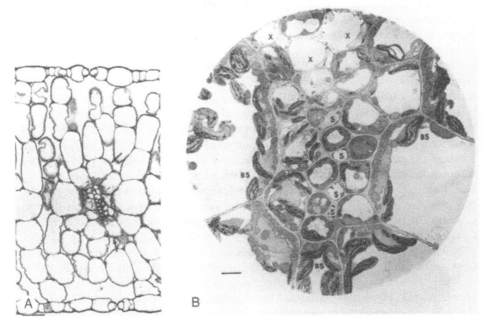

Figure 9 Photomicrographs of the anatomical and ultrastructural features of *Moricandia arvensis* (C_3–C_4) leaves. (A) Transverse section showing centripetal organization of organelles in bundle-sheath cells. As indicated by the unlabeled arrows, however, some chloroplasts are found on the outer bundle-sheath walls, especially in those areas bounded by intercellular air spaces. Bar = 52.63 μm magnified 190×. (B) Electron micrograph of vascular bundle and surrounding bundle-sheath cells. Note the extreme centripetal arrangement of mitochondria along the innermost tangential wall of the bundle sheath. BS, bundle-sheath cell; X, xylem element; S, sieve element. Bar = 4.54 μm magnified 2,200×. (C) Electron micrograph showing details of bundle-sheath cell ultrastructure. Bar = 2.33 μm magnified 4,300×. (D) Close-up view of the intimate relationship between chloroplasts and mitochondria in bundle-sheath cells. The unlabeled arrows are pointing to phytoferritin particles, a characteristic that is relatively rare in mature chloroplasts, and may have resulted from growth with NH_4^+ as the only nitrogen source. Bar = 0.38 μm magnified 26,400×. The inset shows phytoferritin particles at higher magnification. Bar = 0.10 μm magnified 100,000×. ([From: Winter, K., Usuda, H., Tsuzuki, M., Schmitt, M., Edwards, G. E., Thomas, R. J., and Evert, R. F. (1982). Influence of nitrate and ammonia on photosynthetic characteristics and leaf anatomy of *Moricandia arvensis*. *Plant Physiol.* **70**, 616–625, with permission].

to bundle-sheath anatomy and expression of glycine decarboxylase can be interpreted through a model that proposes the recycling of photorespired CO_2 by C_3–C_4 intermediate bundle sheath cells (Fig. 10). The processes described in this model bring a new metabolic role to the bundle-sheath tissue of C_3–C_4 intermediates, compared to C_3 plants, and as is discussed

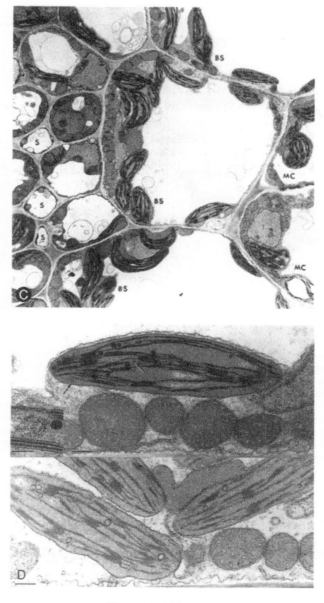

Figure 9—*Continued*

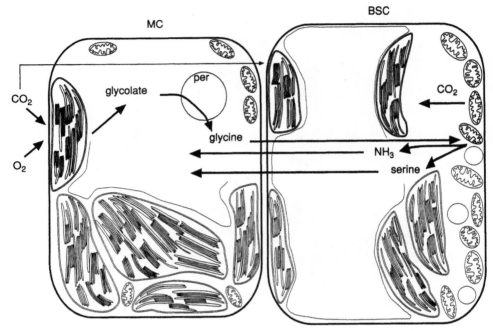

Figure 10 A model showing the recycling of CO_2 through the isolation of glycine decarboxylase to mitochondria in the bundle-sheath cells of C_3–C_4 intermediate species. Glycolate, produced through the oxidation of RuBP is transformed to glycine in mesophyll peroxisomes, and transported via plasmodesmata to the bundle-sheath cell, where it is decarboxylated by mitochondria localized along the innermost cell wall. The released CO_2 is then assimilated by chloroplasts in the bundle-sheath cell before escaping to the mesophyll cell and atmosphere. This results in a reduced photorespiration rate. NH_3 and serine would be assimilated in both cell types, but because of the high activity of glycine decarboxylase in bundle-sheath cells, a considerable fraction of these metabolites would have to be transported back to mesophyll cells. MC, mesophyll cell; BSC, bundle-sheath cell; per, peroxisome.

later in this chapter, may have set the stage for the evolution of fully expressed C_4 photosynthesis.

D. C_4 Metabolism in C_3–C_4 Intermediates of the Genus *Flaveria*

In contrast to other C_3–C_4 species, intermediates in the genus *Flaveria* exhibit C_4 biochemistry, with assimilation of atmospheric CO_2 through the C_4 cycle ranging from 20%–60% (Bassüner *et al.*, 1984; Rumpho *et al.*, 1984; Monson *et al.*, 1986; Moore *et al.*, 1987; Monson, 1996). Despite significant C_4-cycle activity in these species, and evidence that inorganic carbon can be concentrated (Moore *et al.*, 1987), measurements of the quantum yield for CO_2 uptake and $^{13}C/^{12}C$ discrimination patterns indicate

11. The Origins of C_4 Genes and Evolutionary Pattern in the C_4 Metabolic Phenotype

that any CO_2-concentrating activity is ineffective at reducing oxygenase activity and enhancing the catalytic velocity of RuBP carboxylase (Monson et al., 1986, 1988; von Caemmerer and Hubick, 1989). Unlike C_4 plants, the differential distribution of Rubisco and PEP carboxylase between mesophyll and bundle-sheath cells is not complete in the C_3–C_4 intermediate *Flaveria* species examined to date (Bauwe, 1984; Reed and Chollet, 1985). This would result in inefficient coordination of C_3 and C_4 biochemistry, and concomitant futile cycling of CO_2 between carboxylation and decarboxylation events (Monson et al., 1986, 1988).

The lower quantum yield in some C_3–C_4 species at 2% and 21% O_2 compared to C_3 species (Fig. 11) has been attributed to inefficient coupling of the C_3 and C_4 cycles (Monson et al. 1986). The quantum yield for CO_2 uptake is the molar amount of CO_2 assimilated per mole of photosynthetically active photons absorbed by the leaf. The quantum yield is typically measured as the linear slope of the relationship between CO_2 assimilation

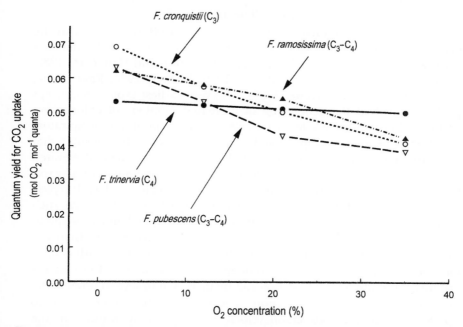

Figure 11 Dependence of the quantum yield for CO_2 uptake in *Flaveria* species with different photosynthetic pathway characteristics on atmospheric O_2 concentration. Note the reduced quantum yield values for the C_3–C_4 species *F. pubescens* at 21% O_2. This is presumably indicative of inefficient coordination of the C_3 and C_4 cycles in this species. [From Monson, R. K., Moore, B. D., Ku, M. S. B., and Edwards, G. E. (1986). Co-function of C_3 and C_4 photosynthetic pathways in C_3, C_4, and, C_3–C_4 intermediate *Flaveria* species. *Planta* **168**, 493–502.]

rate and absorbed photon flux at low photon flux densities. At low photon flux densities, CO_2 assimilation rate is limited by the rate of ATP and NADPH production by the electron transport system of the chloroplast, and thus the quantum yield reflects the energetic cost of assimilating CO_2. Because of the inefficient coupling of the C_3 and C_4 cycles, an ATP cost is imposed for operation of the C_4 cycle, but there is no reduction in the oxygenase activity of Rubisco, and no concomitant reduction in the ATP cost of photorespiration. This would result in a lower quantum yield in C_3–C_4 intermediate species. In the C_3–C_4 intermediate *F. ramosissima*, however, the quantum yield is lower than the C_3 species *F. cronquistii* when measured at 2% O_2, but higher when measured at 21% O_2. Apparently in *F. ramosissima*, there is some reduction in oxygenase activity due an effective CO_2-concentrating function. A reduction in the ATP cost of photorespiration presumably compensates for the extra ATP cost of the C_4 cycle in this species, and provides for a higher quantum yield in 21% O_2.

E. Speculation on an Evolutionary Scenario as Evidenced in C_3–C_4 Intermediate Species

A number of evolutionary scenarios have been offered to explain the origins of C_4 photosynthesis based on the nature of C_3–C_4 intermediates (Monson *et al.*, 1984; Edwards and Ku, 1987; Brown and Hattersley, 1989; Monson, 1989a; Monson and Moore, 1989; Peisker, 1986; Rawsthorne, 1992; Rawsthorne and Bauwe, 1997). The very earliest evolutionary steps probably involved the acquisition of increased numbers of chloroplasts and mitochondria in bundle-sheath cells, and division of photorespiratory metabolism between mesophyll and bundle-sheath cells. It is possible that these modifications resulted in a cellular microenvironment of elevated CO_2 concentration in the bundle-sheath cells. Past modeling studies by von Caemmerer (1989) have shown that by (1) allocating glycine decarboxylase activity and a small fraction of the leaf's Rubisco to the bundle sheath, such that (2) the rate of glycine production in the mesophyll and subsequent decarboxylation in the bundle sheath occur at a greater rate than the assimilation of CO_2 in the bundle sheath, and if (3) bundle-sheath chloroplasts are located in the near vicinity of mitochondria, then (4) CO_2 concentrations at the site of carboxylation can be increased to several times ambient. This would allow the fraction of Rubisco in the bundle sheath to operate at higher carboxylation velocities, compared to Rubisco in the mesophyll. As long as the fraction of the leaf's Rubisco that is allocated to the bundle sheath is relatively small, the overall rate of CO_2 assimilation for the leaf can be increased (von Caemmerer, 1989), along with photosynthetic water- and nitrogen-use efficiencies (Schuster and Monson, 1990). These modifications would provide the greatest benefit to plants growing in warm environments, where rates of photorespiratory glycine synthesis are relatively high (Fig. 12).

The evolution of C_4-like biochemistry is more difficult to explain. It is possible that the key to understanding the evolution of C_4 biochemistry lies in the elevated activity of PEPc (Monson and Moore, 1989). One of the more consistent biochemical patterns among C_3–C_4 intermediate species is a slightly elevated activity of PEPc compared to C_3 species (Kestler *et al.*, 1975; Sayre *et al.*, 1979; Holaday and Black, 1981; Holaday *et al.*, 1981; Ku *et al.*, 1983; Nakamoto *et al.*, 1983; Bauwe and Chollet, 1986; Hattersley and Stone, 1986; Monson and Moore, 1989). If PEP carboxylase activity increased to fulfill some role in C_3–C_4 intermediate photosynthesis, then it might have triggered evolutionary up-regulation in the expression of other C_4-cycle enzymes, such as pyruvate, Pi dikinase, NADP malate dehydrogenase, and NADP malic enzyme. At some critical level of PEP carboxylase activity, these enzymes would have been required to increase the supply of PEP substrate and metabolize the C_4 products of PEP carboxylase. Once organized, C_4 metabolism would have provided the potential to concentrate inorganic carbon at the chloroplastic site of RuBP carboxylation, depending on the relative rates of carboxylation and decarboxylation, and the physical-phase resistance to the diffusion of CO_2 away from the site of carboxylation.

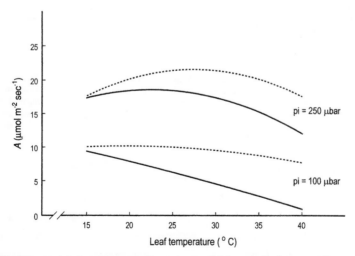

Figure 12 The modeled temperature dependence of CO_2 assimilation rate (A) as a function of leaf temperature in a C_3 species (solid line) and a C_3–C_4 intermediate species (broken line) at two different intercellular CO_2 partial pressures (pi). The model of von Caemmerer (1989) was used for the simulation. Note the higher values for A at higher leaf temperatures. [Redrawn from Schuster, W. S., and Monson, R. K. (1990). An examination of the advantages of C3–C4 intermediate photosynthesis in warm environments. *Plant Cell Environ.* **13,** 903–912.]

A CO_2-concentrating mechanism would have provided the basis for selection for even further increases in photosynthetic water- and nitrogen-use efficiencies, which may have taken over as the principal evolutionary driving forces (Monson, 1989a). A model describing these patterns is presented in Fig. 13.

Two barriers would have to be broached in any explanation of C_4 evolution according to the scenario presented. First, it is not clear what would drive mesophyll PEP carboxylase activities to increase beyond what is found in C_3 species. Second, it is not clear what would drive increases in bundle-sheath RuBP carboxylase activities and decreases in mesophyll RuBP carboxylase activities beyond those explained by the advantages of assimilating CO_2 released during bundle sheath glycine–decarboxylation. In fact, biochemical models have revealed that the advantages of the C_3–C_4 intermediate glycine decarboxylase system are best expressed when RuBP carboxylase

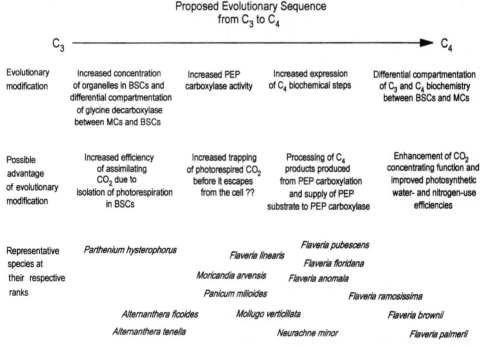

Figure 13 A proposed evolutionary sequence for the progression from fully expressed C_3 to fully expressed C_4 photosynthesis. The model has been constructed on the basis of the distribution of traits in C_3–C_4 intermediate species. [Redrawn from Monson, R. K. (1989a). On the evolutionary pathways resulting in C_4 photosynthesis and Crassulacean acid metabolism. *Adv. Ecol. Res.* **19,** 57–110.]

activity in the bundle sheath is limited to 15% or less of the leaf's total activity (von Caemmerer, 1989).

IV. The Evolution of C_4 Metabolism in Relation to Ecological Factors and Plant Growth Form

A. Why Is C_4 Photosynthesis so Common in Warm Environments?

The more abundant distribution of C_4 species in warm climates has been explained by the observation that photosynthetic quantum yields decrease at warm temperatures in C_3 plants (at least when determined on the basis of CO_2 assimilation), but are not affected in C_4 plants (Ehleringer, 1978). The capacity for C_4 plants to maintain high quantum yields as temperature increases is due to their elimination of photorespiration, a process with strong temperature dependence. Thus, C_4 plants have an important functional advantage in warm climates, but not necessarily in cool climates. In studying C_3–C_4 intermediate species, and the nature of the photorespiratory glycine shuttle, one gets a slightly different perspective. In order to capitalize on the differential compartmentation of glycine decarboxylase and its potential to generate elevated CO_2 concentrations in the bundle-sheath, C_3–C_4 intermediate species require relatively high rates of photorespiration. Without high photorespiration rates, there is little glycine to shuttle to the bundle sheath and little capacity to generate CO_2 through glycine decarboxylation. Thus, although the advantages of fully evolved C_4 photosynthesis obviously include reduced photorespiration rates, the early stages of C_4 evolution may have been driven by those factors that enhance photorespiration rates.

Using the model of von Caemmerer (1989) for C_3–C_4 intermediate species, it can be shown that increases in CO_2 assimilation rate, as well as photosynthetic water- and nitrogen-use efficiencies, are greatest at warm leaf temperatures (Fig. 12; Schuster and Monson, 1989). To date, these advantages have been easier to demonstrate with model simulations than actual measurements. When measured at normal ambient CO_2 concentrations and moderate leaf temperatures, the net CO_2 assimilation rates of C_3 species and C_3–C_4 intermediate congeners are strikingly similar (e.g., Ku et al., 1991). However, when measured at intercellular CO_2 concentrations below those characteristic of normal ambient conditions, C_3–C_4 intermdiates exhibit higher CO_2 assimilation rates and higher photosynthetic water-use efficiencies (Brown and Simmons, 1979; Hattersley et al., 1986; Monson, 1989b). In addition, measurements made at leaf temperatures of 35°C or higher reveal higher rates of photosynthesis per unit of Rubisco activity (an indication of higher photosynthetic nitrogen-use efficiency; Schuster and Monson, 1990), and higher rates of photosynthesis in field-grown plants

(Monson and Jaeger, 1991) for the C_3–C_4 intermediate, *Flaveria floridana*, when compared to C_3 species of similar growth form and ecological habit.

One intriguing analysis that provides some insight into the issue of why C_4 species are most frequently distributed in warm climates, involves consideration of intraspecific variation in the CO_2 compensation point of *Flaveria linearis*, a C_3–C_4 intermediate. Teese (1995) grew *F. linearis* plants with different CO_2 compensation points (reflecting different rates of photorespiratory CO_2 loss) at 25°C and 40°C. At the cooler temperature there was no relationship between growth rate and CO_2 compensation point. However, at the warmer temperature, there was a significant relationship, with those plants exhibiting the lowest CO_2 compensation points (i.e., the lowest rates of photorespiratory CO_2 loss) experiencing the highest growth rates. These results provide a basis for selection based on the efficacy of recycling photorespired CO_2 at high leaf temperatures.

B. Why Is C_4 Photosynthesis so Rare in Trees?

One of the more obvious patterns with respect to the distribution of C_4 photosynthesis is its predominance in herbaceous species and near absence in trees (Ehleringer and Monson, 1993). In the single case in which C_4 trees have been identified (in the genus *Euphorbia*), it is probable that the tree growth habit is derived from C_4 herbaceous ancestors (Pearcy and Troughton, 1975). (C_4 photosynthesis is also present in large members of the Chenopodiaceae from arid and semiarid habitats, and which exhibit some tendency toward woodiness. However, these species are probably best classified as large shrubs rather than trees.) There is no evidence for C_4 photosynthesis evolving *de novo* in a group possessing the tree growth habit. This is a curious observation because there is no obvious reason why the benefits of C_4 photosynthesis would not apply to woody species in the same way that they do for herbaceous species. Likewise, there appear to be no obvious functional disadvantages to possessing the C_4 pathway in those C_4 trees that have been identified (Pearcy and Calkin, 1983).

It is possible that, with respect to the tree growth form, there are "adaptive troughs" during the evolution of C_4 photosynthesis that are not reflected in the final C_4 product. That is, there may be evolutionary phases in which the C_3–C_4 intermediate state is physiologically maladaptive within the tree growth form. As discussed previously, there is evidence from C_3–C_4 *Flaveria* species that intermediate stages of C_4 evolution can be characterized by incomplete compartmentation of the C_3 and C_4 cycles, and concomitant reductions in the quantum yield for CO_2 uptake. Monson (1989a) suggested that this stage of intermediacy represents an adaptive trough during which carbon assimilation rate might be particularly sensitive to low-light environments. Thus, in shaded habitats, such as often exists on the floor of forests, maintenance of a positive carbon balance might be more difficult for early stage biochemical intermediates, compared to habitats with higher incident light intensities.

The minimum quantum yield required to support a positive whole-plant carbon balance in low-light environments should be sensitive to carbon allocation patterns, especially the fractional allocation to leaves that provide a compounded return in the sense of future carbon gains. Thus, the herbaceous growth habit, with its inherent capacity to allocate a greater fraction of assimilated carbon to the production of new leaf area, may be at a distinct advantage in terms of tolerating reduced quantum yields during the intermediary stages of C_4 evolution, and maintaining positive whole-plant carbon balance. Woody species, with their inherited tendency to allocate a greater fraction of carbon to stem growth, may be at a disadvantage in this respect.

In addition to possible barriers posed by quantum-yield constraints, measurements have shown that the metabolic design of NADPme C_4 species compromises their capacity to effectively use short-duration lightflecks, an important energy resource in the dynamic light environment of forest floor habitats (Krall and Pearcy, 1993). In response to a lightfleck, the generation of NADPH in bundle-sheath cells occurs predominantly through the oxidative decarboxylation of malate (NADPme-type C_4 plants typically have low electron transport capacity in the bundle sheath; see Chapter 3). However, the decarboxylation of malate (which frees one molecule of CO_2 to be assimilated by Rubisco) only produces two NADPH molecules. Four NADPH molecules are required to reduce the two PGA molecules produced by the assimilation of one molecule of CO_2. Thus, if all of the bundle-sheath reductant were provided by the decarboxylation of malate, only half of the released CO_2 could be assimilated.

C. Why Is C_4 Photosynthesis so Rare in Dicots?

C_4 photosynthesis has been found in only 16 of approximately 400 dicot families, and within each family it is generally represented in only a few genera (an exception is noted for the Carophyllales, where C_4 photosynthesis is common in some genera) (Chapter 16). In terms of community dominance, C_4 dicots are generally less important than C_4 monocots in those ecosystems where they occur. Ehleringer *et al.* (1997) proposed a model to explain these patterns on the basis of differences in the quantum yield for CO_2 uptake. As discussed previously, the quantum yield for CO_2 uptake is higher for C_3 leaves at moderate to low temperatures, compared to C_4 leaves (Fig. 14). This is because photorespiratory demands for ATP are low at cool temperatures, and the ATP cost of assimilating CO_2 approaches the theoretical maximum of three. At higher temperatures the quantum yield for CO_2 uptake decreases in C_3 leaves (due to the increasing photorespiratory demand for ATP), and eventually drops below that for

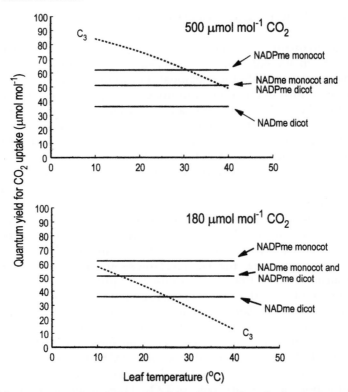

Figure 14 The relationship of quantum yield for CO_2 uptake to leaf temperature in C_3 and C_4 plants. Note the higher quantum yield for C_3 leaves at cooler temperatures, but the higher quantum yield for C_4 leaves at warmer temperatures. The crossover temperature is dependent on which group of C_4 species is being compared to C_3 species. C_4 NADP malic enzyme (NADPme) monocots have higher quantum yields than C_4 NADPme dicots or C_4 NADme dicots, resulting in a lower crossover temperature for the NADPme monocots when compared to C_3 species. Also note that at lower CO_2 concentrations the quantum yield for C_3 plants decreases at all temperatures, resulting in lower crossover temperatures and a lower temperature range where C_4 species have an advantage. The CO_2 concentrations of 500 and 180 μmol mol^{-1} were chosen because these are thought to have been representative of the late Miocene period, and glacial maxima during the Quaternary period, respectively [Adapted and redrawn from Ehleringer, J. R., and Pearcy, R. W. (1983). Variation in quantum yields for CO_2 uptake among C_3 and C_4 plants. *Plant Physiol.* **73,** 555-559, and Ehleringer, J. R., Cerling, T. E., and Helliker, B. R. (1997). C_4 photosynthesis, atmospheric CO_2 and climate. *Oecologia* **112,** 285-299.]

C_4 leaves. (Although the energetics are most easily studied by determining quantum yields at low light intensities, the energetic cost of photorespiration reduces the efficiency of using solar energy to assimilate CO_2 at all light intensities in C_3 compared C_4 leaves.) The "crossover temperature" where C_3 and C_4 leaves would be approximately equal in their energetic costs of assimilating CO_2, varies from 22–30°C (Ehleringer, 1978; Ehleringer and Pearcy, 1983) (Fig. 14). Above the crossover temperature C_4 leaves would be at an advantage, and below the crossover temperature C_3 leaves would be at an advantage. The crossover point occurs at lower temperatures when comparing C_3 monocots to C_4 monocots than when comparing C_3 dicots to C_4 dicots (Ehleringer and Pearcy, 1983). This is because C_4 monocots have higher quantum yields (irrespective of temperature), compared to C_4 dicots. The crossover point is also dependent on atmospheric CO_2 concentration, because at lower CO_2 concentrations the photorespiration rate of C_3 leaves increases, pushing the crossover point to a lower temperature.

On the basis of these differences in quantum yield, C_4 monocots are predicted to have exhibited an advantage over C_3 monocots or dicots in warm-temperature ecosystems during the late Miocene when CO_2 concentrations decreased below approximately 500 μmol mol^{-1} (Ehleringer et al., 1991). During the Miocene, grassland and savanna ecosystems exhibited global expansion, and C_4-like carbon isotope ratios appeared in soil pedogenic carbonates (Chapter 13). Also on the basis of differences in quantum yield, C_4 dicots are predicted to not have gained an advantage over C_3 monocots and dicots until CO_2 concentrations fell below about 250 μmol mol^{-1}, which occurred during glacial maxima in the Quaternary. The predicted late appearance of C_4 dicots has been used to explain their haphazard taxonomic distribution (Ehleringer et al., 1997).

V. Summary

Most evidence collected to date suggests that C_4 genes are homologous with ancestral genes found in C_3 plants, and that most of the modifications required to accommodate C_4 photosynthesis involve changes in gene expression patterns rather than changes to the transcribed regions. Of particular importance to the evolution of C_4-like expression patterns has been changes to *cis*-acting promoter domains, resulting in up-regulation or down-regulation of enzyme activity in response to certain spatial cues that define cells as being located in mesophyll tissue or bundle-sheath tissue. Gene duplication has also been an important process allowing the evolution of gene families, whereby one or more forms of the gene can be modified to take on a C_4 photosynthetic function, while other members of the family retain their original functions. This has been especially important in the evolution of C_4 PEP carboxylase, pyruvate, Pi dikinase, NADP malic enzyme, and cytosolic carbonic anhydrase. The differential C_4-like expression of Rubisco activity between mesophyll and bundle-sheath cells has evolved through the acquisition of complex interactions between a promoter sequence, a sequence within the 3' transcribed region of the gene, and posttranscriptional processes.

Studies of C_3–C_4 intermediate species have provided insight into possible evolutionary steps toward C_4 photosynthesis. In the species examined to date, the initial step appears to be the development of metabolically active bundle-sheath cells, particularly with respect to photorespiratory metabolism. Glycine decarboxylase, the mitochondrial enzyme responsible for the release of CO_2 during photorespiration, is localized to bundle-sheath cells. Models and measurements have shown that at high leaf temperatures or low CO_2 concentrations, when photorespiration rates are high, C_3–C_4 intermediate species exhibit higher CO_2 assimilation rates and higher water- and nitrogen-use efficiencies. C_4-like biochemistry is most highly expressed in the C_3–C_4 intermediate genus *Flaveria*. Studies of the *Flaveria* intermediates have revealed that the initial stages in evolving C_4-like biochemistry are characterized by poor integration between the C_3 and C_4 cycles, low quantum yields for CO_2 uptake, and possible decreased rates of CO_2 assimilation in shaded environments. These results may have significance to explaining the uncommon occurrence of C_4 photosynthesis in species exhibiting the tree growth habit. Differences in the quantum yield for CO_2 uptake among C_4 dicots and monocots, and differences in the temperature dependence of the quantum yield for CO_2 uptake between C_3 versus C_4 species, have been used to explain why C_4 plants tend to be most abundant in warm environments, and why the occurrence of C_4 photosynthesis is so much less frequent among diocots compared to monocots.

Acknowledgments

The author's research discussed in this chapter was supported by grants from the U.S. National Science Foundation. Many thanks are expressed to Professor Gerald Edwards for providing the photomicrographs of *Moricandia*.

References

Aoyagi, K., and Bassham, J. A. (1984). Pyruvate orthophosphate dikinase mRNA organ specificity in wheat and maize. *Plant Physiol.* **76**, 278–280.

Bansal, K. C., Viret, J.-F., Haley, J., Khan, B. M., Schantz, R., and Bogorad, L. (1992). Transient expression from cab-m1 and rbcS-m3 promoter sequences is different in mesophyll and bundle-sheath cells in maize leaves. *Proc. Natl. Acad. Sci. U.S.A.* **89**, 3654–3658.

Bassüner, B., Keerberg, O., Bauwe, H., Pyarnik, T., and Keerberg, H. (1984). Photosynthetic CO_2 metabolism in C_3–C_4 intermediate and C_4 species of *Flaveria* (Asteraceae). *Biochem. Physiol. Pflanzen* **172**, 547–552.

11. The Origins of C_4 Genes and Evolutionary Pattern in the C_4 Metabolic Phenotype

Bauwe, H. (1984). Photosynthetic enzyme activities and immunoflourescence studies on the localization of ribulose-1,5-bisphosphate carboxylase/oxygenase in leaves of C_3, C_4, and C_3–C_4 intermediate species of *Flaveria* (Asteraeae). *Biochem. Physiol. Pflanzen* **179**, 253–268.

Bauwe, H., and Chollet, R. (1986). Kinetic properties of phosphoenolpyruvate carboxylase from C_3, C_4, and C_3–C_4 intermediate species of *Flaveria* (Asteraceae). *Plant Physiol.* **82**, 695–699.

Boinski, J. J., Wang, J.-L., Xu, P., Hotchkiss, T., and Berry, J. O. (1993). Post-transcriptional control of cell type-specific gene expression in bundle sheath and mesophyll chloroplasts of *Amaranthus hypochondriacus*. *Plant Mol. Biol.* **22**, 397–410.

Börsch, D., and Westhoff, P. (1990). Primary structure of NADP-dependent malic enzyme in the dicotyledonous C_4 plant *Flaveria trinervia*. *FEBS Lett.* **273**, 111–115.

Brown, R. H., Bouton, J. H., Rigsby, L. L., and Rigler, M. (1983a). Photosynthesis of grass species differing in carbon dioxide fixation pathways. VIII. Ultrastructural characteristics of *Panicum* species in the *Laxa* group. *Plant Physiol.* **71**, 425–431.

Brown, R. H., and Hattersley, P. W. (1989). Leaf anatomy of C_3–C_4 species as related to evolution of C_4 photosynthesis. *Plant Physiol.* **91**, 1543–1550.

Brown, R. H., Rigsby, L. L., and Akin, D. E. (1983b). Enclosure of mitochondria by chloroplasts. *Plant Physiol.* **71**, 437–439.

Brown R. H., and Simmons, R. E. (1979). Photosynthesis of grass species differing in CO_2 fixation pathways. I. Water-use efficiency. *Crop Sci.* **19**, 375-379

Chollet, R., Vidal, J., and O'Leary, M. H. (1996) Phospho*enol*pyruvate carboxylase: A ubiquitous, highly regulated enzyme in plants. *Annu. Rev. Plant Physiol. Plant Mol. Biol.* **47**, 273–298.

Clark, A. G. (1994). Invasion and maintenance of a gene duplication. *Proc. Natl. Acad. Sci. U.S.A.* **91**, 2950–2954.

Cockburn, W. (1983). Stomatal mechanism as the basis of the evolution of CAM and C_4 photosynthesis. *Plant Cell Environ.* **6**, 275–279.

Dean, C., Pichersky, E., and Dunsmuir, P. (1989). Structure, evolution, and regulation of rbcS genes in higher plants. *Annu. Rev. Plant Physiol.* **40**, 415–439.

Dickinson, W. J. (1980). Evolution of patterns of gene expression in Hawaiian picture-winged Drosophilia. *J. Mol. Evol.* **16**, 73–94.

Dickinson, W. J. (1988). On the architecture of regulatory systems: Evolutionary insights and implications. *BioEssays* **8**, 204–208.

Dobson, D. E., Prager, E. M., and Wilson, A. C. (1984). Stomach lysozymes of ruminants. I. Distribution and catalytic properties. *J. Biol. Chem.* **259**, 11607–11616.

Edwards, G. E., Andreo, C. S. (1992). NADP-malic enzyme from plants. *Phytochemistry* **31**, 1845–1857.

Edwards, G. E., and Ku, M. S. B. (1987). Biochemistry of C_3–C_4 intermediates. In "The Biochemistry of Plants" Vol 10 (M. D. Hatch and N. K. Boardman, eds.), pp 275–325. Academic Press, New York,.

Ehleringer, J. R. (1978). Implications of quantum yield differences to the distributions of C_3 and C_4 grasses. *Oecologia* **31**, 255–267.

Ehleringer, J. R., Cerling, T. E., and Helliker, B. R. (1997). C_4 photosynthesis, atmospheric CO_2, and climate. *Oecologia* **112**, 285–299.

Ehleringer, J. R., and Monson, R. K. (1993). Evolutionary and ecological aspects of photosynthetic pathway variation. *Ann. Rev. Ecol. Syst.* **24**, 411–439.

Ehleringer, J. R., and Pearcy, R. W. (1983). Variation in quantum yields for CO_2 uptake among C_3 and C_4 plants. *Plant Physiol.* **73**, 555–559.

Ehleringer, J. R., Sage, R. F., Flanagan, L. B., and Pearcy, R. W. (1991). Climate change and the evolution of C_4 photosynthesis. *Trends Ecol. Evol.* **6**, 95–99.

Gilmartin, P. M., Sarokin, L., Memelink, J., Chua, N.-H. (1990). Molecular light switches for plant genes. *Plant Cell* **2**, 369–378.

Glackin, C. A., and Grula, J. W. (1990). Organ-specific transcripts of different size and abundance derive from the same pyruvate, orthophosphate dikinase gene in maize. *Proc. Natl. Acad. Sci. U.S.A.* **87**, 3004–3008.

Goldstein, D. J., Rogers, C., and Harris, H. (1982). Evolution of alkaline phosphatase in primates. *Proc. Natl. Acad. Sci. U.S.A.* **79**, 879–883.

Gutteridge, S., and Gatenby, A. A. (1995). Rubisco synthesis, assembly, and regulation. *Plant Cell* **7**, 809–819.

Gutteridge, S., Newman, J., Herrmann, C., and Rhoades, H. (1995). The crystal structures of Rubisco and opportunities for manipulating photosynthesis. *J. Exp. Bot.* **46**, 1261–1267

Hata, S., and Matsuoka, M. (1987). Immunological studies on pyruvate, orthophosphate dikinase in C_3 plants. *Plant Cell Physiol.* **28**, 635–641.

Hatch, M. D., and Burnell, J. N. (1990). Carbonic anhydrase activity in leaves and its role in the first step of C_4 photosynthesis. *Plant Physiol.* **93**, 825–828.

Hattersley, P., Stone, N. E. (1986). Photosynthetic enzyme activities in the C_3–C_4 intermediate *Neurachne minor* S. T. Blake (Poaceae). *Aust. J. Plant. Physiol.* **13**, 399–408.

Hattersley, P. W., Wong, S.-C., Perry, S., Roksandic, Z. (1986). Comparative ultrastructure and gas exchange characteristcs of the C_3–C_4 intermediate *Neurachne minor* S. T. Blake (Poaceae). *Plant Cell Environ.* **9**, 217–233.

Hermans, J., Westhoff, P. (1992). Homologous genes for the C_4 isoform of phosphoenolpyruvate carboxylase in a C_3 and a C_4 *Flaveria* species. *Mol. Gen. Genet.* **224**, 459–468.

Holaday, A. S., and Black, C. C. (1981). Comparative characterization of phosphoenolpyruvate carboxylase in C_3, C_4, and C_3–C_4 intermediate *Panicum* species. *Plant Physiol.* **67**, 330–334.

Holaday, A. S., Lee, K. W., and Chollet, R. (1984). C_3–C_4 intermediate species in the genus *Flaveria*: leaf anatomy, ultrastructure, and the effect of O_2 on the CO_2 compensation point. *Planta* **160**, 25–32.

Holaday, A. S., Shieh, Y. J., Lee, K. W., and Chollet, R. (1981). Anatomical, ultrastructural and enzyme studies of leaves of *Moricandia arvensis*, a C_3–C_4 intermediate species. *Biochem. Biophys. Acta* **637**, 334–341.

Hughes, A. L. (1994). The evolution of functionally novel proteins after gene duplication. *Proc. R. Soc. Lond.* B **256**, 119–124.

Hylton, C. M., Rawsthorne, S., Smith, A. M., Jones, D. A., Woolhouse, H.W. (1988) Glycine decarboxylase is confined to the bundle-sheath cells of leaves of C_3–C_4 intermediate species. *Planta* **175**, 452–459.

Imaizumi, N., Ishihara, K., Samejima, M., Kaneko, S., Matsuoka, M. (1992). Structure and expression of pyruvate, orthophosphate dikinase gene in C_3 plants. *In* "Research in Photosynthesis" (N. Murata, ed.), pp 875–878. Kluwer Academic, Dordrecht, The Netherlands.

Kawamura, T., Shigesada, K., Toh, H., Okumura, S., Yanagisawa, S., Izui, K. (1992). Molecular evolution of phosphoenolpyruvate carboxylase for C_4 photosynthesis in maize: Comparison of its cDNA sequence with a newly isolated cDNA encoding an isozyme involved in the anaplerotic function. *J. Biochem.* **112**, 147–154.

Kestler, D. P., Mayne, B. C., Ray, T. B., Goldstein, L. D., Brown, R. H., and Black, C. C. (1975). Biochemical components of the photosynthetic CO_2 compensation point of higher plants. *Biochem. Biophys. Res. Comm.* **66**, 1439–1446.

Kettler, M. K., Ghent, A. W., and Whitt, G. S. (1986). A comparison of phylogenies based on structural and tissue-expressional differences of enzymes in a family of teleost fishes (Salmoniformes: Umbridae). *Mol. Biol. Evol.* **3**, 485–498.

Kopriva, S., Chu, C.-C., Bauwe, H. (1996). Molecular phylogeny of *Flaveria* as deduced from the analysis of nucleotide sequences encoding the H-protein of the glycine cleavage system. *Plant Cell Environ.* **19**, 1028–1036.

Krall, J. P., Pearcy, R. W. (1993). Concurrent measurements of oxygen and carbon dioxide exchange during lightflecks in maize (*Zea mays* L.). *Plant Physiol.* **103**, 823–828.

Ku, M. S. B., Kano-Murakami, Y., Matsuoka, M. (1996). Evolution and expression of C_4 photosynthesis genes. *Plant Physiol.* **111**, 949–957.

Ku, M. S. B., Monson, R. K., Littlejohn, R. O., Nakamoto, H., Fisher, D. B., Edwards, G. E. (1983). Photosynthetic characteristics of C_3–C_4 intermediate *Flaveria* species. I. Leaf anatomy, photosynthetic responses to CO_2 and O_2, and activities of key enzymes in the C_3 and C_4 pathways. *Plant Physiol.* **71**, 944–948.

Ku, M. S. B., Wu, J. R., Dai, Z. Y., Scott, R. A., Chu, C., Edwards, G. E. (1991). Photosynthetic and photorespiratory characteristics of *Flaveria* species. *Plant Physiol.* **96**, 518–528.

Laetsch, W. M. (1971). Chloroplast structural relationships in leaves of C_4 plants. In "Photosynthesis and Photorespiration" (M. D. Hatch, C. B. Osmond, and R. O. Slatyer, eds.), pp. 323–352. Wiley-Interscience, New York.

Li, W.-H. (1983). Gene duplication. In "Evolution of Genes and Proteins" (M. Nei and R. K. Koehn, eds.), pp. 14–37. Sinauer Associates, Sunderland, Massachusetts.

Lipka, B., Steinmüller, K., Rosche, E., Börsch, D., Westhoff, P. (1994). The C_3 plant *Flaveria pringlei* contains a plastidic NADP-malic enzyme which is orthologous to the C_4 isoform of the C_4 plant *F. trinervia*. *Plant Mol. Biol.* **26**, 1775–1783.

Luchetta, P., Crétin, C., and Gadal, P. (1991). Organization and expression of the two homologous genes encoding the NADP-malate dehydrogenase in *Sorghum vulgare* leaves. *Mol. Gen. Genet.* **228**, 473–481.

Ludwig, M., and Burnell, J. N. (1995). Molecular comparison of carbonic anhydrase from *Flaveria* species demonstrating different photosynthetic pathways. *Plant Mol. Biol.* **29**, 353–365.

Marshall, J. S., Stubbs, J. D., and Taylor, W. C. (1996). Two genes encode highly similar chloroplastic NADP-malic enzymes in *Flaveria*. Implications for the evolution of C_4 photosynthesis. *Plant Physiol.* **111**, 1251–1261.

Martineau, B., Smith, H. J., Dean, C., Dunsmuir, P., Bedbrook, J., and Mets, L. J. (1989). Expression of a C_3 plant rubisco SSU gene in regenerated C_4 *Flaveria* plants. *Plant Mol. Biol.* **13**, 419–426.

Matsuoka, M. (1995). The gene for pyruvate, orthophosphate dikinase in C_4 plants: Structure, regulation and evolution. *Plant Cell Physiol.* **36**, 937–943.

Matsuoka, M., Kyozuka, J., Shimamoto, K., and Kano-Murakami, Y. (1994). The promoters of two carboxylases in a C_4 plant (maize) direct cell-specific, light-regulated expression in a C_3 plant (rice). *Plant J.* **6**, 311–319.

Matsuoka, M., and Numazawa, T. (1991). Cis-acting elements in the pyruvate, orthophosphate dikinase gene in maize. *Mol. Gen. Genet.* **228**, 143–152.

Matsuoka, M., Tada, Y., Fujimura, T., and Kano-Murakami, Y. (1993). Tissue-specific light-regulated expression directed by the promoter of a C_4 gene, maize pyruvate, orthophosphate dikinase, in a C_3 plant, rice. *Proc. Natl. Acad. Sci. U.S.A.* **90**, 9586–9590.

Matsuoka, M., and Yamamoto, N. (1989). Induction of mRNAs for phosphoenolpyruvate carboxylase and pyruvate, orthophosphate dikinase in leaves of a C_3 plant exposed to light. *Plant Cell Physiol.* **30**, 479–486.

Mayfield, S. P., Yohn, C. B., Cohen, A., and Danon, A. (1995) Regulation of chloroplast gene expression. *Annu. Rev. Plant Physiol. Plant Mol. Biol.* **46**, 147–166.

McGonigle, B., and Nelson, T. (1995. C_4 isoform of NADP-malate dehydrogenase: cDNA cloning and expression in leaves of C_4, C_3, and C_3–C_4 intermediate species of *Flaveria*. *Plant Physiol.* **108**, 1119–1126.

Monson, R.K. (1989a). On the evolutionary pathways resulting in C_4 photosynthesis and Crassulacean acid metabolism. *Adv. Ecol. Res.* **19**, 57–110.

Monson, R. K. (1989b). The relative contributions of reduced photorespiration, and improved water- and nitrogen-use efficiencies, to the advantages of C_3–C_4 intermediate photosynthesis in *Flaveria*. *Oecologia* **80**, 215–221.

Monson, R. K. (1996). The use of phylogenetic perspective in comparative plant physiology and developmental biology. *Ann. Missouri Bot. Gard.* **83,** 3–16.

Monson, R. K., Edwards, G. E., and Ku, M. S. B. (1984). C_3–C_4 intermediate photosynthesis in plants. *BioScience* **34,** 563–574.

Monson, R. K., and Jaeger, C. H. (1991). Photosynthetic characteristics of the C_3–C_4 intermediate *Flaveria floridana* (Asteraceae) in natural habitats: Evidence of advantages to C_3–C_4 photosynthesis at high leaf temperatures. *Am. J. Bot.* **78,** 795–800.

Monson, R. K., and Moore, B. D. (1989). On the significance of C_3–C_4 intermediate photosynthesis to the evolution of C_4 photosynthesis. *Plant Cell Environ.* **12,** 689–699.

Monson, R. K., Moore, B. D., Ku, M. S. B., and Edwards, G .E. (1986). Co-function of C_3 and C_4 photosynthetic pathways in C_3, C_4, and C_3–C_4 intermediate *Flaveria* species. *Planta* **168,** 493–502.

Monson, R. K., Teeri, J. A., Ku, M. S. B., Gurevitch, J., Mets, L., and Dudley, S. (1988). Carbon-isotope discrimination by leaves of *Flaveria* species exhibiting different amounts of C_3- and C_4-cycle co-function. *Planta* **174,** 145–151.

Moore, B. D., Ku, M. S. B., and Edwards, G. E. (1987). C_4 photosynthesis and light dependent accumulation of inorganic carbon in leaves of C_3–C_4 and C_4 *Flaveria* species. *Aust. J. Plant Physiol.* **14,** 657–688.

Moore, B. D., Monson, R. K., Ku, M. S. B., and Edwards, G. E. (1988). Activities of principal photosynthetic and photorespiratory enzymes in leaf mesophyll and bundle sheath protoplasts from the C_3–C_4 intermediate *Flaveria ramosissima*. *Plant Cell Physiol.* **29,** 999–1006.

Morgen, C. L., Turner, S. R., and Rawsthorne, S. (1993). Coordination of the cell-specific distribution of the four subunits of glycine decarboxylase and of serine hydroxymethyl transferase in leaves of C_3–C_4 intermediate species from different genera. *Planta* **190,** 468–473.

Nakamoto, H., Ku, M. S. B., and Edwards, G. E. (1983). Photosynthetic characteristics of C_3–C_4 intermediate *Flaveria* species. II. Kinetic properties of phosphoenolpyruvate carboxylase from C_3, C_4 and C_3–C_4 intermediate species. *Plant Cell Physiol.* **24,** 1387–1393.

O'Leary, M. H. (1982). Phosphoenolpyruvate carboxylase: an enzymologist's view. *Annu. Rev. Plant Physiol.* **33,** 297–315.

Ohno, S. (1973). Ancient linkage groups and frozen accidents. *Nature* **244,** 259–262.

Oliver, D. J., Neuberger, M., Bourguignon, J., and Douce, R. (1990). Interaction between the component enzymes of the glycine decarboxylase multienzyme complex. *Plant Physiol.* **94,** 833–839.

Pearcy, R. W., and Calkin, H. W. (1983). Carbon dioxide exchange of C_3 and C_4 tree species in the understory of a Hawaiian forest. *Oecologia* **58,** 26–32.

Pearcy, R. W., and Troughton, J. (1975). C_4 photosynthesis in tree form *Euphorbia* species from Hawaiian rainforest sites. *Plant Physiol.* **55,** 1054–1056.

Peisker, M. (1986). Models of carbon metabolism in C_3–C_4 intermediate plants as applied to the evolution of C_4 photosynthesis. *Plant Cell Environ.* **9,** 627–635.

Poetsch, W., Hermans, J., and Westhoff, P. (1991). Multiple cDNAs of phosphoenolpyruvate carboxylase in the C_4 dicot *Flaveria trinervia*. *FEBS Lett.* **292,** 133–136.

Powell, A. M. (1978). Systematics of *Flaveria* (Flaverinae: Asteraceae). *Ann. Missouri Bot. Gard.* **65,** 590–636.

Rajaveen, M. S., Bassett, C. L., and Hughes, D. W. (1991). Isolation and characterization of cDNA clones for NADP-malic enzyme from leaves of *Flaveria*: Transcript abundance distinguishes C_3, C_3–C_4 and C_4 photosynthetic types. *Plant Mol. Biol.* **17,** 371–383.

Rawsthorne, S. (1992). C_3–C_4 intermediate photosynthesis: Linking physiology to gene expression. *Plant J.* **2,** 267–274.

Rawsthorne, S., and Bauwe, H. (1997). C_3–C_4 photosynthesis. *In* "Photosynthesis, A Comprehensive Treatise" (A. S. Raghavendra, ed), Cambridge University Press, United Kingdom (in press).

Rawsthorne, S., Hylton, C. M., Smith, A. M., and Woolhouse, H. W. (1988). Distribution of photorespiratory enzymes between bundle-sheath and mesophyll cells in leaves of the C_3–C_4 intermediate species *Moricandia arvensis* (L.) DC. *Planta* **176,** 527–532.

Reed, J. E., and Chollet, R. (1985). Immunofluorescent localization of phosphoenolpyruvate carboxylase and ribulose 1,5-bisphosphate carboxylase/oxygenase proteins in leaves of C_3, C_4, and C_3–C_4 *Flaveria* species. *Planta* **165,** 439–445.

Reed, M. L., and Graham, D. (1981). Carbonic anhydrase in plants: Distribution, properties and possible physiological roles. *In* "Progress in Phytochemistry" (L. Reinhold, J. B. Harborne, and T. Swain, eds), pp. 47–94. Pergamon Press, Oxford.

Rosche, E., Streubel, M., and Westhoff, P. (1994). Primary structure of the photosynthetic pyruvate, orthophosphate dikinase of the C_3 plant *Flaveria pringlei* and expression analysis of pyruvate orthophosphate dikinase sequences in C_3, C_3–C_4, and C_4 *Flaveria* species. *Plant Mol. Biol.* **26,** 763–769.

Rosche, E., and Westhoff, P. (1995). Genomic structure and expression of the pyruvate, orthophosphate dikinase gene of the dicotyledonous C_4 plant *Flaveria trinervia* (Asterceae). *Plant Mol. Biol.* **29,** 663–678.

Roth, R., Hall, L. N., Brutnell, T. P., and Langdale, J. A. (1996). *bundle sheath defective 2*, a mutation that disrupts the coordinated development of bundle sheath and mesophyll cells in the maize leaf. *Plant Cell* **8,** 915–927.

Rumpho, M., Ku, M. S. B., Cheng, S.-H., and Edwards, G. E. (1984). Photosynthetic characteristics of C_3–C_4 intermediate *Flaveria* species. III. Reduction of photorespiration by a limited C_4 pathway of photosynthesis in *Flaveria ramosissima*. *Plant Physiol.* **75,** 993–996.

Sayre, R. T., Kennedy, R. A., and Prignitz, D. J. (1979). Photosynthetic enzyme activities and localization in *Mollugo verticillata* populations differing in the levels of C_3 and C_4 cycle operation. *Plant Physiol.* **64,** 293–299.

Schäffner, A. R., and Sheen, J. (1991). Maize rbcS promoter activity depends on sequence elements not found in dicot rbcS promoters. *Plant Cell* **3,** 997–1012.

Schäffner, A. R., and Sheen, J. (1992). Maize C_4 photosynthesis involves differential regulation of phosphoenolpyruvate carboxylase genes. *Plant J.* **2,** 221–232.

Scheibe, R. (1990). Light–dark modulation regulation of chloroplast metabolism in a new light. *Bot. Acta* **103,** 327–334.

Schuster, W. S., and Monson, R. K. (1990). An examination of the advantages of C_3–C_4 intermediate photosynthesis in warm environments. *Plant Cell Environ.* **13,** 903–912.

Seemann, J. R., Badger, M. R., and Berry, J. A. (1984). Variations in the specific activity of ribulose-1,5-bisphosphate carboxylase between species utilizing differing photosynthetic pathways. *Plant Physiol.* **74,** 791–794.

Sheen, J. (1991). Molecular mechanisms underlying the differential expression of maize pyruvate, orthophosphate dikinase genes. *Plant Cell* **3,** 225–245.

Teese, P. (1995). Intraspecific variation for CO_2 compensation point and differential growth among variants in a C_3–C_4 intermediate plant. *Oecologia* **102,** 371–376.

Thorpe, P. A., Loye, J., Rote, C. A., and Dickinson, W. J. (1993). Evolution of regulatory genes and patterns: Relationships to evolutionary rates and to metabolic functions. *J. Mol. Evol.* **37,** 590–599.

Toh, H., Kawamura, T., and Izui, K. (1994). Molecular evolution of phosphoenolpyruvate carboxylase. *Plant Cell Environ.* **17,** 31–43.

Ueda, T., Pichersky, E., Malik, V. S., and Cashmore, A. R. (1989). Level of expression of the tomato rbcS-3A gene is modulated by a far upstream promoter element in a developmentally regulated manner. *Plant Cell* **1,** 217–227.

Viret, J.-F., Mabrouk, Y., Bogorad, L. (1994). Transcriptional photoregulation of cell-type–preferred expression of maize rbcS-m3: 3′ and 5′ sequences are involved. *Proc. Natl. Acad. Sci. U.S.A.* **91,** 8577–8581.

von Caemmerer, S. (1989). A model of photosynthetic CO_2 assimilation and carbon isotope discrimination in leaves of certain C_3–C_4 intermediates. *Planta* **178,** 463–474.

von Caemmerer, S., and Hubick, K. T. (1989). Short-term carbon-isotope discrimination in C_3–C_4 intermediate species. *Planta* **178,** 475–481.

Wedding, R. (1989). Malic enzyme of higher plants: Characteristics, regulation, and physiological function. *Plant Physiol.* **90,** 367–371.

Winter, K., Usuda, H., Tsuzuki, M., Schmitt, M., Edwards, G. E., Thomas, R. J., and Evert, R. F. (1982). Influence of nitrate and ammonia on photosynthetic characteristics and leaf anatomy of *Moricandia arvensis*. *Plant Physiol.* **70,** 616–625.

Yeoh, H. H., Badger, M. R., and Watson, L. (1981) Variations in kinetic properties of ribulose-1,5-bisphosphate carboxylases among plants. *Plant Physiol.* **67,** 1151–1155.

12

Phylogenetic Aspects of the Evolution of C_4 Photosynthesis

Elizabeth A. Kellogg

I. Introduction

One of the great puzzles of C_4 photosynthesis is how it has evolved. The pathway is thought to be an adaptation and is clearly genetically and developmentally complex. It also appears in at least 18 families of flowering plants (Chapter 16). This creates a set of phylogenetic questions: Has the C_4 pathway originated independently in each plant family? Are there multiple origins even within one family? When did C_4 species evolve, and in what lineages of plants? The answers to these questions require a detailed description of the phylogenetic pattern. Fortunately, advances in molecular systematic techniques, coupled with burgeoning numbers of molecular phylogenies, permit unusually precise placement of C_4 species among their C_3 relatives.

The phylogenetic pattern provides clear direction to studies of the physiological and genetic basis of the pathway, by pointing to species outside the standard model systems that might profitably bear investigation. Physiological and genetic studies are best done pairwise, comparing a C_4 taxon with its C_3 sisters. The phylogeny also constrains hypotheses about the sorts of selective pressures that might have led to C_4 photosynthesis, by pointing to particular paleoenvironments in which the pathway must have arisen.

In this chapter, I outline what we know about relationships among C_4 taxa. As I show, the pathway has arisen in many unrelated lineages, and many of the origins must be quite recent. Some of the origins, notably those in the grass family, appear to be more ancient, suggesting that some C_4 lineages originated long before they became ecologically dominant.

II. Phylogenetic Pattern

A. Distribution across Flowering Plants

The C_4 pathway has evolved only in the flowering plants. A simplified and very conservative angiosperm phylogeny is shown in Fig. 1, based on 18S RNA sequences (Soltis *et al.*, 1997). This tree is similar to that generated using *rbcL* sequences (Chase *et al.*, 1993), although not identical. The differences between the trees, however, concern the relationships among the major clades, whereas many of the clades themselves appear to be robust. Inspection of this figure shows immediately that C_4 lineages occur throughout the angiosperms, in both monocots and dicots. Unlike nitrogen fixation, which has evolved repeatedly but within a clade of only 19 families (the "nitrogen fixing clade;" Soltis *et al.*, 1995), C_4 appears in unrelated groups, often in only a handful of species. At least at this level of analysis, there seems to be no particular pattern as to where the pathway has appeared.

In the following sections, I summarize current knowledge of the phylogeny of the various families with C_4 taxa. I illustrate the phylogenetic relationships using "supertrees" created by combining phylogenies from several closely related groups. For example, the phylogeny of the Caryophyllidae has been determined using sequences of ORF2280 in the chloroplast inverted repeat (Downie *et al.*, 1997), with particularly good sampling of the Chenopodiaceae and Amaranthaceae. In a related study of the internal transcribed spacer (ITS) region of the ribosomal RNA genes, Hershkovitz and Zimmer (1997) looked at an overlapping group of taxa, paying particular attention to the Portulaceae. Their phylogenies are grafted together in Fig. 6. The support for the phylogenies varies greatly. For the grass family, for example, the relationships shown in Fig. 11 are supported by eight different molecular data sets (Kellogg, 1997), and may be as close to reliable as anything is in phylogeny reconstruction. For the Heliantheae of the Asteraceae, however (Fig. 3), relationships are based on a single morphological cladogram (Karis and Ryding, 1994b), which has yet to be corroborated by other sources of data. Because many of the trees shown are based on results from independent studies, measures of support such as the bootstrap or decay index are not really meaningful. Nonetheless, it seemed helpful to indicate which branches were strongly supported in the individual analyses, so I have marked with an asterisk branches that received more than 90% bootstrap support in one or more of these analyses. Other relationships are simply the best current estimate, but must be regarded as preliminary. Future studies may or may not suggest other relationships. For each group, therefore, users of the phylogeny will need to evaluate how well it is corroborated by other data.

12. Phylogenetic Aspects of the Evolution of C_4 Photosynthesis   413

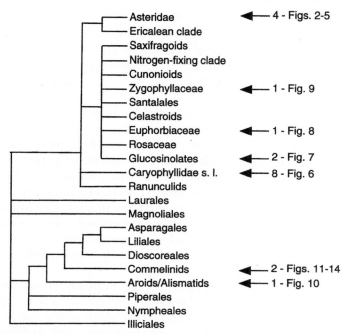

Figure 1 Phylogeny of the flowering plants, showing relationships among clades containing C_4 taxa. Terminal taxa include one (e.g., Zygophyllaceae) to multiple (e.g., Asteridae) families. Relationships among dicot clades based on *rbcL* sequences [Chase, M. W., Soltis, D. E., Olmstead, R. G., Morgan, D., Les, D. H., Mishler, B. D., Duvall, M. R., Price, R. A., Hills, H. G., Qiu, Y.-L., Kron, K. A., Rettig, J. H., Conti, E., Palmer, J. D., Manhart, J. R., Sytsma, K. J., Michaels, H. J., Kress, W. J., Karol, K. G., Clark, W. D., Hedrén, M., Gaut, B. S., Jansen, R. K., Kim, K.-J., Wimpee, C. F., Smith, J. F., Furnier, G. R., Strauss, S. H., Xiang, Q.-Y., Plunkett, G. M., Soltis, P. S., Swensen, S. M., Williams, S. E., Gadek, P. A., Quinn, C. J., Eguiarte, L. E., Golenberg, E., Learn, G. H., Jr., Graham, S. W., Barrett, S. C. H., Dayanandan, S., and Albert, V. A. (1993). Phylogenetics of seed plants: An analysis of nucleotide sequences from the plastid gene *rbcL*. *Ann. Missouri Bot. Gard.* **800,** 528–580.] and on 18S ribosomal RNA sequences [Soltis, D. E., Soltis, P. S., Nickrent, D. L., Johnson, L. A., Hahn, W. J., Hoot, S. B., Sweere, J. A., Kuzoff, R. K., Kron, K. A., Chase, M. W., Swensen, S. M., Zimmer, E. A., Chaw, S.-M., Gillespie, L. J., Kress, W. J., and Sytsma, K. J. (1997). Angiosperm phylogeny inferred from 18S ribosomal DNA sequences. *Ann. Miss. Bot. Gard.* **84,** 1–49.]. Monocot relationships based on combined data from *rbcL* and morphology [Chase, M. W., Stevenson, D. W., Wilkin, P., and Rudall, P. J. (1995). Monocot systematics: A combined analysis. *In* "Monocotyledons: Systematics and Evolution" (P. J. Rudall, P. J. Cribb, D. F. Cutler, and C. J. Humphries, eds.), pp. 685–730. Royal Botanic Gardens, Kew.]. Relationships shown are conservative and attempt to reflect only those groups for which there is some agreement. Not all clades are shown. Distribution of C_4 families indicated by arrows. Numbers indicate numbers of C_4 families in clade, and figures cited illustrate relationships within the clade.

B. Asteridae

The asterid group, as determined by studies of molecular systematics, corresponds roughly to the Sympetalae of Engler and Prantl. A supertree for the group is shown in Fig. 2. Unrelated C_4 genera appear in Acanthaceae, Scrophulariaceae, *Heliotropium* ("Boraginaceae" *sensu lato*), and Asteraceae.

The Acanthaceae is supported as monophyletic by *rbcL* sequences (Scotland *et al.*, 1995). *Blepharis*, the sole C_4 genus, has not been included in any molecular phylogenetic study to date, but based on morphological synapomorphies is placed in the tribe Acantheae (McDade, personal communication, 1997). In molecular phylogenies based on nuclear ITS and chloroplast gene spacers (McDade and Moody, 1997), Mendoncioideae and Thunbergioideae are sister to the core Acanthaceae (the latter including all taxa with four monothecous anthers; McDade, personal communication, 1997). The Acantheae lineage, which includes *Acanthus* and *Crossandra* and presumably *Blepharis* as well, is the next diverging branch, sister to all remaining acanths.

Boraginaceae is clearly polyphyletic (Ferguson, in press). *Heliotropium* is separate from the other subfamilies of Boraginaceae and sister to the genus *Tournefortia*. *Heliotropium* includes 260 species, of which only a few are C_4. The genus is divided into about 20 sections (Johnston, 1928; Al-Shehbaz, 1991), of which one, section *Orthostachys*, contains both C_3 and C_4 species. Based on phylogenies of chloroplast restriction site polymorphisms, the C_4 species are all part of a single lineage in *Heliotropium* section *Orthostachys*, comprising subsections *Bracteata* and *Axillaria*, according to Johnston's taxonomy (M. Frohlich, personal communication, 1997). A few members of these groups have chlorenchyma organization intermediate between C_3 and C_4. One of these intermediates, *H. tenellum*, a winter annual of the central United States, may be mostly C_3, based on its well-developed palisade layer and not many chloroplasts in the bundle sheath (Frohlich, 1978). In the cladogram it attaches exactly between the C_3 and C_4 groups. Therefore, C_4 either arose after *H. tenellum* diverged from the line leading to C_4 species, or the change in photosynthesis was not firmly fixed in that species and it quickly lost the C_4 attributes.

Like Boraginaceae, genera traditionally placed in Scrophulariaceae also fall into unrelated lineages (Olmstead and Reeves, 1995; Wagstaff and Olmstead, 1997). *Anticharis*, the one C_4 genus, has not been included in any phylogenetic study to date. Wettstein (1891), however, places it close to the genus *Aptosimum*, which in unpublished molecular phylogenies is closely related to Myoporacaeae and the clade labelled Scrophulariaceae-1 (Olmstead and Reeves, 1995; R. Olmstead, personal communication, 1997; Fig. 2). The important point here is that *Anticharis* is not at all closely

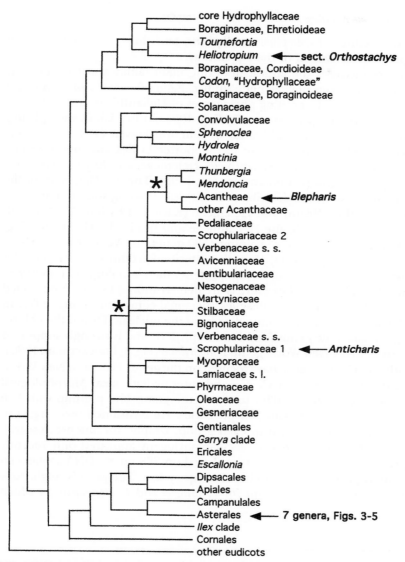

Figure 2 Relationships among members of the Asteridae s.l., depicted by grafting several disparate phylogenies based on different sorts of data. Relationships among Boraginaceae and Hydrophyllaceae based on sequences of *ndhF* (Ferguson, in press). Relationships among Acanthaceae based on combined nuclear ITS and chloroplast gene spacer sequences [McDade, L. A., and and Moody, 1997). Phylogenetic relationships among Acanthaceae: Evidence from nuclear and chloroplast genes. *Amer. J. Bot.* **84**(6, suppl.), 214–215.]. Relationships among other Lamiales based on *rbcL* (Olmstead, R. G., and Reeves, P. A. (1995). Polyphyletic origin of the Scrophulariaceae: Evidence from *rbcL* and *ndhF* sequences. *Ann. Miss. Bot. Gard.* **82**, 176–193; and Wagstaff, S. J., and Olmstead, R. G. (1997). Phylogeny of Labiatae and Verbenaceae inferred from *rbcL* sequences. *Syst. Bot.* **22**, 165–179.]. Asterisks indicate nodes supported by more than 90% of bootstrap replicates in individual gene trees. C_4 genera are indicated in boldface.

related to either *Heliotropium* or *Blepharis,* and thus represents a clearly independent derivation of the C_4 pathway.

The Asteraceae is the largest flowering plant family and has been the subject of numerous phylogenetic studies. Figure 3 shows a phylogeny for the family, created by grafting phylogenies of Heliantheae and Helenieae based on morphology (Karis and Ryding, 1994a,b) to a family wide phylogeny based on the large *ndhF* study of Kim and Jansen (1995). The well-supported nodes of the latter are also supported by *rbcL* and most morphological studies. The tribes Heliantheae, Helenieae, and Eupatorieae form a monophyletic group (Kim and Jansen, 1995). The five C_4 genera of the Asteraceae are all within the tribes Heliantheae (Ryding and Bremer, 1992; Karis and Ryding, 1994b) and Helenieae (Karis and Ryding, 1994a), with apparently one origin in the Heliantheae and two in the Helenieae (Fig. 3). As far as is known, all remaining 1530 genera of Asteraceae are C_3.

Karis and Ryding (1994a) show the subtribe Pectidinae, which includes *Tagetes, Dyssodia,* and *Pectis,* as paraphyletic; their topology is reproduced in Fig. 3. This group, however, is treated as monophyletic (and called Tageteae) by Jansen and colleagues (personal communication, 1997). In a molecular study incorporating data from the nuclear ITS and the chloroplast gene *ndhF,* the genus *Pectis* is monophyletic and is strongly supported as the sister to the New World genus *Porophyllum* (Fig. 4; Loockerman, 1996).

Studies of C_4 Asteraceae have focused on the genus *Flaveria,* a New World genus of 21 species, with C_3, C_4, and intermediate taxa. Morphological distinctions led Powell (1978) to divide the genus into a group with 5–6 phyllaries and a group with 3–4 phyllaries. He envisioned two origins of the C_4 pathway, one origin in each group. Despite the intense research interest in *Flaveria,* the first molecular phylogeny appeared only recently (Kopriva *et al.,* 1996; Fig. 5). This is based on 388 bp of cDNA sequenced from the H-protein of the glycine cleavage system. The gene is single copy in all species examined except for *F. pringlei* and *F. cronquistii,* in which

---

Figure 3 Phylogeny of the Asteraceae, depicted by grafting morphological phylogenies for genera of Heliantheae and Helenieae [Ryding, O., and Bremer, K. (1992). Phylogeny, distribution, and classification of the Coreopsidae (Asteraceae). *Syst. Bot.* **17,** 649–659; Karis, P. O., and Ryding, O. (1994a). Tribe Helenieae. *In* "Asteraceae: Cladistics and Classification" (K. Bremer, ed.), pp. 521–558. Timber Press, Portland, Oregon; and Karis, P. O., and Ryding, O. (1994b). Tribe Heliantheae. *In* "Asteraceae: Cladistics and Classification" (K. Bremer, ed.), pp. 559–624. Timber Press, Portland, Oregon.] to a family level phylogeny based on data from *ndhF* [Kim, K.-J., and Jansen, R. K. (1995). *ndhF* sequence evolution and the major clades in the sunflower family. *Proc. Nat. Acad. Sci. U.S.A.* **92,** 10379–10383.]. Asterisks indicate nodes supported by more than 90% of bootstrap replicates in individual gene trees. C_4 genera are indicated in boldface.

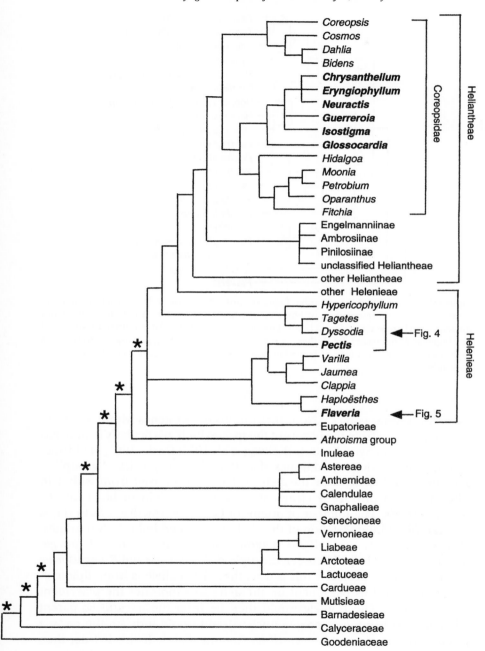

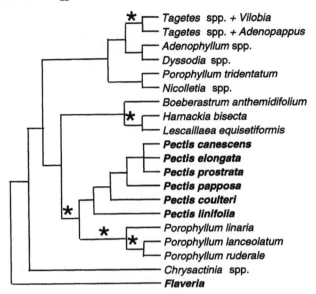

Figure 4 Phylogeny of the subtribe Tageteae, pruned to show only the clade containing *Pectis* and select representatives of other clades; based on combined data from *ndhF* and nuclear ITS [Loockerman, D. J. (1996). Phylogenetic and molecular evolutionary studies of the Tageteae (Asteraceae). Ph.D. thesis. University of Texas, Austin.]. Asterisks indicate nodes supported by more than 90% of bootstrap replicates in individual gene trees. C_4 genera are indicated in boldface.

there are several copies. [Note that the history of genes within a species may be different from the history of the species itself, particularly if there has been a history of hybridization (Kellogg *et al.*, 1996; Doyle, 1997; Maddison, 1997).] The phylogeny indicates that Powell's 3-4 phyllary "group" is actually paraphyletic, with the 5-6 phyllary line derived from within it. The H-protein phylogeny clearly shows a single origin of bona fide C_4 photosynthesis. The number of origins of C_3-C_4 intermediacy, however, is unclear. Strict interpretation of the cladogram would indicate two origins. However, Kopriva *et al.* (1996) suggest that their accession of *F. pringlei* may actually be a hybrid between the diploid C_3 *F. pringlei* and *F. angustifolia*, an intermediate species. If this is true, then the A allele is from diploid *F. pringlei* (C_3) and its placement on the gene tree indicates the placement of the species. The B allele would then be from the *angustifolia* parent (despite now being in a *pringlei* nucleus) and would indicate the correct position of the species *F. angustifolia* (intermediate). If this is true, then optimization of photosynthetic pathway on the cladogram would indicate a single origin of C_3-C_4 intermediacy followed by a single development of

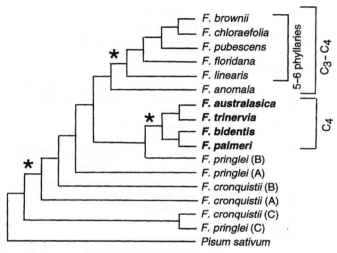

Figure 5 Phylogeny of the genes coding for the H-protein in the glycine cleavage system of species of *Flaveria* [Kopriva, S., Chu, C.-C., and Bauwe, H. (1996). Molecular phylogeny of *Flaveria* as deduced from the analysis of nucleotide sequences encoding the H-protein of the glycine cleavage system. *Plant Cell Environ.* **19**, 1028–1036.]. *F. pringlei* and *F. cronquistii* both have multiple genes encoding the protein. Asterisks indicate nodes supported by more than 90% of bootstrap replicates. C_4 species are indicated in boldface. Letters following species' name indicate different alleles for the H-protein.

true C_4 status. As Kopriva *et al.* (1996) point out, a rigorous answer to the question demands both data from another gene and inclusion of the remaining species of *Flaveria*.

C. Caryophyllidae

The Caryophyllidae has been recognized for some time as a coherent and likely monophyletic group. Within the group, however, the classification is only a poor reflection of the phylogeny, and the phylogeny itself suffers from inadequate sampling. C_4 photosynthesis is common among members of the Caryophyllidae, as is CAM. The number of origins and their relative time, though, are unknown because the phylogeny remains largely unresolved (Fig. 6; P. F. Stevens, personal communication, 1997).

Molecular studies have shown that family delimitations in the Caryophyllaceae will need to be revised, which makes discussion of C_4 origins difficult. As currently circumscribed, a number of the families are para- or polyphyletic (Downie *et al.*, 1997; Hershkovitz and Zimmer, 1997). As shown in Fig. 6, based on ITS sequences (Hershkovitz and Zimmer, 1997), Basellaceae and Didiereaceae, both monophyletic, are sister taxa and are related to genera commonly assigned to Portulacaceae. Three of the four members

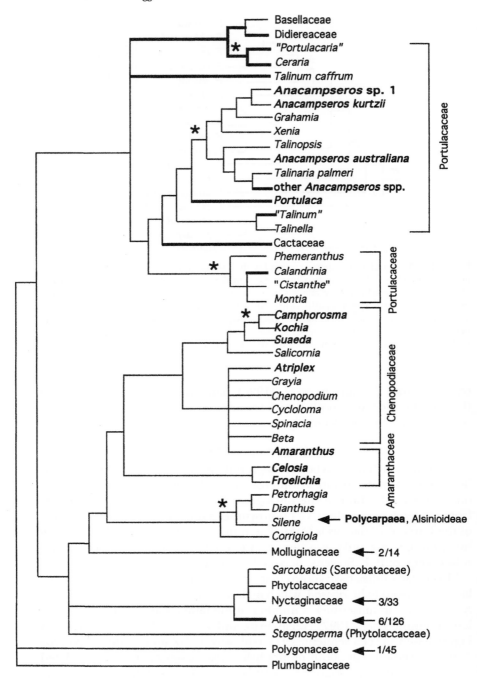

of this clade exhibit Crassulacean Acid Metabolism (CAM; Winter and Smith, 1995). The C_4 species of Portulacaceae are in a separate clade that also includes several C_3 taxa plus the Cactaceae, which exhibits CAM. According to results from Downie et al. (1997), who studied ORF2280 of the chloroplast, this entire clade (Portulacaceae + Cactaceae + Basellaceae + Didieraceae) is an early diverging branch of the Caryophyllales. This conclusion should be accepted with caution, however, as the branching order of the early branches is not strongly supported in the chloroplast trees.

The monophyly of Aizoaceae has never been investigated. The family contains members with CAM and others with the C_4 pathway, but the relationships among the genera are not known. Likewise Molluginaceae, Nyctaginaceae, and Polygonaceae remain largely unstudied. Caryophyllaceae, which was often considered basal in early analyses, appears as a later diverging branch in Downie et al. (1997), although it is monophyletic. The sole C_4 genus, *Polycarpaea*, has not been included in any molecular phylogeny.

Morphological studies of Caryophyllales place Amaranthaceae and Chenopodiaceae together (Rodman, 1994; Downie et al., 1997), a point that is supported by molecular data as well (Manhart and Rettig, 1994; Downie et al., 1997). ORF2280 sequences show a monophyletic Chenopodiaceae, with the exception of *Sarcobatus*, which is apparently unrelated, being included with members of Phytolaccaceae and Nyctaginaceae. *Sarcobatus* has been described as a new family (Behnke, 1997). Amaranthaceae is paraphyletic. Despite much fascinating physiological and morphological variation, the amaranth/chenopod clade has not received the attention it deserves by phylogeneticists. Relationships within it are poorly resolved.

The Portulacaceae is clearly polyphyletic (Fig. 6). The two C_4 genera, *Portulaca* and *Anacampseros*, are apparently closely related, but the latter genus appears polyphyletic (Fig. 6). Both genera also contain CAM species.

Figure 6 Relationships among members of the Caryophyllidae. Relationships in the upper large clade, including Basellaceae, Didieraceae, Cactaceae, and the polyphyletic Portulacaceae are based on sequences of the ITS region of ribosomal RNA [Hershkovitz, M. A., and Zimmer, E. A. (1997). On the evolutionary origins of the cacti. *Taxon*, **46**, 217–232.]. The ITS phylogeny is grafted to a phylogeny based on sequences of the chloroplast encoded ORF2280 [Downie, S. R., Katz-Downie, D. S., and Cho, K.-J. (1997). Relationships in the Caryophyllales as suggested by phylogenetic analyses of partial chloroplast DNA ORF2280 homolog sequences. *Amer. J. Bot.* **84**, 253–273.]. There are many more C_4 genera of Amaranthaceae and Chenopodiaceae that have not been studied phylogenetically. C_4 genera are shown in boldface. For groups with additional C_4 genera, numbers indicate number of C_4 genera in a group/total number of genera. Heavy lines indicate lineages exhibiting CAM metabolism. Nodes supported by more than 90% of bootstrap replicates indicated by asterisks.

The tremendous morphological, ecological, physiological, and biochemical diversity in the Caryophyllidae make it ideal for a phylogenetic study. Not only is there variation for photosynthetic pathway, but many members are also salt- and drought-tolerant. In at least some taxa, there is a direct physiological connection between environmental stimuli such as drought or salt stress and the induction of CAM (see multiple papers in Winter and Smith, 1995). Because much of the photosynthetic machinery is the same in CAM plants as in C_4, this may be relevant to the evolution of the pathway (see later this chapter).

D. Glucosinolate Clade

The glucosinolate clade (Fig. 7) includes all but one of the genera known to produce mustard oils and has been the subject of several comprehensive molecular studies using sequences of *rbcL* (Rodman *et al.*, 1996) and 18S rRNA (Rodman *et al.*, 1997). In all these studies, Brassicaceae and Capparaceae are closely related, a relationship that is strongly supported. A morphological phylogeny of the two families shows that Brassicaceae is actually embedded within a paraphyletic Capparaceae (Judd *et al.*, 1994). Most of the species in the glucosinolate clade are C_3.

Despite the intense biological interest in the family Brassicaceae, there is no published molecular phylogeny for the family. Fortunately, however, Warwick and Black (1997) have produced an extensive study of chloroplast restriction site variation in the tribe Brassiceae, which includes the C_3–C_4 intermediate genus *Moricandia*. The tribe and subtribe are strongly supported as monophyletic (Warwick and Black, 1997; Price, personal communication, 1997), and *Moricandia* itself also is a single lineage (Fig. 7). Sister group relationships of *Moricandia* are unresolved.

Note that the genera *Brassica* and *Diplotaxis* are clearly polyphyletic (Fig. 7). This is consistent with data for other genera (e.g., *Arabidopsis;* O'Kane *et al.*, 1996, Price *et al.*, 1994). Traditional classifications of Brassicaceae have been based largely on characteristics of the fruit, but available molecular data indicate that many of these groups do not reflect phylogenetic history. In this family, we thus see a serious discrepancy between traditional classification and phylogeny, a point that needs to be kept in mind when comparing species in physiological studies. (Note that this is one of the few such examples. It is more common that traditional classification exhibits broad similarity with phylogenetic data.)

Capparaceae are also largely unknown phylogenetically. The genus *Cleome* includes 150 species, of which a few are C_4. Although the taxonomy indicates that the C_4 species, sometimes segregated into the genus *Gynandropsis*, are derived from C_3 ancestors classified in *Cleome*, this has never actually been tested.

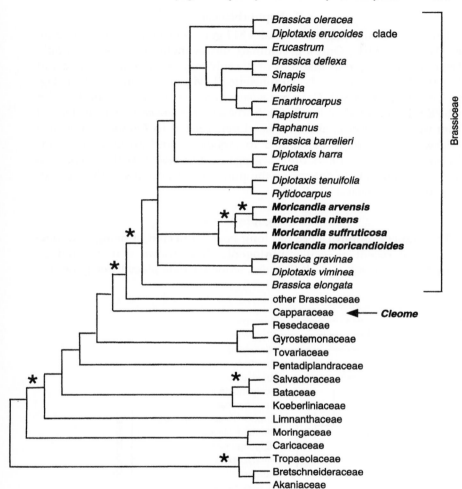

Figure 7 Relationships within the tribe Brassiceae of the family Brassicaceae based on chloroplast restriction site variation [Warwick, S. I., and L. D. Black. (1997). Phylogenetic implications of chloroplast DNA restriction site variation in subtribes Raphaninae and Cakilinae (Brassicaceae, tribe Brassiceae). *Can. J. Bot.* **75,** 960–973.]. The restriction site tree is grafted to a phylogeny for other glucosinolate famlies, based on sequences of *rbcL* and on morphological characters [Rodman, J. E., Karol, K. G., Price, R. A., and Sytsma, K. J. (1996). Molecules, morphology, and Dahlgren's expanded order Capparales. *Syst. Bot.* **21,** 289–307.]. There is no adequate phylogeny within Capparaceae, but the two families are undoubtedly closely related. C_3–C_4 intermediate and C_4 taxa are shown in boldface. Nodes supported by more than 90% of bootstrap replicates indicated by asterisks.

E. Euphorbiaceae

The family Euphorbiaceae includes 317 genera grouped in five subfamilies (Webster, 1994). Only a preliminary phylogenetic study has been published, but this suggests that four of the subfamilies are monophyletic and that the Phyllanthoideae is paraphyletic (Levin and Simpson, 1994; Fig. 8). The two genera with C_4 taxa, *Chamaesyce* and *Euphorbia,* are both in the tribe Euphorbieae of subfamily Euphorbioideae. *Chamaesyce* may be derived from within *Euphorbia,* and may be closely related to *Euphorbia* subgenus *Agaloma. Chamaesyce* is also similar to *Pedilanthus* and possibly the genus *Cubanthus* (Webster, 1994), other segregates of *Euphorbia.* Without a phylogeny, it is impossible to determine the number of C_4 origins or the C_3 sister taxa to the C_4 species.

F. Zygophyllaceae

The family Zygophyllaceae has long been suspected of being heterogeneous, and recent molecular phylogenies have shown that it is indeed polyphyletic (Gadek *et al.,* 1996). *rbcL* sequence data place three genera—

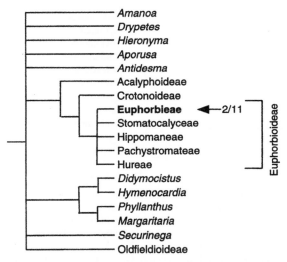

Figure 8 Relationships among members of the Euphorbiaceae, based on morphological characters [Levin, G. A., and Simpson, M. G. (1994). Phylogenetic implications of pollen ultrastructure in the Oldfieldioideae (Euphorbiaceae). *Ann. Miss. Bot. Gard.* **81,** 203–238.] Genera not assigned to a subfamily are included in the polyphyletic Phyllanthoideae. [Tribes of Euphorbioideae follow Webster, G. L. (1994). Synopsis of the genera and suprageneric taxa of Euphorbiaceae. *Ann. Miss. Bot. Gard.* **81,** 33–144.] There is no phylogeny within the tribe Euphorbieae. Numbers following the clade indicate number of C_4 genera and total number of genera respectively.

Peganum, Malacocarpus, and *Nitraria*—in the order Sapindales. The remainder of the family, the Zygophyllaceae *sensu stricto,* including all the C_4 taxa, is more closely related to the Krameriaceae.

A phylogeny of Zygophyllaceae based on *rbcL* sequences and morphological data shows that the family is divided into two major clades, with C_4 species in each (Fig. 9; Sheahan and Chase, 1996). There is at least one origin in the clade including *Kallstroemia* and *Tribulus,* although there may be more depending on the placement of the C_3 species of *Tribulus.* Although the genus *Zygophyllum* contains both C_3 and C_4 members, the phylogeny shows that the genus is polyphyletic, and the species with the different pathways may be unrelated. The photosynthetic pathway is unknown for several genera in the family, which makes it impossible to determine exactly where C_4 originated and how many times.

G. Hydrocharitaceae

The Hydrocharitaceae fall clearly among the alismatid monocots on the basis of analyses of *rbcL* and morphology (Chase *et al.,* 1995; Fig. 1). Based on weighted parsimony analyses of *rbcL* sequences, Les *et al.* (1997) suggest that Hydrocharitaceae is monophyletic (as long as Najadaceae is included

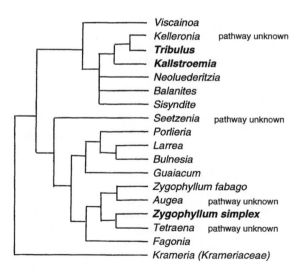

Figure 9 Relationships among members of the Zygophyllaceae s.s., based on *rbcL* sequences combined with morphological data [Sheahan, M. C., and Chase, M. W. (1996). A phylogenetic analysis of Zygophyllaceae R. Br. based on morphological, anatomical and *rbcL* DNA sequence data. *Bot. J. Linn. Soc.* **122,** 279–300.]. No numerical estimate of support was provided in the original study. C_4 genera indicated in boldface. For four genera, the photosynthetic pathway is unknown.

within it), and that *Hydrilla* is sister to a clade that includes *Nechamandra* and *Vallisneria* (Fig. 10). Tanaka *et al.*, (1997) included fewer species, used neighbor-joining algorithms, and sequenced the chloroplast gene *maturase K* (*matK*) in addition to *rbcL*. By combining *rbcL* and *matK* sequences, they found relationships that were similar but not identical to those postulated by Les *et al.*, (1997). The differences between the studies were most likely due to different taxon sampling, different methods of analysis, and inclusion of a second gene by Tanaka *et al.* (1997). Like Les *et al.* (1997), Tanaka *et al.* (1997) found that *Hydrilla* is closely related to *Vallisneria;* they did not

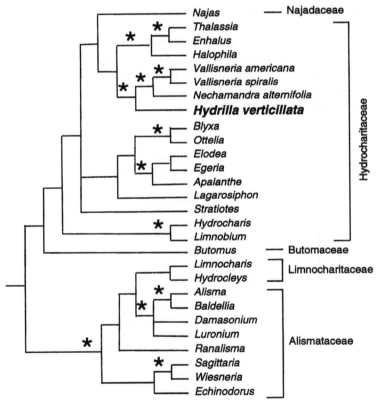

Figure 10 Relationships among members of the Hydrocharitaceae and related alismatid families, based on *rbcL* sequences analyzed with weighted parsimony [Les, D. H., Cleland, M. A., and Waycott, M. (1997). Phylogenetic studies in Alismatidae, II: Evolution of marine angiosperms (seagrasses) and hydrophily. *Syst. Bot.* **22**, 443–463. The sole C₄ genus, *Hydrilla*, is indicated in boldface. Nodes supported by more than 90% of bootstrap replicates indicated by asterisks.

include data on *Nechamandra*. *Hydrilla* thus represents an isolated instance of the C_4 pathway.

H. Cyperaceae

Considering the ubiquity and the diversity of the Cyperaceae, the family has received surprisingly little attention from phylogeneticists. The major work is that of Bruhl (1990, 1995), which includes morphological cladistic and phenetic analyses for all genera of the family, as well as a database of morphological characters. The C_4 genera occur in four tribes, which together form a clade (Fig. 11). The close relationship between Cypereae and Scirpeae is supported by *rbcL* sequence data (Plunkett *et al.*, 1995), but *Rhynchospora* may be more distant. Note, however, that the *rbcL* data set represents only a tiny sample of the family. Different relationships were also found by Simpson (1995), but this study, too, only sampled a handful of genera.

Even among the four tribes, there are apparently several origins of the pathway (Fig. 12). No estimate of support was provided for the morphological phylogenies summarized in Fig. 12, so these results should be interpreted with particular caution.

I. Poaceae

This family has received the lion's share of the attention paid to C_4 photosynthesis. For such a large group, it also has an unusually well studied phylogeny. A semistrict consensus tree for the phylogeny is shown in Fig. 11. The asterisks mark nodes that are well supported, not in a single gene tree, but in two to eight gene trees. They are thus among the strongest phylogenetic statements in this paper. The monophyly of the terminal taxa is also not disputed. There are apparently four C_4 lineages, all members of a single very large clade ("the PACC clade" of Davis and Soreng, 1993) that includes the subfamilies Panicoideae, Arundinoideae, Centothecoideae, and Chloridoideae. Although multiple studies have been carried out using various genes, relationships within this large clade are not well resolved; this is partly due to inadequate sampling and partly to short branches at the base of the clade. Three alternate gene trees for the clade are shown in Fig. 13. Although these differ considerably, they all lead to the conclusion that C_4 arose more than once in the family.

The major limitation of the grass phylogeny for study of C_4 photosynthesis is the lack of a phylogeny for the subfamily Panicoideae. The subfamily is common divided into two very large tribes, the Paniceae and the Andropogoneae, and several smaller tribes such as Arundinelleae, Isachneae, and Neurachneae. Only the Andropogoneae is demonstrably monophyletic. The Arundinelleae are polyphyletic (Mason-Gamer, Weil, and Kellogg, MS submitted, 1998; Spangler *et al.*, unpublished data, 1998), and there are

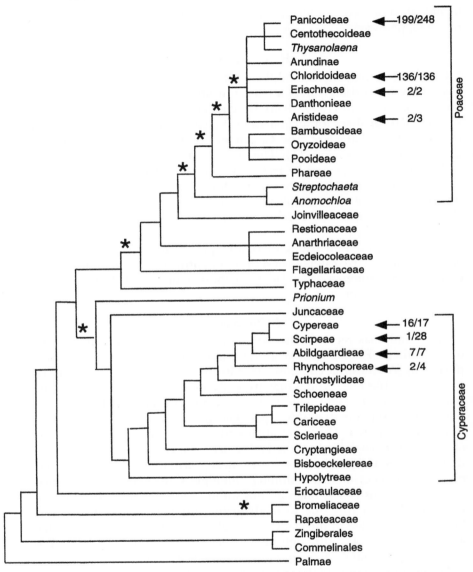

Figure 11 Phylogenies for Poaceae and Cyperaceae grafted to a phylogeny for the commelinid clade, the latter based on sequences of *rbcL* combined with morphology [Chase, M. W., Stevenson, D. W., Wilkin, P., and Rudall, P. J. (1995). Monocot systematics: A combined analysis. *In* "Monocotyledons: Systematics and Evolution" (P. J. Rudall, P. J. Cribb, D. F. Cutler, and C. J. Humphries, eds.), pp. 685–730. Royal Botanic Gardens, Kew; and Linder. H. P., and Kellogg, E. A. (1995). Phylogenetic patterns in the commelinid clade. *In* "Monocotyledons: Systematics and Evolution" (P. J. Rudall, P. J. Cribb, D. F. Cutler, and C. J. Humphries, eds.), pp. 511–542. Royal Botanic Gardens, Kew.]. Relationships among grasses based on

no data on Isachneae, Neurachneae, or Paniceae. The genus *Panicum* is often cited as being unusual because it includes C_3, C_4, and intermediate species, and also includes all C_4 subtypes. It is highly likely, however, that the genus is paraphyletic; we have no evidence that the various photosynthetic types are particularly closely related to each other.

III. Number and Time of C_4 Origins

The phylogenetic data show that there have been at least 31 origins of the C_4 pathway in the flowering plants, with multiple origins in Asteraceae, Poaceae, Cyperaceae, and Zygophyllaceae. The phylogenies in Caryophyllidae are not comprehensive enough, nor are family delimitations clear

---

consensus of data from morphology [Kellogg, E. A., and Campbell, C. S. (1987). Phylogenetic analyses of the Gramineae. *In* "Grass Systematics and Evolution" (T. R. Soderstrom, K. W. Hilu, C. S. Campbell, and M. E. Barkworth, eds.), pp. 217–224. Smithsonian Press, Washington, D.C.], chloroplast restriction sites [Davis, J. I., and Soreng, R. J. (1993). Phylogenetic structure in the grass family (Poaceae), as determined from chloroplast DNA restriction site variation. *Amer. J. Bot.* **80**, 1444–1454.], *rbcL* [Doebley, J., Durbin, M., Golenberg, E. M., Clegg, M. T., and Ma, D. P. (1990). Evolutionary analysis of the large subunit of carboxylase (*rbcL*) nucleotide sequence among the grasses (Gramineae). *Evolution* **44**, 1097–1108; Barker, N. P. (1995). "A molecular phylogeny of the subfamily Arundinoideae (Poaceae)." Unpublished Ph.D. thesis, University of Cape Town; and Duvall, M. R., and Morton, B. R. (1996). Molecular phylogenetics of Poaceae: An expanded analysis of *rbcL* sequence data. *Mol. Phylog. Evol.* **5**, 352–358.], *ndhF* [Clark, L. G., Zhang, W., and Wendel, J. F. (1995). A phylogeny of the grass family (Poaceae) based on *ndhF* sequence data. *Syst. Bot.* **20**, 436–460.], *rpoC2* [Cummings, M. P., King, L. M., and Kellogg, E. A. (1994). Slipped-strand mispairing in a plastid gene: *rpoC2* in grasses (Poaceae). *Mol. Biol. Evol.* **11**, 1–8; and Barker, N. P. (1995). "A molecular phylogeny of the subfamily Arundinoideae (Poaceae)." Unpublished Ph.D. thesis, University of Cape Town.], *rps4* [Nadot, S., Bajon, R., and Lejeune, B. (1994). The chloroplast gene *rps4* as a tool for the study of *Poaceae* phylogeny. *Pl. Syst. Evol.* **191**, 27–38.], rRNA [Hamby, R. K., and Zimmer, E. A. (1988). Ribosomal RNA sequences for inferring phylogeny within the grass family (Poaceae). *Pl. Syst. Evol.* **160**, 29–37.], phytochrome B [Mathews, S., and Sharrock, R. A. (1996). The phytochrome gene family in grasses (Poaceae): A phylogeny and evidence that grasses have a subset of the loci found in dicot angiosperms. *Mol. Biol. Evol.* **13**, 1141–1150; and Mathews, Tsai and Kellogg, MS submitted, 1998), and granule bound starch synthase I (Mason-Gamer Weil, and Kellogg, MS submitted, 1998), as summarized in Kellogg, E. A. [(1998). Relationships of cereal crops and other grasses. *Proc. Natl. Acad. Sci. U.S.A.* **95**, 2005–2010]. Relationships among sedges based on morphological cladograms [Bruhl, J. (1990). "Taxonomic relationships and photosynthetic pathways in the Cyperaceae." Unpublished Ph.D. thesis, Australian National University, Canberra, Australia; and Bruhl, J. J. (1995). Sedge genera of the world: relationships and a new classification of the Cyperaceae. *Aust. Syst. Bot.* **8**, 125–305.]. Clades with C_4 genera indicated by arrows. Numbers following the clade indicate number of C_4 genera and total number of genera respectively. Nodes supported by more than 90% of bootstrap replicates indicated by asterisks.

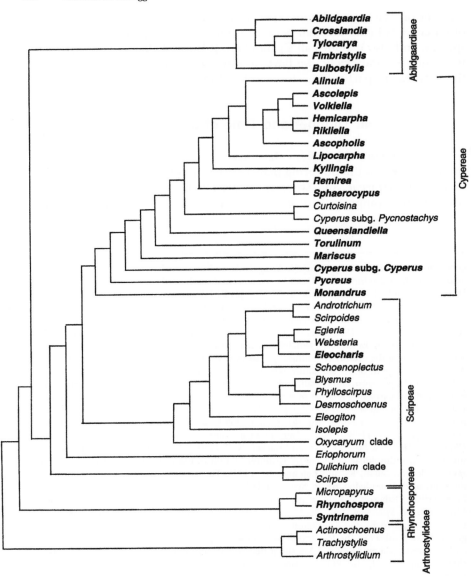

Figure 12 Relationships among C_4 genera of Cyperaceae, based on morphological data [Bruhl, J. (1990). "Taxonomic relationships and photosynthetic pathways in the Cyperaceae." Unpublished Ph.D. thesis, Australian National University, Canberra, Australia; and Bruhl, J.J. (1995). Sedge genera of the world: relationships and a new classification of the Cyperaceae. *Aust. Syst. Bot.* **8**, 125–305.]. Support was not determined in original cladograms. C_4 genera are shown in boldface. Note that *Eleocharis*, *Abildgaardia*, and *Rhynchospora* are variable for photosynthetic pathway. The genus *Cyperus* is also variable, but the C_3 and C_4 species are not closely related to each other.

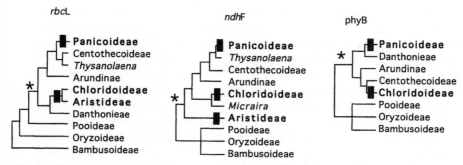

Figure 13 Summary gene trees for three grass genes indicating relationships among the C₄ subfamilies. Discrepancies among the trees lead to the polytomy in Fig. 11, but each individual tree suggests multiple independent origins of the C4 pathway. [*rbcL* tree from Barker, N. P., Linder, H. P., and Harley, E. (1995). Phylogeny of Poaceae based on *rbcL* sequences. *Syst. Bot.* **20**, 423–435, *ndhF* from Clark, L. G., Zhang, W., and Wendel, J. F. (1995). A phylogeny of the grass family (Poaceae) based on *ndhF* sequence data. *Syst. Bot.* **20**, 436–460, phyB from Mathews, S., and Sharrock, R. A. (1996). The phytochrome gene family in grasses (Poaceae). A phylogeny and evidence that grasses have a subset of the loci found in dicot angiosperms. *Mol. Biol. Evol.* **13**, 1141–1150.].

enough, to determine the number of origins per family. Such a comparison also assumes that all families are of comparable evolutionary age, an assumption that is well known to be wrong.

Expansion of C_4 grasslands between 7 and 5 mya has been well documented and has been correlated with lower concentrations of CO_2 worldwide (Cerling *et al.*, 1997; Cerling, Chapter 13). The authors of these studies are careful to note, however, that their data reflect only change in plant communities. They do not claim anything about origin of species.

The phylogenetic data are consistent with the paleoecological data. C_4 species originated well before the expansion of C_4 grasslands and must have persisted as minor components of the vegetation until global climatic conditions permitted their rise to dominance. Our best data on time of origin come from the grass family. We can comfortably date the origin of the grasses at approximately 65 mya based on pollen grains from the Upper Maastrichtian (Linder, 1987). This date can then be used to calibrate molecular clocks of other genes. Although basing any such estimation on a single gene is fraught with difficulties, use of multiple genes gives some confidence in the estimate. Gaut and Doebley (1997) use sequences of malate dehydrogenase and granule-bound starch synthase I to estimate the divergence time of maize and sorghum at approximately 17 mya. Because maize and sorghum are part of the tribe Andropogoneae, a group that is entirely C_4, this must be after the origin the photosynthetic pathway. They also used acohol dehydrogenase sequences to estimate that the time of divergence of maize from *Pennisetum* was more than 25 mya. Because *Pennisetum* is part of the same clade as Andropogoneae, as suggested by preliminary data (Figs. 11 and 13; Kellogg, unpublished 1998), then this date too must be after the origin of the C_4 pathway. This is well before the expansion of C_4 grasslands in the late Miocene. Note also that for the molecular clocks to fit a Miocene origin of the pathway, mutations would have to be accumlating at three times the rate estimated by Gaut and Doebley (1987).

If the C_4 pathway originated well before the expansion of the grasslands, it follows that C_4 taxa must have been present but were relatively minor members of the plant community. Such communities exist today; for example, the Palouse Prairie in the Pacific Northwest is dominated by C_3 grasses, but C_4 species are represented, particularly in drier habitats (Gould and Shaw, 1983). This may affect hypotheses of the adaptive value of the pathway. Although it may have been preadapted to take advantage of lower CO_2 concentrations when they occurred, low CO_2 may not necessarily have caused the pathway to evolve.

It is hazardous to make any inferences about time of origin in other families without some independent calibration point. That said, however, it seems reasonable to infer that the C_4 grass lineages are older than the C_4 species of, for example, *Heliotropium*. Even if one uses the most simple (and simple-minded!) measure of time—uncorrected branch lengths on a parsimony tree for a molecule such as *ndhF*—it is clear that there have been many more mutations since the origin of the chloridoid grasses (22–87 steps; Clark *et al.*, 1995) than there have been since the origin of the genus *Heliotropium* (9 steps; Ferguson, in press). There are two possible inferences from the observed pattern. One is that C_4 was not selectively favored until approximately 25 mya, but since then it has arisen fairly frequently in many disparate groups. The other possible interpretation is that the pathway has been arising and being lost (either by extinction or reversion to C_3) more or less continually in the flowering plants, but has only rarely persisted long enough to lead to large clades of C_4 taxa. It is not clear how these hypotheses could be distinguished on the basis of phylogenetic pattern alone.

How easy is it to lose the C_4 pathway? The phylogenetic pattern is not sufficiently well resolved in most groups to answer this question. There is only one unambiguous case of C_4 loss, and this is in the genus *Eragrostis*, a genus which is comfortably placed in the chloridoid grasses. The subfamily Chloridoideae includes 136 genera, all of which are C_4; it has been demonstrated to be monophyletic in every molecular phylogenetic study to date. *Eragrostis*, with approximately 350 species, is also apparently monophyletic, although this has not been explored as rigorously as one might like. All members of the genus are C_4 except for *E. walteri*, a species native to Namibia (Ellis, 1984). It is morphologically similar to *E. crassinervis*, which is C_4. Unless *E. walteri* is sister to all other species of *Eragrostis* and also sister to all other species of the chloridoids (both unlikely), it must represent a reversion of the C_4 pathway. Other reversals have been suggested in the grass tribe Paniceae and also in the genus *Salsola* (Chenopodiaceae; P'yankov *et al.*, 1997), but in neither case is the supposition tested phylogenetically.

Periodically one finds the suggestion that C_4 has arisen only once and then been repeatedly lost (e.g., Watson *et al.*, 1985). Such an origin would have to have happened very early in angiosperm evolution and have been followed by a very large number of losses. This requires that loss, so far demonstrated only in *E. walteri*, be extremely easy. Certainly in the group for which we have the best phylogeny, the grasses, the pathway appears to be quite stable over evolutionary time. We see very large groups (e.g., the Chloridoideae, or the 100 or so genera of the Andropogoneae) all of which are demonstrably C_4. We do not know, however, whether the pathway might be more labile in other families.

It is also formally possible that C_4 is so complex that it must have arisen only once and then be spread to other lineages via lateral gene transfer. This suggestion is impossible to refute from phylogenetic data alone. Any pattern of parallel or convergent evolution can be explained by lateral transfer. Physiological and genetic data, presented in the next section, are required to address this suggestion.

IV. Evolution of C_4 Photosynthesis

In the following sections, I argue that:

1. The C_4 pathway is not a single pathway, but rather a set of different pathways that share PEPC as the carbon acceptor and a common histological organization.

2. Use of PEPC as carbon acceptor is not uncommon in plants, and is apparently fairly easy to evolve.

3. The central problem of C_4 photosynthesis and the one about which we know the least is leaf anatomy; once leaf anatomy has changed, the physiology becomes stabilized.

A. The C_4 Pathway Is Not Identical in All Lineages

A major assumption underlying much discussion of C_4 photosynthesis is that there are enough correlations among its components that data from one C_4 species can be extrapolated to all other C_4 plants. Both the "repeated loss" hypothesis and the "lateral transfer" hypothesis, outlined previously, in their simplest forms predict that the pathway will be genetically and developmentally identical wherever it occurs. This is clearly not the case. Kranz anatomy is commonly but not universally present (see Chapter 5 by Dengler and Nelson). Between grasses and sedges there are substantial differences in internal leaf structure, despite their relatively close relationship (Fig. 11). It has been known for some years that C_4 plants differ in their decarboxylating enzymes (Hatch *et al.*, 1975), and that NAD-ME and NADP-ME lineages are distinct. Mesophyll-expressed PEPC is controlled by unrelated genes in *Flaveria* and in grasses (Lepiniec *et al.*, 1994; Toh *et al.*, 1994; Chollet *et al.*, 1996; Monson, Chapter 11).

The variability of the C_4 pathway can be seen by comparing the origins within the grass family, the group best studied both genetically and phylogenetically (Fig. 14). Even among the four apparently independent C_4 lineages of the grass family, there are differences in histology (Prendergast *et al.*, 1987), in decarboxylating enzyme (Hattersley and Watson, 1992), and in

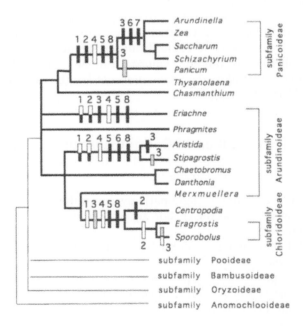

Figure 14 Distribution of different components of the C_4 photosynthetic pathway in the grasses. [Data from the work of Sinha, N. R., and Kellogg, E. A. (1996.) Parallelism and diversity in multiple origins of C_4 photosynthesis in grasses. *Amer. J. Bot.* **83,** 1458–1470, or references therein.] 1, PPDK expression; black, mesophyll localized; white, bundle sheath localized. 2, LHCP expression; black, mesophyll localized; white, bundle sheath localized. 3, C4 subtype; black, NADP-ME; gray, NAD-ME; white, PCK. 4, RuBisCO in mesophyll; white, down-regulated. 5, PEPC in mesophyll; black, up-regulated. 6, loss of grana. 7, bundle sheath reduced to 1. 8, close vein spacing.

cell wall structure (Hattersley and Watson, 1975; Hattersley and Browning, 1981; Hattersley and Perry, 1984; Prendergast et al., 1987). Sinha and Kellogg (1996) compared gene expression across the lineages and found that expression pattern of PPDK and LHCP varied across all lineages. In the panicoids, expression patterns are like maize in that PPDK and LHCP are more strongly expressed in the mesophyll. In the Aristideae and Eriachneae, both enzymes are strongly expressed in both mesophyll and bundlesheath cells. In the Chloridoideae, the expression pattern is exactly backwards from that in maize—both are expressed only in the bundle sheath. This is paradoxical for PPDK, because it implies that PEP must be translocated between cell types rather than pyruvate. The only aspects of the pathway that are consistently present among all grass lineages are 1) close proximity of mesophyll and bundle sheath cells; 2) up-regulation of PEPC in the mesophyll; and 3) down-regulation of RuBisCO in the mesophyll (Sinha and Kellogg, 1996).

The pattern in Fig. 14 argues against a single origin of C_4 even among these closely related lineages. However, there may have been a single origin of the capacity to upregulate PEPC and down-regulate RuBisCO in the mesophyll, and to alter vein spacing. In some lineages, this capacity was fixed, and in others it was lost, thus creating the pattern of closely related C_3 and C_4 lineages. This possibility could be tested by a more detailed molecular phylogeny of PEPC to determine if the C_4 isoform is always the same member of the gene family and if its regulatory sequences are the same.

Thus the only aspects of C_4 photosynthesis that are necessary (if perhaps not sufficient) are those aspects that define the pathway. In other words, if PEPC were not up-regulated in the mesophyll, the plant would not be classed as C_4. However, even among PEPC genes there is variation. PEPC is encoded by a multigene family with 3–5 members (Grula and Hudspeth, 1987; Cushman and Bohnert, 1989 a,b; Poetsch et al., 1991; Schäffner and Sheen, 1992; Lepiniec et al., 1993; Yanai et al., 1994). The isoform coopted for C_4 photosynthesis in maize and sorghum is not the same copy used in *Flaveria* (Hermans and Westhoff, 1992; Chollet et al., 1996; Chapter 11, this book). Furthermore, the maize–sorghum genes appear to have acquired a totally new promoter (Shäffner and Sheen, 1992), whereas the *Flaveria* C_4 gene has a promoter similar to that of the C_3 isozyme (Hermans and Westhoff, 1992).

When one focuses on the variation among C_4 origins, rather than the similarities, it is clear that C_4 is not a unitary phenomenon. This has clear implications for the evolutionary problem. The simplest version of the "single origin" and "lateral transfer" hypotheses is ruled out. If there were a single origin, or if the pathway arose via lateral transfer, then we must also postulate that many of the biochemical and histological details varied with different genetic backgrounds, and then subsequently became fixed. This is a model postulating a single phenomenon with many manifestations (rather like a single actor taking on many roles). Given the complexity of such a model, it is perhaps just as simple to postulate multiple origins.

The C_4 subtypes themselves are also not uniform. In the grasses alone, there are three origins of the NADP-ME subtype, which appears independently in *Aristida*, in *Eriachne*, and in the Panicoideae (Fig. 14). Bundle sheaths are single in NADP-ME panicoids, but double in *Aristida* and *Eriachne*; in *Aristida*, both sheaths are parenchymatous, whereas in *Eriachne* the inner one is a conventional mestome sheath (Prendergast *et al.*, 1987; Hattersley and Watson, 1992; Fig. 14). Bundle-sheath chloroplasts are agranal in NADP-ME Panicoideae and in *Aristida*, but are granal in *Eriachne*. PPDK expression is localized in the mesophyll of the panicoids, is in the mesophyll and outer bundle sheath of *Aristida* and is not localized in *Eriachne* (Sinha and Kellogg, 1996). LHCP expression parallels the structure of the chloroplast; it is reduced in the bundle sheath of NADP-ME panicoids and in *Aristida*, but not in *Eriachne* (Sinha and Kellogg, 1996; Fig. 14). How many other aspects of C_4 subtype are in fact simply idiosyncrasies of a particular lineage or particular species? Ku *et al.* (1996) reviewed literature on several genes involved in C_4 photosynthesis; the reader is referred to their paper for full discussion. In general, though, appropriate comparative studies are lacking.

The NAD-ME subtype has appeared independently in the grass subfamilies Panicoideae and Chloridoideae (Fig. 14). Simply looking at anatomical and ultrastructual characteristics reveals considerable diversity among NAD-ME species (Hattersley and Watson, 1992). For example, chloroplasts in the bundle-sheath cells may be centripetal or centrifugal, a suberized lamella may be present or absent, and the outline of the bundle-sheath cells may be even or uneven. Comparative biochemical and genetic diversity have not been studied in these species. Given the anatomical diversity in a single family, however, it may be that uniformity should not be assumed even within C_4 subtypes.

At least in the grasses, the PCK subtype is always derived from ancestors that had the NAD-ME subtype. (Watson and Dallwitz, 1992; E. A. Kellogg, unpublished data, 1998). Furthermore, all taxa that have been biochemically typed as PCK also have high levels of NAD-ME activity (Guttierrez *et al.*, 1974; Prendergast *et al.*, 1987; Chapter 3). In addition, Burnell and Hatch (1988) showed that in the PCK-type grass *Urochloa panicoides*, decarboxylation of malate by NAD-ME occurred at the same time as decarboxylation of oxaloacetate by PCK. Taken together, these data suggest that PCK decarboxylation is a relatively simple addition to or variant of the NAD-ME pathway. From an evolutionary standpoint, then, it may be more accurate to think about two C_4 subtypes rather than three.

The development of a transformation system in *Flaveria bidentis* (Chitty et al., 1994) is one critical step for understanding the genetic changes necessary for C_4 evolution. This now means that studies can be done on *Flaveria* to compare to those using transient expression systems in maize. For example, Stockhaus et al. (1997) find that *cis*-acting elements of the PEPC gene from *F. trinervia* can direct mesophyll-specific expression, a result similar to that found by Schäffner and Sheen (1992) in isolated maize protoplasts. It will be illuminating when similar experiments are also done with genes and gene constructs from members of other plant families.

Because C_4 photosynthesis has arisen independently multiple times, and because many of the critical photosynthetic genes exist in multiple copies, it may be that different members of each gene family might have been coopted in different lineages. Multiple genes might exist in all angiosperms, and then a different gene deployed (up- or down-regulated) each time C_4 originates. Alternatively, the origin of C_4 photosynthesis might have been immediately preceded by gene duplication. This will lead to different patterns of gene phylogenies. In general the literature on this subject is sparse, but growing. These questions are addressed in detail by Monson (Chapter 11) and also reviewed by Ku et al. (1996).

The physiology of C_3–C_4 intermediates is also reviewed by Monson (Chapter 11). As he points out, whether they shed light on the evolution of C_4 photosynthesis depends in part on whether they are phylogenetically as well as physiologically intermediate. Studies cited previously show that phylogenetic intermediacy has been demonstrated rigorously only in the case of *Flaveria*. It is also possible that the intermediate species of *Panicum*, *Neurachne*, *Eleocharis*, *Mollugo*, and *Alternanthera* will also prove to be part of an evolutionary transformation series, but the phylogenies are not yet sufficiently detailed to be sure. The genus *Moricandia* is composed solely of intermediate species, none of which have given rise to a fully C_4 group; thus it is unclear whether it represents a good model for a group that is "on the way" to C_4 photosynthesis. Given the diversity among C_4 groups, it is possible that *Moricandia* represents a failed attempt, one of many early modifications of C_3 photosynthesis that is not evolving to anything else. Rawsthorne (1992) has suggested that restriction of glycine decarboxylase expression to the bundle sheath, which is found in *Moricandia* and in intermediate species of *Flaveria*, is a necessary early step in evolving C_4. It seems equally possible that it is one of several ways to stabilize the C_3–C_4 condition and evolves later (see also Monson, Chapter 11).

B. Future Directions

C_4 photosynthesis is a complex and diverse set of changes, all of which require activation of a PEPC carbon fixation pathway and its stabilization by altered leaf anatomy. It may be that activation of a PEPC pathway is relatively easy, and that the sporadic appearance of the C_4 pathway has to do with the difficulty of the concomitant anatomical changes. For example, in the grasses, the promoters of both PEPC and PPDK from maize are sufficient for mesophyll-specific expression of the GUS reporter gene in the C_3 species, rice (Matsuoka et al., 1994; Matsuoka, 1995). This implies that the *trans*-acting factors necessary for cell-specific expression are already present in C_3 grasses. Only changes in *cis*-acting regulatory sequences are needed to achieve C_4 levels and distribution of expression. Similarly, in the Caryophyllidae, the widespread occurrence of both C_4 and CAM suggests that genetic changes occurred early in the evolution of the group to allow activation, possibly facultative, of a malate-producing pathway; in some cases this was then fixed as C_4 by histological changes.

The genetic determinants of leaf structure are poorly understood, yet this is the area that is most likely to lead to insights into C_4 evolution. In the grasses, for example, there is a clear distinction between bundle sheath and mesophyll, even in C_3 species. C_4 enzymes must simply respond to the determinants of this prepattern. The basic change is thus one of vein spacing. Once we know what allows changes in vein spacing, we may be better placed to understand the evolution of the C_4 pathway.

In the future, it will be important to pursue phylogenetically informed studies such as that of Sinha and Kellogg (1996). The repeated origin of the C_4 pathway creates a question for evolutionary biologists, but also offers an opportunity for comparative study. Repeated origins can be compared in a manner exactly analogous to the comparison of protein sequences. Proteins are aligned and residues shared by all members of the gene family are inferred to be functionally important. Similarly, one can study pathways by comparing portions of the pathway and determining which aspects are conserved. Hattersley and Watson (1992) outlined some of the questions that require comparative studies:

> . . . we do not yet know . . . , for example, if NADP-ME from an andropogonoid is encoded by the same gene as NADP-ME from an *Eriachne* species, if the genes that determine the leaf structure of a 'classical NAD-ME' type chloridoid are the same as those that determine the leaf structure of a 'classical NAD-ME' type panicoid, or if a modern C_3 panicoid is descended from a C_3 panicoid ancestor or from a C_4 panicoid ancestor. [p. 91]

To this list of unanswered questions, we might add that we also do not know anything about the promoter sequences of the various photosynthetic enzymes, or whether the regulatory networks are identical in all C_4 taxa.

The most powerful study would be one that did pairwise comparisons of C_4 species and their nearest C_3 relatives, determining, for example, orthology of photosynthetic enzymes, similarity of *cis*-regulatory sequences, and identity of leaf morphogenetic mechanisms. Such an approach will determine not only the precise genetic basis of evolutionary change, but will also illuminate the structure of the regulatory networks in the plant cell.

V. Summary

Phylogenetic studies of angiosperms are now sufficiently detailed to show that C_4 photosynthesis has arisen at least 31 times, and multiple origins occur even within single families. With additional phylogenetic work, it should be possible to verify the number of origins. We know more about phylogeny of the grasses than of any other plant family, but there has also been considerable progress in understanding relationships among members of the Asteridae. Major groups still needing investigation are the Caryophyllid clade, the Euphorbiaceae, and the Cyperaceae. Data from the Poaceae suggest that the C_4 pathway originated long before the Miocene expansion of C_4 grasslands; C_4 species therefore must have persisted for millions of years as minor components of the world's flora. Now that organismal phylogenies are available, studies can be designed to compare independent C_4 origins and to find physiological and morphogenetic mechanisms that are shared among all origins of the pathway. Conserved mechanisms are likely to be critical to the origin and/or persistence of the pathway. Comparative data suggest that the C_4 pathway is not a unitary phenomenon, but instead encompasses appreciable morphological, physiological, and genetic diversity, all of which must be accommodated in hypotheses of evolution and adaptation.

Acknowledgments

I thank D. Ferguson, M. Frohlich, R. Jansen, L. McDade for sharing unpublished data, and R. Monson, R. Olmstead, R. Sage, and P. F. Stevens for helpful comments on the manuscript and NSF grant DEB-9419748 for support of some of the work decribed herein.

References

Al-Shehbaz, I. (1991). The genera of Boraginaceae in the southeastern United States. *J. Arnold Arbor., Suppl. Series* **1,** 1–169.
Barker, N. P. (1995). "A molecular phylogeny of the subfamily Arundinoideae (Poaceae)." Unpublished Ph.D. thesis, University of Cape Town.
Barker, N. P., Linder, H. P., and Harley, E. (1995). Phylogeny of Poaceae based on *rbcL* sequences. *Syst. Bot.* **20,** 423–435.
Behnke, H.-D. (1997). Sarcobataceae—a new family of Caryophyllales. *Taxon* **46,** 495–507.
Bruhl, J. (1990). "Taxonomic relationships and photosynthetic pathways in the Cyperaceae." Unpublished Ph. D. thesis, Australian National University, Canberra, Australia.

Bruhl, J. J. (1995). Sedge genera of the world: relationships and a new classification of the Cyperaceae. *Aust. Syst. Bot.* **8**, 125–305.

Burnell, J. N., and Hatch, M. D. (1988). Photosynthesis in phosphoenolpyruvate carboxykinase-type C_4 plants: pathways of C_4 acid decarboxylation in bundle sheath cells of *Urochloa panicoides*. *Arch. Biochem. Biophys.* **260**, 187–199.

Cerling, T. W., Harris, J. M., MacFadden, B. J., Leakey, M. G., Quade, J., Eisenmann, V., and Ehleringer, J. R. (1997). Global vegetation change through the Miocene/Pliocene boundary. *Nature* **389**, 153–158.

Chase, M. W., Soltis, D. E., Olmstead, R. G., Morgan, D., Les, D. H., Mishler, B. D., Duvall, M. R., Price, R. A., Hills, H. G., Qiu, Y.-L., Kron, K. A., Rettig, J. H., Conti, E., Palmer, J. D., Manhart, J. R., Sytsma, K. J., Michaels, H. J., Kress, W. J., Karol, K. G., Clark, W. D., Hedrén, M., Gaut, B. S., Jansen, R. K., Kim, K.-J., Wimpee, C. F., Smith, J. F., Furnier, G. R., Strauss, S. H., Xiang, Q.-Y., Plunkett, G. M., Soltis, P. S., Swensen, S. M., Williams, S. E., Gadek, P. A., Quinn, C. J. , Eguiarte, L. E. , Golenberg, E., Learn, G. H., Jr., Graham, S. W., Barrett, S. C. H., Dayanandan, S., and Albert, V. A. (1993). Phylogenetics of seed plants: An analysis of nucleotide sequences from the plastid gene *rbcL*. *Ann. Missouri Bot. Gard.* **80**, 528–580.

Chase, M. W., Stevenson, D. W., Wilkin, P., and Rudall, P. J. (1995). Monocot systematics: a combined analysis. *In* "Monocotyledons: Systematics and Evolution", (P. J. Rudall, P. J. Cribb, D. F. Cutler, and C. J. Humphries, eds.) pp. 685–730. Royal Botanic Gardens, Kew.

Chitty, J. A., Furbank, R. T., Marshall, J. S., Chen, Z., and Taylor, W. C. (1994). Genetic transformation of the C_4 plant, *Flaveria bidentis*. *Plant J.* **6**, 949–956.

Chollet, R., Vidal, J., and O'Leary, M. H. (1996). Phospho*enol*pyruvate carboxylase: A ubiquitous, highly regulated enzyme in plants. *Ann. Rev. Plant Physiol. Plant Mol. Biol.* **47**, 273–298.

Clark, L. G., Zhang, W., and Wendel, J. F. (1995). A phylogeny of the grass family (Poaceae) based on *ndhF* sequence data. *Syst. Bot.* **20**, 436–460.

Cummings, M. P., King, L. M., and Kellogg, E. A. (1994). Slipped-strand mispairing in a plastid gene: *rpoC2* in grasses (Poaceae). *Mol. Biol. Evol.* **11**, 1–8.

Cushman, J. C., Bohnert, H. J. (1989a). Nucleotide sequence of the gene encoding a CAM specific isoform of phosphoenolpyruvate carboxylase from *Mesembryanthemum crystallinum*. *Nuc. Acids Res.* **17**, 6745–6746.

Cushman, J. C., Bohnert, H. J. (1989b). Nucleotide sequence of the *Ppc2* gene encoding a housekeeping isoform of phosphoenolpyruvate carboxylase from *Mesembryanthemum crystallinum*. *Nuc. Acids Res.* **17**, 6743–6744.

Davis, J. I., and Soreng, R. J. (1993). Phylogenetic structure in the grass family (Poaceae), as determined from chloroplast DNA restriction site variation. *Amer. J. Bot.* **80**, 1444–1454.

Doebley, J., Durbin, M., Golenberg, E. M., Clegg, M. T., and Ma, D. P. (1990). Evolutionary analysis of the large subunit of carboxylase (*rbcL*) nucleotide sequence among the grasses (Gramineae). *Evolution* **44**, 1097–1108.

Downie, S. R., Katz-Downie, D. S., and Cho, K.-J. (1997). Relationships in the Caryophyllales as suggested by phylogenetic analyses of partial chloroplast DNA ORF2280 homolog sequences. *Amer. J. Bot.* **84**, 253–273.

Doyle, J. J. (1997). Trees within trees: Genes and species, molecules and morphology. *Syst. Biol.* **46**, 537–553.

Duvall, M. R., and Morton, B. R. (1996). Molecular phylogenetics of Poaceae: An expanded analysis of *rbcL* sequence data. *Mol. Phylog. Evol.* **5**, 352–358.

Ellis, R. P. (1984). *Eragrostis walteri*—a first record of non-Kranz leaf anatomy in the sub-family Chloridoideae (Poaceae). *S. Afr. J. Bot.* **3**, 380–386.

Ferguson, D. M. Phylogenetic analysis and circumscription of Hydrophyllaceae based on *ndhF* sequence data. *Syst. Bot.* (in press).

Frohlich, M. W. (1978). "Systematics of *Heliotropium* section *Orthostachys* in Mexico." Unpublished Ph.D. thesis, Harvard University, Cambridge, MA.

Gadek, P. A., Fernando, E. S., Quinn, C. J., Hoot, S. B., Terrazas, T., Sheahan, M. C., and Chase, M. W. (1996). Sapindales: Molecular delimitation and infraordinal groups. *Amer. J. Bot.* **83**, 802–811.

Gaut, B. S., and Doebley, J. F. (1997). DNA sequence evidence for the segmental allotetraploid origin of maize. *Proc. Natl. Acad. Sci. U.S.A.*, In press.

Gould, F. W., and Shaw, R. B. (1983). Grass systematics, 2nd ed. Texas A&M University Press, College Station, Texas.

Grula, J. W., and Hudspeth, R. L. (1987). The phosphoenolpyruvate gene family of maize. *In* "Plant Gene Systems and Their Biology" (J. L. Key and I. McIntosh, eds.), pp. 207–216. Liss, New York.

Guttierez, M., Gracen, V. E., and Edwards, G. E. (1974). Biochemical and cytological relationships in C_4 plants. *Planta* **199**, 279–300.

Hamby, R. K., and Zimmer, E. A. (1988). Ribosomal RNA sequences for inferring phylogeny within the grass family (Poaceae). *Pl. Syst. Evol.* **160**, 29–37.

Hatch, M. D., Kagawa, T., and Craig, S. (1975). Subdivision of C_4-pathway species based on differing C_4 acid decarboxylating systems and ultrastructural features. *Aust. J. Plant Physiol.* **2**, 111–118.

Hattersley, P. W., and Browning, A. J. (1981). Occurrence of the suberized lamella in leaves of grasses of different photosynthetic type. I. In parenchymatous bundle sheath and PCR ('Kranz') sheaths. *Protoplasma* **109**, 371–401.

Hattersley, P. W., and Perry, S. (1984). Occurrence of the suberized lamella in leaves of grasses of different photosynthetic type. II. In herbarium material. *Aust. J. Bot.* **32**, 465–473.

Hattersley, P. W., and Watson, L. (1975). Anatomical parameters for predicting photosynthetic pathways of grass leaves: The 'maximum lateral cell count' and the 'maximum cells distant count.' *Phytomorphology* **25**, 325–333.

Hattersley, P. W., and Watson, L. (1992). Diversification of photosynthesis. *In* "Grass Evolution and Domestication," (G. P. Chapman, ed.), pp. 38–116. Cambridge University Press, Cambridge.

Hermans, J, and Westhoff, P. (1992). Homologous genes for the C_4 isoform of phosphoenolpyruvate carboxylase in a C_3 and a C_4 *Flaveria* species. *Mol. Gen. Genet.* **234**, 275–284.

Hershkovitz, M. A., and Zimmer, E. A. (1997). On the evolutionary origins of the cacti. *Taxon* **46**, 217–232.

Johnston, I. M. (1928). Studies in the Boraginaceae. VII. *Contr. Gray Herb.* **81**, 3–83.

Judd, W. S., Sanders, R. W., and Donoghue, M. J. (1994). Angiosperm family pairs: Preliminary phylogenetic analyses. *Harv. Pap. Bot.* **5**, 1–51.

Karis, P. O., and Ryding, O. (1994a). Tribe Helenieae. *In* "Asteraceae: Cladistics and classification" (K. Bremer, ed.), pp. 521–558. Timber Press, Portland, Oregon.

Karis, P. O., and Ryding, O. (1994b). Tribe Heliantheae. *In* "Asteraceae: Cladistics and classification" (K. Bremer, ed.), pp. 559–624. Timber Press, Portland, Oregon.

Kellogg, E. A. (1998). Relationships of cereal crops and other grasses. *Proc. Nat. Acad. Sci. U.S.A.* **95**, 2005–2010.

Kellogg, E. A., and Campbell, C. S. (1987) Phylogenetic analyses of the Gramineae. *In* "Grass Systematics and Evolution" (T. R. Soderstrom, K. W. Hilu, C. S. Campbell, and M. E. Barkworth, eds.), pp. 217–224. Smithsonian Press, Washington, D.C.

Kellogg, E. A., Appels, R., and Mason-Gamer, R. J. (1996). When genes tell different stories: The diplid genera of Triticeae (Gramineae). *Syst. Bot.* **21**, 321–347.

Kim, K.-J., and Jansen, R. K. (1995). *ndhF* sequence evolution and the major clades in the sunflower family. *Proc. Nat. Acad. Sci. U.S.A.* **92**, 10379–10383.

Kopriva, S., Chu, C.-C., and Bauwe, H. (1996). Molecular phylogeny of *Flaveria* as deduced from the analysis of nucleotide sequences encoding the H-protein of the glycine cleavage system. *Plant Cell Environ.* **19**, 1028–1036.

Ku, M. S. B., Kano-Murakami, Y., and Matsuoka, M. (1996). Evolution and expression of C_4 photosynthesis genes. *Plant Physiol.* **111,** 949–957.

Lepiniec, L., Vidal, J., Chollet, R., Gadal, P., and Crétin, C. (1994). Phosphoenolpyruvate carboxylase: Structure, regulation and evolution. *Plant Sci.* **99,** 111–124.

Les, D. H., Cleland, M. A., and Waycott, M. (1997). Phylogenetic studies in Alismatidae, II: Evolution of marine angiosperms (seagrasses) and hydrophily. *Syst. Bot.* **22,** 443–463.

Levin, G. A., and Simpson, M. G. (1994). Phylogenetic implications of pollen ultrastructure in the Oldfieldioideae (Euphorbiaceae). *Ann. Missouri Bot. Gard.* **81,** 203–238.

Linder, H. P. (1987). The evolutionary history of the Poales/Restionales—a hypothesis. *Kew Bull.* **42**(2), 297–318.

Linder, H. P., and Kellogg, E. A. (1995). Phylogenetic patterns in the commelinid clade. In "Monocotyledons: Systematics and Evolution" (P. J. Rudall, P. J. Cribb, D. F. Cutler, and C. J. Humphries, eds.), pp. 511–542. Royal Botanic Gardens, Kew.

Loockerman, D. J. (1996) Phylogenetic and molecular evolutionary studies of the Tageteae (Asteraceae). Ph.D. thesis. University of Texas, Austin.

Ludwig, M., and Burnell, J. N. (1996). Molecular comparison of carbonic anhydrase from *Flaveria* demonstrating different photosynthetic pathways. *Plant Mol. Biol.* **29,** 353–365.

Maddison, W. P. (1997). Gene trees in species trees. *Syst. Biol.* **46,** 523–536.

Manhart, J. R., and Rettig, J. H. (1994). Gene sequence data. In "Caryophyllales: Evolution and Systematics" (H.-D. Behnke and T. J. Mabry, eds.), pp. 235–246. Springer-Verlag, Berlin.

Mathews, S., and Sharrock, R. A. (1996). The phytochrome gene family in grasses (Poaceae): A phylogeny and evidence that grasses have a subset of the loci found in dicot angiosperms. *Mol. Biol. Evol.* **13,** 1141–1150.

Matsuoka, M. (1995). The gene for pyruvate, orthophosphate dikinase in C_4 plants: Structure, regulation and evolution. *Plant Cell Physiol.* **36,** 937–943.

Matsuoka, M., Kyozuka, J., Shimamoto, K., and Kano-Murakami, Y. (1994). The promoters of two carboxylases in a C_4 plant (maize) direct cell-specific, light-regulated expression in a C_3 plant (rice). *Plant J.* **6,** 311–319.

McDade, L. A., and Moody, M. L. (1997). Phylogenetic relationships among Acanthaceae: Evidence from nuclear and chloroplast genes. *Amer. J. Bot.* **84**(6, suppl.), 214–215.

Nadot, S., Bajon, R., and Lejeune, B. (1994). The chloroplast gene *rps4* as a tool for the study of Poaceae phylogeny. *Pl. Syst. Evol.* **191,** 27–38.

O'Kane, S. L. J., Schaal, B. A., and Al-Shehbaz, I. A. (1996). The origins of *Arabidopsis suecica* (Brassicaceae) as indicated by nuclear rDNA sequences. *Syst. Bot.* **21,** 559–566.

Olmstead, R. G., and Reeves, P. A. (1995). Polyphyletic origin of the Scrophulariaceae: Evidence from *rbcL* and *ndhF* sequences. *Ann. Miss. Bot. Gard.* **82,** 176–193.

P'yankov, V. I., Voznesenskaya, E. V., Kondratschuk, A. V., and C. C. Black, Jr. (1997). A comparative anatomical and biochemical analysis of *Salsola* (Chenopodiaceae) species with and without a Kranz type leaf anatomy: A possible reversion of C_4 to C_3 photosynthesis. *Amer. J. Bot.* **84,** 597–606.

Plunkett, G. M., Soltis, D. E., Soltis, P. S., and Brooks, R. E. (1995). Phylogenetic relationships between Juncaceae and Cyperaceae: Insights from *rbcL* sequence data. *Amer. J. Bot.* **82,** 520–525.

Poetsch, W., Hermans, J., and Westhoff, P. (1991). Multiple cDNAs of phosphoenolpyruvate carboxylase in the C_4 dicot *Flaveria trinervia*. *FEBS Lett.* **292,** 133–136.

Powell, A. M. (1978). Systematics of *Flaveria* (Flaveriinae—Asteraceae). *Ann. Miss. Bot. Gard.* **65,** 590–636.

Prendergast, H. D. V., Hattersley, P. W., and Stone, N. E. (1987). New structural/biochemical associations in leaf blades of C_4 grasses (Poaceae). *Aust. J. Plant Physiol.* **14,** 403–420.

Price, R. A., Palmer, J. D., and Al-Shehbaz, I. A. (1994). Systematic relationships of *Arabidopsis*: A molecular and morphological perspective. In "*Arabidopsis*" (E. M. Meyerowitz and C. R. Somerville, eds.), pp. 7–19. Cold Spring Harbor Laboratory Press, Plainview, NY.

Rawsthorne, S. (1992). C_3–C_4 intermediate photosynthesis: Linking physiology to gene expression. *Plant J.* **2**, 267–274.

Rodman, J., Soltis, P., Soltis, D., Sytsma, K., and Karol, K. (1997). Plastid *rbcL* shouts and nuclear 18S-ribosomal DNA whispers, but the message is the same: Dahlgren cuts the mustard. *Amer. J. Bot.* **84**(6, suppl.), 226.

Rodman, J. E. (1994). Cladistic and phenetic studies. *In* "Caryophyllales: Evolution and Systematics" (H.-D. Behnke and T. J. Mabry, eds.), pp. 279–301. Springer-Verlag, Berlin.

Rodman, J. E., Karol, K. G., Price, R. A., and Sytsma, K. J. (1996). Molecules, morphology, and Dahlgren's expanded order Capparales. *Syst. Bot.* **21**, 289–307.

Ryding, O., and Bremer, K. (1992). Phylogeny, distribution, and classification of the Coreopsidae (Asteraceae). *Syst. Bot.* **17**, 649–659.

Schäffner, A. R., and Sheen, J. (1992). Maize C_4 photosynthesis involves differential regulation of phosphoenolpyruvate carboxylase genes. *Plant J.* **2**, 221–232

Scotland, R. W., Sweere, J. S., Reeves, P. A., and Olmstead, R. G. (1995) Higher level systematics of Acanthaceae determined by chloroplast DNA sequences. *Amer. J. Bot.* **82**, 266–275.

Sheahan, M. C., and Chase, M. W. (1996). A phylogenetic analysis of Zygophyllaceae R. Br. based on morphological, anatomical and *rbcL* DNA sequence data. *Bot. J. Linn. Soc.* **122**, 279–300.

Simpson, D. (1995). Relationships within Cyperales. *In* "Monocotyledons: Systematics and Evolution" (P. J. Rudall, P. J. Cribb, D. F. Cutler, and C. J. Humphries, eds.), pp. 497–509. Royal Botanic Gardens, Kew.

Sinha, N. R., and Kellogg, E. A. (1996). Parallelism and diversity in multiple origins of C_4 photosynthesis in grasses. *Amer. J. Bot.* **83**, 1458–1470.

Soltis, D. E., Soltis, P. S., Morgan, D. R., Swensen, S. M., Mullin, B. C., Dowd, J. M., and Martin, P. G. (1995). Chloroplast gene sequence data suggest a single origin of the predisposition for symbiotic nitrogen fixation in angiosperms. *Proc. Nat. Acad. Sci. U.S.A.* **92**, 2647–2651.

Soltis, D. E., Soltis, P. S., Nickrent, D. L., Johnson, L. A., Hahn, W. J., Hoot, S. B., Sweere, J. A., Kuzoff, R. K., Kron, K. A., Chase, M. W., Swensen, S. M., Zimmer, E. A., Chaw, S.-M., Gillespie, L. J., Kress, W. J., and Sytsma, K. J. (1997). Angiosperm phylogeny inferred from 18S ribosomal DNA sequences. *Ann. Miss. Bot. Gard.* **84**, 1–49.

Stockhaus, J., Schlue, U., Koczor, M., Chitty, J. A., Taylor, W. C., and Westhoff, P. (1997). The promoter of the gene encoding the C_4 form of phosphoenolpyruvate carboxylase directs mesophyll-specific expression in transgenic C_4 *Flaveria* spp. *Plant Cell* **9**, 479–489.

Tanaka, N., Setoguchi, H., and Murata, J. (1997). Phylogeny of the family Hydrocharitaceae inferred from *rbcL* and *matK* gene sequence data. *J. Plant Res.* **110**, 329–337.

Toh, H., Kawamura, T., and Izui, K. (1994). Molecular evolution of phosphoenolpyruvate carboxylase. *Plant Cell Environ.* **17**, 31–43.

Wagstaff, S. J., and Olmstead, R. G. (1997). Phylogeny of Labiatae and Verbenaceae inferred from *rbcL* sequences. *Syst. Bot.* **22**, 165–179.

Warwick, S. I., and L. D. Black. (1997). Phylogenetic implications of chloroplast DNA restriction site variation in subtribes Raphaninae and Cakilinae (Brassicaceae, tribe Brassiceae). *Can. J. Bot.* **75**, 960–973.

Watson, L., Clifford, H. T., and Dallwitz, M. J. (1985). The classification of Poaceae: subfamilies and supertribes. *Aust. J. Bot.* **33**, 433–484.

Watson, L., and Dallwitz, M. J. (1992). "The Grass Genera of the World." CAB International, Wallingford.

Webster, G. L. (1994). Synopsis of the genera and suprageneric taxa of Euphorbiaceae. *Ann. Miss. Bot. Gard.* **81**, 33–144.

Wettstein, R. von (1891). Scrophulariaceae. *In* "Die Natürlichen Pflanzenfamilien" Vol. IV. Teil. 3. Abteilung b., (A. Engler and K. Prantl, eds.), pp. 39–107. Wilhelm Englemann, Leipzig.

Winter, K., and Smith, J. A. C. (eds.). (1995). "Crassulacean acid metabolism: Biochemistry, ecophysiology and evolution." Springer-Verlag, Berlin.

Yanai, Y, Okumura, S., and Shimada, H. (1994). Structure of *Brassica napus* phospho*enol*pyruvate carboxylase genes: Missing introns causing polymorphisms among gene family members. *Biosci. Biotech. Biochem.* **58,** 950–953.

13

Paleorecords of C_4 Plants and Ecosystems

Thure E. Cerling

I. Introduction

C_4 plants are different from C_3 plants in anatomy (Hatch and Slack, 1970) and in their $\delta^{13}C$ values (Bender, 1968; Smith and Epstein, 1971). These differences can be used in the geological record to determine the presence of C_4 plants. C_4 plants have Kranz anatomy that has both mesophyll cells, in which CO_2 is fixed by C_4 acids, and bundle-sheath cells, where RuBP carboxylase fixes CO_2 derived from the C_4 acids of the mesophyll cells. The different photosynthetic pathways have different $\delta^{13}C$ values, which result from their different biochemical pathways of CO_2 fixation, averaging about $-26‰$ to $-27‰$ for C_3 plants and about $-12‰$ for C_4 plants. Verification of Kranz anatomy or unequivocal stable isotope evidence can therefore be used to identify C_4 plants or C_4 ecosystems in the geological record.

The C_4 pathway is now thought to be an adaptation to low atmospheric CO_2 levels. Plant metabolism responds directly to atmospheric CO_2 concentrations (Ehleringer *et al.*, 1991; Bowes, 1993). C_4 plants are more productive than C_3 plants at low atmospheric CO_2 levels (Chapters 2 and 5). Atmospheric CO_2 levels have been greater than about 500 ppmV for most of the geological record (Fig. 1; see also Berner, 1991, 1994). Above 500 ppmV C_3 plants have higher efficiency than C_4 plants under most climate conditions (Cerling *et al.*, 1997c; Ehleringer *et al.*, 1997).

It is well known that C_3 plants photorespire when they are under temperature stress and when atmospheric CO_2 levels are low (Ehleringer *et al.*, 1991). A fundamental difference between C_3 and C_4 plants is the

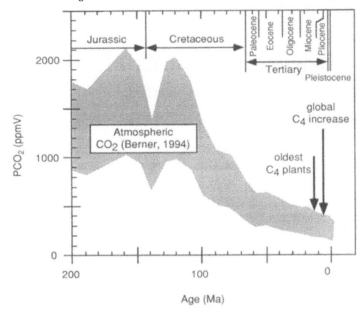

Figure 1 Geological modeling estimates of the history of atmospheric CO_2 concentrations [modified from Berner, R. A. (1991). A model for atmospheric CO_2 over Phanerozoic time. *Amer. J. Sci.* **291**, 339–376; and from Berner, R. A. (1994). GEOCARB II: A revised model of atmospheric CO_2 over Phanerozoic time. *Amer. J. Sci.* **294**, 56–91.] Oldest known C_4 plants and global C_4 expansion as discussed in the text.

quantum yield for CO_2 uptake (Ehleringer and Björkman, 1977). Although under optimal conditions it is expected that C_4 plants should have a lower quantum yield than C_3 plants because of the additional energy expense of the C_4 cycle, under current atmospheric conditions the quantum yield of C_3 plants is significantly reduced because of photorespiration. Ehleringer *et al.* (1997) and Cerling *et al.* (1997b) model the crossover for C_3 plants versus C_4 plants based on which has the greater quantum yield (Fig. 2). This model is based on the equations from Farquhar and von Caemmerer (1982) using the constants determined by Jordan and Ogren (1984), and is discussed in detail in Ehleringer *et al.* (1997). The crossover at higher CO_2 levels is at higher temperatures, such that at growing season temperatures of about 35°C the upper limit appears to be between about 400 and 600 ppmV. This means that atmospheric CO_2 concentrations would have to decrease to at least this range before C_4 plants exhibited an advantage over C_3 plants.

The C_4 photosynthetic pathway is found in many families of plants, but it is particularly prevalent in the monocots, especially the grasses and sedges.

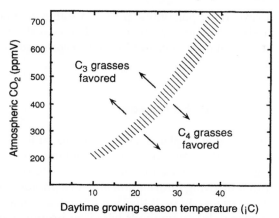

Figure 2 Crossover for C_3–C_4 photosynthesis based on relative quantum yield of grasses [from Cerling, T. E., Harris, J. M., MacFadden, B. J., Leakey, M. G., Quade, J., Eisenmann, V., and Ehleringer, J. R. (1997). Global vegetation change through the Miocene and Pliocene. *Nature* **389**, 153–158.] Shaded area represents the crossover for the different C_4 subpathways that have different quantum yields.

Grasses are relative latecomers in the geological record, with the oldest unequivocal large remains being Oligocene (*ca.* 30 million years ago) in age (Thomasson, 1986) and with possible fragments of grasses being found in Eocene deposits. Grass pollen has been positively identified in Paleocene sediments (*ca.* 55–65 Ma; Muller, 1981). Grasses are rarely preserved as fossils, although grass pollen and siliceous phytoliths are more often found. Unfortunately, except in rare cases, assigning a C_3 or C_4 pathway based on pollen or on phytolith morphologies is not possible.

C_4 dicots are not as abundant as C_4 monocots. Whereas C_4 photosynthesis occurs in perhaps 50‰ of the approximately 10,000 species of grasses (monocots), it is estimated that less than 0.5‰ of the dicots use the C_4 pathway. Ehleringer *et al.* (1997) speculate that C_4 dicots are not abundant because they have significantly lower photosynthetic quantum yields than do the C_4 monocots. Hence, C_4 dicots would be favored only in conditions of extremely low atmospheric CO_2 values such as those found during full Glacial conditions. If so then C_4 dicots would have only short periods where they were able to diversify before unfavorable conditions returned during the Interglacial periods. Therefore, whereas C_4 monocots have had a continuous period of 6–8 million years in which they were favored over C_3 monocots and dicots in tropical regions, C_4 dicots experienced only short intervals of 30,000–50,000 years in which they had an inherent advantage. Ehleringer *et al.* (1997) speculate that this is the reason for the relative paucity of C_4 dicots.

In this paper, I review the paleontological and stable isotope evidence for C_4 photosynthesis in the geological record. I discuss the problems in preservation of the critical evidence, and in the interpretation of measurements or observations.

II. Paleorecords

It has been postulated that C_4 plants have existed for at least 100 million years based on the global distribution of some of the important C_4 families and tribes of the grasses (Brown and Smith, 1972). However, the fossil record is parsimonious in giving information about C_4 photosynthesis. Most C_4 plants grow in ecosystems where, if they are buried, are oxidized so that the fossil record is sparse. Proxy methods based on isotopic compositions of associated phases can be used, but so far the evidence indicates C_4 plants were uncommon before 8 million years ago. There is no unequivocal evidence for C_4 plants older than 12.5 million years ago.

A. Morphology: Kranz Anatomy

C_4 plants have a different anatomy than C_3 plants, an anatomy recognizable in certain well-preserved fossil plants. Kranz anatomy is a specialized leaf anatomy in which vascular bundles are surrounded by radiating mesophyll cells (Chapter 5). Two important fossil occurrences of Kranz anatomy are noted, one from the Miocene Ricardo Formation in California and one from the Miocene Ogallala Formation in Kansas. A third possible occurrence in Kenya is also discussed.

The Miocene Ricardo Formation in southern California is well known for its preservation of flora and fauna (Webber, 1933; Axelrod, 1939) and is reported to have grasses having Kranz anatomy. This locality was originally reported to be Pliocene in age (Nambudiri et al., 1978) but was later reported as being Miocene in age (Tidwell and Nambudiri, 1989). (Part of the confusion of the ages was due to the definition of the Miocene–Pliocene boundary.) The Last Chance Canyon locality in the Dove Spring Formation, the site described by Nambudiri et al. (1978), in the Ricardo Group of southern California is estimated to be about 12.5 Ma (Whistler and Burbank, 1992; Whistler, personal communication, 1996) based on work using radiometric dates and using paleomagnetic records. Fossil grasses from this locality are very well preserved and have been shown to possess the characteristic Kranz anatomy (Nambudiri et al., 1978; Tidwell and Nambudiri, 1989) with low interveinal distances of 3–5 mesophyll cells separating the vascular bundles. This site is the oldest documented locality with Kranz anatomy. Axelrod (1973) interprets the fossil flora of the Last Chance Canyon locality to indicate a thorn scrub to chaparral vegetation like those characteristic of Mediterranean climates, although modern Mediterranean climates, including southern California, have C_3 grasses rather than C_4 grasses. Webber (1933) and Tidwell and Nambudiri (1989) report palms, oaks, and other nongrasses as important components of the fossil flora at the Last Chance Canyon locality. It is possible that this occurrence was associated with perennial springs. (Along similar lines, vernal pools in California today contain C_4 grasses in the genera *Orcuttia* and *Neostapfia*. These grow in Mediterranean settings, with the water in the pools allowing for robust summer growth; R. F. Sage, personal communication, 1997; Keeley 1998). Stable isotope analyses of fossil equids from the Ricardo Formation (discussed later) indicate a C_3-dominated diet. Therefore, we must conclude that C_4 plants, although present, were not abundant at this locality.

A second important Miocene site where Kranz anatomy has been documented is the Minium Quarry in Graham County, Kansas (Thomasson *et al.*, 1986). Grass from this site also shows the presence of vascular bundles that are separated by no more than four mesophyll cells, a feature thought to characterize C_4 anatomy. This site is late Hemphillian in age, between about 5 and 7 million years ago. Proxy isotope evidence (discussed later) suggests that the late Hemphillian was a period of expansion of C_4 biomass in North America.

One other locality that has been put forth as having possible evidence for C_4 grasses is based on cuticle morphology. Fort Ternan (14 million years old) in Kenya has well-preserved grasses that have been studied by Retallack *et al.* (1990), Retallack (1992), and by Dugas and Retallack (1993). Based primarily on cuticle morphology, several different fossil species were recognized that were thought to be most closely related to extant C_4 genera. Unfortunately, Kranz anatomy could not be identified because the interior of the grasses was replaced by calcite or clay, which does not preserve delicate internal features. Silica bodies were preserved, however, and tentative identification and generic assignment was made on that basis. Based on stable isotope evidence of organic carbon and carbonate in fossil soils and on estimates of diets based on the stable isotopic composition of fossil tooth enamel, the ecosystem was dominated by C_3 plants (Cerling *et al.*, 1991, 1992, 1997b). Therefore, the evidence to date for C_4 plants from the 14-million-year-old site of Fort Ternan does not indicate abundant C_4 biomass, but the presence of C_4 plants is not precluded.

Thus, the oldest known specimens with Kranz anatomy are of middle Miocene age, about 12.5 Ma, from the Last Chance Canyon locality in the Ricardo Formation, California. As is shown later, C_4 grasses do not become common anywhere in the world until about 5 million years later.

B. $\delta^{13}C$ of Individual Plants

The $\delta^{13}C$ value of terrestrial C_4 plants is different than that of C_3 plants. The $\delta^{13}C$ values for individual terrestrial plants ranges from about $-10\permil$

to −35‰ (Fig. 3). C_4 plants have a fairly limited range, from about −10 to −14‰, whereas C_3 plants have a much larger range, from about −20 to −35‰. Therefore, studies of the $\delta^{13}C$ value of fossil plant material can be used to help determine whether a plant used the C_4 photosynthetic pathway.

The $\delta^{13}C$ of C_3 plants has a broad range, which is related to their response to water, local variations in the $\delta^{13}C$ of the local atmosphere, and to light levels. In general, the $\delta^{13}C$ of C_3 plants becomes a few ‰ more positive

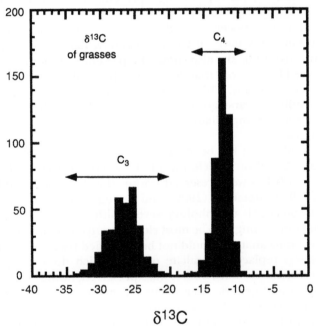

Figure 3 $\delta^{13}C$ values of individual grasses. [Compiled from Smith, B. N., and Epstein, S. (1971). Two categories of $^{13}C/^{12}C$ ratios for higher plants. *Plant Physiol.* **47**, 380–384; Smith, B. N., and Brown, W. V. (1973). The Kranz Syndrome in the Graminae as indicated by carbon isotopic ratios. *Amer. J. Bot.* **60**, 505–513; Winter, K., Troughton, J. H., and Card, K. A. (1976). $\delta^{13}C$ values of grass species collected in the northern Sahara Desert. *Oecologia* **25**, 115–123; Brown, W. V. (1977). The Kranz syndrome and its subtypes in grass systematics. *Mem. Torrey Bot. Club* **23**, 1–97; Vogel, J. C., Fuls, A., and Ellis, R. P. (1978). The geographical distrubution of Kranz grasses in Sought Africa. *S. Afr. J. Sci.* **74**, 209–215; Hattersley, P. W. (1982). ^{13}C values of C_4 types in grasses. *Aust. J. Plant Physiol.* **9**, 139–154; Hattersley, P. W., and Roksandic, Z. (1983). ^{13}C values of C_3 and C_4 species of Australian Neurachne and its allies (Poaceae). *Aust. J. Bot.* **31**, 317–321; Ohsugi, R., Samejima, M., Chonan, N., and Murata, T. (1988). ^{13}C values and the occurrence of suberized lamellae in some *Panicum* species. *Ann. Bot.* **62**, 53–59; and unpublished data.]

under highly water stressed conditions because the isotopic discrimination decreases so that $\delta^{13}C$ values of -21‰ to -22‰ are not uncommon in semiarid and arid regions, and occasional values of -20‰ are observed (Ehleringer, 1993). Under closed canopy conditions, the $\delta^{13}C$ becomes more negative because of changes in the isotopic composition of the atmosphere and because the isotopic discrimination increases at very low light levels. Van der Merwe and Medina (1989) report $\delta^{13}C$ values as low as -35‰ for plants growing at the base of the canopy in the Amazon rain forest.

The $\delta^{13}C$ of C_4 plants is limited, ranging from about -10‰ to -14‰. Much of this variation is encompassed in the slight but significant differences between the subpathways, with the NADP-ME type being the most enriched in ^{13}C and the NAD-ME being the most depleted in ^{13}C with average values about -11.4‰ and -12.7‰, respectively (Hattersley, 1982; Ohsugi et al., 1988). The PCK subpathway has intermediate values.

Therefore, organic matter preserved in the geological record, even if fractionated, has the potential to give an indication of the presence of significant C_4 biomass. For the purposes of identification of terrestrial plants using stable isotopes, a $\delta^{13}C$ "cutoff" value for biomass greater than about -20‰ probably indicates C_4 biomass. A more positive $\delta^{13}C$ "cutoff" value, however, may be needed for geological periods when the isotopic composition of the atmosphere was significantly higher than the pre-Industrial value of approximately -6.5‰. For example, it is thought that the $\delta^{13}C$ of the late Paleozoic atmosphere may have been as positive as -1‰ (Berner, 1994), so that C_3 plants with a discrimination of 15‰ would have a $\delta^{13}C$ of -16‰. As noted later, algae and other aquatic C_3 plants can have $\delta^{13}C$ values in the range of C_4 plants. Thus, high $\delta^{13}C$ values of plants alone is not definitive of C_4 biomass.

Terrestrial and aquatic plants must be considered separately because some aquatic plants have the ability to fix CO_2 from HCO_3^- dissolved in water (Raven, 1970). Dissolved HCO_3^- is enriched in $\delta^{13}C$ by about 8‰ compared to dissolved CO_2 (Mook et al., 1974). Aquatic plants that derive a significant fraction of carbon from HCO_3^- are enriched in ^{13}C compared to those that use only dissolved CO_2 during photosynthesis (Keeley et al., 1986; Keeley, 1990; Keeley and Sandquist, 1992). This generally occurs when CO_2 content in the water column is very low during intense photosynthesis. Therefore, $\delta^{13}C$ values cannot be used to indicate the photosynthetic pathway of aquatic plants (Keeley, 1990). However, this isotope effect is very important in studying productivity in marine and other aquatic ecosystems because high productivity significantly lowers CO_2 concentration (e.g., Rau and Takahashi, 1989; Hollander and McKenzie, 1991). Therefore, the following remarks are directed to terrestrial ecosystems where the bicarbonate isotope effect is not important.

Only a few studies have used carbon isotopes to search for, or to identify, fossil C_4 plants. Nambudiri *et al.* (1978) and Tidwell and Nambudiri (1989) showed that a fossil grass, *Tomlinsonia thomassonii*, from the Ricardo Formation in California had Kranz anatomy (discussed previously) and a $\delta^{13}C$ value of $-13.7‰$, indicating that they used the C_4 photosynthetic pathway. This plant also has Kranz anatomy and is the oldest documented C_4 plant so far. Cerling *et al.* (1998a) analyzed three equids from this locality between 12 and 13 Ma age and found that there was little, if any, C_4 biomass in the diets of those equids. Thus it seems that the oldest documented C_4 plant, a fossil grass from southern California, was a minor part of the local biomass. It was not until about 6–8 million years ago, several million years after *T. thomassonii* made its appearance, that C_4 plants made up a large fraction of any ecosystem anywhere in the world (Cerling *et al.*, 1997b).

Bocherens *et al.* (1994) conducted a systematic search for C_4 plants using carbon isotopes in fossil plants ranging in age from middle Triassic (*ca.* 225 million years ago) to late Tertiary (*ca.* 2 million years ago). They attempted to sample fossil plants from environments ranging from humid to xeric. Individual plant $\delta^{13}C$ values ranged from $-19‰$ to $-28‰$. Modern C_3 plants have $\delta^{13}C$ values as positive as $-20‰$, with such positive $\delta^{13}C$ values being from plants undergoing light and water stress. Allowing for a permil or more variation in the isotopic composition of the atmosphere, and a permil or more due to diagenesis, the values reported by Bocherens *et al.* (1994) are compatible with all of the plants studied having used the C_3 photosynthetic pathway. He also reports a lignite fragment with a $\delta^{13}C$ of $-17.3‰$ and speculates that it is possible that this sample used the CAM photosynthetic pathway.

In summary, the systematic search for plants using stable isotopes to identify the C_4 photosynthetic pathway has not yet identified any plants using the C_4 pathway older than about 12 million years, although not many analyses have been carried out in this type of search. No C_4 dicots or C_4 gymnosperms, have been identified in the fossil record in Tertiary or older sediments (i.e., older than *ca.* 2.5 Ma). Unfortunately, the type of environment where one expects to find C_4 plants (i.e., one that is hot and dry) is not conducive to the preservation of organic matter.

C. $\delta^{13}C$ of Terrestrial Ecosystems

The fraction of C_4 biomass in an ecosystem can be estimated by measuring the $\delta^{13}C$ value of the average carbon in the ecosystem. There are several ways that carbon can be preserved in the geological record, including plant organic carbon, collagen or amino acids derived from animal remains, calcium carbonate, in iron hydroxide minerals, and as biological apatite. For the different phases, the transfer function relating carbon in plants to that preserved in soils or animal fossils has to be determined. Figure 3 showed the wide range of $\delta^{13}C$ values for C_3 plants and the narrower range of $\delta^{13}C$ values for C_4 plants. For the average biomass, unaltered from its original isotopic composition, a simple mixing curve describes the fraction of C_4 biomass in an ecosystem (Fig. 4).

The wide range in the $\delta^{13}C$ of C_3 plants means that interpretation of $\delta^{13}C$ values based on bulk biomass requires some care. For example, a $\delta^{13}C$ value of -23‰ could be either the pure C_3 endmember, or it could be a mixture of 71‰ C_3 biomass with a $\delta^{13}C$ value of -27‰ and 29‰ C_4 biomass with a $\delta^{13}C$ value of -13‰. Unequivocal evidence for C_4 biomass must have $\delta^{13}C$ values that are significantly outside the range of C_3 vegetation, and some precaution must be taken to allow for CAM plants that have $\delta^{13}C$ values that are intermediate to C_3 and C_4 plants. Some discussion of the reasons for the wide range in $\delta^{13}C$ values is in order, because it affects interpretations of $\delta^{13}C$ values and their related ecosystems. Although the average $\delta^{13}C$ of C_3 plants is about -26‰ to -27‰, different ecosystems have systematic differences in the $\delta^{13}C$ of plants growing in those ecosystems. Two of the carbon isotope endmembers are of particular interest. In areas of high temperature and high light $\delta^{13}C$ values of individual plants tend to be shifted toward more positive values, because plants in these environments tend to close their stomata to prevent water loss and, in so doing, inhibit CO_2 exchange. Thus, arid ecosystems tend to have C_3 plants with $\delta^{13}C$ values that are on the order of about -24 to -21‰ (e.g., Ehleringer and Cooper, 1988; Farquhar *et al.*, 1989; Koch *et al.* 1991; Ehleringer, 1993; William and Ehleringer, 1996). At the other extreme, C_3 plants grown under very low light levels, such as closed canopy forests, are very depleted in ^{13}C with $-^{13}C$ values often between -30‰ and -35‰ (e.g., Medina and Minchin, 1980; van der Merwe and Medina, 1989; van der Merwe and Medina, 1991). Because C_4 plants do not grow in closed canopy environments, no mixing line is shown in Fig. 4 for closed-canopy-end members with C_4-end members, although it could be argued (and is an important issue to pursue in dietary studies) that glades of C_4 grasses are present near closed canopies.

Some phases that preserve an ecosystem isotope signal include soil organic matter, peats, soil carbonate or soil goethite, and dietary carbon preserved as collagen, calcium carbonate, or apatite (calcium phosphate). These are addressed in the following paragraphs.

1. Reduced Organic Carbon: Soils, Paleosols, and Peats Organic carbon can be preserved in geological environments that are relatively reducing: highly oxidized environments rarely preserve organic matter over long periods of geological time. Studies of the $\delta^{13}C$ of organic matter in modern soils has been used to estimate the fraction of C_4 biomass in modern ecosystems (e.g., Cerling *et al.*, 1989; Bird *et al.*, 1994). Bird *et al.* (1994) studied the

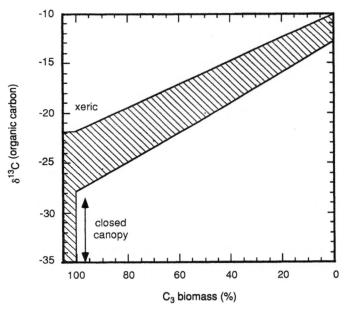

Figure 4 Mixing lines for relative biomass between the C_3 and C_4 ecosystem endmembers. The C_3 endmember for mixing was taken to be $-23‰$ for xeric environments and $-28‰$ for mesic environments. The C_4 endmember was taken to be between -10 and $-13‰$. No mixing with the closed canopy endmember for C_3 plants (ca. $-35‰$) is shown because C_4 plants are extremely rare under closed canopy conditions.

$\delta^{13}C$ of soils in an altitude transect in Papua New Guinea; they point out some of the potential problems that must be considered in using $\delta^{13}C$ values as indicators of the fraction of C_4 biomass in modern soils. In fossil soils, these same problems must be evaluated. Bird et al. (1994) show that forest soils with no C_4 plants (i.e., 100‰ C_3) have a considerable range in the $\delta^{13}C$ values because discrimination in C_3 plants varies in different environments. They found the discrimination in C_3 forest soils to be well correlated with altitude because of changes in the rainfall and temperature. Several other studies have used the $\delta^{13}C$ in modern soils to study Holocene or late Pleistocene vegetation (Schwartz et al., 1986; Guillet et al., 1988; Ambrose and Sikes, 1991; Wang et al., 1993). Wang et al. (1993) studied forest-prairie transitional soils in the midwestern United States. They found the organic matter was in equilibrium with the modern ecosystem (except in agriculturally disturbed soils) but the carbonate component recorded an earlier history of vegetation for the soils, when forest was more dominant than at present.

Peat records the isotopic composition of surrounding vegetation and is a very promising avenue to study ecological changes in the Holocene and late Pleistocene. Changes in the fraction of C_4 biomass should be preserved in the carbon isotopic record of the organic matter. Likewise, lake sediments and sometimes marine sediments (e.g., the Bengal Fan studied by France-Lanord and Derry, 1994) have their organic carbon dominated by terrestrial, rather than lacustrine or marine, input. Using compound specific isotope analysis (e.g., alkenones that are characteristic components of leaf waxes) it may be possible to separate the terrestrial component from the lacustrine (or marine) component in lacustrine (or marine) deposits. For lakes it is particularly important that the algal component be minimized because of problems associated with HCO_3^- assimilation and associated isotope shifts (mentioned previously).

As discussed previously, the theoretical crossover for C_4/C_3 grasses is about 20°C (daytime temperature) for the modern atmospheric level of CO_2 (280 ppmV for the preindustrial value). Figure 2 has some very interesting implications with respect to changes in vegetation between glacial and interglacial conditions. If the temperature change is relatively small and the CO_2 change is large, then cooler temperatures could lead to the expansion of C_4 biomass if the change in CO_2 level was large enough. Glacial atmospheric CO_2 levels were on the order of 180 ppmV (Jouzel et al., 1987; Sowers and Bender, 1995) and the estimated temperature changes were on the order of 5°C lower in the tropics compared to 10–15°C lower at high latitudes (Jouze, et al., 1987; Cuffey et al., 1995; Stute et al., 1995). Therefore it is expected that in tropical regions C_4 plants would have been at a greater advantage over C_3 plants than under modern conditions.

C_4 plants were more abundant at both high and low elevations in tropical Africa during the last full glacial conditions compared to the present (Fig. 5). Tropical forests and woodlands contracted and grasslands or savannas expanded under glacial conditions and that the cores of the tropical rainforests were likely refugia during glacial conditions (Cerling et al., 1998b in review). Bird and Cali (1997) characterized charcoal in sediments cores off the northwest coast of Africa and found that charcoal maxima coincided with glacial conditions, and that C_4 plants also reached higher proportions in each of the last three glacial maxima (lower CO_2) returning to C_3 dominated biomass during interglacial conditions (higher CO_2).

These observations and speculations of Cerling et al. (1997b) and Ehleringer et al (1997) provide a different model, "the CO_2-starvation model," for vegetation changes at the end of the Pleistocene than is generally used. The traditional model of vegetation change for glacial conditions compared to interglacial conditions has temperature differences and precipitation differences as the most important parameters for ecological control. In the "CO_2-starvation model," an important driver for ecological change is the

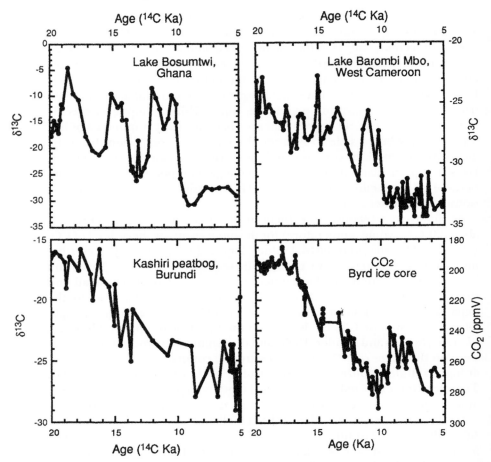

Figure 5 Chronological profiles of the $\delta^{13}C$ of organic matter in lakes and peat bogs from equatorial Africa. Lake Bosumtwi [Talbot, M. R., and Johanessen, T. (1992). A high resolution palaeoclimate record for the last 27,500 years in tropical west Africa from the carbon and nitrogen isotopic composition of lacustrine organic matter. *Earth Planet. Sci. Lett.* **110**, 23–37.] and Lake Barombi Mbo [Giresse, P., Maley, J., and Brenac, P. (1994). Late Quaternary palaeoenvironments in the Lake Barombi Mbo (West Cameroon) deduced from pollen and carbon isotopes of organic matter. *Paleogeog. Palaeoclim. Palaeoecol.* **107**, 65–78.] are low elevation sites (<500 msl), whereas Kashiri peat bog [Aucour, A. M., Hillaire, M. C., and Bonnefil, R. (1993). A 30,000 year record of 13C and 18O changes in organic matter from an equatorial peatbob. *In* "Climate Change in Continental Isotopic Records" (Swart, P. K., Lohmann, K. C., McKenzie, J. A., and Savin, S., eds.), *Am. Geophys. Un. Geophys. Mono.* **78**, 343–351.] is a high elevation site (>2000 msl). [Modified from Ehleringer, J. R., Cerling, T. E., and Helliker, B. R. (1997). C_4 photosynthesis, atmospheric CO_2, and climate. *Oecologia* **112**, 285–299.] pCO_2 in the atmosphere from [Sowers, T., and Bender, M. (1995). Climate records covering the last deglaciation. *Science* **269**, 210–214.]

increased photorespiration of C_3 plants accompanied by decreased water use efficiency under lower atmospheric CO_2 levels. Decreased water use efficiency would appear to have the same result as aridity but it would not require changes in aridity *per se* and changes in the relative rates of photorespiration would change the net photosynthetic rates putting other plants, especially C_4 plants, at an ecological advantage. If this model is valid, it has important implications about the competitiveness of different ecosystems in the modern world where CO_2 levels are increasing because of human activities.

2. Oxidized Carbon Derived from Organic Matter: Soils The oxidation of organic matter produces CO_2 whose isotopic composition is related to the isotopic composition of the original material. The isotopic composition of CO_2 derived from the oxidation of soil organic matter, or from CO_2 respired by plants, can be recorded in soils as pedogenic carbonate. Pedogenic carbonate is often preserved in sedimentary rocks in the form of paleosols (fossil soils), leaving a record of the $\delta^{13}C$ at the ecosystem level. The distribution of $\delta^{13}C$ in soil CO_2 has been described using a diffusion–production model (Cerling, 1984) and has been tested against a variety of modern soils and ecosystems (Cerling *et al.*, 1989; Quade *et al.*, 1989; Cerling and Quade, 1993). There is a 14–17‰ enrichment in pedogenic carbonate compared to primary organic matter or soil-respired CO_2. This is due to the combined effects of diffusion (4–5‰) and fractionation between CO_2 and calcite (10–12‰, depending on temperature). Pedogenic carbonate forms in soils where precipitation is less than about 0.5 m/yr in boreal climates, less than about 1 m/year in temperate climates, and less than about 1.5 m/yr in tropical climates. It is not found under modern closed canopy conditions where ^{13}C depletion in ground level C_3 plants has been observed (e.g., Medina and Minchin, 1980; van der Merwe and Medina, 1991). Therefore, pedogenic carbonate formed in a pure C_3 ecosystems has $\delta^{13}C$ values from about −9‰ to −12‰ and those formed under a pure C_4 vegetation have $\delta^{13}C$ values from about 1‰ to 3‰.

Quade *et al.* (1989a) and Quade and Cerling (1995) identified paleosols by recognizing the pattern of paleosol development: the originally bedded calcareous overbank mudstones show evidence of extensive bioturbation leading to the destruction of the original bedding; the development of a zone in which the matrix was leached of carbonate from the top of the bioturbated zone to many tens of cm depth; prominent color horizons due to translocation of iron and manganese minerals showing the leaching and accumulation of different elements; characteristic geochemical profiles; recognition of ped structure; and the observation of micromorphological features common in modern soils.

A number of studies have used the $\delta^{13}C$ of pedogenic carbonate to estimate the fraction of C_4 biomass in paleoecosystems. Research of soils and on the paleodiets shows that C_4 plants expanded globally between about 6 and 8 million years ago (Cerling *et al.*, 1993, 1997c). An example of that expansion is shown in the data from the Siwaliks in Pakistan, which were deposited between 20 million years ago and the present as a result of the erosion of the rising Himalaya mountain chain (Quade *et al.*, 1989; Quade and Cerling, 1995). Figure 6 shows the abrupt transition from a C_3-dominated biomass to a C_4-dominated biomass between about 8 and 6 million years ago. This figure also shows the isotopic composition of organic carbon from the Bengal Fan, deposited by the combined flows of the

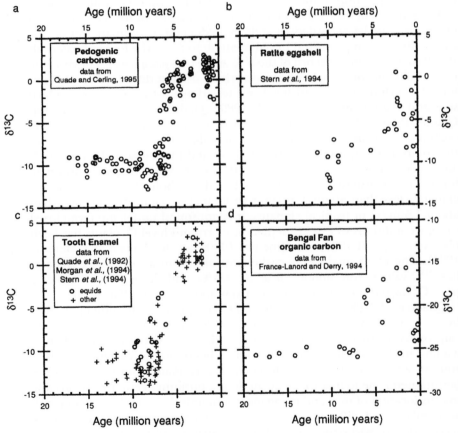

Figure 6 Carbon isotope shifts in southeast Asia during the Neogene are found in many phases, including pedogenic carbonate (a), ratite eggshell (b), fossil tooth enamel (c), and organic matter in the Bengal Fan (d). There is an enrichment of about 14‰ from organic matter to soil carbonate, eggshell, or to bioapatite. Data compiled from various sources.

Ganges and Brahmaputra Rivers, whose carbon is derived primarily by terrestrial input from the Indian subcontinent (France-Lanord and Derry, 1994). The data from the Bengal Fan shows a change from about −25‰ for organic carbon in the early and middle Miocene, to values as high as −17‰ starting about 7 million years ago. This means that the abrupt increase in C_4 biomass recorded by the paleosols from Pakistan is part of a regional change in C_4 biomass for the entire Indian subcontinent starting between 8 and 7 million years ago.

Several other papers have used $\delta^{13}C$ in paleosols to suggest that there was significant C_4 biomass before 8 million years ago (e.g., Wright and Vanstone, 1991; Kingston et al., 1994). However, each of these studies also has contradictory evidence concerning the interpretation of significant C_4 biomass. Kingston et al. (1994) studied the Neogene sequence of sediments in the Baringo Basin, Kenya. Sediments in the Baringo Basin are primarily mudflows with interbedded lacustrine and fluvial sediments. They interpret $\delta^{13}C$ values in carbonates and organic matter associated with the sediments to indicate a heterogeneous environment that included C_4-dominated ecosystems by 15 million years ago. In the Muruyur sediments with an age of 15 million years ago, they suggest that pedogenic carbonates ranges in $\delta^{13}C$ from −5‰ to −8‰ and organic matter ranges from −6‰ to −26‰. However, paleodietary studies using stable isotopes show no evidence of significant C_4 biomass in the diets of mammals at 15 million years ago, but show a significant shift to more positive values at about 7 million years ago (Morgan et al., 1994; Hill, 1995). This apparent contradiction, that significant C_4 biomass was present but was not part of the diet of mammals, is attributed to the immigration of animals more suited to a C_4 diet at about 7 million years ago or some other factor (Hill, 1995). Equids were present in the region by 10 million years ago and had a C_3-dominated diet (Morgan et al., 1994; Cerling et al., 1997c) . Therefore, the dietary evidence (Morgan et al., 1994) and the interpretation of the paleosol evidence (Kingston et al. 1994; Hill, 1995) are at odds for the earlier part of the Baringo Basin. It would be very useful to reevaluate the paleosol evidence, which is much more open to field observations and interpretations than is the fossil material. Kingston et al. (1994) note that many of the characteristics of paleosols are not found (e.g., leached zones, bioturbation, etc.); however, because detrital carbonates are not ubiquitous in the section, they conclude that it is likely that the carbonate is of pedogenic origin. Carbonate formation in the sediments by other means (e.g., during diagenesis) was not evaluated by them.

Wright and Vanstone (1991) studied the $\delta^{13}C$ in late Mississippian (ca. 300 Ma) exposure surfaces that are present as partings [very thin (mm to 10s of cm thick) clay beds] in thick sequences of limestones. These partings are often several cm to several tens of cm thick clays intercalated in thick (tens to hundreds of meters) of marine limestones. The partings have been interpreted to be exposure surfaces in the Mississippian and are interpreted as being related to the calcification of soils into rhizolith crusts, or calcified root mats. These postulated pedogenic carbonates are generally very close to the postulated paleo soil–air interface and are in close proximity to marine carbonate fossils (often within a centimeter or less). The measured $\delta^{13}C$ values, between -1‰ and -4‰, have been interpreted by Wright and Vanstone (1991) to indicate the presence of C_4 or CAM plants in the Mississipian. They have not demonstrated that these sediments meet the requirements of the diffusion model for soil CO_2: namely, that mass transport is diffusion controlled and that all the oxidized carbon species were in isotopic equilibrium (Cerling, 1992). In view of their occurrence in thick sequences of unquestioned marine limestones, it is likely that these carbonates do not fit the criteria for soils formed in a thick unsaturated zone where mass transport is primarily by gaseous diffusion. The data and its interpretation by Wright and Vanstone (1991), that there were abundant C_4 plants in the Mississipian, is equivocal and needs to be corroborated by other evidence.

3. Diet: Reduced and Oxidized Carbon The diet of animals is preserved in carbon-bearing tissues, such as chitin, collagen, or bioapatites (e.g., bone, dentine, tooth enamel). Of these, only the bioapatites are abundant in the fossil record, and of the bioapatites only enamel faithfully preserves dietary information. Collagen, chitin, or other organic phases are useful for very young fossils, generally less than about 100,000 years in age. Fossil bone and dentine, which in modern animals is poorly crystalline, recrystallize and exchange isotopes during diagenesis (Wang and Cerling, 1994). Fossil tooth enamel, however, has been shown to have isotopic fidelity for periods up to 200 million years (Thackery *et al.*, 1990). Using the carbon isotopic composition of fossil tooth enamel as an indicator of dietary preferences is a growing field (e.g., Lee-Thorp and van der Merwe 1989; Cerling *et al.*, 1994; MacFadden, 1996). The enrichment in ^{13}C compared to diet for large herbivore mammals is about 14‰ (Wang *et al.*, 1994; Cerling *et al.*, 1997c). This enrichment is several ‰ greater than that reported by DeNiro and Epstein (1978) for mice. Thus, the $\delta^{13}C$ values for enamel from a pure C_3 diet ranges from about -20‰ for closed canopy habitats to -8‰ for open xeric C_3 habitats, and $\delta^{13}C$ values for pure C_4 diets ranges from about $+2$ to $+4\text{‰}$.

Isotopic measurements of fossil tooth enamel are particularly useful in the search for global changes in the fraction of C_4 biomass. This is because of the selectivity of mammals in their diets which enhances of the C_4 isotope signal; for example, modern wild mammals in East Africa show either a C_3- or a C_4-dominated diet, with very few intermediate values (Cerling *et al.*,

1997b). The Athi Plains in Kenya provides an excellent example of herbivore selectivity for either C_3 or C_4 plants, and so the $\delta^{13}C$ value of tooth enamel can distinguish between browsers and grazers, respectively. Tragelaphine antelope and giraffes from the Athi Plains have $\delta^{13}C$ values between $-13\permil$ and $-8\permil$ whereas coexisting zebras and alcelaphine antelope have $\delta^{13}C$ values between $+1$ and $+4\permil$ (Cerling et al., 1997b; Cerling and Harris, unpublished data, 1998).

Cerling et al. (1997b) completed a survey of hundreds of modern and fossil mammals from Africa, Asia, North America, South America, and Europe. Figure 7 shows that before 8 million years ago mammals had a C_3-dominated diet ($-8\permil$ is taken to be the "cut-off" value for a pure C_3 diet), and that by 6 million years ago C_4-dominated diets are found in North and South America, Asia, and Africa. In many regions, mammals with C_3-dominated diets coexist with those having C_4-dominated diets indicating that there was an abundance of both types of vegetation. In the sequences studied, mammals with hypsodont (high-crowned) teeth (e.g., equids, some bovids, proboscideans) tend to have C_4-dominated diets, whereas low-crowned mammals (e.g., giraffids, cervids) have C_3-dominated diets. However, in high latitude regions ($>ca.$ 35°) even mammals with hypsodont dentition have C_3-dominated diets because C_3 grasses are present at high latitudes.

An intriguing finding of the dietary studies has been that some mammals were preadapted for grazing C_4-grasses when C_4 plants became more abundant. The classic example of animal evolution in response to the evolution of grasses is in the evolution of the dentition of horses (equids). From the Eocene (>50 Ma) to the early Miocene equids had low-crowned teeth. In the middle Miocene (20–15 Ma) equid teeth evolved higher crowns, supposedly for grazing (MacFadden, 1992). It is in the middle Miocene that grasses become more abundant in the geological record, although C_4 grasses are not found until about 12 Ma (see previous discussion). However, it is not until between 8 and 6 Ma that C_4 grasses become an important part of the diets of equids. Equids in Pakistan, North America, and East Africa all adapted to the C_4 diet very quickly. Presumably this means that they were already adapted to grasses or other herbaceous plants. Thus it is possible that C_3 grasses were relatively abundant in the middle Miocene in these regions, which include the tropical through temperate latitudes.

The period from 8 to 6 Ma is one of the most important periods of mammalian turnover in many parts of the world (Cerling et al., 1997b, and references therein). In North America, equids declined in diversity from 12 genera to 3 genera by 5 Ma, and eventually to the single genus *Equus* by the Pleistocene (MacFadden, 1992). The middle Miocene in East Africa had a diversity of large-bodied hominoids that were replaced by cercopithecoid monkeys. Likewise, forest herbivores were replaced by open-adapted

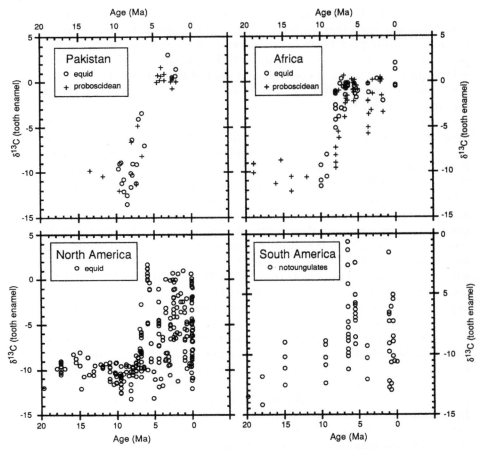

Figure 7 $\delta^{13}C$ values for fossil mammals compared to age. Note the widespread change in C_4 abundance starting between 8 and 6 million years ago (Ma). [Modified from Cerling, T. E., Harris, J. M. Leakey, M. G., and Mudida, N. (ms). Preliminary report on the isotope ecology of the East Africa, with emphasis on the Koobi Fora region, Turkana Basin, Kenya. *In* "Lothagam" (M. G. Leakey and J. M. Harris, eds.).]

bovids that underwent great radiation in the Pliocene. In Asia, the same pattern was observed: large hominoids were replaced by monkeys, and forest-adapted herbivores were replaced by arid-adapted mammals. In some regions, the late Miocene to Pliocene was a time of the largest local mammalian extinctions except for that brought on by humans in the past few thousand years (Webb *et al.*, 1995).

Cerling *et al.* (1997b, 1998b) speculate that the establishment of significant C_4 biomass was responsible for trends in mammalian evolution in the

Pliocene and Pleistocene, including the evolution of hominids. Previously, it was recognized that important faunal changes occurred at the end of the Miocene, and it was generally interpreted that these changes were due to changes in aridity. However, because of problems in correlation it was not clear whether these were regional events unrelated to each other or part of a global pattern. It is now clearly established that there was a global increase in C_4 biomass at the end of the Miocene (Fig. 7) and that the changes in flora were responsible for the changes in faunal composition (Cerling et al., 1998b). If gradually decreasing levels of atmospheric CO_2 was responsible for the changing fraction of C_3 and C_4 biomass, then it is also likely that the change in flora was in response to different water use efficiencies of different plants. The interpretation of global aridity at the end of the Miocene cannot be aridity *per se*, but rather "apparent aridity" where C_3 plant's water-use efficiency changes because of changes in atmospheric CO_2. Therefore, gradual changes in atmospheric CO_2, with increased photorespiration and decreased water use efficiency in C_3 plants but not C_4 plants, would have an impact on global ecosystems (Ehleringer et al., 1997; Cerling et al., 1998).

The evolution of hominids has taken place over the last 5 or so million years, during the time when C_4 plants made up a significant fraction of the global terrestrial biomass. The "C_4-world" was fundamentally different than the previously existing "C_3-world." In the "C_4-world," C_3 plants are under significant stress due to the low CO_2 levels, causing them to have higher rates of photorespiration and decreased water-use efficiency. In such a global environment C_4 ecosystems such as savannas have flourished, and it is in such environments that hominids evolved to become humans. In this way C_4 plants have a unique, and perhaps critical, role in the evolution of humans.

Other phases also record dietary preferences. Modern and fossil eggshell records the fraction of C_4 biomass (von Schirnding et al., 1982). Stern et al. (1994) studied the Siwalik sequence and showed that fossil ratites (e.g., ostrich, emu, rhea) also changed their diets after about 4 million years ago in Pakistan (samples between 7 and 4 Ma are very rare). However, they generally showed a mixed diet after 4 million years ago rather than changing entirely to a C_4 diet. This provides important environmental information because paleosols and fossil mammals are very strongly C_4, and shows that some C_3 biomass was present.

Collagen is also a good recorder of fossil diets. Although it is generally preserved only for several tens of thousands of years, it has provided very important information of the history of diets in human cultures (e.g., van der Merwe and Vogel, 1978) and dietary preferences of other mammals (Lee-Thorp et al., 1989b). It may prove to be particularly useful in studying dietary changes between the late Pleistocene and the Holocene in tropical and subtropical regions, where C_4 plants may have been more abundant in the glacial periods than in the interglacial periods (see previous discussion).

III. Summary

The geological record of C_4 plants is sparse. Individual plants with preserved Kranz anatomy or that have $\delta^{13}C$ values characteristic of C_4 plants, have rarely been found in sediments older than about 1 Ma. This is due, in part, to their preference for warm and well-drained sites, which means that their remains are usually oxidized during early burial. However, C_4 plants with preserved Kranz anatomy and with $\delta^{13}C$ values of about $-13‰$ have been found in 12.5 Ma sediments in California, making them the oldest confirmed C_4 plants. Fossilized cuticles of 14 Ma plants from Africa have been postulated to use the C_4 pathway, but this identification of C_4 photosynthesis has not been confirmed anatomically or isotopically.

A number of proxy records of C_4 plants using stable isotopes show the presence of significant C_4 biomass. These include peat and lake sediments, fossil soils, and fossil mammals (paleodiet studies). Using proxy records, two significant periods are notable. First, between 8 and 6 Ma (latest Miocene), C_4 biomass expanded in North and South America, in Asia, and in Africa. Before 8 Ma, there is no evidence of C_4 biomass in the diets of mammals so far studied on any continent indicating that they were a minor part of global ecosystems, at least in terms of their contribution to the total biomass. The latest Miocene was also a period of major faunal turnover on each of these continents, with elements of the modern fauna being emplaced at this time. It is possible that gradually declining levels of atmospheric CO_2 was largely responsible for this change in global ecology.

A second significant record of C_4 expansion and contraction is found in lakes and peat bogs in central Africa. During the last full Glacial period (ca. 20,000–30,000 years ago) C_4 biomass was more abundant on the fringes of the Zaire (Congo) Basin and surrounding Amazonia. C_4 biomass declined in the tropics during the increase in atmospheric CO_2 between 20,000 and 12,000 years ago. Glacial periods have long been thought to be times of the establishment of refugia, traditionally attributed to aridity. The connection of increased C_4 biomass with decreasing levels of atmospheric CO_2 means that establishment of refugia in the Pleistocene Glacial intervals may be related to atmospheric chemistry as much, if not more, than to aridity. The presence of a "C_4-world" for the last 6 to 8 million years, where C_4 plants are an important component of global ecosystems, unlike any previous period of Earth's history, has important significance for the evolution of mammals, including humans.

Acknowledgments

I thank J. R. Ehleringer, J. M. Harris, and B. J. MacFadden for many discussions over the past decade. Comments by R. F. Sage and R. K. Monson improved the manuscript.

References

Ambrose, S. H., and Sikes, N. E. (1991). Soil carbon isotope evidence for Holocene habitat change in the Kenya Rift valley. *Science* **253**, 1402–1405.
Aucour, A. M., Hillaire, M. C., and Bonnefille, R. (1993). A 30,000 year record of ^{13}C and ^{18}O changes in organic matter from an equatorial peatbog. *In* "Climate Change in Continental Isotopic Records" (Swart, P. K., Lohmann, K. C., McKenzie, J. A., Savin, S. eds.), *Am. Geophys. Un. Geophys. Mono.* **78**, p. 343–351.
Axelrod, D. I. (1939) A Miocene flora from the western border of the Mojave Desert. *Carn. Inst. Wash. Publ.* **516**, 1–128.
Bender, M. M. (1968). Mass spectrometric studies of carbon-13 variations in corn and other grasses. *Radiocarbon* **10**, 468–472.
Berner, R. A. (1991). A model for atmospheric CO_2 over Phanerozoic time. *Am. J. Sci.* **291**, 339–376.
Berner, R. A. (1994). GEOCARB II: A revised model of atmospheric CO_2 over Phanerozoic time. *Am. J. Sci.* **294**, 56–91.
Bird, M. I., and Cali, J. (1997). A million year fire history for the African Sahel from ODP-668-B, Sierra Leone Rise, in *5th. Int. Conf. on CO_2*, Cairns, Australia. (Abstract).
Bird, M. I., Haberle, S. G., and Chivas, A. R. (1994). Effect of altitude on the carbon–isotope composition of forest and grassland soils from Papua New Guinea. *Glob. Biogeo. Cyc.* **8**, 13–22.
Bocherens, H., Friis, E. M., Mariotti, A., and Pedersen, K. R. (1994). Carbon isotopic abundances in Mesozoic and Cenozoic fossil plants: Palaeontological implications. *Lethaia* **26**, 347–358.
Bowes, G. (1993). Facing the inevitable: Plants and increasing atmospheric CO_2. *Ann. Rev. Plant Physiol. Plant Mol. Biol.* **44**, 309–332.
Brown, W. V. (1977). The Kranz syndrome and its subtypes in grass systematics. *Mem. Torrey Botan. Club* **23**, 1–97.
Brown, W. V., and Smith, B. N. (1972). Grass evolution, the Kranz syndrome, $^{13}C/^{12}C$ ratios, and continental drift. *Nature* **239**, 345–346.
Cerling, T. E. (1984). The stable isotopic composition of modern soil carbonate and its relationship to climate. *Earth Planet. Sci. Lett.* **71**, 229–240.
Cerling, T. E. (1992), Further comments on using carbon isotopes in paleosols to estimate the CO_2 content of the atmosphere. *J. Geol. Soc. London* **149**, 673–675.
Cerling, T. E., and Quade, J. (1993). Stable carbon and oxygen isotopes in soil carbonates. *In* "Climate Change in Continental Isotopic Records" (Swart, P. K., Lohmann, K. C., McKenzie, J. A., Savin, S., eds.), American Geophysical Union Geophysical Monograph **78**, Washington, DC, pp. 217–231.
Cerling, T. E., Quade, J., Wang, Y., and Bowman, J. R. (1989). Carbon isotopes in soils and palaeosols as ecology and palaeoecology indicators. *Nature* **341**, 138–139.
Cerling, T. E., Quade, J., Ambrose, S. H., and Sikes, N. E. (1991). Fossil soils from Fort Ternan, Kenya: Grassland or woodland? *J. Hum. Evol.* **21**, 295–306.
Cerling, T. E., Kappelman, J., Quade, J., Ambrose, S. H., Sikes, N. E., and Andrews, P. (1992). Reply to comment on the paleoenvironment of Kenyapithecus at Fort Ternan, *J. Hum. Evol.* **23**, 371–377.

Cerling, T. E., Wang, Y., and Quade, J. (1993). Expansion of C_4 ecosystems as an indicator of global ecological change in the late Miocene. *Nature* **361**, 344–345.

Cerling, T. E., Ehleringer, J. R., and Harris, J. M. (1998b). Carbon dioxide starvation, the development of C_4 ecosystems, and mammalian evolution. *Proc. Royal Soc. London. B.* **353**, 159–171.

Cerling, T. E., Harris, J. M., Ambrose, S. H., Leakey, M. G., Solounias, N. (1997a). Dietary and environmental reconstruction with stable isotope analyses of herbivore tooth enamel from the Miocene locality of Fort Ternan, Kenya. *J. Hum. Evol.* **33**, 635–650.

Cerling, T. E., Harris, J. M., MacFadden, B. J., Leakey, M. G., Quade, J., Eisenmann, V., and Ehleringer, J. R. (1997b). Global vegetation change through the Miocene and Pliocene. *Nature* **389**, 153–158.

Cerling, T. E., Harris, J. M., and MacFadden, B. J. (1998a). Carbon isotopes, diets of North American equids, and the evolution of North American C_4 grasslands. *In* "Stable Isotopes and the Integration of Biological, Ecological, and Geochemical Processes" (H. Griffiths, D. Robinson, and P. Van Gardingen, eds.). BIOS Scientific Publishers, Oxford. pp. 363–379.

Cerling, T. E., Harris, J. M., Leakey, M. G., and Mudida, N. [in review]. Preliminary report on the isotope ecology of the East Africa, with emphasis on the Koobi Fora region, Turkana Basin, Kenya. *In* "Lothagam" (M. G. Leakey and J. M. Harris, eds.).

Cuffey, K. M., Clow, G. D., Alley, R. B., Stuiver, M., Waddington, E. D., and Saltus, R. W. (1995). Large Arctic temperature change at the Wisconsin–Holocene glacial transition. *Science* **270**, 455–458.

DeNiro, M. J., and Epstein, S. (1978) Influence of diet on the distribution of carbon isotopes in animals. *Geochim. Cosmo. Acta* **42**, 495–506.

Dugas, D. P., and Retallack, G. J. (1993). Middle Miocene fossil grasses from Fort Ternan, Kenya. *J. Paleon.* **67**, 113–128.

Ehleringer, J. R. (1993). Carbon and water relations in desert plants: An isotopic perspective. *In* "Stable Isotopes and Plant Carbon/Water Relations" (J. R. Ehleringer, A. E. Hall, and G. D. Farquhar, eds.), pp. 155–172. Academic Press, San Diego.

Ehleringer, J. R., and Björkman, O. (1977). Quantum yields for CO_2 uptake in C_3 and C_4 plants. Dependence on temperature, CO_2 and O_2 concentrations. *Plant Physiol.* **59**, 86–90.

Ehleringer, J. R., and Cooper, T. A. (1988). Correlations between carbon isotope ratio and microhabitat in desert plants. *Oecologia* **76**, 562–566.

Ehleringer, J. R., Sage, R. F., Flanagan, L. B., and Pearcy, R. W. (1991). Climate change and the evolution of C_4 photosynthesis. *Trends Ecol. Evol* **6**, 95–99.

Ehleringer, J. R., Cerling, T. E., and Helliker, B. R. (1997). C_4 photosynthesis, atmospheric CO_2, and climate. *Oecologia* **112**, 285–299.

Farquhar, G. D., and von Caemmerer, S. (1982). Modeling of photosynthetic response to environmental conditions, *Encycl. Plant Physiol.* **12D**, 549–587.

Farquhar, G. D., Ehleringer, J. R., and Hubick, K. T. (1989). Carbon isotope discrimination and photosynthesis. *Ann Rev Plant Physiol. Plant Mol. Biol.* **40**, 503–537.

France-Lanord, C., and Derry, L. A. (1994). $\delta^{13}C$ of organic carbon in the Bengal Fan: Source evolution and transport of C_3 and C_4 plant carbon to marine sediments. *Geochim. Cosmo. Acta* **58**, 4809–4814.

Giresse, P., Maley, J., and Brenac, P. (1994). Late Quaternary palaeoenvironments in the Lake Barombi Mbo (West Cameroon) deduced from pollen and carbon isotopes of organic matter. *Palaeogeog. Palaeoclim. Palaeoecol.* **107**, 65–78.

Guillet, B., Faivre, P., Mariotti, A., and Khobzi, J. (1988). The ^{14}C dates and $^{13}C/^{12}C$ ratios of soil organic matter as a means of studying the past vegetation in intertropical regions: Examples from Columbia (South America). *Palaeogeog. Palaeoclim. Palaeoecol.* **65**, 51–58.

Hatch, M. D., and Slack, C. R. (1970). Photosynthetic CO_2-fixation pathways. *Ann. Rev. Plant Physiol.* **21**, 141–162.

Hattersley, P. W. (1982). ^{13}C values of C_4 types in grasses. *Aust. J. Plant Physiol.* **9**, 139–154.

Hattersley, P. W., and Roksandic, Z. (1983). ^{13}C values of C_3 and C_4 species of Australian Neurachne and its allies (Poaceae). *Aust. J. Bot.* **31,** 317–321.

Hill, A. (1995), Faunal and environmental change in the Neogene of East Africa: Evidence from the Tugen Hills sequence, Baringo District, Kenya. *In* "Paleoclimate and Evolution with Emphasis on Human Origins" (Vrba, E. S., Denton, G. H., Partridge, T. C., and Burckle, L. H., eds.), Yale University Press, pp. 178–193.

Hollander, D. J., and McKenzie, J. A. (1991). CO_2 control on carbon-isotope fractionation during aqueous photosynthesis; a paleo-pCO_2 barometer. *Geology* **19,** 929–932.

Jordan, D. B., and Ogren, W. L. (1984).The CO_2/O_2 specificity of ribulose1,5-biphosphate carboxylase/oxygenase. *Planta* **161,** 308–313.

Jouzel, J., Lorius, C., Petit, J. R., Genthon, C., Barkov, N. I., Kotlyakov, V. M., and Petrov, V. M. (1987). Vostok ice core: A continuous isotope temperature record over the last climatic cycle (160,000 years). *Nature* **329,** 403–408.

Keeley, J. E. (1990). Photosynthetic pathways in freshwater aquatic plants. *Trends Ecol. Evol.* **5,** 330–333.

Keeley, J. E. (1998) C_4 photosynthetic modifications in the evolutionary transition from land to water in aquatic grasses. *Oecologia,* in press.

Keeley, J. E., and Sandquist, D. R. (1992). Carbon: Freshwater plants. *Plant Cell Environ.* **15,** 1021–1035.

Keeley, J. E., Sternberg, L. O., and DeNiro, M. J. (1986). The use of stable isotopes in the study of photosynthesis in freshwater plants. *Aquatic Bot.* **26,** 213–223.

Kingston, J. D., Marino, B. D., and Hill, A. (1994). Isotopic evidence for Neogene hominid paleoenvironments in the Kenya Rift Valley. *Science* **264,** 955–959.

Koch, P. L., Behrensmeyer, A. K., and Fogel, M. L. (1991). The isotopic ecology of plants and animals in Amboseli National Park, Kenya. *Ann. Rep. Dir. Geophys. Lab. Carn. Inst.* **1991,** 105–110.

Lee-Thorp, J. A., Sealy, J. C., and van der Merwe, N. J. (1989b). Stable carbon isotope ratio differences between bone collagen and bone apatite and their relationship to diet. *J. Arch. Sci.* **16,** 585–599.

MacFadden, B. J. (1992). "Fossil Horses: Systematics, Paleobiology, and Evolution of the Family Equidae." Cambridge University Press, Cambridge.

MacFadden, B. J., Cerling, T. E., and Prado, J. (1996). Cenozoic terrestrial ecosystem evolution in Argentina: Evidence from carbon isotopes of fossil mammal teeth. *Palaios* **11,** 319–327.

Medina, E., and Minchin, P. (1980). Stratification of $\delta^{13}C$ values of leaves in Amazonian rain forests. *Oecologia* **45,** 377–378.

Mook, W. G., Bommerson, J. C., and Stavermen, W. H. (1974). Carbon isotope fractionation between dissolved bicarbonate and gaseous carbon dioxide. *Earth Planet. Sci. Lett.* **22,** 169–176.

Muller, J. (1981). Fossil pollen records of extant angiosperms. *Bot. Rev.* **47,** 1–145.

Nambudiri, E. M. V., Tidwell, W. D., Smith, B. N., and Hebbert, N. P. (1978). A C_4 plant from the Pliocene. *Nature* **276,** 816–817.

Ohsugi, R., Samejima, M., Chonan, N., and Murata, T. (1988). ^{13}C values and the occurrence of suberized lamellae in some *Panicum* species. *Ann. Bot.* **62,** 53–59.

Quade, J., and Cerling, T. E. (1995). Expansion of C_4 grasses in the Late Miocene of Northern Pakistan: Evidence from stable isotopes in paleosols. *Palaeogeog. Palaeoclim. Palaeoecol.* **115,** 91–116.

Quade, J., Cerling, T. E., and Bowman, J. R. (1989a). Development of Asian monsoon revealed by marked ecological shift during the latest Miocene in northern Pakistan. *Nature* **342,** 163–166.

Quade, J., Cerling, T. E., and Bowman, J. R. (1989b). Systematic variations in the carbon and oxygen isotopic composition of pedogenic carbonate along elevation transects in the southern Great Basin, United States. *Geolog. Soc. Am. Bull.* **101,** 464–475.

Quade, J., Solounias, N., and Cerling, T. E. (1994). Stable isotopic evidence from Paleosol carbonates and fossil teeth in Greece for forest or woodlands over the past 11 Ma. *Palaeogeog. Palaeoclim. Palaeoecol.* **108**, 41–53.

Raven, J. A. (1970). Exogenous inorganic carbon sources in plant photosynthesis. *Biol. Rev.* **45**, 167–221.

Rau, G. H., and Takahashi, T. (1989). Latitudinal variations in plankton delta ^{13}C; implications for CO_2 and productivity in past oceans. *Nature* **341**, 516–518.

Retallack, G. J. (1992). Middle Miocene fossil plants from Fort Ternan (Kenya) and evolution of African grasslands. *Paleobiology* **18**, 383–400.

Retallack, G. J., Dugas, D. P., and Bestland, E. A. (1990). Fossil soils and grasses of middle Miocene East African grassland. *Science* **247**, 1325–1328.

Schwartz, D., Mariotti, A., Lanfranchi, R., and Guillet, B. (1986). $^{13}C/^{12}C$ ratios of soil organic matter as indicators of vegetation changes in the Congo. *Geoderma* **39**, 97–103.

Smith, B. N., and Brown, W. V. (1973). The Kranz Syndrome in the Graminae as indicated by carbon isotopic ratios. *Am. J. Bot.* **60**, 505–513.

Smith, B. N., and Epstein, S. (1971). Two categories of $^{13}C/^{12}C$ ratios for higher plants. *Plant Physiol.* **47**, 380–384.

Sowers, T., and Bender, M. (1995). Climate records covering the last deglaciation. *Science* **269**, 210–214.

Stern, L. A., Johnson, G. D., and Chamberlain, C. P. (1994). Carbon isotope signature of environmental change found in fossil ratite eggshells from a South Asian Neogene sequence. *Geology* **22**, 419–422.

Stute, M., Forster, M., Frischkorn, H., Serejo, A., Clark, J. F., Schlosser, P., Broecker, W. S., and Bonani, G. (1995). Cooling of tropical Brazil (5°C) during the last glacial maximum. *Science* **269**, 379–383.

Talbot, M. R., and Johannessen, T. (1992). A high resolution palaeoclimate record for the last 27,500 years in tropical west Africa from the carbon and nitrogen isotopic composition of lacustrine organic matter. *Earth Planet. Sci. Lett.* **110**, 23–37.

Thackeray, J. F., van der Merwe, N. J., Lee-Thorp, J. A., Sillen, A., Lanham, J. L., Smith, R., Keyser, A., and Monteiro, P. M. S. (1990). Changes in carbon isotope ratios in the late Permian recorded in therapsid tooth apatite. *Nature* **347**, 751–753.

Thomasson, J. R. (1987). Fossil Grasses: 1820–1986 and Beyond. In "Grass Systematics and Evolution" (T. R. Soderstrom, K. W. Hilu, C. S. Campbell, and M. E. Barkworth, eds.), Smithsonian Institution Press, Washington, DC, pp. 159–167.

Thomasson, J. R., Nelson, M. E., Zakrzewski, R. J. (1986). A fossil grass (Gramineae: Chloridoideae) from the Miocene with Kranz anatomy. *Science* **233**, 876–878.

Tidwell, W. D., and Nambudiri, E. M. V. (1989). *Tomlinsonia thomassonii*, gen. et sp. nov., a permineralized grass from the Upper Miocene Ricardo Formation, California. *Rev. Palaeobot. Palynol.* **60**, 165–177.

Vogel, J. C., Fuls, A., and Ellis, R. P. (1978). The geographical distribution of Kranz grasses in South Africa. *Sou. Afric. J. Sci.* **74**, 209–215.

van der Merwe, N. J., and Vogel, J. C. (1978). ^{13}C content of human collagen as a measure of prehistoric diet in woodland North America. *Nature* **276**, 815–186.

van der Merwe, N. J., and Medina, E. (1989). Photosynthesis and $^{13}C/^{12}C$ ratios in Amazonian rain forests. *Geochim. Cosmo. Acta* **53**, 1091–1094.

van der Merwe, N. J., and Medina, E. (1991). The canopy effect, carbon isotope ratios and foodwebs in Amazonia. *J. Archaeol. Sci.* **18**, 249–260.

von Schirnding, Y., van der Merwe, N. J., and Vogel, J. C. (1982). Influence of diet and age on carbon isotope ratios in ostrich eggshell. *Archaeometry* **24**, 3–20.

Wang, Y., and Cerling, T. E. (1994). A model of fossil tooth and bone diagenesis: Implications for paleodiet reconstruction from stable isotopes. *Palaeogeog. Palaeoclim. Palaeoecol.* 281–289.

Wang, Y., Cerling, T. E., and Effland, W. R. (1993). Stable isotope ratios of soil carbonate and soil organic matter as indicators of forest invasion of prairie near Ames, Iowa. *Oecologia* **95,** 365–369.

Wang, Y., Cerling, T. E., MacFadden, B. J. (1994). Fossil horses and carbon isotopes: new evidence for Cenozoic dietary, habitat, and ecosystem changes in North America. *Palaeogeog. Palaeoclim. Palaeoecol.* **107,** 269–279.

Webb, S. D., Hulbert, R. C., and Lambert, W. D. (1995). Climatic implications of large-herbivore distributions in the Miocene of North America. *In* "Paleoclimate and Evolution, with Emphasis on Human Origins" (Vrba, E. S., Denton, G. H., Partridge, T. C., and Burckle, L. H., eds.), New Haven, Yale University Press, pp. 91–108.

Webber, I. E. (1933). Woods from the Ricardo Pliocene of Last Chance Gulch, California. *Carn. Inst. Wash. Publ.* **412,** 113–134.

Whistler, D. P., and Burbank, D. W. (1992). Miocene biostratigraphy and biochronology of the Dove Spring Formation, Mojave Desert, California, and characterization of the Clarendonian mammal age (late Miocene) in California. *Geolog. Soc. Amer. Bull.* **104,** 644–658.

Williams, D. G., and Ehleringer, J. R. (1996). Carbon isotope discrimination in three semi-arid woodland species along a monsoon gradient. *Oecologia* **106,** 455–460.

Winter, K., Troughton, J. H., and Card, K. A. (1976). $\delta^{13}C$ values of grass species collected in the northern Sahara Desert. *Oecologia* **25,** 115–123.

Wright, V. P., and Vanstone, S. D. (1991). Assessing the carbon dioxide content of ancient atmospheres using palaeocalcretes: theoretical and empirical constraints. *J. Geol. Soc. London* **148,** 945–947.

V

C_4 Plants and Humanity

14

Agronomic Implications of C_4 Photosynthesis

R. Harold Brown

I. Introduction

Crop plants must have characteristics that make them vigorous, competitive, efficient in the use of available resources, and able to adapt to conditions of cultivation. One of the major evolutionary changes that affects these characteristics in plants was development of the C_4 cycle of CO_2 metabolism. In full sunlight, C_4 plants have higher CO_2 assimilation rates above 30°C than C_3 species and thus are better adapted photosynthetically to nonshaded tropical habitats. Whereas crop plants, including those with C_3 photosynthesis, have other traits that allow them to grow in tropical conditions, it has become clear that C_4 species have advantages in crop production in hot climates.

The impact of C_4 crops on the world food supply is enormous. The food supply in many tropical regions is largely based on C_4 plants, including grasses that supply the grains of many tropical diets, and pastures and rangelands that supply forage for animals kept for meat and milk. Maize (*Zea mays*), sorghum (*Sorghum bicolor*), and millet (several species) are the most widely grown C_4 crops. On a land area basis, these species account for 30% of the cereal grains grown worldwide, 70% of the cereals grown in Africa, 46% of those in North America, and 55% of those in South America (FAO, 1994). Sugarcane (*Saccharum officinarum*) is an important crop in tropical and subtropical regions, and many other C_4 species are used for grain, vegetables, and forage. Negative impacts on production are exerted by C_4 plants classified as weeds, which occur from the tropics to the higher latitudes of the temperate agricultural regions.

Table I Domesticated C_4 Crops, Their Region of Domestication, and Hectares Grown[a]

Species	C_4 subtype[b]	Common name	Domesticated in	Millions of hectares[c]
Major cereals				
Zea mays	NADPme(1)	Maize	Mexico	131(1)
Sorghum bicolor	NADPme(1)	Sorghum	Africa	44(1)
Pennesitum glaucum	NADPme(1)	Pearl millet	Africa	28[d](1)
Minor cereals				
Eleusine coracana	NADme(1)	Finger millet	Africa	4(3)
Setaria italica	NADPme(1)	Foxtail millet	China	4.5(3)
Panicum miliaceum	NADme(1)	Proso millet	Egypt or Arabian Peninsula	1.5(3)
Eragrostis tef	NADme(3)	Tef	Africa	1.4(2)
Digitaria exilis	NADPme(3)	Fonio	Africa	0.3(2)
Bracharia deflexa	PEPck(3)	Animal fonio	Africa	—(4)
Digitaria iburua	NADPme(3)	Black fonio	Africa	—(4)
Digitaria sanguinalis	NADPme(1)	Manna	Europe	—(4)
Echinochloa crus-galli	NADPme(1)	Japanese millet	China	—(4)
Coix lacrymi-jobi	NADPme(5)	Adlay	India	—(4)
Digitaria cruciata	NADPMe(3)	Raishan	India	—(4)
Bracharia ramosa	PEPck(1)	—	India	—(4)
Echinochloa colona	NADPme(1)	Sawa	India	—(4)
Paspalum scrobiculatum	NADPme(3)	Khodo	India	—(4)
Setaria pumila	NADPme(3)	Korali	India	—(4)
Panicum sumatrense	NADme(6)	Sama	India	—(4)
Panicum sonorum	NADme(4)	Sauwi	Mexico	—(4)
Miscellanous crops				
Saccharum officinalis	NADPme(1)	Sugarcane	New Guinea	18(1)
Amaranthus edulis	NADme(3)	Grain amaranth	S. America	—(4)
Amaranthus tricolor	NADme(3)	Vegetable amaranth[e]	S. America	—(4)
Gynandropsis gynandra		Spider flower[e]	Africa and S. E. Asia	—(4)
Portulaca oleracea	NADme(2)	Purslane[e]	S. E. Asia	—(4)

[a] Cereals adapted from de Wet, J. M. J. (1992). The three phases of cereal domestication. *In* "Grass Evolution and Domestication" (G. P. Chapman, ed.), pp. 176–198. Cambridge University Press, Cambridge. Miscellaneous crops from several sources.

[b] Subtype classification based on (1) Hattersley, P. W. (1987). Variations in the photosynthetic pathway. *In* "Grass Systematics and Evolution" (T. R. Soderstrom, K. W. Hilu, C. S. Campbell, and M. E. Barkworth, eds.), pp. 40–64. Smithsonian Institution Press, Washington, DC; (2) Elmore, C. D., and Paul, R. N. (1983). Composite list of C_4 weeds. *Weed Sci.* **31,** 686–692. (3) on observations of only this subtype in other species of this genus, (4) other species of this section of *Panicum,* and (5) other species in this tribe of *Gramineae,* and (6) the close relationship to *P. miliaceum* Wanous, M. K. (1990). Origin, taxonomy and ploidy of the millets and minor cereals. *Plant Var. Seeds* **3,** 99–112.

[c] Sources for hectareage shown in parenthesis: (1) FAO (1994). "FAO Production Yearbook." Vol. 48. Food and Agriculture Organization of the United Nations, Rome; (2) National Research Council (1996).

Table I (Continued)

"Lost Crops of Africa. I. Grains." National Academy Press, Washington, DC; (3) de Wet, J. M. J. (1995). Finger millet, pp. 137–140; Foxtail millet, pp. 170–172; Minor cereals, 202–208. *In* "Evolution of Crop Plants" (J. Smart and N. W. Simmonds, eds.), Longman Scientific and Technical, Essex, United Kingdom. (4) hectareages unavailable, probably because they are small and/or scattered among several regions.

[d] Areas of finger millet, foxtail millet, and proso millet from de Wet, J. M. J. (1995). Finger millet, pp. 137–140; Foxtail millet, pp. 170–172; Minor cereals, 202–208. *In* "Evolution of Crop Plants" (J. Smart and N. W. Simmonds, eds.), Longman Scientific and Technical, Essen, United Kingdom were subtracted to obtain this number because FAO reports all millets together.

[e] Used as fresh vegetables.

C_4 crops appear to have been as important in ancient agriculture as they are today. Of the 33 grass species domesticated as cereals, 20 are C_4 species (Table I) (de Wet, 1992). Proso millet (*Panicum miliaceum*) and foxtail millet (*Setaria italica*) are the oldest known cultivated cereals in China, and the first cultivated cereal in the Americas was apparently a species of *Setaria*. Finger millet (*Eleusine coracana*) may have been the first indigenous grass domesticated in Africa. The widely distributed weed, crabgrass (*Digitaria sanguinalis*), was important as a cereal in Europe in Roman times (de Wet, 1992). Maize is the second most important cereal in the world today, after wheat (*Triticum aestivum*), and was domesticated in the Americas 5,000 to 8,000 years ago. The species discussed previously (except wheat) are all C_4 grasses, and the plant grown for paper manufacture in ancient times (*Cyperus papyrus*) is a C_4 sedge.

This chapter deals with characteristics of C_4 crop species that make them suitable for productivity in tropical and temperate cropping systems. The presentation is necessarily comparative, with emphasis on differences between C_3 and C_4 crops. Little consideration is given to management practices because they differ little for C_3 and C_4 species.

II. Adaptation and Characterization of C_4 Crop Plants

Three crops—maize, sorghum, and pearl millet—account for about 95% of the C_4 cereal production in the world (Table I). They are the only ones listed in the annual FAO Agricultural Production Yearbook (1994). Tef (*Eragrostis tef*) is a C_4 grain crop grown mostly in Ethiopia, where it is planted on more hectarage than any other cereal (Wanous, 1990). Many other species, some of which are listed in Table I, are also used for grain, but are of only local importance in most cases and total areas devoted to these species are not well documented.

Africa and Asia produce the majority of C_4 cereals, although Asia and North America have similar areas planted to maize (Table II). In the case

Table II The Land Area of Three C_4 Crops on Four Continents as a Percentage of the World Total for Each Crop[a]

Crop	Period	Percent				
		Asia	Africa	North America	South America	Total
Maize	1978–1981	29.5	14.0	31.4	13.3	88.2
	1992–1994	29.7	16.3	29.7	13.7	89.4
Millet	1978–1981	61.0	30.7	0.2	0.5	92.4
	1992–1994	47.5	45.6	0.4	0.1	93.6
Sorghum	1978–1981	46.2	30.3	16.1	5.5	98.1
	1992–1994	35.6	46.3	13.2	3.1	98.2

[a] Calculated from data in FAO (1994). "FAO Production Yearbook." Vol. 48. Food and Agriculture Organization of the United Nations, Rome.

of millet, Asia and Africa account for more than 90% of the world hectarage. Several species are referred to as millet (Wanous, 1990), and in Table II, millet includes pearl millet (*Pennisetum glaucum*), finger millet, proso millet, barnyard millet (*Echinochloa frumentacea*), and foxtail millet. The total area planted to cereals increased by 29% in Africa from 1978–1981 to 1992–1994, and 82% of that increase was C_4 cereals, mostly sorghum and millet (FAO, 1994). During the same period, the hectarage of both sorghum and millet decreased in Asia by amounts similar to the increase in Africa as indicated by the percentages in Table II. These two crops were domesticated along the southern fringes of the Sahara Desert (de Wet, 1992), and according to Stoskopf (1985), they supply 90% of the food energy requirements for rural populations of the hot, dry Sahelian region.

Several C_4 crops other than cereals are used throughout the tropics (Table I). Sugarcane is of worldwide importance, and is planted on about 18 million hectares (about 8% of the C_4 crops) in tropical and subtropical areas of many countries. Species of *Amaranthus* were domesticated in Central and South America and are cultivated on limited hectarage in various parts of the tropics, although their importance in the region of domestication has decreased greatly since pre-Columbian times (National Research Council, 1984). *Amaranthus edulis* is cultivated for grain and *A. tricolor* is used as a vegetable (Table I). Two other species, *Gynandropsis* (*Cleome*) *gynandra* and *Portulaca oleracea*, are also used as vegetables in limited areas of the tropics (Yamaguchi, 1983; Waithaka and Chweya, 1991). Although *Zea mays* and *Amaranthus* species were domesticated in America, nearly all of the other C_4 crops were domesticated in Africa or Asia.

There are many C_4 species used as forage for livestock throughout tropical and subtropical regions. Table III lists the most widely cultivated perennials of low latitudes (Humphreys, 1981), although there are many other important cultivated and noncultivated forage species. The range of C_4 forages extends beyond the subtropics into temperate regions as indicated by a mean latitude of 31.4 ± 7.5° for 15 stations reporting commercialization of *Cynodon dactylon* (Humphreys, 1981), one of the most widely used grasses in subtropical regions. One of the main limiting factors in the use of

Table III Subtypes and Adaptive Characteristics of Common Cultivated C_4 Pasture Grasses[a]

Species	C_4 subtype[b]	Resistance[c] to			Soil fertility response[d]
		Drought	Cold	Defoliation	
Paspalum dilatatum	NADPme	3	4	4	3
Paspalum notatum	NADPme	3	3	4	3
Paspalum plicatum	NADPme	3	3	4	3
Paspalum commersonii	NADPme	2	3	2	3
Pennisetum clandestinum	NADPme	2	4	5	5
Pennisetum purpureum	NADPme	2	1	3	5
Cenchrus ciliaris	NADPme	5	2	4	4
Digitaria decumbens	NADPme	3	2	4	4
Setaria anceps	NADPme	2	4	3	5
Sorghum almum	NADPme	4	1	3	5
Cynodon dactylon	NADme	3	3	5	4
Panicum coloratum	NADme	3	3	4	3
Panicum maximum	PEPck	3	2	3	5
Chloris gayana	PEPck	3	4	4	4
Bracharia mutica	PEPck	1	1	4	2
Melinis minutiflora	PEPck	2	2	1	1

[a] Adapted from Tables 2.1 and 10.1 of Humphreys, L. R. (1981). "Environmental Adaptation of Tropical Pasture Plants." MacMillan Publishers Ltd., London.
[b] Subtype classification after Hattersley, P. W. (1987). Variations in the photosynthetic pathway. *In* "Grass Systematics and Evolution" (T. R. Soderstrom, K. W. Hilu, C. S. Campbell, and M. E. Barkworth, eds.), pp. 40–64. Smithsonian Institution Press, Washington, DC; and Hattersley, P. W., and Watson L. (1992). Diversification of photosynthesis. *In* "Grass Evolution and Domestication" (G. P. Chapman, ed.), pp. 38–116. Cambridge University Press, Cambridge.
[c] Relative resistance: 1, least resistant; 5, most resistant.
[d] Relative growth response to high soil fertility: 1, least; 5, most.

perennial forages in higher latitudes is susceptibility to low temperatures. Although less resistant to cold than temperate region C_3 grasses, there is considerable variation in the capacity of C_4 species to survive low temperatures, as indicated in Table III. In addition to the species in Table III (which are only those commercialized below approximately 32°C), there are many cultivated and native C_4 forage grasses grown in warmer portions of the temperate world. The C_4 grasses *Panicum virgatum*, *Sorgastrum nutans*, and species of *Bouteloua*, for example, are important components of native North American rangelands and extend into the north central United States and southern Canada (Voigt and MacLauchlan, 1985). Cultivated C_4 species are also used for summer grazing in the north central and northeastern United States (Soil Conservation Society of America, 1986).

The division of C_4 species into three main subtypes based on biochemistry and anatomy may have some implications for adaptation and management, but the evidence is sparse. Most of the agronomic C_4 plants are of the NADPme subtype (Table IV). About 70% of the C_4 grain crops listed by Hattersley and Watson (1992) and de Wet (1992) are of the NADPme subtype, and more than 98% of the grain produced by C_4 cereals is from this subtype [calculated from data for 1978–1981 from FAO (1994) and the production of individual millet species for 1981–1985 from Dendy (1995)]. Twenty of the 33 cultivated C_4 pasture species of Australia are of

Table IV Distribution of C_4 Subtypes among Domesticated C_4 Crops and Weeds

Group	Number of species			
	NADPme	NADme	PEPck	Ref.[a]
Domesticated cereals[b]	9	2	2	1[b]
	15	3	2	2[c]
Australian forages	20	4	9	1
Weeds	22	5	2	3[c]

[a] (1) Hattersley, P. W., and Watson, L. (1992). Diversification of photosynthesis. *In* "Grass Evolution and Domestication" (G. P. Chapman, ed.), pp. 38–116. Cambridge University Press, Cambridge; (2) de Wet, J. M. J. (1992). The three phases of cereal domestication. *In* "Grass Evolution and Domestication" (G. P. Chapman, ed.), pp. 176–198. Cambridge University Press, Cambridge; (3) Holm, L., Plucknett, D. L., Pancho, J. V., and Herberger, J. P. (1977). "The World's Worst Weeds. Distribution and Biology." University of Hawaii Press, Honolulu, Hawaii.

[b] Actually listed as genera, some of which include more than one species.

[c] Reference is for listing of species; classification of subtypes is from Hattersley, P. W. (1987). Variations in the photosynthetic pathway. *In* "Grass Systematics and Evolution" (T. R. Soderstrom, K. W. Hilu, C. S. Campbell, and M. E. Barkworth, eds.), pp. 40–64. Smithsonian Institution Press, Washington, DC; and Elmore, C. D., and Paul, R. N. (1983). Composite list of C_4 weeds. *Weed Sci.* **31**, 686–692.

the NADPme subtype (Table IV), as are 10 of the 16 forage species in Table III. The C_4 cereal and forage crops in Table IV are all grasses, and among the 152 C_4 grasses that have been typed, Hattersley (1987) listed 48 as NADPme, 55 as NADme, and 48 as PEPck subtypes (one had a mixture of PEPck and NADme decarboxylases). Thus, although the grasses that have been typed are nearly evenly divided among the three C_4 subtypes, cultivated species are predominantly NADPme.

Hattersley and Watson (1992) speculated that the nearly exclusive use of NADPme and PEPck subtypes as commercial forages in Australia was due to the occurrence of these subtypes in the wetter, more productive savannas of Africa and Asia that were first exploited during colonization, and later to the attention these subtypes received in attempts at adaptation to more arid cultivated pastures. NADPme and PEPck subtypes may not include the best species for adaptation to arid areas because of the strong correlation (-0.94) between the percentage of native C_4 grasses in Australia made up by NADme subtypes and median annual rainfall (Hattersley and Watson, 1992). This negative correlation indicates that NADme subtype C_4 plants are more drought resistant than the NADPme and PEPck subtypes. However, there does not appear to be a correlation between C_4 subtype and drought resistance of the more commonly grown tropical grasses as indicated by Humphreys (1981) (Table III).

Whatever the reason for the preponderance of NADPme subtype among C_4 crop and pasture species, the same pattern is found for C_4 weeds (Table IV). Of the 29 C_4 species listed among the world's worst weeds (Table V), 21 are of the NADPme subtype, possibly because similar selection factors have operated in the evolution of weeds and cultivated species. Certainly most weeds depend on similar conditions of soil disturbance, soil fertility, and moisture as those of the cultivated crops with which they grow.

III. Productivity of C_4 Crop Plants

A. Maximum Growth Rates and Yield

Much has been written about the comparative productivity of C_3 and C_4 species. There has been controversy in the literature about the advantage of C_4 photosynthesis for plant productivity and Gifford (1974) argued that, although C_4 plants have a great advantage over C_3 at the biochemical level, the advantage diminishes at the leaf level, and disappears at the level of crop growth rate. However, Ludlow (1985) showed that C_4 grasses have twice the leaf photosynthetic rate and twice the crop growth rate of tropical legumes. It is fairly clear now that C_4 plants are more productive if the two types are compared under their respective optimum conditions. This is shown by maximum short-term crop growth rates, under near optimum

Table V C_4 Species among the World's Most Serious Weeds[a]

Ranking	Species	C_4 subtype[b]	No. of countries	No. of crops
Group 1 (most serious in order of importance)				
1	*Cyperus rotundus*	NADPme	92	52
2	*Cynodon dactylon*	NADme	80	40
3	*Echinochloa crusgalli*	NADPme	61	36
4	*Echinochloa colonum*	NADPme	60	35
5	*Eleusine indica*	NADme	60	46
6	*Sorghum halepense*	NADPme	53	30
7	*Imperata cylindrica*	NADPme	73	35
9	*Portulaca oleracea*	NADme	81	45
11	*Digitaria sanguinalis*	NADPme	56	33
14	*Amaranthus hybridus*	NADme	27	27
15	*Amaranthus spinosus*	NADme	44	28
16	*Cyperus esculentus*	NADPme	30	21
17	*Paspalum conjugatum*	NADPme	30	25
18	*Rottboellia cochinchinensis*	NADPme	18	28
Group 2 (Secondary; not listed in order of importance)				
	Axonopus compressus	NADPme	27	13
	Bracharia mutica	PEPck	34	23
	Cenchrus spp.	NADPme	35	18
	Digitaria adscendens	NADPme	22	19
	Digitaria scalarum	NADPme	—	—
	Euphorbia hirta	NADPme	47	—
	Heliotropium indicum	NADPme[c]	28	15
	Ischaemum rugosum	NADPme	26	2
	Leptochloa panicea (*L. chinensis*)	NADme[c]	15	11
	Panicum maximum	PEPck	42	20
	Panicum repens	NADme	27	19
	Paspalum dilatatum	NADPme	28	14
	Pennisetum clandestimum	NADPme	36	14
	Pennisetum purpureum (*P. polystachyon, P. pedicellayum*)	NADPme	25	9
	Setaria viridis	NADPme	35	29
	Tribulus terrestris (*T. cistoides*)	NADPme	37	21

[a] After Holm, L., Plucknett, D. L., Pancho, J. V., and Herberger, J. P. (1977). "The World's Worst Weeds. Distribution and Biology." University of Hawaii Press, Honolulu, Hawaii.

[b] Subtype classification after Hattersley, P. W. (1987). Variations in the photosynthetic pathway. *In* "Grass Systematics and Evolution" (T. R. Soderstrom, K. W. Hilu, C. S. Campbell, and M. E. Barkworth, eds.), pp. 40–64. Smithsonian Institution Press, Washington, DC; and Elmore, C. D., and Paul, R. N. (1983). Composite list of C_4 weeds. *Weed Sci.* **31**, 686–692.

[c] Based on classification of other species in the genus.

field conditions, of 50–55 g m^{-2} day^{-1} for C$_4$ species compared to about 40 g m^{-2} day^{-1} for C$_3$ species (Ludlow, 1985). Seasonal crop growth rates derived by plotting annual yields against length of growing season are about twice as high for C$_4$ as for C$_3$ crop plants (Monteith, 1978).

Higher growth rates of C$_4$ plants are also evident when normalized to three of the resources used in plant growth: nitrogen, solar radiation, and water. Nitrogen is a resource that is usually in short supply for plant growth and is taken up in nearly equal quantities by C$_3$ and C$_4$ plants. The dry matter produced per unit of N taken up, however, is higher for C$_4$ plants (Brown, 1978, 1985; Sage and Pearcy, 1987). This higher N use efficiency (NUE) may result from faster dry matter accumulation in C$_4$ than C$_3$ plants and thus greater dilution of N (see Section III.B). Likewise, if growth of C$_3$ and C$_4$ plants is normalized to the solar radiation intercepted or absorbed by their canopies, C$_4$ plants exhibit maximum radiation use efficiency (RUE) values of about 2.5 g MJ^{-1} (absorbed) compared to 1.7 to 1.9 g MJ^{-1} (absorbed) for C$_3$ species (Gosse *et al.*, 1986). Finally, water-use efficiency (WUE) defined as dry matter produced per unit of water transpired is about twice as high for C$_4$ as for C$_3$ species. These expressions of growth based on N, radiation, and water serve to remove some of the variability associated with comparing growth of plants or plant communities.

Nitrogen, radiation, and water-use efficiencies are integrated measures of the effectiveness of acquired resources in the production of dry matter during the growth of a crop. In crop cultivation, a more practical consideration is the utilization efficiency of cultural energy inputs. Cultivation energy (including energy inputs in fuel, fertilizers, pesticides, etc.) is also used more efficiently by C$_4$ than by C$_3$ plants, probably, as in the case of radiation, N, and water, because of the higher yields of C$_4$ plants. A study by Heichel (1976) showed that maize, sorghum, and sugarcane produced more food energy per unit of cultivation energy (4-5 cal cal^{-1}) than the C$_3$ species, wheat, oats (*Avena sativa*), and soybean (*Glycine max*) (2–3 cal cal^{-1}), grown under similar energy investments.

Does the higher productivity of C$_4$ crops result entirely from their higher leaf photosynthesis rates in spite of the many reports of no relationship between leaf photosynthesis and growth rate or yield (Evans, 1975)? Ludlow (1985) observed a C$_4$–C$_3$ photosynthetic rate ratio of 2.0 for tropical grasses and legumes, and a ratio of 2.1 for crop growth rate. He concluded that the only characteristic of the C$_4$ syndrome that could be clearly linked to the higher crop growth rates was the higher maximum leaf net photosynthesis rates. It is likely, however, that the higher leaf photosynthesis rates of C$_4$ species are accompanied by other physiological differences that contribute to higher yields. For example, higher leaf photosynthesis and sucrose synthesis rates of C$_4$ plants would require higher photosynthate transport rates, as have been observed in C$_4$ leaves (Hofstra and Nelson, 1969; Stephenson et al., 1976). In addition, C_4 plants often have higher relative leaf expansion rates, a trait that can be a powerful determinant of early plant growth (Potter and Jones, 1977). The consistency of differences between C_3 and C_4 species in leaf export rate of photosynthate and relative leaf expansion rate is not well known, but higher growth rates and yield of C_4 plants may involve both.

B. Responses to Nitrogen Nutrition

Although for most mineral nutrients the response of C_3 and C_4 crops differ little, C_4 plants usually use N more efficiently than C_3 plants. The more efficient use of N results at least partially from the localization of the enzyme Rubisco in bundle-sheath cells (BSC) and the concentration of CO_2 in those cells (Brown, 1978). In C_3 leaves, Rubisco makes up a large proportion of the soluble protein, and thus the CO_2 concentrating function of C_4 leaves allows for a reduction of the amount of Rubisco required to fix a given amount of CO_2. Leaf photosynthesis per unit of leaf N is about twice as high in C_4 as in C_3 leaves. This higher efficiency of C_4 species is also expressed in dry matter accumulation; that is, C_4 plants produce more dry matter per unit of N than C_3 plants (Brown, 1978, 1985; Sage and Pearcy, 1987).

The concentration of N in dry matter of whole plants decreases as plants increase in weight (Greenwood et al., 1990), but at any given dry weight, C_4 plants exhibit lower plant N concentration than C_3 plants (Fig. 1). Greenwood et al. (1990) concluded that C_4 plants required only about 72% as much N as C_3 plants of equivalent weight. They also concluded that because C_4 plants produce approximately 32% more dry matter at equal light absorption values (2.5 compared to 1.9 g MJ^{-1}), that N taken up per unit of light intercepted was approximately equal between the two photosynthetic types. Actually, C_3 and C_4 plants grown with adequate N in the field (Kephart and Buxton, 1993) or at very low or adequate N supply in pots (Colman and Lazenby, 1970; Brown, 1985) do not differ in N uptake. Figure 2 shows data for two C_4 and two C_3 grasses grown in pots at two temperature ranges and different rates of N application (Colman and Lazenby, 1970). There is a linear increase in N taken up as applied N increases, but there is no consistent difference between C_3 and C_4 grasses at either temperature. So, the lower N concentration in C_4 plants is not due to lower N uptake, but is likely due to the higher rate of carbon accumulation, which dilutes the N to a greater extent than in C_3 plants.

Most comparisons of nitrogen use efficiency in C_3 versus C_4 plants have been under well-fertilized conditions, but C_4 forages are usually grown on soils deficient in N. In a 3-year experiment at very low soil N, Brown (1985) found that NUE of two C_4 grasses was twice that of two tropical C_3 species and one tropical C_3–C_4 intermediate grass. Dry

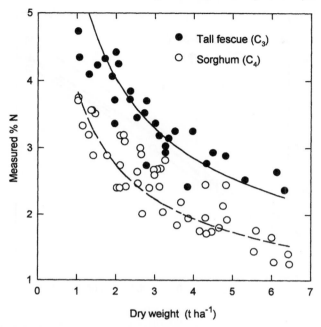

Figure 1 Relationships between plant dry weight and N concentration in foliage of tall fescue (C_3) and sorghum (C_4). [Redrawn with permission from Greenwood, D. J., Lemaire, G., Gosse, G., Cruz, P., Draycott, A., and Neeteson, J. J. (1990). Decline in percentage N of C_3 and C_4 crops with increasing plant mass. *Ann. Bot.* **66,** 425–436, using data points from their Fig. 3 A,B. Lines were drawn using their Eqs. 3 and 5 for C_3 and C_4 species, respectively.]

mass of shoot regrowth decreased with time for all of the species during the 3-year period and N concentration in shoots increased, resulting in an inverse relationship between N concentration and shoot size as observed by Greenwood *et al.* (1990). The lower N concentration of C_4 plants is a benefit to the efficient production of dry matter under limited N, but a disadvantage for the animals that depend on foliage for protein. Although C_4 species as a group have higher NUE than C_3 species, there appears to be considerable variation in the responsiveness of C_4 grasses to soil fertility (mainly N) (Table III).

C. Water Use Efficiency

One of the first clear-cut physiological differences observed between C_3 and C_4 plants was in WUE (Björkman, 1971). The WUE of C_4 crop and weed species is about twice as high as for C_3 species (Table VI) because of higher leaf photosynthesis rates and, under some conditions, lower transpiration (due to lower stomatal conductance). Field WUE (FWUE) is

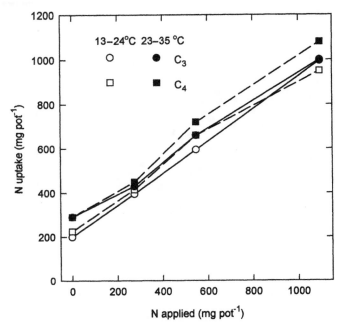

Figure 2 The response of N uptake to N applied for C_4 and C_3 grasses grown in two temperature regimes. [Redrawn with permission from Colman, R. L., and Lazenby, A. (1970). Factors affecting the response of tropical and temperate grasses to fertilizer nitrogen. *Proc. XI Inter. Grassland Cong.* Surfers Paradise, Australia. pp. 393–397, by averaging data for two C_4 grasses (*Digitaria macroglossa* and *Paspalum dilatatum*) and two C_3 grasses (*Lolium perenne* and *Phalaris tuberosa*).]

much more variable than WUE probably because of greater variation in cultural conditions, water supply, and the proportion of water lost directly from the soil. Therefore, differences in FWUE between C_3 and C_4 crops are less consistent than those based only on transpiration and under more controlled conditions. In the experiments represented in Table VI, however, FWUE was 1.7 times higher for the average of two C_4 species than for three C_3 species, almost as great as the advantage for C_4 plants grown in containers. The higher FWUE of C_4 plants was due to their faster growth rates, because evapotranspiration rates were similar to the C_3 crops.

Although C_4 species have higher WUE, they may be no more drought resistant as a group than C_3 species. They may be more appropriate, however, for use in hot environments with limited rainfall, because they produce more yield per unit of water and especially because the WUE advantage of C_4 over C_3 species increases with temperature (Björkman, 1971). C_4 grasses outcompete C_3 grasses in tropical and subtropical grasslands where mean temperatures are greater than 25°C and rainfall is mostly in summer, especially if rainfall is limited (Hattersley and Watson, 1992). If the high FWUE of C_4 photosynthesis is combined with short times from planting to maturity and reasonable drought tolerance, as is the case in some C_4 species, then grain crops can be produced with a minimum of water. These features are apparently responsible for the use of sorghum and millet species on large areas of the arid tropics. *Sorghum bicolor* and the millets *P. miliaceum* and *S. italica* had the highest WUE of 40 crop species tested by Shantz and Piemeisel (1927). Pearl millet is described by Spedding *et al.* (1981) as the "Most productive grain in extremely dry or infertile soils of India and Africa" (p. 75).

D. Influence of Predicted Changes in CO_2 and Temperature

There have been many evaluations of the comparative responses of C_3 and C_4 crop and weed species to increased ambient CO_2 concentrations and the results have been reviewed extensively (Poorter, 1993; Patterson, 1995). The results commonly show that photosynthesis and growth responses to CO_2 are greater for C_3 than C_4 species, at least in well-fertilized soils (Bowes, 1993; Patterson, 1995). In warmer conditions, maximum leaf photosynthesis rates of C_3 and C_4 species approach each other as CO_2 concentration increases above present levels (Pearcy and Ehleringer, 1984), indicating increases in CO_2 concentration should decrease the differences

Table VI Water Use Efficiency of Plants and Crops of Two C_4 and Three C_3 Species

Species	Plant[a] WUE[c] (g kg^{-1})	Crop[b]			
		Growth period (days)	Growth rate (g ha day^{-1})	Mean ET[d] (mm d^{-1})	FWUE[c] (g kg^{-1})
Zea mays (C_4)	2.86	135	126	4.9	2.58
Sorghum bicolor (C_4)	3.28	110	132	5.3	2.49
Solanum tuberosum (C_3)	1.84	128	78	4.2	1.88
Triticum aestivum (C_3)	1.83	112	69	4.2	1.63
Medicago sativa (C_3)	1.18	195	57	5.7	1.01
Means (C_4/C_3)	1.9	—	1.9	0.9	1.7

[a] Plants grown outdoors in large containers (calculated from Table 33 of Shantz, H. L., and Piemeisel, L. N. (1927). The water requirement of plants at Akron, Colo. *J. Agric. Res.* **34**, 1093–1190).
[b] Crops grown in the field (adapted from Table 4.3 of Gardner, F. B., Pearce, R. B., and Mitchell, R. L. (1985). "Physiology of Crop Plants." Iowa State University Press, Ames, Iowa).
[c] WUE, dry matter/transpiration; FWUE, dry matter/evapotranspiration.
[d] Evapotranspiration.

in growth rate and yield and competitive advantage. There is, however, considerable variability in the responses of C_3 and C_4 species, and in some cases there have been substantial CO_2-induced growth increases in C_4 species (Poorter, 1993; Patterson, 1995).

The effects of predicted atmospheric changes in CO_2 concentration and temperature on crop selection and production practices may be much less than physiologically based predictions would indicate. Even on the basis of physiology, C_4 plants may not be disadvantaged by high CO_2 (Henderson et al., 1995). Many plant traits that have large effects on yield, including sink size, photosynthate partitioning, and length of the vegetative and fruiting periods (Evans, 1975), are little affected by CO_2 concentration. Furthermore, nonplant factors such soil type and fertility, length of growing season, and disease and insect pressures exert strong influences on selection and production of crops. Selection of crop species by farmers is based more on economics and tradition than on biological traits. The choice between maize and soybean in the midwestern United States or central India, is not based on differences in yield, but on economics, food value, or tradition. Large areas of the world are committed to specific crops because of soil, climatic characteristics, and traditions of agriculture, such that changes are not likely to be large because of increased CO_2 concentrations.

Moderate increases in temperature and changes in other factors may tend to offset or mask effects of CO_2 on crop distribution. Although the great wheat regions, for example, may be shifted or their areas enlarged slightly by increased yield potential due to increased atmospheric CO_2, increased temperature (if, and to what extent, it occurs) is likely to have a more negative effect on the extent of the crop (Paulsen, 1994). Changes in precipitation patterns with increased atmospheric CO_2 concentration are largely unknown, but precipitation is one of the most important factors in delineation of wheat regions (Loomis and Conner, 1992). Market forces and technology, such as costs and returns, governmental restrictions or inducements, and improved cultivars and pest control are likely to have a more profound effect on crop species distribution than predicted climate changes. Great changes in geographic distribution of crop species have occurred since the last part of the 19th century, largely because of such factors and independent of climatic changes (Wittwer, 1995). Thus, although generalizations can be made that increased atmospheric CO_2 will increase growth of C_3 crops more than C_4 crops, the variability of species response and interactions with economic, cultural, and environmental factors make predictions of CO_2 effects on future production and cropping patterns very tenuous.

IV. Crop Quality

A. Grain Crops

Because nearly all of the C_4 grain crops are grasses, they have the general grain quality characteristic of other cereals. All are low in oil and protein in comparison to other crops grown for those constituents, and humans and domestic animals require supplemental protein for subsistence on diets of grain. C_4 cereals are a major dietary component in tropical areas subject to serious protein deficiency. In addition, protein quality for monogastric animals is deficient in some C_4 cereals, notably maize. The amino acid lysine is below requirements for some animals, but high-lysine lines have been developed through plant breeding (Lambert et al., 1969). The vitamin niacin is chemically bound in a less available form in maize than most cereal grains and its deficiency can result in pellagra in humans (Brouk, 1977).

The seeds of C_4 cereals commonly range from approximately 10% to 13.5% protein and from 2.5% to 6% oil (Table VII), but not all C_4 species

Table VII Protein and Oil Concentration in Seeds of Selected C_4 Grain and Forage Crops and C_4 Weeds[a]

		Percent dry matter		
	Crop	Protein	Oil	Number[b]
Grain crops	Amaranthus spp.	16.4 ± 0.9	6.8 ± 0.9	10
	Eleusine coracana	7.8 ± 0.7	1.5 ± 0.1	3
	Eragrostis tef	10.7 ± 0.8	2.5 ± 0.4	3
	Pennisetum glaucum	13.3 ± 2.5	5.4 ± 0.6	9
	Sorghum bicolor	13.8 ± 3.1	3.8 ± 1.3	17
	Zea mays	11.2 ± 1.1	4.7 ± 0.6	9
Forage crops	Alopecurus spp.	21.6 ± 1.5	14.8 ± 1.3	2
	Angropogon spp.	29.0 ± 4.0	7.1 ± 1.6	2
	Chloris gayana	22.1 ± 1.0	4.9 ± 0.5	3
	Panicum antidotale	12.1 ± 4.1	5.6 ± 0.3	2
	Paspalum notatum	7.4	3.1	1
Weed species	Echinochloa crusgalli	11.5 ± 3.1	4.9 ± 1.6	5
	Kochia scoparia	23.6 ± 3.6	12.0 ± 3.7	3
	Salsola spp.	42.7 ± 4.0	21.5 ± 5.9	3
	Setaria spp.	18.9 ± 3.2	6.7 ± 1.3	3
	Tribulus terrestris	42.3 ± 5.4	33.5 ± 6.1	2

[a] Duke, J. A., and Atchley, A. A. (1986). "Handbook of Proximate Analysis Tables of Higher Plants." CRC Press, Boca Raton, Florida.
[b] The number of accessions or species analyzed.

are low in seed protein or oil. *Amaranthus* spp. are higher with 16.4 ± 0.9% protein and 6.8 ± 0.9% oil. The seeds of some other C_4 grasses are considerably higher in protein than those of the cereal grains; *Andropogon* species have protein concentrations near 30%, and *Alopecurus, Chloris,* and *Setaria* species have concentrations of approximately 20%. These species (except *Chloris gayana*) also have oil concentrations considerably higher than cereal grains. *Eleusine coracana* has very low protein and low oil concentrations. The retention of associated bracts after harvest reduce the seed protein and oil concentration of some of the nongrain grasses in Table VII. Seeds of some C_4 dicots have extremely high protein and oil concentrations. Three *Salsola* weed species average greater than 40% protein and 20% oil, and the weed *Tribulus terrestris,* has 42.3% protein and 33.5% oil (Table VII).

The possibilities of utilizing the seeds of most weeds or noncereal grasses for food are minimal because of their small size and low seed yields, and in some cases their objectionable taste or cooking qualities. It has been demonstrated, however, that protein and oil concentration can be increased through selective breeding in some of the currently used C_4 cereal crops. Oil concentration in maize has been increased from less than 5% to 20.4% in 85 generations by mass selection, although such selection results in reduction in grain size and yield (Alexander, 1989). It does appear possible to increase the concentration from approximately 5% in current maize cultivars to 8% or 10% without significant sacrifice of yield. The increase in oil concentration resulted from both an increase in size of the embryo relative to the endosperm and increased oil concentration in the endosperm. Protein concentrations have also been increased in maize from 10.9% to 26.6% in 70 generations of selection (Dudley *et al.,* 1974). The increased concentration of protein resulted from increased protein concentration in the endosperm and, to a lesser extent, higher concentration in the germ. As in the case of oil, increased protein concentration was associated with decreased kernel weight and yield.

Although the possibilities for other C_4 cereals have barely been investigated, it may be possible to increase oil and protein concentration in situations in which these food components are of primary importance. There are large variations in grain protein concentration in the germplasm of sorghum and pearl millet, with a range of 5% to 20% in both species (FAO, 1995). However, there exist negative correlations between protein concentration and kernel size and grain yield (FAO, 1995; Dudley *et al.,* 1974).

B. Forage Crops

Productivity of animals (meat, milk, wool, or work) consuming mostly forage is directly related to the quality of the forage and the amount consumed. The quality of forages consumed by animals is accounted for, in large part, by their digestibility; the fraction of the plant matter or energy that remains in the body on passage through the gut tract. The intake or consumption of forages is usually positively related with digestibility and is also a reflection of quality. C_4 grasses are lower in digestibility, and usually in intake, than C_3 species, mainly because of their higher fiber concentrations and the structure of the tissues (Minson, 1981; Reid et al., 1988; Wilson, 1994).

Leaf anatomical characteristics associated with C_4 photosynthesis reduce leaf digestibility. Tissues with thickened secondary cell walls, such as vascular bundles, sclerenchyma strands, epidermis, and parenchyma bundle sheaths are the least digestible in grass leaves. Because C_4 grasses have approximately twice as many veins per unit of leaf width as C_3 grasses (Crookston and Moss, 1974; Morgan and Brown, 1979), they have larger percentages of the leaf cross-section composed of these tissues (Table VIII). Epidermal tissues may not be causally associated with photosynthetic type, but C_4 grasses also have higher proportions of epidermis (Wilson et al., 1983). If the three vascular associated tissues in Table VIII are added together, they make up 29.7% and 13.5% of the cross-sections in C_4 and C_3 species, respectively. The most digestible tissue in the leaf is mesophyll (Akin and Burdick, 1975), and it comprises 43% of the cross-section of C_4 leaves in Table VIII compared to 66% of C_3 leaves.

The thickened outer walls of vascular BSC of C_4 grasses are resistant to attack by rumen microorganisms. The resistance of BSC to microbial breakdown prevents digestion of the cell contents, which are rich in protein and carbohydrates, until the walls are degraded (Wilson, 1994). It has been shown that Rubisco protein, which is restricted to BSC in C_4 grasses, remains

Table VIII Mean *in Vitro* Dry Matter Digestibility, Cell Wall Content, and Tissue Percentages in Cross-Sections of Leaves of Seven C_3 and 18 C_4 *Panicum* Species[a]

	Percent						
					Tissues[b]		
Photo type	Dry matter digestibility	Cell wall content	BS	Vas	Scl	Sum[c]	Mes
C_4	69	50	20	8	1.7	29.7	43
C_3	76	33	10	3	0.5	13.5	66

[a] Adapted from Wilson, J. R., Brown, R. H., and Windham, W. R. (1983). Influence of leaf anatomy on the dry matter digestibility of C_3, C_4, and C_3/C_4 intermediate types of *Panicum* species. *Crop Sci.* **23,** 141–146. *In vitro* digestibility is the percentage difference between the initial sample weight and weight of residual sample after treatment with rumen fluid, relative to initial sample weight.
[b] BS, vascular bundle sheath; Vas, vascular tissue; Scl, sclerenchyma; Mes, mesophyll.
[c] Sum of the means for bundle sheath, vascular tissue, and sclerenchyma.

undigested for long periods in the rumen, and some Rubisco passes through the ruminant digestive tract undigested (Miller et al., 1996). In the NADPme and PEPck C_4 grasses, a suberized lamella is present in the peripheral portions of the BSC walls that prevents attack of the secondary walls by cellulose degrading microorganisms, thus limiting use of the cell wall components themselves until the cells are broken (Wilson, 1994).

The difference in quality of C_3 and C_4 grasses has been shown in many studies, some of which are summarized in Table IX. In this summary, *in vivo* digestibility averaged from 7 to 11 percentage units higher in C_3 than in C_4 grasses and intake was 27% higher for the means given by Minson (1981). However, intake was not different in the report of Reid et al. (1988). Although cell walls are variable in composition, the percentage of tissues made up of cell walls and the lignification of cell walls are good indicators of the indigestibility of forages. C_4 grasses have a higher proportion of thick-walled cells associated with indigestible fiber in their leaves than C_3 grasses, a trend that is lessened in leaf sheaths and disappears in stems (Wilson, 1994). The higher fiber concentration in C_4 compared to C_3 grasses is shown in Table VIII (cell wall content) and Table IX. Although Minson (1981) attributed the lower digestibility of tropical grasses to higher growth temperatures, which increase the fiber content and reduces digestibility of all species, an important contributing factor is the Kranz anatomy associated with C_4 photosynthesis.

Another characteristic of C_4 grasses that contributes to their lower nutritive value than C_3 grasses is their lower N concentration. Microorganisms

Table IX Dry Matter Intake, Dry Matter Digestibility, and Fiber Concentration of C_4 and C_3 Grasses[a]

Grass type	No. of samples	Dry matter intake[b] (g day^{-1} kg$^{0.75}$)	Dry matter digestibility (%)	Acid detergent fiber (% of dry matter)	Ref.
C_4 (Sheep)[c]	28	56	62	—	Minson, 1981
C_3 (Sheep)	23[d]	71	71	—	Minson, 1981
C_4 (Sheep)	61	65.7	54.5	42.5	Reid et al., 1988
C_3 (Sheep)	184	66.2	65.5	35.8	Reid et al., 1988
C_4 (Cattle)	78	89.8	60.0	42.7	Reid et al., 1988
C_3 (Cattle)	38	89.5	67.0	38.3	Reid et al., 1988

[a] Dry matter digestibility is the percentage difference between forage intake and fecal output, relative to initial intake.
[b] Consumption based on animal weight.
[c] Animals used in experiments.
[d] Dry matter digestibility based on 58 samples.

in the rumen of ruminant animals can convert nonprotein N into protein and thus crude protein (N × 6.25) is commonly used to express the protein value of ruminant diets. A minimum crude protein concentration of about 9% to 10% of dry weight (1.44–1.92% N) is required in forage consumed by beef cattle (McDowell, 1985). C_4 grasses have lower N concentrations in dry matter than C_3 grasses at equal levels of N fertilization (Brown, 1978; 1985) and at equal yields (Fig. 1), and thus are more often deficient in crude protein than C_3 grasses. As a result, features of C_4 grasses, including lower N concentration and compartmentation of much of that N in BSC, that contribute to higher NUE also contribute to lower nutritive quality.

The lower digestibility and intake of C_4 compared to C_3 grasses results in lower productivity of animals consuming tropical grasses. Although tropical grasses are higher yielding than temperate ones, beef or milk production is usually lower in tropical pastures, in part because of the lower forage quality. Nitrogen fertilized temperate grasses were estimated to produce from 700 to 1400 kg ha^{-1} of beef compared to 400 to 1200 kg ha^{-1} for tropical grasses (Simpson and Stobbs, 1981). Summaries of several experiments of responses of beef production (kg ha^{-1} year^{-1}) to N applied (kg ha^{-1} year^{-1}) resulted in regressions with similar slopes, but quite different intercepts (for tropical grasses, Y = 255 + 1.50 N; for temperate grasses, Y = 783 + 1.125 N), indicating higher beef production from temperate grasses at all N levels in the 0 to 500 kg ha^{-1} range. Expected milk yields from N-fertilized grass were estimated to be in the range of 8,000–17,000 L ha^{-1} for temperate grasses and 4,500–9,500 L ha^{-1} for tropical grasses. These differences in tropical and temperate beef and milk production may also reflect other factors such as direct environmental effects on the animals and differences in management of the experiments, but further illustrate the lower productivity of animals in the tropics.

V. Weeds

The characteristics that make C_4 species productive also contribute to their competitiveness as weeds. Because weeds are similar in many ways to crops, and often are closely related to the species with which they compete, they respond similarly to environmental and edaphic factors. C_4 plants are the major weeds in tropical countries, especially at low altitudes. However, even on a worldwide basis, C_4 species constitute 14 of the 18 worst weeds (Table V; Holm et al., 1977). Individual species of the 14 worst C_4 weeds are serious problems in from 18 to 92 countries and from 21 to 52 crops in those countries. In contrast, only 3 of the 15 major crops are C_4. In surveys of crops in the United States, it was found that C_4 weeds made up more than 50% of the species in maize, cotton (*Gossypium hirsutum*), sorghm, tobacco (*Nicotiana tobaccum*), sugarcane, sugarbeet (*Beta vulgaris*), and peanut (*Arachis hypogaea*) in all regions in which those crops were grown (Table X). In rice (*Oryza* sativa), 40% of the weeds in the southern region and 60% of weeds in the western region were C_4. These are crops that grow only in summer, except sugarcane, which is perennial but produces most of its growth during the warmest seasons. In wheat and other cool season cereals, C_4 weeds made up a small percentage of the total, as they did in hay crops and pastures, except in annual pasture crops in the western states (62%).

Morphology and growth habits are not uniform among the most troublesome weeds. Of the world's 14 worst C_4 weeds, 8 are perennial, 3 are dicots, 2 are sedges that reproduce by tubers as well as seed, and 8 genera of grasses are represented. Six of the species possess rhizomes or stolons. For many of the tropical weeds fresh seeds germinate readily, but for some, seeds may remain dormant for months or years (Maillet, 1991). Some tropical weeds have very short life cycles that allow regeneration before competition from the crop becomes too severe. *Echinochloa colona, Setaria*

Table X Percentage of Weeds That Are C_4 in Various Crops and Regions of the United States[a]

Crop	Northeast	North Central	Southern	Western
Maize	64	59	68	66
Cotton	—	60	59	53
Peanut	—	—	55	80
Rice	—	—	40	60
Sorghum	53	64	56	63
Soybean	47	38	46	—
Sugarbeet	59	55	—	54
Sugarcane	—	—	100	60
Tobacco	60	67	61	—
Wheat	4	18	2	15
Other cereal grain	32	19	2	19
Hay	5	21	34	16
Annual pasture	40	31	20	62
Perennial pasture	8	21	19	14

[a] Reprinted with permission from "Weed Physiology. Vol. 1. Reproduction and Ecophysiology" (S. O. Duke, ed.), pp. 101–129. CRC Press, Boca Raton, FL. Copyright CRC Press, Boca Raton, Florida. ©1985.

verticillata, and *Dactyloctenium aegyptium* may produce mature seeds within 30 to 45 days under short photoperiods (Maillet, 1991). Early maturity under longer photoperiods also allows C_4 weeds to produce seeds and persist at latitudes where summers are short.

For most crops, the detrimental effects of weeds are well known, as shown in Fig. 3. Increasing density of *Setaria faberii* had similar depressive effects on yields of soybean (C_3) and maize (C_4) (Moolani *et al.,* 1964), but *Amaranthus hybridus* reduced yield of soybean more than that of maize, probably in part because *Amaranthus* was taller than soybean but not maize (Knake and Slife, 1962). Despite the competitive capacity of C_4 plants, other factors such as greater plant height and more favorable leaf display reduce the competitive advantage of C_4 photosynthesis in some cases. The yields of pea (*Pisum sativum*) and soybean were decreased more by C_3 dicot weeds than C_4 grasses (Aldrich, 1984), and the likely reason was more horizontal display of leaves and thus greater shading by the dicot weeds. In the case of soybean, yield reduction by the dicot, *Abutilon theophrasti* (C_3), was twice

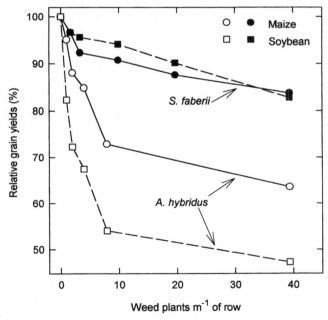

Figure 3 Reduction in grain yield of maize and soybean with increasing density of two C_4 weeds (*Setaria faberii* and *Amaranthus hybridus*). [Drawn from data of Knake, E. L., and Slife, F. W. (1962). Competition of *Setaria faberii* with corn and soybeans. *Weeds* **10,** 26–29; Moolani, M. K., Knake, E. L., and Slife, F. W. (1964). Competition of smooth pigweed with corn and soybeans. *Weeds* **12,** 126–128 (*A. hybridus*).]

as great per unit of weed weight as for *Setaria lutescens* or *Setaria viridis* (C_4). It is also possible that low temperature may have reduced the competitive effects of the C_4 weeds in this Iowa experiment, although temperatures were not reported. However, the dicot weed *Xanthium pensylvanicum* (C_3) reduced yield of soybean 70% (six cultivars over 3 years) compared to 34% for the monocot weed *Sorghum halepense* (C_4) in the southern Mississippi delta, where temperatures would be expected to favor *S. halepense* (McWhorter and Hartwig, 1972).

The competitive advantage of C_4 plants is altered by environmental factors that differentially influence CO_2 assimilation rates of C_3 and C_4 plants. In replacement series experiments, C_4 weeds were shown to be more competitive than C_3 weeds at daytime temperatures near 30°C (Pearcy *et al.*, 1981; Roush and Radosevich, 1985). However, at 17/14°C day/night temperatures, *Chenopodium album* (C_3) was more competitive than *Amaranthus retroflexus* (C_4) (Pearcy *et al.*, 1981). The greater competitive capacity of C_4 plants is also diminished at higher CO_2 concentrations because of the greater response of photosynthesis rate and growth of C_3 species to increased CO_2 (Patterson and Flint, 1980; Patterson, 1995). These observations have led to predictions that C_4 weeds and crops may become less competitive toward C_3 species in the future (Patterson and Flint, 1980).

The extent of future decreases in competitiveness of C_4 weeds may depend on the increase in temperature that accompanies the rise in CO_2. Relative competitiveness between *C. album* (C_3) and *A. hybridus* (C_4) did not change when CO_2 was increased from 350 to 1,000 μmol mol^{-1} at daytime temperatures of 34°C (Grisé 1997). A rise in global temperature may actually enlarge the area in which C_4 weeds compete vigorously with crops. Patterson (1995) suggests that global warming may cause northward expansion of aggressive species of tropical and subtropical origin currently restricted to the southern United States, and Henderson *et al.* (1995) predicted a southward shift of native C_4 grasses in Australia.

VI. C_3–C_4 Mixtures

A. Intercropping of Grain Crops

The cultivation of more than one crop in the same space at the same time is known as *intercropping*. Intercropping involves the planting of two or sometimes more species in a mixture at the same time or with one species interplanted during the growth of another species. It is commonly practiced in developing countries in the tropics, especially in subsistence farming. Although intercropping may involve mixtures of C_3 or C_4 crops, a large proportion of intercropping systems in the tropics are C_3–C_4 mixtures involving C_4 cereals (maize, sorghum, or pearl millet) and C_3 legumes

(Keating and Carberry, 1993). Spedding *et al.* (1981) listed 14 examples of crop mixtures giving substantially higher yields or greater financial returns than pure stands of component species, and 12 were C_3–C_4 mixtures. The land equivalency ratio is the relative land area that would be needed in sole crops to obtain the same total yield as with the crop mixture (Willey, 1979), and for most intercrops (C_3, C_4, or mixed) the land equivalency ratio is greater than one (Fukai and Trenbath, 1993). This is especially true for crop mixtures in which one species has a longer growth duration than another.

The C_4 species is usually dominant in an intercrop of mixed types, and its advantage in these mixtures is shown by the relatively small change in its yield with increasing density of the companion species. The increase in density of bean (*Phaseolus vulgaris*) plants had little effect on yield of maize, but with increases in maize density, yield of both bush (Fig. 4) and climbing beans was decreased drastically (Francis *et al.*, 1982a,b). The most important factor in the dominance of the C_4 species, however, is likely to be its greater relative height, rather than the CO_2 fixation cycle, because the most

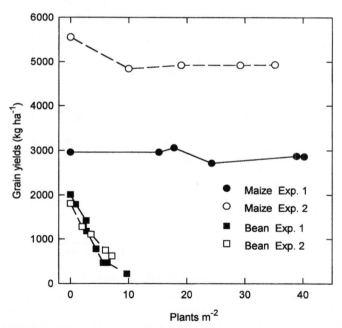

Figure 4 Grain yields of maize and bush bean intercrops as influenced by plant density of the other intercrop species. From data of Francis, C. A., Prager, M., and Tejada, G. (1982). Density interactions in tropical intercropping. II. Maize (*Zea mays* L.) and bush beans (*Phaseolus vulgaris* L.) *Field Crops Res.* **5,** 253–264.

common intercropped C_4 species are usually taller than their companion crops.

The taller C_4 species is more efficient in competition for solar radiation. In addition, the dominated species is often a legume and the shading can have a significant effect on its N_2 fixation capacity. The higher RUE of C_4 crops is evident in these C_3–C_4 mixtures (Fig. 5), and may also be a factor in the dominance by C_4 species. C_3–C_4 mixtures usually exhibit higher RUE than the C_4 sole crop, apparently because partial shading of the C_3 species in the mixture enhances RUE compared the unshaded C_3 sole crop (Keating and Carberry, 1993). The increased RUE in partial shade may result from an increase in the proportion of diffuse radiation on the shaded component as was indicated by model analysis by Sinclair et al.(1992). Increased RUE may be responsible, in part, for the higher land equivalency ratio of the mixtures compared to sole crops.

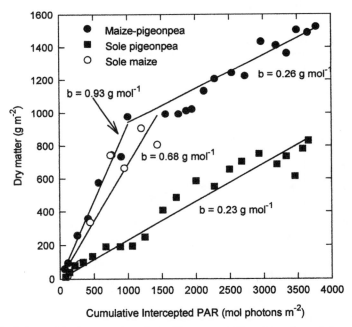

Figure 5 Relationships between cumulative dry matter yields and intercepted photosynthetically active radiation (PAR) for maize, pigeonpea (*Cajanus cajan*), and the mixture of the two species. Steep slopes for maize and the maize–pigeonpea mixture end at about 1,000 g m^{-2} because maize was harvested earlier than pigeonpea. [Redrawn with permission from Sivakumar, M. V. K., and Virmani, S. M. (1980). Growth and resource use of maize, pigeonpea and maize/pigeonpea intercrop in an operational research watershed. *Expl. Agric.* **16**, 377–386. The b-values are the slopes of the linear regressions taken from their Table 3.]

The use of intercrops is often effective in suppressing weeds, because weeds usually become a problem when a crop intercepts a small fraction of the incident radiation. Repression of weed growth in C_4 cereals is most effective when a low growing dicot, such as cowpea (*Vigna unguiculata*), bean, melon (*Cucumis* spp.), or sweet potato (*Ipomoea batatas*) is interplanted (van Rijn, 1991) because of the horizontal leaf display and/or spreading nature of the interplanted crops. However, C_4 cereals are often interplanted in crops that develop more slowly such as cassava (*Manihot esculenta*) and help control weeds until the later developing crop is large enough to intercept most of the incoming radiation.

B. Intercropping of Forages

Mixtures of species are more common than pure stands in grazed pastures and rangeland. The mixtures in cultivated pastures are most frequently grasses and legumes. Such mixtures have advantages over monospecific pastures because legumes have higher nutritive value for ruminants and fix atmospheric N_2 for the mixture. In addition, species in the mixture frequently have different resource requirements or responses to the environment and thus the mixture allows for broader resource exploitation in changing environments. Grasses are more persistent than legumes, thus giving stability and longevity to the sward.

Legumes are C_3 and the interactions of temperate legumes in mixtures with temperate (C_3) grasses have been extensively studied. Temperate legumes are also grown in association with C_4 grasses in subtropical and warm temperate regions. When temperate legumes are grown in mixture with C_4 grasses, the grass and legumes grow rapidly during different seasons, thus minimizing competition. Low temperature early in the growing season in temperate and subtropical regions is an advantage for the temperate legume, and allows them to persist in some mixtures with C_4 grasses. In a study of production and competition in a mixture of the temperate legume alfalfa (*Medicago sativa*) (C_3) and C_4 bermudagrass (*Cynodon dactylon*), it was found that alfalfa tended to dominate the mixture because it gained an initial advantage in the cool spring, but also because it regrew faster following defoliation in summer (Brown and Byrd, 1990).

Coexistence of tropical legumes and grasses is more problematical, however, because their seasonal temperature optima for growth are more similar, and the potential for interspecific competition is greater. Maintenance of high percentages of legume is difficult in most tropical pastures. This difficulty is often attributed to the aggressiveness of C_4 grasses (Humphreys, 1991), although grazing preference for legumes may also contribute. However, it is possible to maintain tropical legumes in mixtures with C_4 grasses if the appropriate species are chosen, if the mixture is not overgrazed, and if no N or only a moderate amount is applied. Ludlow and Charles-Edwards

(1980) reported that a *Setaria anceps/Desmodium intortum* mixture maintained 40–50% legume under 3- and 7-week cutting frequencies for 5 years. Competition between C_4 grasses and legumes is determined by many factors other than their CO_2 assimilation cycles, such as height, rapidity of leaf area expansion after defoliation, and N availability. Under some conditions, the advantage of the C_4 cycle may be negated, as in the example of alfalfa/bermudagrass mixture cited previously (Brown and Byrd, 1990).

A special case of C_4 forages and C_3 plants is the use of pasture among trees. C_3 trees occur in savanna grasslands, and in many tropical and subtropical areas pasture is sown in plantations of tropical trees (Shelton *et al.*, 1987). Reynolds (1988) suggested that a large percentage of the 8.91 million hectares of coconut (*Cocos nucifera*) worldwide could be intercropped with forages with little or no reduction in coconut yields, except where water or nutrients are limiting. The main limitation of grass productivity under trees is low solar radiation, and most research shows a near linear decrease in forage yield under trees with decreases in transmission of solar radiation (Reynolds, 1988). Because C_4 grasses respond strongly to solar radiation intensity and generally grow in sunny environments, they are not considered to be very shade tolerant. There is, however, a considerable range of shade tolerance in C_4 grasses that in some cases can be striking (see Section VII). Differences exist at the species level, with species of *Bracharia* occurring in the low, medium, and high shade tolerance classes (Humphreys, 1991). Shelton *et al.* (1987) noted that few of the high yielding forage grasses have shown genuine shade tolerance and suggested that species be collected from forest margins and shaded habitats for evaluation in tree plantations.

Even for those C_4 grasses that do not exhibit shade tolerance under otherwise optimal conditions, research shows that shading is often beneficial (Wong and Wilson, 1980; Wilson *et al.*, 1990; Wilson, 1996). The increase in yield of several grass species in shade compared to full sun has been attributed to the greater N supply in shade. The greater N supply occurs because organic debris decays faster in the shade than in the sun, probably due to the greater moisture retention in the shaded litter layers (Wilson *et al.*, 1986, 1990). The example in Table XI shows the increase in forage and N yield of a *Paspalum notatum* sward growing in the shade of *Eucalyptus grandis* and in full sun. The lower percentage of dead matter in the sward in shade probably reflects the faster decay of litter, resulting in higher percentages of N and K in the grass shoots. Similarly, Wong and Wilson (1980) observed higher leaf photosynthesis rates in *Panicum maximum* grown in shade, probably because of higher N content in leaves of shaded plants.

Nutritive quality for ruminants is apparently increased by growing C_4 grasses in shade (Fig. 6) (Deinum *et al.*, 1996; Samarakoon *et al.*, 1990a,b). Protein appears to be higher in plants grown at low irradiance and cell

Table XI Yield and Composition of *Paspalum notatum* Herbage after Growth over Spring and Summer, Either in the Open in Full Sunlight or Under the Canopy of a Plantation of *Eucalyptus grandis*[a]

Plant attribute	Open	Under trees	Significance[b]
Dry weight of herbage[c] (kg ha^{-1})	2542	3431	*
Dry weight of green leaf (kg ha^{-1})	2045	3079	***
Nitrogen yield of herbage (kg ha^{-1})	23.1	38.6	***
Leaf in total herbage (%)	80.7	90.2	**
Stem in total herbage (%)	2.8	3.5	ns
Dead in total herbage (%)	16.5	6.3	**
N concentration in shoots (%)	1.01	1.28	***
K concentration in shoots (%)	0.84	1.55	***
P concentration in shoots (%)	0.12	0.12	ns

[a] From Wilson, J. R., Hill, K., Cameron, D. M., and Shelton, H. M. (1990). The growth of *Pasplatum notatum* under shade of a *Eucalyptus grandis* plantation canopy or in full sun. *Trop. Grasslands* **24**, 24–28. With permission.

[b] Significance of difference between means with "*t*" test; ns, not significant; *P <0.05; **P<0.01; ***P<0.001.

[c] Herbage = shoots + stubble

wall concentration is lower (Samarakoon *et al.*, 1990b; Deinum *et al.*, 1996). Digestion of tissues by rumen microorganisms is concomitantly greater for plants grown at low irradiance. Thickness of sclerenchyma cell walls was also observed to be less in plants grown in shade (Deinum *et al.*, 1996). The positive effect of shading on digestibility appears to be greater in pot-grown (Deinum *et al.*, 1996) than field-grown plants (Samarakoon *et al.*, 1990a,b), perhaps because of the higher N concentration of pot-grown plants. The possibility of both higher quality and yield of C_4 grasses in shade (Wilson, 1996) makes the culture of pastures in tree plantations quite attractive.

VII. Turf

In tropical, subtropical, and to a considerable extent in temperate regions, C_4 grasses are used for aesthetic or recreational purposes. Athletic fields, playgrounds, public grounds, roadsides, and home lawns in tropical and subtropical regions are covered almost exclusively by C_4 grasses. Turfgrass is maintained on 14.2 million hectares in the United States and 4 million hectares in the southern states (Duble, 1996), where the grasses

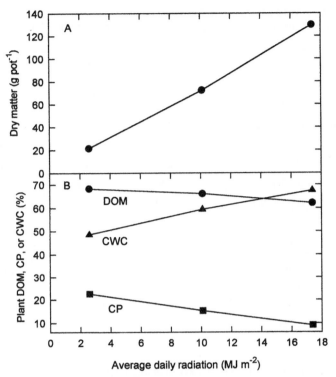

Figure 6 (A) Effect of daily radiation on dry matter yield and (B) digestibility of organic matter (DOM), cell wall concentration (CWC), and crude protein concentration (CP). [Drawn from averages for two C_4 plants (*Bracharia brizantha* and *Panicum maximum* var. *trichoglume*) from Deinum, B., Sulastri, R. D., Zeinab, M. H. J., and Maassen, A. (1996). Effects of light intensity on growth, anatomy and forage quality of two tropical grasses. (*Bracharia brizantha* and *Panicum maximum* var. *trichoglume*). *Neth. J. Agric. Sci.* **44**, 111–124.]

used are mostly C_4. Culture of much of the turfgrass is intense in developed countries, with high inputs of fertilizer and pesticides, especially on golf courses and home lawns. It has been estimated that $45 billion were spent on maintenance of turf areas in the United States in 1993 (Duble, 1996). This is nearly twice the average yearly value of maize grain produced in the United States over the years 1992 to 1994 [198 million tons (FAO, 1994) times an assumed price of $120 ton = $23.8 billion].

There are about 14 warm-season (C_4) species used for turfgrass purposes throughout the world. They originated in several regions, including Africa (*Cynodon* spp. and *Pennisetum clandestinum*), South America (*Paspalum* spp.), Southeastern Asia (*Zoysia* spp.), the West Indies (*Stenotaphrum secundatum* and *Axonopus* spp.), and North America (*Buchloe dactyloides*, and *Bouteloua* spp.). According to Beard (1973), the warm season (C_4) turfgrasses as a group are substantially more tolerant of close mowing, are deeper rooted; and are more drought, heat, and wear tolerant than the cool season (C_3) turfgrasses. In a list of 22 turfgrasses, however, Beard places only C_4 species (*Buchloe dactyloides, Cynodon dactylon, Zoysia* spp., and *Paspalum notatum*) in the group classified as excellent for drought resistance. There were also two C_4 turfgrasses (*Eremochloa ophiuroides* and *Axonopus* spp.) listed among the six poorest in drought resistance. C_4 turfgrasses also tend to be more coarse in texture, and this is related to their greater wear tolerance. Beard (1973) stated that, "The more wear tolerant turfgrasses possess a tough, coarse stem and leaf structure as well as a high shoot density and an increased lignin content" (p. 369). Of the four species he ranked as very fine, none were C_4, and of the five ranked very coarse, only one was C_3.

There are wide differences in adaptation of the C_4 grass species to the geographic locations and to the uses for which the turf is intended. Winter hardy species are needed for the transition zone between temperate and subtropical regions. Tolerance is needed for trampling in high traffic areas such as playgrounds and for shade under trees in the landscape. Beard (1973) ranked several C_4 grasses for their tolerance of these and other factors that influence adaptation (Table XII). As expected, they have excellent tolerance of high temperatures, but poor to good tolerance of low temperatures.

The last three species listed in Table XII are semiarid grasses that have received little improvement through breeding; no turf-type cultivars have been developed, and their use has diminished in the semiarid areas of the United States as irrigation of turf has become more widespread (Beard, 1973). They are tolerant of drought and are of the NADme subtype, which is consistent with the correlation of the frequency of this subtype with aridity in Australian and African grasslands (Hattersley, 1987). *Bouteloua* and *Buchloe* may increase in importance in semiarid areas if water supplies become limiting enough to require use of more drought tolerant grasses. *Cynodon* (NADme) is also characterized as having excellent drought tolerance, as are the *Zoysia* species (PEPck), and the remaining grasses (NADPme) are variable in this trait (Table XII).

VIII. Summary

Crops with the C_4 cycle of CO_2 fixation are well adapted to tropical, subtropical, and even temperate regions. They supply staple foods for a large proportion of people in tropical regions of the world. They also constitute a large proportion of the weed species worldwide. The high potential yields and competitiveness of C_4 plants are traceable to their

Table XII Rankings of Species of C_4 Turfgrasses for Tolerance to Five Factors Related to Adaptation[a]

Species[b]	Tolerance to				
	Treading	High temperature	Low temperature	Shade	Drought
Cynodon species	Very good	Excellent	Poor	Very poor	Excellent
Zoysia species	Excellent	Excellent	Poor to intermediate	Good	Excellent
Stenotaphrum secundatum	Intermediate	Excellent	Extremely poor	Excellent	Fair
Eremochloa ophiuroides	Poor	Excellent	Very poor	Intermediate to fair	Poor
Axonopus species	Intermediate to poor	Excellent	Very poor	Intermediate to fair	Poor
Paspalum notatum	Good	Excellent	Poor	Good	Excellent
Pennisetum clandestinum	Good	Excellent	Very poor	Intermediate	Good
Buchloe dactyloides[c]	NR[d]	Excellent	Good	Poor	Excellent
Bouteloua gracilis	NR	Good	Good	NR	Excellent
Bouteloua curtipendula	NR	Excellent	NR	NR	Excellent

[a] Adapted from Beard, J. B. (1973). "Turfgrass: Science and Culture." Prentice-Hall, Inc., Englewood Cliffs, New Jersey.
[b] Four species are included in *Cynodon* (*C. dactylon*, *C. transvaalensis*, *C. incompletus* var. *hirsutus*, and *Cynodon* x *magennisii*), three in *Zoysia* (*Z. japonica*, *Z. matrella*, and *Z. tenuifolia*), and two in *Axonopus* (*A. affinis* and *A. compressus*).
[c] The last three species were not included in Beard's Table 4-1, but were evaluated based on descriptions in the text.
[d] NR, not ranked.

higher photosynthetic rates, although in many instances other factors are more important in determining adaptation, yield, and the outcome of competition among species. C_4 crops are also more efficient than C_3 crops in the use of N, water, and solar radiation in the production of dry matter and in the conversion of cultivation energy inputs to food energy outputs. Of the three subtypes of C_4 species, the NADPme subtype accounts for the majority of species in the C_4 cereals, pasture grasses, and weeds. The basis for predominance of this subtype among cultivated crops and weeds is unknown, and research on this aspect of photosynthesis could have significant impacts on cultivation of crops. The C_4 grains are low in protein compared to many food stocks, and the quality of the protein is low in these grains because some amino acids, particularly lysine, are deficient for human and domestic animal diets. In addition, digestibility of C_4 grass herbage by ruminants is less than for C_3 grasses and legumes, and the lower digestibility is partially a result of the leaf anatomy required for C_4 photosynthesis. Intercropping of C_3 species (usually legumes) with C_4 grain and pasture crops in tropical regions helps balance the diet, stabilize yields, and increase the efficiency of resource use. Intercropping of C_4 species is also common in tree plantations in the tropics. C_4 grasses are widely used as turf in tropical and subtropical regions. Survival under trees, whether in plantations or as turf, requires some degree of shade tolerance, and there are C_4 species with considerable tolerance of shade. The response of photosynthesis and growth of C_3 species to increasing CO_2 concentrations is much greater than that of C_4 species, and this has led to speculation that the productivity and competitiveness of C_4 plants will decrease in the future with increases in atmospheric CO_2 concentrations. However, if increases in CO_2 concentration are accompanied by increased temperatures, the C_4 species may retain their competitive advantage and the areas of their adaptation may even increase.

References

Akin, D. E., and Burdick, D. (1975). Percentage of tissue types in tropical and temperate grass leaf blades and degradation of tissues by rumen microorganisms. *Crop Sci.* 15, 661–668.

Aldrich, R. J. (1984). "Weed-Crop Ecology. Principles in Weed Management" Breton Publishers, North Scituate, Massachusetts.

Alexander, D. E. (1989). Maize. *In* "Oil Crops of the World. Their Breeding and Utilization" (G. Röbbelen, R. K. Downey, and A. Ashri, eds.), pp. 431–437. McGraw-Hill Publishing Company, New York.

Beard, J. B. (1973) "Turfgrass: Science and Culture." Prentice-Hall Inc., Englewood Cliffs, New Jersey.

Björkman, O. (1971). Comparative photosynthetic CO_2 exchange in higher plants. *In* "Photosynthesis and Photorespiration" (M. D. Hatch, C. B. Osmond, and R. O. Slatyer, eds.), pp. 18–32. John Wiley & Sons, New York.

Bowes, G. (1993). Facing the inevitable: Plants and increasing atmospheric CO_2. *Annu. Rev. Plant Physiol. Plant Mol. Biol.* **44**, 309–332.

Brouk, B. (1977). "Plants Consumed by Man." Academic Press, London.

Brown, R. H. (1978). A difference in N use efficiency in C_3 and C_4 plants and its implications in adaptation and evolution. *Crop Sci.* **18**, 93–98.

Brown, R. H. (1985). Growth of C_3 and C_4 grasses under low N levels. *Crop Sci.* **25**, 954–957.

Brown, R. H., and Byrd, G. T. (1990). Yield and botanical composition of alfalfa–bermudagrass mixtures. *Agron. J.* **82**, 1074–1079.

Colman, R. L., and Lazenby, A. (1970). Factors affecting the response of tropical and temperate grasses to fertilizer nitrogen. *Proc. XI Inter. Grassland Cong.* Surfers Paradise, Australia. pp. 393–397.

Crookston, R. K., and Moss, D. N. (1974). Interveinal distance for carbohydrate in leaves of C_3 and C_4 grasses. *Crop Sci.* **14**, 123–125.

Deinum, B., Sulastri, R. D., Zeinab, M. H. J., and Maassen, A. (1996). Effects of light intensity on growth, anatomy and forage quality of two tropical grasses (*Bracharia brizantha* and *Panicum maximum* var. *trichoglume*). *Neth. J. Agric. Sci.* **44**, 111–124.

Dendy, D. A. V. (1995). Sorghum and the millets. *In* "Sorghum and Millets. Chemistry and Technology" (D. A. V. Dendy, ed.), pp. 11–26. American Association of Cereal Chemists, Inc., St. Paul.

de Wet, J. M. J. (1992). The three phases of cereal domestication. *In* "Grass Evolution and Domestication" (G. P. Chapman, ed.), pp. 176–198. Cambridge University Press, Cambridge.

de Wet, J. M. J. (1995). Finger millet, pp. 137–140; Foxtail millet, pp. 170–172; Minor cereals, 202–208. *In* "Evolution of Crop Plants" (J. Smart, and N. W. Simmonds, eds.) Longman Scientific and Technical, Essex, United Kingdom.

Duble, R. L. (1996). "Turfgrasses. Their Management and Use in the Southern Zone." Texas A&M University Press, College Station, Texas

Dudley, J. W., Lambert, R. J., and Alexander, D. E. (1974). Seventy generations of selection for oil and protein concentration in the maize kernel. *In* "Seventy Generations of Selection for Oil and Protein in Maize" (J. W. Dudley, ed.) pp. 181–212. Crop Science Society of America, Madison, Wisconsin.

Duke, J. A., and Atchley, A. A. (1986). "Handbook of Proximate Analysis Tables of Higher Plants." CRC Press, Boca Raton.

Elmore, C. D., and Paul, R. N. (1983). Composite list of C_4 weeds. *Weed Sci.* **31**, 686–692.

Evans, L. T. (1975). The physiological basis of crop yield. *In* "Crop Physiology. Some Case Histories" (L. T. Evans, ed.), pp. 327–355. Cambridge University Press. London.

FAO (1994). "FAO Production Yearbook." Vol. 48. Food and Agriculture Organization of the United Nations, Rome.

FAO (1995). "Sorghum and Millets in Human Nutrition." FAO Food and Nutrition Series, No. 27. Food and Agriculture Organization of the United Nations, Rome.

Francis, C. A., Prager, M., and Tejada, G. (1982a). Density interactions in tropical intercropping. I. Maize (*Zea mays* L.) and climbing beans (*Phaseolus vulgaris* L.). *Field Crops Res.* **5**, 163–176.

Francis, C. A., Prager, M., and Tejada, G. (1982b). Density interactions in tropical intercropping. II. Maize (*Zea mays* L.) and bush beans (*Phaseolus vulgaris* L.). *Field Crops Res.* **5**, 253–264.

Fukai, S., and Trenbath, B. R. (1993). Processes determining intercrop productivity and yields of component crops. *Field Crops Res.* **34**, 247–271.

Gardner, F. P., Pearce, R. B., and Mitchell, R. L. (1985). "Physiology of Crop Plants" Iowa State University Press, Ames, Iowa.

Gifford, R. M. (1974). A comparison of potential photosynthesis, productivity and yield of plant species with different photosynthetic metabolism. *Aust. J. Plant Physiol.* **1**, 107–117.

Gosse, G., Varlet-Grancher, C., Bonhomme, R., Chartier, M., Allirand, J.-M., and Lemaire, G. (1986). Production maximale de matière sèche et rayonnement soliare intercepté par un couvert végétal. *Agronomie* **6**, 47–56.
Greenwood, D. J., Lemaire, G., Gosse, G., Cruz, P., Draycott, A., and Neeteson, J. J. (1990). Decline in percentage N of C_3 and C_4 crops with increasing plant mass. *Ann. Bot.* **66**, 425–436.
Grisé, D. J., (1997). Elevated CO_2 and temperature interactions on a C_3 annual (*Chenopodium album*) and a C_4 annual (*Amaranthus hybridus*). Ph.D Dissertation. University of Georgia, Athens, Georgia.
Hattersley, P. W. (1987). Variations in the photosynthetic pathway. *In* "Grass Systematics and Evolution" (T. R. Soderstrom, K. W. Hilu, C. S. Campbell, and M. E. Barkworth, eds.), pp. 40–64. Smithsonian Institution Press, Washington.
Hattersley, P. W., and Watson, L. (1992). Diversification of photosynthesis. *In* "Grass Evolution and Domestication" (G. P. Chapman, ed.), pp. 38–116. Cambridge University Press, Cambridge.
Heichel, G. H. (1976). Agricultural production and energy resources. *Am. Sci.* **64**, 64–72.
Henderson, S., Hattersley, P., von Caemmerer, S., and Osmond, C. B. (1995). Are C_4 plants threatened by global climatic change? *In* "Ecophysiology of Photosynthesis" (E. D. Schulze and M. M. Caldwell, eds.) pp. 529–549. Springer-Verlag, Berlin.
Hofstra, G., and Nelson, C. D. (1969). A comparative study of translocation of assimilated ^{14}C from leaves of different species. *Planta* **88**, 103–112.
Holm, L., Plucknett, D. L, Pancho, J. V., and Herberger, J. P. (1977). "The World's Worst Weeds. Distribution and Biology." University of Hawaii Press, Honolulu, Hawaii.
Humphreys, L. R. (1981). "Environmental Adaptation of Tropical Pasture Plants." MacMillan Publishers Ltd., London.
Humphreys, L. R. (1991). "Tropical Pasture Utilization." Cambridge University Press, Cambridge.
Keating, B. A., and Carberry, P. S. (1993). Resource capture and use in intercropping: Solar radiation. *Field Crops Res.* **34**, 273–301.
Kephart, K. D., and Buxton, D. R. (1993). Forage quality responses of C_3 and C_4 perennial grasses to shade. *Crop Sci.* **33**, 831–837.
Knake, E. L., and Slife, F. W. (1962). Competition of *Setaria faberii* with corn and soybeans. *Weeds* **10**, 26–29.
Lambert, R. J., Alexander, D. E., and Dudley, J. W. (1969). Relative performance of normal and modified protein (Opaque-2) maize hybrids. *Crop Sci.* **9**, 242–243.
Loomis, R. S., and Conner, D. J. (1992). "Crop Ecology: Productivity and Management in Agricultural Systems." Cambridge University Press, Cambridge.
Ludlow, M. M. (1985). Photosynthesis and dry matter production in C_3 and C_4 pasture plants, with emphasis on tropical C_3 legumes and C_4 grasses. *Aust. J. Plant Physiol.* **12**, 557–572.
Ludlow, M. M., and Charles-Edwards, D. A. (1980). Analysis of the regrowth of a tropical grass/legume sward subjected to different frequencies and intensities of defoliation. *Aust. J. Agric. Res.* **31**, 673–692.
Maillet, J. (1991). Control of grassy weeds in tropical cereals. *In* "Tropical Grassy Weeds" (F. W. G. Baker, and P. J. Terry, eds.), pp. 112–143. CAB International, Wallingford, United Kingdom.
McDowell, L. R. (1985). Nutrient requirement of ruminants. *In* "Nutrition of Grazing Ruminants in Warm Climates" (McDowell, L. R., ed.), pp. 21–36. Academic Press Inc., Orlando, Florida.
McWhorter, C. G., and Hartwig, E. E. (1972). Competition of johnsongrass and cocklebur with six soybean varieties. *Weed Sci.* **20**, 56–59.
Miller, M. S., Moser, L. E., Waller, S. S., Klopfenstein, T. J., and Kirch, B. H. (1996). Immunofluorescent localization of RuBPCase in degraded C_4 grass tissue. *Crop Sci.* **36**, 169–175.

Minson. D. J. (1981). Nutritional differences between tropical and temperate grasses. In "World Animal Science, B 1; Grazing Animals" (F. H. W. Morley, ed.), pp. 143–157. Elsevier Scientific Publishing Company, Amsterdam.

Monteith, J. L. (1978). Reassessment of growth rates for C_3 and C_4 crops. Expl. Agric. **14,** 1–5.

Moolani, M. K., Knake, E. L., and Slife, F. W. (1964). Competition of smooth pigweed with corn and soybeans. Weeds **12,** 126–128.

Morgan, J. A., and Brown, R. H. (1979). Photosynthesis in grass species differing in carbon dioxide fixation pathways. II. A search for species with intermediate gas exchange and anatomical characteristics. Plant Physiol. **64,** 257–262.

National Research Council. (1984). "Amaranth: Modern Prospects for an Ancient Crop." National Academy Press, Washington, DC.

National Research Council. (1996). "Lost Crops of Africa. I. Grains." National Academy Press, Washington, DC.

Patterson, D. T. (1985). Comparative ecophysiology of weeds and crops. In "Weed Physiology. Vol. 1. Reproduction and Ecophysiology" (S. O. Duke, Ed.), pp. 101–129. CRC Press, Boca Raton, Florida.

Patterson, D. T. (1995). Weeds in a changing climate. Weed Sci. **43,** 685–701.

Patterson, D. J., and Flint, E. P. (1980). Potential effects of global atmospheric CO_2 enrichment on the growth and competitiveness of C_3 and C_4 weed and crop plants. Weed Sci. **28,** 71–75.

Paulsen, G. M. (1994). High temperature responses of crop plants. In "Physiology and Determination of Crop Yield" (Boote, K. J., Bennett, J. M., Sinclair, T. M., and Paulsen, G. M., eds.), pp. 365–389. American Society of Agronomy, Inc., Crop Science Society of America, Inc., and Soil Science Society of America, Inc., Madison, Wisconsin.

Pearcy, R. W., and Ehleringer, J. (1984). Comparative ecophysiology of C_3 and C_4 plants. Plant Cell Environ. **7,** 1–13.

Pearcy, R. W., Tumosa, N., and Williams, K. (1981). Relationships between growth, photosynthesis and competitive interactions for a C_3 and a C_4 plant. Oecologia **48,** 371–376.

Poorter, H. (1993). Interspecific variation in the growth response of plants to an elevated ambient CO_2 concentration. Vegetatio **104/105,** 77–97.

Potter, J. R., and Jones, J. W. (1977). Leaf area partitioning as an important factor in growth. Plant Physiol. **59,** 10–14.

Reid, R. L., Jung, G. A., and Thayne, W. V. (1988). Relationships between nutritive quality and fiber components of cool season and warm season forages: A retrospective study. J. Anim. Sci. **66,** 1275–1291.

Reynolds, S. G. (1988). "Pastures and Cattle Under Coconuts." FAO Plant Production and Protection Paper. No. 91. FAO, Rome.

Roush, M. L., and Radosevich, S. R. (1985). Relationships between growth and competitiveness of four annual weeds. J. Appl. Ecol. **22,** 895–905.

Sage, R. F., and Pearcy, R. W. (1987). The nitrogen use efficiency of C_3 and C_4 plants. I. Leaf nitrogen, growth, and biomass partitioning in Chenopodium album (L.) and Amaranthus retroflexus (L.). Plant Physiol. **84,** 954–958.

Samarakoon, S. P., Wilson, J. R., and Shelton, H. M. (1990a). Voluntary feed intake by sheep and digestibility of shaded Stenotaphrum secundatum and Pennisetum clandestinum herbage. J. Agric. Sci. (Camb.) **114,** 143–150.

Samarakoon, S. P., Wilson, J. R., and Shelton, H. M. (1990b). Growth, morphology and nutritive quality of shaded Stenotaphrum secundatum, Axonopus compressus and Pennisetum clandestinum. J. Agric. Sci. (Camb.) **114,** 161–169.

Shantz, H. L., and Piemeisel, L. N. (1927). The water requirement of plants at Akron, Colo. J. Agric. Res. **34,** 1093–1190.

Shelton, H. M., Humphreys, L. R., and Batello, C. (1987). Pastures in the plantations of Asia and the Pacific: Performance and prospect. Trop. Grasslands **21,** 159–168.

Simpson, J. R. and Stobbs, T. H. (1981). Nitrogen supply and animal production from pastures. *In* "World Animal Science, B 1; Grazing Animals" (F. H. W. Morley, ed.), pp. 143–157. Elsevier Scientific Publishing Company, Amsterdam.

Sinclair, T. R., Shiraiwa, T., and Hammer, G. L. (1992). Variation in crop radiation use efficiency with increased diffuse radiation. *Crop Sci.* **32,** 1281–1284.

Sivakumar, M. V. K., and Virmani, S. M. (1980). Growth and resource use of maize, pigeonpea and maize/pigeonpea intercrop in an operational research watershed. *Expl. Agric.* **16,** 377–386.

Soil Conservation Society of America. (1986). "Warm Season Grasses: Balancing Forage Programs in the Northeast and Southern Corn Belt." Soil Conservation Society of America. Ankeny, Iowa.

Spedding, C. R. W., Walsingham, J. M., and Hoxey, A. M. (1981). "Biological Efficiency in Agriculture." Academic Press, New York.

Stephenson, R. A., Brown, R. H., and Ashley, D. A. (1976). Translocation of ^{14}C-labeled assimilate and photosynthesis in C_3 and C_4 species. *Crop Sci.* **16,** 285–288.

Stoskopf, N. C. (1985). "Cereal Grain Crops." Reston Publishing Company, Reston, Virginia.

van Rijn, P. J. (1991). Integrated control of grassy weeds in mixed crops. *In* "Tropical Grassy Weeds" (F. W. G. Baker and P. J. Terry, eds.), pp. 153–163. CAB International, Wallingford, United Kingdom.

Voigt, P. W., and MacLauchlan, R. S. 1985. Native and other western grasses. *In* "Forages: The Science of Grassland Agriculture" (M. E. Heath, R. F. Barnes, and D. S. Metcalfe, eds.), pp. 177–187. Iowa State University Press. Ames, Iowa.

Waithaka, K., and Chweya, J. D. (1991). "*Gynandropsis gynandra* (L.) Briq.—A Tropical Leafy Vegetable." FAO Plant Production and Protection Paper No. 107. Food and Agriculture Organization of the United Nations, Rome.

Wanous, M. K. (1990). Origin, taxonomy and ploidy of the millets and minor cereals. *Plant Var. Seeds* **3,** 99–112.

Willey, R. W. (1979). Intercropping—its importance and research needs. Part 1. Competition and yield advantages. *Field Crops Abst.* **32,** 1–10.

Wilson, J. R. (1994). Cell wall characteristics in relation to forage digestion by ruminants. *J. Agric. Sci.* (*Camb.*) **122,** 173–182.

Wilson, J. R. (1996). Shade-stimulated growth and nitrogen uptake by pasture grasses in a subtropical environment. *Aust. J. Agric. Res.* **47,** 1075–1093.

Wilson, J. R., Brown, R. H., and Windham, W. R. (1983). Influence of leaf anatomy on the dry matter digestibility of C_3, C_4, and C_3/C_4 intermediate types of *Panicum* species. *Crop Sci.* **23,** 141–146.

Wilson, J. R., Catchpoole, V. R., and Weier, K. L. (1986). Stimulation of growth and nitrogen uptake by shading a rundown green panic pasture on Brigalow clay soil. *Trop. Grasslands* **20,** 134–143.

Wilson, J. R., Hill, K., Cameron, D. M., and Shelton, H. M. (1990). The growth of *Paspalum notatum* under shade of a *Eucalyptus grandis* plantation canopy or in full sun. *Trop. Grasslands* **24,** 24–28.

Wittwer, S. H. (1995). "Food, Climate, and Carbon Dioxide: The Global Environment and World Food Production." CRC Press, Inc., Boca Raton, Florida

Wong, C. C., and Wilson, J. R. (1980). Effects of shading on the growth and nitrogen content of green panic and siratro in pure and mixed swards defoliated at two frequencies. *Aust. J. Agric. Res.* **31,** 269–285.

Yamaguchi, M. (1983). "World Vegetables—Principles, Production, and Nutritive Values." AVI Publishing Company, Westport, Connecticut.

15

C_4 Plants and the Development of Human Societies

Nikolaas J. van der Merwe and Hartmut Tschauner

I. Introduction

The earliest recognizable ancestor of humans lived more than 4 million years ago and the earliest fossil representatives of the genus *Homo* are 2 million years old. Whether one reckons the age of humanity to be 4 million years or 2 million years, the amount of time that humans have spent on Earth as hunter–gatherers still works out to more than 99%. During this period, humans apparently made little or no direct use of grasses for food, whether C_4 or C_3. We know that modern hunter–gatherers like the San of the Kalahari use grass seeds that are accumulated and stored by harvester ants, grinding them into flour to make unleavened bread. This procedure requires stone mortars and pestles, implements that appeared no more than 15,000 years ago in, for example, the Near East (K. I. Wright, 1991, 1993, 1994). The intensive harvesting of grasses by people started at the end of the Pleistocene; it led to the domestication of C_3 barley and wheat in the Near East about 10,000 years ago. The domestication of two C_4 millet species occurred soon thereafter in northern China, followed in turn by the domestication of sorghum and several other millets in Africa and of maize in Mexico. These C_4 plants became the staple crops of urban societies in Asia, Africa and the Americas.

Although the direct use of C_4 plants as food by humans is a very recent phenomenon, the spread of C_4 grasses around the world may have played an important role in the biological evolution of our human ancestors. This part of the story starts more than 7 million years ago, before the emergence of the earliest hominids.

II. C₄ Plants, Carbon Isotopes, and Human Evolution

In the Late Miocene, between about 8 Ma and 5 Ma (million years ago), C_4 plants expanded rapidly in both the Old and New World (Cerling et al., 1997; Cerling, Chapter 13, this volume). The evidence for this expansion is to be found in the stable carbon isotope ratios ($\delta^{13}C$ values) of fossil tooth enamel: no change occurred in the $\delta^{13}C$ values of browsing animals (whose diets are focused on C_3 trees and shrubs), whereas the $\delta^{13}C$ values of grazing animals changed from C_3 to C_4 diets in those parts of the world where C_4 grasses became established. This shift in the carbon isotope signatures of grazers was first documented in Pakistan (Quade et al., 1989, 1992) and North America (Cerling et al., 1993; Wang et al., 1994; MacFadden and Cerling, 1996) and has now also been observed in South America (MacFadden et al., 1996) and East Africa (Cerling et al., 1997). No isotopic shift took place in Western Europe, where C_4 grasses did not become established. Cerling and co-workers (1997) attribute these changes to a decline in atmospheric CO_2 content in the late Miocene below a critical threshold where C_3 plants are photosynthetically less competitive than C_4 plants, especially when subject to warm temperatures. The result was a widespread expansion of C_4 plants at the expense of C_3, leading to a landscape resembling the modern distribution of these photosynthetic types. It is suggested that the threshold in atmospheric CO_2 content lies around 500 ppm (Ehleringer et al., 1991, 1997; Ehleringer and Björkman, 1977). In comparison, atmospheric CO_2 content in the early Miocene may have been 800 ppm (Cerling, Chapter 13, this volume). During the most recent glacial event it was as low as 180 ppm, when C_4 plants were particularly successful in warmer regions. The current value for the open atmosphere is about 350 ppm, but local concentrations as high as 500–1000 ppm can be found at ground level in rain forests where CO_2 from rotting leaf litter accumulates (Medina and Minchin, 1980; Medina et al., 1986).

Stable carbon isotope ratios have been used since the late 1970s to measure the relative contributions of C_3 and C_4 plants to prehistoric foodwebs (Vogel and van der Merwe, 1977; van der Merwe and Vogel, 1978; for reviews, see van der Merwe, 1982; DeNiro, 1987; Schoeninger and Moore, 1992; Katzenberg and Harrison, 1997). In the arguments presented in this chapter, carbon isotopes are frequently used as evidence for C_4 plants in the diets of prehistoric animals and humans, so an explanation of the procedure is in order. The preferred sample material for archaeological applications has been bone collagen, a protein that is resistant to isotopic alteration, but that is preserved for at most 100 kyrs under the best of circumstances (e.g., cold, dry caves). Carbonates in biological apatite also record a dietary isotopic signal, but bone mineral can be contaminated by carbonates from the burial environment. Tooth enamel has been shown to retain a reliable dietary signal as well, lasting many millions of years (Lee-Thorp and van der Merwe, 1987; Thackeray et al., 1990). Measuring the carbon isotopes in tooth enamel for the purpose of reconstructing the dietary habits of fossil animals is now a widely used method (e.g., Lee-Thorp et al., 1989b; Wang et al., 1994; MacFadden et al., 1996).

In order to interpret a dietary isotope signal, it is necessary to know the $\delta^{13}C$ of the source CO_2 (usually that of the open atmosphere) at a given time; the isotopic fractionation effects of the relevant photosynthetic pathways (C_3 and C_4 in this context, but there is also CAM) in a given environment; and the isotopic shift between diet and a given consumer tissue. Figure 1 outlines $\delta^{13}C$ values in the modern atmosphere, the C_3 and C_4 plants that use this atmosphere for photosynthesis, and the herbivores that eat the plants. The current atmosphere has a CO_2 content of about 350 ppm and a $\delta^{13}C$ value of -8‰. Any disturbance of the equilibrium between the different carbon reservoirs (lithosphere, biosphere, oceans, atmosphere) may alter the CO_2 content and $\delta^{13}C$ content of the atmosphere and produce plants and animals with altered $\delta^{13}C$ values. An example of such a disturbance is the burning of fossil fuel, which is of C_3 origin and depleted in ^{13}C. The CO_2 content of the atmosphere has been rising for some time, and its $\delta^{13}C$ value has become more negative. Carbon isotope measurements on ice core air

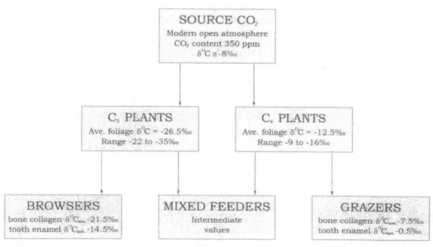

Figure 1 $\delta^{13}C$ values for the modern atmosphere, C_3 and C_4 plants, and large herbivores that feed on them. The $\delta^{13}C$ value of the atmosphere became more negative by 1.5‰ during the Industrial Revolution; in the mid-1800s, all the $\delta^{13}C$ values in the figure were 1.5‰ more positive. The preindustrial atmosphere ($\delta^{13}C \cong -6.5‰$) may have become established during the Miocene. For details and references, see text.

bubbles and archaeological maize kernels (Friedli *et al.*, 1986; Tieszen and Fagre, 1993) show that the $\delta^{13}C$ value of the atmosphere was about $-6.5‰$ between 1000 and 150 BP (years Before Present). The $\delta^{13}C$ value has been changing in a negative direction since 150 BP and this trend accelerated in the 20th century (Marino and McElroy, 1991); it will continue to do so until the burning of fossil fuel ceases.

The isotopic effect of the Industrial Revolution is an unusual example of human origin, but the $\delta^{13}C$ values of past atmospheres have also varied substantially in response to climatic and geological events. For example, the $\delta^{13}C$ value of the atmosphere declined by $8‰$ during the last 20 million years of the Permian (at about 250 Ma), as measured in the tooth enamel of mammal-like reptiles (Thackeray *et al.*, 1990). Atmospheric changes during the past 10 million years, which are of concern here, were not nearly as dramatic. The earliest known C_4 plant fossil (from California, at 12.5 Ma) has a $\delta^{13}C$ value of $-13.7‰$ (Nambudiri *et al.*, 1978; Tidwell and Nambudiri, 1989), within the range of modern C_4 plants (at the negative end). Similarly, tooth enamel $\delta^{13}C$ values of late Miocene animals are much the same as those for their modern counterparts, once the Industrial Effect of $1.5‰$ has been taken into account. Small changes in atmospheric $\delta^{13}C$ values probably accompanied changes in atmospheric CO_2 concentrations—during glacial events, for example, when the CO_2 content fell as low as 180 ppm—but these have not yet been measured with accuracy.

The connection between atmospheric $\delta^{13}C$ values and herbivore tooth enamel is, of course, the plants eaten by the animals. The average $\delta^{13}C$ value for modern C_3 foliage is about $-26.5‰$; the values range between about $-22‰$ in desert conditions and about $-35‰$ under closed forest canopies, but most C_3 plants have values near the average (Vogel *et al.*, 1978; van der Merwe and Medina, 1989; Ehleringer *et al.*, 1993; Cerling *et al.*, 1997). C_4 plants have a restricted range of $\delta^{13}C$ values from about $-9‰$ to $-16‰$, with an average value of about $-12.5‰$; the NADP-me and NAD-me subtypes have average values of $-11.4‰$ and $-12.7‰$, respectively (Vogel *et al.*, 1978; Hattersley, 1982; Cerling *et al.*, 1997). When large herbivores eat plant foliage, they produce bone collagen with $\delta^{13}C$ values about $5‰$ more positive than the dietary average (van der Merwe and Vogel, 1978; Vogel, 1978) and tooth enamel that is about $12‰$ more positive than the diet (Lee-Thorp and van der Merwe, 1987; Lee-Thorp *et al.*, 1989a; but compare Cerling *et al.*, 1997, who use $14‰$). Browsing animals (C_3-plant eaters) in temperate, open habitats have average $\delta^{13}C$ of $-21.5‰$ in their bone collagen and $-14.5‰$ in their tooth enamel, whereas grazers in C_4 grass areas have average $\delta^{13}C$ values of $-7.5‰$ in bone collagen and $-0.5‰$ in tooth enamel. For collagen, further isotopic shifts of about $+1.5‰$ take place on each trophic level. This is not the case for tooth enamel: a carnivore has approximately the same $\delta^{13}C$ value in its enamel as the herbivores it preys on.

To compare diets and environmental events across long time scales by means of carbon isotope measurements, it is necessary to establish the C_3 and C_4 end members in each case. This is done by measuring the $\delta^{13}C$ values of tooth enamel from obligate browsers and grazers, insofar as these dietary habits can be established. In East Africa, where much of the evidence for hominid evolution is to be found, Cerling (Chapter 13, this volume) and Cerling and co-workers (1997) measured carbon isotope ratios in tooth enamel of large mammals over the past 20 million years. They found that all the animals had C_3-dominated diets before 8 Ma, as if C_4 plants were not available in the region. Between 8 Ma and 7 Ma, grazing proboscideans (elephant-like animals) and equids (horses) changed to C_4-dominated diets, whereas the C_3-dominated isotopic signals of browsing deinotheres remained unchanged. Cerling and co-workers (1997) conclude that C_4 plants were scarce or absent in East Africa before 8 Ma, but had become widespread by 7 Ma. A conflicting conclusion has been reached on the basis of carbon isotopic measurements in soil carbonates and organic matter, indicating that mosaics of C_4 and C_3 plants existed without significant change in East Africa over the past 10 million years (Kingston et al., 1994; Morgan et al., 1994; Hill, 1995). Cerling (Chapter 13) makes a compelling case that the latter conclusion is in error.

Of importance here is that East Africa had become a "C_4 world" by 7 Ma (Cerling et al., 1997), with a landscape of savannas and grasslands. It follows that similar plant communities emerged in central and southern Africa at the same time. It is widely argued that these changes were of critical importance to the evolutionary steps that saw the divergence of the earliest hominids from the great apes and the emergence of bipedalism. This "savanna hypothesis" has come to be questioned in the light of fossil finds in Ethiopia (White et al., 1994, 1995) and Kenya (Leakey et al., 1995), where hominids more than 4 million years old may have lived in wooded habitats (for a review of the debate, see Shreeve, 1996). This debate remains to be settled, but we are still left with the evidence that Africa cooled and dried over the last 5 million years (de Menocal, 1995), with concomitant changes in faunal communities (Vrba, 1995). The climate oscillated between warm, humid episodes and cooler, dry periods, with a distinct shift toward more arid conditions after 2.8 Ma (de Menocal, 1995). The evolution of such early hominids as the Australopithecines and the genus *Homo* itself took place in this context, with the progressive exploitation of savanna plants and animals as a central theme. In short, C_4 plants played a part (details as yet unclear) in the evolution of early hominids.

Direct evidence for the role of C_4 plants in human evolution has been found in tooth enamel $\delta^{13}C$ values of early hominids in South Africa (Lee-

Thorp and van der Merwe, 1993; van der Merwe *et al.*, n. d.). The results show that *Australopithecus africanus* from Sterkfontein (*ca.* 3.0–2.5 Ma), as well as *A. robustus* and *Homo* sp. from Swartkrans (*ca.* 1.8 Ma), had diets with a C_4 contribution of 25% or more. This does not mean that these early human ancestors ate grass: it is more likely that the C_4 component came from animals and insects that ate grass. A grass-eating baboon, *Theropithecus* (ancestral to the modern gelada of Ethiopia), was present at the same time, but its teeth were specially adapted to eat grass, unlike those of the hominids; contemporaneous *Parapapio* (an extinct ancestor of the chacma baboon) had a diet based exclusively on C_3 plants (Lee-Thorp *et al.*, 1989b). The C_4 component in the tooth enamel of the early hominids suggests that meat, whether scavenged or hunted, was an important element in the diets of all hominid species after 3 Ma. It is not known how far back in time this dietary adaptation emerged because East African hominid fossils have not been analyzed isotopically. Observations of chimpanzees in Tanzania and Uganda, however, show that they hunt for meat on a regular basis and that it makes up about 8% of their diets (Stanford, 1995). The hunting behavior of chimpanzees, genetically the closest relatives of humans, coupled with the isotopic evidence for meat in early hominid diets, suggest that the hunting adaptation may have been present before the evolutionary divergence of chimps and hominids.

III. The Emergence of Food Production

After more than 4 million years of a hunter–gatherer lifestyle, humans in several parts of the world domesticated plants and animals within a relatively short time and emerged as food producers. These events took place between about 10,000 and 5,000 years ago and laid the foundations for the development of complex societies and urban life (see, for example, Clark and Brandt, 1984; Heiser, 1990; Cowan and Watson, 1992; Zohary and Hopf, 1994; Price and Gebauer, 1995; Smith, 1995). In at least five parts of the world, independent episodes of plant domestication produced agricultural societies that depended on cereal crops (Fig. 2). These were western Asia (wheat and barley, plus noncereals like lentils and other pulses); eastern Asia (millets and Asian rice); the African savannas (millets, sorghum, and African rice); Mexico (maize, amaranth, plus beans and squashes); and northeastern America (several chenopods, plus sunflower, and gourds). Root crops are less well documented in archeology, but their centers of domestication are known to include Peru for potatoes (plus cereals like quinoa) and tropical America for cassava. Other species of root crops were domesticated in southeastern Asia and tropical Africa. Of the

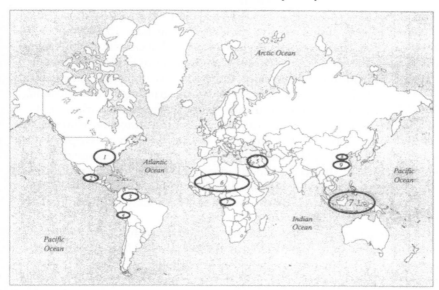

Figure 2 Map of the world showing the location of centers of early plant domestication. (1) Eastern Agricultural Complex (sunflower, sumpweed, goosefoot, maygrass, etc.); (2) maize; (3) cassava (yuca, manioc; location approximate); (4) potatoes and Andean tubers; (5) wheat, barley, pulses; (6) African sorghum and millets; (7) yams (location speculative); (8) Chinese millets; (9) Chinese rice. For details and references, see text.

plants mentioned only millets, sorghum, maize, and amaranth are C_4 plants (Table I).

Archeological evidence and DNA studies of modern populations suggest that anatomically modern *Homo sapiens sapiens* evolved in Africa some

Table I Estimated Annual World Production of Selected Food Crops

Crop	Metric tons × 10^6	Crop	Metric tons × 10^6
Sugarcane[a]	932	Sweet potato	110
Wheat	536	Soybean	95
Maize	481	Sorghum[a]	71
Rice (Asian)	476	Oat	48
Potato	309	Rye	32
Sugar beet	286	Millet (all species)[a]	31
Barley	180	Bean (dry)	15

Note: Sugarcane appears to be the most productive crop, but it is represented by its cane weight, which has a high water content; cereals are represented by their dry kernel weight.

[a] C_4 plant. "FAO Production Yearbook," vol. 40, Rome, 1986, quoted by C. B. Heiser, Jr. (1990). "Seed to Civilization: The Story of Food." Harvard University Press, Cambridge, p. 63.

200,000 years ago, spread through the Near East to Europe and Asia after 100,000 years ago, and arrived in the Americas at least 15,000 years ago (Gamble, 1994). The retreat of the glaciers and global climatic amelioration at this time set the stage for new subsistence adaptations by modern humans. By 15,000 BP, hunter–gatherers of the Near East were harvesting wild grasses intensively, and processing it by stone grinding (Hillman *et al.*, 1989). These grasses were C_3 and their task became easier with time: between 15,000 and 12,000 BP, the CO_2 content of the atmosphere increased from 200 to 270 ppm, which improved the productivity of C_3 plants by 25–50% (Sage, 1995). By 10,000 BP, wheat and barley had become domesticated in the Levant (Bar-Yosef and Meadow, 1995). By 9,000 BP, two species of C_4 grasses were probably already domesticated in China, although the record there is a lot less detailed than in the Near East (Underhill, 1997; Cohen, n. d.).

The process by which plants became domesticated has been best documented in the Levant (Bar-Yosef and Belfer-Cohen, 1991; Bar-Yosef and Meadow, 1995), where environmental and archeological research has been intense. Between about 13,000 and 10,000 years ago, the people of the Near East developed a sedentary lifestyle, living in small, permanent settlements instead of moving around on a seasonal basis to find food. Such sedentism was made possible by an environment rich in natural resources: they harvested the abundant stands of wild wheat and barley with flint-bladed sickles, stored the grain in storage pits for use in the off-season, and also hunted gazelle in large numbers. In archeological terminology, this culture is known as the Natufian. During the late Natufian, between about 10,700 and 10,000 BP, global climates deteriorated with the glacial readvance of the Younger Dryas, resulting in a decline of the natural resource base. During this relatively short period, people of the Levant increasingly manipulated their plant resources to maintain food supplies, apparently saving part of the cereal harvest and planting it during the next growing season. This manipulation included, for example, bringing einkorn wheat seed from the hilly interior of the Near East, the natural habitat of the wild plant, to the narrow coastal strip of the Levant, where it did not occur naturally, but could be grown with more success. By 10,000 BP, such manipulation of wild C_3 cereals had resulted in their domestication. Local barley had changed morphologically from its wild to a domestic form, whereas einkorn was now grown in an area where it had not occurred in the wild.

Based on the botanical and archeological evidence from the Near East, plus somewhat less satisfactory information from other centers of domestication, archeologists can agree on some of the conditions that led to plant (and animal) domestication (Price and Gebauer, 1995; Watson, 1995). These include complex hunter–gatherer organization (as opposed to small, egalitarian, mobile bands), sedentism, and an abundance of resources.

These conditions are interrelated: domestication usually occurred in coastal or riverine areas with a diversity of resources, which made sedentism and more complex social organization possible. It is tempting to add to these conditions the notion of a crisis, such as the environmental deterioration that the Younger Dryas apparently occasioned in the Near East, but the archeological and chronological records in other areas of the world are simply not fine-grained enough.

IV. Millets in China

It is natural to think of Chinese civilization as rice-based, but the classic Bronze Age dynasties (Hsia, Shang, and Chou, *ca.* 4,000–2,700 BP) were built on millet agriculture (Chang, 1980, 1986). The domestication of millet (*Setaria italica* and *Panicum miliaceum,* both C_4) took place along the Yellow River and in the loess plains of North China (see Fig. 3) after about 9000 BP. At the same time, rice (*Oryza sativa*) was domesticated further

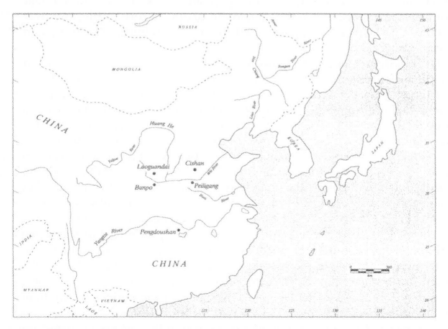

Figure 3 Map of China showing the location of sites mentioned in the text. [From: Chang, K. C. (1986). "The Archaeology of Ancient China," 4th ed. Yale Univeristy, Press, New Haven and London; and Barnes, G. (1993). "China, Korea, and Japan: The Rise of Civilization in East Asia." Thames and Hudson, London.]

south, along the Yangtze River. Rice apparently did not play a subsistence role in the power centers of early China.

Archeological evidence for early plant domestication in China is still sparse (Chang, 1986; Cohen, n. d.; Crawford, 1992a; Underhill, 1997), although it is rapidly being filled in; the same is true of radiocarbon dates. About 10,000 years ago, China was populated by small groups of hunter–fisher–gathering folk. By about 8,500 or 8,000 BP (depending on how one interprets the radiocarbon dates), sedentary villages of agriculturalists had become established along the Yellow River (or Huanghe), from where it descends from the western highlands and cuts westward through the North China plain. This plain of loess was a relatively arid steppe with C_4 grasses in the west, shading to a deciduous woodland in the east (Ho, 1984; Crawford, 1992a). The villages of North China during the period 8,500–7,500 BP are grouped together as the Peiligang culture, after a site by that name in Henan province (see Fig. 3); regional variants are called Cishan (or Tz'u-shan) in Hebei, or Laoguandai (or Lao-kuan-t'ai) in Shanxi (Zhou, 1981; Rodwell, 1984). The villages were fairly large, as much as two hectares in extent (Smith, 1995), with round houses some 2 to 3 m in diameter. Domesticated animals included dogs, pigs, and chickens, but hunted deer contributed importantly to the diet, as did pheasant, carp, turtle, and river shellfish. Wild plants were gathered, including walnuts, hazelnuts, and hackberry (Rodwell, 1984).

Of importance here, of course, is the evidence for early agriculture in the Peiligang culture. Most of this is indirect (Smith, 1995): the villages are larger than those of even complex hunter–gatherers, and the grave goods include stone axes for tree-felling; stone hoes for tilling; mortars with feet, on which to grind grain with pestles; and serrated sickles for harvesting. Numerous storage pits occurred between the houses, some with the remains of grain. At Cishan, 80 storage pits were found, some as deep as 5 m and with heaps of rotting grain up to 1 m thick in them (Rodwell, 1984). Some of the seeds at Cishan have been identified to be foxtail millet (*S. italica*) and those at Peiligang as broomcorn millet (*P. miliaceum*) (Ren, 1995; but see Crawford, 1992a, for problems of identification). Clear botanical proof that these millets were domesticated is lacking, but the sheer volume of stored grain suggests domestication. Both species of millet were extensively cultivated in the succeeding Yangshao culture (after 7,000 BP), when the population of North China was fully agricultural (Chang, 1986), and they continued as staple cereals into the historic period. By the time of the Shang dynasty (after 3,700 BP), when China had entered the Bronze Age, separate Chinese ideographs for the two millet species occurred on oracle bones (Chang, 1980).

Although millets were being domesticated in the Yellow River valley, the same thing was happening with rice in the Middle Yangtze valley, about

800 km to the south. Here the site of Pengdoushan (or Peng-tou-shan, Fig. 3) gives its name to an archeological culture much like that of Peiligang (Chang, 1986; Yan, 1991; Higham, 1995). Pengdoushan is actually a few centuries earlier than Peiligang, but it is likely that the earliest evidence for millet and rice domestication has yet to be found. Although the two cultures were clearly distinctive local developments, they are sufficiently similar to be considered as part of a broader whole. The two agricultural systems did not blend together, however, because the two environments are so different. Carbon isotope ratios for human collagen illustrate this clearly (Table II; Cai and Qin, 1984). This is the only dietary isotope study done in China until now, and the results show the great potential of such analysis. In reading this table, one can take the C_3 and C_4 end members for collagen to be at approximately -20‰ and -6‰. The authors note in their discussion of the results that the modern bone from Hebei reflects mixed agriculture of wheat, millet, and maize, whereas the Warring States specimen from Sichuan reflects a culture based entirely on rice (and a C_3 biome). For Shanxi Province, the heartland of the Peiligang culture, the authors conclude that millet agriculture was very well established during the Yangshao and Longshan cultures at the sites of Banpo (or Pan-p'o), Beishouling (or Pei-shou-ling), and Taosi (or T'ao-ssu), as it was also in Gansu at the Zhuanglang (or Chuang-lang) site (Fig. 3). They interpret the data from Shandong province to mean that millet agriculture arrived there late and did not take hold quickly. In contrast, they conclude that millet made up 90% of the diet in Jilin by 2,900 BP. These interpretations

Table II $\delta^{13}C$ Values for Human Bone Collagen (Unless Otherwise Indicated) from Northern China[a]

Province	Archeological context	Average $\delta^{13}C$ (no. of specimens)
Hebei	Modern	−16.8 (1)
Jilin	Yangtun, Tang Dynasty, ca. 1,300 BP	− 8.5 (1)
Sichuan	Warring States–Han Dynasty, Puge county, ca. 2,500–1,800 BP	−20.4 (1)
Gansu	Xujianian site, 2,745 ± 130 BP	−11.1 (1)
Shandong	Lingyanghe site, 3630 ± 145 BP	−16.8 (1)
Shandong	Baishicun site, ca. 5500 BP	−19.9 (1)
Shaanxi	Taosi site, Longshan culture, ca. 5,000–4,000 BP	−11.3 (3)
Shaanxi	Taosi site, Longshan culture, ca. 5,000–4,000 BP, pig bone	−11.8 (1)
Shaanxi	Huxizhuang site, Longshan culture, ca. 5,000–4,000 BP	−13.7 (3)
Shaanxi	Beishouling site, Yangshao culture, ca. 7,000–6,500 BP	−13.8 (3)
Shaanxi	Banpo site, Yangshao culture, ca. 7,000–6,500 BP	−13.7 (3)

[a] See Fig. 3 for site locations. From: L. C. Cai and S. H. Qiu. (1984). Carbon-13 evidence for ancient diet in China. *Kaogu* **10,** 949–955. Translated by Susan R. Weld.

are not entirely correct: the authors do not consider the dietary contribution of animal food and the differences in plant cover between the different provinces. The pig bone from Taosi in Shanxi indicates that the animal was fed substantially on millet, whereas it is likely that Shandong had more C_3 plants in the environment than Shanxi.

Chinese millets spread far and wide in prehistoric times. Both broomcorn (*P. miliaceum*) and foxtail millet (*S. italica*) were cultivated in Japan by 3,000 BP (Crawford, 1992b). Of particular interest is the occurrence of these two species of millet in early European contexts. During the Neolithic and early Bronze Age (*ca.* 6,000–3,500 B. P.) millets were found in a broad band across central Eurasia from East Asia to the Alps (Kroll, 1981, 1983). At the extensively excavated site of Kastanas in Macedonia, broomcorn millet was an important cultivar throughout the Late Bronze and early Iron Ages (around 3,500–3,000 BP). It is mentioned in literary sources from classical Greece, but it was considered an exotic cereal, a characterization confirmed by archeological finds from Kastanas. Foxtail millet was also found at Kastanas, but in very small numbers and not in storage contexts. It may have been cultivated for a brief period around the beginning of the Iron Age (Kroll, 1983). Finally, at the Hallstatt Iron Age (*ca.* 2,700 BP) site of Magdalenska gora in Slovenia, the average $\delta^{13}C$ for human collagen measures $-16‰$, evidently from a substantial millet component in the diet of the people or their domestic animals (Murray and Schoeninger, 1988).

The spread of Chinese millets to Europe still leaves an interesting puzzle that has not been fully addressed. *Setaria italica* or foxtail millet is generally considered to be the domesticated form of the wild *Setaria viridis* (green bristle grass), with which it is interfertile. The wild form occurs in Europe and may have been independently domesticated there (J. M. J. de Wet, pers. comm., quoted in Rodwell, 1984). The same is not true for *P. miliaceum*, for which no wild ancestor has been identified. It has been suggested that it was domesticated in areas other than China, for example, central Asia (see Hancock, 1992) or southwest Asia (Brown, Chapter 14, this volume, Table I), but China is its center of diversity.

V. C_4 Plants in Africa

Africa provides an illustration of the contrasting environmental requirements of domesticated C_3 and C_4 cereals. The North African coast has winter rainfall like the Near East, so that wheat and barley spread quickly into this region from their Levantine source. The Nile Valley has its own regular water supply and participated in these developments. South of the Atlas mountains, however, the climate of the African continent is dominated by summer rainfall—except for a small area around Cape Town, which is as far from the equator as Morocco and has a Mediterranean climate to match. For agriculture to take hold south of the Sahara required the domestication of African C_4 grasses. The best known C_4 African domesticates are sorghum (*Sorghum bicolor*) and several millets: finger millet (*Eleusine coracana*), pearl millet (*Pennisetum glaucum*), fonio (*Digitaria* sp.), and tef (*Eragrostis tef*). Africa also produced a C_3 cereal: African rice (*Oryza glaberrima*), which is still grown in the West African tropics.

The archeological evidence for African cereal domestication is sparse, nearly nonexistent: it can be cataloged in a paragraph. At the site of Nabta Playa in the Egyptian western desert (Fig. 4), several wild cereals and legumes were harvested some 8,000 years ago (Wasylikowa *et al.*, 1995). One of these was sorghum, which occurred in every house in the village. The seeds have been compared both morphologically and chemically with modern sorghum races, with inconclusive results. They are smaller than modern seeds and it is not clear if they were cultivated or merely harvested. The next piece of evidence is from Adrar Bous, in the central Sahara, where a domesticated sorghum seed left an impression in the wet clay of

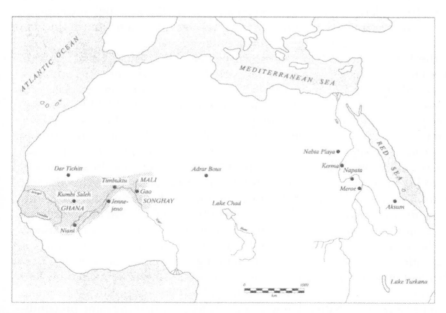

Figure 4 Map of northern Africa showing the location of sites mentioned in the text. Shaded areas represent the territories of the West African kingdoms of ancient Ghana (1,600–800 BP), Mali (800–500 BP), and Songhay (500–400 BP). [From: Shaw, T., Sinclair, P., Andah, B., and Okpoko, A. (1995, eds.). "The Archaeology of Africa: Foods; Metals and Towns." Routledge, London and New York.]

a pot before it was fired some 4,000 years ago (Clark et al., 1973). At the same time, domesticated sorghum had already made its way from Africa to India, apparently by way of Arabia (Kajale, 1988). For pearl millet, the earliest archeological evidence in Africa comes from the site of Dar Tichitt in Mauritania, at the southwestern edge of the Sahara, some 3,000 years ago (Munson, 1976). By that time, however, pearl millet was also grown in India (Vishnu-Mittre, 1974). Finger millet, which probably originated in Uganda and Ethiopia, has also been identified in India at 3,000 BP by Vishnu-Mittre, but this identification has been questioned (Harlan, 1992). By 1,250 BP, pearl millet had reached the southernmost extent of its range in South Africa, the result of a southward movement of Bantu-speaking farmers (Maggs and Ward, 1984). These archeological finds do not provide much of a picture of when African cereals were domesticated or where. Fine sieving and water screening, the archeological techniques used to recover seeds, are fairly new in Africa. Nabta Playa is an example of what can be accomplished with such techniques; improvements can be expected.

Climatic records, linguistic evidence, and botanical distributions provide some avenues of approach to the reconstruction of African domestication. These lines of evidence have received considerable attention (see Harlan et al., 1976; Clark and Brandt, 1984; Shaw et al., 1995), although the conclusions have remained unconfirmed by archeological evidence since the late 1970s. The wild ancestors of sorghum and millets are currently to be found in the Sahel, a broad band across Africa between about 5° and 15° north (Harlan, 1992, 1995). During wetter periods in the past, this band probably occurred further north, in the Sahara itself. Such wet phases occurred between 9,500 and 8,000 BP and also between 6,500 and 4,500 BP (Grove, 1995; Muzzolini, 1995). The latter period is called the Neolithic Wet Phase, when agriculture apparently became established in the northern savannas of Africa. During the wet phases, lakes and watercourses occurred in the Sahara, and Lake Chad was substantially larger than it is now. In the absence of archeological evidence, the question remains open as to whether domestication took place in the Sahara during one of the wet phases, or in the Sahel during the dry phase between 8,000 and 6,500 BP. We do know from archeological evidence that sedentary communities had become established at the margins of water sources by 8,000 BP, and that the inhabitants depended for food on fishing, with hunting a secondary activity, and that they gathered and processed grasses (Muzzolini, 1995). Nabta Playa is one example of such a sedentary community, but they also occurred in many other places in the Sahel. These communities provided the same preconditions for plant domestication as those exhibited by sites of incipient domestication in the Near East and China.

Sorghum (*S. bicolor*) has been in cultivation long enough for five distinct, regional races to be developed; these are associated with particular language groups (Harlan, 1995). The *guinea* race is mostly West African, adapted to high rainfall, and grown by people of the Niger–Congo language family. The *caudatum* race is grown from Lake Chad to the Ethiopian border by people of the Chari–Nile language family. A southern race is grown by people of southern Africa who speak Bantu languages, a subgroup of Niger–Congo. The most primitive race is *bicolor*, which occurs nearly everywhere that sorghum is grown; it closely resembles wild sorghum, *S. verticilliform*, which grows with the density of planted fields in the Chad–Sudan region. The *bicolor* race is the earliest domestic form, from which the other races were derived. *Durra* sorghum is the most highly derived; it was developed in India (probably from *bicolor*) and returned to Africa, where it is grown along the edge of the Sahara by Islamic speakers. Sorghum is grown in 30 countries today, producing more than 70 million tons of grain annually and providing the staple food for 500 million people (National Research Council, 1996; Brown, Chapter 14).

Pearl millet (*Pennisetum glaucum*, which includes *P. typhoides* and *P. americanum*) is the most modified domesticate of all cereals, notwithstanding similar claims for maize. The wild form has seed heads up to 100 mm in length, whereas the cultivated form may produce seed heads as long as 2 m. The morphology of the inflorescence and the size and color of the grains were also altered by the domestication process (Harlan, 1995). The wild forms of pearl millet occur from the Sahel to deep in the Sahara. The most drought resistant of all cereals, it was probably domesticated in the Sahara and then moved to the Sahel with the onset of seriously dry conditions. The current global production is 10 million tons per year, of which half is grown in India (National Research Council, 1996). Its Indian name is *bajra* and its flour is used to make the unfermented bread *roti* and the fermented *kisra*. In Africa, it grows from Senegal to Somalia and in drought-stricken parts of southern Africa, including the edges of the Kalahari desert (Harlan, 1995). It is commonly used to make porridge or beer and can also be used to make couscous. In mountain Nigeria, the staple food is a mixture of pearl millet flour, dried dates, and dried goat cheese; this concentrated energy food is taken on long trips across the Sahara (National Research Council, 1996).

Finger millet (*E. coracana*) is a crop plant of East and southern Africa. Its wild form, *E. africana*, can be found in the highlands of Uganda, Ethiopia, and Kenya (Harlan, 1995). The seed head looks like a hand with fingers, whence the name. It was probably domesticated in Uganda–Ethiopia: today it is grown across East and southern Africa, and it forms the major crop plant in Uganda (National Research Council, 1996). It is also important in the Indian hill country, where it is called *ragi*. Its uses include the making of porridge, malt, beer (it ferments very well), distilled *arake* (in Ethiopia), and fodder. Current global production is 4.5 million tons per year, split about evenly between Africa and India, with a small amount grown in the United States as birdseed.

Fonio is perhaps the oldest African cereal; it was formerly the major food crop of West Africa, where it was presumably domesticated (National Research Council, 1996). It occurs in two forms: white fonio (*D. exilis*), which is widely distributed in West Africa, and black fonio (*D. iburna*), with a more restricted range. Two other species of this millet have been domesticated at some unknown time in the past, in India and Europe (Harlan, 1995). The European domesticate, *D. sanguinalis,* was harvested until quite recently. Fonio is the world's fastest maturing cereal and also one of the best tasting. It is reputed to make better couscous than wheat: the latter may have come to be processed into small grains to imitate fonio for this purpose. It can also be cooked into porridge, popped, or made into bread and beer. Although it has lost much of its former importance, fonio is still the staple food in parts of Mali, Burkina Faso, Guinea, and plateau Nigeria (National Research Council, 1996).

Tef (*E. tef*) is a peculiarly Ethiopian cereal, where it accounts for half of the acreage under cereal cultivation, and fetches the highest price (National Research Council, 1996). Ethiopia's staple grain, it is primarily used to make *injera*, a fermented flat bread like a pancake, more than half a meter in diameter, on which fiery meat stews (*wat*) are presented to the diners. The stickiness of the dough made from tef provides it with large, bubbly air spaces, which give it a spongy texture. Tef is also used to make a gruel (*muk*), a sweet unleavened bread (*kita*), and fermented beverages. It provides good straw for animal fodder and is grown commercially for this purpose in South Africa and Australia. The history of domestication of this small-seeded millet is unknown, and there is no wild form in Ethiopia. Its wild progenitor is probably *E. pilosa,* one of the harvested wild grasses of the Sahel (Harlan, 1995), which suggests that is was brought to Ethiopia and then domesticated.

The lack of archeological evidence for African crops makes it difficult to be specific about the role of particular C_4 plants in the development of African societies. The most readily recognized urban societies of Africa are those of the Nile Valley, particularly of Dynastic Egypt during the Bronze Age (5,000–2,500 BP). Egyptian civilization was built on the cereals of the eastern Mediterranean, wheat and barley, with an admixture of African crops. Less well known are the kingdoms of the Nubian Nile (modern Sudan) during the Bronze and Iron Ages: Kerma (3,500–2,900 BP), Napata (2,900–2,400 BP), and Meroe (2,400–1,800 BP) (see Shaw *et al.*, 1995). The Nubian kingdoms had close connections with Egypt, but we do not know to what extent Mediterranean or African grains provided their diet staples. Ethiopia's first urban society, the kingdom of Aksum (1,800–

1,400 BP), presumably relied exclusively on African C_4 cereals, unlike the irrigation cultures of the Nile.

In West Africa, urban societies emerged during the past 2,000 years. The earliest documented case is Jenne-jeno (2,000–600 BP), which developed in the resource-rich inland delta of the Niger in Mali (McIntosh and McIntosh, 1995). African rice, sorghum, and millet (species not given) are reported as the important cereals here, whereas C_3 root crops such as African yams probably played an important role as well. Other West African Iron Age states include the kingdom of ancient Ghana (1,600–800 BP), the empire of Mali (800–500 BP), and the short-lived state of Songhay (500–400 BP) (Shillington, 1995). The principal source of income of these states was the lucrative trans-Sahara gold and salt trade. The rise and fall of Ghana, located in the dry Sahel, is particularly closely correlated with the trans-Sahara trade. The ensuing empire of Mali, however, was better situated for agricultural purposes, stretching across the southern savanna, where rainfall was adequate for the production of a regular food surplus. The main savanna crops were sorghum and millet, although rice was grown in the Gambia Valley and on the upper Niger floodplain, and camels, sheep, and goats were herded in the Sahel. Foodstuffs were exchanged between these specialized regions. This "dull [yet secure] diet of pounded millet, honey and milk" (Shillington, 1995, p. 99), and the wealth produced by the Sahara trade, sustained sophisticated Islamic societies, which were very much in contact with the rest of the Islamic world and could afford to turn Timbuktu into a great center of Islamic learning.

Finally, sorghum and millet (species unclear) were the staple cereals that accompanied Bantu-speaking agriculturalists as they spread across the southern subcontinent of Africa after 2,000 BP. Urban societies developed in this region during the past 1,000 years, the best known being the kingdom of Great Zimbabwe (800–550 BP), which exercised political control over Zimbabwe and parts of Botswana, South Africa, and Mozambique.

VI. Maize in the Americas

A. Origins and Dispersal

After decades of vigorous debate, botanists seem to have settled on teosinte (*Zea mexicana* ssp. *parviglumis* or *mexicana*) as the wild ancestor of maize (Iltis, 1983; Galinat, 1985a,b; Benz and Iltis, 1990; Dobley, 1990; see also Eubanks, 1995). Its original domestication probably took place somewhere in the west-central highlands of Mexico, where teosinte naturally occurs; this domestication process remains to be documented archeologically. The earliest remains of Mesoamerican maize still are those from the Tehuacán Valley, an arid, marginal valley in southwestern Mexico.

Tehuacan was selected for study because it promised good preservation of macrobotanical remains (MacNeish, 1967, 1981), but it is located outside the natural range of teosinte. Moreover, the Tehuacán maize has been accelerator dated to no more than 4,700 BP (Long et al., 1989), more recent than other early maize evidences in South America. Clearly then, Tehuacan was not a primary center of maize domestication. Models of this process and of the origins of Mesoamerican agriculture have therefore run into serious problems (see MacNeish, 1967; Flannery, 1968, 1986; DeNiro and Epstein, 1981; Stark, 1981; Rindos, 1984; Farnsworth et al., 1985; Schoenwetter, 1990).

Judging from phytolith and pollen core evidence, maize had spread to western Panama by 5,000 BP (Piperno et al., 1985, 1990), Colombia by 6,700 BP (Bray et al., 1987), and eastern Ecuador by 5,300 BP (Bush et al., 1989). Isotope data for human bone collagen confirm the presence of maize in Panama for the period 7,000–4,500 BP (Norr, 1995). High carbon concentrations in the pollen cores suggest intentional burning of the tropical forest even before the introduction of maize (Piperno, 1990). Thus, maize may have been adopted into tropical forest economies based on slash-and-burn agriculture (Lathrap, 1987). The archeological Valdivia culture on the southern coast of Ecuador—one of the earliest ceramic, "Formative," cultures of the Americas—probably represents such a tropical forest adaptation (Damp, 1988; Pearsall, 1988; Raymond, 1988; Marcos, 1993). Maize phytoliths have been identified here in deposits dating to at least 5,300 BP and perhaps even to the preceding Preceramic period (8,000–7,000 BP; Pearsall and Piperno, 1990; Piperno, 1991). Isotope analysis of human skeletons, however, does not support a maize diet for the early Valdivia periods, but points to a focus on forest and riverine resources (van der Merwe et al., 1993; cf. Burleigh and Brothwell, 1978). Both isotope signatures and macrobotanical remains (Lippi et al., 1984) provide more solid evidence of maize for the subsequent Middle Formative (ca. 3,500 BP) period on the coast of Ecuador. Maize reached the late Preceramic Peruvian coast between 4,500 and 3,500 BP (Bonavia and Grobman, 1989; but cf. Bird, 1990), probably from Ecuador (Sánchez González, 1994).

In North America, maize agriculture was established in the Southwest between 3,000 and 2,000 BP (Wills, 1988). In the eastern United States, maize first appeared around 1800 BP in Tennessee and Ohio (e.g., Fritz, 1990; Wagner, 1994), but did not become an important crop until after 1,200 BP (van der Merwe and Vogel, 1978).

Its spread across the Americas makes maize not only the most important crop, but the most widespread human artifact of the entire pre-Columbian New World (Hastorf, 1994). This also implies that the story of maize is largely one of adoption of a new, "foreign" crop into widely differing subsistence systems.

B. Patterns of Maize Adoption and the Role of Maize in pre-Columbian Societies

In the Tehuacán Valley, isotope data indicate that maize quickly established itself as a staple food, and from there on was consumed at about the same level throughout the pre-Hispanic sequence (DeNiro and Epstein, 1981; Farnsworth et al., 1985). For Mesoamerica's forested, humid lowlands, where cassava (yuca, manioc) agriculture may predate the introduction of maize (Davies, 1975; Lowe, 1975; but see Lewenstein and Walker, 1984), the process of maize adoption during the early Formative period (4,200–3,100 BP) seems to have been more gradual (Blake et al., 1992a,b). Subsequently, a dramatic increase in maize use accompanied the rise of La Venta on the Mexican Gulf Coast as a major center with abundant evidence of elite culture and ceremonial activities (Rust and Leyden, 1994). For the Maya Lowlands, isotope data show a temporal and spatial trend of increasing reliance on maize from a low of approximately 30% at Preclassic Cuello in Belize (starting 3,200 BP) to an almost total dependence on maize and beans at Classic (1,700–1,100 BP) Copán in western Honduras (White and Schwarcz, 1989; White et al.,. 1993; Gerry, 1994; Reed 1994; L. E. Wright, 1994; Tykot et al., 1996). Mesoamerican societies ultimately grew so dependent on maize as a staple that the food complex centered on it—maize, beans, squash, and chile peppers—and the associated maize processing technology became a defining feature of the Mesoamerican culture area (Kirchhoff, 1943). Archaic hunter–gatherers in the North American Southwest adopted the Mesoamerican food complex as a package, known to archeologists as the Upper Sonoran Complex (Ford 1985; Wills, 1988; Minnis, 1992). As in Mesoamerica, maize rose to become the unrivaled staple crop of the Southwest.

In contrast to Mesoamerica, maize was adopted as a nonlocal domesticate into food-producing, agricultural economies in eastern North America, as well as tropical lowland and Andean South America. In eastern North America the benign climate of the Hypsithermal (8000–4000 BP) allowed human populations a more sedentary way of life, based on abundant river-valley resources. Extensive human disturbance of floodplain environments got a coevolutionary process under way (O'Brien, 1987; Smith, 1987b), which resulted in the domestication of several starchy C_3 weeds, such as sunflower (*Helianthus annuus*), sumpweed (*Iva annua* var. *macrocarpa*), goosefoot (*Chenopodium bushianum*), maygrass (*Phalaris caroliniana*), and giant ragweed (*Ambrosia trifida*), collectively known as the Eastern Agricultural Complex (Ford, 1985). This is the context in which maize arrived around 1800 BP. For centuries it remained a minor crop, appearing in minute quantities in the archeobotanical record and hardly registered in human isotope signatures (van der Merwe and Vogel, 1978; Bender *et al.*, 1981; Lynott *et al.*, 1986; Rose *et al.*, 1991; Scarry, 1993a; Tuross *et al.*, 1994). Multiple lines of evidence indicate that maize was a special crop consumed in community ceremonies and social gatherings (Johannessen, 1993a,b; Wymer, 1994).

Several centuries after its introduction, an explosion in maize production and consumption took place between 1,200 and 1,000 BP, correlated with the emergence of more complex forms of social organization—late prehistoric, Mississippian chiefdoms, in archeological terminology. The timing of this change varied across the area, and it was not related to the introduction of an improved maize race. In fact, there was a great deal of variation in the maize strains grown and also in the degree of maize reliance—both between and within Late Prehistoric communities (Buikstra *et al.*, 1987, 1988; Stothers and Bechtel, 1987; Fritz, 1990; Schurr, 1992; Scarry, 1993a, 1994; Schurr and Schoeninger, 1995). Maize was grown in fixed fields—including raised fields (Riley, 1987; Woods, 1987)—and people lived close to their fields in scattered hamlets that were dependent upon mound centers under the control of newly emerged elites. Food sharing continued to be important at these mound centers (Blitz, 1993; Johannessen, 1993b).

A similar pattern of maize adoption is observed in tropical lowland South America. Human occupation of the tropical lowlands dates to the 10th millennium BP (Roosevelt *et al.*, 1991), probably based on the tropical-forest subsistence pattern of cassava agriculture, supplemented by fishing and hunting (Lathrap, 1970). For several centuries or even millennia after its introduction to the area, maize was apparently seen as an exotic vegetable (Pearsall, 1994). At Parmana on the Venezuelan Orinoco, maize is first found by 2,800–2,300 BP; due to a lack of skeletal specimens, human isotope signatures (van der Merwe *et al.*, 1981) do not register its consumption until about 1,600 BP, when maize had become a staple. At the same time site sizes increased substantially, indicating a large population increase (Roosevelt, 1980). In light of ethnohistoric sources from the contact period, these changes may be indicative of the rise of complex chiefdoms practicing maize agriculture on the floodplains. Paramount chiefs reportedly collected maize contributions from their subjects, hosting community religious ceremonies involving music, dances, and consumption of copious amounts of maize beer (Roosevelt, 1980, 1987).

An intimate association between maize, social relations, and power is particularly evident in Andean South America. Maize was present on the Peruvian coast from at least the late Preceramic (4,500–3,800 BP) and is depicted in artworks of the Early Horizon (4,500–3,800 BP) Chavín style (Lathrap, 1973). Isotope analyses have shown, however, that is was not the staple crop in the highlands (Burger and van der Merwe, 1990). Similarly, on the coast maize is found as a grave offering in burials dated between 2,200 and 1,400 BP, whereas the food of the living consisted mostly of seafood (Gumerman, 1994). These findings suggest a pattern well documented for Inca society, in which maize was a prestige and state crop, while local Andean crops were low-status foods (Murra, 1973). Maize beer (*chicha*) was a substance of enormous prestige, the chief vehicle for building and manipulating social and political alliances and gaining access to labor (Morris, 1979; Hastorf, 1990a,b, 1991; Gero, 1992; Hastorf and Johannessen, 1993). Early Colonial documents (Netherly, 1978) make it clear that people would refuse to participate in public works projects unless these took place in pseudo-reciprocal, festive settings, hosted and sponsored by the authorities and inevitably involving the consumption of maize beer. The same mechanism of labor recruitment has been postulated for previous Andean states and local elites (Hastorf and DeNiro, 1985; Johannessen and Hastorf, 1989; Moore, 1989; Hastorf, 1993; Hastorf and Johannessen, 1993).

The cases summarized previously establish a clear developmental pattern. In its origin center in Mesoamerica and in the Southwest, maize was the first major domesticated plant food in hunter–gatherer economies and it went on to become the unrivaled staple crop of both areas. Elsewhere, maize was incorporated as a foreign plant into existing agricultural economies, where it lingered on for centuries as an economically insignificant, but ritually and socially charged crop. When its production started in earnest, it was correlated with the rise of elites and social inequality and it played a key role in the negotiation of power. Why did maize consistently assume this special role although there were better adapted, productive, storable, and sometimes even more nutritive indigenous alternatives?

C. Maize and Indigenous Crop Complexes in Eastern North America, Lowland South America, and the Andes

Hastorf and Johannessen (1993; 1994) offer a purely cultural, symbolic explanation of the special role of maize; for them, the feature that sets maize apart is its enormous responsiveness to human manipulation. Maize is indeed an artifact, a species or subspecies created through human selection (Galinat, 1985a,b). Because it has no wild counterpart outside of its Mexican homeland and is completely unable to survive without human care, it is conceptually associated with humans and culture rather than with nature. In the Andes, where great importance is attached to serving cooked, culturally transformed foods in social contexts, maize beer is a highly regarded substance because it entails a triple transformation: its main ingredient—maize—is a cultural artifact, a substantial amount of labor is invested in its preparation and cooking, and, being an alcoholic beverage, it intoxicates and transforms its consumers.

There is another, more mundane, feature unique to maize: it is an exacting crop that, under the right conditions, has the highest grain yield potential of all major cereals (Doswell *et al.*, 1996). The productive conditions are considerably more narrowly defined than those of the indigenous crops it eventually replaced (Olsen and Sander, 1988; Shaw, 1988; Norman *et al.*, 1995; Doswell *et al.*, 1996). Maize is sensitive to both cold and excessive heat and low solar radiation, to both drought and excessive humidity, and makes high demands on soil fertility. Rainfall and temperature are the major limiting factors for maize agriculture. Maize does best in temperate environments with warm days and cool nights (Doswell *et al.*, 1996). There is an acute risk of frost in these environments, however, and the relative timing of rains and frosts requires a precise planning of the agricultural season before the year's rainfall pattern has been established (Kirkby, 1973; Sanders *et al.*, 1979; Wills, 1988).

In contrast to maize, the indigenous crops of North America, apart from squashes and gourds (both C_3), were starchy and oily annual C_3 weeds [sunflower (*Helianthus annuus*), sumpweed (*Iva annua* var. *macrocarpa*), goosefoot (*Chenopodium bushianum*), maygrass (*Phalaris caroliniana*), giant ragweed (*Ambrosia trifida*)], tolerant of a broad range of conditions (Asch and Asch, 1985; Cowan, 1985; Ford, 1985; Heiser, 1985; Keegan and Butler, 1987). Although their productivity is considerable (Asch and Asch, 1978; Smith, 1987a), it is substantially lower than that of maize. Their cultivation probably involved a low-intensity, fairly inefficient horticulture (Riley, 1987). However, they were easily stored and are found in storage bags, gourds, and pits by 2,000 BP (Fritz, 1990).

Maize field agriculture had a greater efficiency and a substantially higher yield potential, particularly with double-cropping regimes; at the same time the emphasis on a single crop in fixed fields raised the risk of crop losses dramatically (Keegan and Butler, 1987; Nassaney, 1987). Moreover, given that maize is a far more demanding and delicate crop than the indigenous plants, maize agriculture brought with it increased local variation, both in yield potential and risk.

In the tropical lowlands of South America, cassava (*Manihot esculenta*, C_3, yuca, manioc) was the primary crop before the adoption of maize. Not unlike the Eastern Agricultural Complex, cassava requires only moderate soil fertility and does not deplete the soil. For a tropical crop it is relatively drought resistant and withstands the acidity of Amazonian soils, but its sensitivity to waterlogging and its long growth cycle make it inappropriate for the nutrient-rich, seasonally flooded floodplain soils (Lathrap, 1970; Roosevelt, 1980). The bitter variety is storable as flour or bread, and cassava roots can be stored in the ground for about 2 years (Lathrap, 1970; Roosevelt, 1980). Because cassava contains essentially no protein and the availability of riverine fish is restricted to the low-water season, protein is the limiting factor of the cassava-based tropical forest economy (Roosevelt, 1980).

Maize fills this protein gap—even though its deficient amino-acid balance results in health problems in maize-dependent populations (Doswell *et al.*, 1996). Maize also makes more efficient use of the fertile floodplain soils and the relatively short season during which these are not flooded (Lathrap, 1970; Roosevelt, 1980). In fact, in the Amazon forest maize will only grow on the scarce floodplain soils (but *cf.* Moran, 1983; Roosevelt, 1989). Maize is itself sensitive to flooding; relatively short periods of waterlogging cause drastic yield reductions. Therefore, given that floodplains in the Amazon are not uniformly affected by flooding, the less frequently flooded plots are more desirable for farming (Lathrap, 1970). In general, the tropical forest with its excessive heat and humidity, hot nights, relatively low solar radiation, and pests and diseases is a high-stress environment for maize.

The salient characteristic of native Andean crops, in sharp contrast to maize, is their resiliency in the highland environment (National Research Council, 1989). These crops include a wide variety of potatoes, tubers, grains, and legumes [such as oca, ullucu, mashwa, kañiwa, kiwicha (*Amaranthus caudatus,* a C_4 plant), quinoa, tarwi]. They adapt to a wide altitudinal range and most tolerate frost, heat, and drought. They are not demanding of soil fertility and their nutritive value is exceptionally high. The grains are as storable as maize, and even potatoes may be stored as freeze-dried *chuño* (Troll, 1931) or fermented *tokush*. The periods to maturity are highly variable for the grains, however, and some shatter their seeds, making harvests unpredictable and labor intensive.

Maize harvesting is clearly more efficient than the harvesting of both Andean grains and potatoes (Knapp, 1991). According to Knapp's data from the northern Ecuadorian highlands, maize requires only one-half to one-third of the labor input of potatoes, and maize performance disproportionately improves under irrigation and at higher levels of intensity. Hastorf (1993) similarly found maize to outperform other crops in the central Peruvian Mantaro Valley when grown in its optimum altitudinal belt, this is, in the valleys below 3250 m.

The superiority of maize applies strictly to its optimum zone. Maize in the highlands is a "vulnerable, handicapped plant" (Murra, 1973, p. 379), far more precarious than native Andean crops. Water is generally scarce in the Central Andes (Farrington, 1985; Earls, 1989: Doswell *et al.*, 1996) and frosts are an obvious threat in the highlands. Also, maize suffers from a chronic shortage of fertilizer in traditional Andean agriculture: significant herds of camelids, the only large domestic animals of the Andes, live at a higher altitude where maize does not grow, and their dung is used for fuel there.

In sum, relative to indigenous crop complexes, maize may be characterized as an exotic, high-status, labor-efficient and high-yielding, spatially concentrated resource; at the same time, it is a riskier, more demanding crop that does well only in select, favorable niches. Maize is the diva of pre-Columbian crops: capricious, but potentially quite rewarding.

A final important feature follows from this "split personality" of maize: it responds exceptionally well to measures of agricultural intensification and rationalization. These include the breeding of numerous, highly specialized races adapted to specific ecological niches, and a precise scheduling of the agricultural cycle with its requisite time control by means of calendars (Forde, 1931; Earls, 1986, 1989, 1991); irrigation and preplanting irrigation (Sanders *et al.*, 1979; Hastorf, 1993); terraces in the highlands (Earls, 1989); and ridged fields, found in eastern North America, tropical lowland South America, and the Andes (Erickson, 1986; Riley, 1987; Woods, 1987; Knapp, 1991). The crucial point about these intensification measures is not only that they substantially raised the yields and efficiency of maize, but that by controlling the principal risk factors, they also made maize agriculture practicable in the first place in many instances, thus transforming a capricious diva of a plant into one of the most predictable engineered ancient crops.

D. The Adoption of Maize and the Emergence of Social Inequality

A dependable, storable grain surplus has played an important role in theories of the evolution of social complexity. The characteristics of maize discussed here point to a different, more direct link between maize and the rise of elites; they suggest a plausible scenario in which maize agriculture itself creates opportunities and selects for social inequality. The key to understanding this role of maize agriculture is a shift of perspective from the group to the individual level. Drawing on numerous ethnographic examples, Hayden (1990, 1995; see also Bender, 1978; Hastorf, 1993; Price, 1995) proposes that in communities with a dependable resource base, where resource sharing is not critical to people's livelihoods, sanctions against openly competitive behavior are likely to be relaxed. This allows ambitious "aggrandizers" to use competitive feasts and other mechanisms as a means for creating debt and alliance relationships and enhancing their status and power in the community, control of labor, and access to desirable mates and resources. Cultural evolutionary theory (Boyd and Richerson, 1985; *cf.* Aldenderfer, 1993; Spencer, 1993) predicts that aggrandizers, whose feasts are imposing displays of their success, should be particularly attractive cultural models to emulate. If one is unable to become an aggrandizer oneself, however, the second-best strategy is to join a proved successful alliance, contributing as much as necessary to its promotion and continuance, while benefiting as much as possible from the network and its privileged access to resources, labor, and mates. Thus the followers of a successful aggrandizer may soon find themselves engaged in corporate projects headed by the Big Man.

The adoption of maize fits handsomely into this scenario. That the intensification of maize agriculture was not a community-level phenomenon finds some empirical support in the intracommunity variability of isotope signatures observed during the emergence of more complex social formations in Late Prehistoric eastern North America (Bender *et al.*, 1981; Lynott *et al.*, 1986; Buikstra *et al.*, 1987; Buikstra and Milner, 1991; Schurr, 1992). The special status that accrued to the exotic, human-made plant and its place in community ceremonies provided the initial impetus for ambitious individuals to grow maize. On a symbolic level, the extreme genetic malleability of maize and its lack of a local wild relative may have suggested that successful maize cultivators had special powers over nature, as aggrandizers in many societies were quick to assert and rulers in pre-Columbian stratified societies invariably claimed. If an aggrandizer happened to have access to a suitable piece of land, maize agriculture was capable of yielding high returns of a high-status item on the limited labor input of a single household. The risk associated with maize agriculture was acceptable if, as seems highly likely, early innovators initially planted only a portion of their lands in the new crop; the plants of the Eastern Agricultural Complex continued to be cultivated after the maize explosion (Fritz, 1990), and both cassava and native Andean crops have been planted until today. By growing the new, special crop, wealth could be accumulated even if, in theory, the community owned the land and individuals were only granted temporary land-use rights.

Because maize was a risky, initially poorly adapted and poorly known crop and quite discriminating in terms of the niches where it would do well, many maize innovators were bound to fail. Some people would have produced potentially large amounts of a special, high-status crop, whereas others could not produce it at all (*cf.* Coe and Diehl, 1980). Risk factors such as flooding, drought, and frost would have had highly localized effects (Lathrap, 1970; Scarry, 1993a,b). Moreover, success with the demanding and delicate new crop depended on a considerable amount of knowledge regarding soil productivity, crop selection, cultivation techniques, and the precise scheduling required (Nassaney, 1987). Early innovators thus had a competitive advantage over latecomers because of their greater experience with the crop. As a result of these ample opportunities for differential success afforded by maize agriculture, many people settled for the second-best of the two predicted responses to competitive feasting, teaming up with a successful aggrandizer and seeking to profit from the alliance.

Because maize responds so well to intensification measures and its labor efficiency increases at higher levels of intensification (Knapp, 1991), luring followers into corporate projects should have been fairly easy and most rewarding to an aggrandizer. What is more, irrigation, terraces, and ridged fields, by curbing the principal risk factors, would have allowed the cultivation of maize in areas where individual maize projects were destined to fail; this gave the alliances a distinct, qualitative advantage. Hastorf (1993; also D'Altroy, 1994) envisages such a scenario for the pre-Inca Mantaro Valley, resulting in a number of fairly large alliances competing for the maize-growing lands. The Inca state ultimately emerged victorious out of such rivalries and made intensive maize agriculture its principal state project. Turned into a highly rationalized Andean agrotechnology, it allowed a precise planning of the allocation of labor forces. Thus maize agriculture structured the agricultural and ceremonial calendar imposed by the Inca over their territory and constituted one of the primary integrative mechanisms of the empire (Earls, 1986, 1989, 1991).

Both aspects of the diva nature of maize have important implications for social evolution because they affect different people differently and thus foster inequality. This may be exploited by aspiring elites when a stable economic system, based on a resource other than maize, has sufficiently lessened the dependence on reciprocal sharing and opened the door to unconcealed competition. In this manner, a plant species created by human selection became itself a powerful selective agent in human social evolution.

VII. The Past 500 Years

Explorers and traders of the 16th century AD were responsible for the spread of maize beyond the Americas. High yield potential quickly turned maize from a botanical curiosity and kitchen garden vegetable into a commercial field crop that was widely grown in Spain, France, and Italy by the end of the 16th century. Less than a century after the European conquest of the Americas, maize had conquered virtually the entire world. Portuguese traders introduced it to Africa in the early 1500s; by 1700 it had become a prominent crop in the western and central parts of the continent. Portuguese and Arab sea trade took maize to the west coast of India, and via the Silk Route it reached Pakistan. By 1550 it had arrived in Southeast Asia; it was well established in Indonesia, the Philippines, and Thailand by 1650. In 17th-century China, maize allowed the opening-up of new agricultural lands outside of the rice-growing plains, required by rapid population growth (Norman *et al.*, 1995; Doswell *et al.*, 1996).

Sugarcane (*Saccharum officinarum* L., C_4), originally domesticated in New Guinea (Artschwager and Brandes, 1958; Warner, 1962), was known in Europe since the times of Alexander the Great, but the spread of sugar production on a significant scale was associated with the expansion of Islam

(Watson, 1974). Sugar was the only one of the prized Oriental "spices" whose production Europeans learned to master in the wake of the crusades. Reacting to the pressure of Islamic expansion, Portugal and Spain introduced sugarcane to their newly acquired islands in the Atlantic (the Canaries, Azores, Cape Verdes, and São Tomé) by the end of the 15th century. Slaves had already been used in the Mediterranean to cultivate the extremely labor-intensive crop (Galloway, 1977). On the previously uninhabited Atlantic islands, this link between sugarcane and slavery was cemented (Greenfield, 1977, 1979).

The Atlantic islands were the stepping stones by which sugarcane reached the New World (Mintz, 1985). Spaniards unsuccessfully attempted to start plantations in the Antilles, but a highly successful transfer to the New World was achieved by the Portuguese in northeastern Brazil—along with plantations, slavery, and settlers from the Atlantic islands. Sugarcane allowed them to settle economically otherwise worthless territories and to turn them into highly profitable colonies (Greenfield, 1979). On the flat Brazilian coast with its adequate rainfall, sugarcane thrived, and the Brazilian sugar plantations grew more profitable than any others before them. By 1600, Brazil was one of the most prosperous colonies on Earth thanks to sugar. In competition with Portugal, the British, Dutch, and French introduced sugarcane and the associated racially mixed slave-plantation society to their Caribbean possessions and all across the Americas. Demand for sugar in Europe increased steadily as sugar underwent a transformation from a curiosity, medicine, and spice for the rich to an item of daily consumption by everyone, making it the most important agricultural import from the colonies. Owing to its close link with slavery and its enormous economic significance (Table I), sugarcane has been one of the massive demographic forces in human history, causing both forced and unforced migrations of millions of people around the globe (Mintz, 1959). Due to its capital-intensive nature and the factory-like mills on the premises of sugar plantations, the sugarcane agroindustry was a major precursor of European industrialization (Mintz, 1985).

To bring this 7-million-year saga of people and C_4 plants to a conclusion, we provide two footnotes from the present.

Maize, the most cultivated C_4 cereal of all (Table I), has become the grain staple of many parts of the world, including east, central, and southern Africa. Its introduction was resisted until recently, however, in an isolated corner of the Transvaal in South Africa. In the village of Modjadji the Rain Queen, perched on the side of the Drakensberg escarpment near Tzaneen, it was forbidden until the 1950s to process maize by stamping and grinding (Davison 1984). It was believed that this foreign grain could bring foreign influences into the village and interfere with the rain-making process. Modjadji V, the current incumbent, has seen her power wane in the face of agribusiness, irrigation, and the availability of mealiemeal (maize flour) at the trading store. She is the descendant of generations of rain queens by the same name, whose lineage derives from the royal dynasties of Zimbabwe kingdoms (Krige, 1949). So powerful were the Modjadjis in the past, that a Victorian predecessor is said to have inspired the fictional character of She-who-must-be-obeyed, a Sheba-like African queen invented by novelist H. Rider Haggard (Haggard, 1887).

At about the same time that Haggard was traveling in South Africa, hearing stories about a powerful and reclusive queen, the least cultivated of all C_4 cereals, tef, was introduced in the Transvaal. In 1886, the Royal Botanic Gardens at Kew distributed Ethiopian tef seeds to the British colonies for trial, with some of it getting as far as the Zuid-Afrikaanse Republiek, or independent Transvaal. Here it grew well, but found no market as a grain (National Research Council, 1996). The story goes that a Transvaal farmer sent some tef hay to the market in Johannesburg, the new boomtown that started with the discovery of gold in 1886. The hay was sold as bedding material to a racing stable. The horses then ate the bedding instead of their feed (C_3 oats?) and won races. Soon everyone wanted tef for their horses, with the result that this prized Ethiopian cereal was used to feed the horses of both sides during the Anglo–Boer War of 1899–1902. After the war, it put the dairy industry of Johannesburg on its feet.

Tef continued through the 20th century as the most important food grain in Ethiopia, but nowhere else. It is still grown commercially in South Africa for animal feed, with some experimentation in Australia. The migration of Ethiopian Jews to Israel has produced a small demand for it in that country. At the end of the 2nd millennium AD, however, the health food demand for gluten-free grain has created a potential market for tef in the First World, whereas the expansion of Ethiopian restaurants across the United States and Europe has made it a necessity. The centerpiece of an Ethiopian meal is *injera*, the flat round bread on which the rest of the meal is served. It cannot be made without tef flour: imitations made from wheat go stale and hard in 1 day instead of 3 days and lack the requisite sour sponginess. An American farmer has responded to this need: most of the current demand for tef flour in the United States is supplied, on 200 acres, by Wayne Carlson of Caldwell, Idaho (National Research Council, 1996). Competition has developed in Oklahoma (W. Carlson, 1997, personal communication).

At the end of the first millennium AD, the Northern Flint variety of maize swept across the northeastern United States and became the staple cereal of most Native Americans in the woodlands. Today, demand for this maize variety is limited to enthusiasts of stone milling, who use it to make traditional jonnycakes. Essentially all the Northern Flint maize grown in New England (where it is known as White Cap) is supplied by Paul Pieri of Little Compton, Rhode Island, and a few friends on fewer than 100 acres (Fussell, 1992; Pieri. 1997, personal communication).

What might the third millennium have in store for Northern Flint maize and tef? In the spirit of scientific inquiry, we investigated these two grains and their food products. This involved kitchen trials, interviews with farmers Pieri and Carlson, and frequent lunchtime consultations with Mrs. Lettensa Afework of the Asmara Restaurant in Central Square, Cambridge, Massachusetts. Jonnycakes, we discovered, are pan-fried fritters of coarse maize flour mixed with water or milk, depending on which side of Narragansett Bay one comes from. It appears unlikely that they will sweep the world as a rediscovered taste sensation, but a minor upsurge in demand for stone-ground maize flour may occur to satisfy the need for polenta in upscale restaurants. Injera we found to be an extraordinarily tasty dish, but it is difficult to make unless one is a professional chef or an Ethiopian. Mrs. Afework is both, and was willing to part with some of her sourdough starter, but this did not guarantee success when we tried it at home. Finally, our investigation also resulted in Northern Flint maize being shipped to Caldwell, Idaho, for trial, and quinoa seeds being sprouted in Little Compton, Rhode Island. The maize did very well in Idaho, unlike the quinoa in Rhode Island.

In idle moments, we speculate about the future of C_4 plants in the third millennium. Will the continuing rise in atmospheric CO_2 content favor the productivity of C_3 plants over C_4? Perhaps, but the migration of rainfall belts across national boundaries as a result of global warming is far more likely to disrupt the political economy of food production. The history of C_4 plants we have related in this chapter prompts us to wager a small sum on two predictions. First, the future of C_4 plants will be subject to serendipity, as suggested by the unintended consequences of our investigation of tef and Northern Flint maize. Second, developments will be shaped by human fickleness. The rest is an open question and the odds are six to five against.

VIII. Summary

The worldwide spread of C_4 grasses in the late Miocene and the formation of savannas created environmental conditions in Africa in which bipedal hominids evolved. Early hominids and their human descendants were foragers for 4 million years; they included grass-eating animals in their diets, but made little or no direct use of C_4 plants as food. Harvesting of grasses by humans started in the late Pleistocene, leading to the domestication of C_3 grasses in the Near East. The domestication of C_4 grasses in China, Africa, and the Americas followed, providing the staple crops for the development of urban societies in those continents. During the past 500 years,

C_4 plants like maize and sugarcane became important worldwide; their large-scale production beyond their natural range has had a major impact on food production and human demography.

References

Aldenderfer, M. (1993). Ritual, hierarchy, and change in foraging societies. *J. Anthropol. Arch.* **12,** 1–40.
Artschwager, E., and Brandes, E. W. (1958). "Sugar Cane: Origin, Classification Characteristics, and Descriptions of Representative Clones." U. S. Department of Agriculture Handbook No. 122. Government Printing Office, Washington, D. C.
Asch, D. L., and Asch, N. E. (1978). The economic potential of *Iva annua* and its prehistoric importance in the lower Illinois Valley. *In* "The Nature and Status of Ethnobotany" (R. I. Ford, ed.), pp. 301–341. Anthropological Papers No. 67. Museum of Anthropology, University of Michigan, Ann Arbor.
Asch, D. L., and Asch, N. E. (1985). Prehistoric plant cultivation in west-central Illinois. *In* "Prehistoric Food Production in North America" (R. I. Ford, ed.), pp. 149–203. Anthropological Papers No. 75. Museum of Anthropology, University of Michigan, Ann Arbor.
Barnes, G. (1993). "China, Korea and Japan: The Rise of Civilization in East Asia." Thames and Hudson, London.
Bar-Yosef, O., and Belfer-Cohen, A. (1991). From sedentary hunter–gatherers to territorial farmers in the Levant. *In* "Between Bands and States" (S. A. Gregg, ed.), pp. 181–202. Occasional Paper, Center for Archaeological Investigations, no. 9. Southern Illinois University at Carbondale.
Bar-Yosef, O., and Meadow, R. H. (1995). The origins of agriculture in the Near East. *In* "Last Hunters, First Farmers: New Perspectives on the Prehistoric Transition to Agriculture" (T. D. Price and A. B. Gebauer, eds.), pp. 39–94. School of American Research, Santa Fe, New Mexico.
Bender, B. (1978). From gatherer–hunter to farmer: A social perspective. *World Arch.* **10,** 204–222.
Bender, M. M., Baereis, D. A., and Stevenson, A. L. (1981). Further light on carbon isotopes and Hopewell agriculture. *Am. Antiquity* **46,** 346–353.
Benz, B. F., and Iltis, H. H. (1990). Studies in archaeological maize I: The 'wild' maize from San Marcos cave reexamined. *Amer. Antiquity* **55,** 500–511.
Bird, R. McK. (1990). What are the chances of finding maize in Peru dating before 1000 B. C.? Reply to Bonavia and Grobman. *Am. Antiquity* **55,** 828–840.
Blake, M., Chisholm, B., Clark, J. E., and Mudar, K. (1992a). Non-agricultural staples and agricultural supplements: Early Formative subsistence in the Soconusco region, Mexico. *In* "Transitions to Agriculture in Prehistory" (A. B. Gebauer and T. D. Price, eds.), pp. 133–151. Prehistory Press, Madison, Wisconsin.
Blake, M., Chisholm, B., Clark, J. E., Voorhies, B., and Love, M. W. (1992b). Prehistoric subsistence in the Soconusco region. *Curr. Anthro.* **33,** 83–94.
Blitz, John H. (1993). Big pots for big shots: feasting and storage in a Mississippian community. *Am. Antiquity* **58,** 80–96.
Bonavia, D., and Grobman, A. (1989). Andean maize: Its origin and domestication. *In* "Foraging and Farming: The Evolution of Plant Exploitation" (D. C. Harris and G. C. Hillman, eds.), pp. 456–470. Unwin Hyman, London.
Boyd, R., and Richerson, P. L. (1985). "Culture and the Evolutionary Process." University of Chicago Press, Chicago.

Bray, W., Herrera, L., Schrimpff, C., and Monsalve, J. G. (1987). The ancient agricultural landscape of Calima, Colombia. In "Pre-Hispanic Agricultural Fields in the Andean Region" (W. M. Denevan, K. Matthewson, and G. Knapp, eds.), vol. 2, pp. 443–481. BAR International Series 359. Oxford.

Buikstra, J. E., Bullington, J., Charles, D., Cook, D., Frankenburg, S., Konigsberg, L., Lambert, J., and Xue, L. (1987). Diet, demography, and the development of horticulture. In "Emergent Horticultural Economies of the Eastern Woodlands" (W. W. Keegan, ed.), pp. 67–85. Occasional Paper 7. Center for Archaeological Investigations, Southern Illinois University, Carbondale.

Buikstra, J. E., Autry, W., Breitburg, E., Eisenberg, L., and van der Merwe, N. J. (1988). Diet and health in the Nashville Basin: human adaptation and maize agriculture in middle Tennessee. In "Diet and Subsistence: Current Archaeological Perspectives." (B. V. Kennedy and G. M. LeMoine, eds.), pp. 243–259. Proceedings of the 19th Annual Chacmool Conference. Archaeological Association of the University of Calgary, Calgary.

Buikstra, J. E., and Milner, G. R. (1991). Isotopic and archaeological interpretations of diet in the central Mississippi Valley. *J. Archaeol. Sci.* **18,** 319–329.

Burger, R. L., and van der Merwe, N. J. (1990). Maize and the origin of highland Chavín civilization: an isotopic perspective. *Am. Anthropol.* **92,** 85–95.

Burleigh, R., and Brothwell, D. (1978). Studies on Amerindian dogs, 1: Carbon isotopes in relation to maize in the diet of domestic dogs from early Peru and Ecuador. *J. Archaeol. Sci.* **5,** 355–362.

Bush, M. B., Piperno, D. R. and Colinvaux, P. A. (1989). A 6,000 year history of Amazonian maize cultivation. *Nature* **340,** 303–305.

Cai, L. Z., and Qiu, S. H. (1984). Carbon-13 evidence for ancient diet in China. *Kaogu* **10,** 949–955.

Cerling, T. E., Wang, Y., and Quade, J. (1993). Expansion of C_4 ecosystems as an indicator of global ecological change in the Miocene. *Nature* **361,** 344–345.

Cerling, T. E., Harris, J. M., MacFadden, B. J., Leakey, M. G., Quade, J., Eisenmann, V., and Ehleringer, J. R. (1997). Global vegetation change through the Miocene and Pliocene. *Nature* **389,** 153–158.

Chang, K. C. (1980). "Shang Civilization." Yale University Press, New Haven and London.

Chang, K. C. (1986). "The Archaeology of Ancient China," 4th ed. Yale University, Press, New Haven and London.

Clark, J. D., and Brandt, S. A. (1984, eds.). "From Hunters to Farmers: The Causes and Consequences of Food Production in Africa." University of California Press, Berkeley.

Clark, J. D., Williams, M .A. J., and Smith, A. B. (1973). The geomorphology and archaeology of Adrar Bous, central Sahara: Preliminary report. *Quaternaria* **18,** 245–297.

Coe, M. D., and Diehl, R. A. (1980). "In the Land of the Olmec," Vol. 2. "The People of the River." University of Texas Press, Austin.

Cohen, D. J. (n. d.). The origins of domesticated cereals and the Pleistocene–Holocene transition in East Asia. *Rev. Archaeol.* in press.

Cowan, C. W. (1985). Understanding the evolution of plant husbandry in eastern North America: Lessons from botany, ethnography, and archaeology. In "Prehistoric Food Production in North America" (R. I. Ford, ed.), pp. 205–243. Anthropological Papers No. 75. Museum of Anthropology, University of Michigan, Ann Arbor.

Cowan, C. W., and Watson, P. J. (1992, eds.). "The Origins of Agriculture: An International Perspective." Smithsonian Institution Press, Washington, D. C.

Crawford, G. W. (1992a). Prehistoric plant domestication in east Asia. In "The Origins of Agriculture" (C. W. Cowan and P. J. Watson, eds.), pp. 7–38. Smithsonian Institution Press, Washington, D. C.

Crawford, G. W. (1992b). The transition to agriculture in Japan. In "Transitions to Agriculture in Prehistory" (A. B. Gebauer and T. D. Price, eds.), pp. 117–132. Prehistory Press, Madison, Wisconsin.

D'Altroy, T. N. (1994). Factions and political development in the central Andes. *In* "Factional Competition and Political Development in the New World" (E. M. Brumfiel and J. W. Fox, eds.), pp. 171-187. Cambridge University Press, Cambridge.

Damp, J. E. (1988). "La primera ocupación Valdivia de Real Alto: Patrones económicos, arquitectónicos e ideológicos." Centro de Estudios Arqueológicos y Antropológicos, ESPOL, Guayaquil.

Davison, P. (1984). "Lobedu Material Culture." *Ann. South Afr. Mus.* vol. 94(3).

Davies, D. D. (1975). Patterns of Early Formative subsistence in southern Mesoamerica. *Man* (N. S.) **10,** 41-59.

de Menocal, P. B. (1995). Plio-Pleistocene African climate. *Science* **270,** 53-59.

DeNiro, M. J. (1987). Stable isotopy and archaeology. *Am. Sci.* **75,** 182-191.

DeNiro, M. J., and Epstein, S. (1981). Influence of diet on the distribution of nitrogen isotopes. *Geo. Cosmo. Acta* **45,** 341-351.

Dobley, J. F. (1990). Molecular evidence and the evolution of maize. *Econ. Bot.* **44,** 6-27.

Doswell, C. R., Paliwal, R. L., and Cantrell, R. P. (1996). "Maize in the Third World." Westview Press, Boulder.

Earls, J. (1986). Evolución de la administración ecológica Inca. *In* "Andenes y camellones en el Perú andino: historia, presente y futuro" (C. de la Torre and M. Burga, eds.), pp. 23-57. CONCYTEC, Lima.

Earls, J. (1989). "Planificación agrícola andina: Bases para un manejo cibernético de sistemas de andenes." Universidad del Pacífico, Centro de Investigación, and Ediciones COFIDE, Lima.

Earls, J. (1991). "Ecología y agronomía en los Andes." Hisbol, La Paz.

Ehleringer, J. R., and Björkman, O. (1977). Quantum yields for CO_2 uptake in C_3 and C_4 plants. Dependence on temperature, CO_2 and O_2 concentrations. *Plant Physiol.* **59,** 86-90.

Ehleringer, J. R., Sage, R. F., Flanagan, L. B., and Pearcy, R. W. (1991). Climate change and the evolution of C_4 photosynthesis. *Trends Ecol. Evol.* **6,** 95-99.

Ehleringer, J. R., Hall, A. E., and Farquhar, G. D. (1993, eds.). "Stable Isotopes and Plant Carbon/Water Relations." Academic Press, San Diego.

Ehleringer, J. R., Cerling, T. E., and Helliker, B. R. (1997). C_4 photosynthesis, atmospheric CO_2, and climate. *Oecologica* **112,** 285-299.

Erickson, C. L. (1986). Waru-waru: Una tecnología agrícola del altiplano prehispánico. *In* "Andenes y camellones en el Perú andino: historia, presente y futuro" (C. de la Torre and M. Burga, Eds.), pp. 59-84. CONCYTEC, Lima.

Eubanks, M. (1995). A cross between two maize relatives: *Tripsacum dactyloides* and *Zea diploperennis* (Poaceae). *Econ. Bot.* **49,** 172-182.

Farnsworth, P., Brady, J. E., DeNiro, M. J. and MacNeish, R. S. (1985). A re-evaluation of the isotopic and archaeological reconstruction of diet in the Tehuacán Valley. *Am. Antiquity* **50,** 102-116.

Farrington, I. S. (1985). Operational strategies, expansion and intensification within the prehistoric irrigation agricultural system of the Moche Valley, Peru. *In* "Prehistoric Intensive Agriculture in the Tropics" (I. S. Farrington, ed.), vol. 2, pp. 621-651. BAR International Series 232. Oxford.

Flannery, K. V. (1968). Archaeological systems theory and early Mesoamerica. *In* "Anthropological Archaeology in the Americas" (B. J. Meggers, ed.), pp. 67-87. Anthropological Society of Washington, Washington, D. C.

Flannery, K. V. (1986, ed.). "Guilá Naquitz: Archaic Foraging and Early Agriculture in Oaxaca, Mexico." Academic Press, Orlando.

Ford, R. I. (1985). Patterns of prehistoric food production in North America. *In* "Prehistoric Food Production in North America" (R. I. Ford, ed.), pp. 341-364. Anthropological Papers No. 75. Museum of Anthropology, University of Michigan, Ann Arbor.

Forde, C. D. (1931). Hopi agriculture and land ownership. *J. Roy. Anthropol. Inst.* **61,** 357–405.
Friedli, H., Lotscher, H., Oeschger, H., Siegenthaler, U., and Stauffer, B. (1986). Ice core record of the $^{13}C/^{12}C$ ratio of atmospheric CO_2 in the past two centuries. *Nature* **314,** 237–238.
Fritz, G. J. (1990). Multiple pathways to farming in precontact eastern North America. *J. Prehist.* **4,** 387–435.
Fussel, B. (1992). "The Story of Corn." New York, Alfed A. Knopf.
Galinat, W. C. (1985a). The missing links between teosinte and maize: A review. *Maydica* **30,** 137–160.
Galinat, W. C. (1985b). Domestication and diffusion of maize. *In* "Prehistoric Food Production in North America" (R. I. Ford, ed.), pp. 245–278. Anthropological Papers No. 75. Museum of Anthropology, University of Michigan, Ann Arbor.
Galloway, J. H. (1977). The Mediterranean sugar industry. *Geograph. Rev.* **67,** 177–192.
Gamble, C. (1994). "Timewalkers: The Prehistory of Global Colonization." Harvard University Press, Cambridge.
Gero, J. M. (1992). Feasts and females: Gender ideology and political meals in the Andes. *Nor. Arch. Rev.* **25:**15–30.
Gerry, J. P. (1994). "Diet and Status Among the Classic Maya: An Isotopic Perspective." Doctoral dissertation, Department of Anthropology, Harvard University. University Microfilms, Ann Arbor.
Greenfield, S. M. (1977). Madeira and the beginnings of New World sugar cultivation and plantation slavery: A study in institution building. *In* "Comparative Perspectives on Slavery in New World Plantation Societies" (V. Ruben and A. Tuden, eds.). Annals of the New York Academy of Sciences No. 292. New York Academy of Sciences, New York.
Greenfield, S. M. (1979). Plantations, sugar cane and slavery. *In* "Roots and Branches: Current Directions in Slave Studies" (M. Craton, ed.), pp. 85–119. Pergamon Press, Toronto.
Grove, A. T. (1995). Africa's climate in the Holocene. *In* "The Archaeology of Africa: Food, Metals and Towns" (T. Shaw, P. Sinclair, B. Andah, and A. Okpoko, eds.), pp. 32–42. Routledge, London and New York.
Gumerman, G., IV. (1994). Corn for the dead: The significance of *Zea mays* in Moche burial offerings. *In* "Corn and Culture" (S. Johannessen and C. A. Hastorf, Eds.), pp. 399–410. University of Minnesota Publications in Anthropology 5. Westview Press, Boulder.
Haggard, H. R. (1887). "She, a History of Adventure." London, Longmans Green.
Hancock, J. F. (1992). "Plant Evolution and the Origin of Crop Species." Prentice Hall, Englewood Cliffs.
Harlan, J. R. (1992). Indigenous African agriculture. *In* "The Origins of Agriculture: An International Perspective." (W. Cowan and P. J. Watson, eds.), pp. 59–70. Smithsonian Institution Press, Washington, D. C.
Harlan, J. R. (1995). The tropical African cereals. *In* "The Archaeology of Africa: Food, Metals and Towns" (T. Shaw, P. Sinclair, B. Andah, and A. Okpoko, eds.), pp. 53–60. Routledge, London and New York.
Harlan, J. R., de Wet, J. M. J., and Stemmler, A. B. L. (1976, eds.). "Origins of African Plant Domestication." Mouton, The Hague.
Hastorf, C. A. (1990a). The effect of the Inka state on Sausa agricultural production and crop consumption. *Am. Antiquity* **55,** 262–290.
Hastorf, C. A. (1990b). One path to the heights: negotiating political inequality in the Sausa of Peru. *In* "The Evolution of Political Systems" (S. Upham, ed.), pp. 146–176. Cambridge University Press, Cambridge.
Hastorf, C. A. (1993). "Agriculture and the Onset of Political Inequality Before the Inka." Cambridge University Press, Cambridge.
Hastorf, C. A. (1994). Introduction to part four. *In* "Corn and Culture" (S. Johannessen and C. A. Hastorf, eds.), pp. 395–398. University of Minnesota Publications in Anthropology 5. Westview Press, Boulder.

Hastorf, C. A., and DeNiro, M. J. (1985). Reconstruction of prehistoric plant production and cooking practices by a new isotopic method. *Nature* **315,** 489–491.

Hastorf, C. A., and Johannessen, S. (1993). Pre-Hispanic political change and the role of maize in the Central Andes of Peru. *Am. Anthropol.* **95,** 115–138.

Hastorf, C. A., and Johannessen, S. (1994). Becoming corn-eaters in prehistoric America. In "Corn and Culture" (S. Johannessen and C. A. Hastorf, eds.), pp. 427–443. University of Minnesota Publications in Anthropology 5. Westview Press, Boulder.

Hattersley, P. W. (1982). ^{13}C values of C_4 types in grasses. *Aust. J. Plant Physiol.* **9,** 139–154.

Hayden, B. (1990). Nimrods, piscators, pluckers, and planters: The emergence of food production. *J. Anthropol. Arch.* **9,** 31–69.

Hayden, B. (1995). Pathways to power: principles for creating socioeconomic inequalities. In "Foundations of Social Inequality" (T. D. Price and G. M. Feinman, eds.), pp. 15–86. Plenum Press, New York.

Heiser, C. B., Jr. (1985). Some botanical considerations of the early domesticated plants north of Mexico. In "Prehistoric Food Production in North America" (R. I. Ford, ed.), pp. 57–72. Anthropological Papers No. 75. Museum of Anthropology, University of Michigan, Ann Arbor.

Heiser, C. B., Jr. (1990). "Seed to Civilization: The Story of Food." Harvard University Press, Cambridge.

Higham, C. (1995). The transition to rice cultivation in Southeast Asia. In "Last Hunters, First Farmers: New Perspectives on the Prehistoric Transition to Agriculture" (T. D. Price and A. B. Gebauer, eds.), pp. 127–155. School of American Research, Santa Fe.

Hill, A. (1995). Faunal and environmental change in the Neogene of East Africa: evidence from the Tugan Hills sequence, Baringo district, Kenya. In "Paleoclimate and Evolution" (E. S. Vrba, G. H. Denton, T. C. Partidge, and L. H. Burckle, eds.), pp. 178–193. Yale University Press, New Haven.

Hillman, G. C., Colledge, S., and Harris, D. R. (1989). Plant food economy during the epi-Paleolithic period at Tell Abu Hureyra, Syria: Dietary diversity, seasonality and modes of exploitation. In "Foraging and Farming: The Evolution of Plant Exploitation" (G. C. Hillman and D. R. Harris, eds.), pp. 240–266. Unwin Hyman, London.

Ho, P.-T. (1984). The paleoenvironment of North China—a review article. *J. Asian Stud.* **43,** 723–733.

Iltis, H. H. (1983). From teosinte to maize: The catastrophic sexual transmutation. *Science* **222,** 886–894.

Johannessen, S. (1993a). Farmers of the Late Woodland. In "Foraging and Farming in the Eastern Woodlands" (C. M. Scarry, Ed.), pp. 57–77. University Press of Florida, Gainesville.

Johannessen, S. (1993b). Food, dishes, and society in the Mississippi Valley. In "Foraging and Farming in the Eastern Woodlands" (C. M. Scarry, ed.), pp. 182–205. University Press of Florida, Gainesville.

Johannessen, S., and Hastorf, C. A. (1989). Corn and culture in central Andean prehistory. *Science* **244,** 690–692.

Kajale, M. D. (1988). Ancient plant economy at Chalcolithic Juljapur Crarhi, district Amraoti, Maharashtra. *Current Science* **57,** 377–379.

Katzenberg, M. A., and Harrison, R. G. (1997). What's in a bone? Recent advances in archaeological bone chemistry. *J. Arch. Res.* **5,** 265–293.

Keegan, W. F., and Butler, B. M. (1987). The microeconomic logic of horticultural intensification. In "Emergent Horticultural Economies of the Eastern Woodlands" (W. W. Keegan, ed.), pp. 109–127. Occasional Paper 7. Center for Archaeological Investigations, Southern Illinois University, Carbondale.

Kingston, J. D., Marino, B. D., and Hill, A. (1994). Isotopic evidence for Neogene hominid palaeoenvironment in the Kenya Rift Valley. *Science* **264,** 955–959.

Kirchhoff, P. (1943). Mesoamérica. *Acta Americana* **1**, 92–107.
Kirkby, A. V. T. (1973). "The Use of Land and Water Resources in the Past and Present Valley of Oaxaca." Memoirs of the Museum of Anthropology, University of Michigan, No. 5. Ann Arbor.
Knapp, G. (1991). "Andean Ecology: Adaptive Dynamics in Ecuador." Dellplain Latin American Studies, No. 27. Westview Press, Boulder.
Krige, E. J. (1943). "Realm of a Rain Queen: A Study of the Pattern of Lovedu Society." Oxford University Press, London.
Kroll, H.(1981). Thessalische Kulturpflanzen. *Zeitschrift für Archäologie* **15**, 97–103.
Kroll, H. (1983). "Kastanas: Ausgrabungen in einem Siedlungshügel der Bronze- und Eisenzeit Makedoniens 1975–1979. Die Pflanzenfunde." Prähistorische Archäologie in Südosteuropa 2. Spiess, Berlin.
Lathrap, D. W. (1970). "The Upper Amazon." Thames and Hudson, London.
Lathrap, D. W. (1973). Gifts of the caiman: some thoughts on the subsistence base of Chavín. *In* "Variation in Anthropology" (D. W. Lathrap and J. Douglas, eds.), pp. 91–105. Illinois Archaeological Survey, Urbana.
Lathrap, D. W. (1987). The introduction of maize in prehistoric eastern North America: the view from Amazonia and the Santa Elena peninsula. *In* "Emergent Horticultural Economies of the Eastern Woodlands" (W. F. Keegan, ed.), pp. 345–371. Occasional Paper No. 7. Center for Archaeological Investigations, Southern Illinois University, Carbondale.
Leakey, M. G., Feibel, C. S., McDougall, I., and Walker, A. (1995). New four-million-year-old hominid species from Kanapoi and Allia Bay, Kenya. *Nature* **376**, 565–571.
Lee-Thorp, J. A., and van der Merwe, N. J. (1987). Carbon isotope analysis of fossil bone apatite. *South Afric. J. Sci.* **83**, 712–713.
Lee-Thorp, J. A., and van der Merwe, N. J. (1993). Stable carbon isotope studies of Swartkrans fossils. *In* "Swartkrans: A Cave's Chronicle of Early Man," pp. 251–256. Transvaal Museum Monographs, Pretoria.
Lee-Thorp, J. A., Sealy, J. C., and van der Merwe, N. J. (1989a). Stable carbon isotope differences between bone collagen and bone apatite and their relationship to diet. *J. Arch. Sci.* **16**, 585–599.
Lee-Thorp, J. A., van der Merwe, N. J., and Brain, C. K. (1989b). Isotopic evidence for dietary differences between two baboon species from Swartkrans. *J. Hum. Evol.* **18**, 183–190.
Lewenstein, S. M., and Walker, J. (1984). The obsidian chip/manioc grating hypothesis and the Mesoamerican Preclassic. *J. New World Arch.* **6**, 25–38.
Lippi, R. D., Bird, R. McK., and Stemper, D. M. (1984). Maize recovered at La Ponga, an early Ecuadorian site. *Am. Antiquity* **49**, 118–124.
Long, A. B., Benz, B. F., Donahue, D. J., Tull, A. J. T. and Toolin, L. J. (1989). First direct AMS dates on early maize from Tehuacán, Mexico. *Radiocarbon* **31**, 1030–1035.
Lowe, G. W. (1975). "The Early Preclassic Barra Phase of Altamira, Chiapas: A Review With New Data." New World Archaeological Foundation Paper No. 38. Provo.
Lynott M. J., Boutton, T. W., Price, J. E., and Nelson, D. E. (1986). Stable carbon isotopic evidence for maize agriculture in Southeast Missouri and Northeast Arkansas. *American Antiquity* **51**, 51–65.
MacFadden, B. J., and Cerling, T. E. (1996). Mammalian herbivore communities, ancient feeding ecology, and carbon isotopes: a 10 million-year sequence from the Neogene of Florida. *Journal of Vertebrate Paleontology* **16**, 103–115.
MacFadden, B. J., Cerling, T. E., and Prado, J (1996). Cenozoic terrestrial ecosystem evolution in Argentina: evidence from carbon isotopes of fossil mammal teeth. *Palaios* **11**, 319–327.
MacNeish, R. S. (1967). A summary of the subsistence. *In* "The Prehistory of the Tehuacán Valley," vol. 1: "Environment and Subsistence" (D. S. Byers, Ed.), pp. 290–309. The University of Texas Press, Austin.

MacNeish, R. S. (1981). Tehuacán's accomplishments. In "Supplement to the Handbook of Middle American Indians," vol. 1: "Archaeology" (J. A. Sabloff, Vol. Ed., V. R. Bricker, Gen. Ed.), pp. 31–47. University of Texas Press, Austin.

Maggs, T., and Ward, V. (1984). Early Iron Age sites in the Muden area of Natal. *Natal Mus. J. Hum.* **5,** 109–151.

Marcos, J. P. (1993). Los agroalfareros Valdivia de Real Alto, en el antiguo Ecuador: Un modelo para la 'revolución neolítica' en el Nuevo Mundo. *Gaceta Arqueológica Andina* **7**(23), 11–31.

Marino, B. D., and McElroy, M. B. (1991). Isotopic composition of atmospheric CO_2 inferred from carbon in C_4 plant cellulose. *Nature* **349,** 127–131.

McIntosh, S. K., and McIntosh. R. J. (1995). Cities without citadels: understanding urban origins along the middle Niger. In "The Archaeology of Africa: Food, Metals and Towns" (T. Shaw, P. Sinclair, B. Andah, and A. Okpoko, eds.), pp. 622–641. Routledge, London and New York.

Medina, E., and Minchin, P. (1980). Stratification of $\delta^{13}C$ values of leaves in Amazonian rainforests. *Oecologica* **45,** 377–378.

Medina, E., Montes, G., Cueves, E., and Rokzandic, Z. (1986). Profiles of CO_2 concentration and $\delta^{13}C$ values in tropical rain forests of the Upper Rio Negro basin, Venezuela. *J. Trop. Ecol.* **2,** 207–217.

Minnis, P. (1992). Earliest plant cultivation in the desert borderlands of North America. In "The Origins of Agriculture: An International Perspective" (C. W. Cowan and P. J. Watson, eds.), pp. 121–141. Smithsonian Institution Press, Washington, D. C.

Mintz, S. W. (1959). The plantation as a sociocultural type. In "Plantation Systems of the New World," pp. 42–50. Social Science Monographs 7. Pan American Union, Washington, D. C.

Mintz, S. W. (1985). "Sweetness and Power: The Place of Sugar in Modern History." Elisabeth Sifton Books, New York.

Morgan, M. E., Kingston, J. D., and Marino, B. D. (1994). Carbon isotope evidence for the emergence of C_4 plants in the Neogene from Pakistan and Kenya. *Nature* **367,** 162–165.

Moore, J. (1989). Pre-Hispanic beer in coastal Peru: Technology and social context of prehistoric production. *Am. Anthropol.* **91,** 682–695.

Moran, E. (1983). Mobility as a negative factor in human adaptability: the case of South American tropical forest populations. In "Rethinking Human Adaptation: Biological and Cultural Models" (R. Dyson-Hudson and M. A. Little, eds.), pp. 117–135. Westview Press, Boulder.

Morris, C. (1979). Maize beer in the economics, politics, and religion of the Inca empire. In "Fermented Food Beverages in Nutrition" (C. F. Gastineau, W. J. Darby, and T. B. Turner, eds.), pp. 21–34. Academic Press, New York.

Munson, P. J. (1976). Archaeological data on the origins of cultivation in the southwestern Sahara and their implications for West Africa. In "Origins of African Plant Domestication" (J. R. Harlan, J. M. J. de Wet, and A. B. L. Stemmler, eds.), pp. 187–209. Mouton, The Hague.

Murra, J. V. (1973). Rite and crop in the Inca state. In "Peoples and Cultures of Native South America" (D. R. Gross, ed.), pp. 377–389. Natural History Press, Garden City.

Murray, M., and Schoeninger, M. J. (1988). Diet, status, and complex social structure in Iron Age central Europe: some contributions of bone chemistry. In "Tribe and Polity in Late Prehistoric Europe" (D. B. Gibson and M. N. Geselowitz, eds.), pp. 155–176. Plenum Press, New York.

Muzzolini, A. (1995). The emergence of food-producing economy in the Sahara. In "The Archaeology of Africa: Food, Metals and Towns" (T. Shaw, P. Sinclair, B. Andah, and A. Okpoko, eds.), pp. 227–239. Routledge, London and New York.

Nambudiri, E. M. V., Tidwell, W. D., Smith, B. N., and Hebbert, N. P. (1978). A C_4 plant from the Pliocene. *Nature* **276,** 816–817.

Nassaney, M. S. (1987). On the causes and consequences of subsistence intensification in the Mississippi alluvial valley. *In* "Emergent Horticultural Economies of the Eastern Woodlands" (W. W. Keegan, ed.), pp. 129–151. Occasional Paper 7. Center for Archaeological Investigations, Southern Illinois University, Carbondale.

National Research Council (1989). "Lost Crops of the Incas: Little-Known Plants of the Andes with Promise for Worldwide Cultivation." National Academy Press, Washington, D. C.

National Research Council (1996). "Lost Crops of Africa, Vol. 1: "Grains." National Academy Press, Washington, D. C.

Netherly, P. J. (1978). "Local Level Lords on the North Coast of Peru." Doctoral dissertation, Department of Anthropology, Cornell University. University Microfilms, Ann Arbor.

Norman, M. J. T., Pearson, C. J., and Searle, P .G. E. (1995). "Ecology of Tropical Food Crops." Cambridge University Press, Cambridge.

Norr, L. (1995). Interpreting dietary maize from bone stable isotopes in the American tropics: the state of the art. *In* "Archaeology in the Lowland American Tropics" (P. W. Stahl, ed.), pp. 198–223. Cambridge University Press, Cambridge.

O'Brien, M. J. (1987). Sedentism, population growth, and resource selection in the Woodland Midwest: a review of coevolutionary developments. *Curr. Anthropol.* **28,** 177–197.

Olsen, R. A., and Sander, D. H. (1988). Corn production. *In* "Corn and Corn Improvement," 3rd ed. (G. F. Sprague and J. W. Dudley, eds.), pp. 639–686. Agronomy No. 18. American Society of Agronomy, Crop Science Society of America, and Soil Science Society of America, Madison.

Pearsall, D. M. (1988). "La producción de alimentos en Real Alto: La aplicación de las técnicas etnobotánicas al problema de la subsistencia en el período formativo ecuatoriano." Centro de Estudios Arqueológicos y Antropológicos, ESPOL, Guayaquil.

Pearsall, D. M. and Piperno, D. R. (1990). Antiquity of maize cultivation in Ecuador: summary and re-evaluation of the evidence. *Am. Antiquity* **55,** 324–337.

Piperno, D. R. (1990). Aboriginal agriculture and land usage in the Amazon basin, Ecuador. *J. Arch. Sci.* **17,** 665–677.

Piperno, D. R. (1991). The status of phytolith analysis in the American tropics. *J. World Prehist.* **5,** 155–191.

Piperno, D. R., Clary, K. H., Cooke, R. G., Ranere, A. J. and Weiland, D. (1985). Preceramic maize in central Panama: phytolith and pollen evidence. *Am. Anthropol.* **87,** 871–878.

Piperno, D. R., Bush, M. B., Colinvaux, P. A. (1990). Paeloenvironments and human occupation in Late Glacial Panama. *Quat. Res.* **33,** 108–116.

Price, T. D. (1995). Social inequality at the origins of agriculture. *In* "Foundations of Social Inequality" (T. D. Price and G. M. Feinman, eds.), pp. 129–151. Plenum Press, New York.

Price, T. D., and A. B. Gebauer (1995, eds.). "Last Hunters, First Farmers: New Perspectives on the Prehistoric Transition to Agriculture." School of American Research, Santa Fe.

Quade, J., Cerling, T. E., and Bowman, J. R. (1989). Development of Asian monsoon revealed by marked ecological shift during the late Miocene in northern Pakistan. *Nature* **342,** 163–166.

Quade, J., Cerling, T. E., Barry, J. C., Morgan, M. E., Pilbeam, D. R., Chivas, J. A., Lee-Thorp, J. A. and van der Merwe, N. J. (1992). A 16-Ma record of paleodiet using carbon and oxygen isotopes in fossil teeth from Pakistan. *Chem. Geol.* **94,** 183–192.

Raymond, J. S. (1988). Subsistence patterns during the early Formative in the Valdivia Valley, Ecuador. *In* "Diet and Subsistence: Current Archaeological Perspectives" (B. V. Kennedy and G. M. LeMoine, eds.), pp. 159–164. Proceedings of the 19th Annual Chacmool Conference. Archaeological Association of the University of Calgary, Calgary.

Reed, D. M. (1994). Ancient Maya diet at Copán, Honduras, as determined through the analysis of stable carbon and nitrogen isotopes. *In* "Paleonutrition: The Diet and Health of Prehistoric Americans" (K. D. Sobolik, ed.), pp. 210–221. Occasional Paper 22. Southern Illinois University, Center for Archaeological Investigations, Carbondale.

Ren, S. (1995). Several important achievements in Neolithic culture before 5000 B. C. *Kaogu* 1995(1), 37–49.
Riley, T. J. (1987). Ridged-field agriculture and the Mississippian agricultural pattern. *In* "Emergent Horticultural Economies of the Eastern Woodlands" (W. W. Keegan, ed.), pp. 295–304. Occasional Paper 7. Center for Archaeological Investigations, Southern Illinois University, Carbondale.
Rindos, D. (1984). "The Origins of Agriculture: An Evolutionary Perspective." Academic Press, Orlando.
Rodwell, S. (1984). China's earliest farmers: the evidence from Cishan. *Bull. Indo-Pacific Prehist. Assoc.* **5**, 55–63.
Roosevelt, A. C. (1980). "Parmana: Prehistoric Maize and Manioc Subsistence Along the Amazon and Orinoco." Academic Press, New York.
Roosevelt, A. C. (1987). Chiefdoms in the Amazon and Orinoco. *In* "Chiefdoms in the Americas" (R. D. Drennan and C. A. Uribe, Eds.), pp. 153–185. University Press of America, Lanham.
Roosevelt, A. C. (1989). Resource management in Amazonia before the conquest: Beyond ethnographic projection. *In* "Resource Management in Amazonia: Indigenous and Folk Strategies" (D. A. Posey and W. Balée, eds.), pp. 30–62. Advances in Economic Botany 7. New York Botanical Garden, Bronx, New York.
Roosevelt, A. C_3, Housley, R. A., Imazio da Silveira, M., Maranca, S., and Johnson, R. (1991). Eighth millennium pottery from a prehistoric shell midden in the Brazilian Amazon. *Science* **254**, 1621–1624.
Rose, J. C., Marks, M. K., and Tieszen, L. L. (1991). Bioarchaeology and subsistence in the central and lower portions of the Mississippi Valley. *In* "What Mean These Bones" (M. L. Powell, P. S. Bridges, and A. M. Mires, eds.), pp. 7–21. University of Alabama Press, Tuscaloosa.
Rust, W. F., and Leyden, B. W. (1994). Evidence of maize use at Early and Middle Preclassic La Venta Olmec sites. *In* "Corn and Culture" (S. Johannessen and C. A. Hastorf, eds.), pp. 181–201. University of Minnesota Publications in Anthropology 5. Westview Press, Boulder.
Sage, R. F. (1995). Was low atmospheric CO_2 during the Pleistocene a limiting factor for the origin of agriculture? *Glob. Change Biol.* **1**, 93–106.
Sánchez González, J. J. (1994). Modern variability and patterns of maize movement in Mesoamerica. *In* "Corn and Culture" (S. Johannessen and C. A. Hastorf, eds.), pp. 135–156. University of Minnesota Publications in Anthropology 5. Westview Press, Boulder.
Sanders, W. T,. Parsons, J. R., and Santley, R. S. (1979). "The Basin of Mexico: Ecological Processes in the Evolution of a Civilization." Academic Press, New York.
Scarry, C. M. (1993a). Variability in Mississippian crop production strategies. *In* "Foraging and Farming in the Eastern Woodlands" (C. M. Scarry, ed.), pp. 78–90. University Press of Florida, Gainesville.
Scarry, C. M. (1993b). Agricultural risk and the beginning of the Moundville chiefdom. *In* "Foraging and Farming in the Eastern Woodlands" (C. M. Scarry, ed.), pp. 157–181. University Press of Florida Gainesville.
Schoeninger, M. and Moore, K. (1992). Bone stable isotope studies in archaeology. *J. World Prehist.* **6**, 247–296.
Schoenwetter, J. (1990). Lessons from an alternative view. *In* "Powers of Observation: Alternative Views in Archaeology" (S. M. Nelson and A. B. Kehoe, eds.), pp. 103–112. Archaeological Papers of the Am. Anthropological Association, No. 2. Washington, D. C.
Schurr, M. R. (1992). Isotopic and mortuary variability in a Middle Mississippian population. *Am. Antiquity* **57**, 300–320.
Schurr, M. R., and Schoeninger, M. J. (1995). Associations between agricultural intensification and social complexity: an example from the prehistoric Ohio Valley. *J. Anthropological Archaeol.* **14**, 315–339.

Shaw, R. H. (1988). Climate requirement. In "Corn and Corn Improvement," 3rd ed. (G. F. Sprague and J. W. Dudley, eds.), pp. 609–638. Agronomy No. 18. American Society of Agronomy, Crop Science Society of America, and Soil Science Society of America, Madison.

Shaw, T., Sinclair, P., Andah, B., and Okpoko, A. (1995, eds.). "The Archaeology of Africa: Food, Metals and Towns." Routledge, London and New York.

Shillington, K. (1995). "History of Africa." St. Martin's Press, New York.

Shreeve, J. (1996). Sunset on the savanna. *Discover,* July, 116–125.

Smith, B. D. (1987a). The economic potential of *Chenopodium berlandiari* in prehistoric eastern North America. *J. Ethnobiol.* **7,** 29–54.

Smith, B. D. (1987b). The independent domestication of indigenous seed-bearing plants in eastern North America. In "Emergent Horticultural Economies of the Eastern Woodlands" (W. W. Keegan, ed.), pp. 3–47. Occasional Paper 7. Center for Archaeological Investigations, Southern Illinois University, Carbondale.

Smith, B. D. (1995). "The Emergence of Agriculture." Scientific American Library, W. H. Freeman, New York.

Spencer, C. S. (1993). Human agency, biased transmission, and the cultural evolution of chiefly authority. *J. Anthropol. Arch.* **12,** 41–74.

Stanford, C. B. (1995). Chimpanzee hunting behavior. *Am. Sci.* **83,** 256–261.

Stark, B. L. (1981). The rise of sedentary life. In "Supplement to the Handbook of Middle American Indians," vol. 1: "Archaeology" (J. A. Sabloff, Vol. Ed., V. R. Bricker, Gen. Ed.), pp. 345–372. The University of Texas Press, Austin.

Stothers, D. M., and Bechtel, S. K. (1987). Stable carbon isotope analysis: an interregional perspective. *Arch. East. North Am.* **15,** 137–154.

Thackeray, J. F., van der Merwe, N. J., Lee-Thorp, J. A., Sillen, A., Lanham, J. L., Smith, R., Keyser, A., and Monteiro, P. M. S. (1990). Changes in carbon isotope ratios recorded in therapsid tooth apatite. *Nature* **347,** 751–753.

Tidwell, W. D., and Nambudiri, E. M. V. (1989). *Tomlinsoniá thomasonii*, gen. et sp. nov., a permineralized grass from the Upper Miocene Ricardo Formation, California. *Rev. Palaeobot. Palynol.* **60,** 165–177.

Tieszen, L. L., and Fagre, T. (1993). Carbon isotopic variability in modern and archaeological maize *J. Arch. Sci.* **20,** 25–40.

Troll, C. (1931). Die geographischen Grundlagen der Andinen Kulturen und des Inkareiches. *Iberoamerikanisches Archiv* **5,** 258–294.

Tuross, N., Fogel, M. L., Newsom, L., and Doran, G. H. (1994). Subsistence in the Florida Archaic: The stable-isotope and archaeobotanical evidence from the Windover Site. *Am. Antiquity* **59,** 288–303.

Tykot, R. H., van der Merwe, N. J., and Hammond, N. (1996). Stable isotope analysis of bone collagen, bone apatite, and tooth enamel in the reconstruction of human diet: a case study from Cuello, Belize. In "Archaeological Chemistry: Organic: Inorganic, and Biochemical Analysis" (M. V. Orna, ed.), pp. 355–365. American Chemical Society, Washington, D. C.

Underhill, A. P. (1997). Current issues in Chinese Neolithic archaeology. *J. World Prehis.* **11,** 103–161.

van der Merwe, N. J. (1982). Carbon isotopes, photosynthesis, and archaeology. *Am. Sci.* **70,** 596–606.

van der Merwe, N. J., and Medina, E. (1989). Photosynthesis and $^{13}C/^{12}C$ ratios in Amazonian rain forests. *Geochim. Cosmochim. Acta* **53,** 1091–1094.

van der Merwe, N. J., Roosevelt, A. C., and Vogel, J. C. (1981). Isotopic evidence for prehistoric subsistence at Parmana, Venezuela. *Nature* **292,** 536–538.

van der Merwe, N. J., Lee-Thorp, J. A., and Raymond, J. S. (1993). Light stable isotopes and the subsistence of Formative cultures at Valdivia, Ecuador. In "Prehistoric Human Bone: Archaeology at the Molecular Level" (J. B. Lambert and G. Grupe, eds.), pp. 63–98. Springer-Verlag, Berlin.

van der Merwe, N. J., Thackeray, J. F., and Lee-Thorp, J. A. (n. d.). Diets of South African australopithecines compared by stable carbon isotopes in their tooth enamel. In prep.

van der Merwe, N. J., and Vogel, J. C. (1978). ^{13}C content of human collagen as a measure of prehistoric diet in woodland North America. *Nature* **276,** 815–816.

Vishnu-Mittre (1974). Palaebotanical evidence in India. In "Evolutionary Studies in World Crops: Diversity and Change in the Indian Subcontinent" (J. B. Hutchinson, ed.), pp. 3–30. Cambridge University Press, Cambridge.

Vogel, J. C. (1978). Isotopic assessment of the dietary habits of ungulates. *South Af. J. Sci.* **74,** 298–301.

Vogel, J. C., and van der Merwe, N. J. (1977). Isotopic evidence for early maize cultivation in New York state. *American Antiquity* **42**:238–242.

Vogel, J. C., Fuls, A., and Elis, R. P. (1978). The geographic distribution of Kranz grasses in southern Africa. *South Af. J. Sci.* **75,** 209–215.

Vrba, E. (1995). Fossil record of African antelopes (Mammalia, Bovidae) in relation to human evolution and paleoclimate. In "Paleoclimate and Evolution, with Emphasis on Human Origins." (E. Vrba, ed.), pp. 385–424. Yale University Press, New Haven.

Wagner, G. E. (1994). Corn in eastern Woodlands late prehistory. *In* "Corn and Culture" (S. Johannessen and C. A. Hastorf, eds.), pp. 335–346. University of Minnesota Publications in Anthropology 5. Westview Press, Boulder.

Wang, Y, Cerling, T. E., and MacFadden, B. J. (1994). Fossil horses and carbon isotopes: new evidence for Cenozoic dietary, habitat and ecosystem changes in North America. *Palaeog. Palaeoclim. Palaeoecol.* **107,** 269–279.

Warner, J. N. (1962). Sugar cane: An indigenous Papuan cultigen. *Ethnology* **1,** 405–411.

Wasylikowa, K., Harlan, J. R., Evans, J., Wendorf, F., Schild, R., Close, A. E., Krolik, H., and Housley, R. A. (1995). Examination of botanical remains from early Neolithic houses at Nabta Playa, western desert, Egypt, with special references to sorghum grains. *In* "The Archaeology of Africa: Food, Metals and Towns" (T. Shaw, P. Sinclair, B. Andah, and A. Okpoko, eds.), pp. 154–164. Routledge, London and New York.

Watson, A. M. (1974). The Arab agricultural revolution and its diffusion, 700–1100. *J. Econ. His.* **34,** 8–35.

Watson, P. J. (1995). Explaining the transition to agriculture. *In* "Last Hunters, First Farmers: New Perspectives on the Prehistoric Transition to Agriculture" (T. D. Price and A. B. Gebauer, eds.), pp. 21–37. School of American Research, Santa Fe.

White, C. D., and Schwarcz, H. P. (1989). Ancient Maya diet: as inferred from isotopic and elemental analysis of human bone. *J. Arch. Sci.* **16,** 451–474.

White, C. D., Healy, P. F., and Schwarcz, H. P. (1993). Intensive agriculture, social status, and Maya diet at Pacbitun, Belize. *J. Anthropol. Res.* **49,** 347–375.

White, T. D., Suwa, G., and Asfaw, B. (1994). *Australopithecus ramidus,* a new species of early hominid from Aramis, Ethiopia. *Nature* **371,** 306–312.

White, T. D., Suwa, G., and Asfaw, B. (1995). *Australopithecus ramidus,* a new species of early hominid from Aramis, Ethiopia. *Nature* **375,** 388.

Wills, W. H. (1988). "Early Prehistoric Agriculture in American Southwest." School of American Research Press, Santa Fe.

Woods, W. I. (1987). Maize agriculture and the Late Prehistoric: a characterization of settlement location strategies. *In* "Emergent Horticultural Economies of the Eastern Woodlands" (W. W. Keegan, ed.), pp. 275–294. Occasional Paper 7. Center for Archaeological Investigations, Southern Illinois University, Carbondale.

Wright, K. I. (1991). The origin and development of ground stone assemblages in Late Pleistocene Southwest Asia. *Paléorient* **17,** 19–45.

Wright, K. I. (1993). Early Holocene ground stone assemblages in the Levant. *Levant,* **XXV,** 93–111.

Wright, K. I. (1994). Ground-stone tools and hunter–gatherer subsistence in Southwest Asia: Implications for the transition to farming. *American Antiquity* **59,** 238–263.

Wright, L. E. (1994). "The Sacrifice of the Earth? Diet, Health, and Inequality in the Pasión Maya Lowlands." Doctoral dissertation, Department of Anthropology, University of Chicago. Ann Arbor: University Microfilms.

Wymer, D. A. (1994). The social context of early maize in the mid-Ohio Valley. *In* 'Corn and Culture" (S. Johannessen and C. A. Hastorf, Eds.), pp. 411–426. University of Minnesota Publications in Anthropology 5. Westview Press, Boulder.

Yan, W. (1991). China's earliest rice agriculture remains. *Bull. Indo-Pac. Prehist. Assoc.* **10,** 118–126.

Zhou, B. X. (1981). Animal remains from Cishan village, Wuan, Hebie Province, China. *Kaogu Xuebao* **3,** 339–347.

Zohary, D., and Hopf, M. (1994). "Domestication of Plants in the Old World." Clarendon Press, Oxford.

16

The Taxonomic Distribution of C_4 Photosynthesis

Rowan F. Sage, Meirong Li, and Russell K. Monson

I. Introduction

The lack of comprehensive and readily available listings of C_4 taxa has delayed the widespread appreciation of C_4 photosynthesis outside of the plant sciences. The earliest comprehensive lists date from the mid-1970s (Troughton *et al.*, 1974; Downton, 1975; Raghavendra and Das, 1978), but these lacked extensive coverage of remote regions such as the tropics and central Asia. Subsequent listings were primarily regional (for example, Winter and Troughton, 1978; Watson and Dallwitz, 1980; Winter, 1981; Ziegler *et al.*, 1981; Collins and Jones, 1985), or focused on select groups, most notably grasses (Brown, 1977; Hattersley, 1987). Within the systematics community, the C_4 syndrome has often been ignored, such that very few taxonomic guides and monographs consider C_4 photosynthesis or even Kranz anatomy. Thus, extensive literature searches are often required to determine whether a given plant is C_4. This situation has improved, particularly for grasses. Brown (1977) provides a comprehensive treatise of grass systematics, including the distribution of the Kranz syndrome in the Poaceae. Hattersley and co-workers built on this work throughout the 1980s, developing a robust set of anatomical characteristics that enabled screening of most grass species for C_4 photosynthesis, often to the level of biochemical subtype (see Chapter 5, Dengler and Nelson, for description). This resulted in two detailed lists: Hattersley (1987) and Hattersley and Watson (1992). The list in Hattersley and Watson (1992) represents a near-complete photosynthetic typing of grass genera and is now available in updated versions in the DELTA database available on the world wide web (Watson and

Dallwitz, 1998; website: http://www.biodiversity.uno.edu/delta/). Similarly, the C_4 genera of the Cyperaceae are listed in the DELTA database, largely as a result of the work of Bruhl and co-workers (Bruhl et al., 1992; Bruhl, 1995). Dicots have been more haphazardly assessed than graminoids for a variety of reasons. C_4 dicots have less economic significance and are often from inaccessible and politically less stable regions. Pyankov's work in central Asia has proved valuable in adding taxa to the list of C_4 dicots (Pyankov and Mokronosov, 1993; Pyankov et al., 1997), and Akhani et al. (1997) published a comprehensive list of C_4 species of the Chenopodiaceae from Asia, Africa, and Europe.

In this chapter, we build on the work of many previous studies (see tables for references) to estimate the numbers of genera and species in the families with C_4 members (Tables I and II). For grasses, we describe the distribution of C_4 photosynthesis in subfamilies and tribes (Table III), and list the known 372 C_4 grass genera (Table IV). We also list all known C_3–C_4 intermediate species (Table V). As much as possible, we summarize the biochemical subtype for C_4 taxa, based on biochemical (Guttierez et al., 1974; Edwards and Walker, 1983; Glagoleva et al., 1992; Pyankov et al., 1997; Pyankov and Black, 1998) and anatomical data (Hattersley and Watson, 1992; Watson and Dallwitz, 1998). It is hoped this comprehensive list will provide a convenient reference for those requiring photosynthetic pathway identifications, and will encourage further ecological and floristic assessments of C_4 distributions across the globe.

II. Methodology

Discrepancies exist between previous lists of C_4 taxa, and numerous sources of error are now recognized in early C_4 screening methods. These problems lead us to review the various criteria that have been used to identify C_4 taxa and classify their biochemical subtype.

A. Photosynthetic Pathway

Four general approaches have been employed to identify C_4 photosynthesis in terrestrial plants. These are:

1. Anatomical, namely the occurrence of Kranz anatomy. Also, in C_4 plants, mesophyll to bundle sheath cell ratios are less than two, whereas C_3 plants generally have ratios above three (Dengler and Nelson, Chapter 5).

2. Carbon isotope ratios ($\delta^{13}C$): in C_4 plants, $\delta^{13}C$ ratios are generally between $-10‰$ and $-15‰$, which contrast with the common C_3 range of $-21‰$ to $-30‰$ (Cerling, Chapter 13);

3. Gas exchange: photosynthetic CO_2 compensation points in C_4 plants are between 0 and 5 μmol CO_2 mol^{-1} air, whereas in C_3 species, CO_2

compensation points are typically above 30 μmol mol^{-1} (Osmond et al., 1982);

4. Biochemistry: C$_4$ plants have a ratio of PEPCase to Rubisco activity above two, whereas in C$_3$ plants the PEPCase:Rubisco ratio is well below one (Hatch, 1987; Sage et al., 1987). Also, unlike C$_3$ photosynthesis, C$_4$ photosynthesis initially produces high levels of C$_4$ acids such as malate and aspartate (Edwards and Walker, 1983). Thus, characterization of initial photosynthetic products can identify C$_4$ metabolism so long as the species are not CAM (Glagoleva et al., 1992; Keeley, 1998).

Potential difficulties are associated with each screening method, so two or more should be used in searching for new C$_4$ taxa. Problems with the anatomical screens arise in C$_3$ species that have green Kranz-like bundle sheaths, or if C$_4$ plants have weakly developed Kranz patterns. In addition, plants from harsh habitats can produce leaves that lack laminate anatomies with easily identifiable Kranz cells. An example of an apparent anatomical error occurs with *Evolvulus* spp. in the Convolvulaceae. Welkie and Caldwell (1970) and Meinzer (1978) identify Kranz-like anatomy (green bundle sheaths) in *Evolvulus alsinoides*, leading to inclusion in numerous C$_4$ lists (for example, Downton, 1975). Subsequent analysis using δ^{13}C shows *E. alsinoides* is C$_3$ (Winter, 1981).

Carbon isotope ratios can be problematic because non-C$_4$ plants can exhibit C$_4$-like δ^{13}C signatures. CAM plants and algae commonly have C$_4$-like δ^{13}C ratios, and in arid situations, some C$_3$ species can exhibit high δ^{13}C ratios approaching C$_4$ values (Cerling, Chapter 13; Williams and Ehleringer, 1996). *Lithops* spp. included in the C$_4$ list of Raghavendra and Das (1978) based on high δ^{13}C are now known CAM plants. An example of problems with reliance on just the carbon isotope method is illustrated with *Bienertia cycloptera* in the Chenopodiaceae. Winter (1981) and Akhani et al. (1997) determined C$_4$-like δ^{13}C ratios for this species, whereas Glagoleva et al. (1992) show C$_3$-like enzyme activities and metabolite fluxes. *Bienertia cycloptera* has fleshy leaves, indicating potential for CAM activity. No Kranz anatomy is reported for the species. Given the conflicting results, further evaluation is needed before *Bienertia* can be safely considered C$_4$. Similar problems may reside with the Chenopodiaceae genus *Axyris*, which has C$_4$-like δ^{13}C ratios but C$_3$-like leaf anatomy (Akhani et al. 1997).

Evaluation of CO$_2$ compensation points requires exceptionally well calibrated and stable gas exchange instruments, because near the CO$_2$ compensation point, CO$_2$ flux rates are low, allowing leaks, slight calibration errors, and gas adhesion problems to have undue significance. Finally, PEPCase:Rubisco ratios require that both these enzymes are extracted and assayed without changing the *in vivo* stoichiometry; this is problematic in wild plants with abundant secondary compounds and tough tissues. Reliance on this method alone will likely result in the greatest number of false C_4 identifications, and thus it should be used as a secondary screen to back up one of the other three.

Although not used as a primary screen, *in situ* labeling of Rubisco has been used successfully to supplement anatomical observations and is particularly useful where Kranz anatomy is weakly developed (Hattersley *et al.*, 1977; Hattersley and Watson, 1992; Hudson *et al.*, 1992). In fully developed C_4 plants, labeling compounds tagged to Rubisco or Rubisco mRNAs are localized to functional bundle sheath–like tissue, whereas in C_3 plants, the label predominantly occurs in mesophyll tissue. *In situ* labeling is often inappropriate for large-scale screens because of the cost and time involved.

B. Biochemical Subtype

C_4 plants can be generally classified into one of three biochemical subtypes based on the enzyme used to decarboxylate C_4 acids in the bundle sheath compartment (Kanai and Edwards, Chapter 3). The best screen for C_4 subtype is direct biochemical assay of the decarboxylating enzyme (either NADP-malic enzyme, NADP-ME; NAD-malic enzyme, NAD-ME; or PEP carboxykinase, PCK) and immediate photosynthetic fixation products (Hatch, 1987). This approach is expensive and time consuming, and requires living plant material that can be biochemically characterized. Many species of wild plants are difficult to assay due to phenolics and other compounds that inhibit enzyme activity and/or the presence of fiber bundles that prevent enzyme extraction. In addition, because C_4 plants are often from undeveloped regions, transporting live material to an appropriate lab can be difficult. As a result, relatively few C_4 taxa have been classified to subtype based on biochemical assays.

In the grasses, biochemical subtype correlates with leaf anatomy and cellular ultrastructure, so that screens based on leaf properties can be used to subtype C_4 taxa. Eight major anatomical types occur in the grasses, and these generally are correlated with one of the three decarboxylation pathways. Most grasses fall into one of three "classical" anatomical types that differ in number of characteristics (Dengler and Nelson, Chapter 5). In the classical group, asparatate forming species (NAD-ME or PCK subtypes) typically have two sheath layers; non-aspartate forming species (NADP-ME subtype) have one. Delineation between NAD-ME and PCK types is possible by studying chloroplast position (NAD-ME is centripetal, whereas PCK is centrifugal or scattered); suberization of the sheath lamellae (none in NAD-ME, extensive in PCK); and eveness of the chloroplast outline (smooth in NAD-ME and uneven in PCK) (Denger and Nelson, Chapter 5). The other five "nonclassical" anatomical types also exhibit specific features that are generally indicative of decarboxylation pathways (Hattersley and Watson, 1992; Dengler and Nelson, Chapter 5). Caution should be exercised when using anatomical data for subtype determinations, however, because in some cases, biochemical activities do not match anatomical patterns. For example, numerous species of *Eragrostis, Enneapogon,* and *Triraphis* are anatomically PCK but biochemically NAD-ME (Watson and Dallwitz, 1998). Also, numerous taxa have high activities of both NADP-ME and NAD-ME (for example, *Neostapfia colusana;* Keeley, 1998) indicating a mixed subtype. These exceptions appear to be infrequent, so that the chance of error is relatively small in most situations (Hattersley and Watson, 1992). For groups lacking anatomical or biochemical descriptions, delineation of subtype is still possible on taxonomic grounds if the taxa in question are classified into tribes exhibiting only one mode of decarboxylation. For example, species in the Andropogoneae, Arundinelleae, and Maydeae have a high probability of being NADP-ME, whereas species in one of the main-assemblage tribes of the Chloridoideae are most likely NAD-ME or PCK (Table III).

III. Lists of C$_4$ Taxa

C$_4$ photosynthesis occurs in 18 families of flowering plants, 15 of which are dicot and 3 monocot (Table I). The syndrome is apparently restricted to just the angiosperms. Most C$_4$ species are found in the grass family (approximately 4600 species), with the sedges (Cyperaceae) containing 1330 C$_4$ species. We estimated there are 1600 C$_4$ dicot species. The dicot estimates are likely low, because in the genera with both C$_4$ and C$_3$ species, we often relied on direct observations of C$_4$ species occurrence. These typically do not include all the C$_4$ species in the genus. Furthermore, many of the dicot groups need revision to account for new phylogenetic analyses and the improved availability of specimens from remote areas. As surveys become more complete and additional dicot groups are identified, the species of C$_4$ dicots will likely rise to near 2000, giving a total of approximately 8000 C$_4$ species.

Early estimates place total C$_4$ species numbers near 1000 (Raghavendra and Das, 1978). Textbooks commonly list about 2000 C$_4$ species (Larcher, 1995). Successive estimates have risen over the years such that 6000 to 10,000 are now given (Ehleringer *et al.,* 1997), which is in agreement with our approximation. A total of 8000 C$_4$ species would represent about 3% of the 250,000 to 300,000 terrestrial plant species.

By comparison, there are 16,000 to 20,000+ CAM species in 33 families (Larcher, 1995; Winter and Smith, 1996). Four divisions of terrestrial plants contain CAM species, including nonseed plants in the Lycophyta (Isoetaceae) and Pterophyta (Polypodiaceae and Vittariaceae), nonflowering seed plants in the Gnetophyta (Welwitchiaceae), 8 monocot families,

Table I Occurrence of C_4 Photosynthesis in Angiosperm Families[a]

		Genera statistics			Species statistics			Percent of all C_4
		C_4 (no.)	Total	C_4%	C_4 (no.)	Total	C_4%	
Dicotyledoneae (subclass)								
Asterales (order)	Asteraceae	8	1500	1	150	13,000	1.3	2.0
Brassicales	Capparidaceae	1	34	3	2+	925	<1	<0.2
Caryophyllales	Aizoaceae	5	126	4	~30	2500	1.2	0.4
	Amaranthaceae	11	74	14	~250	1000	25	3
	Caryophyllaceae	1	88	1	50	2200	2	0.6
	Chenopodiaceae	45	105	43	550	1400	39	7
	Molluginaceae	2	14	14	4+	120	3	<0.2
	Nyctaginaceae	3	33	9	5+	400	1	0.3
	Portulacaceae	2	20	10	70	450	16	0.9
Euphorbiales	Euphorbiaceae	1	300	<1	250	5000	5	3
Linnales	Zygophyllaceae	3	30	10	~50	240	21	0.7
Polygonales	Polygonaceae	1	45	2	80	1100	7	1
Solanales	Boraginaceae	1	250	<1	6+	2000	<1	<0.2
Acanthaceae		1	250	<1	80	2400	3	1
Scrophulariales	Scrophulariaceae	1	280	<1	14	3000	1	<0.2
Total Dicot 15		86	3150	3	~1600	35,735	4.5	21
Monocotyledoneae								
Alismatidae	Hydrocharitaceae	1	16	6	1+	100	1	<0.1
Juncales	Cyperaceae	28	131	21	1330	5000	27	18
Poales	Poaceae	372	~800	47	4600	10,000	46	61
Total monocot 3		401	947	42	~6000	15,000	40	79
Total C_4 18		487			~7600			100

[a] Genera statistics refer to number of genera having C_4 species, total number of genera in the family, and percent of those genera with C_4 members. Species statistics refer to total number of C_4 species estimated for the family, total species in the family, and percent of those species that are C_4. Percent of all C_4 refers to the number of C_4 species in a taxonomic group divided by 7600, the estimated total number of C_4 species. Species numbers estimated from Table II and IV and references there-in. A "+" after the species number indicates a probable underestimate.

and 22 dicot families (Smith and Winter, 1996). Thus, relative to CAM, C_4 photosynthesis apparently evolved much later and is more limited in taxonomic distribution.

A. Dicots

In the dicots, C_4 photosynthesis is taxonomically diverse, indicating numerous independent origins (Table II). At least 86 dicot genera contain C_4 species, with the Chenopodiaceae (~45 genera; ~550 species), Amaranthaceae (11 genera; 250 species) and Asteraceae (8 genera; 150 species) having the largest number. Major C_4 genera are *Gomphrena* (120 species), *Atriplex* (111 species), *Salsola* (~100 species), *Chamaesyce* (=*Euphorbia*) (250 species), and *Pectis* (100 species). These five genera account for more than 40% of all known C_4 dicots. C_3 species are reported in at least 16 of the 86 dicot genera that contain C_4 species. In these mixed genera, extensive species by species characterization may be required to completely distinguish the C_3 from the C_4 representatives.

No dicot family is strictly C_4, and in all cases, the number of C_4 genera and species present in a family are in the minority, usually by a large degree. Only 4 dicot families (Amaranthaceae, Chenopodiaceae, Molluginaceae, Zygophyllaceae) have more than 10% C_4 representation at the generic and species level, whereas seven families have only one identified genus with C_4 species. The broad taxonomic occurrence of C_4 photosynthesis in the Dicotyledonae, and its low proportion within families, indicate that most C_4 dicots appeared recently in geological time (in the past 2 million years), possibly in response to the depleted atmospheric CO_2 levels of the Quaternary (Ehleringer *et al.*, 1997).

C_4 dicots are NADP-ME or NAD-ME, with no known PCK types (Table II); however, subtype evaluations are limited in number and scope, and except for the Chenopodiaceae, have not been extensively pursued for dicots since the 1970s. Three families (Amaranthaceae, Chenopodiaceae, Portulaceae) have both subtypes, three families (Asteraceae, Euphorbiaceae, Zygophyllaceae) are NADP-ME only, and one family (Molluginaceae) is specific for NAD-ME. In dicots, NADP-ME forms are more common (20 NADP-ME versus 12 NAD-ME dicot genera). In the Chenopodiaceae, the most extensively subtyped dicot genus, 12 genera are known to have NADP-ME species whereas 9 have NAD-ME representatives. Two genera contain both subtypes: *Salsola* in the Chenopodiaceae and *Portulaca* in the Portulacaceae. *Salsola* also contains a minority of C_3 species (<10% of all *Salsola* spp.; Akhani *et al.*, 1997), which may be revertants from C_4 ancestors (Pyankov *et al.*, 1997).

The list of dicot species will likely grow in coming years, as floras of more remote regions are studied and dicot groups are systematically revised. In the Chenopodiaceae, most genera of the tribe Salsolaeae are C_4. However,

Table II Families and Genera Containing Identified C_4 Species[a]

Subclass	Family[b]	Genus[c]	Notes and refs.[d]
Dicotyledoneae			
	Acanthaceae (1/250)	*Blepharis*, 80	(5)
	Aizoaceae (5/126)	*Cypselea*, 1	(1) Welkie and Caldwell, 1970
		Gisekia, 5	(1)
		Sesuvium, 1+/12	Also C_3, Hartmann, 1993
		Trianthema, 17	
		Zaleya, 6	
		Kranz anatomy reported in *Lithops*, a known CAM group	
	Amaranthaceae (11/74)	[*Acanthochiton*]	= *Amaranthus*, Kew genlist 1998
		[*Acnidia*]	= *Amaranthus*, Kew genlist, 1998
		Achyranthes, 1/7	Kranz anatomy, also C_3
		Aerva, 1+/10	Also C_3
		Alternanthera, 6+/100	NADP-ME, also C_3 and C_3–C_4 intermediate
		Amaranthus, 60	NAD-ME
		Blutaparon, 4	Kranz anatomy, $\delta^{13}C$ (14)
		[*Brayulinea*]	= *Guillemenia*, Townsend, 1993
		Celosia, 1+/45	also C_3; Kellogg, Chapter 12, this volume; Kranz anatomy (14)
		Froelichia, 18	NADP-ME
		Gomphrena, 120	NADP-ME
		[*Gossypianthus*]	= *Guillemenia*, Townsend, 1993
		Guillemenia, 5	
		Lithophila, 2	
		[*Philoxerus*]	= *Gomphrena* and *Blutaparon*, Kew genlist 1998
		Tidestromia, 7	NADP-ME, Mabberley, 1987, for spp. number

Asteraceae (8/1500)	*Chrysanthellum*, 13	= *Chrysanthellum* by Bremer 1994 NADP-ME, also C_3 and C_3–C_4 intermediate
	Eryngiophyllum, 1	
	Flaveria, 4	
	Glossocardia, 12	= *Glossocardia*, Bremer, 1994
	[*Glossogyne*]	
	Guerreroia, 1	Kranz anatomy, $\delta^{13}C$ (14)
	Isostigma, 11	
	Neuractis, 1+	Kellogg, Chapter 12, this volume
	Pectis 100	
Boraginaceae (1/120)	*Heliotropium*, 6+/250	Also C_3
Capparidaceae (1/34)	*Cleome*, 2+/150	Also C_3 (C_4 species often put in *Gynandropsis*)
Caryophyllaceae (1/88)	*Polycarpaea*, 50	
Chenopodiaceae (~45/105)	[*Aellenia*]	= *Halothamnus*, Kuhn, 1993
	Agathophora, 1+	
	Anabasis, 42	
	Arthrophytum, 9	
	Atriplex, 111/180	NAD-ME, Osmond *et al.*, 1980
	? *Axyris*, 6	C_4 $\delta^{13}C$ (12); Kranz lacking (Carolin *et al.*, 1975)
	Bassia, 5+/36	NADP-ME, also C_3 (Kuhn, 1993)
	? *Bienertia*, 1	C_4 $\delta^{13}C$(6,12), C_3 metabolite flux (2), see text
	Camphorosma, 10	NADP-ME
	Chenolea, 4	Shomer-Ilan *et al.*, 1981
	Climacoptera, 33	NAD-ME; spp. number from Czerepanov, 1995
	Cornulaca, 6	(6)
	Cyathobasis, 1	= *Bassia*, Kuhn 1993
	[*Echinopsilon*]	(12)
	Fadenia, 1	NAD-ME
	Gamanthus, 5	(6) NAD-ME
	Girgensohnia, 3	

(*continues*)

Table II (Continued)

Subclass	Family	Genus	Notes and refs.
		Halanthium, 3	
		Halarchon, 1	
		Halimocnemis, 13	NAD-ME; spp. number from Czerepanov, 1995
		Halocharis, 13	NAD-ME
		Halogeton, 9	NADP-ME
		? *Halosarcia*, 1/23	(12), largely C_3 genus in otherwise pure C_3 tribe
		[*Halostigmeria*]	(6)
		Halothamnus, 23	NADP-ME
		Halotis, 2	
		Haloxylon, ≈25	NADP-ME
		Hammada, 6+	NADP-ME, = *Haloxylon* by Kuhn, 1993
		Horaninowia, 7	NADP-ME
		[*Hypocyclix*]	= *Salsola*, Kuhn 1993
		Iljinia, 1	NADP-ME (14)
		Kirilowia, 2	(3)
		Kochia, 10	NADP-ME, = *Bassia* by Kuhn, 1993
		? *Lagenantha*, 1	Probable C_4, confirmation needed
		Londesia, 1	NADP-ME (13), = *Bassia* by Kuhn, 1993
		Nanophyton, 7	(12); spp. number by Czerepanov, 1995
		Noaea, 3	
		?*Nucularia*, 1	Probable C_4, confirmation needed
		Ofaiston, 1	(12)
		Panderia, 1	(12)
		Petrosimonia, 11	NAD-ME (2)
		Physandra, 1	(12)
		Piptotera, 1	(12)
		Salsola, ~100/113	NADP-ME, NAD-ME, ~7% C_3 (12)
		Seidlitzia, 7	NADP-ME
		Sevada, 1	(12)

		Suaeda, ~58	NAD-ME and C$_3$; 100 spp. total, ~58% C$_4$
		Theleophyton, 1	
		Traganopsis, 1	(12)
		Traganum, 2	(12)
	Euphorbiaceae (1/300)	*Chamaesyce*, 250 or *Euphorbia*, 250/1500	NADP-ME; spp. number from Webster 1994 Euphorbia also C$_3$ and CAM; Webster (1994) places all C$_4$ *Euphorbia* in *Chamaesyce*, not universally accepted.
	Molluginaceae (2/14)	*Glinus*, 1+/6	Also C$_3$
		Mollugo, 3+/35	NAD-ME; also C$_3$, C$_3$–C$_4$; Kennedy et al., 1981
	Nyctaginaceae (3/33)	*Allionia*, 2	Also C$_3$
		Boerhavia, 1+/20	Kranz anatomy only
		Okenia, 2	
	Polygonaceae (1/45)	*Calligonum*, 80	
	Portulacaceae (2/20)	*Anacampseros*, 30	NADP-ME; NAD-ME; also facultative CAM in *P. oleracea*
		Portulaca, 40	
	Scrophulariaceae (1/280)	*Anticharis*, 14	
	Zygophyllaceae (3/30)	*Kallstroemia*, 18	
		Tribulus, 25	NADP-ME
		Zygophyllum, 1+/80	Also C$_3$
Monocotyledoneae	Cyperaceae: (28/131)		All NADP-ME except *Eleocharis* (NAD-ME); Bruhl et al., 1992; Bruhl 1995 (see Watson and Dallwitz, 1998)
		Abildgaardia 15/17	Also C$_3$
		Alinula, 5	
		Ascolepis, 15	
		Ascopholis, 1	

(*continues*)

Table II (*Continued*)

Subclass	Family	Genus	Notes and refs.[a]
		Bulbostylis, 100	
		Crosslandia, 1	
		Cyperus subgenus *Cyperus*, ~500;	Subgenus *Pycnostachys* is C_3, ~150 spp.
		Eleocharis, ~50/200	NAD-ME; also C_3 and C_3–C_4
		Fimbristylis, 200	
		Hemicarpha, 4	
		Kyllinga, 60	
		Lipocarpha, 22	
		Mariscus, 200	
		Monandrus, 5	
		Nelmesia, 1	
		Nemum, 10	
		Oreobolopsis, 1	May be C_3–C_4 intermediate
		Pycreus, 100	
		Queenslandiella, 1	
		Remirea, 1	
		Rhynchocladium, 1	May be C_3–C_4 intermediate
		Rhynchospora, 21/221	Also C_3
		Rikliella, 4	
		Spearocyperus, 1	
		Syntrinema, 1	
		Torulinium, 6	
		Tylocarya, 1	
		Volkiella, 1	
	Hydrocharitaceae	*Hydrilla* 1	NADP-ME, C_3 with inducible C_4; Reiskind *et al.*, 1997
	Poaceae (Gramineae)	C_4 in 372 genera (see Table IV)	
		C_3 in 428 genera	

[a] Developed from L. Watson and M. J. Dallwitz (1998). "Grass Genera of the World: Descriptions, Illustrations, Identification, and Information Retrieval; Including Synonyms, Morphology, Anatomy, Physiology, Cytology, Classification, Pathogens, World and Local Distribution, and References." URL http://biodiversity.uno.edu/delta/; and from the references listed below[d].

[b] Numbers in parentheses indicate number of C_4 genera/total number of genera in the family based on L. Watson and M. J. Dallwitz (1998). "Grass Genera of the World: Descriptions, Illustrations, Identification, and Information Retrieval; Including Synonyms, Morphology, Anatomy, Physiology, Cytology, Classification, Pathogens, World and Local Distribution, and References." URL http://biodiversity.uno.edu/delta/

[c] Numbers following generic name indicate: (1) Estimated number of C_4 species in the genus primarily based on published genera descriptions as follows: For Asteraceae: K. Bremer (1994). "Asteraceae: Cladistics and Classification." Timber Press, Portland. For Caryophyllales: K. Kubitzki (1993). "The Families and Genera of Vascular Plants. Vol. II. Flowering Plants—Dicotyledons—Magnolid, Hamamelid and Caryophyllid Families" (K. Kubitzki, J. G. Rohwer, and V. Bitrich, eds.). Springer-Verlag, New York. For other dicot groups: D. J. Mabberley. (1987). "The Plant Book." Cambridge University Press. Cambridge. For Cyperaceae (except *Cyperus* which is from M.-R. Li, unpublished, 1998): J. J. Bruhl. (1995). Sedge genera of the world: Relationships and a new classification of the Cyperaceae. *Aust. Syst. Bot.* 8, 125–305; or (2) If followed by a "+", the number of C_4 species directly observed for the genus. In genera with C_3 species, the total species number in the genus follows the C_4 species number and a slash. A "?" indicates a genus of uncertain status because of conflicting results, or where anatomy and taxonomic affinity indicate a high probability of C_4 members, but direct confirmation is needed. Generic names enclosed by [ ] indicates that the name is no longer in use.

[d] References: 1: W. J. S. Downton. (1975). The occurrence of C_4 photosynthesis among plants. *Photosynthetica* 9, 96–195. 2: T. A. Glagoleva, M. V. Chulanovskaya, M. V. Pakhomova, E. V. Voznesenskaya, and Y. V. Gamalei. (1992). Effect of salinity on the structure of assimilating organs and ^{14}C labelling patterns in C_3 and C_4 plants of Ararat Plain. *Photosynthetica* 26, 363–369. 3: V. I. Pyankov and A. T. Mokronosov. (1993). General trends in changes of the earth's vegetation related to global warming. *Russian J. Plant Physiol.* (Engl. Translation) 40, 515–531. 4: B. N. Smith and B. L. Turner. (1975). Distribution of Kranz syndrome among Asteraceae. *Am. J. Bot.* 62, 541–545. 5: A. S. Raghavendra and V. S. R. Das. (1978). The occurrence of C_4-photosynthesis: A supplementary list of C_4 plants reported during late 1974–mid 1977. *Photosynthetica* 12, 200–208. 6: K. Winter. (1981). C_4 plants of high biomass in arid regions of Asia—occurrence of C_4 photosynthesis in Chenopodiaceae and Polygonaceae from the Middle East and USSR. *Oecologia* 48, 100–106. 7: H. Ziegler, K. H. Batanouny, N. Sankhla, O. P. Vyas, and W. Stichler. (1981). The photosynthetic pathway types of some desert plants from India, Saudi Arabia, Egypt and Iraq. *Oecologia* 48, 93–921. 8: L. Watson and M. J. Dallwitz (1992, eds). "The Grass Genera of the World." CAB International, Wallingford. 9: V. I. Pyankov, E. V. Voznesenskaya, A. V. Kondratschuk, and C. C. Black, Jr. (1997). A comparative anatomical and biochemical analysis in *Salsola* (Chenopodiaceae) species with and without a Kranz type leaf anatomy: A possible reversion of C_4 to C_3 photosynthesis. *Am. J. Bot.* 84, 597–606. 10: V. I. Pyankov and D. V. Vakhrusheva. (1989). The pathways of primary CO_2 fixation in Chenopodiaceae C_4-plants of central Asian arid zone. *Soviet Plant Physiol.* 36, 178–187. 11: V. I. Pyankov, A. N. Kuzmin, E. D. Demidov, and A. I. Maslov. (1992). Diversity of biochemical pathways of CO_2 fixation in plants of the families Poaceae and Chenopodiaceae from arid zone of central Asia. *Soviet Plant Physiol.* 39, 411–420. 12: H. Akhani, P. Trimborn, and H. Ziegler. (1997). Photosynthetic pathways in Chenopodiaceae from Africa, Asia, and Europe with their ecological, phytogeographical, and taxonomical importance. *Plant Syst. Evol.* 206, 187–221 (for Chenopodiaceae). 13: V. I. Pyankov, Gunin, and C. C. Black, unpublished, 1998. 14: R. F. Sage and N. Dengler, personal observation (1998) from herbarium specimens of the indicated genus. Specific references for a group listed in the table.

not all Salsoleae genera have been typed, including the genera *Lagenantha* and *Nucularia*. These two genera are probably C_4 because they are closely related to other C_4 taxa in the Salsolaeae, and they occur in hot, arid, open habitats of northern Africa. In the Zygophyllaceae, the untyped genus *Kelleronia* is probably C_4 because of its close relationship with the C_4 genera *Kallstroemia* and *Tribulus*, and it occurs in hot, arid areas of Somalia, where 95% of the grass flora is C_4 (Sage and Wedin, Chapter 10; Sheahan and Chase, 1996). Similarly, Acanthaceae, Aizoaceae, and Amaranthaceae appear to be good possibilities for additional C_4 taxa given their large species number and distribution in arid regions of the tropics and subtropics. Based on Rubisco:PEPcase activities of Chinese taxa, Yin and Li (1997) list 20 new C_4 genera and 5 new families. We have not included these genera because of previously mentioned problems with the activity ratio and the lack of confirmation with other screening methods.

Table II excludes numerous genera previously noted as being C_4. Many are no longer accepted, having been reclassified into other genera (for example, *Acnidia* into *Amaranthus*). *Moricandia* spp. are now accepted as C_3-C_4 intermediate (Table V; Apel *et al.*, 1997). Errors involving CAM species occur on a number of early lists. One CAM genus mentioned in Table II is *Lithops* in the Aizoaceae. This inclusion is based on a report in Watson and Dallwitz (1998) that some *Lithops* have Kranz anatomy. *Evolvulus* in the Convolvulaceae (numerous lists such as Downton, 1975) is excluded because carbon isotope ratios indicate it is C_3 (Winter, 1981). *Biernertia* in the Chenopodiaceae is retained because of its of C_4-like $\delta^{13}C$ ratios (Winter, 1981; Akhani *et al.*, 1997), although a question mark is placed beside its name because it reportedly has a C_3 pattern of metabolite production (Glagoleva *et al.*, 1992).

B. Monocots

Together, the Poaceae and Cyperaceae account for 79% of all C_4 species. C_4 photosynthesis occurs in 28 genera in the Cyperaceae, and about 372 genera in the Poaceae. (Many of these genera have single species and are likely to be revised as phylogenetic details are resolved). In the Cyperaceae, 27% of the estimated 5000 species are C_4, compared to 46% of the estimated 10,000 grass species (Table I). Major sedge genera are *Cyperus* (with 500 species in the C_4 subgenus *Cyperus*), *Fimbristylis* with 200 species, and *Bulbostylis* and *Pycreus* with about 100 species each (Table II). Four genera (*Abildgardia, Cyperus, Eleocharis*, and *Rhynchospora*) also contain C_3 species. Except for *Eleocharis*, C_3 and C_4 lines of these genera segregate into distinct subgenera, so precise estimates of C_4 species number are feasible. Fifteen C_4 species have been identified in *Eleocharis*, but we estimate that 50 of the 200 *Eleocharis* species worldwide may be C_4. This is based on the ratio of C_4 to C_3 species in *Eleocharis* from Australia and eastern Asia (Bruhl *et al.*,

1987; Ueno et al., 1989). All C_4 sedges are NADP-ME except for *Eleocharis*, which is NAD-ME.

In the grasses, C_4 photosynthesis occurs in three of the six subfamilies, and 24 of the estimated 58 tribes (Table III). The Arundinoideae has both NADP-ME and NAD-ME subtypes, though in separate tribes. The Chloridoideae is the only group that is completely C_4, with the one exception of *Eragrostis walteri*, a C_3 species thought to have reverted from C_4 (Ellis, 1984). Most C_4 Chloroideae species are either NAD-ME or PCK, with the Orcuttieae tribe containing the known exceptions. The Panicoideae are biochemically the most diverse subfamily, with C_3, C_3–C_4 intermediate, and C_4 species of all three subtypes. NADP-ME is present in all five C_4 tribes in this subfamily, with only the Paniceae having NAD-ME and PCK taxa. The three tribes that are almost exclusively C_4 (Arundinelleae, Andropogoneae, and Maydeae; all NADP-ME), contain many of the major C_4 crops, such as maize, sorghum, and sugarcane (Brown, Chapter 14).

In contrast to the low latitude and warm growing conditions associated with the C_4 tribes, the three exclusively C_3 subfamilies stand out for their association with cool growing season (Pooideae and Stipoideae) and shaded or flooded habitats of low latitude (Bambusoideae). In general, the C_3 tribes are not associated with heat and aridity. This contrast in habitat between C_3 and C_4 tribes likely reflects evolutionary specialization driven in large part by photosynthetic pathway.

Table IV lists all the grass genera known to be C_4, and where possible, the associated subtype. Seven genera have both C_3 and C_4 members, these being *Alloteropsis, Chaetobromus, Eragrostis, Neurachne, Panicum, Steinchisma,* and *Streptostachys*. *Neurachne* and *Steinchisma* are predominantly C_3 or C_3–C_4 intermediate, whereas *Chaetobromus* and *Streptostachys* contain one and two C_3 species, respectively. The two C_3 representatives in *Alloteropsis* and *Eragrostis* are particularly interesting because one (*Alloteropsis semialata* ssp. *ecklonianua*) is a subspecies of a predominantly C_4 species, whereas the other (*Eragrostis walterii*) is a reversion from a C_4 species (Hattersley and Watson, 1992). *Panicum*, with about 600 species, has hundreds of C_3 and C_4 species, and at a minimum, dozens of representative of each C_4 subtype. It is the only major genus to have all three C_4 subtypes. Photosynthetic types and subtypes neatly segregate along subgenus lines in *Panicum;* unfortunately, the systematics are not clear enough to estimate worldwide species numbers for each group.

The grass list in Table IV is reasonably complete, so that it can easily be used to estimate photosynthetic pathway of any grass flora in the world (as is done by Sage *et al.* in Chapter 10 for island floras). Only in *Panicum* should there be many questions about photosynthetic pathway. This problem can be minimized by consulting Brown (1977), who lists many C_3 and C_4 *Panicum* species, and Zuloaga (1987) who lists New World *Panicum* species by subge-

Table III Distribution of C_4 Photosynthesis in the Grass Subfamilies and Tribes[a]

Subfamily[b]	Notes and tribes
Stipoideae Tribes (5)	20 genera, all C_3, mainly temperate, often in cold, arid regions
Stipeae	Common on north temperate steppes and cold deserts
Four minor tribes	Ampelodesmeae, Anisopogoneae, Lygeae, Nardeae
Arundinoideae Tribes (9)	Cosmopolitan; 5 C_4 genera of ~60 total genera (8% C_4)
Aristideae	2 C_4 genera (*Aristida*, *Stipagrostis*, both NAD-ME), one C_3 genus (*Sartidia*)
Danthonieae	All C_3 but *Centropodia* (NAD-ME)
Eriachneae	2 C_4 (NADP-ME), *Eriachne* and *Pheidochloa*
6 minor C_3 tribes	Amphipogoneae, Arundineae, Cyperochloeae, Micrairieae, Spartochloeae, Steyermarkochloeae
Bambusoideae Tribes (15)	125 genera, all C_3; low latitudes, forests, shade and wet places
Bambuseae	Forest grasses with woody stems, shade adapted
Centotheceae	Tropical forest shade grasses
Olyreae	Semiwoody forest grasses, shade
Oryzeae	Marshes, mainly low latitudes
11 minor C_3 tribes	Anomochloeae, Brachyelytreae, Diarrheneae, Ehrharteae, Guaduelleae, Phaenospermateae, Phareae, Phyllorhachideae, Puelieae, Streptochaeteae, Streptogyneae

Chloridoideae Tribes (16)		~165 genera, all C_4 except *Eragrostis walteri*, a revertant to C_3 (Ellis, 1984); mainly low latitudes and dry climates Main assemblage (13 tribes)—Aeluropodeae, Chlorideae, Cynodonteae, Eragrosteae, Jouveae, Lappagineae, Leptureae, Perotideae, Pommereulleae, Spartineae, Sporoboleae, Trageae, Unioleae. All NAD-ME and/or PCK
	Orcuttieae	NADP-ME (Keeley, 1998), and NAD-ME
	Pappophoreae	NAD-ME and PCK
	Triodieae	NAD-ME
Panicoideae		C_3, C_3–C_4 intermediate, and all three C_4 subtypes. Tropical and warm temperate regions; ~200 C_4 genera of ~250 total (80% C_4)
Tribes (6)	Andropogoneae	C_4, NADP-ME
	Arundinelleae	C_4 (NADP-ME) and rare C_3
	Isachneae	C_3
	Maydeae	C_4, NADP-ME
	Neurachneae	C_3,C_3–C_4 intermediate and C_4 (NADP-ME)
	Paniceae	C_3, intermediate and C_4 (all 3 biochemical subtypes)
Pooideae Tribes (7)		~180 genera, all C_3; cool temperate to polar regions, tropical mountains Triticeae, Brachypodieae, Bromeae, Aveneae, Poeae, Seslerieae, Meliceae

[a] Compiled from L. Watson and M. J. Dallwitz (1998). "Grass Genera of the World: Descriptions, Illustrations, Identification, and Information Retrieval; Including Synonyms, Morphology, Anatomy, Physiology, Cytology, Classification, Pathogens, World and Local Distribution, and References." URL http://biodiversity.uno.edu/delta/

[b] Number of tribes in a subfamily are given in parentheses.

Table IV Known Distribution of C_4 Photosynthesis in the Genera of the Poaceae[a]

Genus	C_4 species number[b]	C_4 biochemical subtype and notes[c]
Acamptoclados Nash	1	NAD-ME or PCK (**A**)
Achlaena Grieseb.	1	NADP-ME (**A**)
Acrachne Wright & Arn. ex Chiov.	3	NAD-ME (**A**)
Aegopogon Humb. & Bonp. ex Willd.	3	NAD-ME or PCK (**A**)
Aeluropus Trin.	5	NAD-ME (**A**)
Afrotrichloris Chiov.	2	NAD-ME or PCK (**A**)
Agenium Nees	4	NADP-ME (**A**)
Alexfloydia B. K. Simon	1	NADP-ME (**A**)
Allolepis Soder. & Deck.	1	NAD-ME or PCK (**A**)
Alloteropsis Presl	7	NADP-ME & PCK (**A, B**); also a C_3 subspecies
Anadelphia Hackel	13	NADP-ME (**A**)
Andropogon L.	100	NADP-ME (**A, B**)
Andropterum Stapf	1	NADP-ME (**A**)
Anthaenantiopsis Pilger	4	NADP-ME (**A**)
Anthenantia P. Beauv.	2	NADP-ME (**A**)
Anthephora Schreber	12	NADP-ME (**A**)
Apluda L.	1	NADP-ME (**A**)
Apochiton C. E. Hubb.	1	NAD-ME or PCK (**A**)
Apocopis Nees	15	NADP-ME (**A**)
Aristida L.	290	NADP-ME (**A, B**)
Arthragrostis Lazarides	3	NAD-ME or PCK (**A**)
Arthraxon P. Beauv.	7	NADP-ME (**A**)
Arthropogon Nees	5	NADP-ME, NAD-ME or PCK (**A**)
Arundinella Raddi	55	NADP-ME (**A, B**)
Asthenochloa Buese	1	NADP-ME (**Tax**)
Astrebla F. Muell.	4	NAD-ME (**A, B**)
Austrochloris Lazarides	1	NAD-ME (**A**)
Axonopus P. Beauv.	114	NADP-ME (**A, B**)
Baptorhachis Clayt. & Renv.	1	NADP-ME (**A**)
Bealia Scribner	1	NAD-ME (**A**)
Beckeropsis Figari & de Not.	6	NADP-ME (**A**)
Bewsia Goossens	1	NAD-ME (**A**)
Bhidea Staph. ex Bor	3	NADP-ME (**A**)
Blepharidachne Hackel	3	NAD-ME (**A**)
Blepharoneuron Nash	2	NAD-ME (**A**)
Bothriochloa Kuntze	35	NADP-ME (**A**)
Bouteloua Lag.	40	NAD-ME (**A, B**) & PCK (**A, B**)
Brachiaria (Trin.) Grieseb.	3	PCK (**A, B**)
Brachyachne (Benth.) Stapf	10	NAD-ME & PCK (**A**)
Brachychloa Phillips	2	NAD-ME (**A**)
Buchloë Engelm.	1	NAD-ME (**A, B**)
Buchlomimus Reed., Reed. & Rzed.	1	NAD-ME or PCK (**A**)
Anthaenantiopsis Pilger	4	NADP-ME (**A**)
Calamovilfa Hackel	5	NAD-ME or PCK (**A**)
Camusiella Bosser	2	NADP-ME (**A**)

Table IV (Continued)

Genus	C₄ species number[b]	C₄ biochemical subtype and notes[c]
Capillipedium Stapf	14	NADP-ME (A)
Catalepis Stapf & Stent	1	PCK (A)
Cathestechum J. Presl	6	NAD-ME or PCK (A)
Cenchrus L.	22	NADP-ME (A, B)
Centrochloa Swallen	1	NADP-ME (A)
Centropodia Reichenb.	4	NAD-ME (A)
Chaboissaea Fourn,	4	NAD-ME or PCK (A)
Chaetium Nees	3	NADP-ME, & NAD-ME or PCK (A)
Chaetobromus Nees	1(2)	C₄ by δ^{13}C (*C. involucratus*), C₃ (*C. dregeanus*)
Chaetopoa C. E. Hubb	1	NADP-ME (A)
Chaetostichium (Hochst.) Hubb.	1	NAD-ME or PCK (A)
Chamaeraphis R. Br.	1	NADP-ME (A)
Chasmopodium Stapf.	2	NADP-ME (Tax)
Chionachne R. Br.	7	NADP-ME (A)
Chloris O. Swarz	55	PCK (A, B)
Chlorocalymma W. Clayton	1	NADP-ME (Tax)
Chrysochloa Swallen	5	NAD-ME or PCK (A)
Chrysopogon Trin.	25	NADP-ME (A)
Chumsriella Bor.	1	NADP-ME (Tax)
Cladoraphis Franch.	2	NAD-ME or PCK (A)
Clausospicula Lazarides	1	NADP-ME (A)
Cleistachne Benth.	1	NADP-ME (A)
Coelachyropsis Bor	1	NAD-ME or PCK (A)
Coelachyrum Hochst. & Nees	6	NAD-ME or PCK (A)
Coelorachis Brongn.	20	NADP-ME (A)
Coix L.	5	NADP-ME or PCK (A)
Cottea Kunth	1	NAD-ME or PCK (A)
Craspedorhachis Benth.	6	NAD-ME (A)
Crypsis Aiton	8	NAD-ME or PCK (A)
Ctenium Panzer	20	NAD-ME (A)
Cyclostachya J. & C. Reeder	1	NAD-ME or PCK (A)
Cymbopogon Spreng.	40	NADP-ME (A, B)
Cymbosetaria Schweick.	1	NADP-ME (A)
Cynodon Rich.	10	NAD-ME (A, B)
Cypholepis Chiov.	1	NAD-ME or PCK (A)
Dactyloctenium Willd.	13	PCK (A)
Daknopholis W. Clayton	1	NAD-ME or PCK (A)
Danthoniopsis Stapf	20	NADP-ME (A)
Dasyochloa Willd. ex Rydberg	1	NAD-ME (A)
Decaryella A. Camus	1	NAD-ME or PCK (A)
Desmostachya (Hook f.) Stapf.	6	NAD-ME or PCK (A)
Diandrochloa de Winter	7	PCK (A)
Diandrostachya Jacq.-Fel.	5	NADP-ME (A)
Dichanthium Willem.	16	NADP-ME (A)

(*continues*)

Table IV (Continued)

Genus	C₄ species number[b]	C₄ biochemical subtype and notes[c]
Diectomis Kunth	1	NADP-ME (A)
Digastrium (Hackel) A. Camus	2	NADP-ME (A)
Digitaria Haller	220	NADP-ME (A, B)
Digitariopsis C. E. Hubb.	2	NADP-ME (A)
Dignathia Stapf	5	NAD-ME or PCK (A)
Diheteropogon (Hack.) Stapf	5	NADP-ME (A)
Dilophotriche Jacq.-Fel.	3	NADP-ME (Tax)
Dimeria R. Br.	40	NADP-ME (A)
Dinebra Jacq.	3	PCK (A)
Diplachne P. Beauv.	18	NAD-ME (A)
Dissochondrus (Hillebr.) Kuntze	1	NADP-ME (A)
Distichlis Raf.	6	NAD-ME (A, B)
Drake-Brockmania Stapf	1	NAD-ME or PCK (A)
Dybowskia Stapf.	1	NADP-ME (A)
Eccoilopus Steud.	4	NADP-ME (Tax)
Eccoptocarpha Launert	1	NAD-ME or PCK (A)
Echinochloa P. Beauv.	35	NADP-ME (A, B)
Ectrosia R. Br.	12	NAD-ME or PCK (A)
Ectrosiopsis (Ohwi) Jansen	1	PCK (A)
Eleusine Gaertn.	9	NAD-ME (A, B)
Eliomurus Humb. & Bonpl.	15	NADP-ME (A)
Elymandra Stapf	4	NADP-ME (A)
Enneapogon Desv. ex P. Beauv.	30	NAD-ME (B)
Enteropogon Nees	11	NAD-ME (A)
Entoplocamia Stapf	1	NAD-ME (A)
Eragrostiella Bor	5	NAD-ME or PCK (A)
Eragrostis N. M. Wolf	350	NAD-ME, also 1 C₃ (*E. walteri*) (B)
Eremochloa Buese	9	NADP-ME (A)
Eremopogon Stapf	4	NADP-ME (A)
Eriachne R. Br.	40	NADP-ME (A, B)
Erianthus Michx.	28	NADP-ME (A)
Eriochloa Kunth	30	PCK (A, B)
Eriochrysis P. Beauv.	7	NADP-ME (A)
Erioneuron Nash	5	NAD-ME (A)
Euchlaena Schrad.	4	NADP-ME (A)
Euclasta Franch.	2	NADP-ME (A)
Eulalia Kunth	30	NADP-ME (A)
Eulaliopsis Honda	2	NADP-ME (A)
Eustachys Desf.	10	NAD-ME (A, B) & PCK (A)
Exotheca Anderss.	1	NADP-ME (A)
Farrago W. Clayton	1	NAD-ME or PCK (A)
Fingerhuthia Nees	2	NAD-ME (A)
Garnotia Brongn.	30	NADP-ME (A)
Germainia Bal. & Poitr.	8	NADP-ME (A)
Gilgiochloa Pilger	1	NADP-ME (A)
Glyphochloa W. Clayton	8	NADP-ME (A)

Table IV (Continued)

Genus	C₄ species number[b]	C₄ biochemical subtype and notes[c]
Gouinia Fourn.	9	NAD-ME or PCK (**A**)
Griffithsochloa G. J. Pierce	1	NAD-ME or PCK (**Tax**)
Gymnopogon P. Beauv.	15	NAD-ME (**A**)
Hackelochloa Kuntze	2	NADP-ME (**A**)
Halopyrum Stapf	1	NAD-ME or PCK (**A**)
Harpachne Hochst	2	NAD-ME or PCK (**A**)
Harpochloa Kunth	1	NAD-ME (**A**)
Hemarthria R. Br.	12	NADP-ME (**A**)
Hemisorghum C. E. Hubb.	2	NADP-ME (**Tax**)
Heterachne Benth.	3	NAD-ME or PCK (**A**)
Heterocarpha Stapf & Hubb.	1	NAD-ME or PCK (**Tax**)
Heteropholis C. E. Hubb	5	NADP-ME (**A**)
Heteropogon Pers.	7	NADP-ME (**A**)
Hilaria Kunth	9	NAD-ME or PCK (**A, B**)
Homozeugos Stapf	5	NADP-ME (**A**)
Hubbardochloa Auquier	1	NAD-ME or PCK (**A**)
Hygrochloa Lazarides	2	NADP-ME (**A**)
Hyparrhenia Anderss.	55	NADP-ME (**A, B**)
Hyperthelia W. Clayton	6	NADP-ME (**A**)
Hypogynium Nees	2	NADP-ME (**A**)
Imperata Cyr.	8	NADP-ME (**A**)
Indopoa Bor	1	NAD-ME or PCK (**A**)
Isalus J. Phipps	3	NADP-ME (**A**)
Ischaemum L.	60	NADP-ME (**A**)
Ischnochloa J. D. Hook	1	NADP-ME (**Tax**)
Ischnurus Balf.	1	NAD-ME or PCK (**A**)
Iseilema Anderss.	20	NADP-ME (**A**)
Ixophorus Schlechtd.	3	NADP-ME (**A**)
Jardinea Steud.	3	NADP-ME (**A**)
Jouvea Fourn.	2	NAD-ME or PCK (**A**)
Kampochloa W. Clayton	1	NAD-ME or PCK (**A**)
Kaokochloa de Winter	1	PCK (**A**)
Kengia Packer	10	NAD-ME or PCK (**A**)
Kerriochloa C. E. Hubb.	1	NADP-ME (**Tax**)
Lasiorachis (Hack.) Stapf	1	NADP-ME (**A**)
Lasiurus Boiss.	3	NADP-ME (**A**)
Lepargochloa Launert	1	NADP-ME (**Tax**)
Leptocarydion Stapf	1	NAD-ME (**A**)
Leptochloa P. Beauv.	27	NAD-ME (**B**) & PCK (**B**)
Leptochloöpsis Yates	2	NAD-ME & PCK (**A**)
Leptocoryphium Nees	1	Subtype unknown
Leptoloma Chase	10	NADP-ME (**A**)
Leptosaccharum (Hack.) A. Camus	1	NADP-ME (**A**)
Leptothrium Kunth	2	NAD-ME or PCK (**A**)
Lepturella Stapf	2	NAD-ME or PCK (**A**)
Lepturidium Hitchc. & Ekman	1	NAD-ME or PCK (**A**)

(*continues*)

Table IV (Continued)

Genus	C$_4$ species number[b]	C$_4$ biochemical subtype and notes[c]
Lepturopetium Morat	2	NAD-ME or PCK (A)
Lepturus R. Br.	8	NAD-ME or PCK (A)
Leucophrys Rendle	1	PCK (A)
Lintonia Stapf	2	PCK (A)
Lophacme Staph	2	NAD-ME (A)
Lopholepis Decne.	1	NAD-ME or PCK (A)
Lophopogon Hackel	2	NADP-ME (**Tax**)
Loudetia Hochst.	26	NADP-ME (A)
Loudetiopsis Conert	11	NADP-ME and possibly PCK (A)
Louisiella Hubb. & Leonard	1	NAD-ME or PCK (A)
Loxodera Launert	3	NADP-ME (A)
Lycurus Kunth	3	NAD-ME (A)
Manisuris L.	1	NADP-ME (A)
Megaloprotachne C. E. Hubb.	1	NADP-ME (A)
Melanocenchris Nees	3	NAD-ME or PCK (A)
Melinis P. Beauv.	12	PCK (**A, B**)
Mesosetum Steud.	35	NADP-ME (A)
Microchloa R. Br.	4	NAD-ME & PCK (A)
Microstegium Nees	15	NADP-ME (**A, B**)
Mildbraediochloa Butzin	1	NAD-ME or PCK (A)
Miscanthidium Stapf	7	NADP-ME (A)
Miscanthus Anderss.	20	NADP-ME (A)
Mnesithea Kunth	5	NADP-ME (A)
Monanthochloe Engelm.	3	NAD-ME (A)
Monelytrum Hackel	2	NAD-ME (A)
Monium Stapf	7	NADP-ME (A)
Monocymbium Stapf	4	NADP-ME (A)
Monodia S. W. L. Jacobs	1	NAD-ME (A)
Mosdenia Stent	1	NAD-ME (A)
Muhlenbergia Schreber	160	NAD-ME (A) & PCK (**A, B**)
Munroa J. Torr.	5	NAD-ME or PCK (A)
Myriostachya J. D. Hook.	1	NAD-ME or PCK (A)
Narenga Bor	2	NADP-ME (A)[c]
Neeragrostis Bush	1	NAD-ME or PCK (A)
Neesiochloa Pilger	1	NAD-ME or PCK (A)
Neobouteloua Gould	1	NAD-ME or PCK (A)
Neostapfia Davy	1	NADP-ME & NAD-ME (**B**) Keeley 1998 for subtype
Neostapfiella A. Camus	3	NAD-ME & PCK (A)
Neurachne R. Br.	1(6)	NADP-ME (**A, B**) also C$_3$, C$_3$–C$_4$
Neyraudia Hook. f.	2	NAD-ME or PCK (A)
Ochthochloa Edgwe.	1	NAD-ME or PCK (A)
Odontelytrum Hackel	1	PCK (A)
Odyssea Stapf	2	NAD-ME (A)
Ophiochloa Filg., Dav. & Zuloaga	1	NADP-ME (A)
Ophiuros Gaertn. f.	4	NADP-ME (A)

Table IV (Continued)

Genus	C$_4$ species number[b]	C$_4$ biochemical subtype and notes[c]
Opizia J. & C. Presl	1	NAD-ME or PCK (A)
Orcuttia Vasey	5	NADP-ME (B; Keeley 1998); NAD-ME (A)[d]
Orinus A. Hitchc.	2	NAD-ME or PCK (A)
Oropetium Trin.	4	NAD-ME (A)
Oryzidium Hubb. & Schw.	1	NAD-ME (A)
Oxychloris Lazarides	1	NAD-ME (A, B)
Oxyrhachis Pilger	1	NADP-ME (A)
Panicum L.	~260(370)	NADP-ME, NAD-ME, PCK; also C$_3$, C$_3$–C$_4$ (B)
P. subgenus *Agrostoides*		C$_4$, NADP-ME
P. subgenus *Magathyrsus*		C$_4$, PCK
P. subgenus *Panicum*		C$_4$, NAD-ME
P. subgenus *Steinchisma*		C$_3$ & C$_3$–C$_4$ intermediates (= *Steinchisma*)
Pappophorum Schreber	8	NAD-ME (A)
Parahyparrhenia A. Camus	5	NADP-ME (A)
Paraneurachne S. T. Blake	1	NADP-ME (B)
Paratheria Griseb.	2	NADP-ME (A)
Parectenium P. Beauv. corr. Stapf	1	NADP-ME (A)
Paspalidium Stapf	27	NADP-ME (A)
Paspalum L.	320	NADP-ME (A, B)
Pennisetum Rich.	80	NADP-ME (A, B)
Pentarrhaphis Kunth	3	NAD-ME or PCK (A)
Pereilema J. & C. Presl	3	NAD-ME or PCK (A)
Perotis Aiton	10	NAD-ME (A)
Phacelurus Griseb.	7	NADP-ME (A)
Pheidochloa S. T. Blake	2	NADP-ME (A, B)
Plagiosetum Benth.	1	NADP-ME (A)
Planichloa B. Simon	1	NAD-ME or PCK (A)
Plectrachne Henrard	17	NAD-ME (A)
Pleiadelphia Stapf	2	NADP-ME (A)
Pobeguinea Jacq.-Fel.	4	NADP-ME (A)
Pogonachne Bor	1	NADP-ME (Tax)
Pogonarthria Stapf	4	NAD-ME (A)
Pogonatherum P. Beauv.	3	NADP-ME (A)
Pogoneura Napper	1	NAD-ME or PCK (A)
Pogonochloa C. E. Hubb.	1	NAD-ME or PCK (A)
Polevansia de Winter	1	NAD-ME (A)
Polliniopsis Hayata	1	NADP-ME (Tax)
Polytoca R. Br.	2	NADP-ME (A)
Polytrias Hackel	2	NADP-ME (A)
Pommereulla L. f.	1	NAD-ME or PCK (A)
Pringleochloa Scribner	1	NAD-ME or PCK (A)
Psammagrostis Gard. & Hubb.	1	NAD-ME (A)
Pseudanthistiria (Hackel) Hook. f.	4	NADP-ME (A)

(*continues*)

Table IV (Continued)

Genus	C$_4$ species number[b]	C$_4$ biochemical subtype and notes[c]
Pseudochaetochloa A. Hitchc.	1	NADP-ME (**A**)
Pseudodichanthium Bor	1	NADP-ME (**Tax**)
Pseudopogonatherum A. Camus	2	NADP-ME (**A**)
Pseudoraphis Griff.	6	NADP-ME (**A**)
Pseudosorghum A. Camus	2	NADP-ME (**Tax**)
Pseudovossia A. Camus	1	NADP-ME (**Tax**)
Pseudozoysia Chiov.	1	NAD-ME or PCK (**A**)
Psilolemma Phillips	1	NAD-ME or PCK (**A**)
Pterochloris A. Camus	1	NAD-ME or PCK (**A**)
Ratzeburgia Kunth	1	NADP-ME (**Tax**)
Redfieldia Vasey	1	NAD-ME (**A**)
Reedrochloa Soder. & Decker	1	NAD-ME or PCK (**A**)
Reimarochloa A. Hitchc	4	NADP-ME (**A**)
Rendlia Chiov.	1	NAD-ME (**A**)
Reynaudia Kunth	1	NADP-ME (**A**)
Rhynchelytrum Nees	14	PCK (**A, B**)
Rhytachne Desv.	12	NADP-ME (**A**)
Richardsiella Elf. & O'Byrne	1	PCK (**A**)
Robynsiochloa Jacq.-Fel.	1	NADP-ME (**A**)
Rottboellia L. f.	4	NADP-ME (**A**)
Saccharum L.	5	NADP-ME (**A, B**)
Saugetia A. Hitchc. & Chase	2	NAD-ME or PCK (**A**)
Schaffnerella Nash	1	NAD-ME or PCK (**A**)
Schedonnardus Steud.	1	NAD-ME (**A**)
Schenckochloa J. Ortiz	1	NAD-ME or PCK (**A**)
Schizachyrium Nees	60	NADP-ME (**A**)
Schmidtia Steud.	2	PCK (**A**)
Schoenefeldia Kunth	2	NAD-ME or PCK (**A**)
Sclerachne R. Br.	1	NADP-ME (**A**)
Sclerodactylon Stapf	1	NAD-ME (**A**)
Scleropogon Phil.	1	NAD-ME or PCK (**A**)
Sclerostachya A. Camus	3	NADP-ME (**Tax**)
Scutachne Hitchc. & Chase	2	NAD-ME or PCK (**A**)
Sehima Forssk.	5	NADP-ME (**A**)
Setaria P. Beauv.	110	NADP-ME (**A, B**)
Setariopsis Scribner ex Millsp.	2	NADP-ME (**A**)
Silentvalleya Nair, Sreek, Vag. & Bharg.	1	NAD-ME or PCK (**A**)
Snowdenia C. E. Hubb.	4	NADP-ME (**A**)
Soderstromia Morton	1	NAD-ME or PCK (**A**)
Sohnsia Airy Shaw	1	NAD-ME (**A**)
Sorghastrum Nash	20	NADP-ME (**A, B**)
Sorghum Moench	30	NADP-ME (**A, B**)
Spartina Schreber	16	PCK (**A, B**)
Spathia Ewart	1	NADP-ME (**A**)
Spheneria Kuhlm	1	NADP-ME (**A**)
Spinifex L.	4	NADP-ME (**A, B**)

Table IV (Continued)

Genus	C₄ species number[b]	C₄ biochemical subtype and notes[c]
Spodiopogon Trin.	10	NADP-ME (A)
Sporobolus R. Br.	160	NAD-ME (A, B) & PCK (A, B)
Steinchisma Raf.	1(4)	NAD-ME or PCK, C₃, C₃–C₄; = *Panicum* (Zuloaga 1987)
Steirachne Ekman	2	NAD-ME or PCK (A)
Stenotaphrum Trin.	7	NADP-ME (A)
Stereochlaena Hackel	5	NADP-ME (A)
Stiburus Stapf	2	PCK (A)
Stipagrostis Nees	50	NAD-ME or PCK (A) (but *S. paridisea* not Kranz)
Streptolophus Hughes	1	NADP-ME (A)
Streptostachys Desv.	2(3)	C₄ (*S. macrantha, S. ramosa*) C₃ (*S. asperifolia*)
Swallenia Soder. & Decker	1	NAD-ME or PCK (A)
Symplectrodia Lazarides	2	NAD-ME (A)
Tarigidia Stent	1	NADP-ME (A)
Tatianyx Zuloaga & Soder.	1	NAD-ME or PCK (A)
Tetrachaete Chiov.	1	NAD-ME or PCK (A)
Tetrachne Nees	1	NAD-ME (A)
Tetrapogon Desf.	6	NAD-ME (A)
Thaumastochloa C. E. Hubb.	7	NADP-ME (A)
Thelepogon Roth.	1	NADP-ME (A)
Thellungia Stapf	1	NAD-ME (A)
Themeda Forssk.	18	NADP-ME (A)
Thrasya Kunth	20	NADP-ME (A)
Thrasyopsis L. Parodi	2	NADP-ME (A)
Thuarea Pers.	2	PCK (A)
Thyrsia Stapf	4	NADP-ME (A)
Trachypogon Nees	13	NADP-ME (A)
Trachys Pers.	1	NADP-ME (A)
Tragus Haller	7	NAD-ME (A)
Tricholaena Schrad.	12	PCK (A)
Trichoneura Anderss.	7	NAD-ME (A)
Trichopteryx Nees	5	NADP-ME (A)
Tridens Roem. & Schult.	18	PCK (A, B)
Trilobachne Schenk ex Henrard	1	NADP-ME (Tax)
Triodia R. Br.	35	NAD-ME (A, B)
Triplasis P. Beauv.	2	NAD-ME (A)
Triplopogon Bor.	1	NADP-ME (A)
Tripogon Roem. & Schult.	30	NAD-ME (A)
Tripsacum L.	12	NADP-ME (A)
Triraphis R. Br.	7	NAD-ME (B)
Tristachya Nees	20	NADP-ME (A)
Tuctoria J. Reeder	3	NADP-ME (B) Keeley 1998 for subtype
Uniola L.	2	NAD-ME or PCK (A)

(*continues*)

Table IV (Continued)

Genus	C_4 species number[b]	C_4 biochemical subtype and notes[c]
Uranthoecium Stapf	1	NADP-ME (A)
Urelytrum Hackel	7	NADP-ME (A)
Urochloa P. Beauv.	120	PCK (A, B)
Urochondra C. E. Hubb.	1	NAD-ME or PCK (A)
Vaseyochloa A. Hitchc.	1	NAD-ME (A)
Vetiveria Bory	10	NADP-ME (A)
Vietnamochloa Veldkamp & Nowack	1	NAD-ME (A)
Viguierella A. Camus	1	NAD-ME & PCK (A)
Vossia Wall. & Griff.	1	NADP-ME (A)
Whiteochloa C. E. Hubb.	6	NADP-ME (A)
Willkommia Hackel	2	NAD-ME or PCK (A)
Xerochloa R. Br.	4	NADP-ME (A)
Yakirra Lazarides & Web.	7	NAD-ME (A)
Ystia P. Compere	1	NADP-ME (**Tax**)
Yvesia A. Camus	1	NAD-ME (A)
Zea L.	1	NADP-ME (A, B)
Zonotriche Phipps	3	NAD-ME or PCK (A)
Zoysia Willd.	10	PCK (A, B)
Zygochloa S. T. Blake	1	NADP-ME (A)
Total 372 genera	~4600 species	

[a] Updated from P. W. Hattersley and L. Watson (1992). Diversification of photosynthesis. *In* "Grass Evolution and Domestication." (G. P. Chapman, ed.), pp. 38–116. Cambridge University Press, New York; and L. Watson and M. J. Dallwitz (1998). "Grass Genera of the World: Descriptions, Illustrations, Identification, and Information Retrieval; Including Synonyms, Morphology, Anatomy, Physiology, Cytology, Classification, Pathogens, World and Local Distribution, and References." URL http://biodiversity.uno.edu/delta/

[b] In genera with C_3 and C_3–C_4 species, the total species number for the genus (in parentheses) follows the C_4 species number.

[c] Assignment to C_4 biochemical subtype is based either on anatomical descriptions (**A**), biochemical determinations (**B**), or taxonomic assignment to a tribe containing only C_4 members of that subtype (**Tax**). "NAD-ME or PCK" indicates members of the genera are one of the two subtypes, but anatomical data is insufficient to allow delineation to specific subtype. "NAD-ME & PCK" indicates genera with species of each subtype. "NADP-ME & NAD-ME" indicates genera with species expressing high activities of both of these enzymes, indicating a mixed subtype.

[d] Sectioned material from *Orcuttia tenuis* shows centripedal chloroplast arrangement, indicative of NAD-ME subtype (R. Sage and T. Sage, unpublished, 1998).

[e] Anatomical observations for *Narenga* from V. S. Rama Das and S. K. Vats (1993). A Himalayan monsoonal location exhibiting unusually high preponderance of C_4 grasses. *Photosynthetica* **28**, 91–97.

nus and photosynthetic pathway. Habitat descriptions also help, because tropical C_3 *Panicum* spp. tend to be found in high elevation or moist, shaded locations, whereas C_4 *Panicum* spp. are mainly low elevation in open habitats (Zuloaga, 1987).

C. C_3–C_4 Intermediates

More than 30 species in 7 families are listed as intermediate between fully expressed C_3 and C_4 physiology (Table V). Although photosynthetic pathway intermediates were initially recognized as strictly intermediate in form and physiology, that opinion has changed as a wider variety of patterns have become recognized. Here, we identify three types of intermediates:

1. C_3–C_4 and C_4-like types exhibit intermediate patterns of a. leaf anatomy, b. initial carbon flow into C_4 acids, and/or c. patterns of enzyme expression between fully developed C_3 and C_4 types. Most intermediates belong to this group, and some may be in the process of evolving from one photosynthetic type to the other (Monson, Chapter 11, this volume; 1989).

2. C_3/C_4 types can exhibit complete C_3 and C_4 metabolism on the same plant, although in spatially or temporally segregated sets of leaves. *Eleocharis vivipara*, for example, is C_4 in the terrestrial state, but produces C_3 tissues when submerged (Ueno *et al.*, 1988; Ueno, 1996). Other *Eleocharis* species also exhibit mixtures of C_3 and C_4 modes in different leaves (for example *E. spegazzinii, E. quinquangularis, E. reverchonii*), indicating the patterns expressed by *E. vivipara* might not be isolated to one species (Ueno *et al.*, 1989). These possibilities remain to be confirmed at the biochemical and gas exchange level.

3. $C_3 \times C_4$ species have normal C_3 and C_4 modes in different subspecies, but these can form hybrids having intermediate patterns of C_4 enzyme expression (Hattersley and Watson, 1992) This pattern has been identified in one African species, *Alloteropsis semialata* (Ellis, 1981; Hattersley and Watson, 1992).

Of the C_3–C_4 types of intermediate species, most have a C_4-like anatomy and C_4-like CO_2 compensation points, but they express C_3-like carbon isotope ratios. Most also lack C_4-like biochemistry and assimilate CO_2 through C_3 photosynthesis in mesophyll cells, with a unique recycling of photorespired CO_2 occurring in the bundle sheath cells (Monson, Chapter 11). Within the dicot genus *Flaveria*, however, expression of C_4-like biochemistry is apparent. Most of the C_3–C_4 intermediate species in *Flaveria* (for example, *F. anomala, F. floridana, F. linearis,* and *F. pubescens*) exhibit C_4-like biochemistry, but it is poorly integrated between the mesophyll and bundle-sheath cells. In these species, the majority of CO_2 assimilation occurs in mesophyll cells (Monson *et al.*, 1986). In *F. brownii* and *F. palmerii*, however, the C_4-like biochemistry results in 70–95% of assimilated CO_2 passing through the C_4 cycle, with concomitant enhancement of CO_2 level in the bundle sheath cells and a significant reduction in carbon isotope discrimination (Moore *et al.*, 1987; Monson *et al.*, 1988; Monson, 1989).

Aquatic species in the Hydrocharitaceae present an interesting case. Keeley (1990) notes that *Vallisneria spiralis* incorporates 20% of it initial

Table V Families and Genera Known to Contain C_3–C_4 Intermediate and C_4-like Photosynthesis, with the Methods Used to Evaluate the Degree of Intermediacy

Family	Species	Intermediate type and trait[a]	Refs.
Amaranthaceae	*Alternanthera ficoides* L.R.Br. R.	C_3–C_4: K,Γ,P	Rajendrudu *et al.*, 1986
	A. tenella Colla.	C_3–C_4: K,Γ,P	Rajendrudu *et al.*, 1986; Devi and Raghavendra, 1993
Asteraceae	*Flaveria angustifolia* (Cav.) Pers.	C_3–C_4: K,O,Γ	Monson and Moore, 1989; Ku *et al.*, 1991
	F. anomala B. L. Robinson	C_3–C_4: K,O,Γ,P	Monson and Moore, 1989; Ku *et al.*, 1991
	F. linearis Lag.	C_3–C_4: K,O,Γ,P	Monson and Moore, 1989; Ku *et al.*, 1991
	F. oppositifolia (DC.) Rydb.	C_3–C_4: K,O,Γ	Ku *et al.*, 1991
	F. palmeri J. R. Johnston	C_4-like: K,O,Γ,P,Δ	Ku *et al.*, 1991
	F. pubescens Rydb.	C_3–C_4: K,O,Γ,P	Monson and Moore, 1989; Ku *et al.*, 1991
	F. ramosissima Klatt.	C_3–C_4: K,O,Γ,P	Monson and Moore, 1989; Ku *et al.*, 1991
	F. sonorensis A. M. Powell	C_3–C_4: K,O,Γ	Ku *et al.*, 1991
	F. vaginata Robinson & Greenm.	C_4-like: K,Γ,P	Monson and Moore, 1989; Ku *et al.*, 1991
	Parthenium hysterophorus L.	C_3–C_4: K,O,Γ,Δ	Hedge and Patil, 1981; Moore *et al.* 1987
Brassicaceae	*Diplotaxis tenuifolia* (L.) DC.	C_3–C_4: K,Γ	Apel *et al.*, 1980; Apel *et al.*, 1997
	Moricandia arvensis (L.) DC.	C_3–C_4: P,O,K	Apel, Ticha, and Peisker, 1978; Apel *et al.*, 1997
	M. nitens (Viv.) Durd. & Barr.	C_3–C_4: Γ	Apel *et al.*, 1997
	M. sinaica Pomel.	C_3–C_4: K,Γ	Apel and Ohle, 1979; Apel *et al.*, 1997
	M. spinosa Boiss.	C_3–C_4: K,Γ	Apel, 1980; Apel *et al.*, 1997
	M. suffruticosa (Desf.) Coss. & Durr.	C_3–C_4: Γ	Apel *et al.*, 1997

Family	Species	Traits	References
Cyperaceae	Eleocharis. acicularis R. Br.	C_3–C_4: C,Δ	Keeley, 1990; Keeley and Sundquist, 1990
	E. vivipara Link	C_3/C_4: Δ,K	Ueno et al., 1988, 1996
	Other Eleocharis spp.	C_3–C_4: K	See text
Molluginaceae	Mollugo nudicaulis Lam.	C_3–C_4: K,Γ,O,P	Kennedy et al., 1980
	M. verticillata L.	C_3–C_4: K,Γ,O,P	Kennedy et al., 1980
Hydrocharitaceae	Vallisneria spiralis	C_3–C_4, C	Keeley, 1990
Poaceae	Alloteropsis semialata (R.Br.) Hitchc.	$C_3 \times C_4$: Δ,K	Ellis, 1974; Hattersley and Watson, 1992
	Neurachne minor S. T. Blake	C_3–C_4: K,O,Γ,P	Hattersley and Stone, 1986; Hattersley et al., 1986
	Panicum subgenus steinchisma	Entire subgenus reported as C_3–C_4	Zuloaga, 1987 (= Steinchisma, Watson and Dallwitz, 1998)
	P. cupreum	Requires confirmation	
	P. decipiens Nees & Trin.	C_3–C_4: K,Γ	Morgan and Brown, 1979
	P. exiguiflorum	Requires confirmation	
	P. milioides Nees ex Trin.	C_3–C_4: K,O,P	Morgan and Brown, 1979
	P. schenckii Hack.	C_3–C_4: K,O,Γ,P	Morgan and Brown, 1979
	P. stenophyllum	Requires confirmation	

[a] Methods abbreviations: Δ = stable carbon isotope ratio; Γ = low CO_2 compensation point; C, C_4 metabolite production; K = Kranz anatomy; O = oxygen inhibition of photosynthesis; P = PEP carboxylase activity. Species listed exhibit traits that are either 1) intermediate between normal C_3 and C_4 physiology (indicated as C_3–C_4, or C_4-like when close to full C_4 expression), 2) simultaneously C_3 and C_4 tissues in different regions of the plant (C_3/C_4), or 3) produce hybrids with intermediate characteristics between C_3 and C_4 subspecies ($C_3 \times C_4$).

photosynthetic fixation products into C_4 acids, indicating it is intermediate between C_3 and C_4 modes. *Hydrilla verticillata*, by contrast, incoporates more than 70% of its initial fixation products into C_4 acids. This aquatic monocot operates in both a C_3 and C_4 mode, with high temperature or low CO_2 inducing the C_4 from the C_3 form (Reiskind *et al.*, 1997). When fully induced, a full C_4 cycle operates within individual cells, apparently shuttling CO_2 between cytoplasm and cytosol. Because the C_4 cycle effectively concentrates CO_2 and suppresses photorespiration, we have included it on the list of C_4 taxa (Table II), rather than the intermediate list. Although C_4, *H. verticillata* represents a substantial variation from the pattern of terrestrial C_4 photosynthesis, where both biochemistry and anatomy are modifed to allow for CO_2 concentration.

IV. Summary

In this chapter, we present comprehensive lists of known C_4 taxa, which allow a wide range of scientists to easily identify photosynthetic pathways in terrestrial plants. From the lists, we estimate there are 7600 known C_4 species in 487 genera of 18 angiosperm families. Most are grasses (4600 species), followed by sedges (1330 species), whereas all dicot groups account for about 1600 C_4 species. These estimates are rough, and will change as more C_4 species are found and systematic revisions occur in response to phylogenetic data and better access to specimens. We predict the C_4 dicot numbers will eventually approach 2000 species in more than 100 genera, and a final C_4 tally of between 8000 to 10,000 species in 500 genera is realistic. Grass and sedge estimates should be less dynamic as these groups have been examined in much greater detail than dicot groups. In the final analysis, about 60% of all C_4 plants will be grasses, sedges will account for about 18%, and dicots between 20% to 23%. Of the dicot C_4 species, the Chenopodiaceae account for a more than one-third, and the Amaranthaceae, 16%, such that half of all C_4 dicots occur in these two families. When considering that 93% of all C_4 plants fall into just five plant families (Poaceae, Cyperaceae, Chenopodiaceae, Amaranthaceae, and Euphorbiaceae), and that C_3 members of these five families are commonly found in hot, arid, or saline habitats, it appears that adaptation to these harsh conditions may have predisposed these groups to evolve the C_4 pathway once declining atmospheric CO_2 content favored plants with CO_2 concentrating systems.

Acknowledgments

We wish to thank Jim Ehleringer for comments and sharing information from his list of C_4 taxa prior to publication (Ehleringer *et al.*, 1997), Vladimir Pyankov for his helpful comments on Chenopodiaceae, Leslie Watson for comments and information on the DELTA database, and Elizabeth Kellogg and Jarmila Pittermann for valuable manuscript reviews. This work was prepared with financial assistance from National Science and Research Council (Canada) grant number 91-37100-6619 to R.F.S.

References

Akhani, H., Trimborn, P. and Ziegler, H. (1997). Photosynthetic pathways in *Chenopodiaceae* from Africa, Asia, and Europe with their ecological, phytogeographical, and taxonomical importance. *Plant Syst. Evol.* **206**, 187–221.

Apel, P. (1980). CO_2 compensation concentration and its O_2 dependence in *Moricandia spinosa* and *Moricandia moricandioides* (Cruciferae). *Biochemie und Physiologie der Pflanzen* **175**, 386–388.

Apel, P. and Ohle, H. (1979). CO_2 compensation point and leaf anatomy in species of the genus *Moricandia* (Cruciferae). *Biochemie und Physiologie der Pflanzen* **174**: 68–75.

Apel, P., Hillmer, S., Pfeffer, M. and Muhle, K. (1980). Carbon metabolism type of *Diplotaxis tenuifolia* (L.) DC. (Brassicaceae). *Photosynthetica* **32**, 237–243.

Apel, P., Horstmann, C. and Peffer, M. (1997). The *Moricandia* syndrome in species of the Brassicaceae evolutionary aspects. *Photosynthetica* **33**, 205–215.

Apel, P., Ticha, I. and Peisker, M. (1978). CO_2 compensation concentrations in leaves of *Moricandia arvensis* (L.) DC. at different insertion levels and O_2 concentrations. *Physiologie der Pflanzen* **179**, 631–634.

Bremer, K. (1994) "Asteraceae: Cladistics and Classification." Timber Press, Portland.

Brown, W. V. (1977). The Kranz syndrome and its subtypes in grass systematics. *Mem. Torrey Bot. Club.* **23**, 1–97.

Bruhl, J. J. (1995). Sedge genera of the world: Relationships and a new classification of the Cyperaceae. *Aust. Syst. Bot.* **8**, 125–305.

Bruhl, J. J., Stone, N. E. and Hattersley, P. W. (1987). C_4 acid decarboxylation enzymes and anatomy in sedges (Cyperaceae): First record of NAD-malic enzyme species. *Aust. J. Plant Physiol.* **14**, 719–728.

Bruhl, J. J., Watson, L., and Dallwitz, M. J. (1992). Genera of Cyperaceae: Interactive identification and information retrieval. *Taxon* **41**, 225–235.

Carolin, R. C., Jacobs, S. W. L. and Vesk, M. (1975). Leaf structure in Chenopodiaceae. *Bot. Jahrb. Syst.* **95**, 226–255.

Collins, R. P. and Jones, M. B. (1985). The influence of climatic factors on the distribution of C_4 species in Europe. *Vegetatio* **64**, 121–129.

Czerepanov, S. K. (1995). "Vascular Plants of Russia and Adjacent States (the Former USSR)." Cambridge University Press, Cambridge.

Devi, M. T. and Raghavendra, A. S. (1993). Partial reduction in activities of photorespiratory enzymes in C_3–C_4 intermediates of *Alternanthera* and *Parthenium*. *J. Exp. Bot.* **44**, 779–784.

Downton, W. J. S. (1975). The occurrence of C_4 photosynthesis among plants. *Photosynthetica* **9**, 96–195.

Edwards, G. E. and Walker, D. A. (1983). "C_3, C_4: Mechanism, and Cellular and Environmental Regulation, of Photosynthesis." Blackwell Scientific Publications, Oxford.

Ehleringer, J. R., Cerling, T. E. and Helliker, B. R. (1997). C_4 photosynthesis, atmospheric CO_2, and climate. *Oecologia* **112**, 285–299.

Ellis, R. P. (1974). The significance of the occurrence of both Kranz and non-Kranz leaf anatomy in the grass species *Alloteropsis semialata*. *S. Afr. J. Sci.* **70**, 169–173.

Ellis, R. P. (1984). *Eragrostis walteri*—a first record of non-Kranz leaf anatomy in the sub-family Chloridoideae (Poaceae). *S. Afr. J. Bot.* **3**, 380–386.

Glagoleva, T. A., Chulanovskaya, M. V. Pakhomova, M. V., Voznesenskaya, E. V. and Ganaileii, Y. V. (1992). Effect of salinity on the structure of assimlating organs and ^{14}C labelling patterns in C_3 and C_4 plants of Ararat Plain. *Photosynthetica* **26,** 363–369.

Gutierrez, M., Gracen, V. E. and Edwards, G. E. (1974). Biochemical and cytological relationships in C_4 plants. *Planta* **119,** 279–300.

Hartmann, H. E. K. (1993). Aizoaceae. In: "The Families and Genera of Vascular Plants. Vol. II. Flowering Plants—Dicotyledons—Magnoliid, Hamamelid and Caryophyllid Families." (Kubitzki, K., Rohwer, J. G. and V. Bittrich, eds.), pp. 37–69. Springer-Verlag, New York.

Hatch, M. D. (1987). C_4 photosynthesis: a unique blend of modified biochemistry, anatomy and ultrastructure. *Biochem. Biophys. Acta* **895,** 81–106.

Hattersley, P. W. (1987).Variations in photosynthetic pathways. *In* "Grass Systematics and Evolution" (T. R. Soderstrom, K. W. Hilu, C. S. Campbell and M. E. Barkworth, eds.), pp. 49–64. Smithsonian Institute Press, Washington DC.

Hattersley, P. W. and Stone, N. E. (1986). Photosynthetic enzyme activities in the C_3–C_4 intermediate *Neurachne minor* S. T. Blake (Poaceae). *Aust. J. Plant Physiol.* **4,** 523–539.

Hattersley, P. W. and Watson, L. (1992). Diversification of photosynthesis. *In* "Grass Evolution and Domestication." (G. P. Chapman, ed.), pp. 38–116. Cambridge University Press, New York.

Hattersley, P. W., Watson, L. and Osmond, C. B. (1977). *In situ* immunofluorescent labelling of ribulose-1,5-bisphosphate carboxylase in leaves of C_3 and C_4 plants. *Aust. J. Plant Physiol.* **4,** 523–539.

Hattersley, P. W., Wong, S.-C., Perry, S. and Roksandic, Z. (1986). Comparative ultrastructure and gas-exchange characteristics of the C_3–C_4 intermediate *Neurachne minor* S. T. Blake. *Plant, Cell and Environ.* **9,** 217–233.

Hedge, B. A. and Patil, T. M. (1981). *Parthenium hysterophorus* (L.), a C_3 plant with Kranz syndrome. *Photosynthetica* **15,** 1–4.

Hudson, G. S., Dengler, R. E., Hattersley, P. W. and Dengler, N. G. (1992). Cell-specific expression of Rubisco small subunit and Rubisco activase genes in C_3 and C_4 species of *Atriplex. Aust. J. Plant Physiol.* **19,** 89–96.

Keeley, J. E. (1990). Photosynthetic pathways in freshwater aquatic plants. *Trends Ecol. Evol.* **5,** 330–333.

Keeley, J. E. (1998). C_4 photosynthesis modification in the evolutionary transition from land to water in aquatic grasses. *Oecologia* **116,** 85–97.

Keeley, J. E. and Sandquist, D. R. (1990). Carbon: freshwater plants. *Plant, Cell and Environ.* **15,** 1022–1035.

Kennedy, R. A., Eastburn, J. L. and Jensen, K. G. (1980). C_3–C_4 photosynthesis in the genus *Mollugo*: Structure, physiology and evolution of intermediate characteristics. *Am. J. Bot.* **67,** 1207–1217.

Kew Genlist (1998). List of genera in Gramineae. Royal Botanic Gardens, Kew. URL http://www.rbgkew.org.uk/cgi-bin/web.dbs/.

Koch, K. E. and Kennedy, R. A. (1982). Crassulacean acid metabolism in the succulent C_4 dicot, *Portulaca oleracea* L. under natural environmental conditions. *Plant Physiol.* **69,** 757–761.

Ku, M. S. B., Wu, J., Dai, Z., Scott, R. A., Chu, C. and Edwards, G. E. (1991). Photosynthetic and photorespiratory characteristics of *Flaveria* species. *Plant Physiol.* **96,** 518–528.

Kubitzki, K. (1993). "The Families and Genera of Vascular Plants. Vol. II. Flowering Plants—Dicotyledons—Magnoliid, Hamamelid and Caryophyllid Families." (K. Kubitzki, J. G. Rohwer and V. Bittrich, eds.). Springer-Verlag, New York.

Kuhn, U. (1993). Chenopodium. *In* "The Families and Genera of Vascular Plants. Vol. II. Flowering Plants—Dicotyledons—Magnoliid, Hamamelid and Caryophyllid Families." (K. Kubitzki, J. G. Rohwer and V. Bittrich, eds.). pp. 253–281. Springer-Verlag, New York.

Larcher, W. (1995). "Physiological Plant Ecology." 3rd ed., Springer, New York.

Mabberley, D. J. (1987). "The Plant Book." Cambridge University Press. Cambridge.
Meinzer, F. C. (1978). Observaciones sobre la distribucion taxonomica y ecologica de la fatosintesis C_4 en la vegetacion del noroneste de centroamerica. *Rev. Biol. Trop.* **16,** 359–369.
Monson, R. K. (1989). On the evolutionary pathways resulting in C_4 photosynthesis and crassulacean acid metabolism (CAM). *Adv. Ecol. Res.* **9,** 57–110.
Monson, R. K. and Moore, B. D. (1989). On the significance of C_3–C_4 intermediate photosynthesis to the evolution of C_4 photosynthesis. *Plant, Cell Environ.* **12,** 689–699.
Monson, R. K. Moore, B. D., Ku, M. S. B., Edwards, G. E. (1986). Co-function of C_3 and C_4 photosynthetic pathways in C_3, C_4 and C_3–C_4 intermediate *Flaveria* species. *Planta* **168,** 493–502.
Monson, R. K., Terri, J. A., Ku, M. S. B., Gurevitch, J., Mets, L. J. and Dudley, S. (1988). Carbon-isotope discrimination by leaves of *Flaveria* species exhibiting different amounts of C_3- and C_4-cycle co-function. *Planta* **174,** 145–151.
Moore, B. D., Franchesci, V. R., Cheng, S.–H., Wu, J. and Ku, M. S. B. (1987). Photosynthetic characteristics of the C_3–C_4 intermediate *Parthenium hysterophorus*. *Plant Physiol.* **85,** 984–989.
Morgan, J. A., Brown, R. H. (1979). Photosynthesis in grass species differing in carbon dioxide fixation pathways. II. A search for species with intermediate gas exchange and anatomical characteristics. *Plant Physiol.* **64,** 257–262.
Osmond, C.B., Björkman, O. and Anderson, D. J. (1980). "Physiological processes in Plant Ecology—Toward a Synthesis with *Atriplex*." Springer-Verlag, Berlin.
Osmond, C. B., Winter, K. and Ziegler H. (1982). Functional significance of different pathways of CO_2 fixation in photosynthesis. *In:* "Encyclopedia of Plant Physiology, New Series, Physiological Plant Ecology, II." Vol. 12B. (O. L. Lange, P. S. Nobel, C. B. Osmond and H. Zeigler, eds.), pp. 480–547, Springer-Verlag, Berlin.
Pyankov, V. I. and Vakhrusheva, D. V. (1989). The pathways of primary CO_2 fixation in Chenopodiaceae C_4-plants of central Asian arid zone. *Soviet Plant Physiol.* **36,** 178–187.
Pyankov, V. I., Kuzmin, A. N., Demidov, E. D. and Maslov, A. I. (1992). Diversity of biochemical pathways of CO_2 fixation in plants of the families Poaceae and Chenopodiaceae from arid zone of central Asia. *Soviet Plant Physiol.* **39,** 411–420.
Pyankov, V. I., Voznesenskaya, E. V., Kondratschuk, A. V. and Black Jr., C. C. (1997). A comparative anatomical and biochemical analysis in *Salsola* (Chenopodiaceae) species with and without a Kranz type leaf anatomy: A possible reversion of C_4 to C_3 photosynthesis. *Am. J. Bot.* **84,** 597–606.
Pyankov, V., and Black, C. C. (1998). C_4 photosynthesis in Mongolia vegetation and their physiological roles. Abstract 670 to 1998 American Society of Plant Physiology Conference.
Pyankov, V. I. and Mokronosov, A. T. (1993). General trends in changes of the earth's vegetation related to global warming. *Russian. J. Plant Physiol.* (Engl. Translation). **40,** 515–531.
Raghavendra, A. S., and Das, V. S. R. (1978). The occurrence of C_4-photosynthesis: A supplementary list of C_4 plants reported during late 1974–mid 1977. *Photosynthetica* **12,** 200–208.
Rajendrudu, G., Prasad, J. S. R. and Das, V. S. R. (1986). C_3–C_4 intermediate species in *Alternanthera* (Amaranthaceae). *Plant Physiol.* **80,** 409–414.
Reiskind, J. B., Berg, R. H., Salvucci, M. E. and Bowes, G. (1989). Immunogold localization of primary carboxylases in leaves of aquatic and a C_3–C_4 intermediate species. *Plant Science* **61,** 43–52.
Reiskind, J. B., Madsen, T. V. Van Ginkel, L. C. and Bowes, G. (1997). Evidence that inducible C_4-type photosynthesis is a chloroplastic CO_2-concentrating mechanism in *Hydrilla*, a submersed monocot. *Plant Cell Environ.* **20,** 211–220.
Sage, R. F., Pearcy, R. W. and Seemann, J. R. (1987). The nitrogen use efficiency of C_3 and C_4 plants. III. Leaf nitrogen effects on the activity of carboxylation enzymes in *Chenopodium album* L. and *Amaranthus retroflexus* L. *Plant Phys.* **84,** 355–359.

Sheahan, M. C. and Chase, M. W. (1996). A phylogenetic analysis of Zygophyllaceae R. Br. based on morphological, anatomical and *rbcL* DNA sequence data. *Bot. J. Linn. Soc.* **122**, 279–300.

Smith, B. N. and Turner, B. L. (1975). Distribution of Kranz syndrome among Asteraceae. *Am. J. Bot.* **62**, 541–545.

Smith, J. A. C. and Winter, K. (1996). Taxonomic distribution of Crassulacean acid metabolism. *In:* "Crassulacean Acid Metabolism—Biochemistry, Ecophysiology and Evolution," (K. Winter and J. A. C. Smith, eds.), pp. 427–436. Springer, New York.

Townsend, C. C. (1993). Amaranthaceae *in* "The Families and Genera of Vascular Plants. Vol. II. Flowering Plants—Dicotyledons—Magnoliid, Hamamelid and Caryophyllid Families." (K. Kubitzki, J. G. Rohwer and V. Bittrich, eds.). pp. 70–91. Springer-Verlag, New York.

Troughton, J. H., Card, K. A. and Hendy, C. H. (1974). Photosynthetic pathway and carbon isotope discrimination by plants. *Carnegie Inst. Washington Yrbk.* **73**, 768–780.

Ueno, O. (1996). Immunocytochemical localization of enzymes involved in the C_3 and C_4 pathways in the photosynthetic cells of an amphibious sedge, *Eleocharis vivipara*. *Planta* **199**, 394–403.

Ueno, O., Samejima, M. and Koyama, T. (1989). Distribution and evolution of C_4 syndrome in *Eleocharis*, a sedge group inhabiting wet and aquatic environments, based on culm anatomy and carbon isotope ratios. *Ann. Bot.* **64**, 425–438.

Ueno, O., Samejima, M., Muto, S. and Miyachi, S. (1988). Photosynthetic characteristics of an amphibious plant, *Eleocharis vivipara*: expression of C_4 and C_3 modes in contrasting environments. *Proc. Natl. Acad. Sci. USA* **85**, 6733–6737.

Watson, L., and Dallwitz, M. J. (1980). "Australian Grass Genera, Anatomy: Morphology and Keys." Research School of Biological Sciences, The Australian National University, Canberra.

Watson, L., and Dallwitz, M. J. (1998). "Grass Genera of the World: Descriptions, Illustrations, Identification, and Information Retrieval; Including Synonyms, Morphology, Anatomy, Physiology, Cytology, Classification, Pathogens, World and Local Distribution, and References." URL http://biodiversity.uno.edu/delta/.

Watson, L., and Dallwitz, M. J. (1992, eds.). "The Grass Genera of the World." CAB International, Wallingford.

Webster, G. L. (1994). Synopsis of the genera and suprageneric taxa of Euphorbiaceae. *Ann. Missouri Bot. Gard.*, **81**, 33–144.

Welkie, G. W. and Caldwell, M. (1970). Leaf anatomy of species in some dicotyledon families as related to the C_3 and C_4 pathways of carbon fixation. *Can. J. Bot.* **48**, 2135–2146.

Williams, D. G. and Ehleringer, J. R. (1996). Carbon isotope discrimination in three semi-arid woodland species along a monsoon gradient. *Oecologia* **106**, 455–460.

Winter, K. (1981). C_4 plants of high biomass in arid regions of Asia—occurrence of C_4 photosynthesis in Chenopodiaceae and Polygonaceae from the Middle East and USSR. *Oecologia* **48**, 100–106.

Winter, K. and Smith, J. A. C. (1996). An introduction to crassulacean acid metabolism: biochemical principles and ecological diversity. *In:* "Crassulacean Acid Metabolism—Biochemistry, Ecophysiology and Evolution." (K. Winter and J. A. C. Smith, eds.), pp. 1–16. Springer, New York.

Winter, K. and Troughton, J. H. (1978). Photosynthetic pathways in plants of coastal and inland habitats of Israel and the Sinai. *Flora* **167**, 1–34.

Yin, L., and Li, M.-R. (1997). A study on the geographic distribution and ecology of C_4 plants in China. I. C_4 plant distribution in China and their relation with regional climatic condition. *Acta Ecologica Sinica* **17**, 350–363.

Ziegler, H., Batanouny, K. H., Sankhla, N., Vyas, O. P. and Stichler, W. (1981). The photosynthetic pathway types of some desert plants from India, Saudi Arabia, Egypt and Iraq. *Oecologia* **48**, 93–921.

Zuloaga, F. O. (1987). Systematics of New World species of *Panicum* (Poaceae: Paniceae). *In* "Grass Systematics and Evolution." (T. R. Sodenstrom, K. W. Hilu, C. S. Campbell and M. E. Barkworth, eds.), pp. 287–306. Smithsonian Institute Press, Washington DC.

Index

Acanthaceae, evolution, 414
Africa, human society development, C_4 plant agriculture role, 520–525
Agriculture
　C_4 photosynthesis implications
　　adaptation, 475–479
　　C_3–C_4 mixtures, 494–499
　　　forage crop intercropping, 497–499
　　　grain crop intercropping, 494–497
　　crop plant characterization, 475–479
　　crop quality, 487–491
　　　forage crops, 488–491
　　　grain crops, 487–488
　　overview, 473–475, 501–503
　　productivity, 479–486
　　　CO_2 change predictions, 485–486
　　　maximum growth rates and yield, 479–482
　　　nitrogen nutrition responses, 481–483
　　　temperature change predictions, 485–486
　　　water use efficiency, 483–485
　　turf, 499–501
　　weeds, 491–494
　human society development, 509–538
　　African C_4 plants, 520–525
　　America's maize, 525–534
　　　adoption patterns, 527–529, 532
　　　dispersal, 525–527
　　　indigenous crop complexes, 529–532
　　　origins, 525–527
　　　pre-Columbian societies, 527–529
　　　social inequality emergence, 532–534
　　carbon dating, 510–514
　　Chinese millets, 517–520
　　food production emergence, 514–517
　　human evolution, 510–514

　　overview, 509, 537
　　past 500 years, 534–537
Altitude, biogeographical distribution effects, 326–327
Aminotransferases
　biochemistry, 58–59
　C_4 pathway regulation, 107–109
Angiosperms, *see also specific types*
　taxonomic distribution, 553
Arthropods, herbivory, selectivity and distribution, 300–301
Aspartate aminotransferase
　biochemistry, 58–59
　C_4 pathway regulation, 107–109
Asteridae, evolution, 414–419
ATP, consumption rate modeling equation, 179
Autoradiography
　bundle sheath cell–mesophyll cell transport study, 72
　historical perspectives, 19–22

Benson–Calvin cycle, C_4 pathway regulation
　glycerate-3-P reduction in mesophyll cells, 113–114
　overview, 120–121
　Rubisco role, 111–112
Biochemistry
　bundle sheath cells, 60–62
　CO_2 concentration, 60–62
　compartmentation, 148–150, 162
　C_4 pathway enzymes, 53–60
　　aminotransferases, 58–59
　　C_4 acid decarboxylation enzymes, 56–58
　　carbonic anhydrase, 60
　　NADP-malate dehydrogenase, 55
　　phosphoenolpyruvate carboxylase, 54–55
　　pyruvate, P_i dikinase, 55–56
　　RuBP carboxylase–oxygenase, 59–60

585

C_4 subgroups, 50–53
  minimum energy requirements, 62–63
  NAD-malic enzyme, 53
  NADP-malic enzyme, 51–53
  phosphoenolpyruvate carboxykinase, 53
energetics, 62–68
  in vivo energy requirements, 63–65
  maximum quantum yield versus energy requirements, 65–68
  minimum energy requirements, 62–63
intercellular metabolite transport
  bundle sheath chloroplasts, 76–78
  mesophyll cell–bundle sheath cell coordination, 69
  mesophyll chloroplasts, 74–76
    C_4 acid translocators, 74
    phosphoenolpyruvate translocator, 74–75
    pyruvate transport, 75–76
    triosephosphate translocator, 74–75
  mesophyll cell–bundle sheath cell coordination, 68–74
    cell separation, 68–69
    enzyme distribution, 68–69
    evidence, 69–74
    intercellular metabolite transport, 69
  overview, 49–50, 78–80
  Rubisco activity, 60–62
  taxonomic distribution, 553, 554–555
Biogeography, *see also* Taxonomy
  controlling factors, 331–343
    aridity, 341
    C_4 grass–C_3 woodlands, 335–337
    disturbance, 334–339
    human disturbance, 337–339
    light availability, 334–339
    nitrogen, 341–343
    salinity, 340–341
    temperature, 332–334
    water, 339–340
  future of C_4 plants
    climate warming, 12–14, 346–347
    CO_2 enrichment effects, 344–346
    disturbance regimes, 348–350
    human land use management, 350
    invasive species, 348–350
    terrestrial eutrophication, 348
  global distribution
    altitudinal distribution, 326–327
    C_4-dominated biomes, 327–331
    flowering plant evolution, 412–429
    latitudinal distribution, 314–326
    C_4–C_3-dominated grassland transitions, 324–326
    floristic approach limitations, 322–323
    floristic assessments, 314–322
    temperature correlations, 324
    vegetation assessments, 323
  overview, 350–351
Biomes, *see also* Biogeography; *specific biomes*
  categorization, 329
  C_4-dominated biomes, 327–331
Boraginaceae, evolution, 414
Brassicaceae, evolution, 422–423
Bundle sheath cells
  biochemistry
    C_4 subgroups, 50–53
      minimum energy requirements, 62–63
      NAD-malic enzyme, 53
      NADP-malic enzyme, 51–53
      phosphoenolpyruvate carboxykinase, 53
    intercellular metabolite transport, 76–78
    mesophyll cell coordination, 68–74
      cell separation, 68–69
      enzyme distribution, 68–69
      evidence, 69–74
      intercellular metabolite transport, 69
    overview, 60–62
  carbon dioxide concentration, historical perspectives, 27–30
  herbivore food resources
    C_4 plant responses and features, 290–291
    tolerance and resistance, 302
  mesophyll cell communication, clonal relationships, 157–158
  modeling equations
    decarboxylation types
      conductance, 202–203
      O_2 evolution, 203
    enzyme limited photosynthesis
      CO_2 assimilation rate, 176
      CO_2 concentration, 177
      O_2 concentration, 177
    high irradiance model, conductance, 187–188

Calvin cycle, *see* Photosynthetic carbon reduction cycle

Capparaceae, evolution, 422
Carbon dating
  $\delta^{13}C$ paleorecords
    individual plants, 449–452
    terrestrial ecosystems, 452–464
      dietary carbon, 460–464, 513
      oxidized carbon, derived from organic matter, 457–460
      reduced organic carbon, 453–457
  human society development, 510–514
Carbon dioxide
  atmospheric increase
    crop plant productivity implications, 485–486
    distribution effects, 344–346
    grassland effects, 276–277
  concentration in bundle sheath cells
    biochemistry, 60–62
    historical perspectives, 27–30
  diffusion
    barriers, 143–146
    intercellular airspaces to mesophyll cytosol, 194–195
  enrichment, future biogeographical effects, 344–346
  fixation
    assimilation equation, 175
    energetics, 62–68
      in vivo energy requirements, 63–65
      maximum quantum yield versus energy requirements, 65–68
      minimum energy requirements, 62–63
    modeling equations
      bundle sheath cells
        assimilation rate, 176
        concentrations, 177
      fixation
        electron transport partition optimization, 195–196
        leakiness, 197–199
        quantum yield, 199–202
      geologic estimates, 445–446
      intercellular diffusion, 194–195
      leakage, 175, 197–199
      response curves, 188–191
  photorespiration, Rubisco role, 5–10
Carbonic anhydrase
  biochemistry, 60
  gene evolution, 384–386
Carbon isotopes, discrimination, historical perspectives, 34–35

Carboxylase, Rubisco carboxylation, 4–5
Caryophyllidae, evolution, 419–422
Cell–cell communication
  bundle sheath cell–mesophyll cell clonal relationships, 157–158
  differential effects of mutant cells, 160–161
  intercellular signalling during differentiation, 158–160
Cell wall, CO_2 permeability, historical perspectives, 27–30
C_4 acid decarboxylation enzymes, biochemistry, 56–58
C_4 acid translocator, transport in mesophyll cells, 74
C_4 pathway, see Biochemistry; Regulation
C_4 photosynthesis, see specific aspects
Chemistry, see Biochemistry
Chinese millets, human society development, 517–520
Chloroplasts, Kranz anatomy, dimorphism, 150–153
Climate warming, see also Temperature
  crop plant productivity implications, 485–486
  future biogeographical effects, 12–14, 346–347
Compartmentation, see also Bundle sheath cells; Mesophyll cells
  C_4 plant development, 148–150, 162
Competition, see Success of C_4 photosynthesis
Crop plants, see Agriculture
C_3–C_4 intermediates
  evolutionary patterns, 389–399
    anatomical traits, 389–390
    characteristics, 389–390
    evolutionary scenarios, 396–399
    Flaveria, 391–394
    glycine decarboxylase, differential expression, 391–394
    historical perspectives, 33–34
    light-dependent concentration changes, 90–92
    taxonomic distribution, 578–580
C_3 photosynthesis
  agricultural implications, C_3–C_4 mixtures, 494–499
    forage crop intercropping, 497–499
    grain crop intercropping, 494–497

C_4 photosynthesis compared
  bundle sheath cell chloroplast
      transport mechanisms, malate
      transport, 78
  paleorecords, 447
  Rubisco role, 7–9
  temperate subhumid grasslands, 254
  tropical savannas, 265–267
  historical perspectives, C_3–C_4
      intermediates, 33–34
Cyperaceae
  evolution, 427
  Kranz anatomy, 133, 140

Decarboxylation
  C_4 acid decarboxylation enzyme
      biochemistry, 56–58
  modeling analysis, 202–205
      bundle sheath conductance, 202–203
      bundle sheath O_2 evolution, 203
      energy requirements, 203–205
DELTA database, 551–552
Development
  bundle sheath cell mutants, 160–161
  cell–cell communication role, 157–160
  cell differentiation, 156–161
  cell-specific enzyme expression, 156–157
  clonal relationships, 157–158
  differential effects, 160–161
  environmental signals, 161–162
  intercellular signalling, 158–160
  mesophyll cell mutants, 160–161
  overview, 133–134, 163–164
  physiological signals, 162–163
  vascular pattern
      formation regulation, 154–156
      ontogeny, 153–154
$\delta^{13}C$ paleorecords
  human society development, 510–514
  individual plants, 449–452
  methodology, 449–455, 510–513
  terrestrial ecosystems, 452–464
      dietary carbon, 460–464, 513
      oxidized carbon, derived from organic
          matter, 457–460
      reduced organic carbon, 453–457
Dicotyledons, see also specific types
  evolutionary patterns, 401–403
  Kranz anatomy, 140–143
  taxonomic distribution, 557, 564
Diffusion
  carbon dioxide barriers, 143–146
  intercellular metabolites, 146–148

Distribution, see Biogeography
Disturbance
  controlling factors, 334–339
  future of C_4 plants, 348–350
  human disturbance, 337–339
Diversity, see Biogeography
Dominance, see Success of C_4
      photosynthesis
Drought, tallgrass prairie success, 260–261

Electron transport
  limiting light modeling analysis, CO_2
      fixation, partition optimization,
      195–196
  modeling equations, 179–184
      ATP consumption rate, 179
      C_3 and C_4 electron transport rate
          partitioning, 179–181
      CO_2 assimilation rate quadratic
          expression, 182–183
      light-dependent electron transport
          rate, 182
      NADPH consumption rate, 179
Energetics
  decarboxylation modeling, 203–205
  herbivory, 295
  in vivo energy requirements, 63–65
  minimum energy requirements, 62–63
  quantum yield
      carbon dioxide fixation
          energetics, 65–68
          modeling equations, 199–202
      temperature effects, 232–235
      theoretical efficiencies, 216–217
Environment
  biomes, see Biomes
  C_4 photosynthesis responses
      herbivory, 295–298
      light, 216–221
          equatorial floodplain emergent
              vegetation, 219–221
          production potential, 218–221
          stand level efficiency, 217–218
          temperate coastal saltmarshes, 221
          theory, 216–217
      nitrogen, 221–226
          efficient use theory, 221–223
          leaf and plant efficiencies, 223–226
      overview, 215, 240–241
      salinity, 230–231
      temperature, 231–240
          evolutionary patterns, 399–400

low temperature impairment, 235–238
  Miscanthus, 238–240
  theory, 232–235
 water, 226–231
  theory, 227–228
  whole plant efficiency, 228–230
C_4 success, *see* Success of C_4 photosynthesis
C_4 syndrome development
 evolutionary patterns, 399–403
  monocots versus dicots, 401–403
  warm environments, 399–400
 signals, 161–162
 geologic records, *see* Paleorecords
Enzymes, *see also specific types*
 cell-specific expression, 156–157
 C_4 pathway regulation, activity modulation, 93–95
 enzyme limited photosynthesis modeling equations, 176–179
  bundle sheath CO_2 assimilation rate, 176
  bundle sheath CO_2 concentration, 177
  bundle sheath O_2 concentration, 177
  enzyme limited CO_2 assimilation rate quadratic expression, 178–179
  phosphoenolpyruvate carboxylation rate, 177–178
Equations, *see* Modeling
Equatorial floodplains, ecological significance, 219–221
Euphorbiaceae, evolution, 424
Eutrophication, future biogeographical effects, 348
Evolution
 argument for C_4 evolution, 433–439
  future directions, 437–439
  lineage differences, 433–437
  overview, 439
 C_3–C_4 intermediates, 389–399
  anatomical traits, 389–390
  characteristics, 389–390
  evolutionary scenarios, 396–399
  Flaveria, 391–394
  glycine decarboxylase, differential expression, 391–394
 C_4 genes, 378–389
  carbonic anhydrase gene, 384–386
  evolutionary patterns, 389
  NADP-malate dehydrogenase gene, 386

  NADP-malic enzyme gene, 383–384
  phosphoenolpyruvate carboxylase gene, 378–380
  pyruvate, P_i dikinase gene, 380–383
 crop plants
  adaptation, 475–479
  human society development, 510–514
 ecological factors, 399–403
  monocots versus dicots, 401–403
  warm environments, 399–400
 geologic records, *see* Paleorecords
 overview, 50, 377–378, 403–404
 phylogenetic aspects, 411–439
  Asteridae, 414–419
  Caryophyllidae, 419–422
  C_4 origin number and times, 429–433
  Cyperaceae, 427
  distribution in flowering plants, 412–429
  Euphorbiaceae, 424
  glucosinolate clade, 422–423
  Hydrocharitaceae, 425–427
  overview, 411
  Poaceae, 427–429
  Zygophyllaceae, 424–425
 Rubisco, 3–6, 10–12, 386–388

Fertilization, *see also specific types*
 composition changes and growth induction, neotropical savannas, 271–272
 temperate grasslands, 341–343
Fire, C_4 photosynthesis success
 tallgrass prairies, 256–260, 262
 tropical savannas, 274–275, 337–338
Flaveria, C_3–C_4 intermediate metabolism, 391–394, 571, 579
Floodplains, ecological significance, 219–221
Floristic assessments, latitude distribution analysis
 limitations, 322–323
 procedures, 314–322
Foliage, *see* Leaves
Forage crops, *see* Agriculture
Fossil records, *see* Paleorecords
Fructose-1,6-bisphosphatase, C_4 pathway regulation in leaves, 116–117
Fungi, mycorrhizal fungi symbiosis, 293–294

Genes, evolutionary origins, 378–389
 carbonic anhydrase gene, 384–386

evolutionary patterns, 389
NADP-malate dehydrogenase gene, 386
NADP-malic enzyme gene, 383–384
overview, 377–378, 403
phosphoenolpyruvate carboxylase gene, 378–380
pyruvate, P_i dikinase gene, 380–383
Geologic records, see Paleorecords
Global warming, see also Temperature
 crop plant productivity implications, 485–486
 future biogeographical effects, 12–14, 346–347
Glucosinolate clade, evolution, 422–423
Glycerate-3-P, Benson–Calvin cycle regulation, reduction in mesophyll, 113–114
Glycine decarboxylase, differential expression, 391–394
Grasslands
 C_4 photosynthesis success
  increasing CO_2 effects, 276–277
  neotropical savannas
   CO_2 assimilation, 274
   CO_2 concentrating mechanisms, 275–276
   common species, 320–321
   competitors, 265–267
   crop species, 477–478
   distribution, 264–265, 320, 327–331
   dominance enhancing factors, 267–272
   fire tolerance, 274–275
   growth, 271
   growth-induced fertilization, 271–272
   nitrogen requirements, 271, 275
   organic matter production rates, 268–271
   photosynthetic characteristics, 267–268
   species composition changes, 271–272
   temperate grasslands compared, 276
   tropical savannas, 272–276
   water use efficiency, 274
  overview, 251–252, 276–277
  Rubisco role, 11–12
  tallgrass prairies
   competition, 253–256
   distribution, 324–331
   dominance enhancing factors, 256–261
   dominance reducing factors, 261–262
   drought, 260–261
   fire, 256–260, 262
   grazing, 261–262
   ideal environment, 262–264
   latitude effects, 324–326
   leaf-level gas exchange, 254–256
   leaf nitrogen role, 256
   plant water relations, 253–254
   tropical grasslands compared, 276
 crop plants, see Agriculture
 grass subfamilies, taxonomic distribution, 566–567
Grazing, see Herbivory

Herbivory
 C_4 characteristics effects on tolerance and resistance, 295–306
  bundle sheath food resources, 302
  ecosystem contexts, 295–298
  herbivore selectivity and distribution, 298–301
   arthropods, 300–301
   mammals, 298–299
  herbivory defenses, 305–306
  light responses, 302–305
  temperature responses, 287, 291–292, 302–305
  vascular tissue food resources, 302
  water use patterns, 305
 C_4 plant responses and features, 286–294
  biomass allocation, 292–293
  bundle sheath, 290–291
  fiber content, 286–290
  habitat selection, 294
  mycorrhizal fungi symbiosis, 293–294
  nutrient allocation, 292–293
  nutrient content, 286–290
  photosynthetic characteristics, 291–292
  vascular tissue food resources, 290–291
 overview, 285–286, 306–307
 paleorecords, diets, 460–464
 tallgrass prairie grazing, 261–262
Historical perspectives, 17–40
 biochemical subgroups, 25–27
 carbon isotope discrimination, 34–35
 C_3–C_4 intermediates, 33–34
 cell permeability, 27–30
 CO_2-concentrating function, 27–30

controversies, 35–37
discovery (1954-1967), 19–22
light–dark regulated enzymes, 30–32
   NADP-malate dehydrogenase, 31–32
   phosphoenolpyruvate carboxylase, 22, 32
   pyruvate, P_i dikinase, 30–31
mechanisms (1965-1970), 22–24
metabolite transport, 32–33
overview, 17–18, 39–40
people involved, 20, 23–24, 37–39
prediscovery scene (before 1965), 18–19
Human disturbance
  biogeographical control, 337–339
  land use management, 350
Human society development, C_4 plants role, 509–538
  African C_4 plants, 520–525
  American maize, 525–534
    adoption patterns, 527–529, 532
    dispersal, 525–527
    indigenous crop complexes, 529–532
    origins, 525–527
    pre-Columbian societies, 527–529
    social inequality emergence, 532–534
  carbon dating, 510–514
  Chinese millets, 517–520
  food production emergence, 514–517
  human evolution, 510–514
  overview, 509, 537
  past 500 years, 534–537
Hydrocharitaceae, evolution, 425–427

Intercellular transport, *see* Transport
Invasive species, future biogeographical effects, 348–350

Kranz anatomy
  biochemical compartmentation, 148–150
  chloroplast dimorphism, 150–153
  CO_2 diffusion barriers, 143–146
  Cyperaceae, 140
  dicotyledons, 140–143
  features, 134–135
  metabolite diffusion, 146–148
  overview, 133–134, 163–164
  paleorecords, 448–449
  peripheral reticulum, 151
  photosynthetic carbon oxidation cycle features, 148–153
  plant–herbivore interactions, 287–288

Poaceae, 135–140
variation, 135–143

Land use management, future biogeographical effects, 350
Latitude, biogeographical distribution effects, 314–326
  C_4–C_3-dominated grassland transitions, 324–326
  floristic approach limitations, 322–323
  floristic assessments, 314–322
  temperature correlations, 324
  vegetation assessments, 323
Leaves
  fiber content, 286–290
  gas exchange, tallgrass prairie species, 254–256
  nitrogen use
    efficiencies, 223–226
    tallgrass prairie species, 256
  nutrients
    allocation, 292–293
    content, 286–290
  structure and development, 133–164
    C_4 syndrome development, 153–163
      bundle sheath cell mutants, 160–161
      cell–cell communication role, 157–160
      cell differentiation, 156–161
      cell-specific enzyme expression, 156–157
      clonal relationships, 157–158
      differential effects, 160–161
      environmental signals, 161–162
      intercellular signalling, 158–160
      mesophyll cell mutants, 160–161
      physiological signals, 162–163
      vascular pattern formation regulation, 154–156
      vascular pattern ontogeny, 153–154
    Kranz anatomy, 134–153
      biochemical compartmentation, 148–150, 162
      chloroplast dimorphism, 150–153
      CO_2 diffusion barriers, 143–146
      Cyperaceae, 140
      dicotyledons, 140–143
      features, 134–135
      metabolite diffusion, 146–148
      peripheral reticulum, 151
      photosynthetic carbon oxidation cycle features, 148–153

Poaceae, 135–140
variation, 135–143
overview, 133–134, 163–164
Light
biogeographical control, 334–339
high irradiance modeling analysis, 186–195
bundle sheath conductance, 187–188
CO_2 diffusion from intercellular airspaces to mesophyll cytosol, 194–195
CO_2 response curves, 188–191
C_4 oxygen sensitivity, 191–194
phosphoenolpyruvate carboxylase activity, 186–187
Rubisco activity, 186–187
intermediate concentration changes, 90–92
limiting light modeling analysis, CO_2 fixation, 195–202
electron transport partition optimization, 195–196
leakiness, 197–199
quantum yield, 199–202
modeling equations, light and electron transport limited photosynthesis, 179–184
ATP consumption rate, 179
C_3 and C_4 electron transport rate partitioning, 179–181
CO_2 assimilation rate quadratic expression, 182–183
light-dependent electron transport rate, 182
NADPH consumption rate, 179
physiological responses, 216–221
equatorial floodplain emergent vegetation, 219–221
production potential, 218–221
stand level efficiency, 217–218
temperate coastal saltmarshes, 221
theory, 216–217
plant–herbivore interactions, 287, 302–305
Light–dark regulated enzymes, *see specific types*

Maize, human society development, 525–534
adoption patterns, 527–529, 532
dispersal, 525–527
indigenous crop complexes, 529–532
origins, 525–527
pre-Columbian societies, 527–529
social inequality emergence, 532–534
Mammals, herbivory, selectivity and distribution, 298–299
Mathematical modeling, *see* Modeling
Mesophyll cells
bundle sheath cell coordination
cell separation, 68–69
clonal relationships, 157–158
enzyme distribution, 68–69
evidence, 69–74
intercellular metabolite transport, 69
C_4 subgroup biochemistry, 50–53
minimum energy requirements, 62–63
NAD-malic enzyme, 53
NADP-malic enzyme, 51–53
phosphoenolpyruvate carboxykinase, 53
high irradiance modeling analysis, CO_2 diffusion from intercellular airspaces to mesophyll cytosol, 194–195
intercellular metabolite transport
active pyruvate transport, 75–76
bundle sheath cell coordination, 69
C_4 acid translocator, 74
phosphoenolpyruvate translocator, 74–75
P_i/triose phosphate translocator, 74–75
Metabolites, *see* Biochemistry; *specific types*
Millet, human society development, 517–520
Miscanthus, temperature responses, 238–240
Mitochondria, C_4 pathway interactions, 109–111
Modeling, 173–205
analysis, 184–205
CO_2 fixation at limiting light, 195–202
electron transport partition optimization, 195–196
leakiness, 197–199
quantum yield, 199–202
decarboxylation types, 202–205
bundle sheath conductance, 202–203
bundle sheath O_2 evolution, 203
energy requirements, 203–205
high irradiance model, 186–195
bundle sheath conductance, 187–188

CO_2 diffusion from intercellular
    airspaces to mesophyll cytosol,
    194–195
CO_2 response curves, 188–191
C_4 oxygen sensitivity, 191–194
phosphoenolpyruvate carboxylase
    activity, 186–187
Rubisco activity, 186–187
parameterization, 184–186
equations, 174–183
    enzyme limited photosynthesis
        equations, 176–179
        bundle sheath CO_2 assimilation rate,
            176
        bundle sheath CO_2 concentration,
            177
        bundle sheath O_2 concentration, 177
        enzyme limited CO_2 assimilation rate
            quadratic expression, 178–179
        phosphoenolpyruvate carboxylation
            rate, 177–178
    light and electron transport limited
        photosynthesis equations, 179–184
        ATP consumption rate, 179
        CO_2 assimilation rate quadratic
            expression, 182–183
        electron transport rate partitioning,
            179–181
        light-dependent electron transport
            rate, 182
        NADPH consumption rate, 179
    overview, 173–174, 205
Monocots, *see also specific types*
    evolutionary patterns, 401–403
    taxonomic distribution, 564–565, 576
Mycorrhizal fungi, symbiosis, 293–294

NAD-malic enzyme
    biochemistry, 53, 56, 62–63
    C_4 pathway regulation, 105–106, 109
NADPH, consumption rate modeling
    equation, 179
NADP-malate dehydrogenase
    biochemistry, 55
    C_4 pathway regulation, 103–104
    gene evolution, 386
    historical perspectives, 31–32
NADP-malic enzyme
    biochemistry, 51–53, 56, 62–63
    bundle sheath cell chloroplast
        transport mechanisms, C_3–C_4 plants
        compared, 78

C_4 pathway regulation, 104–105
gene evolution, 383–384, 434–436
historical perspectives, 22, 24–27
Nitrogen
    biogeographical control, 341–343
    C_4 photosynthesis success
        neotropical savannas, 271, 275
        tallgrass prairies, 256
    crop plant productivity, 481–483
    physiological responses, 221–226
        efficient use theory, 221–223
        leaf and plant efficiencies, 223–226
    plant–herbivore interactions, 287

Overcycling, modeling equation, 176
Oxaloacetic acid
    biochemistry, 57–58
    historical perspectives, 23
Oxygen, modeling equations
    bundle sheath oxygen
        concentration, 177
    evolution, 203
    decarboxylation, 203
    enzyme limited photosynthesis equations,
        177
    sensitivity, 191–194
Oxygenase, *see* Rubisco

Paleorecords, 445–464
    $\delta^{13}C$ values
        individual plants, 449–452
        terrestrial ecosystems, 452–464
            dietary carbon, 460–464, 513
            oxidized carbon, derived from
                organic matter, 457–460
            reduced organic carbon, 453–457
    Kranz anatomy morphology, 448–449
    overview, 445–448, 464
Pathways, *see* Biochemistry; Regulation;
    specific types
Peripheral reticulum, Kranz anatomy, 151
Peroxisomes, photosynthetic carbon
    reduction cycle, 152–153
Phosphate translocator
    C_4 pathway regulation in leaves,
        115–116
    transport in mesophyll cells, 74–75
Phosphoenolpyruvate, transport
    mechanisms
    historical perspectives, 33
    translocators in mesophyll cells,
        74–75

Phosphoenolpyruvate carboxykinase
  biochemistry, 53, 56–57, 62–63
  C_4 pathway regulation, 106–107, 109
Phosphoenolpyruvate carboxylase
  biochemistry, 54–55
  carboxylation rate modeling equation, 177–178
  C_4 pathway regulation, 95–100
  gene evolution, 378–380, 435–437
  high irradiance activity model, 186–187
  historical perspectives, 22, 32
3-Phosphoglycerate, transport mechanisms, historical perspectives, 33
Photorespiration, Rubisco role, 5–10
Photosynthesis, see C_3 photosynthesis; *specific aspects*
Photosynthetic carbon oxidation cycle
  biochemistry, 50, 60–65, 69
  Kranz anatomy, 148–153
Photosynthetic carbon reduction cycle
  biochemistry, 49–51, 63, 69
  historical perspectives, 18–22
  specialized cell features, 148–153
    biochemical compartmentation, 148–150
    chloroplast dimorphism, 150–153
    peroxisomes, 152–153
Physiology, see also Biochemistry
  C_4 leaf development signals, 162–163
P_i/triose phosphate translocator, transport in mesophyll cells, 74–75
Poaceae
  evolution, 427–429
  Kranz anatomy, 133, 135–140
  taxonomic distribution, 568–576
Portulacaceae, evolution, 421
Prairies, see Grasslands
Productivity
  agricultural implications, 479–486
    CO_2 change predictions, 485–486
    maximum growth rates and yield, 479–482
    nitrogen nutrition responses, 482–483
    temperature change predictions, 485–486
    water use efficiency, 483–485
  light responses, 218–221
  nitrogen use, see Nitrogen
  organic matter production rates
    neotropical savannas, 268–271
    tallgrass prairies, 256

quantum yields
  carbon dioxide fixation
    energetics, 65–68
    modeling equations, 199–202
    temperature effects, 232–235
    theoretical efficiencies, 216–217
Pyruvate, transport mechanisms
  historical perspectives, 32–33
  mesophyll chloroplasts, 75–76
Pyruvate, P_i dikinase
  biochemistry, 55–56
  C_4 pathway regulation, 100–103
  gene evolution, 380–383, 435–436
  historical perspectives, 30–31

Quantum yield
  carbon dioxide fixation
    energetics, 65–68
    modeling equations, 199–202
    temperature effects, 232–235
    theoretical efficiencies, 216–217

Radiotracers
  bundle sheath cell–mesophyll cell transport study, 72
  historical perspectives, 19–22
Regulation, 89–121
  aminotransferase role, 107–109
  Benson–Calvin cycle regulation, 111–114
    glycerate-3-P reduction in mesophyll cells, 113–114
    Rubisco role, 111–112
  enzyme activity modulation, 93–95
  intermediates, light-dependent concentration changes, 90–92
  intracellular metabolite transport, 92–93
  mitochondrial metabolism interactions, 109–111
  NAD-malic enzyme role, 105–106
  NADP-malate dehydrogenase role, 103–104
  NADP-malic enzyme role, 104–105
  overview, 89, 120–121
  phosphoenolpyruvate carboxykinase role, 106–107
  phosphoenolpyruvate carboxylase role, 95–100
  product synthesis regulation in leaves, 114–120
    fructose-1,6-bisphosphatase, 116–117
    phosphate translocator, 115–116

starch synthesis enzymes, 119–120
sucrose phosphate synthase, 117–119
pyruvate, P_i dikinase role, 100–103
Rubisco
  Benson–Calvin cycle regulation, 111–112
  CO_2 assimilation equation, 175
  C_4 photosynthesis role
    activity in bundle sheath cells, 60–62
    carboxylase activity, 4–5
    C_3 plants compared, 7–9
    efficiency, 10
    evolution, 3–6, 10–12, 386–388
    future prospects, 12–14
    mechanisms, 3–9
    overview, 3, 14
    oxygenase activity, 4–6
    photorespiration, 5–10
    success, 11–12
    taxonomic variation, 10–11
  high irradiance activity model, 186–187
  light effects, 215, 240
RuBP carboxylase–oxygenase, biochemistry, 59–60

Salinity
  biogeographical control, 340–341
  physiological responses, 230–231
  temperate saltmarshes, 221
Savannas, see Grasslands
Scrophulariaceae, evolution, 414
Societal interactions, see Agriculture
Starch synthesis enzymes, C_4 pathway regulation in leaves, 119–120
Stomata, conductance, plant–herbivore interactions, 287
Success of C_4 photosynthesis
  increasing CO_2 effects, 276–277
  neotropical savannas, 264–276
    competitors, 265–267
    distribution in tropical South America, 264–265
    dominance enhancing factors, 267–272
      growth, 271
      growth-induced fertilization, 271–272
      nitrogen requirements, 271, 275
      organic matter production rates, 268–271
      photosynthetic characteristics, 267–268
      species composition changes, 271–272
  tropical savannas, 272–276
    CO_2 assimilation, 274
    CO_2 concentrating mechanisms, 275–276
    fire tolerance, 274–275
    temperate grasslands compared, 276
    water use efficiency, 274
  overview, 251–252, 276–277
  Rubisco role, 11–12
  tallgrass prairies, 252–264
    competition, 253–256
      leaf-level gas exchange, 254–256
      leaf nitrogen, 256
      plant water relations, 253–254
    dominance enhancing factors, 256–261
      drought, 260–261
      fire, 256–260, 262
    dominance reducing factors, 261–262
    grazing, 261–262
    ideal environment, 262–264
    tropical grasslands compared, 276
Sucrose phosphate synthase, C_4 pathway regulation in leaves, 117–119
Symbiosis
  mutualistic symbiosis, 288
  mycorrhizal fungi symbiosis, 293–294

Tallgrass prairies, see Grasslands
Taxonomy, see also Biogeography
  biochemical subgroups, historical perspectives, 25–27
  C_4 success, see Success of C_4 photosynthesis
  distribution, 551–580
    angiosperm families, 556
    biochemical subtypes, 554–555
    grass subfamilies, 566–567
    identified C_4 species, 558–563
    overview, 551–552, 580
    photosynthetic pathway, 552–554
    Poaceae, 568–576
    taxa lists, 555, 557, 564–565, 576–577, 580
    C_3–C_4 intermediates, 577, 580
    dicots, 557, 564
    monocots, 564–565, 576
  evolution, see Evolution
  Kranz anatomy, 135–143
  Rubisco variations, 10–11
Temperature
  atmospheric increase
    crop plant productivity implications, 485–486

future biogeographical effects, 12–14, 346–347
biogeographical control, 332–334
evolutionary patterns, 399–400
latitude distribution correlations, 324
physiological responses, 231–240
  low temperature impairment, 235–238
  miscanthus, 238–240
  theory, 232–235
plant–herbivore interactions, 287, 291–292, 302–305
Terrestrial eutrophication, future biogeographical effects, 348
Transport
  electron transport limited photosynthesis modeling analysis, CO_2 fixation partition optimization, 195–196
  modeling equations, 179–184
    ATP consumption rate, 179
    C_3–C_4 partitioning, 179–181
    CO_2 assimilation, 182–183
    light dependence, 182
    NADPH consumption rate, 179
  historical perspectives, 32–33
  intracellular metabolites
    biochemistry
      bundle sheath chloroplasts, 76–78
      C_4 acid translocators, 74
      mesophyll cell–bundle sheath cell coordination, 69
      mesophyll chloroplasts, 74–76
      phosphoenolpyruvate translocator, 74–75
      pyruvate transport, 75–76
      triosephosphate translocator, 74–75
    C_4 pathway regulation, 92–93
Triosephosphates, transport mechanisms
  historical perspectives, 33
  translocator in mesophyll cells, 74–75
Tropical savannas, *see* Grasslands
Turf, C_4 photosynthesis implications, 499–501

Variability, *see* Taxonomy
Vascular system
  C_4 leaf development
    ontogeny, 153–154
    pattern formation regulation, 154–156
  herbivore food resources
    C_4 plant responses and features, 290–291
    tolerance and resistance, 302
Vegetation assessments, latitude distribution analysis, 323

Water
  biogeographical control, 339–341
  C_4 photosynthesis success
    tallgrass prairies, 253–254
    tropical savannas, 274
  crop plant productivity, 483–485
  herbivory effects, use patterns, 305
  physiological responses, 226–231
    theory, 227–228
    whole plant efficiency, 228–230
Weeds, agricultural implications, 491–494
Woodlands
  biogeography controlling factors, 335–337
  evolutionary patterns, 400–401

Zygophyllaceae, evolution, 424–425

Physiological Ecology
A Series of Monographs, Texts, and Treatises

Series Editor
Harold A. Mooney
Stanford University, Stanford, California

Editorial Board
Fakhri A. Bazzaz F. Stuart Chapin James R. Ehleringer
Robert W. Pearcy Martyn M. Caldwell E.-D. Schulze

T. T. KOZLOWSKI. Growth and Development of Trees, Volumes I and II, 1971

D. HILLEL. Soil and Water: Physical Principles and Processes, 1971

V. B. YOUNGER and C. M. McKELL (Eds.). The Biology and Utilization of Grasses, 1972

J. B. MUDD and T. T. KOZLOWSKI (Eds.). Responses of Plants to Air Pollution, 1975

R. DAUBENMIRE. Plant Geography, 1978

J. LEVITT. Responses of Plants to Environmental Stresses, Second Edition
Volume I: Chilling, Freezing, and High Temperature Stresses, 1980
Volume II: Water, Radiation, Salt, and Other Stresses, 1980

J. A. LARSEN (Ed.). The Boreal Ecosystem, 1980

S. A. GAUTHREAUX, JR. (Ed.). Animal Migration, Orientation, and Navigation, 1981

F. J. VERNBERG and W. B. VERNBERG (Eds.). Functional Adaptations of Marine Organisms, 1981

R. D. DURBIN (Ed.). Toxins in Plant Disease, 1981

C. P. LYMAN, J. S. WILLIS, A. MALAN, and L. C. H. WANG. Hibernation and Torpor in Mammals and Birds, 1982

T. T. KOZLOWSKI (Ed.). Flooding and Plant Growth, 1984

E. L. RICE. Allelopathy, Second Edition, 1984

M. L. CODY (Ed.). Habitat Selection in Birds, 1985

R. J. HAYNES, K. C. CAMERON, K. M. GOH, and R. R. SHERLOCK (Eds.). Mineral Nitrogen in the Plant–Soil System, 1986

T. T. KOZLOWSKI, P. J. KRAMER, and S. G. PALLARDY. The Physiological Ecology of Woody Plants, 1991

H. A. MOONEY, W. E. WINNER, and E. J. PELL (Eds.). Response of Plants to Multiple Stresses, 1991

F. S. CHAPIN III, R. L. JEFFERIES, J. F. REYNOLDS, G. R. SHAVER, and J. SVOBODA (Eds.). Arctic Ecosystems in a Changing Climate: An Ecophysiological Perspective, 1991

T. D. SHARKEY, E. A. HOLLAND, and H. A. MOONEY (Eds.). Trace Gas Emissions by Plants, 1991

U. SEELIGER (Ed.). Coastal Plant Communities of Latin America, 1992

JAMES R. EHLERINGER and CHRISTOPHER B. FIELD (Eds.). Scaling Physiological Processes: Leaf to Globe, 1993

JAMES R. EHLERINGER, ANTHONY E. HALL, and GRAHAM D. FARQUHAR (Eds.). Stable Isotopes and Plant Carbon–Water Relations, 1993

E.-D. SCHULZE (Ed.). Flux Control in Biological Systems, 1993

MARTYN M. CALDWELL and ROBERT W. PEARCY (Eds.). Exploitation of Environmental Heterogeneity by Plants: Ecophysiological Processes Above- and Belowground, 1994

WILLIAM K. SMITH and THOMAS M. HINCKLEY (Eds.). Resource Physiology of Conifers: Acquisition, Allocation, and Utilization, 1995

WILLIAM K. SMITH and THOMAS M. HINCKLEY (Eds.). Ecophysiology of Coniferous Forests, 1995

MARGARET D. LOWMAN and NALINI M. NADKHARNI (Eds.). Forest Canopies, 1995

BARBARA L. GARTNER (Ed.). Plant Stems: Physiology and Functional Morphology, 1995

GEORGE W. KOCH and HAROLD A. MOONEY (Eds.). Carbon Dioxide and Terrestrial Ecosystems, 1996

CHRISTIAN KÖRNER and FAKHRI A. BAZZAZ (Eds.). Carbon Dioxide, Populations, and Communities, 1996

THEODORE T. KOZLOWSKI and STEPHEN G. PALLARDY. Growth Control in Woody Plants, 1997

J. J. LANDSBERG and S. T. GOWER. Application of Physiological Ecology to Forest Management, 1997

FAKHRI A. BAZZAZ and JOHN GRACE (Eds.). Plant Resource Allocation, 1997

LOUISE E. JACKSON (Ed.). Ecology in Agriculture, 1997

ROWAN F. SAGE and RUSSELL K. MONSON (Eds.). C_4 Plant Biology, 1999